Recent Trends in Hydrogeology

Edited by

T. N. Narasimhan
Lawrence Berkeley Laboratory
University of California
Berkeley, California 94720

SPECIAL PAPER
189

THE GEOLOGICAL SOCIETY OF AMERICA
P.O. Box 9140 · 3300 Penrose Place · Boulder, Colorado 80301

P.O. Box 9140, 3300 Penrose Place
Boulder, Colorado 80301

Printed in U.S.A. by Malloy Lithographing, Inc.,
Ann Arbor, Michigan.

Printed 1982. Reprinted 1983.

The publishing of this volume was generously supported by the donation of editorial and word–processor time by Lawrence Berkeley Laboratory, University of California, Berkeley, California 94720

Library of Congress Cataloging in Publication Data
Main entry under title:

Recent trends in hydrogeology.

(Special paper / The Geological Society of America; 189)
Papers presented at a symposium sponsored by the Earth Sciences Division of Lawrence Berkeley Laboratory on Feb. 8–9, 1979.
Includes bibliographical references.
1. Hydrogeology—Congresses. I. Narasimhan, T. N. (Thiruppudaimarudhur N.), 1935– . II. Lawrence Berkeley Laboratory. Earth Sciences Division. III. Series: Special paper (Geological Society of America); 189.
GB1001.2.R43 551.48 82-2872
ISBN 0-8137-2189-X AACR2

Contents

Preface

Hydrogeology may be broadly considered to be the discipline concerned with those geologic processes that are influenced by water. Until recently, the subject has been considered to be synonymous with ground-water hydrology, dealing mainly with the dynamics of subsurface fluid-flow systems with minor influence, if any, on variations in temperatures. Historically, such a limited scope of definition was adequate, since man's interest in ground water was largely governed by its role as a source of fresh water. However, over the past few decades, the rapid growth of technology and the desire to improve the standard of living the world over has driven man to look deeper into the Earth's crust for more resources, to perturb his environment physically and chemically in ever-increasing magnitudes, and to seek ways of protecting himself from natural calamities such as earthquakes and landslides. In this expanded context, field experience indicates that the role of the ubiquitous subsurface water in controlling geologic processes is far more significant than what was inherent in the historical scope of hydrogeology. To cite a few examples: meteoric waters may circulate down to depths as great as tens of kilometres and take part in hydrothermal convection mechanisms over geologically small time periods; in tectonically critical stress areas relatively small changes in fluid pressures may trigger significant earthquakes; even in areas of igneous-metamorphic or ancient argillaceous stratigraphy it is extremely difficult to find a truly impermeable environment down to a few kilometres below the land surface; and large-scale withdrawal of fluids can cause ground subsidence, fault rejuvenation, and earth fissuring. Many more examples could be added to the list.

Indeed, the subject of hydrogeology encompasses all geologic processes that involve water, either near the land surface (surface-water/ground-water interaction) or at depths of several kilometres or more (geopressured systems; hydrothermal mineralization). The physical phenomena of interest range from the mere motion of water due to mechanical, chemical, or other gradients of potential to the interaction between fluid potential and tectonic stresses, the chemical interactions between the fluid and rock matrix, the transport of the chemical constituents, and the transport of energy by the dynamic water. Because of the variety of the processes and scales involved, studies coming within the folds of hydrogeology are carried out in widely varying fields of science and engineering. The conceptual framework of hydrogeology has been enriched significantly through related research in many disciplines, such as soil mechanics and soil physics.

In any field of the arts or sciences, knowledge continues to accumulate with time through the contribution of a multitude of efforts. As this accumulation continues to take place, it behooves us to step back periodically and take an overall look at our accomplishments and failures. This provides us with an opportunity to synthesize seemingly unrelated pieces of knowledge and to identify simpler patterns in nature that were not obvious earlier. As science progresses at a rapid rate, such an overview becomes an increasingly difficult task for two reasons. First, the scope of the subject is too vast for any single worker to undertake an overview of the entire field. A reasonable success in the venture actually demands a coordinated endeavor of researchers with different backgrounds. Second, even if one wanted to assemble a team of active workers to carry out the survey, a strong motivation would be essential to inspire them to take time off from their regular research to collect their thoughts on the subject.

In the case of the present volume, the catalyst for undertaking a survey of the directions in which hydrogeology is proceeding happened to be an important occasion in the life of a distinguished hydrogeologist, Paul A. Witherspoon, Professor of Mineral Engineering at the Berkeley Campus of the University of California as well as an Associate Director of the Lawrence Berkeley Laboratory and leader of its Earth Sciences Division. Through his own insight and research, through the many distinguished students of hydrogeology who have grown under his tutelage, and through his associations with many researchers from the United States and abroad, Professor Witherspoon has made significant contributions in such varied topics as well-testing, multiple-aquifer systems, regional ground-water motion, numerical modeling, fluid flow in fractured rocks, heat and mass transfer in ground-water systems, and so on. It was felt, therefore, by a number of people who know him, that a survey of the current trends in hydrogeology would be a worthy effort for the occasion of his sixtieth birthday, February 9, 1979. Accordingly, a symposium entitled Recent Trends in Hydrogeology was convened by the Earth Sciences Division of Lawrence Berkeley Laboratory at Berkeley, California, on February 8 and 9, 1979. The contributors were invited from among the students and others who had had associations with Professor Witherspoon. The twenty-three papers included in this volume were presented at the symposium.

Although it was recognized that a complete review of the subject would be a very difficult if not impossible task, the purpose of the symposium was to attempt a reasonable coverage of the important facets of hydrogeology as we then knew them. In that attempt, the aim was to present the thoughts in such a manner that the papers would collectively provide a global picture of hydrogeology.

To achieve such a unification here, the papers in this volume have been organized into five groups: Conceptual Aspects, Measurement Aspects, Mathematical Aspects, Regional Aspects, and Engineering Aspects. Within the framework of these five categories, we shall now introduce and relate the various papers that are to follow.

A fundamental aim of science is to draw inferences from empirical observations of cause-effect relationships. An imperative need in this regard is a conceptual framework. In the field of hydrogeology, the conceptual framework includes the principles of hydrodynamics, heat and solute transport, and solid-fluid chemical interactions in subsurface systems. These phenomena are governed by the laws of mass, energy, and momentum conservation as well as the laws of chemical equilibrium. The papers on Saturated-Unsaturated Flow (Narasimhan), Flow in Geothermal Systems (Pinder and Shapiro), and Contaminant Migration (Gillham and Cherry) attempt to cover the overall conceptual foundations of heat and mass transfer in porous subsurface systems. Although the laws of motion and conservation are precisely obeyed in nature, their application to subsurface systems is masked by uncertainty stemming from our limited means of observing and quantifying physical processes that take place in porous media. That is, insofar as porous media are concerned, our observations are of a gross nature and meaningful only in a statistical sense. It is therefore necessary that the physical parameters that are defined in the conceptual framework be properly constrained by the appropriate uncertainties. Over the past decade, many workers have been attempting to include the statistical constraints on the physical model, and research is active in this direction. The paper on Rainfall-Runoff Predictions (Freeze) discusses some developments in the application of statistics with special emphasis on the surface-water/ground-water relationships governing base flow to streams. In a related paper on Aquifer Heterogeneities (Neuman), the statistical concepts related to parameter estimation in heterogeneous aquifer systems are discussed in detail. Subsurface fluid-flow systems are, in general, characterized by a deformable porous matrix. In characterizing the physics of deformation of a fluid-filled porous medium, it is customary to treat the porous solid as deforming in a series of static steps. Although in practice it is sufficient for most purposes, this approach leads to certain conceptual difficulties when one wishes to consider a system subject to general three-dimensional deformation caused by withdrawal of fluids from within the pores. The paper on Conceptual Aspects of Subsidence (Helm) is an attempt to examine our current ideas on deformable fluid-filled systems and to suggest an alternate, improved framework in which the matrix itself is treated as a dynamic fluid with an associated velocity field.

In any scientific endeavor the importance of measurement cannot be overemphasized. In the field of hydrogeology, measurements are made in the laboratory and in the field. These measurements may commonly be indirect in nature (for example, geophysical measurements). In addition, we are increasingly forced to make measurements under hostile in situ conditions (such as in the case of geothermal reservoirs). The subject of hydrogeologic measurement cuts conspicuously across a variety of disciplines--soil mechanics, rock mechanics, petroleum engineering, geophysics, and soil physics, to cite only a few. In the present volume the measurement of physical properties of geologic materials is covered by three related papers. The paper on Physical Properties of Porous Geologic Materials (Houston and Kasim) relates to poorly consolidated materials at room temperatures, whereas the paper on Permeability Characteristics of Fractured Rock (Gale) focuses attention on massive fractured rocks. The paper on Porous Rock-Fluid Systems (Somerton) examines the measurement of petrophysical properties at elevated temperatures and pressures. The field aspects of measurement are discussed in two complementary papers. One, on Instrumentation for Well Tests (Schroeder), discusses instrumentation needs especially for hostile geothermal environments. The other, on Borehole Geophysics (Nelson), summarizes the instrumentation currently available for borehole geophysics in hydrogeology.

Although the conceptual aspects provide the framework in which we can analyze our measurements, the ultimate objective of quantification is to be realized by solving problems with the aid of appropriate mathematical models. Perhaps the most widely used tool for problem solving in physics is the differential equation, which allows the subsurface system to be treated as a statistical continuum at the macroscopic level. Starting with the early investigations of Dupuit, Forchheimer, Slichter, Thiem, and others at the turn of the century, the differential-equation approach received its most marked boost in hydrogeology with the work of Theis in the mid-1930s when he idealized transient ground-water motion in the form of a parabolic differential equation. Contemporaneously, the same concept was introduced in the petroleum industry. Theis's work was later extended to a variety of subsurface conditions, all based on solving the parabolic equations. Indeed, until the early 1960s the attention of hydrogeology was heavily directed toward the interpretation of well tests. The paper on Analytical Solutions in Subsurface Fluid Flow (Javandel) summarizes the various techniques used to solve the parabolic equation and highlights some of the more important solutions relating to porous, heterogeneous aquifer systems. Closely related to this is the paper on Fractured Reservoirs (Gringarten), which examines solutions to the parabolic equation with respect to fractured reservoirs. Although the parabolic equation describes the physical phenomenon within the aquifer, the processes going on within the well are governed by the efficiency of the well itself insofar as it behaves as a hydraulic system. The near-well phenomena have traditionally been treated as a well-loss function in hydrogeology or a skin effect in petroleum engineering. The paper on Well-Loss Function and the Skin Effect (Ramey) examines the historical development of these concepts and provides a bridge between the two fields.

During the mid-1960s, a new direction began to establish itself in the field of hydrogeology. This consisted of an increased emphasis on quantifying generalized fluid flow in natural systems rather than studying the particular case of flow to a well. This general approach required the availability of a more powerful tool for problem solving than that of obtaining analytical solutions for relatively simple systems. The numerical approach, which essentially consists of directly integrating the conservation equations, has since become a very widely used tool for analyzing complex hydrogeologic systems. The paper on Numerical Modeling (Narasimhan) examines the nature and use of numerical models in simulating isothermal ground-water systems.

In the field, the most basic of all hydrogeologic concepts is that of regional ground-water motion. It is now well established that such diverse features as hydrocarbon accumulations, geochemical evolution of ground waters, and the evolution of tectonic stresses in the Earth's crust can be eventually related to the hydrodynamics of the ground-water system on a local or a regional scale. Commencing with the investigations on artesian flow of the late nineteenth century, the existence of ground-water systems spanning several tens or even hundreds of kilometres has been recognized and mapped by many hydrogeologists both in the U. S. and elsewhere. The earliest notions of regional motion were centered on the pattern of fluid flow that can be observed as water moves from areas of recharge to those of discharge. This notion was embellished by several other basic concepts of later workers. These concepts included: the importance of the compressibility of the porous skeleton in releasing water from storage, the influence of flow dynamics on geochemical evolution of ground waters, and, more recently, the role of ground-water movement on the evolution of tectonic effective-stress fields. The paper on Regional Ground-Water Flow (Bredehoeft, Back, Hanshaw) surveys development of the concept of regional ground-water motion in the United States and identifies the relevant physical processes of interest. The paper on Induced Seismicity (Snow) considers the mechanisms governing reservoir-induced earthquakes in two well-known areas and attempts to relate hydrogeologic observations to the evolution of tectonic stresses over geologic time. The paper on Isotope Dating (Fritz) analyzes the status of available geochemical techniques to determine ages of old ground waters. It may be pointed out here that accurate age-dating of waters constitutes one of the best ways of understanding regional ground-water systems that have evolved over long periods of time. Enhanced understanding of regional ground-water systems can be applied to isolate specific sites where new resources could be developed or where specific geological engineering activities (for example, construction of impoundments; disposal of wastes) could be safely carried out. The paper on Geothermal Resource Exploration (Wollenberg) summarizes a variety of geological, geophysical, and geochemical techniques that are currently used to prospect for ground water in general and geothermal resources in particular.

The papers included in the final category of this volume relate to the engineering aspects of hydrogeology. Engineering applications indeed provide the motivation for the generation of new and daring concepts and provide the testing grounds for such ideas. As technology advances rapidly, the related engineering questions demand answers of ever-increasing precision. A dramatic case in point is that of nuclear waste disposal. Because of the high toxicity of nuclear wastes, hydrogeologists are required to predict the consequences of geologically disposed nuclear wastes to the ground-water systems over periods of centuries or even millenia. Considering the complexities of natural geologic systems and the limited database, the predictive problem is extremely difficult if not impossible to assess. Additionally, the problem is complicated by its sensitive sociopolitical nature. The paper on the Hydrogeology of Radioactive Waste Disposal (Davis) provides a provocative perspective on the challenges that lie ahead in a rational assessment of the consequences of radioactive waste disposal. The important role that ground water plays in the development and operations of mines is well known. The paper on Open-Pit and Underground Mines (Cook) summarizes the important field problems associated with the hydrogeology of these environments. An unusual engineering problem of recent times is that of the chemical effects of brine reinjection in depleted reservoirs. In geothermal energy development, brine reinjection is often a necessity both for minimizing water pollution and for providing pressure support to the depleting reservoir. However, the physicochemical interactions between the injected brine, the well, and the formation should be expected to lead to corrosion and plugging of expensive wells, thereby adversely influencing the economic viability of resource development. The paper on Reinjection of Geothermal Brines (Weres and Apps) analyzes the important problems associated with brine reinjection. One of the important field problems in hydrogeology that has arisen in the context of the world's declining energy sources is the use of shallow ground-water systems for energy storage. There is a widespread interest the world over in utilizing these systems as temporary repositories to store heat that would otherwise be wasted. The paper on Aquifer Thermal Energy Storage (Tsang and Hopkins) surveys current activities in this area of research.

As evidenced by the contributions in this volume, it is clear that the scope of hydrogeology is extemely wide. In the future, man no doubt will reach to greater and greater depths in the Earth's crust to expand his resource base to meet his growing needs. The influence of water on the occurrence and exploitation of these resources will certainly be of more than academic interest. At the present time, we do possess a reasonable understanding of the fundamental physical quantities that govern the behavior of hydrogeologic systems. We also know, in a conceptual sense, the mutual relationships among these quantities. The greatest challenge to our knowledge, however, lies in the area of quantification. As any natural scientist knows, subsurface hydrogeologic systems are often characterized by an extreme complexity of geometry, material distributions, and boundary and initial conditions. This complexity is compounded by our inability to acquire quantitative data to characterize these systems even with minimal confidence. There is yet another challenge that is even more disconcerting than those related to system complexity and paucity of data. This is the challenge concerning the scale-dependence of hydrogeologic parameters. Fundamental quantities such as fluid potential, permeability, porosity, compressibility, and concentration have only a statistical significance in the hydrogeologic context. In addition to their variability in space and time, these quantities are also very much dependent on the scale of observation. At a given location, the same parameter may legitimately assume widely different values depending on whether one is interested in a thin layer or in a complex of beds representing a geologic era. Again, in such problems as earthquake prediction, we may not even know the relevant scale of observation for fluid pressure and stresses that may be presumed to control earthquakes. Thus, although technology demands precise answers to questions of ever-increasing complexity, and although the electronic revolution gives us increasingly sensitive measuring devices, we find ourselves in a situation of parameter uncertainty, which takes away much of the quantitative precision from our highly refined measurement and computational tools. The question then arises whether there is a need to search for alternative conceptual frameworks that could lead to more precise answers to important hydrogeologic questions. The hydrogeologist's awareness of this question has resulted in an increased effort to incorporate uncertainty into the equations that govern the behavior of hydrogeologic systems. Although this approach holds promise, it is still far from being a well-developed tool that one could use when precise answers are needed, as in problems of toxic waste disposal. We are now faced with a conflict between the demands of high precision and a basic negation of precision of the governing equation once uncertainty is introduced. A great deal of research activity lies ahead in order to bridge this gap.

While the theoreticians are engaged in developing better methods for quantifying the field problem, hydrogeology is sure also to witness field-oriented research of unprecedented magnitude and breadth. Within the past decade, extemely precise pressure-measuring devices, downhole flow meters, downhole televiewers, and other geophysical instruments have been developed in response to the requirements of geothermal exploration and the problems associated with toxic waste disposal. Simultaneously, better core-recovery equipment, multiple packers, automatic data recorders, and related equipment are being developed to collect, store, and retrieve great volumes of data. Hydrogeology is already at the threshold of a data explosion. Although our ability to do justice to these data is less than one would have desired, we are already beginning to observe subtle physical phenomena that could not be observed in the past. There is promise that many regional ground-water systems, ranging from small watersheds to large geologic basins, will receive systematic study in the near future in response to our needs for better ground-water management and environmental protection. On a more academic level, there is increased research activity on the role of meteoric waters in diagenetic and mineral-forming processes. An important aspect of this activity is to understand the chemical interaction between the interstitial fluid and the host rock, taking into consideration sorption, dissolution, and precipitation mechanisms. The studies of regional ground water systems will be aided by the use of improved tracers of various kinds in order to determine the movement and the age of the in situ ground waters.

It should be quite clear from a glance at the Contents page that an enormous amount of time has been spent by the twenty-nine authors who have contributed to this volume. I would like to thank each one of them for their efforts. The symposium on Recent Trends in Hydrogeology, of which this volume is an outcome, was sponsored by the Lawrence Berkeley Laboratory. I am very thankful to Andrew Sessler, former Director of the Laboratory, and Earl Hyde, former Deputy Director of the Laboratory, who were very generous with help and encouragement in conducting the conference and getting up this volume. As required by the Geological Society of America, all the manuscripts were reviewed by independent reviewers, to each of whom my thanks are due.

For the sake of uniformity and for producing this volume in camera-ready form, all the typing was done at the Lawrence Berkeley Laboratory, a job of no mean proportions. Liberal use was made of the laboratory's facilities and resources, with support drawn from Geothermal, Waste Isolation, and Geosciences projects sponsored by the U.S. Department of Energy. But for such valuable support, this volume cannot have materialized.

The manuscripts were technically edited by Ellen Diamond and Dick Johnson. In addition to doing a lot of typing herself, Cheri Plumlee also cheerfully spent a lot of time in coordinating the typing activities shared by Lois Armetta, Gloria Mitchell, and Jeanette Grant. The highly mathematical nature of some of the papers posed a challenge to the technical editors and the typists. The Conversion Tables appended to this volume were compiled by several colleagues, including Thomas Doe, Keshav Goyal, Brian Kanehiro, Kenji Karasaki, Don Mangold, and Robert Sterrett.

I would like particularly to thank R. A. Freeze for his help and advice in regard to editorial aspects. My thanks are also owed to my wife Vijayalakshmi, who with patience and good humor bore with me during the long hours spent at home in compiling this volume.

Finally, thanks to you, Paul, for providing the motivation for all of this.

January 1981 T. N. Narasimhan

I. Conceptual Aspects

Geological Society of America
Special Paper 189
1982

Physics of Saturated-Unsaturated Subsurface Flow

T. N. NARASIMHAN
Earth Sciences Division
Lawrence Berkeley Laboratory
University of California
Berkeley, California 94720

ABSTRACT

Shallow ground-water systems that communicate with the atmosphere are of great interest in the fields of hydrogeology, soil physics, and geotechnical engineering. Under certain circumstances, such multiphase systems can be approximated as saturated-unsaturated complexes in which the prime focus is the water. In general, saturated-unsaturated porous media are characterized by transient fluid motion in conjunction with deformation and desaturation phenomena. This paper identifies, to the extent of present knowledge, the observable physical parameters that characterize saturated-unsaturated systems, and places the parameters in mutual cause-effect relationships. The paper then examines the adequacy of the conceptual framework for studying field problems.

INTRODUCTION

Our interest in the science of hydrogeology stems from the followng facts: ground water is a valuable source of fresh water for biological consumption; ground-water reservoirs can provide valuable space for storing fluids and energy; and flowing ground water profoundly alters our environment by influencing the stability of geologic systems and transporting dissolved chemical constituents. In studying the physics of saturated-unsaturated ground-water flow, we have two goals: to identify those physical quantities that characterize the ground-water system as both a resource and a modifier of the environment, and to establish the cause-effect relationships between such quantities.

This paper considers saturated-unsaturated ground-water flow and evaluates our current knowledge of its physics. The phrase "saturated-unsaturated" flow is commonly understood to denote a system which has three phases (solid, water, and air), of which the air phase is continuous and at constant pressure. The system is assumed to be under isothermal conditions. In nature, such a system may exist only in an approximate sense. In reality, the natural shallow ground-water system should be considered as a multiphase system, with each phase coexisting with the others subject to the physical laws of their interaction. Yet, over the past several decades soil physicists have found that the shallow ground-water system can often be studied as a saturated-unsaturated complex. In this paper we will take this approximate approach. Broadly, we will examine the physics of transient ground-water movement in saturated-unsaturated systems. The discussion will begin with a description of the field problem and the relevant parameters that define it, along with the identification of a conceptual framework in which the parameters can be mutually interrelated. We will then proceed to examine how these parameters are actually defined and measured and evaluate how useful they are as currently defined and used. The study will be restricted to the movement of water. No consideration will be given to solute transport.

THE FIELD PROBLEM

In the most general sense, the field problem is concerned with the nature, occurrence, and movement of ground water in a geologic system which has an interface with the atmosphere. Depending on the specific questions studied, the field may vary in size from a few metres to a few kilometres. The interior of the flow region may be composed of more than one material with either well-defined individual properties or properties grading from one material to another.

To understand these complex systems, we must ask the following questions: How much ground water is stored in the system at any given time? How will this quantity of stored water change with time due to natural causes and due to specific exploitation activities? What is the spatial distribution of fluid potentials and how will the distribution change with time? How will the temporal variation of potentials influence the existing state of in situ rock stresses, and what will be the consequent displacements? And, what is the direction and magnitude of ground-water movement in different parts of the system and how will it vary with time? Note that the direction and magnitude of ground-water movement governs the geochemical evolution of the ground water system.

The answers to these questions must be gleaned from the few physical quantities that can be measured in the field. From what we already know, it is obvious that we must be able to measure the following properties in the field: moisture content θ (usually volumetric); fluid potential ϕ; in-situ effective rock stresses σ'; and groundwater flux $\vec{q}$ usually volumetric). Of these, θ can sometimes be measured in the field using borehole-geophysics or other indirect techniques. The in-situ rock stresses can occasionally be measured with some effort; and the direction and magnitude of ground-water flow $\vec{q}$ can only be indirectly estimated in the field with carefully controlled experiments. At present, the only quantity that can be observed with any degree of precision in the field is fluid potential ϕ, or equivalently, p or ψ. Fluid potential measurements can be made either by measuring water levels in piezometers, tensiometers, or wells, or by measuring fluid pressures in wells using precise pressure gauges. Fluid potential data constitute by far the greatest portion of available ground-water data and thus constitute the key physical information used to express the other fundamental quantities such as ρ, θ, $\vec{q}$, and σ'.

The relationships among these quantities are governed by the size, shape, and material makeup of the system as a whole. The relevant system parameters are the geometry, initial condition at a reference time, and the nature of communication between the system and its surroundings. In order to relate the fundamental parameters and the system parameters, a conceptual framework is needed. The basis of the conceptual framework is the rather intuitive law of mass conservation, which states that matter can neither be created nor destroyed. Specifically, the relevance of the conservation law to the ground-water problem lies in the fact that if we sum the mass of water entering or leaving the ground-water system or a portion thereof, the net difference in mass has to be absorbed by the system or a portion thereof. At this fundamental level, we are indeed concerned with the content of water in the system or, equivalently, the parameter, θ. However, as mentioned previously, the most easily observable physical quantity in the field is ϕ, not θ. Hence, the conceptual framework needs a second basis, namely, a fundamental link between θ and ϕ. These two bases, namely the law of mass conservation and the postulated connection between θ and ϕ, are fundamental to all the following theoretical discussions on the physics of saturated-unsaturated ground-water flow.

STORAGE

The first important question is, what is the mass of water stored in the system at any given time? The easiest, intuitive way of evaluating this is to divide the system into a convenient number of non-overlapping subdomains, evaluate the mass of water contained in each, and sum up all the quantities. The mass of water contained in any such subdomain is governed by three fundamental features: the amount of voids available, the magnitude of water saturation, and the mass density of water. Thus

$$M_{w,\ell} = V_{v,\ell}\, S_{w,\ell}\, \rho_{w,\ell}, \tag{1}$$

where $M_{w,\ell}$ is the mass of water, $V_{v,\ell}$ is the void volume, $S_{w,\ell}$ is the volumetric water saturation of the void, and $\rho_{w,\ell}$ is the water density with reference to ℓ^{th} subdomain of the flow region (Narasimhan and Witherspoon, 1977). Furthermore, $\ell = 1, 2, 3, \ldots, N$, where N is the total number of such subdomains in the system. Obviously,

$$M_{w,sys} = \sum_{\ell=1}^{N} V_{v,\ell}\, S_{w,\ell}\, \rho_{w,\ell}, \tag{2}$$

where $M_{w,sys}$ = the total mass of water contained in the system.

At this point, we must recognize that $V_{v,\ell}$, $S_{w,\ell}$, and $\rho_{w,\ell}$ are extremely difficult to measure directly in the field. Therefore we postulate, based on empirical knowledge, that $V_{v,\ell}$, $S_{w,\ell}$ and $\rho_{w,\ell}$ are all functionally related to fluid pressure p or fluid pressure head ψ. It is customary in hydrogeology and soil physics literature to express potential as energy per unit weight and to deal with potential or pressure in terms of equivalent height of a water column. Therefore, we shall henceforth recognize that $p \approx \rho_w g\psi$ where g is gravitational constant and ψ is the pressure in excess of atmospheric, expressed as equivalent height of water column.

Pressure Head and Density of Water

Of the three fundamental features that govern storage, the easiest to set forth is perhaps the relation between ρ_w and ψ, which is otherwise known as the equation of state for water. It is known from extensive experimental studies that at constant temperature, the density of water can be fairly accurately approximated by

$$\rho_w = \rho_{w,o} \exp\,[\beta(p - p_o)]\ , \tag{3}$$

where $\rho_{w,o}$ is water density at atmospheric pressure, p_o, and β is compressibility of water. It is also known empirically that β is very weakly dependent on p. Considering that $p \approx \rho_{w,o} g\psi$, and assuming that p is gauge pressure, defined as the excess pressure over atmospheric pressure (zero gauge pressure), we can rewrite equation 3 as

$$\rho_w = \rho_{w,o} \exp\,(\beta\rho_{w,o} g\psi)\ . \tag{4}$$

Pressure Head and Water Saturation

Extensive experimental data from the soil physics literature lend credence to the postulate that S_w is functionally related to ψ in naturally occurring geologic materials, although such a dependence has to be established experimentally for each soil. When ψ is greater than atmospheric pressure, the soil is in a state of complete saturation with reference to water. In such a case, ψ can be easily measured as the height to which water rises in a manometer from the elevation of the point of measurement.

As the fluid pressure in the soil drops below atmospheric, air will attempt to enter the pore spaces and displace the water. However, depending on the pore diameter, specific surface, mineralogy and other properties of the soil, air will not displace water until ψ decreases below a critical value called the "air-entry" value. Once that critical value is overcome, air invasion will cause a progressive decline in saturation with declining pressure head. Although we shall treat this relation in detail in a later section, it is sufficient to point out here that $S_w \leq 1$, for $\psi \leq \psi_A$ and $S_w = 1$ otherwise. When ψ is negative, the resulting pressure is usually termed "moisture suction" or "moisture tension." Soil moisture suction is usually measured in moist media by means of a device called a tensiometer. Although the functional dependence of S_w on ψ is quite complex, it is a practically measureable relation and hence, $S_w = S_w(\psi)$.

Pressure Head and Void Volume

The functional relation between ψ and V_v is normally determined in an indirect manner. As we shall presently see, at a given spatial location within the system, we can treat V_v directly as a function of ψ. In order that the ψ versus V_v relation be understood properly, it is essential to study the mechanisms that control the disposition of the rock skeleton. For our present discussion, let us consider a porous medium, although the concepts discussed could be extended to fractured media.

The relation between the void volume and bulk volume of a porous medium is strictly governed by the stresses borne by the matrix skeleton. Consider a small subdomain of the ground-water system. It is subjected to tectonic stresses and overburden stresses, which tend to compress the subdomain. These may be collectively called the total stress and can be symbolically expressed by the total stress tensor, σ_{ij}. Opposing these external stresses within the pores is the pore-fluid pressure, which is a scalar quantity. Since total stress and pore-water pressure oppose each other, the net stress borne by the skeletal matrix (σ'_{ij}) is directly related to the difference between total stress and pore-fluid pressure. Because the skeletal stresses cannot be easily measured in situ in a water-filled medium, and since pore pressure acts only on some portions of the solid surface, various definitions have been proposed in the literature to relate σ_{ij}, σ'_{ij}, and ψ. For example, the following general relationship has been proposed (Garg and Nur, 1973; Robinson and Holland, 1970),

$$\sigma'_{ij} = \sigma_{ij} - \alpha \rho_w g \psi \delta_{ij} \ , \tag{4}$$

where α is a coefficient ≤ 1 and δ_{ij} is kronecker delta. The coefficient α is sometimes termed boundary porosity.

If we restrict our attention to an isotropic elastic porous material, its volume may be treated as a function of its normal stresses only. Then, void volume can be related reasonably uniquely to the mean principal effective stress by the relation

$$\sigma'_m = (\sigma'_{11} + \sigma'_{22} + \sigma'_{33})/3 \ , \tag{5}$$

where σ'_{11}, σ'_{22}, and σ'_{33} denote the three principal effective stress components. Therefore, if the relationship between σ'_m and V_v for a given material is experimentally established, and if we can determine σ'_m in the field using equation 4, then we can indirectly compute V_v. The value of V_v so computed can then be used in equation 1. The aforementioned relationship implies that volume change is caused only by changes in normal stress, and that deviatoric stresses do not cause volume change. This type of behavior may not be valid for deep-seated rocks, which are known to exhibit shear-enhanced volume strains in the laboratory (Ahmed Abou-Sayed, Terra Tek, written commun.).

A useful field approximation widely used in soil engineering is the assumption that in geologic systems with large lateral/vertical dimension ratios, one could conveniently neglect changes in lateral stresses and strains. In this case, V_v can be related uniquely to the vertical stress $\sigma'_{11} = \sigma'_v$. Furthermore, for a fully saturated soil, α in equation 4 is defined as equal to unity. This gives rise to the well-known effective stress law attributed to Terzaghi,

$$\sigma' = \sigma'_v = \sigma_v - \rho_w g \psi \ . \tag{6}$$

In the soil mechanics field it is a commonplace to establish V_v versus σ' relationships through uniaxial compression tests.

The effective stress law becomes increasingly complicated when the soil is partially saturated. In this case, moisture suction acts only over that part of the solid area which is in contact with water. To account for this, Bishop (1960) proposed the relation,

$$\sigma' = \sigma - [p_a - \chi (p_a - p] \ , \tag{7}$$

where p_a is the air pressure. If we neglect the change in the density of water in shallow ground-water systems, and assume $p \approx \rho_w g \psi$, then

$$\sigma' = \sigma - \rho_w g \ [\psi_a - \chi (\psi_a - \psi)], \tag{8}$$

in which $\chi = 1$ for $\psi \geq \psi_a$, and $\chi \leq 1$ for $\psi < \psi_a$. Further, if we assume that p_a and ψ_a denote zero gauge pressure and zero gauge pressure head respectively, then equation 8 simplifies to $\sigma' = \sigma - \chi \rho_w g \psi$ (Narasimhan and Witherspoon, 1977). In general, χ is a strong function of S_w when $S_w < 1$.

The parameter χ conceptually satisfies the need for a coefficient to account for the fact that pore-pressure change is only partly convertible into effective stress in the unsaturated zone. However, in a quantitative sense, considerable uncertainty exists about χ when saturation is less than one (Mitchell, 1976). Recent work by Fredlund and Morgenstern (1977) suggests that χ may have to be treated as a function of ψ as well as σ. Let us assume for simplicity that χ is a function only of ψ.

Computing $M_{w,\ell}$. In light of the above, if we know ψ_ℓ and σ_ℓ over the subdomain ℓ, the equation of state for water, and the empirical relationships between ψ and S_w and between σ' and void volume, then we can independently evaluate $V_{v,\ell}$, $S_{w,\ell}$ and $\rho_{w,\ell}$. Substituting these quantities into equation 1 directly yields $M_{w,\ell}$.

CHANGE IN STORAGE

In nature, there seldom exists a ground-water system under complete equilibrium. Due to the movement of ground water, however slow it might be, the quantity of fluid stored in a system constantly changes. Simultaneously, pore pressures change since pore pressure and storage are related phenomena. Only two possible mechanisms exist by which the mass of water $M_{w,\ell}$ contained in a subdomain ℓ can change. The first of these is the entry or exit of water across the surface Γ_ℓ which bounds and encloses the subdomain ℓ. The second is the "source" phenomenon by which mass is spontaneously removed from or added to the subdomain by causes external to the physical system.

To describe the physics of transient ground-water flow it is essential to introduce a concept describing the capacity of a material to change its stored fluid content. The amount of water mass that will be released from storage from the subdomain ℓ due to a unit change in ψ shall be defined as its fluid mass capacity, $M_{c,\ell}$.

In other words, $M_{c,\ell}$ is the derivative of $M_{w,\ell}$ with reference to ψ. As will become obvious later, $M_{c,\ell}$ is the vital link which relates the rate of change of mass in a volume element to the more readily observable rate of change of potential. Applying the chain rule of differentiation to equation 1, we obtain

$$M_{c,\ell} = \frac{dM_{w,\ell}}{d\psi} = V_{v,\ell} S_{w,\ell} \frac{d\rho_w}{d\psi} + S_{w,\ell}\, \rho_{w,\ell} \frac{dV_{v,\ell}}{d\psi} + V_{v,\ell}\rho_{w,\ell} \frac{dS_{w,\ell}}{d\psi} \quad (9)$$

(Narasimhan and Witherspoon, 1977). In view of equation 4 we can see that $d\rho_w/d\psi = \rho_{wo}\, \rho_{w,\ell}\beta g$, and in view of equation 8,

$$\frac{d\sigma'}{d\psi} = -\rho_{w,\ell} g[\chi + \psi \frac{d\chi}{d\psi}] \equiv -\rho_{w,\ell} g\chi^*, \quad (10)$$

$$\text{where } \chi^* = [\chi + \psi(d\chi/d\psi)] \; . \quad (11)$$

We can now simplify the expression for fluid mass capacity,

$$M_{c,\ell} = \rho_{w,\ell}\, [V_{v,\ell} S_{w,\ell} \rho_{wo} \beta g - S_{w,\ell}\rho_{w,\ell} g\chi^* \frac{dV_{v,\ell}}{ds'} + V_{v,\ell} \frac{dS_{w,\ell}}{d\psi}] \; . \quad (12)$$

It is obvious from equation 12 that to compute $M_{c,\ell}$ we need, among other things, the quantities $V_{v,\ell}$ and $dV_v/d\sigma'$, corresponding to the existing state of effective stress. For simplicity and ease of reporting, laboratory data relating V_v and σ' are, as a rule, recorded in some volume-normalized form. The normalizing volume, however, is not unique and it could be bulk volume (V_b), solid volume (V_s) or, rarely, void volume (V_v). The choice depends on how one elects to define the subdomain ℓ.

Definition of Volume Element

If one defines the subdomain ℓ (synonymous with volume element ℓ) as always having constant bulk volume in the so-called Eulerian fashion, then one has

$$V_{v,\ell} = V_{b,\ell} n_\ell \quad (13)$$

and

$$-(dV_v/d\sigma') = V_{b,\ell} m_v \; , \quad (14)$$

in which n is porosity, defined as $n = V_v/V_b$, and m_v is volumetric compressibility, defined as $m_v = -(dn/d\sigma')$.

On the other hand, if one chooses to define ℓ as a material volume element (the so-called Lagrangian element), then one has

$$V_{v,\ell} = V_{s,\ell} e_\ell \quad (15)$$

and

$$-(dV_v/d\sigma') = V_{s,\ell} a_v \; , \quad (16)$$

in which e is void ratio defined as $e = V_v/V_s$ and a_v coefficient of compressibility, defined as $a_v = -(de/d\sigma')$.

The choice of normalizing volume element is related very much to the way one chooses to handle the equation of ground-water motion, an aspect which we will consider in detail below. Depending on the choice of the normalizing volume, $M_{c,\ell}$ can be written in either of two ways:

with V_b as the normalizing volume,

$$M_{c,\ell} = V_{b,\ell}\rho_{w,\ell}(n_\ell S_{w,\ell}\rho_{w_o} \beta g + S_{w,\ell}\rho_{w,\ell} g\chi^* m_v + n_\ell \frac{dS_{w,\ell}}{d\psi} \quad (17)$$

or with V_s as the normalizing volume,

$$M_{c,\ell} = V_{s,\ell}\rho_{w,\ell}(e_\ell S_{w,\ell}\rho_{w_o}\beta g + S_{w,\ell}\rho_{w,\ell}g\chi^* a_v + e_\ell \frac{dS_{w,\ell}}{dc} . \tag{18}$$

Specific Storage

Although $M_{c,\ell}$ denotes the ability of an arbitrary element to yield water from storage after a unit change in ψ, it is sometimes convenient to normalize $M_{c,\ell}$ with reference to the bulk volume $V_{b,\ell}$ of the volume element. Thus, one could define a material property coefficient, specific fluid mass capacity, by the relation

$$m_c = M_{c,\ell}/V_{b,\ell} \tag{19}$$

The expressions for m_c corresponding to equations 17 and 18 can be written as

$$m_{c,\ell} = \rho_{w,\ell}\left[n_\ell S_{w,\ell}\rho_{w_o} g\chi^* m_v + n_\ell \frac{dS_{w,\ell}}{d\psi}\right] , \tag{20}$$

where V_b is the normalizing volume, or

$$m_{c,\ell} = \rho_{w,\ell}\left[n_\ell S_{w,\ell}\rho_{w_o} g\chi^*(1 - n_\ell)a_v + n_\ell \frac{dS_{w,\ell}}{d\psi}\right] , \tag{21}$$

where V_s is the normalizing volume. In arriving at equation 21, the following relations have been used:

$V_s/V_b = 1/(1 + e)$; $e/(1 + e) = n$;

and $1/(1 + e) = (1 - n)$.

The concept of specific fluid mass capacity is particularly useful when one attempts to determine it directly through transient drainage experiments such as interference tests between wells. Indeed, as we shall see presently, the concept of specific storage and the related concepts of storage coefficient and storativity are directly related to m_c.

Let us now restrict our attention to a porous medium fully saturated with water. In this case $S_w = 1$, $\chi^* = 1$, and $dS_w/d\psi = 0$. Furthermore, let us ignore changes in density of water and consider m_c in a volumetric sense rather than in the sense of mass. Such a volumetric coefficient of specific fluid mass capacity is called specific storage or storativity. Thus, if we consider equation 20, specific storage can be written as,

$$S_s = m_c/\rho_w = \rho_{w_o} g[nb + m_v] . \tag{22}$$

Or, if we consider equation 21,

$$S_s = m_c/\rho_w = \rho_{w_o} g[nb + (n - 1)a_v] . \tag{23}$$

Note that in equation 23, V_s is chosen as a normalizing volume to express void volume, and V_b is chosen as the normalizing volume to express S_s. However, in equation 22, V_b is the normalizing volume in both cases. In rare instances, especially in the petroleum literature (Earlougher, 1977), the quantity $dV_v/d\sigma'$ is normalized with reference to the void volume. Thus the coefficient, pore volume compressibility, c_f, is defined as

$$c_f = -\frac{dV_v/V_v}{d\sigma'} . \tag{24}$$

By simple transformation one can show that $c_f = m_v/n$ and hence S_s given in equation 22 can also be written as

$$S_s = \rho_{w_o} g[n\beta + nc_f] = \rho_{w_o} gn\,[\beta + c_f] . \tag{25}$$

The sum $(\beta + c_f)$ is sometimes designated as total compressibility, c_t (Earlougher, 1977). Note that the water released from storage includes expansion of fluid and contraction of pores or vice versa. Thus the fluid and pores are deforming in opposite directions, although with respect to fluid release, the two effects are additive. Hence, from the point of view of deformation, the term total compressibility has to be viewed with caution.

The above definitions of specific storage differ somewhat from the earliest definition of storage coefficient introduced in the well-testing literature (Ferris and others, 1962). In this restricted definition, coefficient of storage is the volume of water released from a prism of the aquifer of unit base area due to a unit change in ψ normal to the aquifer. This earliest definition was clearly tailored to suit the analysis of an infinite homogeneous aquifer using a partial differential equation.

Nature of Loading

Physically, S_s signifies the volume of water released from storage per unit bulk volume of the material due to a unit change in ψ. The fact that S_s is a measure of the volume of water released from storage directly suggests that the accompanying skeletal deformation is caused by the drainage phenomenon. In other words, the problem of transient ground-water flow is accompained by a "drained" loading of the porous medium as opposed to an "undrained" loading. A fundamental feature of drained loading is that the porous medium is stressed from within (endogenous loading). For this type of loading, the effective stress relation embodied in equation 4 appears especially reasonable. Under conditions of drained loading, a drop in pore pressure causes expansion of water and reduction in pore volume, both of which make up the quantity of water drained.

In undrained loading, there is no change in the mass of water contained in the system. Nevertheless, as the external load seeks to reduce the pore volume, the consequent compression of water causes an opposing hydrostatic stress in the form of increased pore pressure. The net result is that the external load is shared by the skeleton and the water in such a manner that the change in pore volume of the skeleton matches the change in volume of the contained water.

For a detailed discussion of drained and undrained response of soils, the reader is referred to a soil mechanics text (Lambe and Whitman, 1969). A discussion of storage from the hydrogeology viewpoint can be found in Domenico and Mifflin (1965).

The principal differences between drained and undrained loading are as follows. In the former, total stress remains constant; change in effective stress equals change in pore-fluid pressure; change in void volume is not equal to change in water volume; and the combined change in void volume and water volume is equal to the water drained. In undrained loading, total stress is variable; change in effective stress is not equal to change in water pressure, in general; and change in void volume equals change in water volume.

MOTION OF GROUND WATER

In its natural, dynamic state, ground water is constantly in motion. Fundamental to our physical understanding of transient ground-water flow is knowledge of the rate at which water is transferred from one subdomain to its neighbor. In order to study the nature of ground-water motion, we postulate that water moves from a site of "higher" state to a site of "lower" state. Hubbert (1940) identified this "state" as the energy required to move a unit mass of water from a reference site to the site of interest, whereas hydrogeologists and soil-physicists treat it as energy per unit weight of water (Philip, 1969).

In general, ground water possesses a variety of energy forms: gravity, elastic, kinetic, thermal, electrochemical, and so on. The spatial gradients of each of these energy components impel ground water. Not all these impelling forces, however, may cause ground water to move in the same direction. The net ground-water movement is the result of the movements due to each different energy form. In practice, most ground-water systems are dominated by their components of gravitational and elastic energies. Kinetic energy is invariably small and neglected. In certain special systems, the content of thermal and chemical energy may be important. For our present considerations, we will restrict our discussion to the two forms: gravity and elastic energy.

Fluid Potential

Fluid potential, also called force potential (De Wiest, 1969), and energy potential (Nelson, 1966) may be defined as a scalar quantity whose spatial gradient is directly proportional to the velocity of ground-water flow. Following the investigations of Buckingham (1907) and Hubbert (1940), fluid potential for isothermal saturated-unsaturated flow may be defined as the amount of potential energy and elastic energy possessed by a unit mass of water. As described by Hubbert (1940), it is the energy required to move a unit mass of water from a given elevation and pressure to a different elevation and pressure within the flow region. By normalizing potential with reference to gravity, potential can also be defined as energy per unit weight of water (Philip, 1969). However, potential cannot in general be defined as energy per unit volume of water (Nelson, 1966). A basic requirement for the potential to exist is that it be path-independent (De Wiest, 1969; Nelson, 1966). According to Hubbert (1940), fluid potential can be expressed as

$$\Phi = gz + \int_{p_o}^{p} \frac{dp}{\rho_w(p)} + C_\ell \, , \tag{26}$$

where z is elevation, p is pressure, and c_ℓ is a constant. For equation 26 to be true, ρ_w must be a function only of pressure. In other words, equation 26 cannot hold for nonisothermal systems or systems with strong chemical gradients, in which ρ_w may be a significant function of temperature and chemical concentration. For such problems, a unique potential does not exist.

The potential Φ in equation 26 has dimensions of energy per unit mass. Equivalently, one could define a potential that has units of energy per unit weight (length) by dividing both sides of equation 26 by g. Thus, one obtains

$$\phi = z + \psi \, , \tag{27}$$

where ϕ is the usually defined hydraulic head.

Darcy-Buckingham Law

The most widely used equation of motion in transient ground-water flow studies is based on the original experimental observations made by Henry Darcy in 1858. Whereas Darcy studied fully saturated sands, Buckingham (1907) extended the equation of motion to partially saturated soils. Hence it is often called Darcy-Buckingham Law (for example, Swartzendruber, 1969). Darcy's observations were restricted to steady-state flow through a sand filter involving laminar (nonturbulent) flow of a Newtonian fluid. (A Newtonian fluid is one whose viscosity is independent of rate of shear and time.) Under these conditions, volumetric fluid flux across a surface segment in the porous medium is directly related to the gradient of hydraulic head and is given by

$$\vec{Q} = - K \frac{d\phi}{dx} A \, , \tag{28}$$

where Q is volumetric flux rate, x is the direction

normal to the surface, A is the area of the surface segment, and the negative sign denotes flow in decreasing potential. In a saturated, homogeneous, isotropic medium, the quantity K is independent of orientation and is usually called hydraulic conductivity. The quantity $\vec{q} = \vec{Q}/A$ is widely used in the literature and is designated variously as specific flux or Darcy velocity. A flow phenomenon in which $\vec{q}$ is directly proportional to the first power of potential gradient is sometimes known as Darcian flow. Field experience suggests that under certain conditions of turbulent flow (flow in certain fractures, flow in the vicinity of high-production wells) $\vec{q}$ may have to be treated as proportional to some higher power of potential gradient. Such flows have been termed non-Darcian flow (Ramey, 1965). Current evidence suggests that Darcy's law is valid in saturated clays down to very low hydraulic gradients (Olsen, 1965). In saturated clays, however, some deviations from Darcian flow may sometimes be observed due to electrokinetic effects and gradients of chemical concentrations (Mitchell, 1976). It must be pointed out here that the Darcy-Buckingham law is strictly valid only for steady-state flow. In extending the law to the transient problem, it is generally assumed that Darcy velocity is relative to the velocity of the solids (Philip, 1969).

Richards (1931) who recognized the similarity between capillary potential proposed by Buckingham and hydraulic head introduced by Darcy, suggested the general form of the equation of motion, which unifies the saturated and the unsaturated flow regimes,

$$q = - K(\psi)\nabla(z + \psi) \;, \tag{29}$$

where $K(\psi)$ denotes the dependence of K on ψ.

Hydraulic Conductivity and Intrinsic Permeability

The constant of proportionality, K, is actually a lumped parameter involving the properties of the porous medium and the fluid. The component of K that is exclusively related to the porous medium is variously called "permeability," "absolute permeability" or "intrinsic permeability," k. According to Hubbert (1940), the relationship between the two can be shown to be

$$K = k\rho_w g/\mu \,. \tag{30}$$

The quantity k is a measure of the ease with which a saturated porous medium can transmit a fluid regardless of the property of the fluid.

Intrinsic permeability is a function of the mean pore diameter, specific surface, and tortuosity of the porous medium. Additionally, when more than one fluid phase is present, permeability to a given fluid is a function of its saturation. Such a dependence on saturation has given rise to the parameter "relative permeability," k_r, in the petroleum literature. Relative permeability is defined as the ratio of the permeability at a given saturation to the saturated permeability, k. Clearly, $0 \leq k_r \leq 1$. However, in the consideration of saturated-unsaturated flow with the assumption of a continuous air-phase at atmospheric pressure, the term relative permeability is seldom used. Instead, k is simply treated as a function of water saturation, S_w. Inasmuch as pore geometry and structure is a function of in-situ skeletal stresses σ' and since S_w is a function of pressure head, it follows that in a most general sense k is a function of total stress, σ as well as pressure head, ψ. Therefore $k = k(\sigma,\psi)$. However, if we consider a specific location in the flow domain where σ is a constant, then k can be treated purely as a function of ψ.

In view of equation 30, the Darcy-Buckingham law is usually written as

$$\vec{q} = \frac{k\rho_w g}{\mu} \nabla (z + \psi) \;. \tag{31}$$

If one wishes to use potential in the sense of energy units per unit mass, as is frequently done in the petroleum literature, the equation of motion can also be written equivalently as

$$q = (k/\mu)(\rho_w g\nabla Z + \nabla p) \;. \tag{32}$$

The quantity k/μ is sometimes designated "mobility."

Tensorial Nature of Permeability

The Darcy-Buckingham law as given in equation 31 implies that the direction of the specific flux vector coincides with the direction of the gradient of hydraulic head. This is true only under certain conditions. It can be shown (Day, 1965) that, in general, the direction of macroscopic hydraulic gradient and macroscopic ground-water velocity will be different, and connected by the relation

$$F_i = \frac{\partial \phi}{\partial x_i} = A_{ij}\, q_j \;, \tag{33}$$

where F_i is the force component in the ith direction and A_{ij} is a second-order tensor. The nine components represent incremental components of viscous drag in the liquid phase per incremental change in macroscopic velocity. They also incorporate local geometry as influenced by the local boundary conditions.

Inverting equation 33, we write:

$$q_i = K_{ij}\,(\partial \phi/\partial x_j) = K_{ij} F_j \;, \tag{34}$$

where $K_{ij} = A_{ij}^{-1}$. The tensor K_{ij} is the hydraulic conductivity tensor (Maasland, 1957) whereas A_{ij} is the reciprocal hydraulic conductivity tensor. Inasmuch as K_{ij} transforms a vector F_i into a vector q_i, it is by definition a second-order tensor. Nevertheless, there is no indication that K_{ij} is symmetric, although it is generally assumed to be so. Since the directional bias embodied in K_{ij} stems purely from the geometry of the porous medium, it follows that $K_{ij} = k_{ij}\,\rho_w g/\mu$. That is, intrinsic permeability is a tensor.

If it happens that the direction of the q_i coincides with one of the principal components of the permeability tensor, then the intrinsic permeability tensor is diagonal in nature (that is, its off-diagonal terms are zero). Furthermore, if the material is isotropic, not only are the off-diagonal terms zero, but $k_{11} = k_{22} = k_{33}$. Hence, the Darcy-Buckingham law can be expressed either in the form of equation 29 or equation 34.

MASS CONSERVATION

Having established the appropriate definitions and parameters describing storage, change in storage, and motion of ground water, we are now ready to relate these to each other in the form of a fundamental equation. Consider an arbitrarily shaped three-dimensional subdomain (ℓ) of the flow region under consideration. This subdomain is bounded by the closed surface, Γ_ℓ. The chosen subdomain is small enough so that physical quantities such as fluid density, porosity, pressure, and stress do not vary rapidly within it. If so, these quantities can be represented through appropriate average values associated with the volume element. Let the average fluid velocity perpendicular to the mth surface segment of subdomain ℓ during a small interval of time $\Delta t = q_{\ell m}$. Then, the rate at which water mass increases within the element ℓ can be obtained by algebraically summing up the fluxes across all the surface segments. That is,

$$-\sum_{m=1}^{M} \rho_{\ell m} \vec{q}_{\ell m} \Delta\Gamma_{\ell m} = \frac{\Delta M_{w,\ell}}{\Delta t} , \tag{35}$$

where M denotes the total number of segments into which Γ_ℓ is divided, $\Delta\Gamma_{\ell m}$ denotes the area of the mth surface segment, and $\rho_{\ell m}$ is the water density of the surface segment m. Letting $M \to \infty$, $\Delta t \to 0$, noting the $M_{c,\ell} = dM_{w,\ell}/d\psi$, and applying equation 34, we may write for the volume element ℓ,

$$\int_{\Gamma_\ell} \rho \frac{k_{ij}\rho_w g}{\mu} \frac{\partial}{\partial x_j} (z + \psi)\, d\Gamma = M_{c,\ell} \frac{\partial \psi}{\partial t} . \tag{36}$$

In equation 36 note that the Darcy velocity term within the integral does not include the velocity of the grains since $d\Gamma$ is fixed in time. But for transient flow, Darcy velocity denotes fluid velocity relative to the grains. Strictly speaking, therefore, the left-hand side of equation 36 is an approximation although it is practically valid for most subsurface situations. Nevertheless, one could overcome this by choosing a material volume element (Smiles and Rosenthal, 1968; Narasimhan and Witherspoon, 1977) in which the volume of solids in element remain fixed in time. In such an event, the partial derivative on the right-hand side of equation 36 will be replaced by a total derivative, $D\psi/Dt$.

In light of the Darcy-Buckingham law, it is clear that water will flow into the flow region across those boundary segments where the potential on the exterior exceeds that on the interior, and vice versa. Evidently the flow region continuously strives toward an equilibrium with its natural surroundings. It follows that if there are no sources within the flow region, the maximum and minimum potentials under steady-state conditions can occur only on the boundary of the flow region. The only way that the maximum and/or minimum potential can occur in the interior of the flow region is for these quantities to occur at the initial time. Stated generally, the maximum and minimum potentials in a transient ground water flow system can occur only at the boundaries of the flow region or at the initial time. This statement is usually known as the Maximum Principle. Since a steady-state condition can be visualized as a transient problem after a period of infinite time, the maximum principle simplifies, in this case, to the statement that the maximum and minimum potentials can occur only at the boundary.

Functional Dependencies

Before we proceed further, it is pertinent to enumerate the dependencies of the various parameters occurring in equation 36. Thus, K_{ij} is a function of pore geometry as well as S_w; V_v is a function ψ, assuming that the external stresses on the element ℓ remain constant; ρ_w is a function of ψ through the equation of state for water; S_w is a function of ψ; and $M_{c,\ell}$, by definition, is a function of ψ. The geometric quantities, z, $d\Gamma$, and the gradient operator may or may not be functions of time, depending on whether one defines the volume element in a Eulerian or Lagrangian sense.

MATERIAL PROPERTIES

So far the main endeavor has been to formulate the statement about how a small subdomain within a transient ground-water system evolves from a given initial condition over a small interval of time. Presumably, application of this statement to every subdomain in the system over a large number of successive small time intervals will enable us to fully describe the evolution of the transient system in time. Our consideration of material properties has been restricted thus far to an extent that was sufficient to define that statement. However, the practical utility of the fundamental statement of transient ground water flow as embodied in equation 36 is very much dependent on the assurance with which the material properties can be determined, either in the laboratory or in the field, and on the factors that govern the material properties. We will now consider, in some detail, the nature of the different material properties.

Deformation Parameters

The deformation parameter that is most important to fluid flow is the change of void volume per unit change in mean principal stress in a suitably volume-normalized form. Through an enormous body of experimental literature it is known that naturally occurring geologic materials exhibit very complex deformation behavior. For example, the compressibility of most naturally occurring geologic materials is a function of the magnitude of effective stress, and most geologic materials retain in "memory" the maximum stress ever experienced by them (see, for example, Lambe and Whitman, 1969). Figure 1 shows the relation between void ratio, e, and effective vertical stress for a soft clay. In Figure 2, similar data are presented for a consolidated sandstone. The slope of the curve in Figure 1 shows the coefficient of compressibility of the soil, and the change in the slope of the curve as a function of σ' denotes the stress-dependence of compressibility. Both Figure 1 and Figure 2 demonstrate that volume changes non-linearly with stress. The clay in Figure 1 exhibits much higher compressibilities when stressed beyond its past maximum stress (preconsolidation stress) than otherwise. Note also in Figure 1 that the rebound and recompression curves

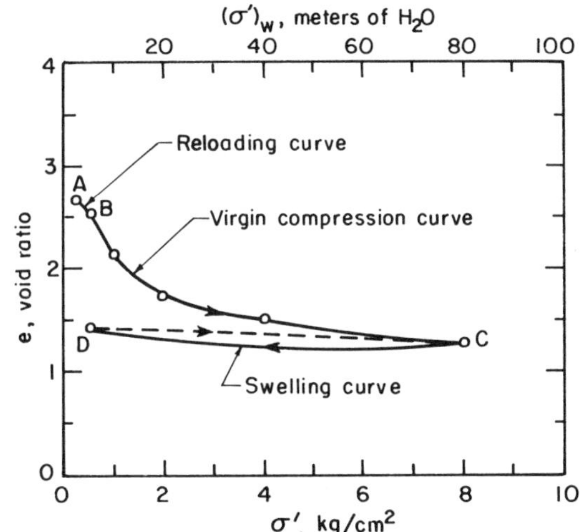

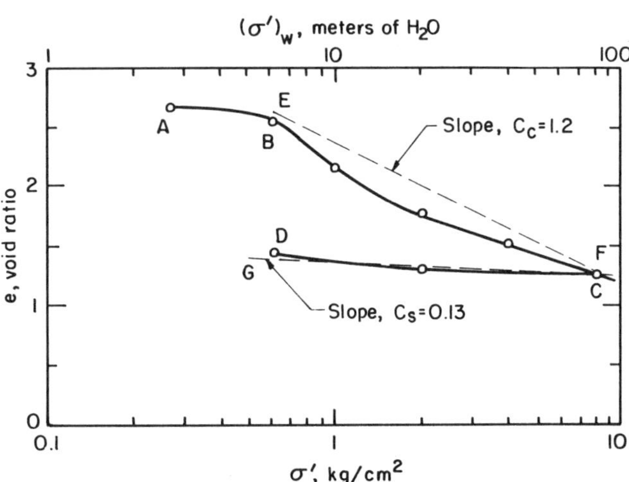

Figure 1. Relation between void ratio and effective stress for a soft clay: Top, cartesian plot; Bottom, semilog plot (data from W. N. Houston, University of California, Berkeley).

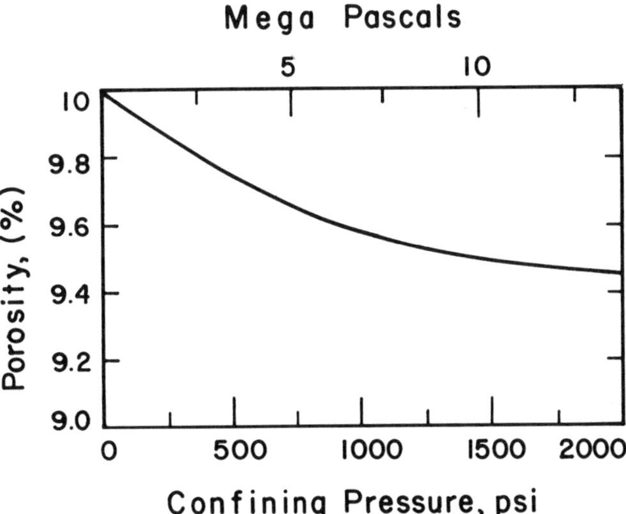

Figure 2. Relation between porosity and hydrostatic stress for a consolidated sandstone (data from Ahmed Abou-Sayed, TerraTek, Salt Lake City, Utah).

follow slightly different paths, exhibiting "hysteresis." It is obvious that between normal consolidation and subsequent rebound, a certain amount of void volume is permanently lost by the material at any given stress, causing plastic deformation. The range of compressibilities of some naturally occurring geologic materials is given in Table 1.

Effective Stress and Pressure Head

In discussing the concept of effective stress we introduced the generally-used expression, equation 8, involving the parameter χ. Intuitively, one would expect that χ is a function of S_w, which in turn is a function of ψ. The variation of χ as a function of S_w for two soils is given in Figure 3. The fact that χ is nonzero when $S_w < 1$ indicates that effective stresses will continue to change gradually during the process of pore desaturation leading to soil

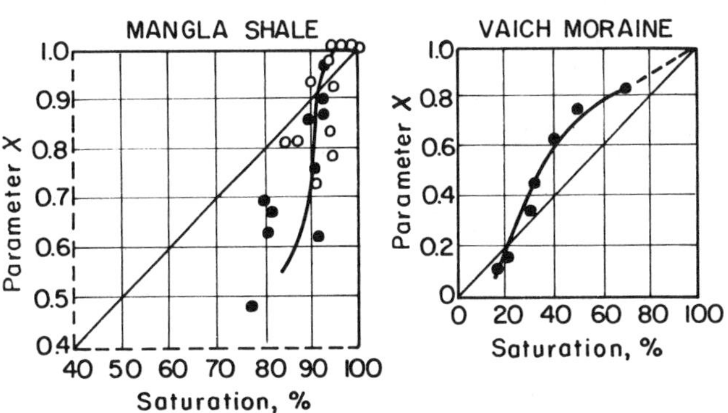

Figure 3. Relation between and saturation for two compacted soils (after Bishop and Blight, 1963).

TABLE 1. COEFFICIENT OF VOLUMETRIC COMPRESSIBILITY FOR SOME ROCKS AND SOILS*

Material	Compressibility, $cm^2/kg \times 10^{-6}$		c_s/m_v
	m_v	c_s†	
Quartzitic sandstone	5.8	2.7	0.46
Quincy granite (100 ft deep)	7.5	1.9	0.25
Vermont marble	17.5	1.4	0.08
Concrete (approx.)	20.0	2.5	0.12
Dense sand	1,800	2.7	0.0015
Loose sand	9,000	2.7	0.0003
London clay (over-consolidated)	7,500	2.0	0.00025
Gospart clay (normally con-solidated)	60,000	2.0	0.00003

*(after Skempton, 1960) Compressibilities measured at a stress of 1 kg/cm^2
Compressibility of water, $c_w = 48 \times 10^{-6}$ cm^2/kg.
† c_s = Compressibility of solid grains.

deformation. For a hypothetical sand with a given S_w versus χ relationship, Figure 4 shows the variation of σ' as a function of ψ for four different assumed χ versus S_w dependencies (Narasimhan, 1979). Because desaturation is initiated in the larger pores, and because larger pores have small specific surfaces, it can be reasoned that the χ versus S_w relationship is most likely to be convex upward, as indicated by Case 1 in Figure 4.

Saturation versus Pressure Head

The variation of S_w versus ψ is known to be a highly path-dependent (hysteretic) relation. As seen in Figure 5 for a typical case of sand, the saturation at a given pore pressure is always higher for the drying curve than for the wetting curve. Moreover, during the drying process, moisture suction has to exceed a critical air-entry or bubble-point value before the desaturation process can begin. The slope of the saturation curve $dS_w/d\psi$ when multiplied by porosity (usually assumed constant) yields $d\theta/d\psi$, which is often called soil moisture capacity. The phenomenon of saturation hysteresis is due to several causes. For example, the contact angle between the meniscus and the solid wall in small pores may change between advancing and receding phases. Also, as water moves through bottlenecks in micropores, excess capillary pressures will be needed to drive the water through the pores.

Hydraulic Conductivity and Pressure Head

Hydraulic conductivity, or equivalently, intrinsic permeability, is a function of pressure head in fully and partially saturated materials. However, since pore geometry does not change rapidly with pore pressure except when effective stresses are extremely low, the hydraulic conductivity versus ψ dependence is not as pronounced as when $\psi < 0$. The variation of hydraulic conductivity with moisture suction is usually very pronounced and characterized by hysteresis. In unsaturated soils,

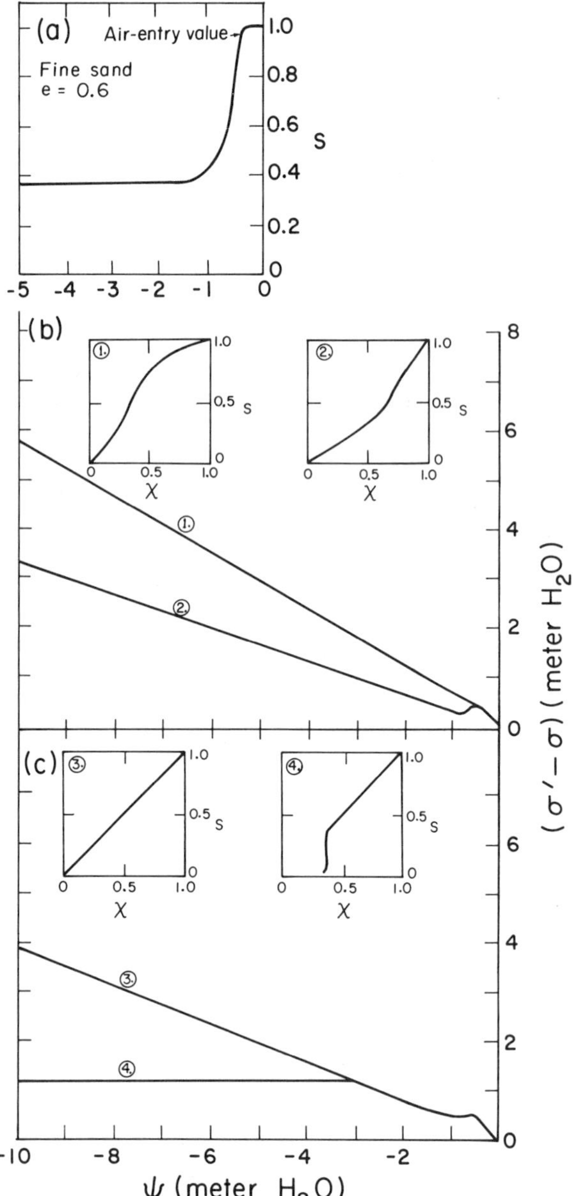

Figure 4. Variation of σ' in the unsaturated zone for a hypothetical soil: a, saturation characteristic; b and c ($\sigma' - \sigma$) versus pressure head for two assumed S_w versus ψ relations.

hydraulic conductivity may vary by several orders of magnitude between full saturation and very low saturations. The relationship between e and logarithm of K for several soils is presented in Figure 6, and the variation of K with ψ for a typical sand is given in Figure 7. The range of intrinsic permeabilities (saturated) of some naturally occurring geologic materials is summarized in Table 2 (after Davis, 1969).

Static Versus Dynamic Parameters

In defining the physical parameters, a tacit assumption was made that the relations between

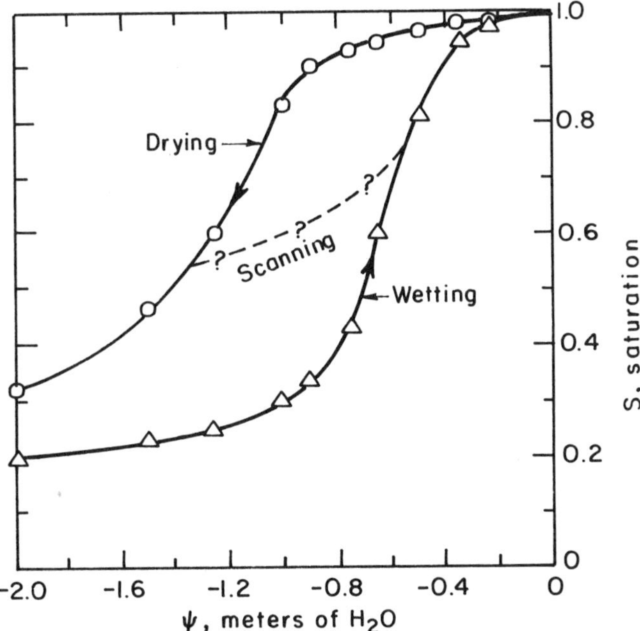

Figure 5. Relation between S_w and ψ for a sand showing hysteresis (after Liakopoulos, 1965).

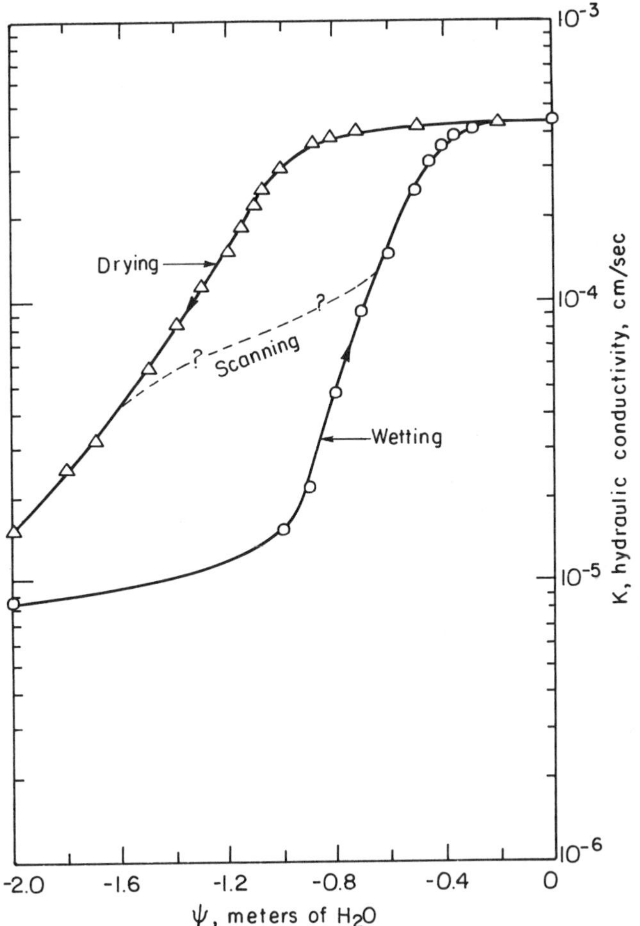

Figure 7. Relation between saturation and hydraulic conductivity for an unsaturated sand (after Liakopoulos, 1965).

saturation and pressure head, or saturation and intrinsic permeability, or void ratio and effective stress, or even between flow and hydraulic gradient, are all independent of the state of flow. Indeed, these parameters are measured in the laboratory under steady-state conditions of flow and are later used to analyze transient flow systems. As Klute (1973) points out, physical properties under unsteady flow conditions may differ from those under steady flow conditions due to several causes, not all of which are clearly understood. Complications are clearly possible in this regard.

Porous and Fractured Media

In our development of the conceptual model of transient ground-water flow, we assumed the existence of a reasonably small subdomain over which such average quantities as e, $M_{c,\ell}$, S_w, ρ_w, and ψ can be defined without difficulty.

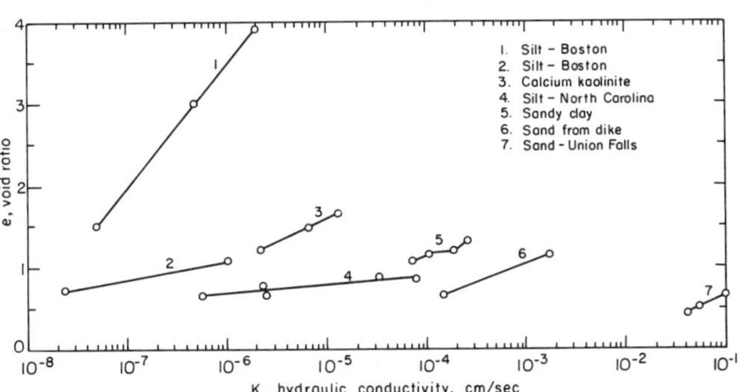

Figure 6. Relation between hydraulic conductivity and void ratio for some saturated soils (after Lambe and Whitman, 1969).

Physically, when dealing with commonplace porous media such as sands and clays, such a subdomain can easily be assumed to exist without any conceptual difficulty. However, the existence of such a subdomain can become practically unrealistic when one considers fractured and fissured media. If one were to treat such media as equivalent porous systems, a reasonable subdomain may have to have dimensions of several tens or even hundreds of metres. In many field situations, therefore, it may be more reasonable to treat such discontinuous media in a more detailed fashion, giving due consideration to the geometry and disposition of individual fractures.

Natural fractures are usually bounded by irregular walls that may be in contact with each other at certain points, called aspirities. The rate of fluid flow in such fractures is strongly a function of the fracture aperture and roughness. By considering the flow of water between two adjoining fracture elements across a common interface, one could define a pseudopermeability for a fracture. Using a simple parallel-plate model to represent a fracture, it has been proposed that the permeability of a fracture may be related to the second power of its aperture. Experimental work (Iwai,

1976) has verified the practical validity of this for plane and natural fractures. In the field, fracture apertures may vary from a few microns for incipient cracks to several metres in the case of lava tubes and limestone caverns.

Most natural fractures are deformable. The deformation characteristics of a fracture are usually expressed in terms of stiffness. Thus the fluid capacity (or specific storage) of a fracture element is directly related to its stiffness. Although fractures are usually treated in practice as elastically deformable openings, there is little reason to doubt that they may also exhibit elastic-plastic properties in much the same manner as porous media.

Uniform, Homogeneous, Heterogeneous, and Anisotropic Media

Intrinsically, all geologic materials are subject to some variability of disposition in space. Yet the fundamental parameters required to define the physics of flow in saturated-unsaturated

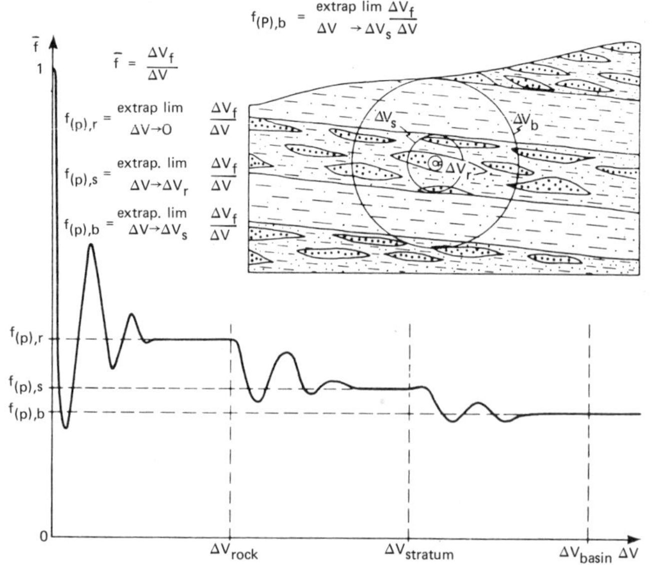

Figure 8. Porosity as a function of sample volume (after Bear, 1972).

systems require a certain minimum critical volume of the material before the parameters can be quantified with reasonable assurance. Coefficients such as porosity, void ratio, saturation, hydraulic conductivity, and permeability fall into this category. That a minimum critical volume exists in a statistical sense is illustrated schematically in Figure 8 (Toth, 1967). Note that several such minimum critical volumes can exist, depending on the scale of interest.

Suppose one were to collect a number of suitable samples from a geologic unit and were to measure a parameter of interest. If all the samples yielded the same magnitude subject only to experimental error, then the geologic unit may be called "uniform." Nevertheless, natural geologic processes seldom create uniformity, even over short distances. Rather, a number of samples collected from a given small region are most likely to show a certain variability about a mean value. Such a variability can be quantified in terms of a mean and a standard deviation, giving rise to the well-known bell-shaped distribution curves. Such being the case, it is natural to define (Freeze, 1975) a homogeneous medium as one whose frequency distribution does not change in space. By implication, therefore, a heterogeneous medium is one in which frequency distribution varies in space. The definitions of uniformity and homogeneity are schematically represented in Figure 9.

Although homogeneity and heterogeneity relate to spatial variation of frequency distribution, it is also conceivable that frequency distribution of certain properties may be dependent on orientation. If the frequency distribution is independent of orientation, the medium is isotropic; otherwise it is anisotropic.

It is obvious from the foregoing that a geologic material which is heterogeneous in one scale may be homogeneous in another. The scale in which one wants to study the physics is therefore of para-

TABLE 2. POROSITY AND PERMEABILITY OF SOME GEOLOGIC MATERIALS (from S. N. Davis, 1969)

Material	Porosity (%)	Permeability (darcies)*
Dense rocks		
Chert		1.9×10^{-7}
Dolomite	6.3	1.0×10^{-3}
Limestone	8.4	1.0×10^{-3}
Quartzite	0.6	1.9×10^{-6}
Rock salt	0.6	7.3×10^{-6}
Slate	3.4	1.3×10^{-6}
Weathered or fractured dense rocks		
Quartz-mica schist	48	3.3×10^{-2}
Metasediments	2.4	3.1×10^{-2}
Schist	3.0	1.4×10^{0}
Volcanic rocks		
Basalt, moderately dense	7.7	1.4×10^{-5}
Tuff, zeolitized	39	4×10^{-5}
Tuff, friable	36	1.4×10^{-3}
Tuff, welded	14	3.3×10^{-4}
Sedimentary rocks		
Arkose, medium grained	25.6	1.1×10^{-3}
Conglomerate	17.3	4.9×10^{-4}
Sandstone, Cretaceous		4.1×10^{-1}
Shale, Cretaceous		4×10^{-6}
Siltstone	9.7	1.2×10^{-4}
Unconsolidated sediments		
Clay		
Kaolinite	58.8	2.38×10^{-3}
Oligocene, subsurface	33.3	1.1×10^{-5}
Silt		
Miocene, subsurface	33.7	3.85×10^{-5}
Overbank deposit	47.5	5.55×10^{-2}
Sand		
Aquifer, alluvium	33.8	1.98 x 10'
Cretaceous formation		3.68 x 10'
Sand dune	35.8	2.8 x 10'
Gravel		
Sand, silt; end moraine		2.2×10^{0}
Silt, clay; colluvium	63.4	5.5×10^{-5}

*For water at 60°F: 1 darcy = 9.86×10^{-9} cm^2
= 9.66×10^{-9} cm^2
= 11.2 gpd (US)/ft^2

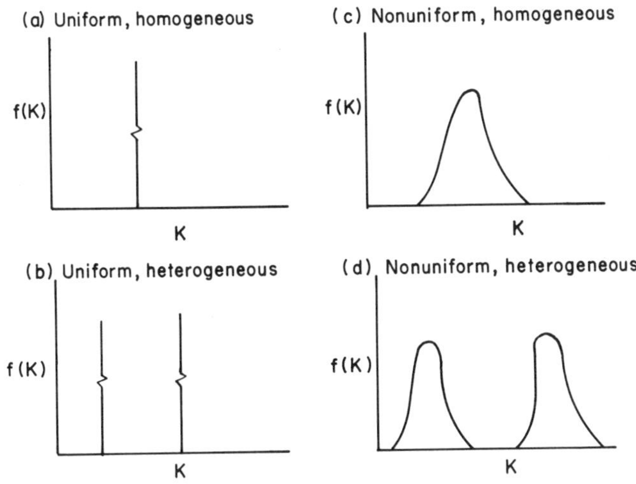

Figure 9. Frequency distribution of hydraulic conductivity illustrating definitions of uniformity and homogeneity (after Greenkorn and Kessler, 1969).

mount importance. Thus, the scale of interest may be pore-to-pore or microscopic within geologic formations, or intraformational or regional and interformational. The larger the scale, the larger the volume over which properties are to be integrated. Because measuring instruments are integrating devices, it would appear that the scale at which observations can be made in the field greatly governs the detail in which physics can be described. This is of special significance when one wants to translate physics into mathematical models to predict system behavior. In such ventures, the nature of sampling or field observation should be given appropriate consideration in empirically obtaining the coefficients that would be input for the model.

BOUNDARY CONDITIONS

The manner in which the flow region communicates with its surroundings is exemplified in the boundary conditions. Such conditions are thus assumed to be known, and may be in the form of information on variation of potential with time, flux with time, or some combination of both. Boundary conditions commonly encountered are described below.

Prescribed Potential

On a boundary segment with prescribed potential, the time-dependent variation of ϕ is assumed to be known. The most realistic physical example of this condition is the communication of the flow region with a large reservoir across the boundary segment. In this case, the prescribed condition will represent the time-dependent variation of water level in the reservoir (Fig. 10A). The prescribed variation may be constant, gradually changing with time (for example, water level in an impounded reservoir), or periodic with time (ocean tides). In the mathematics literature this type of boundary condition is sometimes called the Dirichlet condition.

Prescribed Flux

Across a prescribed flux boundary, the rate at which water is entering or leaving the flow region is known as a function of time. Although exact information on such fluxes may often be hard to obtain due to difficulties associated with flux measurement, certain idealized field situations can be represented in the form of prescribed flux boundary conditions. Perhaps the best example of this is rainfall infiltration, when rainfall intensity is less than the saturated hydraulic conductivity of the soil. In this case, the known rainfall intensity can be treated as known flux into the flow region. A prescribed flux boundary condition is schematically shown in Figure 10B. This boundary condition is generally termed the Neumann condition.

Seepage Face

A seepage face is a boundary condition that is peculiar to saturated-unsaturated ground-water systems. Suppose there is a segment of the boundary across which the flow region communicates with the atmosphere. Consider two points across this boundary that have the same elevation, z. In this case, the point on the exterior side of the boundary has $\phi = z$, since atmospheric pressure is used as a datum for ϕ. If now the flow region around the interior point is saturated ($\psi > 0$), then water can flow out across the boundary segment. On the other hand, if the flow region is unsaturated around the interior point, then no water movement will take place across the boundary since there is no water available on the exterior to maintain flow into the flow region. Such a boundary segment, on the exterior of which $\phi = z$ and across which water can only leave the flow region, is known as a seepage face (Fig. 10C). In the field, seepage faces are often encountered in earth dams, wells draining unconfined aquifers, and other man-made structures. In a natural setting, seepage faces may occur in such locations where the water table intersects a valley slope.

Evaporation and Evapotranspiration

In some field situations, required information at a boundary may only be quantifiable within certain bounds. Such conditions are characteristic of the interface between the saturated flow region and the atmosphere on the one hand (evaporation), and the plant-root system (transpiration) on the other. The amount of moisture that the atmosphere can take from the soil is equal to the sum of potential evaporation and potential evapotranspiration, both of which can presumably be estimated from micrometeorological data (Neuman and others, 1975). In addition, there also exist lower limits to the suctions that can develop at plant roots (wilting pressure of plants) and at the dry soil-atmosphere interface. Thus, at the soil-atmosphere boundary neither potential nor flux is exactly

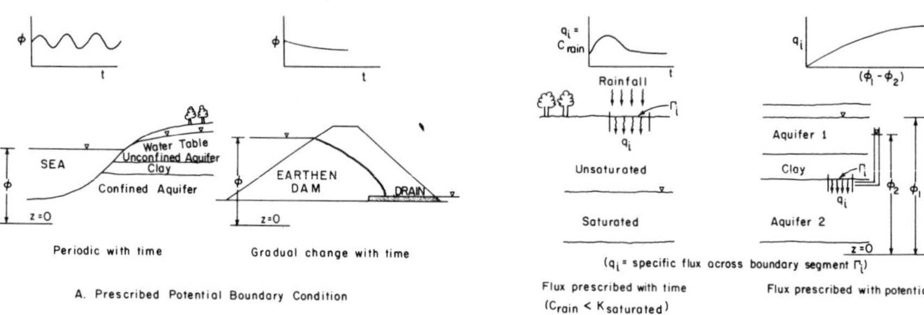

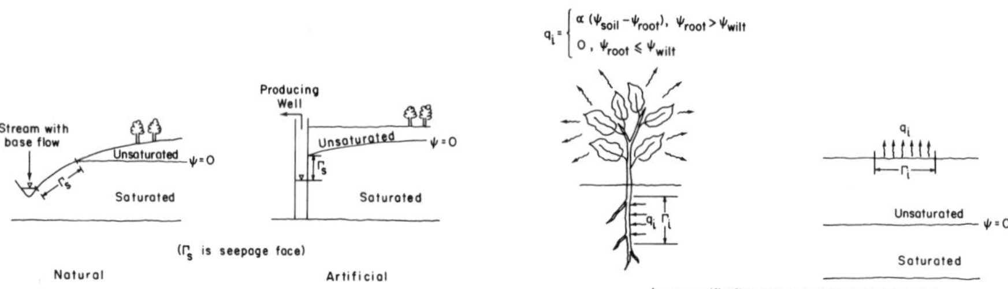

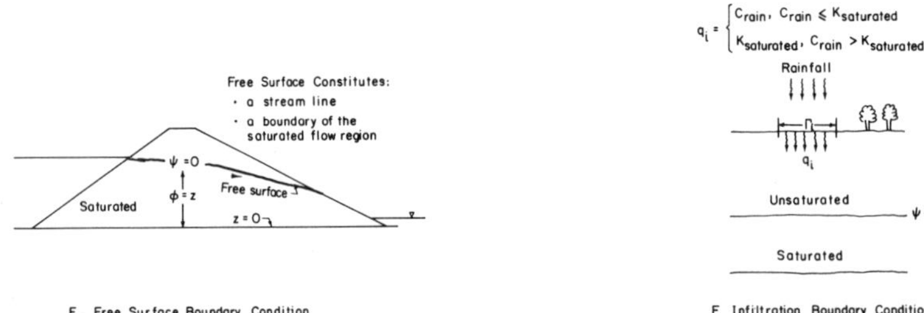

Figure 10. Schematic representation of some typical boundary conditions in saturated-unsaturated flow: a, prescribed potential; b, prescribed flux; c, seepage face; d, evaporation and evapotranspiration; e, free surface; and f, infiltration boundary conditions.

known. Instead, an upper boundary for flux and a lower boundary for ψ are prescribed (Fig. 10D).

Infiltration Boundary

When considering the prescribed flux boundary, I stated that rainfall can be totally absorbed by the flow region across a boundary segment provided that rainfall intensity does not exceed saturated-hydraulic conductivity of water. When this is exceeded, the excess rainfall must be accounted for as runoff. A boundary at which rainfall exceeds saturated hydraulic conductivity is an example of an infiltration boundary (Fig. 10E). At an infiltration boundary, therefore, flux is prescribed subject to an upper boundary.

Free Surface

Simply stated, the free surface is the boundary of the saturated flow region. Thus, at the free surface, potential equals elevation and the pressure head is atmospheric. The concept of a free surface came into existence nearly a century ago with the work of Dupuit, Forrsscheimer, Boussinesq, and others who used it to great advantage in studying many problems of practical interest. In choosing to use the free-surface concept, the unsaturated region is tacitly neglected, and it is assumed that there is a sharp discontinuity at the free surface. By definition, $\phi = z$ at the free surface, flux across the surface is prescribed, and the free surface itself is a streamline. Although the free-surface concept has been of great help as an approximate engineering tool, its physical significance is very limited. For example, although the free surface is by definition a streamline across which no flow can take place, study of saturated-unsaturated flow shows that, in reality, flow invariably takes place across the surface $\psi = 0$. Since ϕ and flux are both known, the free-surface problem is actually one of unknown geometry. The consideration of free surface arises in the study of unconfined ground-water systems and infiltration fronts.

SOURCES

Sources are phenomena by which water can be added to or abstracted from parts of the flow region by external mechanisms. If it is recognized that, in a way, boundary conditions also represent external causes adding or abstracting water from the flow region, it follows that in some situations sources and boundary conditions may become indistinguishable from one another. Nevertheless, there are many cases in which sources have fairly distinct identities.

Fluid Abstraction

Perhaps the most widely used concept of a source is fluid extraction from a portion of the flow region. The best known example of this is production of water from a well, the production being achieved in most cases with some pumping device. Another example of a source especially relevant in saturated-unsaturated flow is that of water uptake by plants (Molz, 1970). By convention, production of fluid from a well constitutes a negative source of a "sink," while injection of fluid is considered a positive source or simple "source." It is useful to bear in mind in this connection that a source or a sink is always associated with a finite subdomain of the flow region. The common practice of using "point" sources or "line" sources are mere mathematical abstractions which are only beneficial to the extent that they aid in obtaining mathematical solutions.

Pore-Pressure Generation, Chemical

Spontaneous generation of pore-pressures constitutes an important class of sources. Although not frequently encountered in conventional hydrogeology, consideration of these sources is beginning to assume importance in the context of our quest to harness geothermal energy and to gain better control of our environment. Spontaneous generation of pore pressures can occur due to chemical as well as mechanical causes.

It is known that clay minerals such as montmorillonite (smectite) are generally chemically unstable at high temperatures and pressures. With progressive burial, therefore, these minerals may break down to less hydrous but more stable minerals such as illite, releasing water in the process. The accumulation of such water in the clays themselves or in the adjoining silty or sandy layers can lead to the generation of pronounced pore-water pressures. The generation of anomalously high pore pressures in the deep sediments of the Gulf Coast in Texas and Louisiana have been attributed by some to such chemical causes. If one were interested in the hydrogeological picture of the Gulf Coast region on a geologic time scale, the generation of pore pressures due to chemical causes should be properly treated as a source term.

Pore-Pressure generation, Mechanical

It is a well-established fact that pore pressures can be generated in saturated media by mechanical loading at the boundary. Such loads can be either gradual or cyclic. Examples of the gradual loading case are pore pressures generated by an increasing overburden in sinking geological basins or by engineering constructions on shallow ground-water systems. For the cyclic case, examples include the low-frequency response (6- to 12-hour) periods of aquifers to earth tides, or the high frequency response of saturated alluvial and other materials to earthquake waves (periods of fractions of a second).

A characteristic of pore-pressure generation due to external mechanical loading of the system is that it represents an instantaneous "undrained" response of the system. Hence, this pore-pressure generation is not accompanied by a change in fluid mass in the system. In contrast, sources due to fluid abstraction and chemical reactions are accompanied by an addition of fluid mass to the flow region. There is also reason to believe that pore pressures could be generated when certain fractures (shear zones, fault zones) are subjected to shearing stresses.

The quantitative nature of chemical and mechanical sources is often extremely hard to determine. For gradual loading of elastic materials, one could compute the source magnitudes by solving a set of stress-strain equations (Poulos and Davis, 1974). These calculations, it must be noted, are external to the physical description of the transient ground-water flow problem. Insofar as high-frequency cyclic loads are concerned, preliminary techniques have been developed (Seed and Lee, 1966) to correlate amplitude and frequency of loading with the magnitude of pore-pressure generation using shaking tables. As an example, the experimentally-observed correlation between the intensity of shaking and pore pressure generated in a sand is given in Figure 11.

INITIAL CONDITIONS

The fundamental problem of hydrogeology is to follow the evolution of the system in time, given a set of known conditions at an initial moment. Therefore, the initial condition of the system must be known before beginning. When dealing with natural ground-water systems, the set of initial conditions will consist of the hydrodynamic state of the system at the particular instant of time at which relevant field observations were made. When implementing or studying controlled field or laboratory experiments, it may sometimes be possible to manipulate the system in such a fashion that at the desired instant of time the initial conditions may have a simple disposition. For example, the initial conditions may be adjusted to be hydrostatic or the distribution of potentials may vary linearly in the domain, and so on.

In problems where the system may be subject to sudden external loads at the initial moment, the initial conditions may first have to be independently determined through a knowledge of the undrained stress-strain response of the system under consideration.

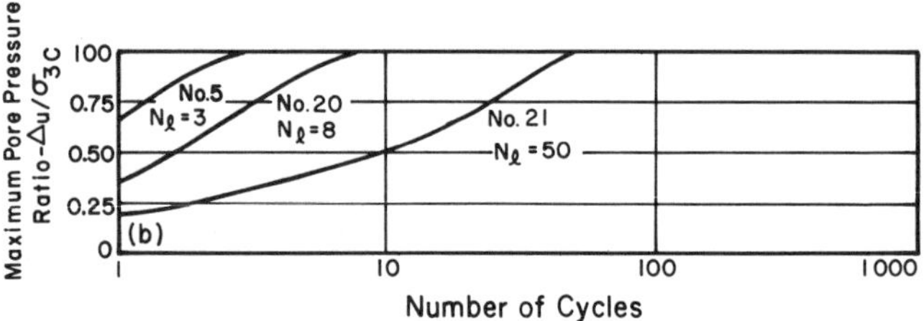

Figure 11. Experimental data on pore-pressure generation due to cyclic loading N_ℓ denotes number of cycles at which liquefaction occurred; u is pore pressure, and σ_{3c} is vertical stress (after Seed and Lee, 1966).

THE GOVERNING EQUATION

We are now in a position to summarize the physics of transient, saturated-unsaturated flow in the form of a single comprehensive statement, the governing equation. Given the basic information on (1) the initial state of the system at an arbitrary moment, (2) the time-wise variation of the state of the surroundings in which the system is embedded, and (3) the time-wise variation of sources or sinks, the governing equation attempts to predict the manner in which the state of the ground-water system will change over an interval of time and attain a new state at the end of that interval. Here, the word "state" implies the spatial distribution of water and its potential over the flow region. The causative phenomena responsible for the dynamic change of the state of the system include the imbalance in potentials existing at any instant either within or at the boundary of the system and the arbitrary sources or sinks. Thus, the governing equation is also a statement of cause and effect.

For a finite subdomain of the flow region whose volume is large enough so that average quantities such as ϕ, n, e, M_c, and others are statistically meaningful, and small enough so that these quantities do not vary rapidly over the subdomain, one could write the governing equation of transient, saturated-unsaturated ground-water flow as follows:

$$\int_{\Gamma_{I,\ell}} \rho_w \frac{k_{ij}\rho_w g}{\mu} \frac{\partial}{\partial x_j}(z+\psi) + \int_{\Gamma_{E,\ell}} \rho_w \frac{k_{ij}\rho_w g}{\mu} \frac{\partial}{\partial x_j}(z+\psi) + G_\ell = M_{c,\ell}\frac{D\psi_\ell}{Dt}, \quad \begin{cases} i = 1, 2, 3 \\ j = 1, 2, 3 \\ \ell = 1, 2, \ldots, L. \end{cases} \tag{37}$$

Here $\Gamma_{I,\ell}$ is that part of the bounding surface Γ of the subdomain interior to the flow region ,and $\Gamma_{E,\ell}$ is the portion coinciding with the external boundary of the flow region. The term G_ℓ in equation 37 represents all the sources or sink terms associated with the subdomain ℓ and the total derivative assumes a material volume element. $M_{c,\ell}$ is as defined in equation 18.

Note that equation 37 is an invariant expression, not associated with any specific coordinate system. The integrals are defined over arbitrary surfaces enclosing a three-dimensional subdomain. Considered explicitly, the three terms on the left-hand side constitute the cause, and the time derivative on the right-hand side denotes the effect. In addition to the material properties and the geometry, the integral over the external surface, $\Gamma_{E,\ell}$ and the source term, G_ℓ are presumed to be known.

The functional dependencies of the parameters in equation 37 are as follows: ρ_w is a function of ψ; k_{ij} and $M_{c,\ell}$ are functions of σ, ψ, and the material contained in element ℓ; G_ℓ and the integral over $\Gamma_{E,\ell}$ are functions of time; and g and μ are constants.

The statement of the governing equation with reference to a material volume element implies that it automatically takes into account that Darcy velocity refers to the velocity of the solid grains. However, due to solid deformation, the geometrical properties of the subdomain, such as its volume, surface area, and location, are functions of time. If one were to choose a Eulerian element to express the governing equation by defining $M_{c,\ell}$ according to equation 18 and replacing D/Dt in equation 37 by the partial derivative $\partial/\partial t$, then the geometric parameters remain fixed in time. But one would have to incorporate the grain velocity in evaluating Darcy velocity within the integrand of equation 37.

Prediction and Inverse Problems

The cause-and-effect logic upon which equation 37 was written assumes that geometry, material properties, boundary conditions, and initial conditions are known. Given this, equation 37 can be used, in principle, to predict the future behavior of the system. Although extremely powerful in theory, the predictive ability of equation 37 is severely limited in practice due to the lack of credible field information to use as input data. Such insufficient input data may be forthcoming from laboratory experiments on small samples or limited field experiments.

One attempt to overcome this problem is to have the time derivative $D\psi_\ell/Dt$ as known input data based on past behavior of the system, and evaluate either material properties, boundary conditions, initial conditions, or sources by a process of history matching. Such a procedure in which the known effect is used to evaluate parts of the causative mechanism may be called the "inverse" problem. Because a given effect can arise from a variety of causes, the inverse application of equation 37 should be expected to lead to nonunique results.

Implementation of Governing Equation

In order to implement equation 37 and advance in the time domain, we must know $M_{c,\ell}$ which includes a coefficient for volume change as a function of pore pressure change. The physical phenomenon of transient flow is such that the change in the effective stress field induced by the time-rate of change of ψ over element ℓ will cause exactly the same volume change in element ℓ as the net volume of water accumulating in ℓ due to Darcy fluxes across its boundaries, subject to density and saturation corrections. Therefore, one cannot independently implement equation 37 without solving a parallel stress-strain equation for the skeleton. The flow equation and the stress-strain equation are related through effective stress laws and coupled through the criterion of volume change compatibility. Fortunately, however, when the medium is isotropic and elastic, the two fundamental equations can be unified into a single equation, as is normally done in all practical situations. In special situations, not commonly encountered in hydrogeology, when the material is nonisotropic to deformation, the two equations have to be solved independently and coupled through iterative techniques.

Determinism Versus Uncertainty

Equation 37 assumes that all the quantities can be measured uniquely at any given time and place. However, we saw that almost all properties of natural geologic systems are statistical in nature and that they may be described in terms of a mean value and the variability about the mean (for example, standard deviation). If this frequency distribution is to be incorporated in equation 37, then the causative factors such as k_{ij}, $M_{c,\ell}$ boundary conditions, and initial conditions would all have to be expressed as frequency distributions, leading to a statistical prediction of system change expresed in terms of a mean charge and probable variability about the mean.

Relation to partial Differential Equation

The governing equation can also be expressed in a volume normalized form as opposed to equation 37. As will be seen presently, this procedure can lead to certain advantages in regard to obtaining mathematical solutions. Combining the two integrals in equation 37 and dividing through by V_ℓ, the bulk volume of the subdomain ℓ may be written

$$\frac{1}{V_\ell} \int_{\Gamma_\ell} \rho_w \frac{k_{ij}\rho_w g}{\mu} \frac{\partial}{\partial x_j} (z + \psi)\, d\Gamma + g_\ell$$
$$= m_{c,\ell} \frac{D\psi_\ell}{Dt} . \tag{38}$$

If we now let $V_\ell \to 0$ in the limit, the integral in equation 38 becomes, by definition, divergence. Thus, ignoring the source term for simplicity,

$$\operatorname{div} \left[\rho_w \frac{k_{ij}\rho_w g}{\mu} \frac{\partial}{\partial x_j} (z + \psi) \right] = m_{c,\ell} \frac{D\psi}{Dt} . \tag{39}$$

Although equations 37 and 39 are conceptually identical, they would lead to different procedures if they were to be practically implemented in order to obtain solutions. The advantage of equation 39 is that if we consider the reference subdomain to be regularly-shaped with the bounding surfaces normal to the principal coordinate axes of symmetry, then the actual evaluation of divergence leads to partial differential equations involving second spatial derivatives of ψ on the left-hand side, and first derivative with reference to time on the right-hand side. In the process, the integral operation disappears from the left-hand side. Thus, whereas one has to evaluate the integrals in equation 37 in order to implement that equation, one simply has to find the appropriate function $\psi(x_i, t)$ which will satisfy both sides of equation 39 in order to implement the partial differential equation.

Implicit in the differential equation approach is that a solution $\psi(x_i, t)$ exists before one sets out to find it. The lack of fast integrating devices, which precluded the general implementation of equation 37, was probably the dominating motivation for the development of the differential equation as a tool for studying mass- and energy-conservation problems. A major philosophical difference between the integral and differential approaches is that the former follows the system as it evolves, whereas the latter assumes the existence of a "solution" before working the problem.

Some Intriguing Concepts

In the foregoing developments we defined and used various quantities such as intrinsic permeability, storage coefficient, density, porosity, and so on with fairly well-defined physical connotation. Nevertheless, these quantities, as well as certain other quantities not hitherto mentioned but still used in the literature, have some intriguing aspects. We will now briefly examine a few of these.

The first among these is the concept of velocity potential, which is the product $K\phi$ where K is hydraulic conductivity and ϕ is hydraulic head. By definition, the gradient of velocity potential should yield Darcy velocity. That is, $\vec{q} = \nabla(K\phi)$. However, this can only be strictly true when K is constant, and hence the concept is useful only for homogeneous systems. Although it does not have any physical meaning, the concept has proved to be of practical use, when applicable, to facilitate the use of complex variable theory to study fluid flow problems.

A second concept of this kind is that of soil moisture diffusivity, or the analogous concept of coefficient of consolidation. For the simple case of a one-dimensional horizontal soil column with an isotropic soil, the differential equation 39 can be simplified as

$$(\partial/\partial x)\ K\ (d\psi/d\theta)\ (\partial\theta/\partial x) = D\theta/Dt.$$

However, one could computationally replace $K\ (d\psi/d\theta\ \partial\theta/\partial x)$ by the relation $D\ (\partial\theta/\partial x)$, where $D = K\ (\partial\psi/\partial\theta)$. The quantity D is called soil moisture diffusivity. The advantage of using this transformation is that the equation is somewhat easier to solve, since D is not as strongly nonlinear with respect to θ as K and $M_{c,\ell}$ are with reference to ψ. Yet, diffusivity by itself has little physical meaning. Where the change in θ is governed dominantly by soil deformation, diffusivity is equivalent to the coefficient of consolidation.

A third concept that may be mentioned here is that of "delayed drainage," which has been used in studying the behavior of unconfined aquifers. In applying the differential equation to unconfined aquifers, the assumption is often made that water moves vertically down into the saturated zone from the region through which the water table moves. Although the flow in the unsaturated domain is neglected, the assumption made above introduces an equivalent, nonlinear source term which gives rise to the delayed drainage coefficient. Here again, the coefficient is a model parameter rather than a physical property.

As the final example, we may cite the case of a point function, such as point potential, point pressure, or point porosity. Potential, pressure, and similar quantities of an intensive nature are but expressions of the spatial distribution of energy or mass. Pressure or potential cannot exist without energy or mass. However, mass and energy require space for their existence. Therefore, intensive quantities do actually represent an average value of said phenomenon over a small region around the point of interest, with the average value being associated with the point of interest. This is not just a semantic exercise. All our conceptualization is intimately related to volume-averaging over finite subdomains of the flow region. The concept of point function entered into the literature with the advent of the concept of a continuum. Although it does not truly simulate a porous medium, the continuum concept, in conjunction with the differentiation of continuous functions, has provided us considerable insights into otherwise intractable problems. Thus, despite its utility, the point function remains an abstraction rather than a physical property.

CONCLUDING REMARKS

As Philip (1973) points out, flow in porous media is approximately an ordered flow in a disordered geometry. The order we observe on the Darcy scale is indeed statistical in nature. For our studies on this scale to be meaningful and for us to be truly aware of the applicability limits of our studies, we must conduct independent studies on a more detailed "Navier-Stokes" scale (Philip, 1973), in which the more precise laws of fluid mechanics are applicable. In constructing the physical theory of saturated-unsaturated flow, therefore, we are confronted with a general problem of compromising between credible volume and quality of information versus detail of prediction. The deterministic approach, which has historically formed the backbone of modern science, has helped us understand in a semiquantitative way the pattern of response of natural systems due to complex interactions between several variables. Nevertheless, the very nature of the field problems in measuring the observable scales creates a gap between the deterministic framework and the real-world situation. Recently, several workers have found that the system response and the uncertainty associated with it is more compatible with reality than the deterministic approach. In seeking to implement this approach, one is naturally led to an algebra in which one is concerned with the fundamental mathematical operations relating to probability distributions rather than real numbers.

An important limitation especially relevant to saturated-unsaturated flow is that the air phase is assumed to be continuous and at constant pressure. This assumption is probably essential in order that large field problems can be handled meaningfully. Nevertheless, on a local scale, one has to bear in mind that the fundamental problem of saturated-unsaturated flow is actually a multiphase flow problem in which mass conservation has to be assured separately for each of the phases. Several workers (for examle, Morel-Seytoux, 1973) have taken the more general two-phase-flow approach to this problem.

The conceptual framework and descriptive parameters of saturated-unsaturated ground-water flow are currently well established. Although the existing parameters may need subtle modifications, it is doubtful that any radically new parameter will be introduced into the literature. If these parameters can be measured accurately, we could handle their complex interactions with considerable computational facility. The greatest challenge, then, is our ability to measure the desired parameters in sufficient quantity and detail in the field. In this regard we are faced with both hardware and interpretation problems. In many instances, parameters are estimated by solving equations subject to specific geometries and boundary conditions. In so doing, we are not certain what has been determined--a physical parameter or a coefficient satisfying a particular geometry and boundary condition. Thus we need to devise means for reliably estimating physical parameters under field conditions so that the well-laid-out physical framework can be credibly implemented to study field problems. Studies of flow in porous media have generally borrowed heavily from the heat-transfer literature. Although the basic similarity between flow through porous media and heat transfer cannot be overlooked, there are certain peculiarities of the porous media problem that do not exist in heat transfer (for example, gravity, deformation). Therefore, there exists at present a very definite need to construct a theory of saturated-unsaturated flow from its own axiomatic foundations.

NOMENCLATURE

Symbol	Definition	Units
a_v	Coefficient of compressibility; $a_v = -(\Delta V_v/V_s)/\Delta\sigma'$	(LT^2/M)
A	Area of cross section	(L^2)
A_{ij}	Reciprocal hydraulic conductivity tensor	(T/L)
c_f	Pore-volume compressibility; $c_f = -(\Delta V_v/V_v)/\Delta\sigma'$	(LT^2/M)
c_t	Total compressibility; $c_t = c_f + \beta$	(LT^2/M)
c_ℓ	A constant	
F_j	A dimensionless force vector	
g	Gravitational constant	(L/T^2)
g_ℓ	Source per unit volume of volume element, ℓ	(M/L^3T)
G_ℓ	Source from volume element, ℓ	(M/T)
k	Permeability; absolute permeability; intrinsic permeability	(L^2)
k_{ij}	Permeability tensor	(L^2)
K	Hydraulic conductivity	(L/T)
K_{ij}	Hydraulic conductivity tensor	(L/T)
ℓ	Subscript denoting a volume element or subdomain	
$m_{c,1}$	Fluid mass capacity of material in ℓ	(M/L)
m_v	Coefficient of volume change	(LT^2/M)
$M_{c,\ell}$	Fluid mass capacity of volume element ℓ	(M/L)
$M_{w,\ell}$	Mass of water contained in volume element	(M)
n	Porosity	
$\vec{n}$	Unit outer normal	
p	Fluid pressure	(M/LT^2)
p_o	Reference fluid pressure (normally atmospheric)	(M/LT^2)
$\vec{q}$	specific flux or Darcy velocity	(L/T)
q_i	Specific flux or Darcy velocity in the ith coordinate direction	(L/T)
Q	Total volumetric flux	(L^3/T)
S_s	Coefficient of specific storage	(L^{-1})
S_w	Water saturation	
$S_{w,\ell}$	Water saturation of element ℓ	
t	Time	
V_ℓ	Bulk volume of element ℓ	(L^3)
V_s	Volume of solids	(L^3)
V_v	Volume of voids	(L^3)
$V_{v,\ell}$	Volume of voids in element ℓ	(L^3)
x_i	Coordinate axes	
α	A coefficient	
β	Compressibility of water; $\beta = -(\Delta V_w/V_w/\Delta\sigma')$	(LT^2/M)
Γ	Surface bounding a subdomain	(L^2)
$\Gamma_{I,\ell}$	Bounding surface segment of element ℓ, interior to the flow region	(L^2)
$\Gamma_{E,\ell}$	Bounding surface segment of element ℓ, coinciding with boundary of flow region	(L^2)
ε_v	Volume strain	
θ	Volumetric moisture content	
ρ_w	Density of water	(M/L^3)
$\rho_{w,\ell}$	Density of water in volume element ℓ	
$\rho_{w,o}$	Reference density of water at $p = p_o$	(M/L^3)
σ	Total stress	(M/LT^2)
σ_{ij}	Total stress tensor	(M/LT^2)
σ_v	Total vertical stress	(M/LT^2)
σ'	Effective stress	(M/LT^2)
σ'_{ij}	Effective stress tensor	(M/LT^2)
σ'_m	Mean principal effective stress	(M/LT^2)

ϕ	Hydraulic head	(L)
Φ	Fluid potential	(L^2/T^2)
χ	A coefficient	
χ^*	$[\chi + d\psi/d\chi]$	
ψ	Pressure head, $(\rho - \rho_o)/\rho_w g$	(L)
ψ_A	Air entry valve	(L)

ACKNOWLEDGEMENTS

I sincerely thank Fred Molz for a critical scrutiny of the manuscript and several constructive suggestions. This work was supported by the Director, Office of Energy Research, Office of Basic Energy Sciences, Division of Engineering, Mathematics, and Geosciences of the U.S. Department of Energy under Contract No. W-7405-ENG-48.

REFERENCES CITED

Bear, J., 1972, Dynamics of fluids in porous media: New York, Elsevier.

Bishop, A. W., 1960, The principle of effective stress: Norwegian Geotechnical Institute Publication, 32, p. 1-5.

Bishop, A. W., and Blight, G. E., 1963, Some aspects of effective stress in saturated and partly saturated soils: Geotechnique, v. 13, no. 1, p. 177-197.

Buckingham, E., 1907, Studies on the movement of soil moisture: Washington, D.C., U.S. Department of Agriculture, Bull. 38.

Darcy, H., 1856, Les fontaines publiques de la ville de Dijon: Paris, Victor Dalmont.

Davis, S. N., 1969, Porosity and permeability of natural materials, in de Wiest, R. J. M., ed., Flow through porous media: New York, Academic Press, p. 54-90.

Domenico, P. A., and Mifflin, M. D., 1965, Water from low permeability sediments and land subsidence: Water Resources Research, v. 1, no. 4, p. 563-576.

Earlougher, R. C., 1977, Advances in well test analysis: Soc. Pet. Eng. AIME.

Ferris, J. G., and others, 1962, Theory of aquifer tests: U.S. Geological Survey Water Supply Paper 1536-E.

Fredlund, D. G., and Morgenstern, N. R., 1977, Stress-state variables for unsaturated soils, in Proceedings, ASCE, Journal of Geotechnical Engineering Division: v. 103, No. GT5, p. 446-447.

Freeze, R. A., 1975, A stochastic-conceptual analysis of one-dimensional ground-water flow in nonuniform homogeneous media: Water Resources Research, v. 11, no. 5, p. 725-741.

Garg, S. K., and Nur, A. M., 1973, Effective stress laws for fluid-saturated rocks: Journal of Geophysical Research, v. 78, no. 26, p. 5911-5921.

Greenkorn, R. A., and Kessler, D. P., 1969, Dispersion in nonuniform anisotropic porous media: Ind. Eng. Chem., v. 61, no. 9, p. 14-32.

Hubbert, M. K., 1940, The theory of groundwater motion: Journal of Geology, v. 48, p. 785-944.

Iwai, K., 1976, Fundamental studies of fluid flow through a single fracture [Ph.D. thesis]: Berkelely, University of California.

Klute, A., 1973, Soil water flow theory and its aplication in field situations: Field Soil Water Regime, Soil Science Society of America, p. 9-35.

Lambe, T. W., and Whitman, R. V., 1969, Soil mechanics: New York, John Wiley.

Liakopoulos, A. C. 1965, Transient flow through unsaturated porous media (Ph.D. thesis): Berkeley, University of California.

Maasland, M., 1977, Soil anisotropy and land drainage, in Luthin, J. N., ed., Drainage of agricultural lands: Madison, American Society of Agronomy, p. 216-285.

Mitchell, J. K., 1976, Principles of soil behavior: New York, John Wiley.

Molz, F., and Remson, I., 1970, Extraction term models of soil moisture use by transpiring plants: Water Resources Research, v. 6, no. 5, p. 1346-1356.

Morel-Seytoux, H. J., 1973, Two-phase flows in porous media, in Chow, V. T., ed., Advances in hydroscience: Academic Press, v. 9, p. 119-202.

Narasimhan, T. N., 1979, The significance of the storage parameter in saturated-unsaturated ground water flow: Water Resources Research, v. 15, no. 3, p. 569-576.

Narasimhan, T. N., and Witherspoon, P. A., 1977, Numerical model for saturated-unsaturated flow in deformable porous media theory: Water Resources Research, v. 13 no. 3, p. 657-664.

Nelson, R. W., 1966, Flow in heterogeneous porous mediums 1, Darcian-type description of two-phase systems: Water Resources Research, v. 2, no. 3, p. 487-495.

Neuman, S. P., Fedder, R. A., and Bresler, E., 1975, Finite element analysis of two-dimensional flow in soils considering water uptake by plant roots 1, theory: Soil Science Society of America Proceedings, v. 39, no. 2, p. 224-237.

Olsen, H. W., 1965, Deviations from Darcy's law in saturated clay: Proceedings, Soil Science Society of America, v. 29, p. 135-140.

Philip, J. R., 1969, Hydrostatics and hydrodynamics of swelling soils: Water Resource Research, v. 5, p. 1070-1077.

Philip, J. R., 1972, Flow in porous media, in Becker, E., and Mikhailov, G. K., eds., Applied mechanics, Proceedings of Thirteenth International Congress of Theoretical and Applied Mechanics, Moscow, August 1972: Berlin, Springer-Verlag.

Poulos, H. G., and Davis, E. H., 1974, Elastic solutions for soil and rock mechanics: New York, John Wiley.

Ramey, H. J., Jr., 1965, Non-Darcy flow and wellbore storage effects in presure buildup and drawdown of gas wells: Journal of Petroleum Technology, v. 17, p. 223-233.

Richards, L. A., 1931, Capillary conduction of liquids through porous mediums: Physics, v. 1, p. 318-333.

Robinson, L. H., and Holland, W. E., 1970, Some interpretation of pore-fluid effects on rock failure, in Rock mechanics theory and practice,

Proceedings, 11th Symposium on Rock Mechanics: New York, Society of Mining Engineers, p. 585-598.

Seed, H. B., and Lee, K. L., 1966, Liquefaction of saturated sands during cyclic loading: American Society Civil Engineering Journal Soil Mechanics and Foundation Division, v. 92(SM5) p. 105-134.

Skempton, A. W., 1961, Effective stress in soils, concrete and rocks, in Pore pressure and suction in soils: London, Butterworths, p. 4-16.

Smiles, D. E., and Rosenthal, M. J., 1968, The movement of water in swelling materials: Australian Journal of Soil Research, v. 6, p. 237-248.

Swartzendruber, D., 1969, The flow of water in unsaturated soils, in De Wiest, R. J. M., ed., Flow through porous media: New York, Academic Press, p. 215-292.

Toth, J., 1967, Groundwater in sedimentary (clastic) rocks, in Proceedings of the National Symposium on Ground-Water Hydrology, San Francisco, November 6-7, 1967: Australian Journal of Soil Research, v. 6, p. 237-248.

Manuscript Accepted by the Society October 14, 1980

Printed in U.S.A.

Geological Society of America
Special Paper 189
1982

Physics of Flow in Geothermal Systems

GEORGE F. PINDER
Department of Civil Engineering
Princeton University
Princeton, New Jersey 08540

ALLEN SHAPIRO
Department of Civil Engineering
Princeton University
Princeton, New Jersey 08540

ABSTRACT

The equations governing the simultaneous flow of steam and water in a geothermal reservoir are developed by treating the porous medium as a continuum. The balance laws for mass, linear momentum, and energy, which are valid for a single continuum, are integrated over each phase (rock, liquid water, and steam) in a representative elementary volume of the medium. The resulting equations are macroscopic point balances, which account entirely for the loss of definition in going from the microscopic to the macroscopic level of observation.

INTRODUCTION

This paper provides a logical development of the equations governing the physics of flow in multiphase geothermal reservoirs. The formulation is entirely general and therefore applicable to a wide range of physical problems. The reservoir is considered to be a porous medium with the presence of two fluid phases, steam and water. No assumption is made about whether the solid matrix is rigid or deformable. However, the porous medium must be regarded as amenable to analysis as a continuum.

The development begins with the integration over the porous medium of the point-balance equations for mass, linear momentum, and energy. The resulting equations, defined in terms of macroscopic variables, describe the behavior of individual phases. These equations are subsequently phase averaged to provide a demonstration of the assumptions and simplifications inherent in this procedure. These equations, in conjunction with suitable constitutive and thermodynamic relationships, form the mathematical basis of most geothermal simulators.

BALANCE EQUATIONS

The following are point equations describing mass, linear momentum, and energy transport:

$$\frac{\partial \rho}{\partial t} + \underset{\sim}{\nabla} \cdot (\rho \underset{\sim}{v}) = 0, \qquad \text{(mass)} \quad (1)$$

$$\frac{\partial (\rho \underset{\sim}{v})}{\partial t} + \underset{\sim}{\nabla} (\rho \underset{\sim\sim}{vv}) - \underset{\sim}{\nabla} \cdot \underset{\approx}{t} - \rho \underset{\sim}{g} = 0, \qquad \text{(momentum)} \quad (2)$$

$$\frac{\partial \left(\rho U + \frac{1}{2}\rho V^2 \right)}{\partial t} + \underset{\sim}{\nabla} \cdot \rho \underset{\sim}{v}\left(U + \frac{1}{2} - \underset{\sim}{v}^2 \right)$$
$$- \underset{\sim}{\nabla} \cdot \underset{\sim}{q} - \underset{\sim}{\nabla} \cdot (\underset{\approx}{t} \cdot \underset{\sim}{v}) - \rho \underset{\sim}{v} \cdot \underset{\sim}{g} = 0, \qquad \text{(energy)} \quad (3)$$

where ρ = mass density,
$\underset{\sim}{v}$ = the mass average velocity,
$\underset{\approx}{t}$ = the stress tensor,
U = internal energy per unit mass,
$\underset{\sim}{q}$ = the heat-flux vector, and
$\underset{\sim}{g}$ = gravitational acceleration.

These equations pertain to any point in a continuum. The intricate geometry of the individual pores and fluid phases makes the solution of these equations an impossible task. One must therefore consider some means of dealing with these physical concepts on a larger scale.

In this paper, the porous-medium balance equations are developed by use of the concepts of mass and volume averages. The conceptual model of a porous medium as used here is presented in Figure 1. There are three immiscible phases pictured in this representative elementary volume (REV): steam (S), liquid water (W), and rock (R). It is assumed that the property ψ_α can be represented at the macroscopic level in a statistically meaningful sense as the volume-phase average.

$$\langle\psi\rangle_\alpha(\chi,t) \equiv \frac{1}{d\forall}\int_{d\forall} \psi_\alpha(\chi + \xi,t)\,dV_\xi, \tag{4}$$

where ψ_α is nonzero only in the α phase and dV_ξ is the microscopic element of volume in $d\forall$. In other words, there exists a macroscopic variable $\langle\psi\rangle_\alpha$ associated with the centroid of the REV, which is related to the microscopic variable ψ_α such that $\langle\psi\rangle_\alpha$ is well behaved in χ and yet maintains the essential physical attributes of ψ_α. However, a requirement of equation 4 is that the product $\psi_\alpha dV_\xi$ is an extensive quantity so that its summation is meaningful.

An alternative and possibly more precise statement of equation 4 is given by Gray and Lee (1977)

$$\langle\psi\rangle_\alpha(\chi,t) \equiv \frac{1}{d\forall}\int_{d\forall} \psi(\underset{\sim}{r},t)\;\gamma_\alpha(\underset{\sim}{r},t)\,dV_\varepsilon, \tag{5}$$

where $r = \chi + \underset{\sim}{\xi}$ and γ_α is defined for the α phase as

$$\gamma_\alpha(\underset{\sim}{r},t) = \begin{cases} 1 & \underset{\sim}{r}\varepsilon d\forall_\alpha \\ 0 & \underset{\sim}{r}\varepsilon d\forall_b \end{cases} \qquad \text{for all } t,\alpha,\beta = S,W,R \tag{6}$$

and ψdV_ξ is defined only as an extensive property. One can now formally define the volume of $d\forall$ as occupied by the α phase. That is,

$$d\forall_\alpha = d\forall_\alpha(\chi,t) = \int_{dV} \gamma_\alpha(\underset{\sim}{r},t)\,dV_\xi. \tag{7}$$

It is apparent from equations 4 through 7 that the integration is carried out over the microscopic local coordinate system defined by ξ. The concept of a macroscopic area element arises naturally from equation 6:

$$dA_{\alpha\alpha} = dA_{\alpha\alpha}(\chi,t) = \int_{dA} q_\alpha(\underset{\sim}{r},t)\,da_\xi. \tag{8}$$

Note in Figure 1 that the surface area of the α phase is made up of two parts: (1) the part that makes up the boundary of dA, that is, $dA_{\alpha\alpha}$; and (2) the part wholly within $d\forall$, that forms the interface of the α-phase with all other phases

$$\sum_{\beta\neq\alpha} dA_{\alpha\beta}.$$

Thus,

$$dA_\alpha = dA_{\alpha\alpha} + \sum_{\beta\neq\alpha} dA_{\alpha\beta}. \tag{9}$$

Although the phase average is unambiguous when applied to extensive variables, in many instances it is not physically meaningful when the integrand involves an intensive variable. To avoid this difficulty, Hassanizadeh and Gray (1978) define a mass average $\bar{\psi}^\alpha$ as

$$\bar{\psi}^\alpha(\underset{\sim}{X},t) \equiv 1/[\langle\rho\rangle_\alpha(\underset{\sim}{X},t)\,d\forall]$$

$$\int_{d\forall} \rho(\underset{\sim}{r},t)\psi(\underset{\sim}{r},t)\gamma_\alpha(r,t)\,dV_\xi, \tag{10}$$

where ψ is now an intensive variable such as energy per unit mass. As the development proceeds, additional relationships are required that relate microscopic and macroscopic variables; they are introduced as the need arises.

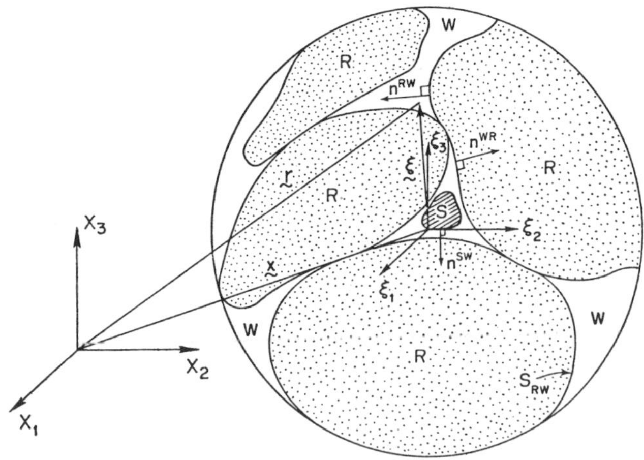

Figure 1. Conceptual model of a representative elementary volume in a multiphase geothermal reservoir composed of steam S, liquid water W, and rock R.

MACROSCOPIC MASS BALANCE

The macroscopic mass balance for the α phase is obtained in a rather straightforward manner from the point-balance equation 1. Multiplication of equation 1 by γ_α, with subsequent integration of the product over $d\forall$, yields

$$\int_{d\forall} \left[\frac{\partial\rho}{\partial t} + \underset{\sim}{\nabla}\cdot(\rho\underset{\sim}{v})\right]\gamma_\alpha dV_\xi = 0. \tag{11}$$

To express equation 11 in terms of macroscopic variables, the time and space derivatives are modified using the following identities (Gray and Lee, 1977):

$$\int_{d\forall} \frac{\partial \rho}{\partial t} \gamma_\alpha dV_\xi = \frac{\partial}{\partial t} \int_{d\forall} \rho \gamma_\alpha dV_\xi - \sum_{\beta \neq \alpha} \int_{dA_{\alpha\beta}} \rho \underset{\sim}{W} \cdot \underset{\sim}{n}^{\alpha\beta} da, \tag{12}$$

where $\underset{\sim}{W}$ is the interface velocity, and

$$\int_{d\underline{V}} [\underset{\sim}{\nabla} \cdot (\underset{\sim}{\rho V})] \gamma_\alpha dV_\xi$$

$$= \underset{\sim}{\nabla} \cdot \int_{d\underline{V}} (\rho V) \gamma_\alpha dV_\xi + \sum_{\beta \neq \alpha} \int_{dA_{\alpha\beta}} (\rho \underset{\sim}{V}) \cdot \underset{\sim}{n}^{\alpha\beta} da, \tag{13}$$

where $n^{\alpha\beta}$ is the outward directed normal from the α phase to the β phase. Combining equations 11 through 13 yields

$$\frac{\partial}{\partial t} \langle \rho \rangle_\alpha + \underset{\sim}{\nabla} \cdot \langle \underset{\sim}{\rho V} \rangle_\alpha$$

$$- \frac{1}{d\forall} \sum_{\beta \neq \alpha} \int_{dA_{\alpha\beta}} \rho (\underset{\sim}{W} - \underset{\sim}{V}) \cdot \underset{\sim}{n}^{\alpha\beta} d\alpha = 0. \tag{14}$$

Application of the mass average operator on the second term in equation 14 yields

$$\frac{\partial}{\partial t} \langle \rho \rangle_\alpha + \underset{\sim}{\nabla} \cdot \left(\langle \rho \rangle_\alpha \bar{\underset{\sim}{V}}^\alpha \right)$$

$$- \frac{1}{\partial V} \sum_{\beta \neq \alpha} \int_{dA_{\alpha\beta}} \rho (\underset{\sim}{W} - \underset{\sim}{V}) \cdot \underset{\sim}{n}^{\alpha\beta} d\alpha = 0, \tag{15}$$

where the final term represents the exchange of mass across interfacial boundaries. Although the development of equation 15 arises from averaging over a discrete volume, it is an expression descriptive of the macroscopic point balance of mass for the α phase. It originates from the reasoning that the REV is the smallest discernible volume from a macroscopic level of observation that is consistent with a statistically meaningful macroscopic variable.

An alternative form of the mass-transport equation is obtained by the summation of equation 15 over all fluid phases. Continuity of mass across the interfaces requires that the surface integrals vanish, whereupon one obtains

$$\sum_{\alpha \neq R} \left\{ \frac{\partial}{\partial t} \langle \rho \rangle_\alpha + \underset{\sim}{\nabla} \cdot \left(\langle \rho \rangle_\alpha \bar{\underset{\sim}{V}}^\alpha \right) \right\} = 0. \tag{16}$$

Equation 16 is further reduced through the assumption of a local average fluid density

$$\bar{\rho}_f = \sum_{\alpha \neq R} \langle \rho \rangle_\alpha, \tag{17}$$

and a mass average fluid velocity

$$\underset{\sim}{\forall}_f = \frac{1}{\bar{\rho}_f} \sum_{\alpha \neq R} \left(\langle \rho \rangle_\alpha \bar{\underset{\sim}{V}}^\alpha \right). \tag{18}$$

Combining equations 16 to 18 yields, for the fluid phases,

$$\frac{\partial}{\partial t} \bar{\rho}_f + \underset{\sim}{\nabla} \cdot \left(\bar{\rho}_f \underset{\sim}{V}_f \right) = 0. \tag{19}$$

The equation for the solid phase remains

$$\frac{\partial}{\partial t} \langle \rho \rangle_R + \underset{\approx}{\nabla} \cdot \left(\langle \rho \rangle_R \bar{\underset{\sim}{V}}^R \right) = 0, \tag{20}$$

because it is assumed that the rock phase is inert, such that

$$\sum_{\beta \neq R} \int_{dA_{R\beta}} \rho (\underset{\sim}{W} - \underset{\sim}{V}) \cdot \underset{\sim}{n}^{R\beta} d\alpha = 0.$$

One may also define a mass-transport equation for the entire continuum inclusive of the rock phase, by the summation of equation 15:

$$(\partial \bar{\rho} / \partial t) + \underset{\sim}{\nabla} \cdot (\bar{\rho} \bar{V}) = 0, \tag{21}$$

where

$$\bar{\rho} = \sum_\alpha \langle \rho \rangle_\alpha \tag{22}$$

and

$$\bar{\underset{\sim}{V}} = \frac{1}{\bar{\rho}} \sum_\alpha \left(\langle \rho \rangle_\alpha \bar{V}^\alpha \right). \tag{23}$$

The use of these relationships will become evident in the manipulation of the macroscopic balance laws that are developed in the following sections.

MACROSCOPIC LINEAR-MOMENTUM BALANCE

Using an approach similar to that of the previous section, the macroscopic linear-momentum balance is obtained through the multiplication of equation 2 by γ_a and integration over $d\forall$:

$$\int_{d\forall} \left[\frac{\partial (\rho \underset{\sim}{V})}{\partial t} + \underset{\sim}{\nabla} \cdot (\rho \underset{\sim}{VV}) - \underset{\sim}{\nabla} \cdot \underset{\approx}{t} - \rho g \right] \gamma_\alpha dV_\xi = 0. \tag{24}$$

Application of equation 12 to the first term and equation 13 to the second and third terms of the integrand in equation 24 yields

$$\frac{\partial \langle\rho\rangle_\alpha \bar{\underset{\sim}{v}}^\alpha}{\partial t} + \frac{1}{dV}\left(\underset{\sim}{\nabla} \cdot \int_{dV} \rho \underset{\sim}{v}\underset{\sim}{v}\gamma_\alpha dV_\xi - v \cdot \int_{dV} \underset{\approx}{t}\gamma_\alpha dV_\xi\right)$$

$$- \langle\rho\rangle_\alpha \underset{\sim}{g} - \frac{1}{dV}\sum_{\beta\neq\alpha}\int_{dA_{\alpha\beta}} \rho\underset{\sim}{v}(\underset{\sim}{w} - \underset{\sim}{v}) \cdot n^{\alpha\beta} da$$

$$- \frac{1}{dV}\sum_{\beta\neq\alpha}\int_{dA_{\alpha\beta}} \underset{\approx}{t} \cdot \underset{\sim}{n}^{\alpha\beta} da = 0. \quad (25)$$

The surface integrals in equation 25 represent the exchange of momentum between the α phase and all other phases as a result of mass exchange and mechanical interaction.

The average of a product of dependent variables, such as appears in the second term of equation 25, can be obtained by using a perturbation expansion:

$$\underset{\sim}{v} = \bar{\underset{\sim}{v}}^\alpha + \tilde{\underset{\sim}{v}}^\alpha. \quad (26)$$

It can be shown (Hassanizadeh and Gray, 1978) that

$$\int_{dV} \rho \underset{\sim}{v}\gamma_\alpha dV_\xi = \int_{dV} \rho \bar{\underset{\sim}{v}}^\alpha \gamma_\alpha dV_\xi + \int_{dV} \rho \tilde{v}^\alpha \gamma_\alpha dV_\xi \,. \quad (27)$$

Applying equation 10 to the integral on the left-hand side of equation 27 and recognizing that $\bar{\underset{\sim}{v}}^\alpha$ is constant over dV, the previous relationship simplifies to

$$\langle\rho\rangle_\alpha \bar{\underset{\sim}{v}}^\alpha = \langle\rho\rangle_\alpha \bar{\underset{\sim}{v}}^\alpha + \int_{dV} \rho \tilde{\underset{\sim}{v}}^\alpha \gamma_\alpha dV_\xi, \quad (28)$$

which implies

$$\int_{dV} \rho \tilde{\underset{\sim}{v}}^\alpha \gamma_\alpha dV_\xi = 0. \quad (29)$$

Returning now to equation 25, the second term may be handled in the following manner:

$$\underset{\sim}{\nabla} \cdot \int_{dV} \rho \underset{\sim}{v}\underset{\sim}{v}\gamma_\alpha dV_\xi = \underset{\sim}{\nabla} \cdot \int_{dV} \rho\left(\bar{\underset{\sim}{v}}^\alpha + \tilde{\underset{\sim}{v}}^\alpha\right)\left(\bar{\underset{\sim}{v}}^\alpha + \tilde{\underset{\sim}{v}}^\alpha\right)\gamma_\alpha dV_\xi.$$

As a consequence of equation 29 this expression reduces to

$$\underset{\sim}{\nabla} \cdot \int_{dV} \rho \underset{\sim}{v}\underset{\sim}{v}\gamma_\alpha dV_\xi = \underset{\sim}{\nabla} \cdot \left(\langle\rho\rangle_\alpha \bar{\underset{\sim}{v}}^\alpha \bar{\underset{\sim}{v}}^\alpha\right)$$

$$+ \underset{\sim}{\nabla} \cdot \int_{d\underline{V}} \rho \tilde{v}^\alpha \tilde{v}^\alpha \gamma_\alpha dV_\xi. \quad (30)$$

Equation 25 can thus be written

$$\frac{\partial\left(\langle\rho\rangle_\alpha \bar{\underset{\sim}{v}}^\alpha\right)}{\partial t} + \underset{\sim}{\nabla} \cdot \left(\langle\rho\rangle_\alpha \bar{\underset{\sim}{v}}^\alpha \bar{\underset{\sim}{v}}^\alpha\right) - \underset{\sim}{\nabla} \cdot \underset{\approx}{\tau}^\alpha - \langle\rho\rangle_\alpha \underset{\sim}{g}$$

$$- \frac{1}{dV}\sum_{\beta\neq\alpha}\int_{dA_{\alpha\beta}} \left[\rho\underset{\sim}{v}(\underset{\sim}{w} - \underset{\sim}{v}) + \underset{\approx}{\tau}\right] \cdot \underset{\sim}{n}^{\alpha\beta} da = 0, \quad (31)$$

where

$$\underset{\approx}{\tau}^\alpha = \frac{1}{dV}\int_{d\underline{V}} \left(\underset{\approx}{t} - \rho\tilde{\underset{\sim}{v}}^\alpha \tilde{\underset{\sim}{v}}^\alpha\right)\gamma_\alpha dV_\xi. \quad (32)$$

Notice that in the loss of definition in going from the microscopic to the macroscopic balance law, one must account for all components of a given macroscopic variable. Consider, for example, the induced stresses that result from microscopic velocity deviations as shown in equation 32. It should be apparent that constitutive relationships must be developed to describe the behavior of such macroscopic quantities in terms of other macroscopic variables.

An alternative form of the linear-momentum balance may be obtained through the summation of equation 31 over all phases. Conservation of momentum across interfaces requires the last term in equation 31 to vanish when summed, resulting in

$$\sum_\alpha \left[\frac{\partial\langle\rho\rangle_\alpha \underset{\sim}{v}^{-\alpha}}{\partial t} + \underset{\sim}{\nabla} \cdot \left(\langle\rho\rangle_\alpha \bar{\underset{\sim}{v}}^\alpha \bar{\underset{\sim}{v}}^\alpha\right) - \underset{\sim}{\nabla} \cdot \underset{\approx}{\tau}^\alpha - \langle\rho\rangle_\alpha \underset{=}{g}\right] = 0. \quad (33)$$

Equation 33 can now be modified by using the mass-averaged continuum-velocity equation 23 in the first term and noting that the gravitational acceleration is a constant in the last term:

$$\frac{\partial(\bar{\rho}\ \bar{\underset{\sim}{v}})}{\partial t} + \sum_\alpha \left[\underset{\sim}{\nabla} \cdot \langle\rho\rangle_\alpha \bar{\underset{\sim}{v}}^\alpha \bar{\underset{\sim}{v}}^\alpha - \underset{\sim}{\nabla} \cdot \underset{\approx}{\tau}^\alpha\right] - \bar{\rho}\underset{=}{g} = 0. \quad (34)$$

In developing a momentum balance that is valid for the continuum as a whole, one must also recognize that once again there is a loss of definition, this time with respect to the individual macroscopic variables of a given phase. Thus, it should not be expected that in all cases the new continuum variables will be merely the summation of the macroscopic phase-averaged components. A perturbation expansion is applied to the second term of equation 34 by defining the mass-averaged phase velocity in terms of the mass-averaged continuum velocity

$$\bar{\underset{\sim}{v}}^\alpha = \bar{\underset{\sim}{v}} + \hat{\underset{\sim}{v}}^\alpha, \quad (35)$$

where we require that $\hat{V}^\alpha$ be defined such that

$$\sum_\alpha \langle\rho\rangle_\alpha \hat{\underset{\sim}{V}}^\alpha = 0. \tag{36}$$

The macroscopic-continuum momentum balance at a given point can now be expressed as

$$\frac{\partial(\bar{\rho}\ \bar{\underset{\sim}{V}})}{\partial_t} + \underset{\sim}{\nabla} \cdot (\overline{\rho \underset{\sim}{V}\underset{\sim}{V}}) - \underset{\sim}{\nabla} \cdot \underset{\approx}{T} - \bar{\rho}\underset{\sim}{g} = 0, \tag{37}$$

where the continuum stress tensor is now defined as

$$\underset{\sim}{T} = \sum_\alpha (\underset{\approx}{\tau}^\alpha - \langle\rho\rangle_\alpha \hat{\underset{\sim}{V}}^\alpha\hat{\underset{\sim}{V}}^\alpha). \tag{38}$$

Equation 37 can be further simplified through application of the mass-conservation relationship for the three-phase continuum given by equation 21:

$$\bar{\rho}\frac{\partial\bar{\underset{\sim}{V}}}{\partial t} + \bar{\rho}\bar{\underset{\sim}{V}} \cdot \underset{\sim}{\nabla}\bar{\underset{\sim}{V}} - \underset{\sim}{\nabla} \cdot \underset{\approx}{T} - \bar{\rho}\underset{\sim}{g} = \underset{\sim}{0}. \tag{39}$$

MACROSCOPIC ENERGY BALANCE

The point of departure in the development of the macroscopic energy balance is the microscopic equation 3. Integration over the REV of this expression multiplied by γ_α yields

$$\int_{d\underline{V}} \left\{ \frac{\partial\left(\rho U + \frac{1}{2} v^2\right)}{\partial t} + \underset{\sim}{\nabla} \cdot \left[\rho \underset{\sim}{V}\left(U + \frac{1}{2} v^2\right)\right] - \underset{\sim}{\nabla} \cdot \underset{\approx}{g} - \underset{\sim}{\nabla} \cdot (\underset{\approx}{\tau} \cdot \underset{\sim}{V}) - \rho\underset{\sim}{V} \cdot \underset{\approx}{g} \right\} \gamma_\alpha dV_\xi = 0. \tag{40}$$

Application of the averaging theorems modifies equation 40 to

$$\frac{\partial\left(\langle\rho\rangle_\alpha \bar{U}^\alpha\right)}{\partial t} + \frac{1}{dV}\left[\frac{\partial}{\partial t}\int_{d\mathbf{V}} \frac{1}{2}\rho v^2 \gamma_\alpha dV_\xi + \underset{\sim}{\nabla} \cdot \int_{d\underline{V}} \rho\underset{\sim}{V}\, U + \frac{1}{2} v^2\ \gamma_\alpha dV_\xi\right]$$

$$- \frac{1}{dV}\underset{\sim}{\nabla} \cdot \int_{d\underline{V}} (\underset{\sim}{q} + \underset{\sim}{t} \cdot \underset{\sim}{V})\gamma_\alpha dV_\xi - \langle\rho\rangle_\alpha \bar{\underset{\sim}{V}}^\alpha \cdot \underset{\sim}{g}$$

$$- \frac{1}{dV}\sum_{\beta\neq\alpha} \int_{DA_{\alpha\beta}} \left[\rho\left(U + \frac{1}{2} v^2\right)(\underset{\sim}{W} - \underset{\sim}{V}) + \underset{\sim}{q} + \underset{\approx}{\tau} \cdot \underset{\sim}{V}\right] \cdot \underset{\sim}{n}^{\alpha\beta} da = 0. \tag{41}$$

In a manner similar to that of preceding sections, the averages of products in the second, third, and fourth terms of equation 41 may be treated using a perturbation expansion for the microscopic velocity, as defined by equation 26. Thus, equation 41 may be rewritten as

$$\frac{\partial\left[\langle\rho\rangle_\alpha e^\alpha + \frac{1}{2}\langle\rho\rangle_\alpha(\bar{\underset{\sim}{V}}^\alpha)^2\right]}{\partial t} + \underset{\sim}{\nabla} \cdot \left\{\langle\rho\rangle_\alpha \bar{\underset{\sim}{V}}^\alpha\left[e^\alpha + \frac{1}{2}(v^\alpha)^2\right]\right\}$$

$$- \nabla \cdot (\underset{\approx}{\tau}^\alpha \cdot \bar{\underset{\sim}{V}}^\alpha + \underset{\sim}{q}^\alpha) - \langle\rho\rangle_\alpha \bar{\underset{\sim}{V}}^\alpha \cdot \underset{\sim}{g}$$

$$- \frac{1}{d\mathbf{V}}\sum_{\beta\neq\alpha}\int_{dA_{d\xi}} \left[\rho\left(U + \frac{1}{2}v^2\right)(\underset{\sim}{W} - \underset{\sim}{V}) + \underset{\sim}{q} + \underset{\approx}{\tau} \cdot \underset{\sim}{V}\right] \cdot n^{\alpha\xi} da = 0, \tag{42}$$

where the macroscopic internal-energy density for the α phase is defined as

$$e^\alpha = \bar{U}^\alpha + \overline{\frac{1}{2}(\tilde{V}^\alpha)^2}^\alpha, \tag{43}$$

and the macroscopic heat flux vector is given by

$$\underset{\sim}{q}^\alpha = \frac{1}{d\mathbf{V}}\int_{d\underline{V}} \left\{\underset{\sim}{q} + \underset{\approx}{\tau} \cdot \tilde{\underset{\sim}{V}}^\alpha - \rho\left[U + \frac{1}{2}(\tilde{V}^\alpha)^2\right]\tilde{\underset{\sim}{V}}\right\}\gamma_\alpha dV_\xi. \tag{44}$$

Here again, in accounting for the descriptive loss in going from the microscopic to macroscopic levels, the macroscopic variables are defined such that they describe the total amount of the microscopic quantity.

The macroscopic energy balance can be summed over all phases in obtaining an expression for the entire continuum. Because equation 42 describes total energy transport, conservation of energy requires the energy transfer between the α phase and all other phases as a result of mass exchange, conduction, and mechanical interaction to sum identically to zero. Therefore, the composite form of equation 42 is written

$$\sum_\alpha \left\{ \frac{\partial\left[\langle\rho\rangle_\alpha e^\alpha + \frac{1}{2}\langle\rho\rangle_\alpha(\bar{V}^\alpha)^2\right]}{\partial t} + \underset{\sim}{\nabla} \cdot \langle\rho\rangle_\alpha \bar{\underset{\sim}{V}}^\alpha\left(e^\alpha + \frac{1}{2}(\bar{V}^\alpha)^2\right) - \nabla \cdot (\underset{\approx}{\tau}^\alpha \cdot \bar{\underset{\sim}{V}}^\alpha + \underset{\sim}{q}^\alpha) - \langle\rho\rangle_\alpha \bar{\underset{\sim}{V}}^\alpha \cdot \underset{\sim}{g} \right\} = 0. \tag{45}$$

Use of a perturbation expansion for the phase-average velocity (equation 35) in the first three terms yields, after some simplification,

$$\frac{\partial}{\partial t}\left[\bar{\rho}\bar{E} + \frac{1}{2}\bar{\rho}(\bar{V})^2\right] + \nabla \cdot \bar{V}\left[\bar{\rho}\bar{E} + \frac{1}{2}\bar{\rho}(\bar{V})^2\right]$$

$$- \nabla \cdot (\tau \cdot \bar{V} + Q) - \bar{\rho}\bar{V} \cdot g = 0, \qquad (46)$$

where continuum variables V and τ are defined by equations 23 and 38, respectively:

$$\bar{E} = \frac{1}{\bar{\rho}}\sum_{\alpha}\left[\langle\rho\rangle_{\alpha} e^{\alpha} + \frac{1}{2}\langle\rho\rangle_{\alpha}(\hat{v}^{\alpha})^2\right] \qquad (47)$$

and

$$Q = \sum_{\alpha}\left\{q^{\alpha} + \tau^{\alpha} \cdot \hat{v}^{\alpha} - \langle\rho\rangle_{\alpha}\left[e^{\alpha} + \frac{1}{2}(\hat{v}^{\alpha})^2\,\hat{v}^{\alpha}\right]\right\}. \qquad (48)$$

Equation 46 can be further simplified through the use of the mass and linear-momentum balances describing the three-phase continuum, equations 21 and 39. Expansion of the first two terms of equation 46, and subsequent combination with equation 21, reduces the continuum energy balance to

$$\bar{\rho}\frac{\partial}{\partial t}\left[\bar{E} + \frac{1}{2}(\bar{V})^2\right] + \bar{\rho}\bar{V} \cdot \nabla\left[\bar{E} + \frac{1}{2}(\bar{V})^2\right] - \nabla \cdot Q$$

$$- \nabla \cdot (\tau \cdot \bar{V}) - \bar{\rho}\bar{V} \cdot g = 0. \qquad (49)$$

The scalar product of equation 39 with the continuum mass averaged velocity results in

$$\bar{\rho}\left(\frac{\partial}{\partial t}\right)\left[\frac{1}{2}(\bar{V})^2\right] + \bar{\rho}\bar{V} \cdot \nabla\left[\frac{1}{2}(\bar{V})^2\right]$$

$$- \bar{V} \cdot (\nabla \cdot T) - \bar{V} \cdot \bar{\rho}\, g = 0 . \qquad (50)$$

The terms that appear in this equation represent a kinetic energy balance, which can now be used to further simplify equation 49 to its final form:

$$\bar{\rho}\frac{\partial \bar{E}}{\partial t} + \bar{\rho}\bar{V}(\nabla\bar{E} - \nabla) \cdot Q - T : \nabla\bar{V} = 0. \qquad (51)$$

Equations 19, 20, 39, and 51 are the phase-averaged mass, linear-momentum, and energy-balance equations for a multiphase geothermal reservoir. They obviously do not represent a closed system of equations and must be augmented by constitutive and thermodynamic information. The course of action followed from this point forward is not unique and depends upon the objectives of the analysis and the experimental information available.

CONCLUSIONS

Three levels of investigation are involved in formulating the equations governing the physics of flow in multiphase geothermal reservoirs: (1) microscopic balance equations; (2) macroscopic balance equations defined for each of the phases; (3) macroscopic continuum balance equations defined for the three phases (rock, water, and steam) as a whole.

The microscopic equations are useful only as a point of departure in obtaining macroscopic equations descriptive of the porous medium. Although these macroscopic equations could be used in reservoir simulation, in practice they are replaced by a single, phase-averaged, continuum equation. In proceeding from one level of observation to another, one loses definition, and constitutive relationships must be invoked to close the system of equations. This paper introduces and demonstrates a logical framework for moving through the various stages of averaging required to formulate the equations generally employed in geothermal-reservoir simulators. Perhaps of greater significance, it provides a physical interpretation for the various macroscopic variables for which constitutive equations are required.

ACKNOWLEDGMENTS

This work was supported by the U. S. Department of Energy under Contract No. W-7405-ENG-48 with Lawrence Berkeley Laboratory through Subcontract No. 3143202.

REFERENCES CITED

Gray, W. G., and Lee, P. C. Y., 1977, On the theorems for local volume averaging of multiphase systems: International Journal of Multiphase Flow, p. 333.

Hassanizadeh, S. M., and Gray, W. G., 1978, General balance equations for multi-phase systems: Princeton, N. J., Princeton University Water Resources Program Report.

Manuscript Accepted by the Society October 14, 1980

Geological Society of America
Special Paper 189
1982

Contaminant Migration in Saturated Unconsolidated Geologic Deposits

ROBERT W. GILLHAM and
JOHN A. CHERRY
Department of Earth Sciences
University of Waterloo
Waterloo, Ontario, N2L 3G1, Canada

ABSTRACT

Models that are most commonly used in analyzing the migration of nonreactive contaminants in ground water are based on the advection-dispersion equation derived by spatial averaging of microscopic processes to represent conditions of advection, dispersion, and diffusion at the macroscopic scale. The advection-dispersion equation provides good representation of the results of tracer experiments in saturated columns of porous, homogeneous geologic materials. In field situations, however, the deposits through which contaminants migrate are invariably heterogeneous at the macroscopic scale; hence the applicability of advection-dispersion models is questionable.

In unfractured silty or clayey deposits, diffusion generally controls the migration of ground-water contaminants. In fractured deposits, where advective transport may occur along the fractures, molecular diffusion acts as a mechanism of attenuation, causing transfer of contaminants from the fractures to the porous but relatively impervious matrix. In contrast to the coefficients of hydraulic conductivity and dispersivity, the diffusion coefficient for nonreactive constituents in unconsolidated deposits varies within a narrow range. For these reasons, diffusion-controlled hydrogeologic zones have become of interest because of the potential they offer for long-term subsurface isolation of toxic wastes.

At higher ground-water velocities, advection and dispersion become important and the influence of heterogeneities becomes dominant. Descriptions of contaminant migration under these conditions are often based on the assumption that the number of heterogeneities is very large relative to the volume of the contaminated zone and that the complex macroscopic velocity field produces dispersion that can be described by the advection-dispersion equation even though the degree of dispersion is orders of magnitude stronger per unit travel distance than in laboratory columns. In field applications, the advection-dispersion model is commonly applied. In deposits that are more discretely heterogeneous, an alternative conceptualization of the transport process involves rapid migration by advection along the more permeable zones. This results in irregular contaminant patterns with fingers or stringers containing relatively undispersed contaminant concentrations that can extend far in advance of contaminant fronts in zones of lesser permeability. Considerable apparent dispersion can result if pumped wells or piezometers for sampling include water from both the contaminated and uncontaminated heterogeneities.

Although some field studies provide evidence that contaminant patterns can be strongly controlled by advection along distinct zones of higher permeability, most documented, intensively monitored occurrences of contaminant zones in heterogeneous sand or gravel aquifers and some field tracer experiments in these deposits show concentration patterns that have regularity, with markedly little or no evidence of appreciable fingering. This is the case even at sites where samples were collected from point samplers yielding water from zones that are small in volume relative to many or most of the distinctive heterogeneities. To account for such smoothly dispersed contaminant zones in distinctly heterogeneous deposits, a third conceptualization of the transport process is suggested. As the contaminants are transported primarily by advection in the more permeable heterogeneities, migration by diffusion occurs into the adjacent heterogeneities of lower permeability, thereby reducing the concentrations in the main zones of advection and increasing the concentrations in the zones of lesser flow. Dispersion at the macroscopic scale and at larger scales within the complex flow system is therefore accomplished primarily by molecular diffusion, which acquires its driving force from the numerous local concentration contrasts continually imposed on the system by advection of contaminants in the more permeable heterogeneities. None of the existing mathematical models for contaminant migration in geologic deposits at the field scale has been developed within this conceptual framework.

To account for the behavior of reactive contaminants in porous geologic materials, a reaction term that can be assigned various forms has been used by numerous investigators. The most common form involves a linear sorption isotherm to describe rapid reversible reactions between the solute and the solids. In contrast to nonreactive solutes, solutes that are well described by linear reversible isotherms in batch experiments provide breakthrough results from column experiments that are poorly represented by the advection-dispersion equation with an incorporated linear isotherm. This indicates that the combined processes of advection, dispersion, and sorption are not well represented even at the laboratory scale by the conventional mathematical formulations. Although many contaminant species in ground water are influenced primarily by adsorption-desorption, many others are controlled by precipitation-dissolution or oxidation-reduction. For many years these processes have been a focus of attention by hydrogeochemists, who have concentrated on developing thermodynamically based models for systems in isolation from the effects of dispersion and diffusion, and by soil scientists, who have focused on the behavior of dynamic systems but at the flow-path scales of a metre or two. There is a necessity to incorporate information from these disciplines into concepts and models of more direct relevance to reactive contaminants migrating along flow paths in dispersive, heterogeneous deposits. Integration of the processes of advection, dispersion, diffusion, and chemical reaction into unified models for analysis of contaminant migration in hydrogeologic systems presents problems of scale in measurement of parameters. Chemical and biochemical processes occur in response to very local reaction conditions which are not necessarily best sampled or described at the scale most suitable for measurements of the physical transport parameters such as hydraulic conductivity or the dispersion coefficient.

Although considerable research on contaminant migration in ground-water flow systems has been conducted during recent decades, this field of endeavor is still in its infancy. Many definitive laboratory and field tests remain to be accomplished to provide a basis for development of mathematical concepts that can be founded on knowledge of the transport processes that exist at the field scale. Without these steps, the tools for assessing the impact of man's activities on the quality of the ground-water environment will remain inadequate.

INTRODUCTION

The 1977 Report to Congress on "Waste Disposal Practices and Their Effect on Ground Water" indicated that at least 17 million waste disposal facilities in the United States place over 1.7 trillion gallons of contaminated liquid into the ground each year (U.S. Environmental Protection Agency, 1977). It is therefore not surprising that during the past 10 to 15 years the emphasis in hydrogeology has been gradually shifting from exploration of aquifers and development of ground-water resources to problems of subsurface contamination. Increased awareness in the scientific community and increased public concern regarding environmental issues will undoubtedly cause the trend to continue for some time.

Locating, monitoring, and, where possible, renovating existing zones of ground-water contamination are important aspects of environmental management. The greatest challenge to hydrogeology and related disciplines at the present time, however, is to develop the ability to predict, with a reasonable degree of certainty, the spatial and temporal distribution of ground-water contaminants that would result from a proposed activity. Nowhere is this challenge more evident than in problems associated with radioactive waste management, where predictions are expected to span time periods of thousands or tens of thousands of years. Decisions involving billions of dollars may very well rest upon the predictive capability of hydrogeologists. Principles of solute migrations through geologic materials, which 25 years ago may have been considered esoteric, are today of immediate importance.

During the past few decades, and particularly during the past ten years, many studies have been directed toward the development of an improved understanding of the processes of solute migration through porous media. These studies have included laboratory and field observations of transport processes, theoretical descriptions of the processes, incorporation of the theoretical descriptions into predictive mathematical models, and testing of the predictive models. Such studies, however, have not yet resulted in reliable methods for predicting the future spatial distribution of contaminants in ground-water systems. Because major uncertainties are inherent in our understanding of the physical and chemical processes that occur in natural or contaminated ground waters, the existing predictive models are suspect.

This paper presents a brief review of the theory of solute migration through porous, nonindurated geologic deposits and examines the laboratory and field evidence upon which transport concepts are currently based. Existing predictive models are considered in relation to the recognized transport processes, including molecular diffusion, mechanical dispersion, and some types of chemical interactions.

THE ADVECTION-DISPERSION MODEL

Development

Various models have been developed in order to arrive at a mathematical description of the solute transport process. Fried (1975) has classified these models into three groups: (1) geometric models, of which the work of Taylor (1953) for solute migration in a single capillary is perhaps the best known example; (2) probabilistic models, such as the random-walk model described by Scheidegger (1954); and (3) statistical geometric models, as described by Bear and Bachmat (1967) and Bear (1972).

All three types of models result in mathematical expressions describing solute transport. The statistical geometric approach, however, has been the most widely used, probably because it is developed directly from what are generally perceived to be the physical processes causing solute migration. Since the parameters used in this approach have deterministic physical representations, comparisons of model performance with experimental results can provide an improved understanding of the physical processes. In recent years this approach has been used extensively in field situations and will be adopted as the basis for the following development.

At the microscopic scale of description, solute transport in porous media is the result of physical and chemical processes occurring within the pores and at solid-solution interfaces. Although physically based equations are available for describing the migration process at this scale, solutions to the equations are not available because of the complexity of the boundary conditions and would not be useful because of our inability to measure the appropriate transport parameters. In the statistical geometric approach, a process of spatial averaging is used to replace the microscopic scale of description with a conceptual macroscopic scale. In the macroscopic description, quantities defined at a point represent the average of microscopic effects in the vicinity of the point. The porous medium is microscopically discontinuous and is thus transformed to a fictitious continuum on the macroscopic scale.

The macroscopic description of a porous medium is represented schematically in Figure 1, which is a graph of porosity at a point in a porous medium versus the volume of the medium about the point. If the medium is not homogeneous, and if the initial volume selected includes more than one stratigraphic layer, then, as the volume is decreased, the measured porosity will change as different layers are included in the averaging process. As the volume is decreased to include only a homogeneous region about the point, the porosity will approach a constant

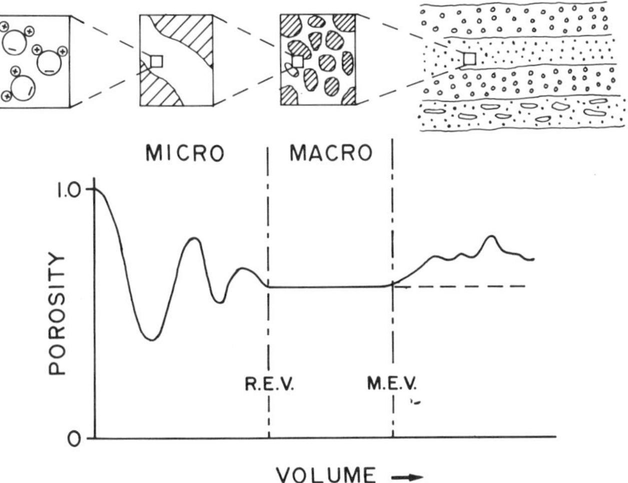

Figure 1. Conceptual representation of the macroscopic scale as applied to the description of a porous medium.

value. With a continued decrease in volume, the porosity will again begin to vary as it becomes dependent upon the proportion of voids and solids in the volume element. In the limit, as the volume approaches zero, the porosity will be zero or one, depending on whether the point is located in a void or in a solid particle. As indicated in Figure 1, the macroscopic description of porosity applies within the region where the porosity is independent of volume. The lower limit of the macroscopic scale has been termed the representative elemental volume (REV) of the medium (Bear, 1972), which for most parameters would include several pores and several particles. In this paper the upper limit of the macroscopic scale is referred to as the maximum elemental volume (MEV). A formal definition and description of the spatial averaging process is given by Bear (1972). In the development that follows, all parameters are defined at the macroscopic scale.

Solute transport is generally viewed as the net effect of two processes, advection and dispersion. Advective transport is attributed to the average motion of the fluid. In processes involving displacement of miscible fluids in one dimension, with advection the only transporting mechanism, a sharp front would be maintained between the initial and displacing fluids, and this front would move at a velocity equal to the average linear pore-water velocity. The average linear pore-water velocity, defined macroscopically at a point in the porous medium, is the microscopic solution velocity averaged over an REV about the point; in the remainder of this paper, it is referred to as the average solution velocity, $\bar{V}$. The advective solute flux is given by

$$\bar{J}_a = \bar{q}C, \tag{1}$$

where $\bar{J}_a$ is the mass of solute crossing a unit area of porous medium oriented normal to the flow direction per unit time $[ML^{-2}T^{-1}]$, $\bar{q}$ is the specific discharge or Darcy flux $[LT^{-1}]$, and C is the solute concentration (mass of solute per unit volume of solution) $[ML^{-3}]$.

Dispersion causes mixing at the interface between two fluids of different composition, and, as defined by Fried (1975), "... is the occurrence and evolution of a transition zone between two domains of the fluid phase with different composition." Bear (1972) further describes dispersion as being nonsteady and irreversible. The effects of advection and dispersion are shown schematically in Figure 2 for one-dimensional displacement of miscible fluids.

Dispersion is the result of two processes, molecular diffusion and mechanical mixing. Diffusion is the result of the thermal kinetic energy of the molecules, which, in the presence of a concentration gradient, results in a net flux of solute toward the low-concentration zone. By modifying Fick's first law to account for the presence of the solid phase, the diffusion flux in a porous medium becomes

$$\bar{J}_d = -\eta D_d \text{ grad } C, \tag{2}$$

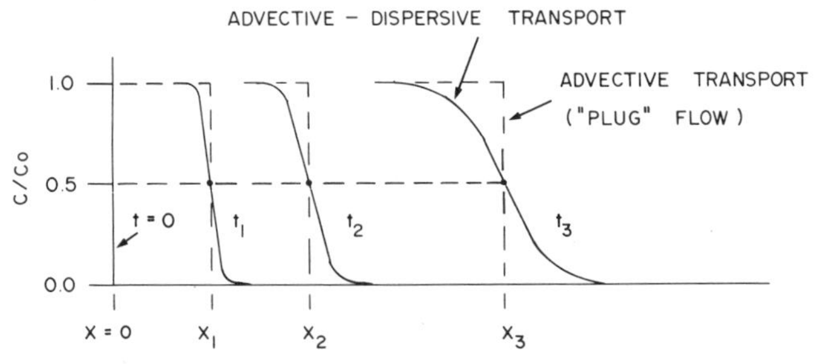

$X_3 > X_2 > X_1 > X$

$t_3 > t_2 > t_1 > t$

$X_3/t_3 = X_2/t_2 = X_1/t_1 = \bar{V}$

Figure 2. Schematic representation of the advective and dispersive solute transport processes.

where $\bar{J}_d$ is the diffusive solute flux (mass of solute per unit area of porous medium per unit time) $[ML^{-2}T^{-1}]$, D_d is the coefficient of molecular diffusion in the porous medium (effective diffusion coefficient) $[L^2T^{-1}]$, η is the porosity of the medium, and C is the solute concentration. The porosity term is introduced because the cross-sectional area available for diffusion is reduced by the presence of the solid phase.

As a consequence of the tortuous diffusion pathway within a porous medium and the presence of the solution-solid interface, the effective diffusion coefficient is generally less than the diffusion coefficient in free solution, D_o. From a review of data on unconsolidated granular media obtained by several investigators, Perkins and Johnston (1963) suggested that the value of D_d/D_o is approximately 0.707. This is in reasonable agreement with the expression given by Bear (1972),

$$D_d = D_o\tau, \tag{3}$$

where τ is the tortuosity of the granular medium, with a value close to 0.67, as suggested by Bear. Further discussions of the effective diffusion coefficient are given in a later section of this paper.

The mechanical-mixing component of the dispersion process is the result of velocity variations within the porous medium. For homogeneous media, there are three microscopic mechanisms that give rise to velocity variations. These mechanisms, shown schematically in Figure 3, include (i) the velocity distribution associated with the flow of viscous fluid through a pore, (ii) variations in velocity as a result of different pore geometries, and (iii) fluctuations in the stream lines with respect to the mean flow direction. The mechanical-mixing component of dispersion is analogous to diffusion in that spreading of the solute is the result of velocity variations across a concentration gradient. Consequently, on the macroscopic scale, the mechanical-mixing component of the solute flux is commonly represented by an equation analogous to equation 2,

$$\bar{J}_m = -\eta D_m \text{ grad } C, \tag{4}$$

where D_m is the coefficient of hydrodynamic (mechanical) dispersion. Adding equations 2 and 4 gives the dispersive flux of the solute as

$$\bar{J} = -\eta D \text{ grad } C, \tag{5}$$

where D is the dispersion coefficient, defined as

$$D = D_d + D_m. \tag{6}$$

Recognizing the advective and dispersive flux components, and applying the principle of conservation of mass gives

$$\frac{\partial(\eta C)}{\partial t} = \text{div}(\eta D \text{ grad } C - \bar{q}\, C). \tag{7}$$

The specific discharge $\bar{q}$ in equation 7 is generally replaced by the average solution velocity according to the relation

$$\bar{V} = (\bar{q}/\eta)^w, \tag{8}$$

where $\bar{V}$ is the average solution velocity $[LT^{-1}]$ and w is an empirical exponent. The work of Ellis and others (1968), as well as many other one-dimensional miscible displacement experiments, indicates that the value of the exponent is very nearly 1 in granular materials.

Substituting equation 8 into equation 7 and assuming the porosity to be constant and the fluid incompressible gives

$$\frac{\partial C}{\partial t} = \text{div}(D \text{ grad } C) - \bar{V} \text{ grad } C. \tag{9}$$

This is frequently referred to as the advective-dispersion equation, or simply the dispersion equation for solute transport. A rigorous development of equation 9 shows the tortuosity and dispersion coefficients to be second-rank tensors (Bear, 1972).

Although equation 9 is in a more general form than is commonly applied to problems of contaminant migration, it is nevertheless subject to several assumptions and limitations:

1. The contaminants are soluble in water.
2. The fluid properties (density and viscosity) are independent of solute concentration.
3. The fluid is incompressible.
4. The coefficients of molecular diffusion and mechanical mixing are additive.

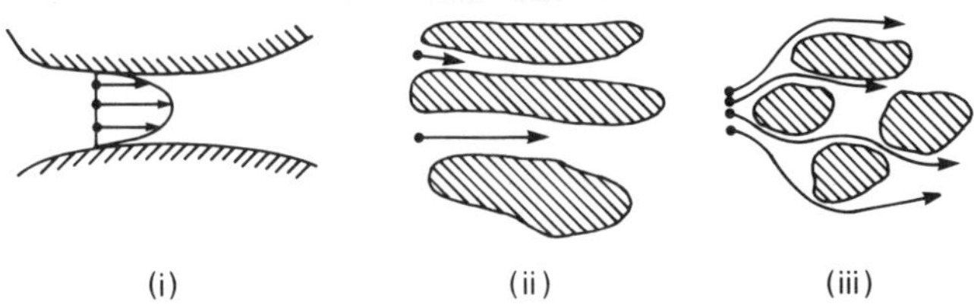

Figure 3. Microscropic components of mechanical mixing.

5. Equation 8 is applicable.
6. The solute is nonreactive.

Laboratory Evaluation of the Model

A large number of laboratory experiments have been conducted to determine the validity of the advection-dispersion model. The majority of these have been one-dimensional miscible-displacement experiments in which the concentration history of the column effluent was compared to the concentrations predicted by the model. Under these conditions, and assuming the medium to be homogeneous, equation 9 reduces to

$$\frac{\partial C}{\partial t} = D_L \frac{\partial^2 C}{\partial x^2} - \bar{V} \frac{\partial C}{\partial x}, \qquad (10)$$

where x is the direction of flow and D_L is the component of the dispersion tensor in the direction of flow (longitudinal dispersion coefficient).

For an experiment in which a column of soil is initially saturated with solution having a tracer concentration of zero and in which the initial solution is displaced with solution of concentration C_o, the concentration history in the column effluent is generally represented by a breakthrough curve: a graph of dimensionless concentration (C/C_o) versus dimensionless time represented by the number of pore volumes. Here C is the measured concentration in the effluent, and pore volume (PV) is the volume of effluent divided by the total volume of voids in the soil sample.

Solutions of equation 10 for a finite column (Brenner, 1962) result in almost straight lines when C/C_o is plotted against $\ln$ PV on probability paper (Passioura and Rose, 1971). Other commonly used solutions of equation 10 are those of Ogata and Banks (1961), for a semi-infinite medium, and of Danckwerts (1953), in which dispersion is treated as diffusion about a moving plane. As discussed by Rose (1973), the Ogata and Banks (1961) and Danckwerts (1953) solutions are good approximations of the Brenner solution, provided values of $\bar{V}L/D$ are large, where L is the column length. Furthermore, again provided $\bar{V}L/D$ is large, the breakthrough curve is symmetrical and passes through the point C/C_o = 0.5 and PV = 1.0.

In comparing predicted and measured breakthrough curves for nonreactive tracers, experimental variability undoubtedly contributes to the observed differences; nevertheless, by considering a large amount of experimental data, Scheidegger (1963) concluded that consistent discrepancies do occur. These discrepancies have been attributed to the presence of dead-end pores (Baker, 1977; Fatt and others, 1966; Coats and Smith, 1964; Deans, 1963) and to solution-solid interface processes not accounted for in the spatial averaging process (Fried, 1975). Fried further noted that discrepancies between measured and predicted results seldom exceed 2 to 3% and concluded that, for practical purposes, the model is a good representation of the dispersion process.

Although the conclusion of Fried appears to be appropriate for coarse-grained geologic materials, a number of situations have been identified for which the advection-dispersion equation is not an adequate model. In fine-grained materials where an anion was used as the tracer, the early appearance of the breakthrough curve has been attributed to anion exclusion (Appelt and others, 1975; Thomas and Swoboda, 1970; Kemper and Rollins, 1966). In aggregated media, an extreme tailing of the breakthrough curve has been attributed to diffusion into or out of the aggregates (Weeks and others, 1976; Passioura, 1971; Passioura and Rose, 1971; Philip, 1968), and similar effects observed for partially saturated media have been attributed to diffusion into relatively immobile zones of water (Biggar and Nielsen, 1960; Orlob and Radhakrishna, 1958). Typical deviations from the model are shown in Figure 4. When the model is extended to include the effects of chemical reactions between the solute and the porous medium, substantial differences between measured and predicted concentration distributions are observed. These differences are further considered below.

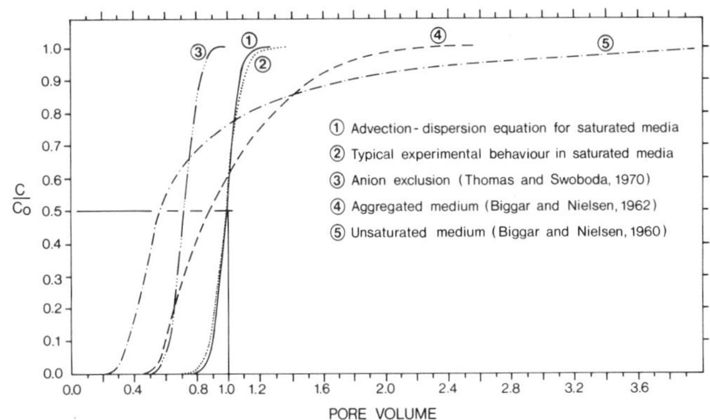

Figure 4. Typical deviations from the breakthrough curve in saturated porous media predicted by the advective-dispersion solute transport model.

In addition to providing tests of the solute transport model, laboratory experiments have provided considerable information on the nature of the dispersion process, such as the effect of velocity on the dispersion coefficient. Much of the available data was compiled by Perkins and Johnston (1963) and later by Rose (1973). Figure 5a is a graph of dimensionless dispersion coefficient D_L/D_o versus Peclet number P_e, defined as $\bar{V}d/D_o$, where d is a characteristic length of the medium, generally taken to be the mean grain diameter. This graph shows the dispersion coefficient to be strongly dependent on the average solution velocity $\bar{V}$ as well as on the geometrical properties of the medium, represented by d. At low Peclet numbers, the dispersion coefficient has a constant value and is equal to the effective diffusion coefficient of the tracer in the porous medium. The dispersion coefficient is observed to increase with increasing Peclet number, becoming a linear function of P_e at Peclet numbers in excess of approximately 1.0. Figure 5a includes a graph of the empirical equation of Perkins and Johnston (1963),

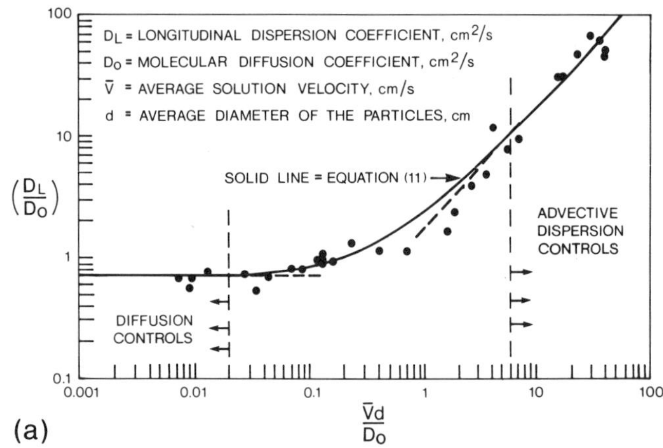

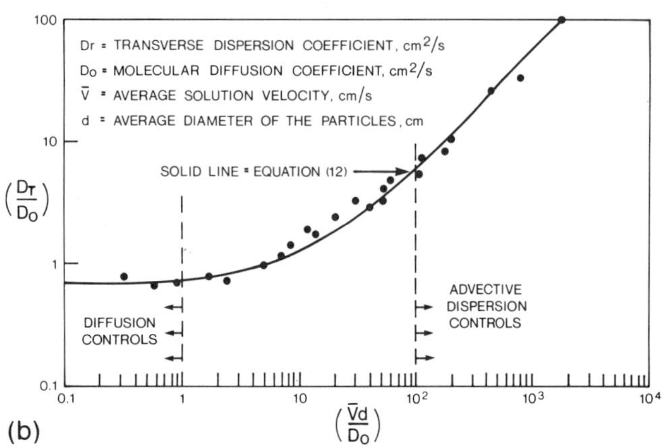

Figure 5. Graphs of the dimensionless dispersion coefficient versus Peclet number, $\bar{V}d/D_O$. (a) D_L/D_O vs P_e. (b) D_T/D_O vs P_e (after Perkins and Johnston, 1963).

$$\frac{D_L}{D_o} = \frac{D_d}{D_o} + 1.75\,\frac{\bar{V}d}{D_o}. \tag{11}$$

A physical interpretation of Figure 5a or equation 11 suggests that at low velocities (low P_e) the dispersion process is predominantly the result of molecular diffusion, whereas at high velocities mechanical mixing processes predominate. At intermediate values of P_e, both diffusion and mechanical mixing play a significant role in the dispersion process.

For steady flow in the x-direction and dispersion in the x- and y-directions, and assuming the medium to be homogeneous and isotropic, equation 9 becomes

$$\frac{\partial C}{\partial t} = D_L\frac{\partial^2 C}{\partial x^2} + D_T\frac{\partial^2 C}{\partial y^2} - \bar{V}\frac{\partial C}{\partial x}, \tag{12}$$

where D_T is the dispersion coefficient in the y-direction, normal to the direction of flow. Figure 5b is a graph of the empirical relationship between D_T and P_e derived by Perkins and Johnston (1963):

$$\frac{D_T}{D_o} = \frac{D_d}{D_o} + 0.055\,\frac{\bar{V}d}{D_o}. \tag{13}$$

Thus the longitudinal and transverse dispersion coefficients appear to be related to the Peclet number in a similar manner; however, diffusion controls the transverse dispersion to Peclet numbers of approximately 1.0, but controls longitudinal dispersion to Peclet numbers of only 10^{-2}. Similarly, mechanical mixing appears to dominate the transverse and longitudinal dispersion coefficients at Peclet numbers in excess of 100 and 1.0, respectively.

Equations 11 and 13 suggest a more general relationship between the dispersion coefficients and $\bar{V}$

$$D_L = D_d + \alpha_L\,|\bar{V}|^m,$$
$$D_T = D_d + \alpha_T\,|\bar{V}|^n, \tag{14}$$

where α_L and α_T are the longitudinal and transverse dispersivities, respectively, and m and n are empirical constants.

Dispersivity, which has dimensions [L] and can be viewed as a characteristic length of the medium, was introduced into the literature on theoretical grounds by Scheidegger (1954, 1961). In situations where mechanical mixing is the dominant dispersive process, m and n have values equal to or close to 1.0, and, for practical purposes (Bear, 1972), equation 14 reduces to

$$D_L = \alpha_L|\bar{V}|,$$
$$D_T = \alpha_T|\bar{V}|. \tag{15}$$

In an appraisal of the validity of this relation for granular materials, Klotz and Moser (1974) conducted 2,500 laboratory dispersion tests on artificial and natural materials with differing grain size, grain shape, grain roughness, and angularity. They concluded that as long as dispersion is primarily due to mechanical mixing, equation 15 is a sufficiently good representation of the experimental results to justify the use of α as an "index property" of the porous medium. Klotz and Moser further observed α_L to be strongly dependent on the median grain size and on the uniformity coefficients of the grain-size distribution and to be only weakly dependent on grain shape, grain roughness, and grain angularity.

Values of α_L determined in the laboratory generally range between 0.01 and 1.0 cm. Fried (1972) suggested 0.1 cm as an average value. Laboratory values of α_L exceed those of α_T by a factor of 10 to 30. In the most general case, dispersivity is a tensor of rank four, being dependent on the directional properties of both the porous medium and the flow field. In practice, it is generally assumed that the dispersivity tensor is symmetrical, that the medium is isotropic, and that the principal direction of the dispersivity tensor is aligned with the velocity vector. This reduces the 81 components of dispersivity to three, the dispersivity in the direction of flow and two mutually perpendicular components transverse to the flow direction. In two-dimensional problems, the dispersivity is further reduced to two components, the longitudinal dispersivity α_L and the transverse dispersivity α_T. The tensorial nature of α is described in detail by Scheidegger (1961) and Bachmat and Bear (1964).

Analytical and Numerical Solutions

Numerous analytical solutions to equation 9 are available, most of which represent one-dimensional cases for steady flow in homogeneous semi-infinite or finite porous media. Most of the solutions of interest to hydrogeologists are included in Ogata and Banks (1961), Brenner (1962), Biggar and Nielsen (1976), and Ogata (1970). Solutions for homogeneous media with uniform steady flow and three-dimensional dispersion are described by Baetslé (1967, 1969) for contaminants entering the systems at point sources, and by Codell and Schreiber (1979) for point-source, line-source, and planar-source inputs. Lenda and Zuber (1970) have presented various analytical solutions, including one for a cuboid source in a uniform, homogeneous flow field. The most comprehensive set of analytical solutions to the advection-dispersion equation was prepared by Cleary (1978), who developed solutions to cases covering a variety of contaminant input conditions, including time-dependent inputs of finite volume and concentration. Analytical methods and solutions are described in the texts by Bear (1972) and Fried (1975). Graphical representations of analytical solutions for analysis of data from field dispersion tests were developed by Lenda and Zuber (1970), Sauty (1977, 1980), and Sudicky and Cherry (1979).

For analysis of contaminant migration in heterogeneous hydrogeologic systems with complex flow patterns and time-dependent boundary conditions typical of many field situations, numerical solution procedures have been developed. These include finite-difference methods, finite-element methods, and the method of characteristics. A summary of the numerical methods used for solutions to the advection-dispersion equation is provided by Marino (1976). A brief description of the numerical models currently available for analysis of ground-water contamination problems is presented by Bachmat and others (1977). The finite-element method is described in detail by Pinder and Gray (1977). A particularly informative review of current models and their application to field problems is given by Anderson (1979). Case histories in which numerical models based on the advection-dispersion equation are used in the analysis of contaminated ground-water systems are summarized by Konikow (1977).

Each solution procedure has limitations when applied to practical field problems. These can include restrictive boundary conditions, numerical dispersion, convergence and stability problems, and excessive computer storage or time requirements. Nevertheless, at the present time the deterministic application of models based on advection-dispersion equation to field contamination problems is generally not limited by an inability to solve the appropriate equations. With the development of more efficient numerical procedures and improved computer hardware, the capability to handle larger and increasingly complex situations will undoubtedly continue to improve. The major obstacle in applying advection-dispersion-based models in a useful manner for the solution of field problems is not the computational difficulty but rather the deficiencies in our capability to recognize and measure the appropriate parameters for model input.

TRANSPORT CONCEPTS FOR HYDROGEOLOGIC SYSTEMS

The good agreement between laboratory measurements and simulations based on the advection-dispersion equation has led, in recent years, to the direct application of the advection-dispersion model to numerous problems of contaminant migration in various field situations. The extension of the model to field problems is problematic because geologic deposits are normally very heterogenous at the macroscopic scale. In contrast, the porous media used in laboratory experiments are relatively homogeneous or include only simple heterogeneities. Our purpose here is to consider the nature of the transport processes in the context of the conditions that exist or appear to exist in field situations and to consider the role of advection, mechanical dispersion, and molecular diffusion.

Contaminant migration in its least complex form occurs when advection and mechanical dispersion have little or no influence relative to diffusion. As a starting point, the following discussion focuses on molecular diffusion.

Molecular Diffusion

Most studies of solute migration in hydrogeologic systems have been concerned with aquifers and the impact of contaminants on ground-water supplies. As a result, hydrogeologists have generally assumed advection to be the primary transport process, and thus assumed the relationship $D = \alpha|V|$ (equation 15) to be applicable. As shown in Figure 5, mechanical mixing dominates the dispersion coefficient only at high Peclet numbers, whereas at low Peclet numbers the dispersion coefficient is approximately equal to the effective diffusion coefficient. For conditions typical of a fine sand aquifer (hydraulic conductivity = 10^{-3} cm/s, d = 0.02 cm, η = 0.35, and hydraulic gradient = 10^{-2}), the Peclet number would be approximately 2 x 10^{-2} for Cl^- with a free-solution molecular diffusion coefficient of 3 x 10^{-5} cm^2/s. From Figure 5a, this is seen to be approximately equal to the Peclet number at which the longitudinal dispersion coefficient is equal to the effective diffusion coefficient. Thus, to the extent that Figure 5a is applicable to natural hydrogeologic environments in uniform, unfractured, fine-grained materials, the dispersion process would be the result of molecular diffusion. Having established that the dispersion coefficient is equal to the effective diffusion coefficient, the relative importance of the advective and diffusive flux components becomes a pertinent question.

An analytical solution to equation 10 for an infinite travel path and a step-function solute input was used as a means of calculating profiles of solute concentration versus distance for specified values of $t > t_o$. In these calculations the range in velocity was such that the Peclet number (assuming d = 0.02 cm) was always in the range where the dispersion process is dominated by molecular diffusion and thus D was set equal to the effective diffusion coefficient (D_d). The set of results shown in Figure 6 indicates that low values of ground-water velocity have little influence on the advance of the solute front. For example, in Figure 6a, at low velocities, a tenfold increase in velocity (0.32 cm/yr to 3.2 cm/yr) results in less than a 50% increase in the front advance at C/C_o = 0.5. In the preparation of these graphs, diffusion coefficients of 1×10^{-5} cm^2/s and 1×10^{-6} cm^2/s were used. These values span the normal range of values reported in the literature for nonreactive solutes in saturated, unconsolidated deposits such as sand, silt, or clay.

A further comparison of the effect of diffusion relative to advection is represented by the graphs

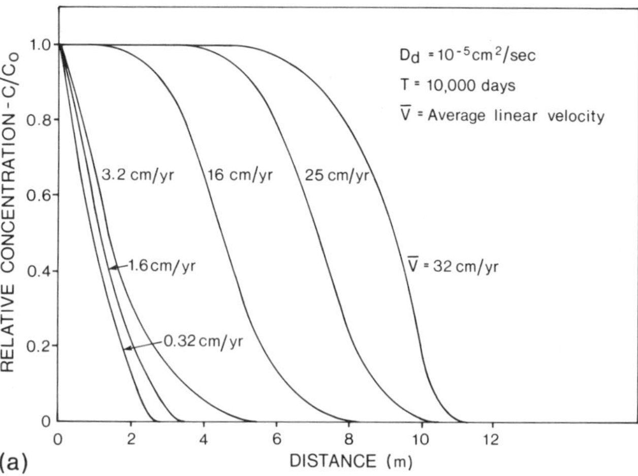

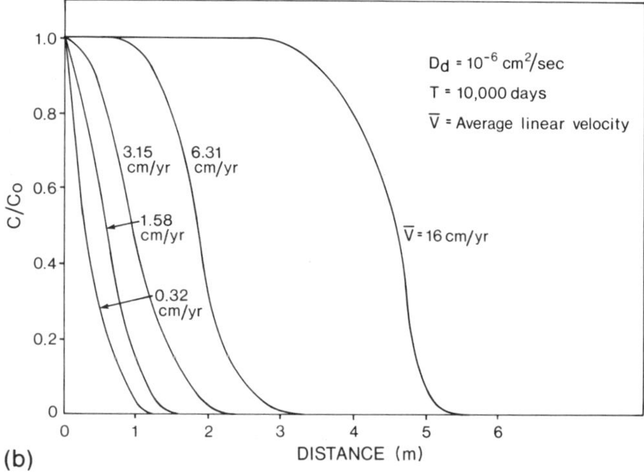

Figure 6. Advance of a solute front along a flow path as a result of advection and molecular diffusion. (a) Effective diffusion coefficient $D_d = 1 \times 10^{-5}$ cm^2/s. (b) Effective diffusion coefficient $D_d = 1 \times 10^{-5}$ cm^2/s.

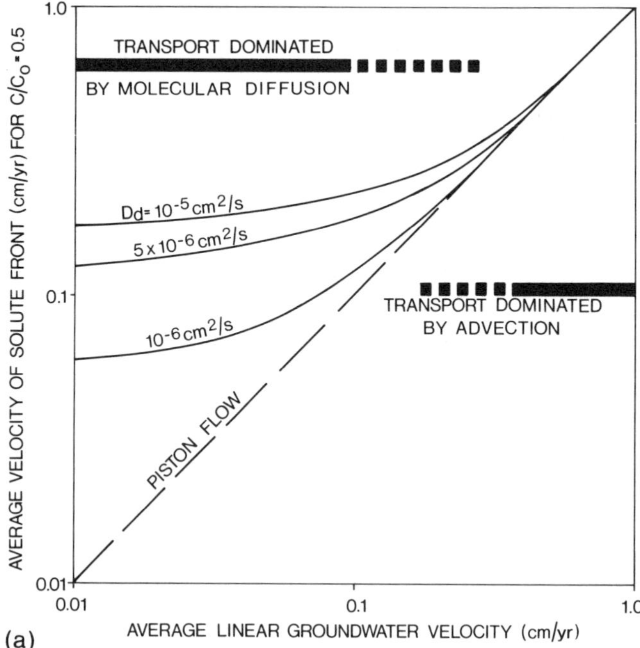

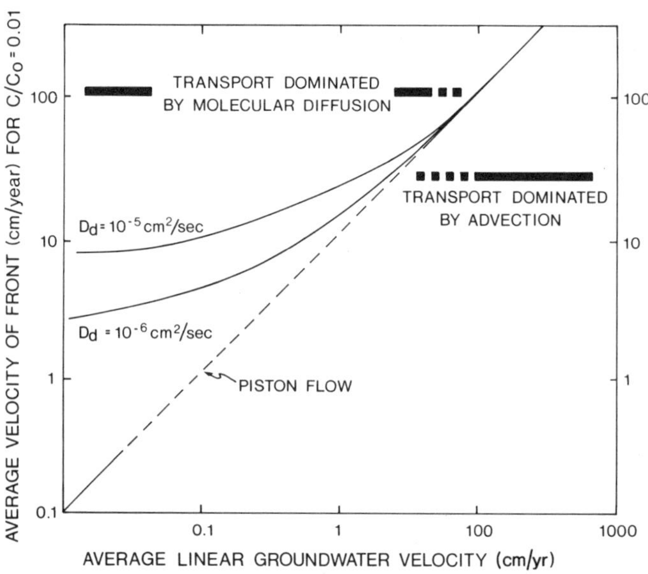

Figure 7. Comparison of the average linear ground-water velocity and rate of solute front advance along a flow path with advection and molecular diffusion. (a) The front is defined by $C/C_0 = 0.5$. (b) The front is defined by $C/C_0 = 0.01$.

in Figure 7, which show the rate of advance of the solute front, calculated in the manner indicated for Figure 6, versus the rate of advance that would occur by advection alone (that is, plug displacement of the solute front). In the development of Figure 7, the rate of front advance was calculated using the C/C_o = 0.01 and 0.5 points on the fronts.

Graphs such as those shown in Figures 6 and 7 are useful in a general way as a means of estimating the conditions under which the various processes are important or unimportant. For example, if we choose 10^{-2} as a representative hydraulic gradient for a particular field setting, 0.35 as a representative porosity, and 1 x 10^{-5} cm^2/s as a representative diffusion coefficient for a nonreactive contaminant, it can be deduced from Figure 7 that in deposits with hydraulic conductivity less than about 10^{-6} cm/s, contaminant migration will be controlled by diffusion and that in deposits with hydraulic conductivity greater than about 10^{-5} cm/s it will be controlled by advection. In the range between these conductivity values, both advection and diffusion would be important. It is reasonable to expect, therefore, that in many hydrogeologic environments consisting of clay, silt, clayey or silty sand or gravel, and very fine sand, the dispersion coefficient is equal to the effective diffusion coefficient, and molecular diffusion is the dominant mode of contaminant migration.

Although hydrogeologists have largely neglected the diffusion process, there is a growing body of field evidence in support of the above conclusion. On the basis of textural properties, field- and laboratory-determined hydraulic conductivity values, vertical hydraulic gradients from piezometer nests, and concentrations of major ions and isotopes obtained from four sites on thick clayey deposits in southwestern Ontario, Desaulniers and others (1980) concluded that the distributions of the natural ^{18}O, ^{2}H, and Cl^- in these deposits are influenced predominantly by molecular diffusion. In profile studies of the ground beneath a sanitary landfill on clayey glacial till near Sarnia, Ontario, Goodall and Quigley (1977) observed that the distributions of major cations from the landfill were consistent with simulated results from a one-dimensional diffusion model. Studies in the Netherlands have identified molecular diffusion as a major process that has contributed during geologic time to the present-day distribution of brackish water within aquifer-aquitard systems (Volker, 1961; Meinardi, 1975). The distribution of the brackish water is a major limiting factor in the development of freshwater zones for water supply. Cherry and others (1979a) and Desaulniers and others (1980) have suggested that thick Quaternary clayey deposits in southern regions of Canada offer potential for burial of toxic wastes to achieve long-term isolation in diffusion-controlled subsurface zones.

Because nonindurated deposits of clay or silt are commonly fractured (Rowe, 1972; Grisak and others, 1976; Burnford and Dixon, 1978), questions commonly arise as to the effect that fractures can exert on the migration of contaminants in these deposits. In an analysis of the migration of tritium in porous, fractured limestone in England, Foster (1975) showed that diffusion of ^{3}H from flowing ground water in the fractures into stagnant pore water in the porous rock matrix could account for rapid loss of tritium from the active ground-water flow system. Day (1977) adopted this "diffusion to the matrix" concept to account for a rapid decline of ^{3}H with depth in fractured clay in the Winnipeg area of Manitoba and developed a combined analytical-numerical model to simulate profiles of ^{3}H versus depth.

Tang and others (1980) presented an analytical model for advective dispersive transport, which is dominant in fractures, and diffusion transport, which is usually dominant in the matrix of low-permeability media. Grisak and Pickens (1980) developed a finite-element model for these transport conditions. With this model, Grisak and others (1980) obtained relatively good agreement between simulations and the results of a laboratory experiment in which a tracer solution was passed through a large column of fractured, clayey glacial till. From these studies, it can be concluded that molecular diffusion can exert a major influence on the rates and patterns of migration of contaminants in fractured argillaceous deposits. Diffusive loss of contaminants from the paths of active flow in the fractures to the matrix can be a dominant mechanism of attenuation.

In many parts of North America, particularly the glaciated regions, clayey or silty deposits of low permeability (and therefore of low velocity) are ubiquitous. Thus it is reasonable to expect that diffusion-controlled hydrogeologic zones are the rule rather than the exception.

In many situations, contaminant-migration distances of only a few tens of centimetres over a decade or two are considered to be of no practical importance. In recent years, however, analysis of the long-term consequences of subsurface disposal of hazardous wastes, particularly radioactive wastes, has required that predictions span many thousands of years or longer. It is therefore necessary to be able to identify diffusion-controlled hydrogeologic environments and to predict the migration of contaminants for these systems (Cherry and others 1979b).

In view of the potential importance of diffusion in the analysis of contaminant transport, it is appropriate to examine further the nature of the diffusion coefficient. The diffusion coefficient in a porous medium is often referred to as the effective or apparent diffusion coefficient. The effective diffusion coefficient, however, is not consistently defined in the literature. In the development presented earlier, the effect of the solid phase in reducing the cross-sectional area available for diffusion (and thus for the diffusion rate) was introduced by including porosity in the diffusion equation (equation 2) as a parameter apart from the effective diffusion coefficient. Under this condition, the relationship between effective diffusion coefficient D_d and the diffusion coefficient in free solution D_o is given by equation 3. Alternatively, porosity is frequently included with the diffusion coefficient giving an effective diffusion coefficient D_d'. The relationship between D_d and D_d' is given by

$$D_d = D_d'/\eta. \tag{16}$$

Measurements of diffusion coefficients in porous media frequently involve the direct fitting of Fick's second law to experimental concentration profiles. The effective diffusion coefficient so obtained is D_d'. Consequently, in using published values of the effective diffusion coefficient in solutions of the transport equation, it is important to determine whether the values represent D_d or D_d' and to make the appropriate adjustments.

From the data summarized by Manheim (1970), Lerman (1978) derived the relation

$$D_d' = D_o \eta^P \quad , \tag{17}$$

where P is an empirical coefficient and all other quantities are as previously defined. Values of P ranged from 1.2 to 2.8, with a mean of approximately 2. Applying equation 16 and recognizing a considerable degree of natural variability in η and τ, the empirical representations of D_d given in equations 3 and 17 are seen not to differ appreciably. Further discussions of the effective diffusion coefficient and alternate expressions for the relationship between D_d and D_o are given by Domenico (1977).

Duursma (1966) reported values of 2×10^{-6} to 6×10^{-6} cm^2/s for experimental diffusion coefficients (D_d') for trivalent and monovalent ions (both positive and negative) in sandy sediments. Manheim (1970) reported that these data agree with other measurements in unconsolidated sediments and stated that the preponderance of experimental evidence indicates a relatively predictable range of diffusion coefficients for ions unaffected by geochemical retardation.

The narrow range of diffusion coefficients suggested by Manheim is encouraging; however, it should be recognized that the majority of published values have come from the disciplines of oceanography and limnology. The applicability of these values to geologic environments is uncertain. Furthermore, in fine-textured materials, the diffusion coefficient can be affected by anion exclusion and the diffusion of ions in the adsorbed phase (Kemper, 1960; van Schaik and others, 1966; van Schaik and Kemper, 1966). In these environments, the relatively simple relationships between effective diffusion coefficient and tortuosity or porosity given by equations 3 and 17 may be inappropriate.

For the reasons presented above, we view molecular diffusion as an important transport process in fine-grained deposits. In a later section of this paper, we suggest that molecular diffusion may also play an important but at present poorly understood role in the process of dispersion in heterogeneous permeable deposits such as sand and gravel. For a detailed evaluation of the nature of the molecular diffusion coefficient and the factors affecting the coefficient values in hydrogeologic systems, further experimental and theoretical investigations are required.

Advective-Dispersive Transport

Although diffusive transport is probably the predominant transport mechanism in unfractured, fine-grained materials and thus of particular importance in the long-term environmental assessment of waste management areas, a wide range of ground-water contamination problems occurs in hydrogeologic regimes where the transport process is controlled by advection. These regimes are of particular importance because, as a result of their relatively high ground-water velocities, they provide the most rapid avenue for contaminant migration from the subsurface to the biosphere. Furthermore, because ground-water supplies in unconsolidated deposits are invariably associated with coarse-textured materials, it is particularly important to understand the mechanism of contaminant transport in these materials.

The preponderance of quantitative information concerning the advection-dispersion process comes from laboratory experiments performed on small homogeneous samples of geologic or artificial porous materials. This information leads to the simplest conceptual model of solute migration in natural hydrogeologic environments--the concept of "homogeneous medium." Applying this concept to contaminants injected continuously as a plane source in a one-dimensional flow field would result in a contaminant front moving in the direction of flow at a velocity $\bar{V}$. The concentration distribution at the front, after a sufficient travel distance from the source, would be Gaussian. Similarly, an instantaneous injection of a contaminant would result in the center of mass of the slug moving at velocity $\bar{V}$, with the dimensions of the slug increasing with travel distance. The concentration distributions would have a Gaussian form both in the direction of flow and in the transverse directions. In applying this concept to field conditions and assuming that the macroscopic scale of description of the medium applies, the actual shape of the concentration profiles is presumed to be determined by dispersion coefficients or dispersivity values comparable to those obtained from laboratory experiments. These concepts are shown schematically in Figures 8a and 8b. Examples of this approach are given by Baetslé (1967, 1969).

Although this concept is appealing in its simplicity, the bulk of the field evidence, together with the results of some laboratory experiments, indicates that the concept of a homogeneous medium is inappropriate for natural hydrogeologic systems. In particular, the observed degree of spreading of a contaminant is generally much greater than would be predicted on the basis of laboratory dispersivity values. This increased spreading is usually attributed to the heterogeneous nature of field environments. An example of the effect of heterogeneities, as perceived by de Josselin de Jong (1958), is shown in Figure 8c. The solute is shown to move most rapidly through a network of high-permeability "fingers" situated in a matrix of lower-permeability materials. The effect of heterogeneities as presented by de Josselin de Jong is supported by the results of several studies using two-dimensional laboratory models. Notable examples are the work of Skibitzke and Robertson (1963), Theis (1967), and Hillier (1975). On the basis of results like those in Figure 8c, Skibitzke and Robertson observed that the presence of small, discrete lenses of high hydraulic conductivity within a sandstone matrix of lower hydraulic conductivity caused much stronger dispersion than when the lenses were not present.

At the laboratory scale of investigation used in the above studies, the distribution of the solute was controlled primarily by the complex velocity field resulting from the heterogeneous nature of the medium. Consequently, we refer to this mechanism of

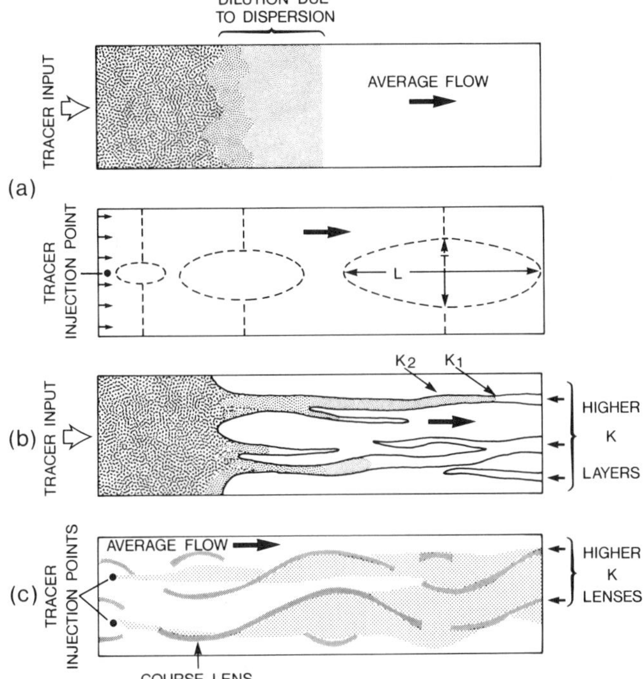

Figure 8. Dispersion in granular materials. (a) Simple dilution front and spread of a slug injection in homogeneous materials. (b) Fingering of tracer front caused by heterogeneities (after de Josselin de Jong, 1958). (c) Spreading caused by high permeability lenses (after Skibitzke and Robertson, 1963).

solute migration as "heterogeneous advection." Heller (1972) suggested that if there are a large number of heterogeneities, if the heterogeneities are randomly distributed, and if the length of the flow path is large with respect to the scale of the heterogeneities, Gaussian concentration distributions will be observed. Similar arguments were presented by de Josselin de Jong (1958). Thus the phenomenological aspects of the heterogeneous advection process would be similar to those predicted by the advection-dispersion equation. This has led to the commonly held view that heterogeneities cause dispersion. It should be noted, however, that Gaussian concentration distributions would be observed only if water samples were collected over some representative number of heterogeneities, in which case the concentrations could not be considered macroscopic. For the advection-dispersion model to rest on a deterministic foundation, the equations would have to be redeveloped with parameters defined at a larger scale. This is considered further below.

Although two-dimensional laboratory models have served a useful function in evaluating solute transport processes, the experiments are generally conducted at spatial and temporal scales that are compressed with respect to the natural processes in hydrogeologic environments. The applicability to natural environments of interpretations based on such experiments is therefore uncertain. However, to determine the specific nature of the transport process in unconsolidated granular deposits, it is necessary to conduct detailed monitoring of the contaminant concentration distributions. Because of the expense and time that is normally associated with this type of endeavor, monitoring at the level of detail required to obtain conclusive evidence of the nature of the contaminant transport process is rarely done. This conclusion was drawn by Theis (1967) and more recently by Palmquist and Sendlein (1975). Although the number of field studies of sites of ground-water contamination has increased markedly during the past few years, the number of intensely documented case histories is still meager relative to the magnitude of the problem.

Childs and others (1974) used a closely spaced, three-dimensional ground-water sampling network to investigate contaminant migration patterns in sandy deposits near tile-field systems that dissipate the water from septic tanks. They concluded that "the migration patterns . . . bifurcate vertically, indicating that the plumes may occupy several levels below the water table with changes in loading rate or water supply, or that the plumes migrate along zones in the regolith that, although they are texturally similar, show subtle differences in fabric that result in slight variations in permeability." They indicated that even where the sand is homogeneous to the eye, complex bifurcating plumes were produced, and concluded that "data collected under the assumption of a single plume model frequently do not allow an accurate characterization of system behavior." The complex migration patterns observed by Childs and others (1974) and the conclusions based on those observations tend to support heterogeneous advection as a useful conceptual model for the transport of contaminants in heterogeneous geologic environments.

During the past four years, the Department of Earth Sciences at the University of Waterloo has conducted detailed monitoring of several zones of ground-water contaminantion. The term "detailed monitoring," as applied to these studies, refers to zones that are monitored areally at numerous sites and at many depths at each site, referred to here as multilevel sampling. In these studies, monitoring was accomplished by sampling from piezometers or tubes with intake zones that are small; in most cases the samples were obtained by drawing water into the piezometers or tubes from domains in the porous media that were generally less than about 0.005 m^3 in volume. The various types of piezometers and multilevel point samplers used in these studies are described by Gillham and others (1980). One of the multilevel samplers is described in detail by Pickens and others (1978a). Examples of the results of these investigations are given in MacFarlane and others (1980), Gillham and Cherry (1978), Trudell (1980), and Blair and others (1980) and described briefly in Pickens and others (1978a).

MacFarlane and others (1980) determined the shape and concentration distribution of a plume of contamination that extends 900 m from a landfill in an unconfined aquifer of glacial outwash. The plume and peripheral zones were sampled at many hundreds of points. The plume has a regular tongue-like shape (Figure 9) in spite of the fact that the

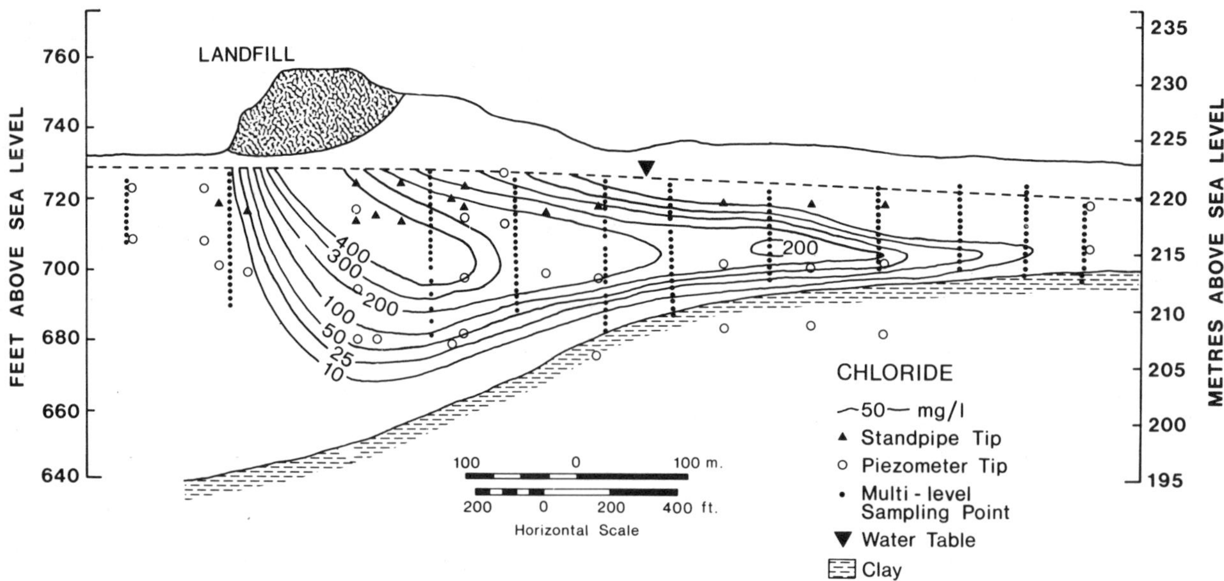

Figure 9. Distribution of a leachate plume (chloride, mg/l) from a landfill situated on an unconfined sandy aquifer in southern Ontario (after Freeze and Cherry, 1980).

outwash deposit is quite heterogeneous because of irregular bedding and lensing of the fine-, medium-, and coarse-grained sand. Vertical concentration profiles of Cl^- and electrical conductance close to and beneath the landfill show considerable irregularity superimposed on the general trend of the plume, but away from the landfill in the down-flow direction, the profiles are smooth, bulge-shaped representations of the plume, indicating that mixing masks the effects of individual heterogeneities even though the contrast in hydraulic conductivity between heterogeneities can commonly be one or two orders of magnitude.

Blair and others (1980) also sampled a large number of points on vertical profiles and delineated a long, tongue-shaped plume of sulfate-rich water from an impoundment of uranium-mill tailings that advanced more than 400 m laterally through a shallow deposit of very heterogeneous sand and fine gravel of glaciofluvial origin. They also observed smooth, bulge-shaped concentration profiles, even close to the contaminant source, which indicates that mixing obliterates any tendency for water of highest sulfate content to preferentially occupy the coarse heterogeneities where advection velocities are largest. Similar conclusions can be drawn from the study by Gillham and Cherry (1978) of nitrate contamination in a shallow, unconfined sandy aquifer in an agricultural area of intensive fertilizer use, although their interpretation is complicated by the occurrence of nitrate loss by denitrification.

Other examples indicating the occurrence of plume-shaped zones with regularity of concentration distributions in heterogeneous granular deposits are described in the literature, although in those studies the monitoring was generally not as detailed as in the examples mentioned above. For example, Kimmel and Braids (1980), Golwer and others (1975), and Palmquist and Sendlein (1975) delineated large, regular, flame- or tongue-shaped plumes of leachate-contaminated ground water in unconfined sandy or gravelly aquifers of outwash or alluvial origin.

Another line of evidence indicating regularity of concentration patterns in heterogeneous sandy deposits is presented by Egboka and others (1980) and Egboka (1980), who used detailed networks of multilevel samplers of the type described by Pickens and others (1978a) to map the distribution of bomb tritium in recharge areas in sandy deposits. Bomb tritium originated in rain and snow as a result of atmospheric tests of nuclear weapons in the period 1953-1967. The tests produced a series of tritium peaks in ground-water recharge during this period. The patterns presented by Egboka and colleagues exhibited relatively smooth tritium profiles, with little or no indication of the peaks and troughs of concentrations that have characterized the occurrence of tritium in rain and snow since 1953.

The conclusion that large-scale contaminant patterns and tritium patterns in heterogeneous sandy aquifers typically exhibit relatively smooth plumes and concentration profiles appears to be inconsistent with the chaotic behavior anticipated on the basis of the heterogeneous advection concept. Furthermore, the highly dispersed nature of the observed plumes is not compatible with the small dispersion coefficient values associated with the homogeneous-medium concept of contaminant migration. As an alternative to these concepts, we propose a third concept, referred to here as the "advection-diffusion" concept.

When contaminants enter the ground-water zone in heterogeneous granular deposits, it is reasonable to expect that they move in complex patterns according to the heterogeneous advection model. As the contaminants are transported, primarily by advection in the more permeable heterogeneities, diffusion causes them to migrate from these heterogeneities into the adjacent heterogeneities of lesser permeability. The net effect is to reduce the concentrations and contaminant flux in the permeable zones and increase the concentrations in the less permeable zones. As time goes on, the diffusion tends to generate increasing uniformity of concentration distribution in the plume and a concentration front that would exhibit increasing spread and a gradual decrease in concentration toward the position of farthest advance. Dispersion at the macroscopic scale and at larger scales within the complex flow system is therefore accomplished primarily by molecular diffusion, which acquires its driving force from transient local concentration gradients imposed on the macroscopic system by preferential advective transport in the more permeable layers or lenses. Figure 10 is a schematic representation of the operation of these processes in a simple heterogeneous system. Dispersion, as it is envisioned in the advection-diffusion concept for heterogeneous systems, is an irreversible process similar to processes of dispersion at the macroscopic scale.

The advection-diffusion representation of the contaminant migration process in heterogeneous unconsolidated deposits is not unlike that proposed for miscible displacement of fluids of different viscosity. In laboratory experiments involving two fluids of different viscosity in homogeneous granular media, Slobod and Thomas (1963) showed that at higher displacement velocities, a distinct fingering of the two fluids was observed, whereas at lower velocities the extent of fingering was greatly reduced. The relative lack of fingering at low velocities was attributed to diffusion from the fingers into the zone of the porous medium between the fingers.

In the advection-diffusion concept of contaminant migration, time and travel distance are necessary to cause plume regularity. In addition, the extent to which the plume is regularized would depend upon the thickness and permeability contrast between adjacent layers of the medium. In this context, a field experiment reported on by Sudicky and Cherry (1979) and by Sudicky and others (1980) is of interest. A slug of chloride tracer was injected into a shallow heterogeneous aquifer and allowed to migrate under the influence of natural ground-water flow. Early in the test, the slug split into two distinct segments, presumably because of small but significant differences in permeability from one side of the test area to the other. Then each of the split segments of the slug rapidly evolved to a classical Gaussian form in spite of the fact that each segment was migrating through materials with numerous distinct heterogeneities.

The advection-diffusion concept of solute migration may have further consequences with regard to the in situ measurement of transport parameters. As represented in the advection-dispersion equation, dispersivity is a characteristic property of the geologic materials through which the contaminants move, and a parameter for which values must be specified so that the equation can be applied in a predictive manner. The test methods can generally be grouped into two categories:

1. Forced advection tests. The tracer is passed rapidly through the aquifer under the influence of a relatively strong hydraulic gradient induced by an injection well and/or pumping well.

2. Natural advection tests. The tracer is pumped rapidly into an aquifer and then allowed to migrate slowly through the aquifer in a natural regime rather than an induced flow regime. In both types of tests, monitoring devices (which might include a well slotted over the thickness of the interval of the aquifer of interest, or a network of small sampling points) are used to determine the distribution of the tracer in the aquifer. The dispersivity for each test is determined by fitting an advection-dispersion-based model of the observed tracer distribution.

The forced advection tests offer considerable convenience because they can be conducted with fewer monitoring devices and in much less time than the natural advection tests. However, if the advection-diffusion concept is a valid representation of the dispersion process in heterogeneous media, we can expect field determinations of dispersivity to be rate dependent because the process of diffusion from permeable lenses to less permeable zones is time dependent. This raises questions with regard to the suitability of field tests of the forced-advection type for deriving dispersivity values for use in predictions of contaminant migration under natural flow conditions.

The possibility of significant rate dependency of field-measured dispersivity values has apparently not been evaluated in field or laboratory experiments. If the results of field dispersion tests prove to be significantly rate dependent, then dispersivity would no longer be a unique geometrically based characteristic of the porous medium. This would further complicate the already difficult problem of field determination of dispersion parameters.

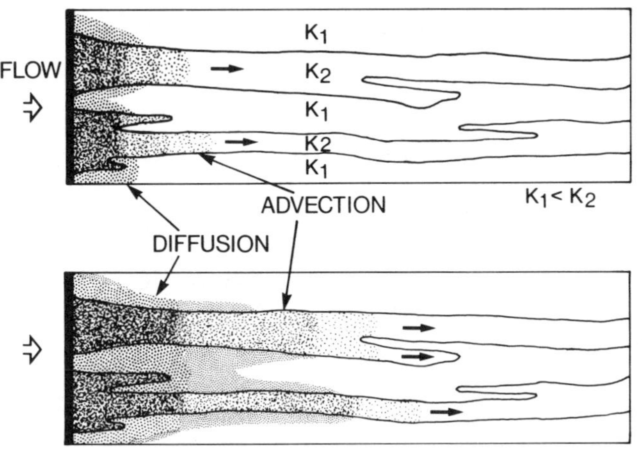

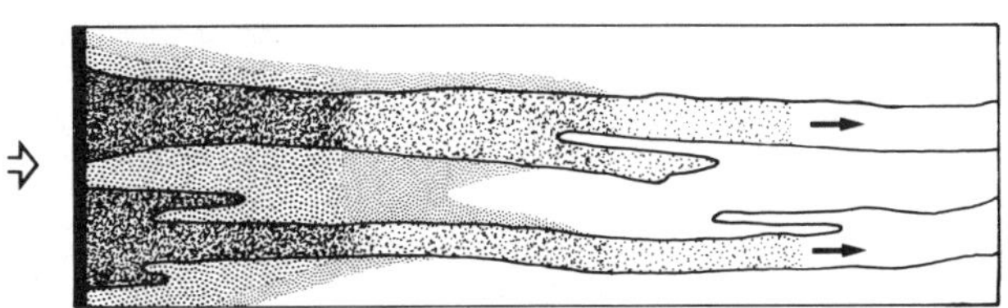

Figure 10. Schematic diagram demonstrating the advection-diffusion transport mechanism.

In summary, there is little evidence in support of the homogeneous-medium concept of contaminant transport in natural hydrogeologic environments. Although considerable evidence indicates that heterogeneities play a major role in the solute transport process, the mechanism through which this role is expressed is uncertain. The heterogeneous-advection model as represented by Figure 8b predicts localized, high-concentration zones of contaminated water extending for considerable distances from the source. On the other hand, the advection-diffusion model (Figure 10) predicts more or less continuous plumes showing a marked decrease in concentration with distance from the source. The contaminants tend to migrate at velocities lower than suggested by the heterogeneous-advection model. The degree of applicability of these two concepts to field situations undoubtedly depends upon such factors as the permeability contrast between adjacent layers, the velocity distribution, and the distance between adjacent high-permeability layers. The type and distribution of heterogeneities in unconsolidated granular deposits varies greatly, depending on the depositional regime in which the sediments were laid down and the source of the sediments. Additional detailed investigations of existing contaminant plumes and of the distribution of environmental isotopes in a broad range of hydrogeologic environments could be instrumental in developing an improved understanding of the solute migration process.

APPLICABILITY OF THE ADVECTION-DISPERSION MODEL

The foregoing discussion leads to the conclusion that the processes of solute migration in natural hydrogeologic systems are not well understood. Formulation of the processes into mathematical expressions for the purpose of making predictions is therefore a tenuous endeavor. The model generally applied to solute migration in natural hydrogeologic systems is the advection-dispersion model. Applications of the model have included the interpretation of field tracer tests, the simulation of existing contaminant plumes, the generation of hypothetical plumes under imposed hydrogeologic conditions, and the prediction of future concentration distributions at waste management facilities. Although there are numerous examples where the advection-dispersion model has been used to simulate existing contaminant distributions, we are not aware of any situation where the predictive capability of the model has been successfully demonstrated. This section considers the applicability of the model in light of our current understanding of the migration process.

The proposed mechanisms of solute migration discussed in the preceding section all involve some combination of advective, dispersive, and diffusive processes. As a result, from a purely theoretical perspective, equation 9 or other representations of the advection-dispersion equation might be expected to provide an adequate predictive model. Major difficulties arise, however, when one considers the scale of the model in contrast with the scale of most problems of hydrogeologic significance.

Equation 9 is based on the principles of spatial averaging, with the quantities in the equation being macroscopic quantities, that is, microscopic quantities averaged over some region of space. As shown schematically in Figure 1, a macroscopic quantity is defined over a region of space having the REV as its lower limit and the MEV as its upper limit. A paramameter can thus be applied to a region of space that is homogeneous with respect to that parameter. Although the REV is conceptually useful, it is of little practical importance because it is generally considered to include several pores and several solid particles and thus occurs at a scale smaller than that at which most field measurements can be made. The MEV is important, however, because it represents the upper boundary of the scale at which the solute transport model (equation 9) is valid. It therefore determines the maximum scale at which the parameters in the model can be specified; in the case of numerical solutions of the transport equation, it determines the maximum scale for discretization of the region of interest. It should also be noted that the macroscopic scale for an equation involving several parameters is the region where the macroscopic definitions of the individual parameters overlap. As a result, the scale at which the equation could be applied would tend to be determined by the parameter exhibiting the greatest spatial variability.

The application of equation 9 to a particular problem requires that the dispersion coefficient and velocity be provided at a macroscopic scale of description throughout the region of interest. Provided that this region is homogeneous and the flow system is relatively simple, specifying the parameters does not constitute a major obstacle. The introduction of a minor degree of heterogeneity, however, results in a complex velocity field, and with the dependence of D on $\bar{V}$ (Figure 5), the problem of parameterization rapidly becomes quite intractable.

Applications of equation 9 to problems of contaminant migration in active ground-water systems have generally considered the problem to be two-dimensional and the dispersion coefficient to be dominated by mechanical dispersion rather than molecular diffusion; that is, it has generally been assumed that $D = \alpha\bar{V}$. With these assumptions, the velocity components in the x- and y-directions are determined and values of the dispersivities in the direction of flow (α_L) and normal to the direction of flow (α_T) are specified. Application of the advection-dispersion model thus requires that every distinct zone within the region of interest be identified and dispersivity of each zone be determined. The velocity distribution can be determined from solutions of the flow equation if the hydraulic conductivity and porosity of each zone are known. It is difficult to conceive of a geologic environment in which these steps could be accomplished with much certainty. For example, sandy environments generally show significant textural changes in both the vertical and horizontal directions and subtle variations or laminations within a region of similar texture. In these situations the macroscopic scale of description may apply to regions of only a few centimetres or less in thickness. Identifying the individual layers and determining their properties would of course be quite impractical. If the advection-diffusion concept of solute migration, as presented in the previous section, is indeed a reasonable representation of the dispersive process at the field scale, a further disadvantage of the model is that the $D = \alpha V$ assumption would not accurately represent the physics of the system. The transverse migration of solutes from one layer to another would be controlled by the diffusion coefficient D_d rather than the mechanical mixing component of the dispersion coefficient D_m.

A possible solution to the impracticalities of the macroscopic scale is to adopt a larger scale of description, which we refer to as the megascopic scale. This scale is conceptualized in Figure 11, an extension of Figure 1, with dispersivity as the dependent variable. As a consequence of encompassing different textural zones, the dispersivity becomes a function of the volume beyond the macroscopic scale. When a sufficient number of layers or lenses are encountered to give a dispersivity value that is statistically representative of the region, it is expected that a further increase in volume would not result in a change in the value of the megascopic dispersivity. The region could be considered "homogeneously heterogeneous." If a further increase in volume were to cause the solute zone to encounter larger-scale heterogeneities, increases in dispersivity beyond the upper limit of this megascopic scale would occur and, by analogy, higher-order megascopic scales controlled by larger scales of heterogeneity could exist.

The megascopic scale, as applied to solute transport, and the mathematical formality of going from the macroscopic scale to the megascopic scale are described by Bear (1977, 1979). In adopting the megascopic scale of description, equation 9 and the concepts used in developing this equation are considered to apply; however, the quantities in the equation are defined on a megascopic scale. That

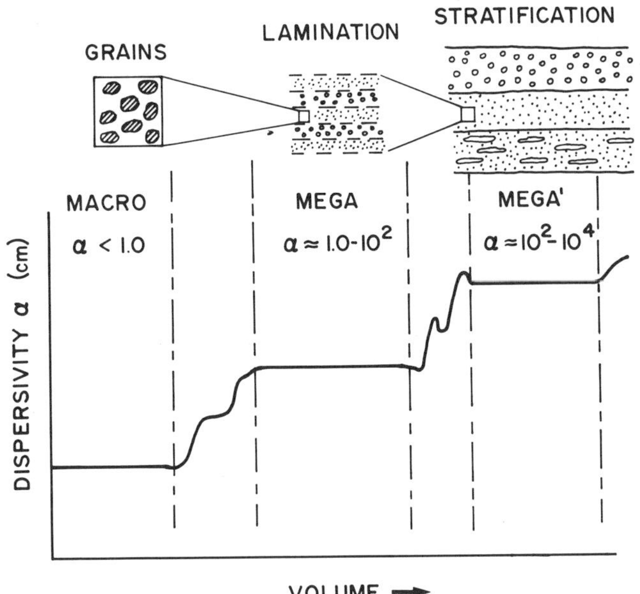

Figure 11. Conceptual representation of megascopic scale as applied to description of a geologic region.

is, the megascopic quantities represent integrated averages of the macroscopic quantities.

Although intuitively attractive, adoption of the megascopic scale presents several difficulties. In order for the megascopic scale to be useful, or indeed to be defined, there must be some range in the scale of description over which the parameter values remain spatially invariant. There is little evidence, either theoretical or experimental, indicating the existence of such a scale range to be a common characteristic of heterogeneous geologic deposits.

The simplest arrangement of heterogeneities is a layering parallel to the direction of flow. From theoretical considerations, Mercado (1967) and Warren and Skiba (1964) showed dispersion in layered systems to be a continuously increasing function of the mean travel distance. In a consideration of similar systems, Gelhar and others (1979) used a stochastic form of the advection-dispersion equation to show that dispersivity is an increasing function of the mean travel distance but that some maximum value is approached as the travel distance becomes large. It should be noted that dispersivity, as used in the above-mentioned studies, is a megascopic parameter, being determined from concentration distributions obtained by integrating macroscopic concentrations along a line normal to the layers.

From simulations of solute transport in media with random heterogeneities, Schwartz (1977) and Smith and Schwartz (1980) indicated that concentration distributions were frequently non-Gaussian and that in many hydrogeologic environments the advection-dispersion equation is an inadequate model. On the basis of computed simulations and laboratory experiments, Heller (1972) arrived at a similar conclusion; he did suggest, however, that a dispersivity value can be defined if there are a large number of randomly distributed heterogeneities and the length of the flow path is large with respect to the size of an individual heterogeneity. This conclusion is supported by the laboratory studies of Martin (1971), as quoted by Fried (1975), which showed that in heterogeneous nonstratified sandstone, the dispersion coefficient approaches a constant value when the distance traveled by the transition zone between two fluids is large with respect to the dimensions of the heterogeneities. In other laboratory studies, the longitudinal dispersivity has been observed to increase continuously with the distance traveled by the tracer (Lawson and Elrick, 1972).

Conclusions regarding the spatial variations of dispersivity based on theoretical considerations or simulation studies are, of course, quite uncertain. This stems from the uncertainty in the applicability to field situations of the conditions and assumptions used in developing the theoretical approaches. Nor are results of laboratory experiments any more conclusive, since the scale of processes observed in the laboratory may have little relevance at the field scale.

Evidence concerning the scale dependence of dispersivity in natural hydrogeologic environments comes principally from two sources: from the trial-and-error fit of solutions of the advection-dispersion equation to existing contaminant plumes and from field tracer tests for the measurement of dispersivity values. Anderson (1979) provides a summary table of dispersivity values reported in the literature for studies in which the trial-and-error-model calibration procedure for large contaminant plumes in alluvial sediments was used. The values for longitudinal dispersivity range from 12 to 200 m and are therefore about four orders of magnitude larger than values normally determined from the use of homogeneous materials in laboratory columns. The studies represented by the values in this table were conducted on different types of monitoring wells in alluvial settings that represented a variety of characteristics and types of contaminant sources; hence comparison of the dispersivity values does not lead to a conclusive statement on the effect of scale apart from the large difference inherent between laboratory values and plume-scale values. There is a tendency, however, for the largest dispersivity values to be associated with the contaminant plumes of largest areal extent.

Table 1 lists the types of field dispersivity tests reported in the literature and includes ranges of reported values from tests conducted in granular unconsolidated materials. The values range from 0.03 to 15 m and in general are intermediate between laboratory values and those derived by fitting the advection-dispersion equation to contaminant plumes.

A small number of field tracer tests have been conducted in which dispersivity values were observed as a function of the mean travel distance of the tracer. On the basis of one such set of experiments, Sauty (1980) concluded that "there is definite evidence that results for dispersivity can vary depending on the distances used in the field" and that "dispersivity increases with distance until a characteristic value is reached where no further increases are observed." In a detailed natural-advection field test in heterogeneous sand, Sudicky and Cherry (1979) and Sudicky and others (1980) observed tracer distributions that were Gaussian in shape in accordance with the classical theory of solute transport and that closely fitted a three-dimensional analytical solution of the advection-dispersion equation. However, the dispersivity values obtained from the analytical solution increased with mean travel distance. The data from this test did not indicate conclusively whether a characteristic dispersivity value was attained at the farthest distance of travel.

TABLE 1. RESULTS OF DISPERSIVITY TESTS PERFORMED ON UNCONSOLIDATED GEOLOGIC MATERIALS

Method	Range of dispersivity values (m)		
	α_L	α_T Horizontal	α_T Vertical
Laboratory (columns and box)	10^{-4} - 10^{-2}	$\approx \alpha_L/30$	---
Tracer slug, natural gradient	10^{-2} - 2	10^{-2} - 3 x 10^{-1}	<0.1
Single-well injection, withdrawal	3 x 10^{-2} 3 x 10^{-1}	---	---
Radial-flow injection test	4 - 11	---	---
Two-well recirculating test	0.5 - 15	---	---
Calibration of digital model to contaminant plumes	3 - 61	1 - 30	0.2 - 0.6

A detailed forced-advection test using a two-well, recirculating, injection-withdrawal test was conducted by Pickens and others (1978b). An apparent dispersivity value of 50 cm was obtained from concentration data from the extraction well that fully penetrated the aquifer, whereas tracer data from within a single layer in the aquifer provided a value of less than 10 cm. A single well test at the same site gave values of about 0.03 m for single layers. By extending the analytical method of Mercado (1967), Pickens (1978) showed the results of the test to be consistent with a model in which apparent dispersivity is a continuous function of scale.

From the field dispersion tests reported in the literature on unconsolidated granular deposits, it can be concluded that increasing dispersivity with scale is probably a normal occurrence. Whether dispersivities in most or all field settings increase with travel distance to some characteristic level, as suggested by Sauty (1980), remains to be determined from field tests at a greater variety of sites.

The concept of continuously increasing values of dispersion parameters as a function of scale or of constant megascopic regions separated by scale-dependent regions, as suggested in Figure 11, is of little practical importance. If one chose to pursue the megascopic approach by assuming the transport parameters to be constant within some range in scale, the experimental difficulties in determining the range applicable for a particular hydrogeologic environment would generally be insurmountable. Moreover, when one considers the expanding nature of a contaminant plume originating from a continuous input at a local source, it becomes clear that different scales of description would be necessary as the plume expands, and thus no single scale could be used to describe the complete evolution of the contaminant plume. Furthermore, as noted by Simpson (1978), the result of simulations in which a large dispersivity value is used to account for the downstream spreading of a large plume in an active flow regime is that contaminants at the source migrate up the hydraulic gradient at an appreciable rate, a condition that is physically unrealistic.

Further limitations of the megascopic scale are demonstrated in Figure 12. This diagram shows a horizontally stratified aquifer with a contaminant that is injected over the entire aquifer thickness. The appropriate megascopic scale, as suggested by Bear (1977), is the entire aquifer thickness. Figure 12 also shows the macroscopic distribution of contaminants in the aquifer at a particular time after start of injection. Megascopic concentrations would be determined from sampling wells slotted over the entire thickness of the aquifer and would thus represent a weighted average concentration over the entire aquifer thickness. The resulting megascopic concentration distribution is shown in Figure 12. Over most of the region, there is no concentration in the aquifer that corresponds to the megascopic concentration. The concentration of contaminants in a well installed in the aquifer would be highly dependent upon the interval over which the well was screened, and this concentration would not be

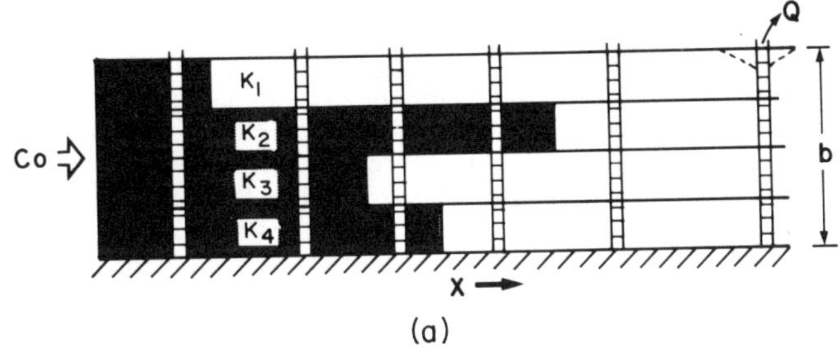

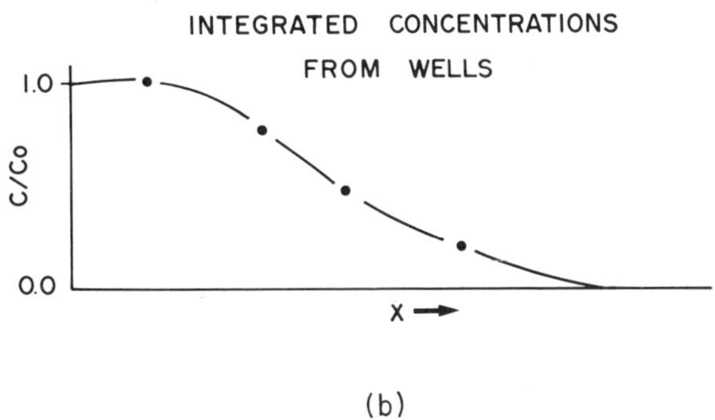

Figure 12. Continuous injection of contaminants into horizontally stratified aquifer. (a) Macroscopic concentration distribution. (b) Corresponding megascopic concentration distribution.

predicted using a megascopic model. The fact that concentrations predicted on the basis of a megascopic model are not necessarily equal to the macroscopic concentrations can also be misleading in cases where some or all of the contaminants of concern undergo chemical reactions as migration occurs. This topic is discussed further in a later section of this paper.

Although the megascopic scale is conceptually attractive, particularly in explaining the large dispersivity values encountered in the field, its occurrence under natural conditions is questionable, and the rigorous application of the advection-dispersion model based on the megascopic scale may be no more practical than the macroscopic approach. Acknowledging the scale dependence of the parameters, there is no justification for predicting migration characteristics beyond the scale or conditions under which the parameters are measured or the model is calibrated. It can be concluded only that in spite of its physical basis and its ability to predict solute migration in uniformly packed laboratory columns, the advective-dispersion model as currently applied to field problems is essentially empirical. No degree of mathematical elegance will overcome the basic limitations in the current knowledge of the physical processes.

As described below, methods other than the one based on the advection-dispersion equation have been developed for the prediction of contaminant migration. These approaches are either very empirical and unrepresentative of the actual field conditions or more sophisticated but in the early stages of development. Because the advection-dispersion approach is well ingrained in the literature and because there is no well-established alternative, we expect that it will remain for some time to come as the principal predictive tool in the analysis of field scale problems. Therefore, it is important that the limitations of the model be appreciated and that an appropriate degree of uncertainty be associated with predictions based on applications of the model.

ALTERNATIVE MODELS FOR CONTAMINANT MIGRATION

There is little doubt that geological heterogeneity is the primary cause of contaminant spreading in natural hydrogeologic systems. The impossible task of mapping the heterogeneities in detail represents the major limitation of the macroscopic advection-dispersion equation, and the fact that the majority of field data suggest the dispersive property of natural geologic materials to be a function of scale presents a major limitation in the application of the advection-dispersion equation with parameters redefined at the megascopic scale. The present challenge is to develop a physically based transport model that incorporates spatially and/or temporally variable dispersion parameters that can be determined in a practical manner and with an acceptable degree of certainty.

Mercado (1967) considered a horizontally stratified aquifer with a normal hydraulic conductivity distribution, and by considering only advection, he concluded that the width of the transition zone is directly proportional to the distance traveled. Warren and Skiba (1964) developed similar expressions for log-normal and arbitrary permeability distributions.

Pickens (1978) extended the work of Mercado (1967) and Warren and Skiba (1964) by deriving expressions relating the longitudinal dispersivity to the distance traveled. In these approaches, the aquifer is considered to be stratified normal to the flow direction, and the tracer is injected over the entire aquifer thickness. In most field situations, these conditions do not exist. Furthermore, to apply the models in a predictive manner, the detailed distribution of hydraulic conductivity would need to be measured in the field and be incorporated into the models. This would be less difficult than attempting a complete macroscopic characterization of the medium but could still prove to be quite impractical for most field problems. The approach suffers from a further difficulty in that it does not allow for dispersion or diffusion between adjacent layers of the medium, and thus the physical basis of the models is questionable.

Schwartz (1977) applied a random-walk model to the analysis of dispersion in hypothetical media in which blocks of lower hydraulic conductivity were inserted into an otherwise homogeneous two-dimensional region. Direct application of the model to field situations would require that each zone of differing hydraulic conductivity be mapped and included in the model. Alternatively, the distribution of heterogeneities could be adjusted until simulated and field-measured contaminant distributions were in reasonable agreement. The model could then be used to predict future concentration distributions. There would be no assurance, however, that the distribution of heterogeneities necessary for a particular contaminant distribution would be unique or that the arrangement of heterogeneities used to calibrate the model could be extrapolated to larger spatial scales. The predictive capability of the model for field conditions is therefore unknown.

Smith and Schwartz (1980) extended the work of Schwartz (1977) by including hydraulic conductivity as a stochastic variable. They use the statistical properties of the hydraulic conductivity distribution determined from field measurements for specification in the model rather than a deterministic description of the medium. Measurement of the required statistical properties is very difficult and raises further questions as to the scale of sampling that should be undertaken for characterization of the medium.

The random-walk approach, as used by Schwartz (1977) and Smith and Schwartz (1980) allows the particles of tracer to move one time step according to the velocity field and then to disperse in a prescribed manner at the end of the time step. After dispersion, the particles are transported through another time step. Although the distribution of particles at the end of any time step can be used to calculate a dispersivity, that parameter is not a necessary input to the model. Although this approach appears to offer an advantage over the advection-dispersion model, it is nevertheless necessary that the distribution of heterogeneities or the statistical properties of the hydraulic conductivity distribution be specified, as well as the manner in which the particles disperse at the end of a time step. In current models, the representation of the process of dispersal is not necessarily a valid representation of the actual process in heterogeneous media.

Recognizing the uncertainty associated with many of the parameters used in both the ground-water flow and solute transport equations, various researchers have introduced stochastic terms into the deterministic differential equations as a means of accounting for this uncertainty. Tang and Pinder (1977) considered the dispersion coefficient to consist of two components, a mean and a variance about the mean. This definition of the dispersion coefficient was introduced into the advection-dispersion equation, and the equation was solved using perturbation theory. A limitation of the approach, as noted by Tang and Pinder, is that the random part of the dispersion parameter must be small compared to the nonrandom part. Furthermore, the stochastic description of dispersion appears to be spatially invariant, and thus the model does not appear to account for the scale dependence in dispersivity.

Gelhar and others (1979) used an approach somewhat different from that of Tang and Pinder in that the variability in hydraulic conductivity and concentration were treated as stochastic processes. The resulting stochastic form of the advection-dispersion equation was applied to a hypothetical horizontally stratified aquifer. The analysis indicated that for large time, the dispersivity approaches a constant value that is dependent on the statistical properties of the medium. Practical limitations of the approach include the problem of determining the statistical properties of hydraulic conductivity for a particular geologic formation, as well as the fact that horizontal stratification is only one of many possible arrangements of heterogeneities. At any particular time, the dispersivity is spatially invariant, and thus the model does not appear to overcome the problem of contaminants moving up the hydraulic gradient from the point of injection.

The simplest alternative to the advection-dispersion equation is to neglect the effects of mechanical dispersion and molecular diffusion and use flow-net or flow-pattern analyses to determine the expected path of contaminant migration. In a series of papers, Nelson (1978a,b,c,d) presents a detailed description of this approach as a means of evaluating the environmental consequences of ground-water contamination. Nelson showed that for some types of problems, the possible effects of dispersion are not sufficiently important to warrant inclusion in the analysis. In cases where the contaminated zone is a slug or cloud arising from a discontinuous source such as a spill, the neglect of dispersion is generally a conservative approach in that the peak concentrations that will arrive at an extraction well or body of surface water will be lower as a result of dispersion than the predicted concentrations from plug-flow analysis. The contaminant will, however, begin to arrive sooner and will cease arriving later. At some field sites the uncertainties in the flow paths and in the advection rates make the inclusion of dispersion unwarranted in the predictive analysis. For some situations, however, it can be shown that dispersion can cause sufficiently strong contaminant attenuation to reduce the hazard to insignificance at some locations along the migration path. In other situations, the dispersive spreading of the zone of contamination can cause the area of significant contamination to become large enough to pose a more severe problem than would be perceived from predictions that do not include dispersion.

Another approach to analyzing the consequences of ground-water contamination is the cell approach, a brief review of which was presented by Bear (1979). A two-dimensional cell model for use in the analysis of the movement of hazardous waste chemicals in and from landfill sites was developed by Elzy and others (1974). In this model the flow domain is represented by an assemblage of cells of rectangular cross section and specified width. Flow through the domain is approximated by vertical flow in specified zones and horizontal flow in others. Since dispersion and diffusion are not accounted for in the model, it is, in effect, a simplified flow net with steady plug flow in uniform materials. The model incorporates the effects of adsorption and rate-dependent biodegradation of contaminants, a simplified approach for vertical moisture movement above the water table, and performs a mass balance on each contaminant in each cell but no mass balance for water. Elzy and colleagues suggested that the model would be useful for evaluation of chemical disposal in landfills.

Another category of cell model is the mixing cell mode, referred to by Gelhar (1977) as a lumped parameter model. In this type of model, a mass balance involving various inputs and outputs of water and solutes is performed on the aquifer or a segment of the aquifer in which complete mixing of all inputs is assumed. The effect of dispersion is therefore implicitly included in the model but as an arbitrary or lumped parameter. An example of this type of model for simple, unconfined aquifers with contaminant sources areally distributed over the aquifer was described by Gelhar and Wilson (1974). The contaminant concentration flowing out of the aquifer is the mixed concentration within the aquifer.

Simpson and Duckstein (1975) described a multi-cell model which they referred to as a finite-state, mixing-cell model. The model has an assemblage of interconnected cells of any specified size through which water and solutes are transported. Each cell can be specified to perform as a cell of complete mixing, a cell of simple piston flow, or a cell that has partial mixing and piston flow.

The mixing-cell and flow-net approaches to predicting contaminant migration do not include dispersion as an integral component of the predictive process. Consequently, although these models may provide useful results in certain situations, their physical bases can indeed be questioned. The layered-aquifer models and stochastic approaches appear to be no less restrictive. Measurement of the necessary statistical properties of the parameters used in the models could prove to be an insurmountable task in most field situations. Furthermore, these approaches, with the exception of those of Schwartz (1977) and Smith and Schwartz (1980), consider the dispersion parameter to be spatially invariant at any particular time. As a result, the large dispersion parameters required to predict downstream contaminant distributions cause the contaminants in the vicinity of the source to move up the hydraulic gradient. The available field evidence suggests that a more realistic approach would be to consider the dispersive parameters to be spatially variable at any particular time, with magnitude increasing with distance from the source.

Extension of the stochastic approach to a consideration of reactive constituents would undoubtedly require a statistical representation of the reactive characteristics of the medium. The fact that there is generally no reason to assume that the dispersive and reactive properties would be correlated could pose a major obstacle to the stochastic approach.

Models other than the advection-dispersion model have seldom been applied to field situations. If these models are to gain acceptance as reliable means of predicting contaminant migration in natural hydrogeologic environments, considerable testing in well-documented field situations will be required.

TRANSPORT OF REACTIVE CONTAMINANTS

The discussion to this point has focused exclusively on the migration of dissolved constituents that do not undergo chemical or biochemical reactions along flow paths. Many solutes of environmental concern, however, are subject to the effects of chemical reactions within the groundwater system, and most problems of practical interest must include consideration of these effects. The chemical processes that can affect the migration of contaminants in ground water include oxidation-reduction (redox processes), precipitation-dissolution, co-precipitation, adsorption-desorption, ion pairing and complexing, and radioactive decay. To incorporate radioactive decay into the advection-dispersion equation (Bear, 1972) requires only the decay constants, which are well established for all radioactive species of relevance. Radioactive decay is the only process that can be accounted for quantitatively without considerable expenditure of effort on laboratory or field measurements. Reviews of the geochemical processes that affect contaminant behavior in ground water are provided by Langmuir (1972) and Back and Cherry (1976).

The one-dimensional form of the advection-dispersion equation for homogeneous media (equation 9), modified to account for the effect of chemical reactions, can be written

$$\frac{\partial C}{\partial t} = D\frac{\partial^2 C}{\partial x^2} - \bar{V}\frac{\partial C}{\partial x} - G, \tag{18}$$

where G is a source-sink term that represents the rate at which the dissolved species is removed from solution (mass of solute per unit volume of solution per unit time, $ML^{-3}T^{-1}$). Expressed in this manner, the reaction term includes whatever specific chemical or biochemical processes are involved in the alteration of the contaminant concentration in the solution phase. To be useful in application, the reaction term must be expressed in a more specific form.

Transport with Sorption

Sorption is the process of mass transfer between the liquid and solid phases that is most commonly incorporated into the advection-dispersion equation. The first step in this incorporation is to express the reaction term in the form

$$G = \frac{\rho}{\eta}\frac{\partial q}{\partial t}, \tag{19}$$

where q is the concentration of the solute in the solid phase (MM^{-1}) and ρ is the dry bulk density of the porous medium (ML^{-3}). If it is assumed that the chemical processes occur rapidly relative to the flow rate and therefore that local chemical equilibrium is achieved, and if it is further assumed that under isothermal conditions the concentration of the contaminant in solution is a function only of the concentration in the solid phase, it follows that

$$\frac{\partial q}{\partial t} = \frac{\partial q}{\partial C} \cdot \frac{\partial C}{\partial t}. \tag{20}$$

Substitution into equation 19 yields

$$\frac{\partial C}{\partial t}\left(1 + \frac{\rho}{n}\frac{\partial q}{\partial C}\right) = D\frac{\partial^2 C}{\partial x^2} - \bar{V}\frac{\partial C}{\partial x}. \tag{21}$$

For chemical species that undergo transfer between the liquid and solid phases as a result of ion-exchange processes, the equilibrium partitioning of the species between the two phases is commonly measured by means of batch tests. In this type of test a specified volume of solution is brought in contact with a known mass of soil for sufficient time to allow equilibrium partitioning of the chemical species to occur between the liquid and solid phases. From these tests, graphs are obtained of the concentration in solution, C, versus the concentration on the solids, q. These graphs are known as isotherms. The slope of the isotherm, $\partial q/\partial C$, is contained in equation 20. Various isotherm forms and their mathematical representations are described by Golubev and Garibyants (1971) and Smith (1970). The most common isotherm reported for contaminants of interest in ground-water studies is the Freundlich isotherm, given by

$$q = K\,C^{a}, \qquad (22)$$

where a and K are empirical coefficients. In some cases, particularly for contaminants at very low concentrations, the isotherm is linear (a = 1) and the proportionality constant is designated as the distribution coefficient, K_d. With these constraints, equation 21 becomes

$$\frac{\partial C}{\partial t} = D' \frac{\partial^2 C}{\partial x^2} - \bar{V}' \frac{\partial C}{\partial x}, \qquad (23)$$

where $D' = D/R$, $\bar{V}' = \bar{V}/R$, and R is the retardation factor (the relative velocity of the reactive constituent), defined as

$$R = \frac{\bar{V}}{\bar{V}'} = 1 + \frac{\rho}{\eta} K_d. \qquad (24)$$

The retardation equation (equation 24) was introduced into the chemical literature by Mayer and Tompkins (1947) and was first incorporated into the advection-dispersion equation for use in chromatographic studies in chemical engineering by Vermeulen and Hiester (1952). This approach to representing the transport of adsorbed contaminants was first reported in the ground-water literature by Higgins (1959) and Nelson (1959). Baetslé (1967, 1969) suggested that the retardation equation can be used to predict the retardation of the position of the center of mass of a contaminant slug or cloud moving from a point source under uniform flow in a homogeneous domain. Contaminant migration in ground water has been analyzed by using numerical solutions to the two-dimensional form of the advection-dispersion equation and incorporating the retardation term represented by equation 24 or equivalent forms; examples of such analyses are provided by Robertson (1974), Pickens and Lennox (1976), Schwartz (1976), and Duguid and Reeves (1976). Before the 1970s, nearly all K_d determinations for contaminants of interest in ground-water studies were for radionuclides. More recently, however, the K_d approach and the form of the transport equation represented by equation 23 have been applied to other contaminants that are hazardous at low concentrations, such as pesticides, PCBs, and other organic compounds.

Although equations 23 and 24 are derived for systems in which the contaminant achieves linear partitioning under local equilibrium along the flow paths and therefore is severely restricted with regard to the types of contaminants and hydrogeologic systems for which it is suitable, these equations commonly appear in the ground-water literature. Equation 23 is convenient because it has the same form as the advection-dispersion equation for nonreactive species and thus can be solved using the same mathematical methods. Analytical solutions for equation 23 and its two- and three-dimensional forms are provided by Baetslé (1967, 1969), Cleary (1978), and Codell and Schreiber (1979).

It should be emphasized that the main appeal of equations 23 and 24 is their mathematical convenience and the fact that simple laboratory tests can be used to obtain K_d values for input to the retardation term. This does not imply, however, that this methodology yields predictions of acceptable accuracy when applied to field problems or, for that matter, to simulation of conditions represented in laboratory experiments in which transport of reactive contaminants is involved. These topics are considered below.

The validity of applying equation 23 to simple hydrogeologic systems can be appraised by comparing simulated results with the results of laboratory experiments in which tracers are passed through columns of saturated granular media under steady flow and other appropriate experimental constraints. A necessary prerequisite for the comparison is that agreement be obtained between simulated and experimental breakthrough curves for nonreactive species (such as Cl^-, ^3H, ^{18}O) in the solution that also contains the reactive species of interest. As indicated earlier in this paper, good agreement is commonly obtained for nonreactive species. Another prerequisite is that the reactive species of interest have a partitioning function, obtained from batch tests, that can be described by a linear isotherm within the concentration range relevant to the column experiment.

The constraints described above were attained by Reynolds (1978) in experiments with a column of fine sand of glacial origin in which strontium was used as the reactive tracer and Cl^- and ^3H as the nonreactive tracers. Figure 13 compares the simulated breakthrough curve for strontium from the column with the measured breakthrough curve. The K used for the reaction term in the simulation was obtained from a batch test on the sand. This test gave a partitioning relation with a high degree of linearity over the entire concentration range of interest. Figure 13 indicates that the measured breakthrough curve is displaced to the right of the simulated breakthrough curve. The displacement indicates retardation that is twice the magnitude of the retardation obtained using the measured K_d. There are very few detailed comparisons between batch-determined retardation factors and column-determined retardation factors reported in the literature, but in the few studies that we have encountered, large differences between the two retardation factors are the rule rather than the

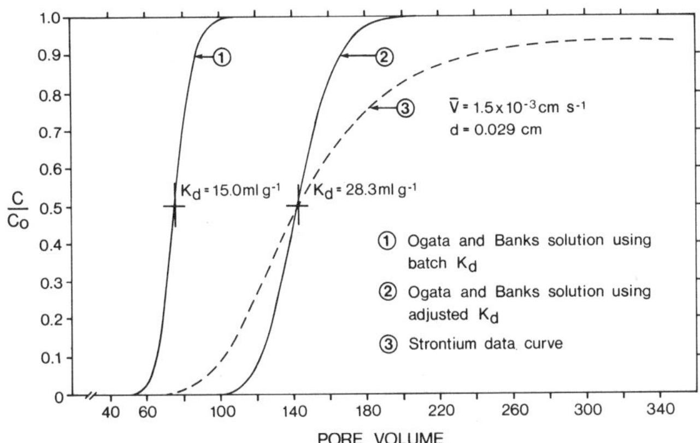

Figure 13. Experimental and simulated column breakthrough curves for strontium (after Reynolds, 1978).

exception. Examples are provided by Nelson (1959) and Schroeder and Jennings (1963). Schroeder and Jennings obtained column K_d values that in many cases were greater than the batch values and that in a significant number of cases were less than the batch values.

Perhaps more important than the differences between the batch and column K_d values is the fact that the shapes of the breakthrough curves differ appreciably from simulated curves based on equation 23. Figure 13 includes a breakthrough curve computed using equation 23 with a K_d value adjusted so that the $C/C_o = 0.5$ point on the simulated curve coincides with the same relative concentration on the measured curve. Inspection of the two curves indicates that although the $C/C_o = 0.5$ points are matched, the measured breakthrough curve is much more spread out than the simulated curve. The laboratory curve has a pronounced asymmetric shape with an extreme extension of the trailing end.

To compare his experimental results with the results of column experiments reported on by other investigators for various reactive constituents with linear isotherms, Reynolds (1978) assembled a group of normalized breakthrough curves. These curves were found to fit within the domain shown in Figure 14. Table 2 lists the sources of data and the experimental conditions for the column experiments. Considering the variety of reactive species used as tracers and the range of experimental conditions, the similarity of the breakthrough curves, which is represented by the narrowness of the envelope curve, is remarkable. The pronounced tailing effect therefore seems to be quite general. This suggests that the tailing effect may result from a common cause. Nelson (1959) also drew attention to the discrepancy between experimental results and simulated results from the advection-dispersion equation with the K_d form of the reaction term and concluded that work should be directed toward developing equations that will adequately define breakthrough curves for long soil columns.

Van Genuchten and Wierenga (1976) matched simulated breakthrough curves to asymmetric experimental curves by adding terms to the advection-dispersion equation to account for the transfer of solute into and out of assumed microzones of immobile fluid in the porous medium. In a more general approach, Cameron and Klute (1977) altered the reaction term in the transport equation by incorporating rate constants for forward and backward reactions. In effect, both these approaches involve the addition of empirical parameters that can produce close matches to experimental data by trial-and-error adjustment of the parameter values. In his column experiments, Reynolds (1978) found that the shape of normalized breakthrough curves for strontium did not vary over a considerable range in V, providing evidence contrary to the kinetic explanations of the tailing effect. As an alternative approach, Reynolds accurately simulated his experimental data using a highly nonlinear empirical isotherm without incorporation of rate-determining parameters.

For calcium migration through laboratory columns, James and Rubin (1978) reported asymmetric breakthrough curves at high transport velocities, but for low velocities they obtained symmetrical curves. The experiments that yielded symmetrical curves were conducted at velocities at which the dispersion process would be expected, on the basis of Figure 5, to be dominated by molecular diffusion.

At the present time there is no generally accepted mechanistic explanation for the pronounced asymmetry of breakthrough curves obtained from column experiments conducted at Peclet numbers above the

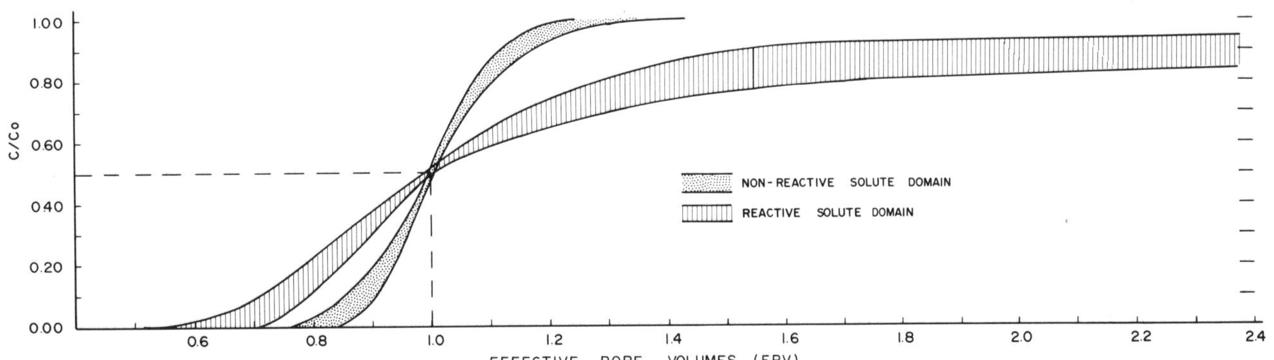

Figure 14. Normalized column breakthrough curves for reactive and nonreactive solutes (after Reynolds, 1978).

diffusion-controlled range. Until it is possible to accurately simulate the results of laboratory column experiments by using equations that are based on a sound understanding of the combined physical and chemical mechanisms that control the transport process and by using partitioning functions that are obtained from experiments that are shown to be appropriate for dynamic conditions, equation 23 or its two- and three-dimensional forms for homogeneous media--or their numerical approximations fcr heterogeneous systems--will continue to provide a questionable means of achieving rigorous analyses of field-scale problems in contaminant hydrogeology.

Transport with Precipitation and Dissolution

The primary mechanisms of transfer between the liquid and solid phases for some ground-water contaminants are precipitation and dissolution. An analysis of the partitioning between liquid and solid phases for the constituents can be carried out using computational models based on mass balance, charge balance, and mass-action relations. Computerized chemical models based on these relations for equilibrium calculations of aqueous systems are commonly used as aids in interpreting data on ground-water quality from natural or contaminanted systems. The models normally serve for computation of saturation indices for the dissolved species with respect to various mineralogical species or amorphous compounds, thereby indicating the degree of disequilibrium of the dissolved species with respect to various solubility constraints. The models also compute the distribution of dissolved species (that is, the concentration of free ions, inorganic ion pairs, and inorganic complexes for each element). These models can provide indications of the thermodynamic controls or lack thereof on various dissolved species. A summary and comparison of computerized hydrochemical models of this type was presented by Nordstrum and others (1978). Two of the most commonly used computer codes, known as WATEQ and SOLMNEQ, were developed by Truesdell and Jones (1974) and Kharaka and Barnes (1973), respectively, for interpreting concentrations of major ions. Ball and others (1978) modified WATEQ for analysis of equilibria involving minor and trace elements in water. Although models such as WATEQ and SOLMNEQ can be useful aids in interpreting the behavior of inorganic contaminants in ground water, they are not designed for predictive purposes.

Predictive-equilibrium chemical models for aqueous systems, which have many of the ingredients of WATEQ, have recently been developed by Plummer and others (1975) and Parkhurst and others (1978). Other models of this type are described by Westall and others (1976) and Mattigod and Sposito (1978). These models are predictive in the sense that they provide computed-equilibrium liquid-phase concentrations for inorganic species, provided that the solid phases (that is, minerals or amorphous compounds)

TABLE 2. NORMALIZED BREAKTHROUGH CURVE CHARACTERISTICS*

						Reactive solutes			Nonreactive solutes		
Researchers	Porous medium	Column length (cm)	Porosity	Bulk density (g/cm^3	Flow velocity (cm/min)	Species	Column† K_d (ml/g)	Flow velocity (cm/min)	Species	Peclet§ number	Dispersivity (cm)
Reynolds (1978)	Chalk R. sand	5.0	0.3461	1.746	0.092	Sr^{2+}	28.3	0.088	^3H	0.95	0.036
Reynolds (1978)	Chalk R. sand	5.0	0.3535	1.726	0.428	Cs^+	167.4	0.435	$C\ell^-$	4.71	0.024
Biggar and Nielsen (1963)	Oakley sand	30.0	0.3967	1.623	0.028	Mg^{2+} (0.05 N)	0.74	0.028	$C\ell^-$	0.61	0.114
Biggar and Nielsen (1963)	Oakley sand	30.0	0.3603	1.721	0.030	Mg^{2+} (0.10 N)	0.31	0.030	$C\ell^-$	0.65	0.216
Elrick and others (1966)	Honeywood silt loam	15.00	0.5130	1.420	0.011	Atrazine	1.43	0.015‡	$C\ell^-$‡	0.32‡	0.107‡
Kay and Elrick (1967)	Honeywood silt loam	7.75	0.4040	1.220	0.019	Lirdane	20.4	0.021	$C\ell^-$	0.45	0.076
								Distribution function parameters			
Model simulation F.D. model	Chalk R. sand	5.0	0.3461	1.746	0.092	Sr^{2+}	0.0 (at $0.5C_0$)	A = -1.15 B = +1.10		P = 0.50	0.035

*After Reynolds (1978).
†Calculated using equation 1.23.
‡Data from Elrick and French (1966).
§$D_0 = 2 \times 10^{-5}$ cm^2/s.

with which the water reacts by way of precipitation or dissolution are assigned and the reaction sequence and thermodynamic data in the form of equilibrium constants or free energies are specified. Although the emphasis in these models is on solubility controls, other liquid-solid phase interactions such as oxidation-reduction and ion exchange can be included (Parkhurst and others, 1978). Comprehensive, predictive-equilibrium water-chemistry models suitable for application to ground-water systems have been available for only a few years, and to date no significant applications to problems of ground-water contamination have appeared in the literature.

The models referred to above are not transport models, in that they do not include velocity or dispersive terms. In the context of contaminant migration in ground water, the question can be raised as to whether it would be appropriate to incorporate relatively sophisticated equilibrium water-chemistry models of the type described by Westall and others (1976), Mattigod and Sposito (1978), or Parkhurst and others (1978) into numerical representations of the advection-dispersion equation as a means of representing all or part of the reactions occurring during transport by advection and dispersion. This would appear to be a fruitful area for research, although difficulties may arise due to excessive computing costs that may be incurred during simulations of complex systems of relevance to field problems. The extent to which most inorganic contaminants in ground water achieve equilibrium with respect to ideal solid phases is at present only in the initial stages of evaluation. The computational capabilities of the existing multicomponent-equilibrium water-chemistry models are in an advanced state relative to the status of studies aimed at evaluation of the importance of equilibrium controls on most of the important contaminant species in ground water.

Although equilibrium chemical routines involving numerous dissolved species and many solid phases have not yet been incorporated into a framework based on the advection-dispersion experiment, equilibrium processes described by the mass-action law were included for a small number of components and reactions by Dutt and others (1972) and Grove and Wood (1979). For example, in a one-dimensional analysis of laboratory column results and of field data, Grove and Wood used a method whereby equation 10 was solved by using the finite-difference method to show the effect of chemical reaction during transport by computing in a cyclic manner new concentrations of dissolved and adsorbed species at each node. The concentrations at the individual nodes were then adjusted for the next time step by applying the advection-dispersion equation to the solute as a nonreactive species and then repeating the reaction computations. Grove and Wood included the activity-coefficient effect and incorporated reactions for complexing of cations and anions, for precipitation and dissolution of gypsum, and for cation exchange (Na^+, Mg^{2+}, Ca^{2+}). They obtained good agreement between model simulations and laboratory and field experiments, and they concluded that the task of adding or deleting chemical reactions to described equilibrium processes involving other minerals or complexes would not be difficult.

Transport with Redox Processes

The discussion of the role of chemical processes on contaminant migration in ground water has to this point focused on sorption and solubility constraints. The literature on ground-water contamination includes studies of contaminants influenced by other chemical processes such as co-precipitation, complexation, and oxidation-reduction (redox processes). Of these, probably the most important are redox processes. Almost all redox reactions of importance in ground-water systems are mediated by bacteria. Dissolved organic contaminants that migrate in ground water commonly decrease in concentration as a result of oxidation: bacteria convert the organic matter to carbon dioxide, biomass, and water. The conversion of nitrate (NO_3^-) to nitrogen (N_2) through the process of denitrification is an example of bacterially mediated reduction.

There are an almost limitless number of contaminants that are influenced by redox processes for which little or no data are available for quantitative representation of their behavior in subsurface systems. An exception to this generalization, however, is NO_3^-, which is the most widespread contaminant in shallow ground water in North America and Europe. The chemical transformations of nitrogen in soil have been the subject of intensive research for several decades. The transformation that is most relevant to ground-water studies is denitrification ($NO_3^- \rightarrow NO_2^- \rightarrow N_2O \rightarrow N_2$) because nitrogen that enters the ground-water system (that is, the nitrogen flux to the water table) is generally in the form of NO_3^-, which is transformed to N_2 as it is transported deeper in the ground-water zone, where reducing conditions typically prevail. On the basis of results of laboratory column studies, Doner and McLaren (1976) developed an expression to describe NO_3^- loss due to steady-state denitrification:

$$G = \frac{d\left[NO_3^-\right]}{dt} = \frac{(\gamma + \beta)_\infty \bar{m}\left[NO_3^-\right]M}{K_m + \left[NO_3^-\right]}, \tag{25}$$

where $\left[NO_3^-\right]$ is the concentration of NO_3^- in solution, γ is N utilized per unit biomass per unit time for maintenance of the bacterial population, β is N utilized per unit biomass in wasted bacterial metabolism, $\bar{m}$ is a steady-state constant, K_m is a saturation constant, and M is the mass of organic matter available to the bacteria that mediate the reaction. γ denotes a maximum for the term in parentheses. Equation 25 contains many parameters that must be known for the equation to be used for simulation of NO_3^- transport in soil water or ground water. Doner and McLaren have successfully used equation 25 as the reaction term for simulation of NO_3^- distributions in laboratory soil columns. Although it is reasonable to expect that this approach is conceptually applicable to NO_3^- transport in ground water, the necessity of obtaining values for the many parameters in the reaction term may render it impractical for analysis of field-scale problems. A similar conclusion can be drawn for other nitrogen transformations, such as nitrification ($NH_4^+ \rightarrow NO_3^-$), which normally occurs above the water table. Reaction terms for this process have also been developed for simulation of nitrogen behavior in laboratory columns (Cho, 1971; Doner and McLaren, 1976). Although the development of relatively deterministic reaction terms for nitrogen transformations in soil columns has not led to equations that can be applied directly to problems of nitrogen migration in ground water, the reaction terms are important as indicators of the parameters that can be expected to exert a significant influence on nitrogen behavior in ground water. The reaction terms may serve as a guide in efforts at developing quantitative methods for analysis of nitrogen transformations in ground-water flow systems. An example of an entirely empirical approach to the incorporation of nitrogen transformations in a model of subsurface movement of nitrogen species has been presented by Duffy and others (1975).

Detailed field studies of the occurrence and behavior of nitrogen species in aquifers are a prerequisite to the development of a methodology to bridge the gap between quantitative laboratory studies of denitrification and the qualitative information currently available on NO_3^- behavior in actual ground-water flow systems. Gillham and Cherry (1978) delineated shallow aquifer zones with abundant dissolved oxygen in which NO_3^- does not undergo reduction and zones with lower but detectable concentrations of dissolved oxygen in which denitrification causes NO_3^- to decline to very low concentrations. Trudell (1980) monitored rates of denitrification following the injection of nitrate-tagged water in a shallow sand aquifer. Because of the importance of microbial factors in redox processes, the transferability of laboratory-derived parameter values to the field is probably minimal; therefore, development of methods such as those described by Trudell for field measurement of process rates seems to be essential.

Description and Sampling Scales

Chemical and biochemical processes in ground water occur in response to very local reaction conditions that, for rigorous description, would require representation on the scale of a single pore or grain surface (the microscopic scale). As in the case of physical transport parameters, measurements of chemical parameters, such as solution concentration, solid-phase concentration, distribution coefficients, selectivity coefficients, or organic-matter content, are made at some macroscopic scale. Use of a macroscopic scale is an unavoidable approach. There is no a priori reason to believe, however, that a macroscopic scale chosen for measurement of the physical transport parameters (hydraulic conductivity, hydraulic head, velocity, dispersivity) is also appropriate for measurement of the chemical parameters that control changes in the contaminant concentration. Determination of scales of description and measurement that are compatible with both the physical and chemical parameters is a task to which almost no analysis has been directed to date.

We will attempt to illustrate some aspects of the scale problem by considering specific situations. The process of denitrification, for example, requires that the redox potential of the pore water (that is, the Eh or pE) be at or below a threshold level that is specified by the thermodynamic constraints on the system and the biochemical constraints on the microbial processes. One of the major thermodynamic constraints is that the water contain not even a trace amount of dissolved oxygen. However, water samples from wells and piezometers in aquifers in which denitrification is known to occur contain appreciable concentrations of dissolved oxygen (Edmunds, 1973; Gillham and Cherry, 1978).

This condition of apparent incompatibility can be attributed to an effect that has become known in the soil science literature as the "microsite" effect. The process of denitrification occurs within the pore space at microsites where the denitrifying bacteria exist and where there are appropriate biogeochemical conditions, such as a lack of dissolved oxygen and an existence of energy-supplying organic matter. The concentrations of the measured dissolved oxygen and dissolved organic matter reflect the concentrations in the pore waters that flow most rapidly to wells or piezometers that are subjected to pumping. These concentrations may bear little relation to the microsite concentrations that actually control the directions and rates of the biologically mediated reactions.

Similarily, there is reason to be concerned about the use of water-chemistry data obtained from field systems at the macroscopic scale (that is, samples from short-interval well screens, piezometers, or "point" samples) in conjunction with models that incorporate reaction terms for processes of precipitation or dissolution described by thermodynamically based equilibrium models. The inputs necessary for computations of equilibrium partitioning between the liquid and solid phases consist primarily of equilibrium constants or free-energy values obtained from laboratory studies using solutions with few components and solid phases in nonflowing conditions. The equilibrium constants pertain to liquid-solid phase systems in which the solution concentration is uniform (that is, fully mixed) and at equilibrium. In porous media in which fluid flow occurs, precipitation-dissolution reactions can take place in each of the many pores within the system and can proceed toward equilibrium at different rates, depending on the ratios of surface to pore volume for the individual pores, diffusion within the pores, and the rates of replacement of pore solution caused by fluid flow. Whether thermodynamically based reaction terms that describe equilibrium partitioning between contaminants in pore water and the solid phases that compose porous media are consistent with the practical scales of ground-water sampling is a question that requires investigation.

In investigations of denitrification in shallow, sandy aquifers, Gillham and Cherry (1978) concluded on the basis of empirical correlations of data obtained from detailed piezometer nests that denitrification can occur when the dissolved oxygen content of the ground water, as represented by piezometer samples, is less than about 2mg/l. Above this level of dissolved oxygen, NO_3 was abundant; below it, little or no NO_3^- was present. Because concentrations of dissolved oxygen declined sharply with depth below the water table, the concentrations of NO_3^- declined sharply with depth, as shown in Figure 15. The 2mg/l threshold has no foundation in thermodynamically based theory, but must bear some relation to the scale of ground-water sampling and the microsite processes.

In studies of ground-water quality in thin, laterally extensive aquifers, it is not uncommon for fully penetrating wells to be used for acquisition of water samples. If this approach had been used in the shallow, sandy aquifers considered in this example, the concentrations, which could be viewed as "megascopic" concentrations, would have been averaged over the aquifer thickness. The data on NO_3^- and dissolved oxygen would therefore have provided little or no evidence of the importance of denitrification on the distribution of NO_3^- in the ground-water flow system and would have rendered futile any attempts at modeling of NO_3^- transport in the aquifer. On the other hand, where vertical components of hydraulic gradients were small, fully penetrating wells in the aquifers would have been adequate for acquisition of hydraulic-head data. Fully penetrating wells also may have provided adequate hydraulic conductivity values from pumping tests or slug tests. In many cases, wells that provide adequate hydrologic data cannot yield adequate hydrochemical data. The problems associated with the acquisition of water samples from wells that have screened intervals across various layers or lenses within stratified aquifers were analyzed by Narasimhan and others (1976), who simulated the concentration yields from wells in several hypothetical layered aquifers. They demonstrated that

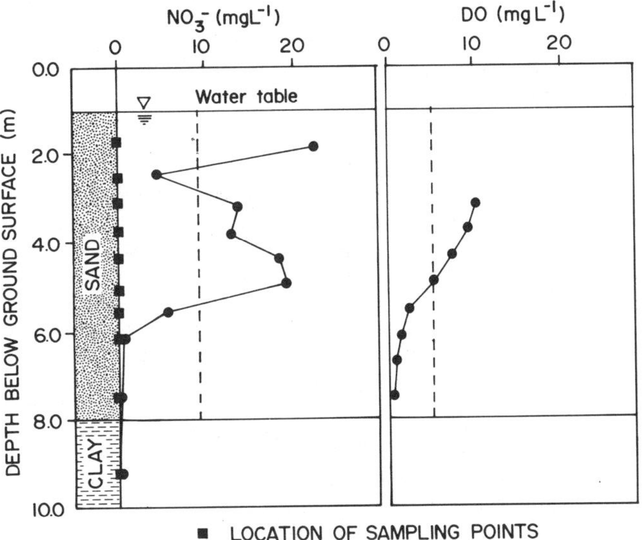

Figure 15. Nitrate and dissolved oxygen profiles as determined from unconfined aquifer in southern Ontario. Dashed lines indicate concentrations that would have been obtained had a "megascopic" approach to sampling been adopted.

wells in a layered aquifer can provide drawdown curves having excellent "match" with Theis' ideal aquifer curves (homogeneous, isotropic case) and therefore that pumping tests can provide no indication of the degree of layering. Simulations of well-yield concentrations, based on the assumption that contaminants migrate in the slightly more permeable layers and not in the other layers in the aquifer, indicated that sampling wells could produce mixed concentrations that approach near steady-state values that represent the contaminant concentrations in none of the aquifer layers. The conclusion (Narasimhan and others, 1976) was that

> water samples from these wells, which are indeed mixtures of waters from different layers, can yield solute concentration data implying that the water meets drinking water standards, while in fact in individual strata, waters may fall far short of acceptable concentrations. Also, from a modelling viewpoint, the mixed samples may falsely suggest a large dispersion coefficient, whereas the actual transport processes may be accompanied by weak dispersion. From a geochemical viewpoint, the water from wells in stratified aquifers can appear to be in a state of thermodynamic disequilibrium when in the aquifer equilibrium exists.

Wells with different lengths of screened intervals and pumped at different rates can, in the same aquifer, yield widely different contaminant concentrations. In many situations, therefore, it is desirable to obtain water samples from wells, piezometers, or other sampling devices that draw in water across small vertical intervals of aquifer and from small lateral distances. Examples of ground-water sampling devices that are consistent with this objective have been described by Edberg (1973), Childs and others (1974), Hansen and Harris (1974), Yare (1975), Pickens and others (1978a), and Sudicky and Cherry (1979). In some situations, nests of piezometers with short-length piezometer tips are well suited for monitoring ground-water quality, although difficulty can be encountered in the choice of depths for the piezometers if prior knowledge of the vertical distribution of contamination is not available.

SUMMARY AND CONCLUSIONS

The development of procedures for modeling the distribution of contaminants in ground-water systems affected by waste management facilities or other changes in land use constitutes a major challenge to the hydrogeologic community. The physical processes contributing to migration of contaminants through granular geologic materials are advection, mechanical mixing, and molecular diffusion. Mathematical expressions for these processes, when combined with the continuity equation, give the advection-dispersion equation. Solutions to this equation, subject to the appropriate initial and boundary conditions, provide predictions of the spatial and temporal distribution of the contaminant.

The physical basis of the advection-dispersion

model and the generally good agreement found between model predictions and the results of experiments with laboratory columns have led to general acceptance of the model and its application to a wide range of ground-water contamination problems. However, fitting of the advection-dispersion equation to contaminant plumes that occur in natural hydrogeologic environments has resulted in two important observations: dispersivity is generally orders of magnitude greater than values obtained from laboratory experiments, and the dispersivity appears to increase as the size of the contaminant plume increases. The large dispersivity values indicate that the macroscopic scale of description used in the development of the advection-dispersion equation is inappropriate for the analysis of field-scale problems. As a result, the physical basis of the model, when applied to field situations, is questionable.

The large dispersivity values associated with natural hydrogeologic environments have been attributed to the effects of geologic heterogeneity. Conceptually, the heterogeneities can cause complex velocity distributions that result in the spread of a contaminant in a manner analagous to the mechanical mixing associated with the intergranular velocity variations of a homogeneous medium. On the basis of this conceptual model, the physical basis of the macroscopic advection-dispersion equation has been extended to field-scale problems by redefining the parameters at a megascopic scale. This approach assumes that the number of heterogeneities encountered by the migrating contaminant is large and that the heterogeneities are distributed such that the medium can be viewed as being homogeneously heterogeneous. The fact that most field evidence suggests dispersivity to be a function of the distance traveled by the contaminant is not consistent with the above assumptions, thus casting doubt on the validity of the continuum approach used in the development of the megascopic advection-dispersion equation. There is little basis for using the equation to predict concentration distributions beyond the scale or conditions under which the parameters are measured. The equation therefore becomes empirical, and predictions based on the equation must be viewed as being highly uncertain.

The introduction of spatially variable or stochastic parameters into the advection-dispersion equation appears to be a constructive approach to the problem; however, models based on this approach are in the early stages of development, and their predictive capability has not been evaluated under field conditions. Furthermore, it should be noted that the adoption of these models would present major practical problems in that the spatial trends or stochastic properties of the parameters would need to be determined experimentally as input to the models.

Although there is little doubt that geologic heterogeneity is a major cause of contaminant spreading in natural hydrogeologic environments, the manner in which the influence is expressed is not clearly understood. From a consideration of velocity variations caused by heterogeneities, one could anticipate the development of chaotic distributions of contaminants. On the other hand, detailed sampling suggests that, although highly dispersed, many contaminant plumes show regular concentration distributions. On the basis of this evidence, an advection-diffusion mechanism of contaminant transport has been proposed. It appears unreasonable to attempt to develop physically based predictive models until the processes of migration are clearly understood. For our predictive capability to advance, there is a need for detailed studies of contaminant migration processes in a wide range of heterogeneous hydrogeologic environments.

In ground-water regimes with low velocity, such as occur in silty or clayey deposits, advection and mechanical dispersion commonly have little or no influence relative to molecular diffusion. Although diffusion-controlled hydrogeologic systems appear to be widespread in nature, there have been few studies of the diffusion process in hydrogeologic systems or of the factors that control the magnitude of the diffusion coefficient. Data from the literature of limnology and oceanography suggest that diffusion coefficients for nonreactive constituents in various types of sediments vary within a range that is smaller than one order of magnitude. This small variability suggests that diffusion-controlled hydrogeologic systems may be more readily amenable to predictive analysis than systems in which advection and dispersion are the dominant contaminant transport processes.

In many investigations of contaminant occurrence and migration in ground water, reactive contaminants rather than unreactive ones are the focus of concern. The problem of predicting contaminant migration in these situations is compounded by combined interactions involving advection, dispersion, diffusion, and chemical reactions such as adsorption, precipitation, coprecipitation, oxidation, reduction, and complexing. In models that include the effects of advection, dispersion, and reaction, the reaction term rarely describes more than the effects of reversible linear sorption represented by linear isotherms. These models have generally provided poor representation of laboratory data. The addition of parameters of a kinetic form to the models provides a means of achieving better fits to laboratory results, but the parameters are little more than an empirical means of achieving the fit. For equilibrium transport of sorbed tracers in laboratory columns, highly nonlinear isotherms also have provided data fits, but the choice of the nonlinearity has no apparent foundation in physicochemical theory. Without resolution of the difficulties encountered in the models applied at the laboratory scale, there can be no firm basis for using the models to predict the behavior of reactive contaminants in field situations that have additional complexity caused by heterogeneities.

Sophisticated thermodynamically based models are available for simulation of the equilibrium behavior of some ionic species that are influenced by precipitation and dissolution and by oxidation and reduction. These hydrogeochemical models are applied to aqueous systems without inclusion of the effects of dispersion and diffusion, and they therefore cannot realistically represent contaminant behavior in advection- or diffusion-controlled groundwater regimes. Combining elements of equilibrium water-chemistry models with advective-dispersive concepts is an endeavor worthy of pursuit.

In the development of transport models for reactive solutes, there is no a priori reason to believe that a scale of measurement chosen for physical transport parameters such as hydraulic conductivity, hydraulic head, and dispersivity is an appropriate scale of description or measurement for chemical parameters such as concentration, adsorption properties, or thermodynamically based equilibrium constants. For example, a well may provide an adequate measurement of the average hydraulic conductivity of the zone intersected by the well screen but when pumped may yield contaminant concentrations that represent a mixture of waters from several zones with different concentrations in each zone. Because chemical reactions are normally described in terms of concentrations at the macroscopic or microscopic scales, concentrations representing mixtures at some megascopic scale may not be useful for predictive purposes. Solutes that are in thermodynamic equilibrium at the macroscopic scale of description can appear to be in disequilibrium when represented by concentrations measured in samples acquired at a megascopic scale.

Although much research on contaminant migration in ground water has been conducted during the past two decades, this field of endeavor is still in its infancy. Many definitive laboratory and field tests remain to be done. It is these tests that will provide much of the framework for development of practical and reliable methods for assessing the impact of man's activities on the quality of the ground-water environment.

ACKNOWLEDGMENTS

During the preparation of this paper, we benefitted from numerous discussions with Emil Frind, Ed Sudicky, and Dan Reynolds, all of the Department of Earth Sciences at the University of Waterloo. Valuable reviews of the manuscript were provided by all of them, as well as by associates at other institutions and agencies, including Mary Anderson, Geology Department, University of Wisconsin; Frank Schwartz, Geology Department, University of Alberta; and Gerry Grisak and John Pickens, National Hydrology Research Institute, Environment Canada. Financial support for some of the work included in this paper was provided by grants from the National Science and Engineering Research Council of Canada.

REFERENCES CITED

Allison, G. B., and Hughes, M. W., 1975, The use of environmental tritium to estimate recharge to a south Australian aquifer: Journal of Hydrology, v. 6, p. 245-254.

Anderson, M. P., 1979, Using models to simulate the movement of contaminants through groundwater flow systems: CRC Critical Reviews in Environmental Control, v. 9, p. 97-156.

Appelt, H., Holtzclaw, K., and Pratt, P. F., 1975, Effect of anion exclusion on the movement of chloride through soils: Soil Science Society of America Proceedings, v. 39, p. 264-267.

Bachmat, Y., and Bear, J., 1964, The general equations of hydrodynamic dispersion: Journal of Geophysical Research, v. 69, p. 2561-2567.

Bachmat, Y., and others, 1978, Utilization of numerical groundwater models for water resource management: U.S. Environmental Protection Agency, EPA-600/8-78-012.

Back, W., and Cherry, J. A., 1976, Chemical aspects of present and future hydrogeological problems, in Proceedings, Advances in groundwater hydrology: American Water Resources Association Symposium, p. 153-172.

Baetslé, L. H., 1967, Computational methods for the prediction of underground movement of radionuclides: Journal of Nuclear Safety, v. 8, p. 576-588.

___1969, Migration of radionuclides in porous media, in Duhamel, A.M.F., ed., Progress in nuclear energy, series 12: Health Physics, v. 2, pt. I.

Baker, L. E., 1977, Effects of dispersion and dead-end pore volume in miscible flooding: Society of Petroleum Engineering Journal, v. 17, p. 219-227.

Ball, J. W., Jenne, E. A., and Nordstrom, D. K., 1978, WATEQ2--A computerized chemical model for trace and major element speciation and mineral equilibria of natural waters, in Jenne, E. A., ed., Chemical modeling in aqueous systems: American Chemical Society Symposium Series 93, p. 815-835.

Banks, R. B., 1974, A mixing cell model for longitudinal dispersion in open channels: Water Resources Research, v. 10, p. 357-358.

Bear, J., 1972, Dynamics of fluids in porous media: New York, American Elsevier, p. 764.

___1977, On the aquifer's integrated balance equations: Advances in Water Resources, v. 1, p. 15-23.

___1979, Hydraulics of groundwater: New York, McGraw-Hill, p. 567.

Bear, J., and Bachmat, U., 1967, A generalized theory of hydrodynamic dispersion in porous media, in Proceedings, Symposium on Artificial Recharge and Management of Aquifers: International Association of Scientific Hydrology, v. 72, p. 7-16.

Biggar, J. W., and Nielsen, D. R., 1960, Diffusion effects in miscible displacement occurring in saturated and unsaturated porous materials: Journal of Geophysical Research, v. 65, p. 2887-2895.

___1962, Miscible displacement: II. Behavior of tracers: Soil Science Society of America Proceedings, v. 26, p. 125-128.

___1973, Miscible displacement: V. Exchange Processes: Soil Science Society of America Proceedings, v. 27, p. 623-627.

___1976, Spatial variability of the leaching characteristics of a field soil: Water Resources Research, v. 12, p. 78-84.

Blair, R. D., and others, 1980, Groundwater monitoring and contaminant occurrence at an abandoned tailings area, in Proceedings, First International conference on Uranium Mine Waste Disposal: Vancouver, Canada, Society of Mining

Engineers of the American Institute of Mining, Metallurgical and Petroleum Engineers, p. 411-444.

Brenner, H., 1962, The diffusion model of longitudinal mixing in beds of finite length: Numerical values: Chemical Engineering Science, v. 17, p. 229-243.

Burnford, A. E., and Dixon, J. M., 1978, Systematic fracturing in young clay of the Cuyahoga River Valley, Ohio, and its relation to bedrock jointing and drainage segments: Geotechnique, v. 28, p. 1201-1206.

Cameron, D. R., and Klute, A., 1977, Convective dispersive solute transport with a combined equilibrium and kinetic adsorption model: Water Resources Research, v. 13, p. 183-188.

Cherry, J. A., and others, 1979a, Hydrogeologic properties and pore water origin and age: Clayey till and clay in south central Canada, in Proceedings, Workshop on Low-Flow Low-Permeability Measurements in Largely Impermeable Rocks: Paris, Organization for Economic Co-operation and Development, Nuclear Energy Agency and International Atomic Energy Agency, p. 31-47.

Cherry, J. A., and others, 1979b, A concept for long-term isolation of solid low-level radioactive waste in fine-grained glacial deposits, in Carter, M. W., ed., Management of low-level radioactive waste, Volume 2: New York, Pergamon Press, p. 1021-1040.

Childs, K. E., Upchurch, S. B., and Ellis, B., 1974, Sampling of variable waste-migration patterns in ground water: Ground Water, v. 12, p. 369-377.

Cho, C. M., 1971, Convective transport of ammonium with nitrification in soil: Canadian Journal of Soil Sciences, v. 51, p. 339-350.

Cleary, R. W., 1978, Analytical models for groundwater pollution and hydrology: Princeton University, Water Resources Program Report No. 78-WR-15.

Coats, K. H., and Smith, B. D., 1964, Dead-end pore volume and dispersion in porous media: Society of Petroleum Engineering Journal, v. 4, p. 73-84.

Codell, R. B., and Schreiber, D. L., 1979, NRC models for evaluating the transport of radionuclides in groundwater, in Carter, M. W., ed., Management of low-level radioactive, Volume 2: New York, Pergamon Press, p. 1193-1212.

Day, M. J., 1977, Analysis of movement and hydrochemistry of groundwater in the fractured clay and till deposits of the Winnipeg area, Manitoba [M.Sc. thesis]: Waterloo, Ontario, University of Waterloo, p. 210.

Deans, H. A., 1963, A mathematical model for dispersion in the direction of flow in porous media: Society of Petroleum Engineering Journal, v. 3, p. 49-52.

de Josselin de Jong, G., 1958, Longitudinal and transverse diffusion in granular deposits: American Geophysical Union Transactions, v. 39, p. 67-74.

Desaulniers, D. E., Cherry, J. A., and Fritz, P., 1981, Origin, age and movement of pore water in argillaceous Quaternary deposits at four sites in southwestern Ontario: Journal of Hydrology (in press).

Domenico, P. A., 1977, Transport phemomena in chemical rate processes in sediments: Earth and Planetary Science, v. 5, p. 287-317.

Doner, H. E., and McLaren, A. D., 1976, Soil nitrogen transformation: A modelling study, in Nriagu, J., ed., Environmental biogeochemistry (Vol. 1): Ann Arbor, Michigan, Ann Arbor Science publishers, Inc., p. 245-258.

Duffy, J., and others, 1975, A simulation model of biophysiochemical transformations of nitrogen in tile-drained corn belt soil: Journal of Environmental Quality, v. 4, p. 477-486.

Duguid, J. O., and Reeves, M., 1976, Material transport in porous media: A finite element Galerkin model: Oak Ridge National Laboratory ORNL-4928, p. 198.

Dutt, G. R. Shaffer, M. J., and Moore, W. J., 1972, Computer simulation model of dynamic biophysiochemical processes in soils: Tucson, University of Arizona Water and Engineering Agricultural Experiment Station Technical Bulletin 196, p. 101.

Duursma, E. K., 1966, Molecular diffusion of radioisotopes in interstitial water in sediments: Vienna International Atomic Energy Agency, SM-72/20, p. 355-371.

Edberg, J., 1973, Design, fabrication and installation of a groundwater sampling device: Los Angeles County, Refuse Department, County Sanitation Districts, Research and Development Division, p. 13.

Edmunds, W. M., 1973, Trace element variations across an oxidation-reduction barrier in a limestone aquifer, in Proceedings, Symposium on Hydrogeochemistry and Biochemistry, Tokyo, 1970: Washington, D.C., The Clark Company, p. 500-526.

Egboka, B.C.E., 1980, Tritium in sandy aquifers [Ph.D. thesis]: Waterloo, Ontario, University of Waterloo, p. 190.

Egboka, B.C.E., Cherry, J. A., and Farvolden, R. N., 1980, Bomb tritium as a tracer of young water in unconfined sandy aquifers [abs.]: EOS (American Geophysical Union Transactions), v. 61, p. 233.

Ellis, R. R., Devi, L., and Wierenga, W. A., 1968, The investigation of water flow through porous mediums by means of radiotracers: Water Resources Research, v. 4, p. 413-416.

Elrick, D. E., and French, L. K., 1966, Miscible displacement patterns on disturbed and undisturbed soil cores: Soil Science Society of America Proceedings, v. 30, p. 153-156.

Elrick, D. E., Erk, K. T., Krupp, H. K., 1966, Applications of miscible displacement techniques to soils: Water Resources Research, v. 2, p. 717-727.

Elzy, E., and others, 1974, Analysis of the movement of hazardous waste chemicals in and from a landfill site, in Disposal of environmentally hazardous wastes: Corvallis, Oregon State University, p. 93.

Fatt, I., Maleki, M., and Upadhyay, R. N., 1966, Detection and estimation of dead-end pore volume in reservoir rock by conventional laboratory tests: Society of Petroleum Engineering Journal, v. 6, p. 206-212.

Foster, S.S.D., 1975, The chalk groundwater tritium anomaly--A possible explanation: Journal of Hydrology, v. 25, p. 159-165.

Freeze, R. A., and Cherry, J. A., 1979, Groundwater: Englewood Cliffs, New Jersey, Prentice-Hall Inc.

Fried, J. J., 1972, Some recent applications of the theory of dispersion in porous media, in Proceedings, Fundamentals of Transport Phenomena in Porous Media: International Association for Hydraulic Research--International Society of Soil Science, Second Symposium, August 7-11, 1972, University of Guelph, Canada, p. 722-731.

___1975, Groundwater pollution: New York, Elsevier, p. 330.

Gelhar, L. W., 1977, Effects of hydraulic conductivity variations on groundwater flows, in Hydraulic problems solved by stochastic methods: Fort Collins, Colorado, Water Resources Publications, p. 409-429.

Gelhar, L. W., and Wilson, J. L., 1974, Ground-water quality modeling: Ground Water, v. 12, p. 399-405.

Gelhar, L. W., Gutjahr, A. L., and Naff, R. L., 1979, Stochastic analysis of macrodispersion in a stratified aquifer: Water Resources Research, v. 15, p. 1387-1397.

Gillham, R. W., and Cherry, J. A., 1978, Field evidence of denitrification in shallow groundwater flow systems, in Proceedings, Thirteenth Canadian symposium on water pollution research: Hamilton, Ontario, McMaster University, Canada, p. 53-71.

Gillham, R. W., Cherry, J. A., and Anderson, G., 1981, Hydrogeological studies of a sandy unconfined aquifer at an abandoned landfill. 2. Contaminant monitoring: Journal of Hydrology (in press).

Golubev, V. S., and Garibyants, A. A., 1971, Heterogeneous processes of geochemical migration (special research report): New York, Consultants Bureau (trans. from Russian), p. 150.

Golwer, A., Matthess, G., and Schneider, W., 1975, Effects of waste deposits on groundwater quality, in Symposium on Groundwater Pollution: International Association for Hydraulic Research Publication No. 103, p. 159-166.

Goodall, D. C., and Quigley, R. M., 1977, Pollutant migration from two sanitary landfill sites near Sarnia, Ontario: Canadian Geotechnical Journal, v. 14, p. 223-236.

Grisak, G. E., and Pickens, J. F., 1980, Solute transport through fractured media. 1. The effect of matrix diffusion: Water Resources Research, v. 16, p. 719-730.

Grisak, G. E., and others, 1976, Hydrogeologic and hydrochemical properties of fractured till in the interior plains region, in Legget, R. F., ed., Glacial till: Royal Society of Canada, Special Publication, p. 304-335.

Grisak, G. E., Pickens, J. F., and Cherry, J. A., 1980, Solute transport through fractured media. 2. Column study of fractured till: Water Resources Research, v. 16, p. 731-739.

Grove, D. B., and Wood, W. W., 1979, Prediction and field verification of subsurface water quality changes during artificial recharge, Lubbock, Texas: Ground Water, v. 17, p. 250-257.

Hansen, E. A., and Harris, A. R., 1974, A ground-water profile sampler: Water Resources Research, v. 10, p. 375.

Heller, J. P., 1972, Observations of mixing and diffusion in porous media, in Proceedings, Fundamentals of transport phenomena in porous media: International Association for Hydraulic Research--International Society of Soil Science, Second Symposium, Guelph, Canada, University of Guelph, August 7-11, 1972, p. 1-26.

Higgins, G. H., 1959, Evaluation of the groundwater contamination hazard from underground nuclear explosions: Journal of Geophysical Research, v. 64, p. 1509-1519.

Hillier, R., 1975, A laboratory investigation of dispersion and groundwater flow through heterogeneous sands [B.Sc. thesis]: Waterloo, Ontario, University of Waterloo, p. 43.

James, R. V., and Rubin, J., 1978, Applicability of the local equilibrium assumption to transport through soils of solutes affected by ion exchange, in Proceedings, American Chemical Society, 176th ACS National Meeting, p. 225-235.

Kay, B. D., and Elrick, D. E., 1967, Adsorption and movement of lindane in soils: Soil Science, v. 104, p. 314-322.

Kemper, W. D. 1960, Water and ion movement in thin films as influenced by the electrostatic charge and diffuse layer of cations associated with clay mineral surfaces: Soil Science Society of America Proceedings, v. 24, p. 10-16.

Kemper, W. D., and Rollins, J. B., 1966, Osmotic efficiency coefficients across compacted clays: Soil Science Society of America Proceedings, v. 30, p. 529-534.

Kharaka, Y. K. and Barnes, I., 1973, SOLMNEQ: Solution-mineral equilibrium computations: Springfield, Virginia, National Technical Information Service Technical Report PB214-899, p. 92.

Kimmel, G. E., and Braids, O. C., 1980, Leachate plumes in ground water from Babylon and Islip landfills, Long Island, New York: U.S. Geological Survey Professional Paper 1085, p. 38.

Klotz, D., and Moser, H., 1974, Hydrodynamic dispersion as aquifer characteristic: Model experiments with radioactive tracers, in Isotope techniques in groundwater hydrology, Volume 2: Vienna, International Atomic Energy Agency, p. 341-355.

Konikow, L. F., 1977, Applications of solute-transport models to groundwater quality problems, in Papers and proceedings of the Water Research Center Conference: Berkshire, University of Reading, Sept. 6-8, 1976, p. 625-643.

Langmuir, D., 1972, Controls on the amounts of pollutants in subsurface waters: Pennsylvania State University, Earth and Mineral Sciences, v. 42, p. 5-13.

Lawson, D. W., and Elrick, D. E., 1972, A new method for determining and interpreting dispersion coefficients in porous media, in Proceedings, Symposium on Fundamentals of Transport Phenomena in Porous Media: International Association for Hydraulic Research--International Society of Soil Science, Guelph, University of Guelph, p. 753-777.

Lenda, A., and Zuber, A., 1970, Tracer dispersion in groundwater experiments: International Atomic Energy Agency, SM-129/37.

Lerman, A., 1978, Chemical exchange across sediment-water interface: Earth and Planetary Science, v. 6, p. 281-303.

MacFarlane, D. S., and others, 1981, Hydrogeological studies of a sandy aquifer at an abandoned landfill. 1. Groundwater flow and contaminant distribution: Journal of Hydrology (in press).

Manheim, F. T., 1970, The diffusion of ions in unconsolidated sediments: Earth and Planetary Science Letters, v. 9, p. 307-309.

Marino, M. A., 1976, Numerical simulation of contaminant transport in subsurface systems: in Saleem, Z. A., ed., Advances in groundwater hydrology: American Water Resources Association, p. 113-129.

Martin, J. M., 1971, Deplacements miscibles dans les milieux poreux naturels de grande extension: Institute France Petrole, Revue, v. 26, p. 1065-1075.

Mattigod, S. V., and Sposito, G., 1978, Chemical modeling of trace metal equilibria in contaminated soil solutions using the computer program GEOCHEM, in Jenne, E. A., ed., Chemical modeling in aqueous systems, American Chemical Society Symposium Series 93, p. 837-856.

Mayer, S. W., and Tompkins, E. R., 1947, Theoretical analysis of the column separations process: Journal of the American Chemical Society, v. 69, p. 2866-2874.

Meinardi, C. R., 1975, Brackish groundwater bodies as a result of geological history and mineralogical conditions: The Hague, Netherlands, Government Institute for Water Supply, r.i.d. mededeling 75-1.

Mercado, A., 1967, The spreading pattern of injected water in a permeability stratified aquifer, in Proceedings, Symposium on Artificial Recharge and Management of Aquifers: International Association of Scientific Hydrology, v. 72, p. 23-36.

Narasimhan, T. N., and others, 1976, Interpretation of water quality data from observation wells in stratified aquifers [abs.]: EOS (American Geophysical Union Transactions), v. 57, p. 912.

Nelson, J. L. 1961, Recent studies at Hanford on soil and mineral reactions in waste disposal, in Proceedings, Symposium on Groundwater Disposal of Radioactive Wastes, Berkeley, California, Aug. 25-27, 1959: Berkeley, Sanitary Engineering Laboratory, University of California.

Nelson, R. W., 1978a, Evaluating the environmental consequences of groundwater contamination. 1. An overview of contaminant animal distributions as general evaluation requirements: Water Resources Research, v. 14, p. 409-415.

___1978b, Evaluating the environmental consequences of groundwater contamination. 2. Obtaining location/arrival time and location/outflow quantity distributions for steady flow systems: Water Resources Research, v. 14, p. 416-428.

___1978c, Evaluating the environmental consequences of groundwater contamination. 3. Obtaining contaminant arrival distributions for steady flow in heterogeneous systems: Water Resources Research, v. 14, p. 429-440.

___1979d, Evaluating the environmental consequences of groundwater contamination. 4. Obtaining and utilizing contaminant arrival distributions in transport flow systems: Water Resources Research, v. 14, p. 441-450.

Nordstrum, D. K. and others, 1978, A comparison of computerized chemical models for equilibrium calculations in aqueous system, in Jenne, E. A., ed., Chemical modeling in aqueous systems: American Chemical Society Symposium Series 93, p. 857-892.

Ogata, A., 1970, Theory of dispersion in a granular medium: U.S. Geological Survey Professional Paper 411-I, p. 36.

Ogata, A., and Banks, R. B., 1961, A solution of the differential equation of longitudinal dispersion in porous media: U.S. Geological Survey Professional Paper 411-A, p. 7.

Orlob, G. T., and Radhakrishna, G. N., 1958, The effects of entrapped gases on the hydraulic characteristics of porous media: American Geophysical Union Transactions, v. 39, p. 648-659.

Palmquist, R., and Sendlein, L.V.A., 1975, The configuration of contaminant enclaves from refuse disposal sites on floodplains: Ground Water, v. 13, p. 167-181.

Parkhurst, D. L., Plummer, L. N., and Thorstenson, D. C., 1978, Chemical models in ground-water systems: Geological Society of America Abstracts with Programs, v. 3, p. 468.

Passioura, J. B., 1971, Hydrodynamic dispersion in aggregated media. 1. Theory: Soil Science, v. 3, p. 339-344.

Passioura, J. B., and Rose, D. A., 1971, Hydrodynamic dispersion in aggregated media. 2. Effects of velocity and aggregate size: Soil Science, v. 3, p. 345-351.

Perkins, T. K., and Johnston, O. C., 1963, A review of diffusion and dispersion in porous media: Society of Petroleum Engineering Journal, v. 3, p. 70-84.

Philip, J. R., 1968, The theory of absorption in aggregated media: Australian Journal of Soil Research, v. 6, p. 1-19.

Pickens, J. F., 1978, The effect of aquifer stratification in the determination of dispersivity: Geological Association of Canada, Mineralogical Association of Canada Abstracts with Programs, v. 3, 472.

Pickens, J. F., and Lennox, W., 1976, Numerical simulation of waste movement in steady groundwater flow systems: Water Resources Research, v. 12, p. 171-180.

Pickens, J. F., and others, 1978a, A multilevel device for ground-water sampling and piezometric monitoring: Ground Water, v. 16, p. 322-327.

Pickens, J. F., Cherry, J. A., and Gillham, R. W., 1978b, Field studies of dispersion in a shallow sandy aquifer, in Proceedings, Symposium on Invitational Well Testing, Berkeley, California, Oct. 19-21, 1977: U.S. Department of Energy, p. 55-62.

Pinder, G. F., and Gray, W. G., 1977, Finite element simulation in surface and subsurface hydrology: New York, Academic Press, p. 295.

Plummer, L. N., Parkhurst, D. L., and Kosiur, D. R., 1975, MIX 2: A computer program for modelling chemical reactions in natural waters: U.S. Geological Survey Water Resources Investigation Report 75-61, p. 73.

Reynolds, W. D., 1978, Column studies of strontium and cesium transport through a granular geologic porous medium [M.S. thesis]: Waterloo, Ontario,

University of Waterloo, p. 149.
Robertson, J. B., 1974, Digital modeling of radioactive and chemical waste transport in the Snake River Plain aquifer at the National Reactor Testing Station, Idaho: U.S. Geological Survey Open File Report IDO-22054.
Rose, D. A., 1973, Some aspects of the hydrodynamic dispersion of solutes in porous materials: Journal of Soil Science, v. 24, p. 284-295.
Rowe, P. W., 1972, The relevance of soil fabric to site investigation practice: Geotechnique, v. 22, p. 195-300.
Sauty, J. P., 1978, Interpretation of tracer tests by means of type curves: Application to uniform and radial flow, in Proceedings, Invitational Well-Testing Symposium, U.S. Department of Energy, Berkeley, California, Oct. 19-21, 1977: Berkeley, Lawrence Berkeley Laboratory, LBL-7027.
___1980, An analysis of hydrodispersive transfer in aquifers: Water Resources Research, v. 16, p. 145-158.
Scheidegger, A. E., 1954, Statistical hydrodynamics in porous media: Journal of Applied Physics, v. 25, p. 997-1001.
___1961, General theory of dispersion in porous media: Journal of Geophysical Research, v. 66, p. 3273-3278.
___1963, The physics of flow through porous media: Toronto, University of Toronto Press.
Schroeder, M. C., and Jennings, A. R., 1963, Laboratory studies of the radioactive contamination of aquifers: Livermore, Lawrence Radiation Laboratory, UCRL-13074.
Schwartz, F. W., 1975, On radioactive waste management: Model analysis of a proposed site: Journal of Hydrology, v. 32, p. 257-277.
___1977, Macroscopic dispersion in porous media: The controlling factors: Water Resources Research, v. 13, p. 743-752.
Simpson, E. S., 1978, A note on the structure of the dispersion coefficient: Geological Association of Canada--Mineralogical Association of Canada Abstracts with Programs, v. 3, p. 493.
Simpson, E. S., and Duckstein, L., 1975, Finite-state mixing cell-models: Proceedings, U.S.-Yugoslavian Symposium on Karst Hydrology and Water Resources, Dubrovnik, p. 489-508.
Skibitzke, H. E. and Robertson, G. M., 1963, Dispersion in groundwater flowing through heterogeneous materials: U.S. Geological Survey Professional Paper 386-B, p. 5.
Slobod, R. L., and Thomas, R. A., 1963, Effect of transverse diffusion on fingering in miscible-phase displacement: Society of Petroleum Engineering Journal, v. 3, p. 9-14.
Smith, J. M., 1970, Chemical engineering kinetics: New York: McGraw-Hill.
Smith, L., and Schwartz, F. W., 1980, A stochastic model of macroscopic dispersion. 1. Justification and methodologies: Water Resources Research, v. 16, p. 303-313.
Sudicky, E. A., 1979, Field observation of non-uniform dispersivity in an unconfined aquifer [M.S. thesis]: Waterloo, Ontario, University of Waterloo, p. 70.
Sudicky, E. A., and Cherry, J. A., 1979, Field observations of tracer dispersion under natural flow conditions in an unconfined sandy aquifer: Water Pollution Research, v. 14, p. 1-17.
___1981, Hydrogeological studies of a sandy unconfined aquifer at an abandoned landfill. 3. A natural gradient dispersion test: Journal of Hydrology (in press).
Tang, D. H., and Pinder, G. F., 1977, Simulation of groundwater flow and mass transport under uncertainty: Advances in Water Resources, v. 1, p. 25-30.
Tang, D. H., Frind, E. O., and Sudicky, E. A., 1981, Contaminant transport in fractured porous media. 1. Analytical solution for a single fracture: Water Resources Research (in press).
Taylor, G., 1953, Dispersion of soluble matter in solvent flowing slowly through a tube: Royal Society Proceedings, ser. A, v. 219, p. 186-203.
Theis, C. V., 1967, Aquifer and models, in Marino, M. A., ed.,Proceedings, National Symposium on Ground-Water Hydrology: American Water Resources Association Proceedings Series No. 4, p. 138-148.
Thomas, G. W. and Swoboda, A. R., 1970, Anion exclusion effects on chloride movement in soils: Soil Science, v. 110, p. 163-166.
Truesdell, A. H., and Jones, B. F., 1974, WATEQ, a computer program for calculating chemical equilibria of natural waters: U.S. Geological Survey Journal of Research, v. 2, p. 233-274.
Trudell, M., 1970, A field-injection study of denitrification in a shallow sandy aquifer [M.S. thesis]: Waterloo, Ontario, University of Waterloo, p. 96.
United States Environmental Protection Services, 1977, The report to Congress: Waste disposal practices and their effects on ground water. U.S. Environmental Protection Agency, EPA-570/9-77-002.
van Genuchten, M.Th., and Wierenga, P. J., 1976, Mass transfer studies in sorbing porous media. 1. Analytical solutions: Soil Science Society of America Journal, v. 40, p. 473-480.
van Schaik, J. C., and Kemper, W. D., 1966, Chloride diffusion in clay-water systems: Soil Science Society of America Proceedings, v. 30, p. 22-25.
van Schaik, J. C., Kemper, W. D., and Olsen, S. R., 1966, Contribution of adsorbed cations to diffusion in clay-water systems: Soil Science Society of America Journal, v. 30, p. 17-22.
Vermeulen, T., and Niester, N. K., 1962, Ion-exchange chromatography of trace components: Industrial Engineering Chemistry, v. 44, p. 636-651.
Volker, A., 1961, Source of brackish groundwater in Pleistocene formations beneath the Dutch polderlands: Economic Geology, v. 56, p. 1045-1061.
Warren, J. E., and Skiba, F. F., 1964, Macroscopic dispersion: Society of Petroleum Engineering Journal, v. 4, p. 215-230.
Weeks, O. L., Stewart, G. L., and Weeks, M. E., 1976, Measurement of non-exchanging pores during miscible displacement in soils: Soil Science, v. 122, p. 139-144.
Westall, J. C., Zachary, J. L., and Morel, F. M. M., 1976, MINEQL, a computer program for the calculation of chemical equilibrium composition of aqueous systems: Keck Lab Tech, Note 18, California Institute of Technology, p. 91.

Yare, B. S., 1975, The use of a specialized drilling and groundwater sampling technique for delineation of hexavalent chromium contamination in an unconfined aquifer, southern New Jersey coastal plain: Ground Water, v. 13, p. 151-154.

Manuscript Accepted by the Society October 14, 1980

Printed in U.S.A.

Geological Society of America
Special Paper 189
1982

Hydrogeological Concepts in Stochastic and Deterministic Rainfall-Runoff Predictions

R. ALLAN FREEZE
Department of Geological Sciences
University of British Columbia
Vancouver, British Columbia V6T 1W5 Canada

ABSTRACT

Two trends are evident in the recent developments in engineering hydrology, one toward integrating surface-water and ground-water concepts, and the other toward integrating stochastic and deterministic techniques. This paper provides a review of these trends with the specific aim of clarifying the role of ground water in the current methodology. Among the interrelationships discussed are (1) the linear-reservoir representation of the instantaneous unit hydrograph and the concepts of ground-water storage; (2) the existence of autocorrelated streamflow sequences and their dependence on baseflow recessions; (3) the conceptual prediction of flood exceedance probabilities and the mechanisms of streamflow generation; and (4) the nature of time-space tradeoffs in assessing the worth of data for rainfall-runoff models.

The process of streamflow generation is controlled in large part by the near-surface saturated-unsaturated hydrogeological regimes that exist on the hillslopes within a watershed. Much remains to be done before the lumped calibration parameters that arise in time-series analysis, frequency analysis, and unit-hydrograph analysis of streamflows can be interpreted in terms of the spatially distributed, physically based parameters that control the runoff-generating processes on hillslopes.

INTRODUCTION

The major application of the science of hydrology in engineering is in the prediction of streamflow regimes for the design and operation of hydraulic structures. At first glance, the techniques developed by engineering hydrologists for this purpose do not appear to offer much scope for input from the hydrogeological community. Such approaches as time-series analysis of streamflow sequences and frequency analysis of flood occurrences come from a stochastic discipline that is not home to most ground-water hydrologists. Even in the more deterministic watershed-simulation models, where subsurface processes do play a role, their treatment often seems cavalier to the hydrologist, and emphasis usually remains on the surface-water components of runoff.

On reflection, however, it seems clear that subsurface hydrology deserves a larger role in assessing streamflow regimes. Streamflow is generated from rainfall or snowmelt on hillslopes, and the apparent mechanisms of generation all require an understanding of the saturated-unsaturated subsurface flow processes in the hillslope soils. In addition, low flows, which often constitute the most critical design variable for reservoirs that must maintain sufficient throughflow to satisfy hydroelectric, navigation, and recreation requirements, are controlled by the baseflow-producing capabilities of the ground-water reservoirs in a watershed.

Two trends are evident in recent developments in engineering hydrology, one toward integrating surface-water and ground-water concepts, and the other toward integrating stochastic and deterministic techniques. The purpose of this paper is to provide a review of these trends with the specific aim of clarifying the role of ground-water in the current methodology. First, a brief review is provided of the four topics that underlie the discussions: (1) time-series analysis of stream-flow sequences, (2) frequency analysis of streamflow extremes, (3) rainfall-runoff models, and (4) mechanisms of streamflow generation on hillslopes. Some of the interrelationships will then be outlined that have been uncovered with respect to, (1) linear reservoirs and ground-water storage, (2) streamflow persistence and baseflow, (3) conceptual prediction of exceedance probabilities, and (4) time-space tradeoffs in assessing the worth of data.

It is first necessary, however, to set the stage with some definitions.

DEFINITIONS

It is common in surface-water hydrology to differentiate between *prediction* and *forecasting*. Following Clarke (1972, 1973), the term *prediction* is reserved for the long-term statistical properties of streamflow. A *design prediction* involves estimating either the frequency with which a rare event will occur at a given site, or the probability that a specific sequence of events will occur at the site in the long run. A prediction provides no estimate of when a particular flow or sequence of flows will occur. *Forecasting*, on the other hand, means estimating the magnitude of specific future runoff events. Forecasting is a short-term process and is sometimes called *one-step-ahead prediction*. Clarke differentiates between two types of forecasting. In a *design forecast*, one is given a hypothetical wet catchment condition and a hypothetical heavy storm and is then asked to estimate the resultant flood hydrograph. In an *operational forecast*, which is the real-time case, one is given the sequence of past flow rates at a site, and perhaps the present flow rates upstream from the site, and asked to estimate future flows.

Table 1 outlines the scope of engineering hydrology in terms of the three main engineering tasks: the design of minor structures, the design of dams and reservoirs, and the operation of water resource systems. In each case, the hydrologic requirements are noted in terms of design predictions, design forecasts, or operational forecasts. The case of spillway design is further considered in Table 2. The design flood for a small dam can be based on a design prediction. Large and intermediate dams require a design forecast.

The various techniques of hydrologic analysis can be classified in a variety of ways. Table 3 lists the hydrologic methods that will be discussed in this paper and classifies them according to input requirements, nature of variables, model structure, spatial and temporal response, and appliccation. The rightmost columns identify that time-series analysis and frequency analysis are the two main techniques used for arriving at design predictions, whereas routing techniques and rainfall-runoff models are the primary methods used to produce operational or design forecasts.

Most of the terms that appear at the top of Table 3 are well established in the hydrologic literature. The terms *lumped* and *distributed* as well as *time-variant* and *time-invariant* should require no discussion. The terms that appear in the columns "nature of variables" and "model structure," however, do require clarification. Once again we will follow Clarke (1972, 1973). Consider a general hydrologic model:

$$q_t = f(p_{t-1}, p_{t-2}, \ldots; q_{t-1} q_{t-2}, \ldots; a_1, a_2, \ldots) + \varepsilon_t. \quad (1)$$

The vectors p_t and q_t are *variables* that assume different numerical values at different times. The vector a_1 contains the system *parameters* that do not change with time. Here p_t is the input variable, q_t is the output variable, and ε_t expresses the lack of fit between model and reality.

A model is termed *stochastic* or *deterministic* according to whether or not it contains random variables. If any of the variables p_t, q_t, or ε_t is thought of as having a distribution in probability, then the model is stochastic; if not, it is deterministic. The terms *conceptual* and *empirical* are defined according to whether the form of the function f in equation 1 is or is not suggested by considering the physical processes acting on the

TABLE 1. SCOPE OF ENGINEERING HYDROLOGY

Engineering task	Hydrologic requirement		
	Design prediction	Design forecast	Operational forecast
Design of minor structures			
Culverts, drainage ditches, urban storm-drainage systems, airport drainage systems, agricultural irrigation and drainage systems, small canals, and aqueducts	X		
Design of dams and reservoirs			
Design of reservoir storage capacity to fulfill reservoir operation requirements listed below	X		
Design of diversion tunnel	X		
Design of spillway (see Table 2)	X	X	
Operational hydrology			
Flood forecasting and drought forecasting in unregulated steams			X
Reservoir operation in regulated steams. Maintain low flows above specified minimum for hydroelectric power generation, navigational requirements, recreation requirements, irrigation abstractions, municipal water supplies, and dilution of pollutant releases. Keep peak flows below specified maximum for flood control.			X

TABLE 2. SPILLWAY DESIGN FLOODS*

	Reservoir storage capacity (acre-ft)	Dam height (ft)	Potential loss of life	Potential damage	Spillway design flood
Large dam	>50,000	>60	Considerable	Excessive	Maximum probable flood (hypothetical flood) derived from maximum possible precipitation event
Intermediate dam	100 to 50,000	40 to 100	Possible but small	Within financial capability of owner	Standard project flood (based on most severe storm of record)
Small dam	<1000	<50	None	Of same magnitude	Frequency analysis: 50- to 100-yr recurrence interval

* Adapted from Dalrymple, 1964.

Table 3. CLASSIFICATION OF HYDROLOGIC METHODOLOGIES

Methodology	Input requirements			Nature of variables		Model structure		Spatial response		Temporal response		Application		
	Streamflow	Precipitation	Watershed parameters	Stochastic	Deterministic	Empirical	Conceptual	Lumped	Distributed	Time-invariant	Time-variant	Design prediction	Design forecast	Operational forecast
Time-series analysis														
Mean annual discharge	X			X		X		X		X		X		
Mean monthly discharge	X			X		X		X		X		X		
Mean daily discharge	X			X		X		X		X		X		
Frequency analysis														
Maximum annual discharge	X			X		X		X		X		X		
Minimum annual discharge	X			X		X		X		X		X		
Streamflow routing														
Hydraulic routing using equations of open channel flow	X				X		X		X		X			X
Hydrologic routing using theory of reservoir storage	X				X	X		X		X				X
Rainfall-runoff models														
Volumes only														
Empirical formulae: rational formulae	X	X		X		X		X		X				X
Rainfall-runoff models														
Volumes and timing														
Unit hydrograph	X	X			X	X		X		X			X	X
Hydrologic watershed routing	X	X			X	X		X		X			X	X
Parametric simulation models: Stanford watershed model	X	X	X		X		X		X		X		X	X
Physics-based simulation models: kinematic cascade	X	X	X		X		X		X		X		X	X

input variables. The terms *conceptual* and *empirical* thus refer to the structure of the model, whereas the terms *deterministic* and *stochastic* refer to the nature of the variables.

There is one term that does not appear in Table 3 that will be used in later sections of the paper. An input-output system is *linear* if the principle of superposition holds. If an input sequence $p_1(t)$ produces an output sequence $q_1(t)$, and if a second input sequence $p_2(t)$ produces an output sequence $q_2(t)$, the system relating the input to the output is linear in the variables p and q if the input sequence $p_1(t) + p_2(t)$ produces the output sequence $q_1(t) + q_2(t)$. The assumption of linearity underlies the theory of the unit hydrograph and most other hydrologic routing techniques. Clarke (1972) refers to systems that are linear according to the above definition as *linear in the system-theory sense (LST)*, to differentiate them from systems that are *linear in the statistical regression sense (LSR)*. There are many input-output relationships that are LST but not LSR, and vice versa.

TIME-SERIES ANALYSIS OF STREAMFLOW SEQUENCES

Consider the hypothetical hydrograph of discharge shown in Figure 1a. For purposes of reservoir design (Table 1), it is sufficient to work with the stochastic properties of such measured streamflows without referring to the nature of the meteorological events that were their source or to the watershed transformation processes that delivered them to the stream. A time-series analysis of the streamflow record will provide an estimate of the probable occurrence of the most critical sequence of wet and dry years, and this is the principal information on which the design of reservoir storage capacity must be based.

Let us abstract the mean annual discharge for each of the years of record from the hydrograph shown in Figure 1a. If these discharge values are plotted, as in Figure 1b, a time series emerges that can be analyzed using the standard methods of time-series analysis (Kisiel, 1969; Box and Jenkins, 1976). If, instead of the mean annual discharges, the mean monthly discharges were abstracted, a time series like that shown in Figure 1c, with a notable seasonal trend, would result. The time series of mean daily flows would also show such a seasonal trend, and the time series of mean hourly flows would show both a seasonal and a diurnal trend.

If one works with mean annual discharge as in Figure 1b, or with monthly, daily, or hourly flows with the trends removed, one is left with a sequence of discharge values ($q_1, q_2, \ldots, q_n$). Such a time series represents one *realization* (or sample function) of the total *ensemble* (or population) of time series that could be realized at that stream gauge. The mathematical model that describes the ensemble is called a *stochastic process.*

For each realization it is possible to calculate the sample mean $\overline{q}$ and the sample variance s_q^2 in the usual way:

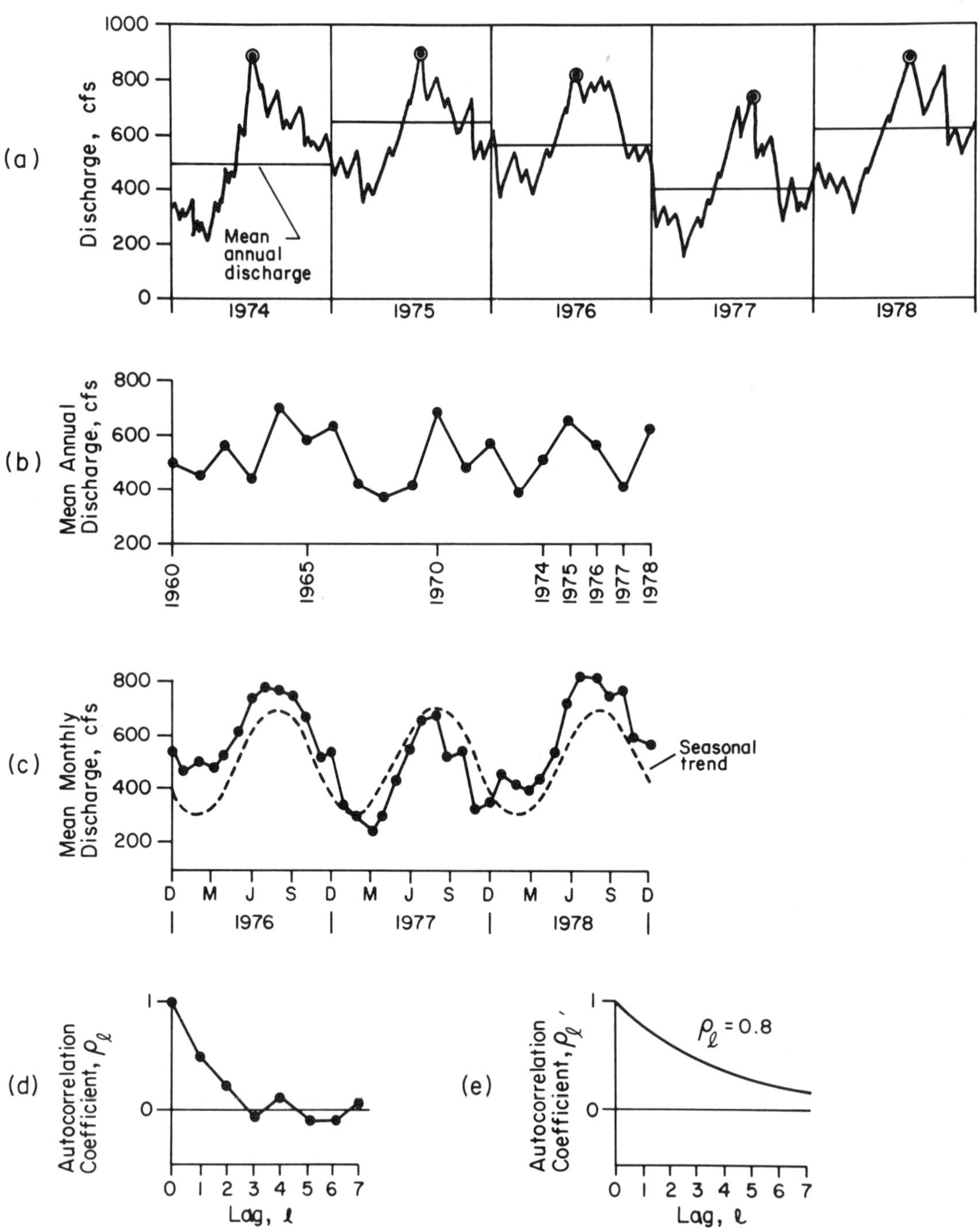

Figure 1. Time-series analysis of streamflow sequences.

$$\bar{q} = \frac{1}{n}\sum_{i=1}^{n} q_i, \tag{2}$$

$$s_q^2 = \frac{1}{n}\sum_{i=1}^{n} (q_i - \bar{q})^2. \tag{3}$$

If we are willing to accept the assumption of ergodicity, on which most methods of time-series analysis rest, we can assume that these time averages, taken over the only available realization, are equal to the ensemble averages that would be obtained if a large number of realizations were available. Under the ergodic assumption, $\hat{\mu}q = \bar{q}$ and $\hat{\sigma}_q^2 = s_q^2$, where $\hat{\mu}q$ and $\hat{\sigma}_q^2$ are estimates of the mean μ_q and variance σ_q^2 of the stochastic process controlling the sequence of q^i. If μ_q and $\sigma_q^2 = s_q^2$ are constant through time, then the time series is said to be stationary. The standard methods of time-series analysis require stationarity. Time series that exhibit trends, either cyclical (as in Fig. 1c) or monotonic, are not stationary. However, one can remove the trend by considering the time series of deviations from the trend (the solid line minus the dashed line in Fig. 1c); such transformed series are often stationary.

The discharges q_i that constitute the time series shown in Figure 1b are each representative of an equal time spacing. The magnitude of the t in such cases is mathematically immaterial, and the time series is totally defined by the series of values q_1, q_2, q_n. The number of time steps ℓ between discharges q_i and $q_{i+\ell}$ is termed the lag. If the stochastic process that describes the sequence q_i is purely random, then there will be no correlation between q_i and $q_{i+\ell}$ for any lag. In the more usual situation, however, the series will show some degree of *autocorrelation* (or serial correlation) among its members. The *autocorrelation function* $\rho(\ell)$ (or ρ_ℓ, as it is usually expressed) describes the degree of correlation between the discharge $q_{i+\ell}$ and an earlier discharge q_i for all ℓ. The lag-1 autocorrelation coefficient ρ_1, which describes the autocorrelation between neighboring values, is usually the most important, with the values of ρ_ℓ decreasing as the lag ℓ increases.

It is possible to calculate the sample autocorrelation function r_ℓ from a single realization, such as in Figure 1b. Under the ergodic assumption and for a stationary series, $\hat{\rho}_p = r_\ell$, where $\hat{\rho}_\ell$ is an estimate of the autocorrelation function, ρ_ℓ, of the stochastic process. The r_ℓ can be calculated according to

$$r_\ell = \frac{c_\ell}{co}, \tag{4}$$

where

$$c_\ell = \frac{1}{n}\sum_{i=1}^{n-\ell} (q_i - \bar{q})\,(q_{i+\ell} - \bar{q}). \tag{5}$$

Note that by comparing equations 5 and 3, $c_o = s_q^2$, so that r_ℓ values are essentially normalized against the variance of the series, and $r_o = 1$. Figure 1d shows a sample autocorrelation function for a hypothetical measured time series.

There are many theoretical stochastic processes that can generate time series with properties similar to those observed in streamflow data and would thus make suitable stochastic models of streamflow. One of the most widely used is the autoregressive model. An *autoregressive process (AR)* of order m is defined as

$$q_i = \phi_1 q_{i-1} + \phi_2 q_{i-2} + \ldots \phi_m q_{i-m} + a_i, \tag{6}$$

where the vector a_i is a normally distributed error term with mean zero and variance σ_a^2. The most commonly used AR process is the first-order AR process (or lag-1 AR process, or first-order Markov process), defined as

$$q_i = \phi_1 q_{i-1} + a_i. \tag{7}$$

For this process it can be shown (Box and Jenkins, 1976) that the autocorrelation function satisfies the relation

$$\rho_\ell = \phi_1 \rho_{\ell-1} \text{ for all } \ell > 0. \tag{8}$$

Note that $\rho_o = 1$ leads to the two relations

$$\phi_1 = p_1 \tag{9}$$

and

$$\rho_\ell = \rho_1^\ell. \tag{10}$$

Substituting equation 9 into equation 7 yields

$$q_i = \rho_1 q_{i-1} + a_i, \tag{11}$$

which is the equation for a first-order AR process with a lag-1 autocorrelation coefficient ρ_1. It can also be shown that the variance of the a_i is given by

$$\sigma_a^2 = (1 - \rho_1^2)\sigma_q^2. \tag{12}$$

The stochastic process is fully defined by the three parameters μ_q, σ_q^2, and ρ_1. Estimates of these parameters can be obtained from the streamflow realization, with $\mu_q = \bar{q}$, $\sigma_q^2 = s_q^2$, and $\rho_1 = r_1$, where $\bar{q}$, s_q^2, and r_1 are given by equations 2, 3, and 4. As long as one works with only the first two moments (μ, σ) and the autoregressive structure, there is no need to specify the nature of the statistical distribution from which the values for q_i are taken. Figure 1e shows a theoretical autocorrelation function for the case $\rho_1 = 0.8$.

Recently, surface-water hydrologists have been examining many alternative stochastic processes as possible models for streamflow. The reasons for this search can be ascribed to Mandelbrot and Wallis (1968), who pointed out that one very important property of streamflow sequences, the so-called Hurst phenomenon, is not preserved by first-order AR models. Consider a sequence q_i of mean annual discharges. Now define

$$W_j = \sum_{i=1}^{j} q_i - \bar{q} \quad \text{for} \quad j = 1, 2, \ldots, n, \tag{13}$$

and

$$R_n = \max |W_j' - W_j''|, \tag{14}$$

where W_j' and W_j'' are the maximum and minimum values of W_j and n is the length of record. The parameter W_j represents the cumulative departure of the flows from the sample mean at the end of the jth year. The parameter R_n is the range of cumulative departures. If we view the q_i as inflows to a reservoir, then R_n is the storage required in the reservoir to assure outflow equal to $\bar{q}$ each year. In general,

$$R_n \approx n^h, \tag{15}$$

where $0.5 \leq h \leq 1.0$. In an empirical study of 690 time series, Hurst (1956) found the value for h to be 0.73. The first-order AR process, however, produces an h value of 0.50.

Mandelbrot and Wallis (1968) pointed out that if first-order AR models are used to produce design predictions of streamflow sequences for reservoir design, such reservoirs will be underdesigned. They referred to the Hurst phenomenon as *long-term persistence*, and suggested an alternative model, *fractional Gaussian noise (FGN)*, that would preserve both short-term and long-term persistence. Later work has shown that integrated autoregressive moving-average models may also satisfy the requirements. O'Connell (1972) provides a clear review of the development of the various concepts and models.

For our purposes, the point to be made is that mean annual-discharge sequences exhibit both a short-term autoregressive persistence and a long-term persistence. Such persistence is presumably controlled by the release of water to the streams from ground-water storage. In a later section we will review the work of Moss and Bryson (1974), who analyzed streamflow autocorrelation in terms of carry-over storage in the ground-water zone.

FREQUENCY ANALYSIS OF STREAMFLOW EXTREMES

As noted in Tables 1, 2, and 3, frequency analysis of extreme flows is the standard approach for the derivation of design predictions for the

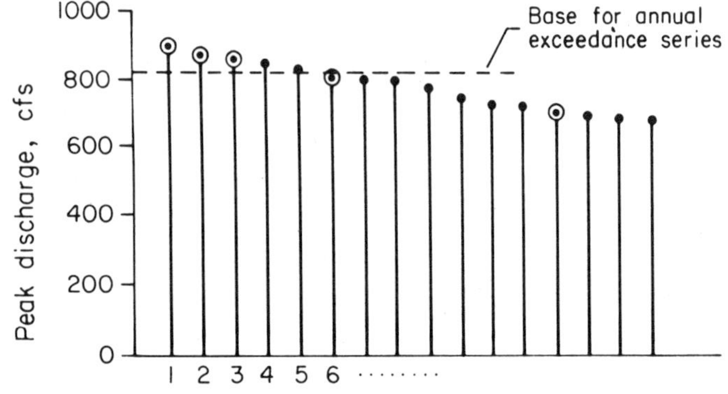

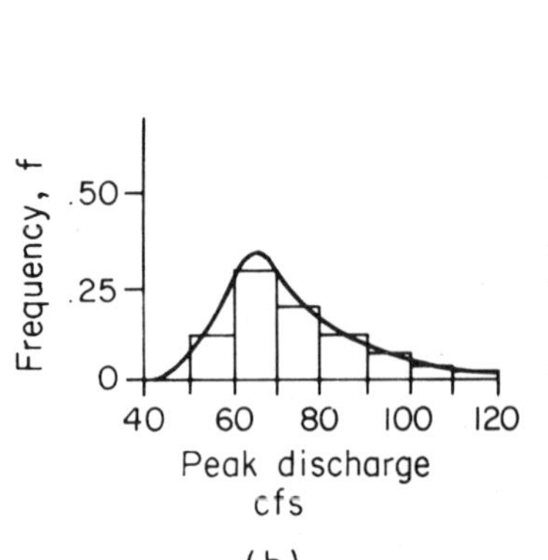

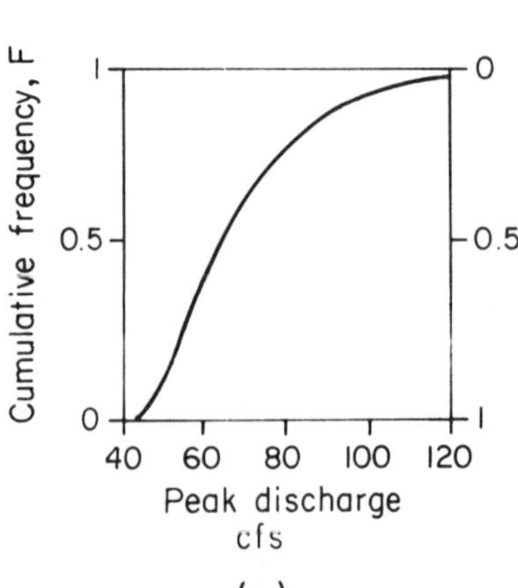

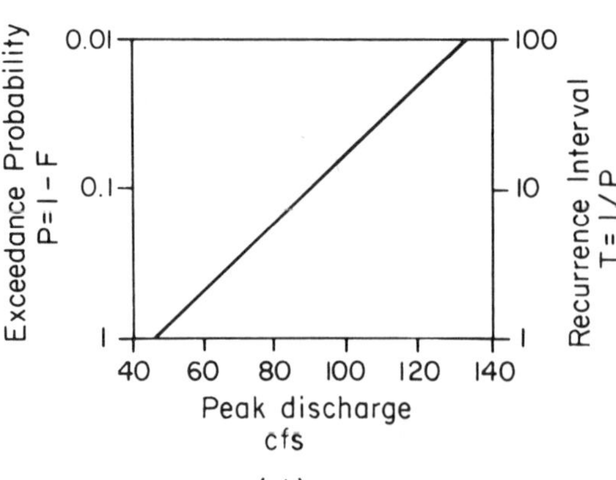

Figure 2. Frequency analysis of streamflow extremes.

design of minor structures and spillways for small dams.

Consider once again the hydrograph shown in Figure 1a. If we wish to carry out a flood-frequency analysis on the basis of a record of this type, the usual approach is to make a plot of peak discharges, arranged in decreasing order, as shown in Figure 2a. If the length of record is n years, there are two ways to extract a sample of size n for flood-frequency analysis (Chow, 1964a). The first method makes use of the annual-maxima series (the discharge values circled in Figs. 1a, 2a). In the second method, one chooses a base discharge (as shown in Fig. 2a), such that it is exceeded by exactly n peaks. This series, which may contain several peaks from one year and none from others, is called the annual-exceedance series (or partial-duration series). The annual-maxima series is used in cases where the design must protect against low-probability events, as in spillway design. The annual-exceedance series is used where only high-probability events must be considered, as in the design of some minor structures.

Figure 2b shows a hypothetical flood-frequency curve for an annual series of peak discharges. Figure 2c shows a cumulative-frequency curve. The vertical axis is labeled on the left side in terms of the cumulative frequency (F), and on the right side in terms of the exceedance probability ($P = 1 - F$). Figure 2d plots the exceedance probability on a logarithmic scale. The recurrence interval T given on the right vertical axis is the inverse of the exceedance probability. For this particular example, it would be expected that a peak discharge of 90 cubic feet per second would be equaled or exceeded once every 10 yr. Such a flood has an exceedance probability of 0.10.

In order to arrive at a theoretically defensible design flood, it is necessary that one consider the measured frequency distribution of the peaks as an estimate of the probability-density function for the population of peaks from which the measured peaks have been drawn. Unlike time-series analysis, frequency analysis requires knowledge of the full distribution, not just the first two moments. There has been considerable controversy in the literature as to which theoretical distribution is most suitable as a model of flood frequency. Among those that have been used are the lognormal, the two-parameter gamma, the log-Pearson Type III, and the Gumbel extreme-value distribution. In 1968, a United States national committee recommended (Benson, 1968) that a uniform model for flood-frequency distribution be adopted by all federal agencies involved in flood prediction and forecasting. The distribution model they recommended was the log-Pearson Type III.

Eagleson (1972) introduced a radical departure from the traditional approach to determining flood frequencies. He developed an equation, based on a simplified catchment model, that translates the probability distribution of rainfall events into a probability distribution of flood events. The catchment model involves an infiltration component that is dependent on the soil-moisture conditions in the watershed. Eagleson's approach thus implicitly recognizes that flood frequencies are controlled in part by watershed hydrogeology. The sole purpose in presenting the material in this section has been to introduce the concept of exceedance probabilities so that we may examine Eagleson's approach in greater detail in a later section.

RAINFALL-RUNOFF MODELS

For the purposes of this paper, discussion will be limited to only two of the rainfall-runoff models listed on Table 3, namely the unit hydrograph and the kinematic cascade.

Unit Hydrographs

A *unit hydrograph* can be defined (Chow, 1964b) as the hydrograph of direct runoff resulting from 1 in. of effective rainfall generated uniformly over a watershed area at a uniform rate for a specified duration. *Direct runoff* is defined as total runoff minus baseflow; *effective rainfall* is defined as total rainfall minus interception. Chow (1964b) discusses the methods of baseflow separation that can be used to determine direct runoff. He also discusses the methods by which a unit hydrograph for any given duration can be developed from a given set of rainfall-runoff data on a watershed.

In recent years, it has become common to use the theoretical concept of an *instantaneous unit hydrograph (IUH)*, which is the hydrograph that would result from the instantaneous (zero-duration) application of a unit rainfall. Chow (1964a) discusses the concept in detail and describes the methods of developing an IUH from field data.

Once an IUH has been developed for a basin, it can be used as a mathematical model for the prediction of runoff from rainfall. Its use is based on two main assumptions. The first is that of time invariance, where it is assumed that any two rainfall events with identical effective-rainfall rates will lead to identical runoff events. The second is linearity, where it is assumed that the stream discharge q(t) at any time t will be twice as great for a 2-in. effective rainfall as for a 1-in. effective rainfall of the same duration.

Nash (1957) and Dooge (1959) cooperated in the development of a general theory of the instantaneous unit hydrograph. They showed that one can view the IUH as an impulse-response function, u(t), and the prediction of runoff as a convolution, as follows:

$$q(t) = \int_o^t u(t - \tau)p(\tau)d\tau, \qquad (16)$$

where p(t) is the effective-rainfall hydrograph, q(t) is the direct-runoff hydrograph, and u(t) is the instantaneous unit hydrograph. Figure 3 is a graphic illustration of the concepts of the IUH as a rainfall-runoff predictor.

Kinematic Cascade

With the increased use of computers has come an increase in the use of conceptual hydrologic simulation models. This approach circumvents the need for

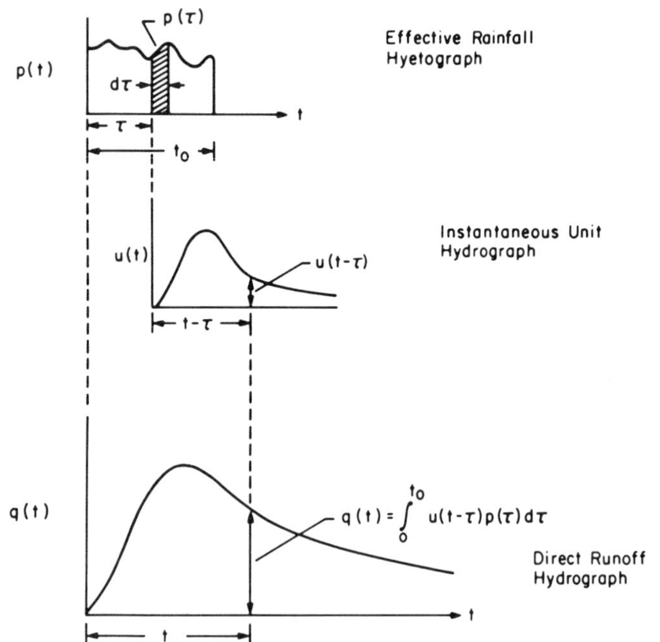

Figure 3. The instantaneous unit hydrograph as a rainfall-runoff model (after Chow, 1964).

the often-unacceptable assumptions of linearity and time invariance inherent in the unit-hydrograph model.

The model that will be described here makes use of the kinematic form of the shallow-water equations, consisting of an equation of continuity and an equation of motion. Both were originally developed for analyzing flow in channels, but they can also be used for sheet-flow representation of overland flow. In both cases it is assumed that the flow is *transient, gradually varied, turbulent,* and *subcritical*. For definition of these terms and for a review of the development of the kinematic form of the shallow-water equations the reader is directed to Eagleson (1970).

The approach outlined here makes use of the simplest geometrical configuration possible. It was first documented by Wooding (1965) and reviewed by Eagleson (1970). Consider a V-shaped pair of hillslope planes feeding a straight, rectangular stream channel (Fig. 4a). The watershed is assumed to be totally impermeable. A rainfall event with a really uniform intensity p(t) generates *overland* flow on the hillslopes, and this overland flow produces *lateral inflows* $q_P(t)$ to the stream along its length. The build-up of lateral inflows is evidenced by a runoff hydrograph q(t) at the outlet.

The one-dimensional equation of continuity for overland sheet flow in the x-direction is given by

$$\frac{\partial h}{\partial t} - p(t) + \frac{\partial}{\partial x}(\alpha h^m) = 0, \qquad (17)$$

where $\alpha = \sqrt{S_o}/n$ and m = 5/3. As shown in Figure 4b, S_o is the slope of the plane, n is Manning's friction coefficient, and h is the depth of flow. Equation 17 can be solved for h(x,t).

(a)

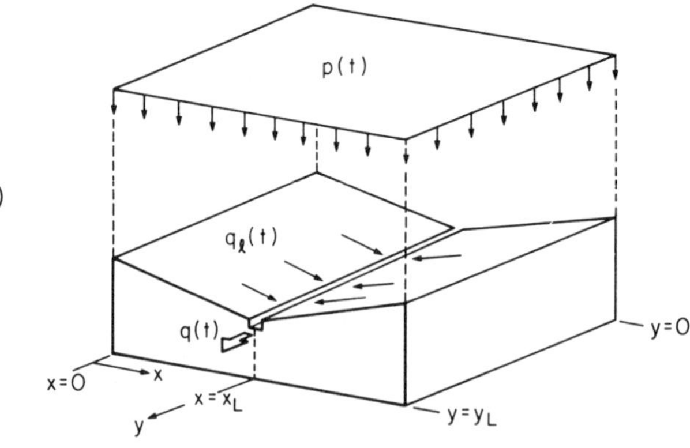

(b)

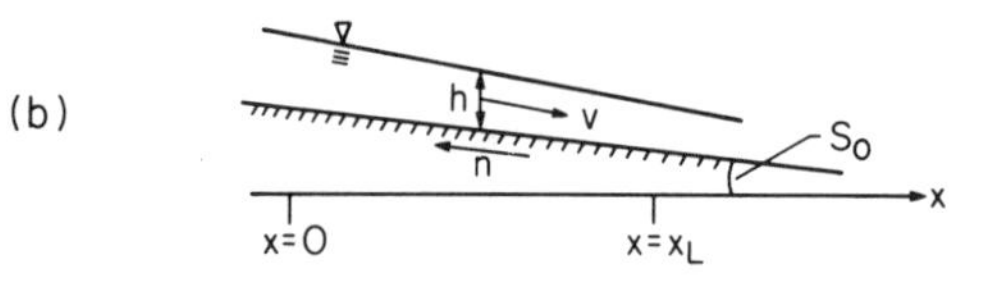

(c)

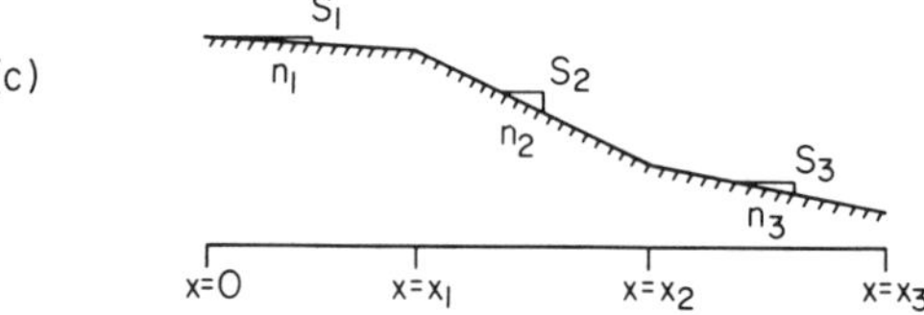

Figure 4. Conceptual hydrologic simulation with a kinematic cascade (part a after Eagleson, 1970).

The velocity v(x,t) can then be determined from the kinematic equation of motion,

$$v = \alpha h^{m-1}, \qquad (18)$$

and the lateral inflow to the stream $q_\ell(t)$ is given by

$$q_\ell(t) = v(x_L,t) \cdot h(x_L,t), \qquad (19)$$

where x_L is the value of x at the lower end of the overland flow plane.

The kinematic equations are then reinvoked for one-dimensional flow in the y-direction (Fig. 4a) in the stream channel. If the channel is assumed to be rectangular with a width B, the continuity equation 17 can be rewritten as

$$\frac{\partial h}{\partial t} - \frac{q_\ell(t)}{B} + \frac{\partial}{\partial y}(\alpha h^m) = 0. \qquad (20)$$

This equation is solved for the depth of flow h(y,t) in the channel. If B >> h(y,t) for all (y,t), the velocity v(y,t) can be obtained from equation 18, as before. The outflow hydrograph q(t) is given by

$$q(t) = B \cdot v(y_L, t) \cdot h(y_L . t), \qquad (21)$$

where y_L is the value of y at the outlet.

In order to bring overland-flow simulations

closer to reality, it is necessary to attempt simulations on planes that are not restricted to a single slope S_o. The most advanced approach of this type is that of a *kinematic cascade* (Fig. 4c), defined as a sequence of n discrete overland flow planes on which the kinematic-wave equations are used to describe the transient flow. The watershed parameters needed to utilize the kinematic cascade as a rainfall-runoff predictor are the Manning friction coefficients $n_1, n_2, \ldots, n_n$ and the surface slopes $S_1, S_2, \ldots, S_n$ for each of the planes of the cascade. Kibler and Woolhiser (1970) describe the numerical methods of solution for the determination of hydrographs with a kinematic cascade.

Without consideration of infiltration, the kinematic cascade as thus far described is somewhat unrealistic. Some authors include infiltration by treating the p(t) term in equation 17 as precipitation minus infiltration. Smith and Woolhiser (1971) have carried out the more complete analysis, wherein infiltration rates are allowed to vary with space as well as time. Their analysis is based on the unsaturated-flow theory outlined in the following section.

MECHANISMS OF STREAMFLOW GENERATION

Detailed descriptions of the mechanisms of streamflow generation can be found in a recent book edited by Kirkby (1978) and in the review by Freeze (1974). Only a brief synopsis of each will be included here.

Although the subject has aroused much controversy, it seems clear that, apart from certain humid, forested environments, subsurface flow mechanisms seldom

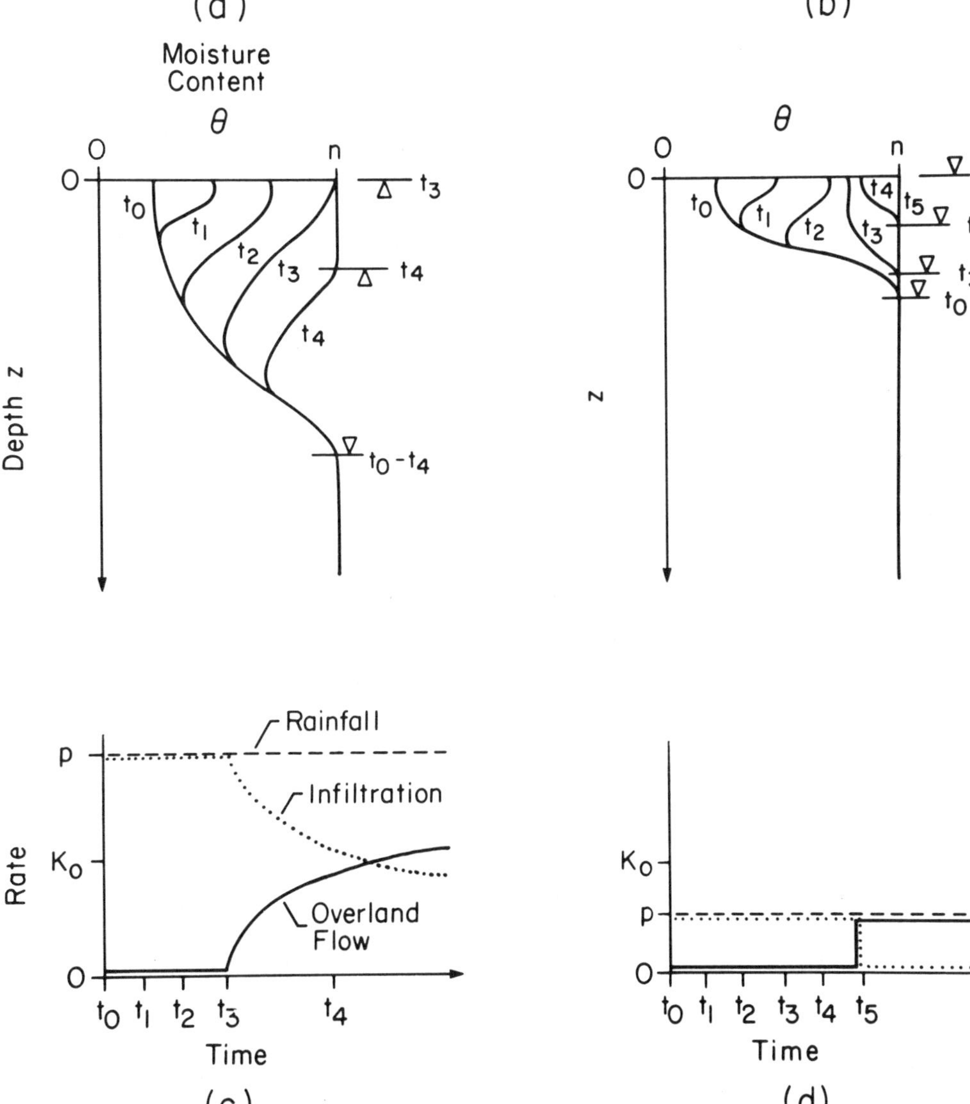

Figure 5. Moisture-content profiles and generation of overland flow under conditions of surface saturation from above and surface saturation from below.

contribute significant lateral inflows to streams during storm-runoff periods. This is not to say that subsurface flow is unimportant. It supplies the lateral inflows that sustain streams during the between-storm periods; and the saturated-unsaturated flow regime in the near-surface soils controls the timing, distribution, and total volume of overland flow that is generated by a given storm on a given hillslope. It is quite logical that a conceptual rainfall-runoff prediction model should calculate the lateral inflows into streams as overland flow, but the model must include a subsurface component to control the generation of the overland flows on the hillslope.

Overland flow cannot occur until surface ponding takes place. Ponding cannot occur until the surface soil layers become saturated. It is now recognized that the great heterogeneity of soil types at the ground surface over a watershed and the very irregular patterns of precipitation in both time and space create a very complex hydrologic response on the land surface. Surface saturation, ponding, and overland flow are generated from partial areas of the watershed that are functions of both time and space. It is also recognized that surface saturation usually occurs as a result of one of two mechanisms. We will discuss each in turn.

The classic mechanism, first espoused by Horton (1933) and placed in a more scientific framework by Rubin and Steinhardt (1963), is for a precipitation rate p that exceeds the saturated hydraulic conductivity K_o of the surface soil. As illustrated in Figure 5a, a moisture-versus-depth profile during such a rainfall event will show moisture contents that increase at the surface as a function of time. At some point in time (t_3 in Fig. 5a) the surface becomes saturated and an inverted zone of saturation begins to propagate downward into the soil. It is at this time (Fig. 5c) that the infiltration rate drops below the rainfall rate and overland flow is generated. The time t_3 is called the ponding time. The conditions necessary for the generation of overland flow under the Horton-Rubin mechanism are (1) a rainfall rate greater than the saturated conductivity of the soil, and (2) a rainfall duration longer than the ponding time required for a given existing initial-moisture profile.

The second mechanism, identified by Dunne and Black (1970), is illustrated in Figures 5b and 5d. In this case, $p < K_o$ and the initial water table is shallow. Surface saturation is the result of a rising water table; ponding and overland flow occur at time t_5, when no further soil-moisture storage is available.

The Horton-Rubin mechanism is soil controlled and is more common on upslope partial areas. The Dunne mechanism is hydrologically controlled and is more common in near-channel wetlands. Both cases lead to variable-source areas that expand and contract through wet and dry periods.

Both mechanisms can be simulated with the aid of unsaturated-flow theory. A conceptual prediction model that can predict ponding times and ponding rates must be based on the equation for unsaturated flow, which for one-dimensional vertical profiles takes the form

$$\frac{\partial}{\partial z}[K(\Psi)(\frac{\partial \Psi}{\partial z} + 1)] = C(\Psi)\frac{\partial \Psi}{\partial t}. \qquad (22)$$

Solutions to this equation are in terms of the pressure-head profile $\Psi(z, t)$ in the soil. One can easily then determine the moisture-content profile $\theta(z, t)$ (as shown in Figs. 5a and 5b), and the infiltration rates (as shown in Figs. 5c and 5d). The model requires as input the unsaturated characteristic curves of the soil $K(\Psi)$ and $C(\Psi)$, where K is the unsaturated hydraulic conductivity and C is the specific moisture capacity. The most important single parameter is the saturated hydraulic conductivity K_o.

The conceptual rainfall-runoff model of Smith and Woolhiser (1971) mentioned in the previous section utilizes a set of one-dimensional, unsaturated-flow models to predict ponding times and overland flow rates on a kinematic cascade. As can be seen in Figure 4a, their model allows areally nonuniform rainfall $p(x,y,t)$ to fall on a heterogeneous slope where S_o, n, and K_o are all functions of x and y. A nonuniform lateral inflow $q_\ell(y,t)$ leads to the outflow hydrograph $q(t)$.

LINEAR RESERVOIRS AND GROUND-WATER STORAGE

The preceding four sections of this paper reviewed a diverse set of hydrologic processes and methodologies: autoregressive models of streamflow sequences, exceedance probabilities for extreme events, unit hydrographs, kinematic cascades, and the theory of unsaturated overland-flow generation. This section and the three that follow will review some of the interrelationships that have been uncovered by other researchers among these diverse concepts. Perhaps more important will be an attempt to show that where this integrated understanding stops, hydrogeological research could lead to further progress.

Let us first look at the concept of a linear reservoir as a lynchpin between the unit-hydrograph and autoregressive models of streamflow. Klemes (1978) provides an excellent review of these concepts and of reservoir theory in general. Of the many authors who have recognized the linkage between IUH models and AR models, Quimpo (1973) has probably examined the question in greatest detail.

The equation of continuity for any reservoir can be written as

$$p - q = \frac{ds}{dt}, \qquad (23)$$

where p is the inflow, q is the outflow, and s is the storage. A linear reservoir is one in which the outflow is a linear function of the storage:

$$q = ks. \qquad (24)$$

Combining equations 23 and 24 leads to the differential equation

$$\frac{dq}{dt} + kq = kp. \qquad (25)$$

The solution to this equation for the initial condition q = 0 at t = 0 is

$$q(t) = \int_0^t ke^{-k(t-\tau)}p(t)d\tau. \quad (26)$$

Equation 26 is identical to equation 16 if

$$u(t) = ke^{-kt}. \quad (27)$$

This analysis shows that if an instantaneous unit hydrograph u(t) can be put in the form of equation 27, the use of the IUH and equation 16 for rainfall-runoff prediction is equivalent to consideration of the watershed as a single linear reservoir with rainfall input p(t), storage set s(t), and runoff q(t). The model calibration would be carried out in terms of the single watershed parameter k. In storage theory, the inverse, $\kappa = 1/k$, is more commonly used. It is known as the storage coefficient. In that case equation 27 becomes

$$u(t) = \frac{1}{\kappa}e^{-t/\kappa}, \quad (28)$$

and calibration would be carried out in terms of κ. Nash (1957) has shown that one may also develop an IUH for a system of n identical linear reservoirs in series. For this case,

$$u(t) = \frac{1}{\kappa\Gamma(h)}\left(\frac{t}{\kappa}\right)^{n-1} e^{-t/\kappa}, \quad (29)$$

where $\Gamma(n)$ is the nth-order gamma function. In this case, model calibration would involve two parameters, κ and n.

Let us now turn to Quimpo's (1973) analysis that relates the linear-reservoir model of an IUH and the autoregressive model for streamflow sequences. First, note that a simple change in variables leads to the following alternative form of equation 16 for an IUH rainfall-runoff prediction:

$$q(t) = \int_0^t u(\tau)p(t - \tau)d\tau. \quad (30)$$

This formulation is based on the continuous functions p(t), u(t), and q(t). As shown schematically on Figure 6, we can formulate equation 30 equally well for the discrete representations p_i, u_i, and q_i. For $\Delta\ell = 1$, we have

$$q_i = \sum_{\ell=o}^{i} u_\ell \, p_{i-\ell}. \quad (31)$$

Now, recall the first-order AR representation for a streamflow sequence as given in equation 11:

$$q_i = \rho_1 q_{i-1} + a_i. \quad (32)$$

Applying equation 32 recursively to itself leads to

$$q_i = a_i + \rho_1 \, q_{i-1} + \rho_1^2 \, a_{i-2} + \ldots \quad (33)$$

or

$$q_i = \sum_{\ell=1}^{i} \rho_1^\ell a_{i-\ell}. \quad (34)$$

Comparing equations 31 and 34, it should be clear that there is a one-to-one correspondence between the IUH rainfall-runoff model and the AR streamflow model with $u_p \leftrightarrow 1$ and $p_i ET a_i$. In other words, if we view the normally distributed uncorrelated error term a_i in equation 32 as the stochastic rainfall input that creates the streamflow sequence, then the autocorrelation in the streamflow sequence, ρ_1, is related to the IUH, u(t). And since we have shown in equations 28 and 29 that u(t) can be represented by one or more linear reservoirs with storage coefficient κ, it is clear that ρ_1 and κ are closely related. In fact, for a single linear reservoir,

$$\rho_1 = e^{-1/\kappa}. \quad (35)$$

Unit hydrographs and time-series analysis of streamflow are two of the most widely used techniques in surface-water hydrology. As models of the rainfall-runoff process, they each involve only a single watershed parameter, κ in the first case and ρ_1 in the second. In addition, these two parameters apparently are closely interrelated. What should be of interest to hydrogeologists is the fact that these parameters are highly dependent on ground-water and soil-moisture conditions in a watershed. The storage coefficient κ must be related to the mechanism of overland-flow generation on the hillslopes of a given watershed, and hence to the distribution of saturated hydraulic-conductivity values and the usual range of moisture-content profiles. These latter two properties also control the delivery of baseflow to streams, and it is the sequence of baseflows that controls, in large part, the autocorrelation properties of the streamflow. There is much work to be done by hydrogeologists on both of these fronts. At this writing, nothing on the hydrogeological implications of linear-reservoir theory appears to have been covered in the literature. Some work has begun on streamflow autocorrelation, and this work is reviewed in the following section.

STREAMFLOW PERSISTENCE AND BASEFLOW

Moss and Bryson (1974), building on the earlier work of Fiering (1967), used a conceptual baseflow model to explain the presence of autocorrelation (or persistence) in streamflow sequences. They worked with the sequence of mean monthly discharges q_i, q_2 . . . , q_n. For month i (Fig. 7a), the volume of flow during the monthly time period is Q_i. Moss and Bryson (1974) envisaged this volume as being the sum of three components (Fig. 7b), such

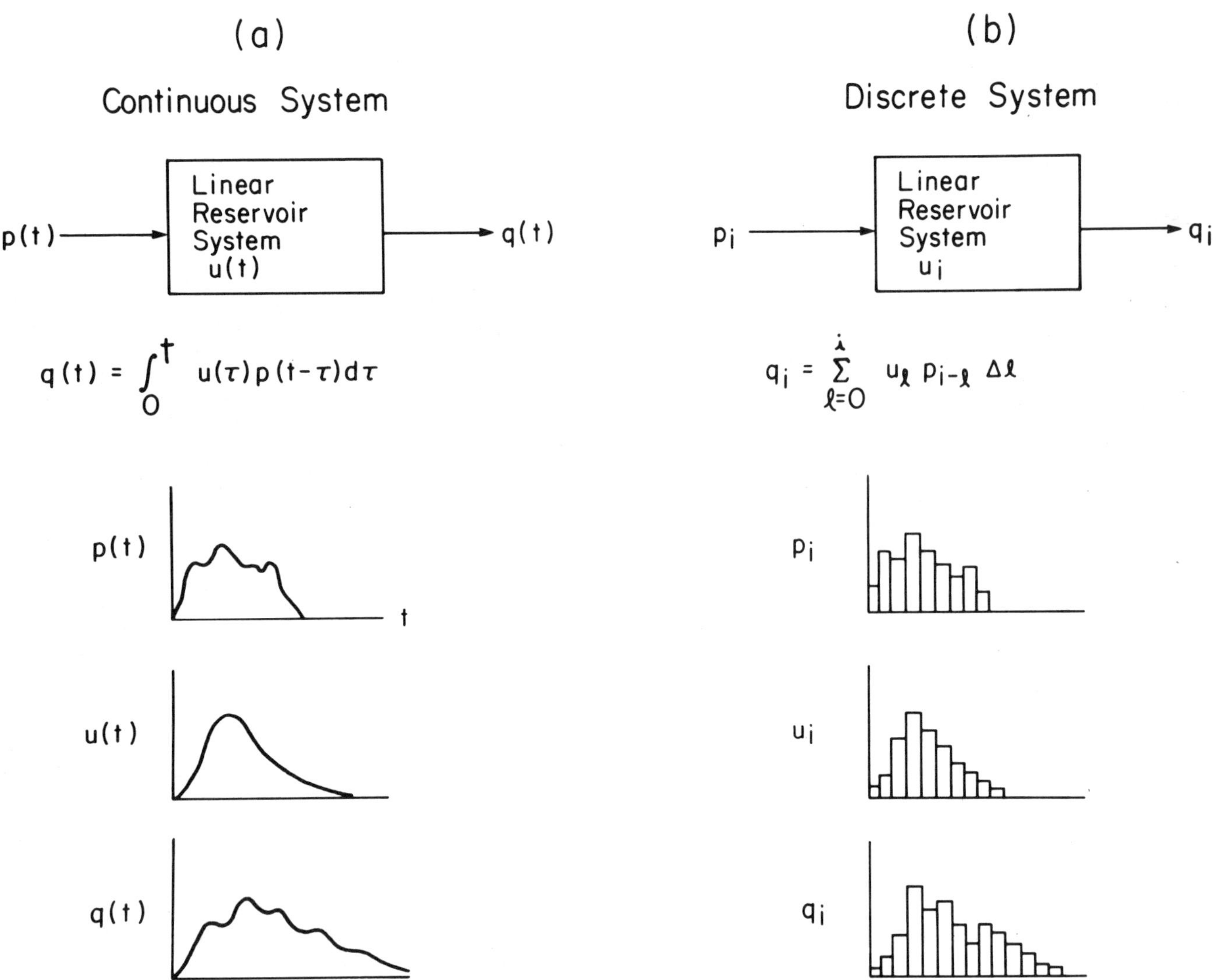

Figure 6. Continuous and discrete representations of the linear-reservoir model for instantaneous unit hydrograph (after Eagleson and others, 1966).

that

$$Q_i = G_i + F_i + (1 - f_i)P_i, \tag{36}$$

where G_i is the volume of baseflow caused by over-period storage, F_i is the volume of baseflow caused by within-period precipitation, and $(1 - f_i)P_i$ is the volume of direct runoff. P_i is the effective precipitation during period i. The sequence of precipitation events that may have occurred during time period i are represented by a single effective-precipitation event at time T_i. The parameter f_i is the percentage of effective precipitation that becomes infiltration.

The model is a lumped-parameter model. The value of f_i is assumed to hold over the entire watershed. If we let s_i represent the volume of ground water in storage at the end of the ith monthly period, the equation of continuity for the watershed is

$$s_i = s_{i-1} + P_i - Q_i. \tag{37}$$

The volumes F_i and G_i in equation 36 are calculated on the assumption that the ground-water storage zone delivering baseflow to the stream operates as a linear reservoir. The linear reservoir must satisfy equations 23 and 24, where $p - p'$ is the inflow due to infiltration and $q - q'$ is the outflow from the reservoir as baseflow. If we assume that the duration of the effective-infiltration event is short with respect to the monthly time periods, we can approximate the system by an instantaneous recharge followed by a long-lived recession. For the recession period, $p' = 0$ in equations 23 and 25. The initial condition at time $t = 0$ is $q' = q_0'$, where, from equation 24, $q_0' = ks_{i-1}$. The solution to equation 25 under these conditions is

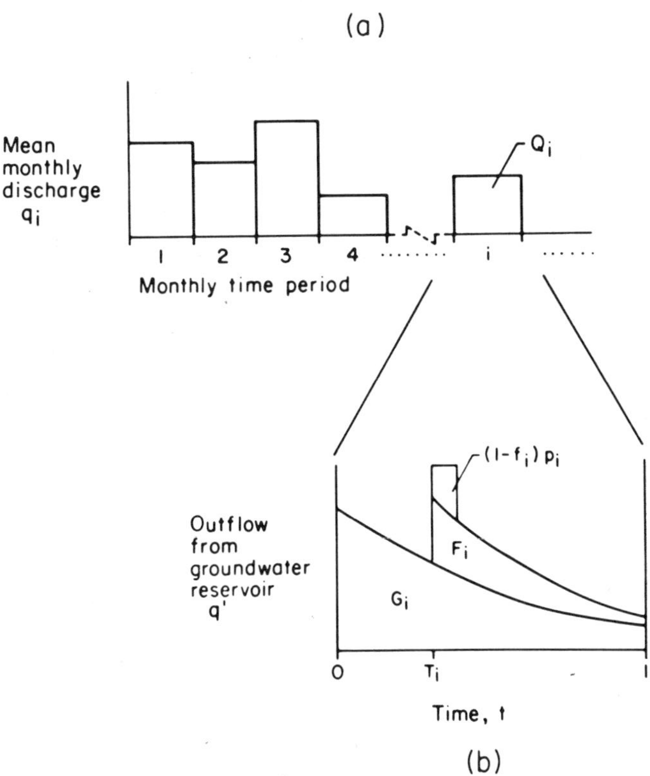

Figure 7. Conceptual monthly baseflow model (after Moss and Bryson, 1974).

$$q'(t) = ks_{i-1} e^{-kt}. \tag{38}$$

Equation 38 is the equation of the line that forms the upper boundary of volume G_i in Figure 7b. Therefore, since

$$G_i = \int_0^1 q'(t)dt, \tag{39}$$

we have

$$G_i = s_{i-1}(1 - e^k). \tag{40}$$

Similarly, integrating over the interval T_i to 1, we have

$$F_i = f_i P_i [1 - e^{k(t_i - 1)}]. \tag{41}$$

Combining equations 36, 37, 40, and 41 leads to

$$Q_i = e^{-k} Q_{i-1} + [f_{i-1} e^{k(t_{i-1})}] P_{i-1} + [1 - f_i e^{k(t_{i-1})}] P_i \tag{42}$$

We can replace the volumes P_i and Q_i by the precipitation rate p_i and mean monthly discharge rate q_i without loss of generality. Setting $c_i = f_i e^{k(t_{i-1})}$, equation 42 becomes

$$q_i = e^{-k} q_{i-1} + c_{i-1} p_{i-1} + c_i p_i. \tag{43}$$

Moss and Bryson (1974) treat the parameters k and c in equation 43 as deterministic parameters, and the variables p and q as random variables. They show that equation 43 can be viewed as the output from a mixed, autoregressive moving-average stochastic process. It is possible to calculate the covariance function for q_i and q_{i-1} and thus estimate the first-order autocorrelation coefficient ρ_1 for the streamflow sequence. They show that values of ρ_1 for an actual watershed calculated through the conceptual baseflow model are in close agreement with the values determined from the measured streamflow sequence.

Equation 43 as developed by Moss and Bryson (1974) bears a striking resemblance to equation 32 as developed by Quimpo (1973), especially in light of equation 35. This is not surprising, as both studies used linear-reservoir theory to link rainfall-runoff processes to the observed properties of streamflow sequences. Quimpo used the unit-hydrograph model. Moss and Bryson utilized a conceptual baseflow model that is probably closer to reality.

The model of Moss and Bryson (1974) is well suited to highly permeable watersheds, where subsurface delivery of streamflow is more important than overland flow. As noted in the earlier section "Mechanisms of Streamflow Generation," this is not thought to be the usual case. There is great scope for the extension of the ideas of Quimpo (1973) and Moss and Bryson (1974) to watershed models that do use the nonlinear, areally variable mechanisms of streamflow generation that rest on an understanding of the saturated-unsaturated subsurface-flow conditions on hillslopes.

CONCEPTUAL PREDICTION OF EXCEEDANCE PROBABILITIES

As noted in Tables 1, 2, and 3, the usual approach to design prediction for culverts, small dams, and other minor structures involves a frequency analysis of peak-discharge events. As shown in Figure 2d, the usual form of the design prediction is a graph of peak discharge versus exceedance probability or recurrence interval.

Eagleson (1972) introduced a new approach to the determination of exceedance probabilities. Rather than trying to work with the set of flood events in a stream, which usually constitutes a small statistical sample under the limited recorded lengths available, he chose to work with the set of rainfall events on the watershed, which usually provides a much larger sample. He then looked for a method for transforming the probability-density function for rainfall into a probability-density function for streamflow. The method he chose utilizes the kinematic rainfall-runoff model of Wooding (1965) as described in an earlier section of this paper in connection with Figure 4a.

A point-rainfall event can be described statistically in terms of two parameters, the average point-rainfall intensity $\bar{p}$, and the storm duration t_p. Eagleson stated that the probability-density functions for both of these parameters take the form of exponential distributions, so that

$$f(\bar{p}) = \beta e^{-\beta\bar{p}}, \tag{44}$$

and

$$t(t_p) = \lambda e^{-\lambda t_p}. \tag{45}$$

He then showed that their areal joint distribution is given by

$$f(\bar{p},t_p) = \frac{\lambda\beta}{K} e^{(-\lambda t_p - \frac{\beta}{K}\bar{p})}, \tag{46}$$

where K is a factor that reduces point-rainfall depths to areal averages.

Eagleson then developed an analytical expression for the probability-density function $f(q_p)$, where q_p is the peak discharge in the stream. As noted in Figures 2b, 2c, and 2d, the probability-density function $f(q_p)$ can easily be converted into the cumulative probability-density function, $F(q_p)$, the exceedance-probability curve $P(q_p)$, or the recurrence-interval relation $T(q_p)$. Eagleson's analytical expression is based on solutions to equations 17 through 21 as applied to the watershed model in Figure 4a. It involves many simplifying assumptions with respect to the geometry of the system and the nature of the watershed parameters. The final form of the equation relating peak discharge q_p and recurrence interval T is very complicated and quoting it here in full would serve little purpose. However, it is worth noting its functional form:

$$q_p = q_p(T,\lambda,\beta,K,\theta,P,A,A_p,y_L,n_o,n_s,\phi_1, \phi_2). \tag{47}$$

The parameters T, λ, β, and K are as defined earlier; θ and P are the average annual number of independent rainfall events and the average annual rainfall depth, respectively; A and A_p are the total catchment area and the partial area respectively that produce direct runoff; y_L is the length of stream reach (Fig. 4a); n_o and n_s are the friction parameters on the overland flow plane and in the stream, respectively; ϕ_1 is the fraction of rainfall occurring as runoff; and ϕ_2 is the fraction of runoff occurring as direct runoff. The parameters λ, β, K, θ, and P are estimated from the available rainfall records. The parameters A and y_L can be gleaned from watershed maps. The parameters A_p, n_o, n_s, ϕ_1 and ϕ_2 are lumped-deterministic parameters. They must be obtained by a combined (and probably iterative) process of estimation and model calibration.

Using Eagleson's analytical expression, it is possible to hold all but one of the watershed parameters constant and examine the influence of any single parameter on the resulting flood-frequency curves. The method can be used to predict how the form of the frequency distribution changes from catchment to catchment or within the same catchment under the influence of man-made changes.

The parameter of greatest interest to hydrogeologists in equation 47 is ϕ_2, which is, in effect, the infiltration parameter. As suggested schematically in Figure 8a, an increase in ϕ should lead to decreased q_p at any given recurrence interval T.

In Eagleson's analysis, ϕ was taken as a lumped-deterministic parameter. In a subsequent study, Wood (1976) repeated Eagleson's analysis with ϕ defined statistically. He assessed the uncertainty that is introduced in the exceedance-probability curves when ϕ is assumed to be randomly distributed on the watershed. He assessed the influence of various statistical distributions and various values of the mean and variance of ϕ. Figure 8b is a sample of his results for a particular set of rainfall and watershed parameters, for a particular distribution of ϕ. The curve labeled $P(q_p \mid \phi_2|$, is the exceedance-probability curve generated by Eagleson's equation with ϕ_2, set deterministically at a value equal to $\bar{\phi}_2$, the mean of the infiltration-parameter distribution. The curve

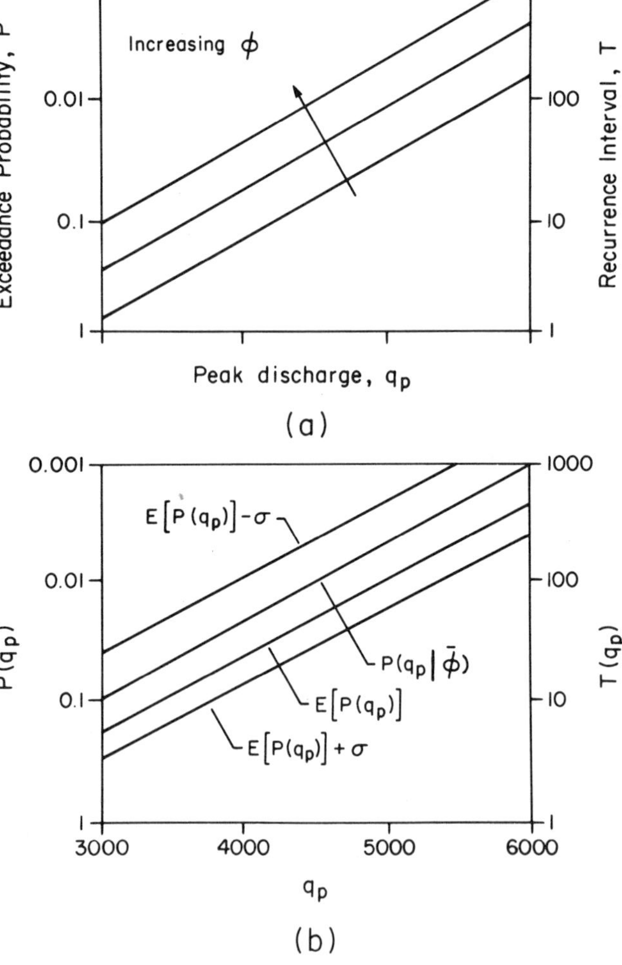

Figure 8. Effect of infiltration rates on conceptual prediction of exceedance probabilities (part b after Wood, 1976).

labeled E $[P(q_p)]$ is the expected value of the exceedance-probability curve as determined by Wood's stochastic approach. It is interesting to note that these two curves are not coincident. The outside curves show the expected exceedance-probability curve plus and minus one standard deviation.

Results like those of Figure 8a show the importance of infiltration in determining exceedance-probability curves. Results like those of Figure 8b show the importance that uncertainties in our knowledge of the infiltration rates play in assessing the uncertainty associated with predicted exceedance probabilities. It is clear that the understanding that exists within the hydrogeological community regarding the infiltration process could be brought to bear on improving the conceptual prediction of flood-frequency curves. The spatial variation in saturated hydraulic conductivity and the spatial and temporal variations in soil-moisture profiles and ponding times must play an important role.

TIME-SPACE TRADEOFFS IN ASSESSING WORTH OF DATA

Applying a rainfall-runoff model to a given watershed requires a calibration. It is best to view this process in terms of three distinct time periods: calibration, verification, and prediction. During the calibration period, the model parameters are estimated on the basis of the available records of rainfall and runoff. During the verification period, the calibration model is applied to the available rainfall record, and computed runoffs are compared with the observed runoff records in order to assess the predictive efficiency of the model. If the efficiency is adequate, the model can then be used for runoff prediction in the prediction period. In the case of operational forecasts, the prediction period is the immediate future; in the case of a design forecast, the prediction period is a hypothetical future.

Let us look at the concepts of calibration and verification with respect to two of the models that have been discussed in this paper: the instantaneous unit hydrograph and Smith and Woolhiser's (1971) kinematic cascade on an infiltrating surface.

If we accept the single linear-reservoir representation of an IUH, then, as shown by equation 28, calibration is carried out to estimate the single parameter κ. If the IUH is represented by a system of n linear reservoirs (equation 29), then calibration must be carried out in terms of both κ and n. For Smith and Woolhiser's model (equations 17 through 22) a general calibration would be in terms of the overland-flow parameters $n_1, n_2, \ldots, n_n$; $S_1, S_2, \ldots, S_n$; and the saturated hydraulic conductivity $K_o(x,y)$. For a single homogeneous plane, the model parameters are reduced to n, S_o, and K_o. The parameter κ in the IUH model is an empirical parameter that cannot be determined in any other way than through calibration. The parameters n, S_o, and K_o in the conceptual hydrologic-simulation model are physically based parameters that can be determined either by calibration or by field measurement or by a combination of these two methods. Eagleson and others (1966) discuss the calibration of instantaneous unit hydrographs; Nash and Sutcliffe (1970) discuss the calibration of conceptual-simulation models.

Nash and Sutcliffe (1970) also suggest a verification procedure based on a sum-of-squares criterion:

$$F^2 = \sum (q' - q)^2, \tag{48}$$

where q and q' are the observed and computed stream discharges at corresponding times, and F^2 is the residual variance. The sum may be taken over all q's at some interval t or at preselected times, such as at peaks or troughs in the hydrograph. The initial variance F_o^2 is defined by

$$F_o^2 = \sum (q - \bar{q})^2, \tag{49}$$

where $\bar{q}$ is the mean of the observed q's. The efficiency of the model can then be defined in terms of the proportion of the initial variance accounted for by the model:

$$R^2 = \frac{F_o^2 - F^2}{F_o^2}. \tag{50}$$

Our desire is to maximize model efficiency by minimizing R^2.

The parameter R^2 can be used to compare the efficiencies of two models, such as the unit hydrograph and the kinematic cascade described above. It can also be used to establish the worth of additional data for either model. The worth-of-data concept rests on the fact that it may be possible to reduce R^2 by taking more data measurements. The cost of the added measurements must then be weighed against the value of the increased model efficiency. In the design of large dams (Table 2), for example, the value may be substantial.

There is another aspect of the worth-of-data question that deserves consideration. For time-dependent data, such as rainfall or runoff, it is possible to increase the database only by waiting for further measurements. There may be a cost associated with the delay. The databases of parameters that are time independent, such as n, S_o, and K_o in the simulation model, can be immediately increased by a measurement program that increases knowledge of the parameter distribution through space. The relative increase in efficiency that may be achieved through an increase in these two types of database is referred to as the question of time-space tradeoffs.

Figure 9 is a schematic illustration for discussing time-space tradeoffs where comparing of the IUH model with the Smith-Woolhiser simulation model for prediction of runoff from rainfall. In the IUH model (Fig. 9a), records of p(t) from a rain gauge and q(t) from a stream gauge are calibrated in terms of the model parameter κ. This parameter is an integrated watershed parameter; it cannot be viewed as having a distribution in space. An increase in efficiency in the IUH model can only be

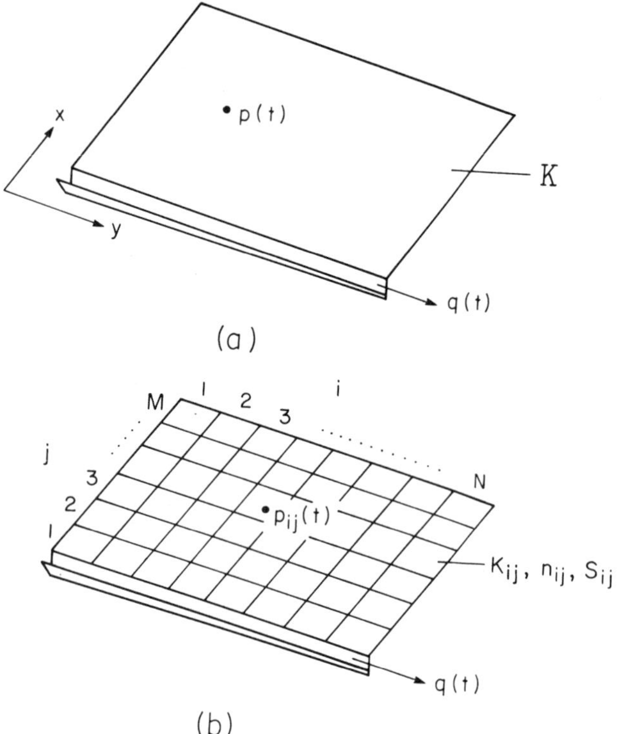

Figure 9. Hillslope representations for assessment of time-space tradeoffs in comparison of two rainfall-runoff models.

attained by extending the p(t) and q(t) records or by adding additional rain gauges to estimate the space-time behavior of p(x,y,t). Either approach requires a waiting period while the new data are generated.

If we view the conceptual model (Fig. 9b) as spatially discretized, we may be able to increase the efficiency of the model either by extending the rainfall record $p_{ij}(t)$ or by taking additional measurements of the physically based model parameters n_{ij}, S_{ij}, and K_{ij}. Our earlier discussion of the mechanisms of streamflow generation should make it clear that the time until ponding is more important in creating streamflow lags than the time of delivery of overland flow. The hydraulic conductivities K_{ij} are therefore likely to be more sensitive than the overland-flow parameters, n_{ij} and S_{ij}, in the runoff predictions. Research is currently under way to examine the space-time tradeoffs between empirical and conceptual rainfall-runoff models. Emphasis will be placed on the role of the distribution of hydraulic conductivities of hillslope soils.

The question of time-space tradeoffs is inherently bound to the question of how best to describe the spatial distribution of physically based watershed parameters such as hydraulic conductivity. It is now widely recognized that the best approach is to view these parameters as spatially stochastic variables. The work of Wood (1976), described earlier, is in this spirit. Wood's infiltration parameter, $\phi 2$, which he used in his assessment of Eagleson's exceedance-probability model, was randomly distributed. Recently, Smith and Hebbert (1980) utilized a simplified form of the model of Smith and Woolhiser to predict runoff from a watershed with a randomly distributed infiltration parameter. Both these studies assumed the infiltration parameters to be independently distributed through space. A more realistic assumption is that these parameters can be described by a spatial stochastic process, such as a first-order autoregressive process. For hydraulic conductivities along a one-dimensional line, a space series would be analogous to the time series of equation 11, so that

$$K_i = \rho_1 K_{i-1} + a_i. \qquad (51)$$

Field measurements of K would then be used to obtain estimates of μ_K, σ_K^2, and ρ_1. Smith and Freeze (1980) followed this approach, using a slightly different stochastic-process model, to assess two-dimensional, steady-state ground-water flow in a bounded domain. I am currently adapting the approach to the specification of spatial variability on a discretized hillslope (Fig. 9b) in order to examine the influence of such variability on the efficiency of various rainfall-runoff models.

ACKNOWLEDGMENTS

This paper was prepared while I held a sabbatical appointment in the Department of Hydrology and Water Resources at the University of Arizona. The stimulating environment provided there is much appreciated. I also wish to acknowledge the helpful reviews of this paper by Francis Hall, Shlomo Neuman, and Roger Smith.

REFERENCES CITED

Benson, M. A., 1968, Uniform flood-frequency estimating methods for federal agencies: Water Resources Research, v. 4, p. 891-908.

Box, G.E.P., and Jenkins, G. M., 1976, Time series analysis, forecasting and control: New York, Holden-Day, 575 p.

Chow, V. T., 1964a, Statistical and probability analysis of hydrologic data; Part I, Frequency analysis, in Chow, V. T., ed., Handbook of applied hydrology: New York, McGraw-Hill, p. 8.1-8.42.

___1964b, Runoff, in Chow, V. T., ed., Handbook of applied hydrology: New York, McGraw-Hill, p. 14.1-14.54.

Clarke, R. T., 1972, General report on stochastic and statistical models, in Proceedings, International Symposium on Uncertainties in Hydrologic and Water Resource Systems: Tucson, Arizona, p. 1325-1356.

___1973, A review of some mathematical models used in hydrology, with observations on their calibration and use: Journal of Hydrology, v. 19, p. 1-20.

Dalrymple, T., 1964, Hydrology of flow control; Part

I, Flood characteristics and flow determination, in Chow, V. T., ed., Handbook of applied hydrology: New York, McGraw-Hill, p. 25.1-25.33.

Dooge, J.C.I., 1959, A general theory of the unit hydrograph: Journal of Geophysical Research, v. 64, p. 241-256.

Dunne, T., and Black, R. D., 1970, An experimental investigation of runoff production in permeable soils: Water Resources Research, v. 6, p. 478-490.

Eagleson, P. S., Meija, R., and March, F., 1966, Computation of optimum realizable unit hydrographs: Water Resources Research, v. 2, p. 755-764.

___1970, Dynamic hydrology: New York, McGraw-Hill, 462 p.

___1972, Dynamics of flood frequency: Water Resources Research, v. 8, p. 878-897.

Fiering, N. B., 1967, Streamflow synthesis: Cambridge, Harvard University Press.

Freeze, R. A., 1974, Streamflow generation: Reviews of Geophysics and Space Physics, v. 12, p. 627-647.

Horton, R. E., 1933, The role of infiltration in the hydrologic cycle: American Geophysical Union Transactions, v. 14, p. 446-460.

Hurst, H. E., 1956, Methods of using long-term storage in reservoirs, in Institute of Civil Engineers, Proceedings: London, v. 5, p. 519-590.

Kibler, D. F., and Woolhiser, D. A., 1970, The kinematic cascade as a hydrologic model: Colorado State University Hydrology Paper 39, 27 p.

Kirkby, M., 1978, Hillslope hydrology: New York, Wiley-Interscience, 389 p.

Kisiel, C. C., 1969, Time series analysis of hydrologic data: Advances in Hydroscience, v. 5, p. 1-119.

Klemes, V., 1978, Physically based stochastic hydrologic analysis: Advances in Hydroscience, v. 11, p. 285-356.

Mandelbrot, B. B., and Wallis, J. R., 1968, Noah, Joseph, and operational hydrology: Water Resources Research, v. 4, p. 909-918.

Moss, M. E., and Bryson, M. C., 1974, Autocorrelation structure of monthly streamflows: Water Resources Research, v. 10, p. 737-744.

Nash, J. E., 1957, The form of the instantaneous unit hydrograph: International Association of Science Hydrology Publication 45, v. 3, p. 114-121.

Nash, J. E., and Sutcliffe, J. V., 1970, River flow forecasting through conceptual models; Part I, A discussion of principles: Journal of Hydrology, v. 10, p. 282-290.

O'Connell, P. E., 1972, General report on stochastic and statistical models, in Proceedings, International Symposium on Uncertainties in Hydrologic and Water Resource Systems: Tucson, Arizona, p. 1363-1402.

Quimpo, R. G., 1973, Link between stochastic and parametric hydrology: American Society of Civil Engineers Journal of Hydrology Division Proceedings, v. 99, no. HY3, p. 461-470.

Rubin, J., and Steinhardt, R., 1963, Soil water relations during rain infiltration, Part 1, Theory: Soil Science Society of America Proceedings, v. 27, p. 246-251.

Smith, R. E., and Woolheiser, D. A., 1971, Overland flow on an infiltrating surface: Water Resources Research, v. 7, p. 899-913.

Smith, L., and Freeze, R. A., 1980, Stochastic analysis of steady-state groundwater flow in a bounded domain, 2, Two-dimensional simulations: Submitted to Water Resources Research.

Smith, R. E., and Hebbert, R.H.B., 1980, A Monte Carlo analysis of the hydrologic effect of spatial variability of infiltration: Submitted to Water Resources Research.

Wood, E. F., 1976, An analysis of the effects of parameter uncertainty in deterministic hydrologic models: Water Resources Research, v. 12, p. 925-932.

Wooding, R. A., 1965, A hydraulic model for the catchment stream problem, 1, Kinematic-wave theory: Journal of Hydrology, v. 3, p. 254-257.

Manuscript Accepted by the Society October 14, 1980

Printed in U.S.A.

Geological Society of America
Special Paper 189
1982

Statistical Characterization of Aquifer Heterogeneities: An Overview

SHLOMO P. NEUMAN
Department of Hydrology and Water Resources
University of Arizona
Tucson, Arizona 85721

ABSTRACT

In recent years, ground-water hydrologists have become increasingly aware of the need to describe the spatial variability of aquifer characteristics in statistical terms. This has led to the development of new theoretical models whose parameters, representing the aquifer characteristics, are treated as stochastic variables rather than deterministic functions of space. In this paper, the state of the art in stochastic modeling is reviewed, and the experience and new knowledge gained with these models are summarized. Many of the stochastic models developed to date allow the parameters to fluctuate with equal amplitude at every point in space, including points at which the material properties have actually been measured. A more recent trend has been to try to reduce the variance of the computed hydraulic head values by conditioning the model on measured values of aquifer transmissivities. It is argued that a further reduction in this variance could be effected by conditioning the model not only upon measurements of the aquifer characteristics, but also upon historical data relating to the prevailing flow regime. This additional conditioning can be achieved by estimating the model parameters with the aid of inverse methods that are compatible with the stochastic interpretation of spatial variability. Two such inverse methods are described in this paper. It is suggested that in the future, the output from inverse models should be used as input into stochastic models of groundwater flow. In addition to the need for interphasing inverse models with stochastic models, additional research is required to improve the reliability, versatility, and computational efficiency of these models.

Dedicated to the author's teacher, colleague, and friend, Paul A. Witherspoon, on the occasion of his 60th birthday

The statisticians believe those who prove points deductively to be dangerously intuitive. But, by their colleagues, those who are controlled by numbers are often thought unduly cautious or even dull.

John Kenneth Galbraith
The New Industrial State
1967

INTRODUCTION

Most quantitative methods in subsurface hydrology are founded on the classical theory of flow through porous media. This theory is concerned with macroscopic physical quantities, which vary in a deterministic manner and obey the principles of continuum mechanics. These principles are usually expressed in the form of partial differential equations, the parameters of which represent various material properties of the porous continuum such as hydraulic conductivity, storativity, and porosity, etc. In order to use the available mathematical tools to solve these partial differential equations, it is usually necessary to assume that their parameters remain uniform throughout the entire flow region of interest. Only in rare situations, such as flow to wells in multi-aquifer systems have hydrologists been able to overcome this restriction to some extent by assuming that the flow region can be subdivided into several layers, each of which can be characterized by a distinct set of uniform parameters (Neuman and Witherspoon, 1969).

In reality, subsurface materials are seldom uniform; on the contrary, their macroscopic properties usually vary from point to point in what appears to be a random manner. Evidence of such random spatial variability is provided by laboratory analyses of soil and rock samples, lithological and geophysical logs of boreholes, well tests, surface geophysical surveys, and direct geological observations. This evidence is in clear contradiction to the classical theoretical approach in which material characteristics are represented by constant parameters. Can this contradiction be resolved by defining an equivalent uniform porous medium that behaves mathematically in a manner analogous to that of the actual medium? Would such a definition depend only on the material itself, or would it be affected by the prevailing flow regime? How can one determine the "effective" properties of such an equivalent medium in the field, and how do these fictitious properties relate to the true ones? If, on the other hand, the concept of an equivalent uniform medium is valid only under limited conditions, what utility, if any, can the practicing hydrologist hope to derive from the wealth of analytical solutions that have been derived on the basis of this concept?

Although these questions have been asked by hydrologists for many years, a serious attempt to answer them has not begun until very recently. Instead, with the advent of digital computers, the emphasis in hydrogeology has shifted away from mathematical analysis toward numerical techniques, which are more flexible in their ability to deal with variable materials and other complexities of nature. In the numerical approach, material properties are represented by parameters that are either assumed to remain uniform over discrete subregions of the flow field, or are allowed to vary slowly in space. In both cases, the rate of spatial variability allowed in the numerical model is limited by practical considerations such as computer time, computer storage, and availability of data. As a result of such limitations, the rate at which the parameters of a numerical model vary from point to point is usually much slower than the rate of variation of the actual material properties. When this is the case, the parameters cannot be regarded as representatives of the true material characteristics but only, at best, of some average quantities related to the latter. Can such average quantities be properly defined, and if so, how and under what conditions? Would such a definition depend only on the material and the degree of numerical discretization, or would it also be affected by the manner in which flow takes place? How could these parameters be determined on the basis of actual field and laboratory data? If the concept of average parameters is invalid under certain conditions, or if these averages cannot be determined in a satisfactory manner from the available data, what utility, if any, can the hydrologist hope to derive from the large number of sophisticated computer models that are presently at his disposal?

The above questions are similar to those that have been raised previously in connection with the concept of an equivalent uniform porous medium, the difference being primarily one of scale. If the parameters of a discrete subregion in a numerical model are assumed to be constant, then this subregion plays the role of an equivalent, locally uniform porous medium. If the parameters are allowed to vary slowly in space, then the subregion can be regarded as an equivalent medium which is neither completely uniform nor heterogeneous to the same extent as the true material. Regardless of which case one is dealing with, the questions posed earlier cannot be answered without postulating a comprehensive theory that could clarify the relationship between the macroscopic properties of subsurface materials as determined with the aid of field or laboratory procedures and the parameters of relevant theoretical models, whether analytical or numerical. In addition, this theory should be able to address itself to the following three questions: How can we estimate the parameters of theoretical models on the basis of field and laboratory data? How can we determine the degree of uncertainty associated with such estimates? How can we decide what effect this uncertainty may have on the reliability of mathematical or numerical results provided by the theoretical model for specific problems?

The experience of recent years has shown that progress toward the development of such a theory can be achieved most efficiently by regarding the macroscopic properties of subsurface materials as random variables. This means that the parameters of theoretical models are also random, and so are the solutions provided by these models for given problems. In addition to accounting for the random nature of material characteristics, some researchers have also begun considering uncertainties in other quantities entering into theoretical models of subsurface fluid flow, such as pumpage and recharge. The result has been a gradual change in emphasis from the traditional deterministic framework of porous media theory toward a modified theoretical approach based on stochastic differential equations and their discrete equivalents.

In this paper, I will attempt to trace briefly the history of this development toward a statistical theory of macroscopic flow through porous media and to provide an overview of the state of the art. Various researchers have been introducing statistical elements in to the established deterministic setup of porous media theory, and their approaches may appear to be unrelated or even contradictory. For example, some workers have viewed the parameters of the governing equations in a continuum sense as random fields, while others have treated them as random vectors in a discrete framework. Some hydrologists have dealt with the effect of random parameter variations on fluctuations in hydraulic head, others have tried to do the same by requiring that the parameters conform to available field or laboratory measurements at specific sites. Still others have been developing statistical methods for estimating these parameters by inverse techniques based on theoretical models of flow through porous media. In the discussion that follows, I will emphasize the relationships among these approaches, stressing that all of them can be viewed as attempts to deal with different aspects of the same problem: the effect of parameter variability and uncertainty on our ability to describe the past, present, and future states of a given hydrological system. Once these relationships are clarified, the road should then be open for closer cooperation among researchers interested in material heterogeneity and, consequently, toward a more rapid development of the theory and practice of statistical subsurface hydrology.

FREQUENCY DISTRIBUTION OF MATERIAL PROPERTIES

The search for a statistical description of aquifer heterogeneities has for many years focused on the frequency distribution of various material properties within given geologic formations. The general approach has been to assume that measured values of a material property (for example, hydraulic conductivity) at various points within the geologic formation can be regarded as belonging to a population having a univariate probability distribution. This is the same as saying that these measurements are independent of each other in a statistical sense and are controlled by the same probability law at each point. When this is true, the governing probability law can be determined from a histogram of the measured values, provided that the sample is sufficiently large. Based on such an approach, researchers have concluded that the distribution of hydraulic conductivities in a given geologic unit is approximately lognormal (Davis, 1969 p. 76). The work leading to this conclusion has been summarized by Freeze (1975) and Bakr (1976). The lognormal distribution was originally discovered by Law (1944) on the basis of core data from a carbonate oil-field reservoir. It was later supported to various degrees by the works of Bulness (1946) on similar data, by Warren and others (1961) on core data from sandy and dolomitic reservoirs, by Bennion and Griffiths (1966) on tens of thousands of core data from reservoir rocks including sand, conglomerate, and limestone, by Ilyin and others (1971) on samples of various rock types, and by Bakr (1976) on core data from unconsolidated aquifer materials. Similar conclusions were reached with regard to the hydraulic conductivity of shallow soils on the basis of laboratory and field tests by Willardson and Hurst (1965), Brust and others (1968), and Nielsen and others (1973). The statistical parameters of the lognormal distribution determined on the basis of some of these data have been summarized by Freeze (1975, Table 1).

Other material properties that are directly related to hydraulic conductivity also appear to have a lognormal distribution. This is true of transmissivities (McMillan, 1966; Farengolts and Kolyada, 1969; Jetel, 1974; Krasny, 1974; Rousselot, 1976); specific capacities (Csallany and Walton, 1963; Walton and Neil, 1963; Seaber and Hollyday, 1968); and grain sizes of porous materials (Krumbein, 1936; Griffiths, 1967), although the work of Bagnold (1968) raises some questions about the last category. On the other hand, the distribution of porosities is usually regarded to be normal (see Law, 1944; Bennion and Griffiths, 1966). If one is willing to assume that hydraulic conductivity is an exponential function of porosity as suggested by some empirical formulas, then a normal distribution of porosities implies that the distribution of hydraulic conductivities must be lognormal.

The recognition that the hydraulic properties of porous materials can be described in statistical terms has led some researchers to investigate the effect of a random distribution of parameters in the governing equations on the flow of water through such materials. The tool required for such an investigation is a stochastic model. The rest of our discussion in this paper will center around the description of various stochastic models that have been developed during recent years for the analysis of flow through porous materials with random parameters.

STOCHASTIC MODELS WITH INDEPENDENT PARAMETERS

Stochastic models differ from deterministic ones in that some of their input functions are uncertain and therefore, the model output must be described in probabilistic rather than deterministic terms. In this paper, we shall be concerned primarily (though not exclusively) with stochastic models of flow through porous media in which all input functions are prescribed (for instance, initial and boundary conditions, sink and source terms) with the exception of the model parameters, which are treated as random functions of space.

Let $\alpha(x)$ be the generic notation for a parameter function that is defined over a continuum and varies with the spatial coordinate vector $x = (x_1, x_2, x_3)$. Assume that $\alpha(x)$ is a random field that can be described by a stochastic process. This implies that in order to obtain a complete description of $\alpha(x)$, one would have to specify its joint probability density function over the entire flow field. However, this joint density function is seldom known in real-life situations and therefore one must usually resort to simplifying assumptions. The most common simplifying assumptions are those of stationarity and ergodicity. Strict stationarity implies that the probability law governing $\alpha(x)$ remains constant in space. Ergodicity means that this law can be derived either from repeated sampling of an ensemble of statistically equivalent media at a given point in space, or from samples collected at different points within a single medium (that is, from a single realization of the stochastic process). Since in reality one always deals with a specific geologic formation rather than an ensemble of such formations, one has no choice but to adopt the assumption of ergodicity as a working hypothesis for the stochastic approach. When this hypothesis fails, as may happen in some situations, the stochastic approach can lead to inaccurate results.

The assumption of strict stationarity can often be relaxed by postulating weaker assumptions related to the first few moments of $\alpha(x)$. The mixed nth moment of $\alpha(x)$ is defined as:

$$E[\alpha^i(\underline{x}_i)\ \alpha^j(\underline{x}_2)] \equiv \iint_{-\infty}^{\infty} \alpha_1^i\ \alpha_2^j\ g(\alpha_1, \alpha_2; \underline{x}_1, \underline{x}_2)\ d\alpha_1\ d\alpha_2, \qquad (1)$$

where $0 < i + j = n$, $\underline{x}_1$ and $\underline{x}_2$ are two arbitrary points in the field, $\alpha_1 \equiv \alpha(\underline{x}_1)$, $\alpha_2 \equiv \alpha(\underline{x}_2)$, and $g(\alpha_1,\alpha_2; \underline{x}_1,\underline{x}_2)$ is the joint density of $\alpha(\underline{x})$ at points x_1 and x_2. The ensemble average of $\alpha(x)$ at a given point is given by its first moment ($i = 1$, $j = 0$),

$$E[\alpha(\underline{x})] = \int_{-\infty}^{\infty} \alpha \; g(\alpha;\underline{x}) \; d\alpha. \tag{2}$$

The spatial covariance (or autocovariance) of (x) is defined in terms of the first and second moments as

$$C_\alpha(\underline{x}_1,\underline{x}_2) \equiv E[\alpha(\underline{x}_1)\;\alpha(\underline{x}_2)] - E[\alpha(\underline{x}_1)]\;E[\alpha(\underline{x}_2)], \tag{3}$$

whereas the variance at a given point is

$$\sigma^2_\alpha(\underline{x}) \equiv C_\alpha(\underline{x},\underline{x}) \tag{4}$$

The spatial correlation (or autocorrelation) function is defined in terms of C_α and σ^2_α as

$$\rho_\alpha(\underline{x}_1,\underline{x}_2) = \frac{C_\alpha(\underline{x}_1,\underline{x}_2)}{\sigma_\alpha(\underline{x}_1)\;\sigma_\alpha(\underline{x}_2)} \;. \tag{5}$$

In the particular case where $E[\alpha(\underline{x})]$ is independent of $\underline{x}$, and $C_\alpha(\underline{x}_1,\underline{x}_2)$ does not depend on the actual choice of x_1 and x_2, but only on the vector $\underline{\zeta} = \underline{x}_1 - x_2$ separating these points, so that

$$C_\alpha(\underline{x}_1,\underline{x}_2) = C_\alpha(\underline{\zeta}); \tag{6}$$

then $\alpha(\underline{x})$ is said to be a second-order or weakly stationary stochastic process. If $C_\alpha(\underline{x}_1,\underline{x}_2)$ does not depend on $\underline{\zeta}$ but only on its modulus (that is, on the scalar distance between x_1 and x_2), the process is said to be not only weakly stationary, but also statistically isotropic (otherwise, it is anisotropic). Similar definitions apply when $\alpha(\underline{x})$ is replaced by a vector, $\underline{\alpha} = (\alpha_1,\alpha_2,\ldots,\alpha_N)$, representing discrete values of the parameter over finite subregions of the field. Different stochastic models of flow through porous media will differ from each other primarily in their manner of defining α (whether over a continuum or a discrete grid), their assumptions with regard to the density function or the first two moments of α, and the method of solution.

This section is limited to stochastic models in which the probability density of α is assumed to be identical at each point in space so that α becomes a (strictly) stationary process. In addition, the values of α at different points are assumed to be statistically independent. We will discuss models in which α is defined in a discrete manner as well as models in which it is defined over a continuum.

The first stochastic model of macroscopic flow through porous media was constructed by Warren and Price (1961). They used a numerical Monte Carlo approach to investigate the effect of heterogeneities in the hydraulic conductivity on steady and transient flows in quasi-uniform and quasi-radial flow fields. In their model, the hydraulic conductivity was represented by discrete values, which were assumed to be statistically independent and were assigned at random according to a known frequency distribution characterized by a relatively small variance. The authors' objective was to examine the influence of various frequency distributions on the definition of effective hydraulic conductivities. One of their important conclusions was that when the distribution is lognormal and flow takes place in three dimensions under steady state, replacing the heterogeneous medium with an equivalent uniform medium will preserve the flow rate through the system, provided that the equivalent (or effective) conductivity is taken to be the geometric mean of the "true" values. Although their transient analysis was less rigorous and considered only small variances, Warren and Price concluded that the geometric mean should be equally representative of the total medium under conditions of nonsteady state.

Following the early work of Warren and Price (1961), the development of stochastic models with statistically independent parameters has been pursued by several other workers. Warren and Skiba (1964) used numerical Monte Carlo simulations of convective mass transport to study chemical dispersion in a three-dimensional system. A numerical Monte Carlo study of the relationship between hydraulic conductivity variations and fluctuations in hydraulic head values in two- and three-dimensional flow regimes was performed by McMillan (1966). Bibby and Sunada (1971) investigated the effect of normally distributed errors in the measurement of hydraulic conductivity, storage coefficient, aquifer thickness, initial head, boundary head, and pumping rate on the numerical solution of transient flow to a well in a confined aquifer. The effect of variable hydraulic conductivity on dispersion in a two-dimensional system was analyzed with the aid of a numerical model by Heller (1972). A different approach was taken by Lippmann (1973) who looked at steady-state flow across two-dimensional systems consisting of blocks generated by a "Poisson lines" model with Markovian properties. His blocks represented fragments of two distinct uniform materials so that the system became a binary mixture of finite fragments, with a certain amount of statistical dependence between them. Finally, Wu and others (1973) studied the problem of flow across porous layers whose parameters were obtained from statistical distributions based on limited field data.

The development of stochastic flow models received a considerable impetus with the publication of Freeze's (1975) paper, "A stochastic-conceptual analysis of one-dimensional groundwater flow in nonuniform homogeneous media." In this paper, Freeze essentially extended the approach of Warren and Price (1961) to the multivariate case in which hydraulic conductivities, compressibilities, and porosities all vary randomly in space and are mutually correlated. Furthermore, Freeze allowed the variance of the hydraulic conductivities to be much larger than in the work of Warren and Price. However, he considered only one-dimensional flow and assumed the medium to consist of discrete blocks of equal length, each block being characterized by uniform parameter values.

Freeze considered two situations: steady-state flow under constant head boundary conditions, and transient consolidation of a clay layer under similar boundary conditions and prescribed head at time zero. The analysis was carried out by performing a large number of Monte Carlo runs, each consisting of two steps: (1) assigning random values to the parameters of each block; and (2) computing the resulting hydraulic head distribution in the system. In order to accomplish the first step, the block parameter values were drawn at random from a population whose joint probability function represented a trivariate normal distribution of log-conductivities, compressibilities, and porosities. However, the parameter values of one block were still statistically independent of those of all the other blocks. The second step was performed with the aid of an analytical formula for the steady-state case, and a numerical method for the transient case. The final stage included a statistical analysis of all Monte Carlo solutions corresponding to a given type of problem, a given number of blocks, and a given choice of statistical parameters such as mean parameter values, variances, and correlation coefficients.

On the basis of this work, Freeze (1975) arrived at some far-reaching conclusions with regard to the validity of the deterministic philosophy traditionally shared by most subsurface hydrologists. His major conclusions can be summarized as follows: (1) there is no simple way to define an equivalent uniform porous medium for transient flow in nonuniform materials; (2) the consideration of spatial variability in a stochastic sense may lead to very large fluctuations in the computed values of hydraulic head; and (3) therefore, the values of hydraulic head computed with the aid of deterministic models, whether analytical or numerical, may often suffer from very large errors. These conclusions caused Freeze to throw into question the usefulness of determining aquifer characteristics by means of pump test analyses, which rely on deterministic analytical solutions based on the assumption that the aquifer is uniform. He expressed similar doubt regarding the utility of determining the spatial variability of aquifer characteristics with the aid of inverse methods, which rely on the measurement of hydraulic head values.

As we shall see later, the one-dimensional aspect of Freeze's analysis has caused him to overestimate the degree to which hydraulic heads will fluctuate in response to random variations in the parameters under realistic two- and three-dimensional conditions. This means that the negative effect of spatial variability on the accuracy of deterministic models is, in reality, less severe than has been implied in Freeze's work. We will also see that the inverse method need not be regarded as an antithesis of the stochastic approach. On the contrary, this method should be considered an essential element of the stochastic procedure, which can help generate the kind of statistical information without which a stochastic model would not be complete. Nevertheless, Freeze's conclusions had a strong positive impact on hydrologists by making them acutely aware of the potential dangers involved in regarding the subsurface environment from a purely deterministic point of view.

The most recent contribution to the development of models with independent parameters has been the work of Dagan and Bresler (1979) and Bresler and Dagan (1979). In this interesting study, the authors consider the problem of vertical fluid flow and solute transport in an unsaturated soil caused by a steady rate of infiltration at the surface. The hydraulic properties of the soil are considered to be known functions of the water content and the saturated hydraulic conductivity, K_s. However, K_s is assumed to depend on a scaling factor, δ, which is distributed lognormally in the horizontal plane, as suggested by Warrick and others (1977). The fluid flow is at steady state due to a steady rate of infiltration at the surface. However, the infiltration rate is random and has a rectangular frequency distribution throughout the field. The problem consists of determining the probability that, at a certain depth z and time t, the solute concentration will take on a given value when the actual magnitude of K_s is unknown.

In order to solve this problem, the authors assume that horizontal hydraulic and concentration gradients are small in comparison to vertical ones, and can be neglected. Since the horizontal dimensions of the field are taken to be much larger than the depth of the soil, this assumption is intuitively appealing and provides formal justification for treating the values of K_s at different points in the field as being statistically independent. The authors rely on an analytical solution for the concentration which disregards hydrodynamic dispersion due to pore-scale effects. Their results show that pore-scale hydrodynamic dispersion is much less important than field-scale dispersion, the latter being due entirely to convective transport coupled with spatial variability of the soil properties.

STOCHASTIC MODELS WITH AUTOCORRELATED PARAMETERS DEFINED OVER A CONTINUUM

In the numerical model of Freeze (1975), the parameter values are taken to be uniform in each block, and the values in neighboring blocks are assumed to be statistically independent. This means that there is perfect positive correlation between the parameter values at different points within a given block, but no correlation between values corresponding to different blocks. In performing his one-dimensional analysis on a system of a given length, Freeze discovered that the variance of the computed head values increases with the size of the blocks in his model (see Freeze, 1975, Fig. 5). Since the size of the block in this case reflects the maximum length over which point values of the parameters are correlated, the implication of Freeze's finding is that his solutions are influenced by the specific correlation structure of the parameters implied by the nature of his model. From this we may conclude that for a stochastic model to yield reliable results, it must take into account the manner in which point values of a given parameter are correlated in space. That such a correlation must exist in nature is evident from the known tendency of geologic deposits to exhibit a definite spatial arrangement which must reflect itself in a spatial structure of the material properties as well.

The importance of considering the spatial structure of hydraulic conductivity in dealing with stochastic models of ground-water flow has been recognized by Gelhar (1976) and his coworkers, including Gelhar and others (1977), Bakr and others (1977), and Gutjhar and others (1978). Their approach is to view the spatial variation of hydraulic conductivities in the continuum sense as a random field characterized by theoretical spatial covariance functions or, equivalently, by spectral density functions in the wave-number domain. When such an hydraulic conductivity function is introduced into the partial differential equations governing groundwater flow, they are transformed into stochastic differential equations. The theory then proceeds by assuming that the random hydraulic conductivity process is weakly stationary and isotropic, and can be represented by a relatively simple covariance function.

To solve a specific problem, the random variables in the appropriate stochastic differential equation are represented by Fourier-Stieltjes integrals (see Lumley and Panofsky, 1964). The analysis produces functional relationships between the wave-number spectrum of the hydraulic conductivity fluctuations and the spectra of flow rate or hydraulic heat variations. From these spectral relationships it is then possible to determine the covariances of the latter flow characteristics based on the known autocorrelation function of the hydraulic conductivity. The method does not require making any specific assumptions about the probability distribution of the model parameters. However, the solution usually requires linearizing the governing stochastic differential equations and therefore, the results become invalid when the variance of the hydraulic conductivity exceeds a certain value. Furthermore, the method has so far only been applied to situations in which the flow field extends to infinity and flow takes place under the influence of a uniform mean hydraulic gradient.

In his original paper on the subject, Gelhar (1976) used the spectral method to investigate the difference between one- and two-dimensional flows in a sloping phreatic aquifer under conditions of steady state. He assumed that in the one-dimensional case, the flow is governed by the following form of the Boussinesq equation (Bear, 1972):

$$\frac{d}{dx}\left(Kh\frac{df}{dx}\right) = 0; \quad f = h + \beta x, \tag{7}$$

where h is hydraulic head, K is hydraulic conductivity, x is horizontal coordinate, and β is a constant. Based on previous work implying that point values of K are lognormally distributed (in a macroscopic sense, core data can be regarded as point data), Gelhar decided to express equation 7 in terms of ln K as

$$\frac{d}{dx}\left(h\frac{df}{dx}\right) + \frac{d(\ln K)}{dx} \quad h\frac{df}{dx} = 0, \tag{8}$$

and to treat ln K as a weakly stationary stochastic process. The variables in this equation are represented as the sum of their mean values (denoted by overbars) and some perturbations (denoted by primes) as $\bar{h} = h + h'$, $\bar{f} = f + f'$, and $Y = \bar{Y} + Y'$ where $Y \equiv \ln K$. Substituting into equation 8 and neglecting all second-order terms including products of perturbations, Gelhar obtained a linearized stochastic differential equation in terms of h' and Y',

$$\bar{h}\frac{d^2h'}{dx^2} + \beta\frac{dh'}{dx} = -\bar{h}\beta\,\frac{dY'}{dx}\ . \tag{9}$$

This equation is limited to small perturbations and since small changes in Y may correspond to large variations in K, there is an obvious advantage in working in terms of ln K instead of K. After representing h' and Y' in the form of Fourier-Stieltjes integrals, the author was able to obtain a closed form expression for the spectrum (or spectral density function) of h' in terms of the spectrum of Y. In order to evaluate the former, the author assumed that the autocovariance of Y, $C_Y(\zeta)$ has the simple exponential form

$$C_Y(\zeta) = E[Y'(x)\ Y'(x + \zeta)] - \sigma_Y^2\ e\ {-\lambda|\zeta|}\ , \tag{10}$$

where ζ is distance from a given point x,σ_Y^2 is the variance of Y', and λ is the so-called "integral scale." The latter is defined by Gelhar in analogy to Lumley and Panofsky (1964) as

$$\lambda = 1/\sigma_Y^2 \int_0^{\infty} C_Y(\zeta)\ d\zeta, \tag{11}$$

where ζ_o is the smallest value of ζ at which $C_Y(\zeta)$ becomes negative [when $C_Y(\zeta)$ is given by equation 10, $\zeta_o = \infty$]. Equation 11 implies that the integral scale represents the average distance over which ln K is correlated in space. With this, Gelhar was able to determine the spectrum of Y' which ultimately led to a closed-form solution for the variance of h', σ_h^2. This solution shows that the standard deviation of the head, σ_h, relative to the saturated thickness of the aquifer, $\bar{h}$, will generally be much smaller than the standard deviation of ln K, σ_Y, unless the saturated thickness is much smaller than the integral scale, λ. The solution also shows that the computed heads are correlated over a larger distance than the conductivity perturbations (Lumley and Panofsky, 1964, Fig.2). Furthermore, the variance of head increases as the mean correlation distance of point hydraulic conductivities (the integral scale λ) increases.

This finding explains why in Freeze's (1975) analysis the standard deviation of h increased as the size of the blocks in his model became larger (Freeze, 1975, Fig. 5). If the discretization interval in a numerical model is chosen to be much larger than the integral scale of the heterogeneities, the effect of these heterogeneities will be greatly overestimated. From this, Gelhar concluded that the spatial correlation structure of hydraulic conductivity variations is a central element in the flow process and must be explicitly taken into account in every stochastic analysis. In particular, the discretization interval in numerical models must never exceed λ, and the parameter values assigned to each interval must be generated on the basis of realistic autocovariance functions. Another conclusion was that the traditional statistical description of spatially-varying data based on probability density functions, which assume that data at different points in space are statistically independent, is inadequate because the spatial structure of the data is ignored.

In a later study, Bakr and others (1978) performed a similar one-dimensional analysis for confined flow with another autocorrelation function adopted from Erdelyi (1954),

$$C_Y(\zeta) = \sigma_Y^2(1 - |\zeta|/\ell)\, e^{-|\zeta|/\ell} , \tag{12}$$

where ℓ is an empirical parameter representing correlation distance. According to Bakr (1976), this latter form of $C_Y(\zeta)$ can be fitted with reasonable accuracy to realistic core data which often exhibit a negative correlation beyond a certain value of $|\zeta|/\ell$ (Bakr, 1976, Chapter 5).

In his study of two-dimensional flow systems, Gelhar (1976) assumed that the average hydraulic gradient remains uniform and therefore essentially one-dimensional; the two-dimensional aspect of the flow is entirely due to local deviations of the gradient from its mean. The principles of the mathematical analysis remain the same as in the one-dimensional case except that the autocovariance of Y is now assumed to have radial symmetry and take the form

$$C_Y(\zeta) = E[Y'(\underline{x})\ Y'(\underline{x} + \underline{\zeta})] = \sigma_Y^2\, b\zeta K_1(b\zeta) , \tag{13}$$

where $\underline{x}$ is the vector of horizontal coordinates, $\underline{\zeta}$ is the vector separating two given points in space, ζ is the modulus of $\underline{\zeta}$, K_1 is the modified first-order Bessel function of the first kind, and b is a constant related to the integral scale, λ, acccording to

$$\lambda = (1/\sigma_Y^2) \int_0^\infty C_Y(\zeta)\ d\zeta = \pi/2b . \tag{14}$$

This form of autocovariance function has been suggested by Whittle (1954) as being more suitable than the exponential form commonly used for one-dimensional processes, and it has also been used by Rodriguez-Iturbe and Mejia (1974) in the design of rain-gauge networks. The fact that C_Y is radially symmetric and depends only on ζ, not on $\underline{\zeta}$, implies that the two-dimensional covariance structure is assumed to be isotropic. By following a procedure similar to that outlined by Whittle, Gelhar was able to obtain an integral expression for the autocovariance of head which showed that it is an anisotropic stochastic process, even though the structure of ln K is isotropic. Gelhar was also able to develop an approximate closed-form expression for the variance of h' which is valid when $\lambda << \bar{h}/\beta$. A comparison of this expression with that for the one-dimensional case showed that for a given variance of the log-conductivity, σ_Y^2, the variance of head fluctuations, σ_h^2, is much smaller for two-dimensional flow than for one-dimensional flow. Furthermore, in the two-dimensional case, σ_h^2 is more strongly dependent on the integral scale, λ. For example, in the particular case where $\lambda = 0.02\ \bar{h}/\beta$, σ_h^2 is reduced by two orders of magnitude when one deals with a two-dimensional instead of a one-dimensional system. This result can be easily understood if we recognize that in the two-dimensional case, most of the flow will generally bypass zones of low conductivity, whereas in the one-dimensional case this is not possible.

Gelhar (1976) pointed out that in reality, the variation of point K values, as determined from core samples, is three-dimensional. When the governing equations are expressed in one- or two-dimensional form, they are suitable only for the description of spatially averaged quantities. For example, in working with two-dimensional models in the horizontal plane, the statistical information should be based on depth-averaged parameters such as transmissivities, whereas in one-dimensional models, this information must be based on parameters averaged over a plane. According to Gelhar and others (1977), such space-averaged inputs will generally have a smaller variance than the original point values. When the averaging distance is large in comparison with the integral scale of the heterogeneities, the variance of the space-averaged quantities may become very small. In dealing with point data, the appropriate form of analysis would be three-dimensional.

A three-dimensional extension of Gelhar's (1976) original work has been published recently by Bakr and others (1978). They considered steady-state flow in an unbounded medium subject to a constant mean hydraulic gradient, J, in the x_1 direction. The governing equation is expressed in terms of ln K as

$$\nabla^2 h + \nabla(\ln K) \cdot \nabla h = 0, \tag{15}$$

where $\nabla \equiv [(\partial/\partial x_1), (\partial/\partial x_2), (\partial/\partial x_3)]$ is the three-dimensional gradient operator. The hydraulic head is expressed as

$$h = - Jx_1 + h'. \tag{16}$$

Substitution of equation 16 into equation 15 under the assumption that $|\partial h'/\partial x_1| << J$ leads to

$$\nabla^2 h' - J(\partial/\partial x_1)(\ln K) = 0, \quad (17)$$

which is a stochastic differential equation for the head perturbations, h', in terms of ln K. Bakr and others show the autocovariance function in equation 12 to be be suitable for three dimensions and assume a new function having a simple exponential form of the kind discussed by Agterberg (1974, p. 341) and Gelhar and others (1974, p. 78),

$$C_Y(\zeta) = \sigma_Y^2 e^{-\zeta/\lambda}, \quad (18)$$

where ζ is again the modulus of $[\underline{\zeta}]$. This form is preferred over the S-shaped cumulative Gaussian function used in an earlier work by Buyevich and others (1969) for porosity fluctuations because the former appears to provide a better fit with available data (Bakr, 1976), whereas the Gaussian function has been considered unrealistic for geophysical processes by Bras and Rodriguez-Iturbe (1976). Equation 18 implies spherical symmetry of $C_Y(\zeta)$, which conforms to the assumption that ln K is a weakly stationary and isotropic process. The assumption of statistical isotropy has been maintained here only for the sake of simplicity. The authors recognized that in layered media, the autocovariance would be different parallel and perpendicular to the bedding and therefore, statistical anisotropy will be the rule rather than the exception. The solution leads to a closed-form expression for the autocovariance of hydraulic head fluctuations that depends not only on the constants J, σ_Y^2, and λ, but also on the angle between the separation vector $\underline{\zeta}$ and the direction of mean flow, x_1. The last fact demonstrates that the hydraulic head fluctuations are statistically anisotropic even though the input function, ln K, is not. The anisotropy is such that, for a given value of ζ/λ, the autocorrelation of fluctuations in hydraulic head is largest in a direction perpendicular to the mean flow and smallest parallel to the mean flow. Furthermore, the head perturbations are correlated over a much larger distance than ln K. Based on these results, the authors raise a question about the applicability of statistical analyses of hydraulic head fluctuations that assume that the water level data are independent (Gates and Kisiel, 1974; Yakowitz, 1976).

In general, the results of Bakr and others (1978) confirm the most important findings of Gelhar's earlier (1976) study. Bakr and others found that when $\ell/\lambda = 2.5$, the one- and three-dimensional expressions for C_Y in equations 12 and 18 have very similar functional forms. When this value of ℓ/λ is substituted into the corresponding expressions for the variance of hydraulic head, it decreases by a factor of almost 20 in passing from one to three dimensions. In addition, Bakr and others were able to show that there is no one-dimensional ln K covariance that would produce finite head variance in a one-dimensional analysis and still be generalized to an isotropic process in three dimensions--thereby exposing another weakness of the one-dimensional approach. Based on all these findings, the authors conclude that we may be justified in holding a more optimistic view about the validity and accuracy of deterministic ground-water flow models than previously held by Freeze (1975).

One of the questions that remained to be answered regarding the generality of the previous findings concerns the conditions under which the linearization in ln K, on which the theory is founded (see equations 8 and 9), is valid. To address this question, Gutjahr and others (1978) solved the one-dimensional problem by an exact method and compared the results with the linearized solution. The comparison is based on the assumption that ln K is a stationary Gaussian process. It shows that under these conditions, the logarithmic approximation is valid as long as $\sigma_Y < 1$, but the errors increase rapidly when $\sigma_Y > 1$. For $\sigma_Y < 0.5$, the logarithmic solution overestimates the head variance by less than 2%; for $\sigma_Y = 1.0$, by about 10%; for $\sigma_Y = 1.4$, by 50%. At $\sigma_Y^2 = 2.0$, the error becomes unbounded. To what extent these results can be generalized to three dimensions has not been made clear. On the other hand, Gutjahr and others showed that K cannot be modeled as a lognormal process in one dimension when $\sigma_Y^2 > 2$. However, the analysis of core data has shown that the magnitude of σ_Y is typically around 1, and may sometimes be as high as 3 (see Freeze, 1975; Bakr, 1976). Thus, we see that the linearized one-dimensional approach is inherently limited in its ability to deal with realistic σ_Y values.

Another important question addressed by Gutjahr and others (1978) relates to the concept of an equivalent uniform porous medium. They showed that under one-dimensional steady-state conditions, the effective hydraulic conductivity, K_e, is given by the harmonic mean,

$$K_e = \{E[K^{-1}]\}^{-1}. \quad (19)$$

This result is well-known from deterministic considerations of flow normal to the bedding of a layered medium (see Bear, 1972, p. 154) and has also been confirmed by the statistical analysis of Freeze (1975). If ln K is normally distributed, K_e takes on the special form

$$K_e = K_g e^{-0.5\sigma_Y^2}, \quad (20)$$

where K_g is the geometric mean,

$$\ln K_g = E[\ln K]. \quad (21)$$

This exact result is compared by the authors with another expression,

$$K_e = K_g(1 - \sigma_Y^2/2), \quad (22)$$

which results from the approximate linearized solution. This solution agrees with equation 20 to first order in σ_Y^2; it underestimates the exact value of K_e by about 18% when $\sigma_Y = 1$, but the error reduces to less than 1% when $\sigma_Y = 0.5$. The linearized analysis also shows that in the case of a uniform mean gradient, the effective hydraulic conductivity in two dimensions is equal to the geometric mean, K_g, whereas in three dimensions, it is given by

$$K_e = K_g(1 + \sigma_Y^2/6), \qquad (23)$$

which is correct to first order in σ_Y^2. A similar result was obtained earlier by Matheron (1967) for an isotropic lognormal conductivity field by using a different approach. The finding that in two-dimensional uniform flow, K_e is equal to K_g has been confirmed in a recent analysis by Dagan (1979), who also assumes K to be lognormal. Equations 20, 22, and 23 show that in one dimension, K_e is smaller than K_g, whereas in three dimensions, it is slightly larger. It will be recalled that Warren and Price (1961) found the geometric mean to provide a good approximation for K_e in the three-dimensional case, which agrees with equation 22 for typical values of σ_Y. The finding that in two-dimensional uniform flows, K_e is equal to K_g when K is lognormal is in agreement with an earlier study based on the use of an electric analogue.

In addition to these theoretical questions, Gelhar (1976) and Gutjahr and others (1978) have examined the implication of spatial heterogeneity on the design of two simple laboratory and field experiments, one for determining the average hydraulic resistivity of a core, the other for evaluating the vertical specific discharge across the aquitard. The analysis considers measurement errors and further emphasizes the importance of the correlation length, ℓ, in one-dimensional problems of this kind.

The most recent contribution of Gelhar and his coworkers has been the analysis of large-scale dispersion phenomena in a stratified aquifer (Gelhar and others 1979). Here the aquifer is assumed to consist of horizontal layers having uniform hydraulic conductivities which vary along the vertical, describing a weakly stationary one-dimensional stochastic process. The porosity of the aquifer is constant, and a uniform (deterministic) hydraulic gradient causes the water to flow in one specified horizontal direction. Because both longitudinal and transverse mixing are allowed, the analysis is related to that of Taylor (1953) for dispersion in tubes. By treating the concentration as a weakly stationary stochastic process and applying spectral analysis to a linearized form of the governing stochastic differential equation, Gelhar and others were able to obtain an explicit expression for the macroscopic longitudinal dispersion coefficient in terms of parameters describing the spatial variability of K. Their results indicate that at sufficiently large values of time, the longitudinal dispersivity approaches a constant value and the dispersion coefficient becomes a product of dispersivity and mean velocity.

However, at early time, the dispersivity is not constant and Fick's law does not apply. Furthermore, the asymptotic approach to Fickian conditions is slow, and deviations from Fick's law may manifest themselves quite strongly over considerable distances from the chemical source. For this reason, the dispersivity may vary with distance from the source and may thus appear to depend on the scale of the experiment, as has indeed been observed by various researchers in the laboratory and in the field. The authors also recommend modifying the classical dispersion-convection equation by adding to it two third-order terms. However, the results have been proven only for the case where $\sigma_K << \bar{K}$, σ_K being the standard deviation of the hydraulic conductivity and $\bar{K}$ being its mean. In spite of this, the authors expect their asymptotic results to be valid even when $\sigma_K/\bar{K}$ is large.

In their work on dispersion, Gelhar and others (1979) used the special autocorrelation function defined in equation 12. It is easy to verify that the integral of this autocorrelation function with respect to $|\zeta|$, between the limits 0 and B, is 0. Furthermore, the Laplace transform of equation 12 is linear in the Laplace parameter near the origin. According to de Marsily (1978), Matheron demonstrated several years ago that both of these conditions (the area under the autocorrelation function being zero and linear behavior of the Laplace transform near the origin) must be satisfied simultaneously if the system is to behave according to Fick's law at large values of time. Matheron's analysis differs slightly from that of Gelhar and others. Mahterson's model treats the pore-scale dispersion coefficient as a constant, whereas Gelhar and others allow it to vary linearly with the velocity.

STOCHASTIC MODELS WITH AUTOCORRELATED PARAMETERS DEFINED OVER A GRID

Models based on the continuum approach are very useful for gaining insight into the effect of spatial variability on certain flow situations. However, just like deterministic analytical methods, their usefulness is restricted to relatively simple problems. In particular, we saw that the spectral method has so far been limited to uniform flow in infinite domains in which the parameter field is weakly stationary, isotropic, and has a variance that does not exceed unity. In order to study realistic boundary-value problems of the kind usually arising in subsurface hydrology, the analytical approach must be supplemented by numerical methods.

A numerical Monte Carlo approach to two-dimensional, steady-state ground-water flow in bounded domains with autocorrelated parameters has been described very recently by Smith (1978) and Smith and Freeze (1979b). In their model, the flow domain is represented by a grid of square blocks. In each block, the hydraulic conductivity is uniform, but is no longer statistically independent of K values in the neighboring blocks. Instead, the discrete values of K are autocorrelated via a first-order autoregressive scheme called the "nearest neighbor" model (Whittle, 1954; Martin, 1974; Bartlett, 1975). To describe this model, let

$\underline{K} = (K_1, K_2, \ldots, K_N)$, where K_n is the value of K in the nth block and N is the total number of blocks in the grid. Let us assume that K_n is lognormally distributed for each n, and let $\underline{\mu}$ be the mean (or trend) of log $\underline{K}$ (that is, μ_n may vary from block to block),

$$\underline{\mu} = E[\log \underline{K}]. \qquad (24)$$

We can then express log $\underline{K}$ as

$$\log \underline{K} = \underline{\mu} + \underline{Y}, \qquad (25)$$

where $\underline{Y}$ is the deviation of log $\underline{K}$ from its mean. This implies that Y_n is normally distributed with zero mean for each n, and log K is not necessarily a stationary process. In the nearest neighbor model, Y is assumed to have a constant variance, σ_Y^2, which, according to equation 25, is equal to the variance of log K. Thus, Y is a strictly stationary process. If we orient the grid so that the positive x coordinate points east and the positive z coordinate points north, we can express the nearest neighbor model of Smith and Freeze for a typical block in the form

$$Y_C = \alpha_x(Y_W + Y_E) + \alpha_z(Y_S + Y_N) + \varepsilon_C. \qquad (26)$$

Here Y_C corresponds to a central block; Y_W, Y_E, Y_S, and Y_N correspond to the immediate neighbors of C on the west, east, south, and north, respectively; ε_C is a normally distributed random variable with mean zero and variance σ_ε^2, uncorrelated with ε_W, ε_E, ε_S, and ε_N in the neighboring blocks; α_x is an autoregressive parameter expressing the degree to which Y_C depends on its neighbors in the x direction, Y_W and Y_E; and α_z is a similar parameter determining the degree to which Y_C depends on its neighbors in the z direction, Y_S and Y_N. If $\alpha_x = \alpha_z$, the Y process has an isotropic covariance structure; otherwise, the structure will be anisotropic, with the covariance between neighboring blocks depending on orientation.

When equations of similar kind are written for each block in the grid, the result is a set of simultaneous linear algebraic equations that can be expressed in matrix form as

$$\underline{Y} = \underline{\underline{W}}\,\underline{Y} + \underline{\varepsilon}, \qquad (27)$$

where $\underline{\underline{W}}$ is an N x N "spatial lag" matrix. The typical term of this matrix is defined as

$$w_{nm} = (\gamma_{nm}/M), \qquad (28)$$

where $\gamma_{nm} = \alpha_x$ if blocks n and m are contiguous in the x direction; $\gamma_{nm} = \alpha_z$ if they are contiguous in the z direction; $\gamma_{nm} = 0$ if n = m, or if blocks n and m are not contiguous; and M is the total number of contiguous blocks surrounding block n. The scaling factor M is required to minimize deviations from stationarity in the generated sequence of Y_n values near the boundaries of the flow region: Equation 26, corresponding to blocks near the boundary, is truncated because the blocks do not extend across the boundary. This means that the statistical dependence between Y_n values in neighboring blocks near the boundary will be different than in the interior of the mesh, and the stochastic process will become nonstationary. In the particular systems treated by Smith and Freeze (1979b), the Y_n values in the interior of the grid appear to be stationary, indicating that the boundary effect does not extend all the way to the interior.

Equation 27 can be solved for $\underline{Y}$ in terms of $\underline{\varepsilon}$,

$$\underline{Y} = (\underline{\underline{I}} - \underline{\underline{W}})^{-1}\underline{\varepsilon}. \qquad (29)$$

Because $\underline{Y}$ has mean zero, its covariance is given simply by

$$\mathrm{Cov}[\underline{Y}] = E[\underline{Y}\underline{Y}^T] = \alpha_\varepsilon^2\,\underline{\underline{V}}, \qquad (30)$$

where $\underline{\underline{V}} = [(\underline{\underline{I}} - \underline{\underline{W}})(\underline{\underline{I}} - \underline{\underline{W}})^T]^{-1}$, and its correlation matrix is given by

$$\underline{\underline{C}} = (\sigma_\varepsilon^2/\sigma_Y^2)\,\underline{\underline{V}}. \qquad (31)$$

In general, $\underline{\underline{V}}$ and $\underline{\underline{C}}$ will be full matrices, so that the Y_n value in each block, n, will be correlated with Y_m in all other blocks, $m \neq n$. Thus, if σ_ε^2 and $\underline{\underline{V}}$ are known, it is possible to obtain a piecewise constant representation of the autocovariance function of Y in any prescribed direction by plotting the values of $\sigma_\varepsilon^2 V_{nm}$ against distance in that direction, starting with the nth block. The shape of this autocovariance function will depend on the particular choice of α_x and α_z, on σ_ε^2 and, to some extent, on the size of the blocks. Based on these functions, it is possible to compute the integral scales, λ_x and λ_z, in the principal directions x and z. In order to generate a sequence of $\underline{Y}$ vectors with a prescribed stepwise autocovariance structure, one must first determine the corresponding values of α_x, α_z, and σ_ε^2. On the other hand, if α_x and α_z are known, $\underline{\underline{V}}$ can be computed on the basis of $\underline{\underline{W}}$ and, since $\mathrm{Cov}[\underline{Y}] = \sigma_Y^2\,\underline{\underline{C}}$ is prescribed, σ_ε^2 can be obtained from equation 31.

Any particular vector, $\underline{Y}$, conforming to the nearest neighbor model, can be viewed as a discrete realization of the stochastic process ln K. Monte Carlo simulation consists of generating a large number of such realizations at random and, for each realization, obtaining a solution to the resulting deterministic flow problem by a suitable numerical method. In the work of Smith and Freeze (1979b), the constants α_x, α_z, and σ_Y are assumed to be known. During each Monte Carlo run, the authors first generate $\underline{\varepsilon}$ from a normal distribution (N) with mean zero and standard deviation σ_ε (denoted by $N[0,\sigma_\varepsilon]$) by means of an appropriate random number generator, and then determine $\underline{Y}$ according to equation 29. Because a linear combination of normally distributed random variables is also normal, Y_n will be $N[0,\sigma_Y]$ for each n. Thus, according to equation 25, log K_n will be $N[\mu_n,\sigma_Y]$. The next step is to compute $\underline{K}$ from the relationship

$$K_n = \exp(2.3026Y_n), \quad (32)$$

and assign these values to the appropriate blocks. In so doing, there is some loss in the strength of the autocorrelation of $\underline{K}$ in relation to that of $\underline{Y}$ which depends on the choice of α_x, α_z, and α_Y (Smith and Freeze, 1979a, Fig. 3). The final step is to solve the resulting deterministic flow problem by a finite element method for hydraulic head at the nodes of the grid, subject to prescribed head or prescribed flux boundary conditions. After completing a sufficiently large number of Monte Carlo runs, the frequency distribution of head, h_n, at any particular node, n, can be analyzed to obtain estimates of the mean, $\bar{h}_n$, and standard deviation, σ_{h_n}, of h_n.

Most of the results presented by Smith and Freeze (1979b) are based on 300 Monte Carlo runs over a rectangular grid with 10 square blocks parallel to the x direction and 20 blocks parallel to the z direction. All the components of $\underline{\mu}$ in equation 25 are set equal to $\mu = -2.00$, which, according to the authors, is reasonable when K is measured in cm/s. As the finite-element method sometimes fails to converge for σ_Y values equal to or exceeding unity, the largest value considered by the authors is $\sigma_Y = 0.91$.

To discuss the effect of unidirectional flow under a uniform mean gradient, let us again orient the positive z and x axes toward the north and east, respectively. In this way, the longer axis of the grid will coincide with the west-east direction. A uniform mean gradient parallel to this direction can be created by imposing no-flow conditions on the northern and southern boundaries, a constant dimensionless head of, say, 100 on the western boundary, and zero head on the eastern boundary. Although the mean flow is unidirectional, the streamlines may deviate locally from the mean direction due to material nonuniformity. The results of the Monte Carlo analysis show that when the conductivity field is isotropic so that $\lambda_x = \lambda_z = \lambda$, σ_h increases as σ_Y becomes larger. The values of σ_h are zero at the constant head boundaries and increase gradually toward a maximum value midway between these boundaries. This clearly indicates that the hydraulic head field is nonstationary, the phenomenon being due entirely to the imposition of constant (deterministic) heads on the inflow and outflow boundaries. For a fixed value of σ_Y, σ_h increases as the integral scale λ increases. Except for the boundary effects, these results confirm the earlier findings of Gelhar (1976) for the case where K is defined over a continuum.

Because the solution of Smith and Freeze (1979b) concerns bounded domains and is expressed in dimensionless quantities, it must be interpreted not in terms of λ, but rather in terms of a dimensionless integral scale, λ/L, where L is a characteristic length of the flow field. In the particular case of one-dimensional flow, L is simply the length of the flow region. For this case, Smith and Freeze (1979a, Fig. 7) are able to demonstrate that when L and λ are varied while the boundary heads and the size of each block remain unchanged, σ_h also stays unchanged as long as λ/L remains fixed. This is true provided that the size of the block remains sufficiently small in comparison to λ to preserve a reasonably accurate stepwise representation of the autocorrelation function of Y. Thus, instead of being dependent on λ alone, σ_h actually increases as λ/L increases. Furthermore, if λ/L is preserved, σ_h shows very little dependence on the block size as long as the shape of the autocorrelation function is represented with reasonable accuracy. In general, preserving the value of λ/L is more important than preserving the exact shape of the autocorrelation function.

In the case of two-dimensional flow under a uniform gradient in the x direction, σ_h will generally depend on λ_x/L_x and λ_z/L_z where L_x and L_z are distances between the constant head and no-flow boundaries, respectively. When λ_z/L_z is decreased by modifying λ_z while keeping everything else in the model fixed, σ_h decreases. Smith and Freeze (1979b) state that the effect of λ_z/L_z on the magnitude of σ_h is less dramatic than that of λ_x/L_x. However, a careful examination of their Figures 5 and 6 reveals that when either λ_z/L_z or λ_x/L_x are reduced by an equal amount, the resulting decrease in σ_h is approximately the same.

One of the conclusions reached by Smith and Freeze (1979b) is that σ_h for the case of two-dimensional uniform flow in a bounded domain is about half as large as for one-dimensional flow. According to Gelhar (1976), the reduction in σ_h in passing from one to two dimensions is much greater in an unbounded domain. The relatively small reduction observed by Smith and Freeze can probably be explained by the fact that their particular autocorrelation functions cause Y_n to be mutually correlated over very large portions of the flow field. The question as to what extent this is realistic will require further investigation.

Smith and Freeze (1979b) have also looked at nonuniform flow fields created over the same grid by a rearrangement of the boundary conditions. The results show that here σ_h is generally greater than in the uniform case, especially in regions of high mean hydraulic gradient which are sufficiently removed from the constant head boundaries. The conclusion is that σ_h depends not only on the nature of the heterogeneities, but also on the shape of the flow region and the boundary conditions. Another problem Smith and Freeze considered is that of nonuniform flow in a system consisting of two different, stationary layers, with cross flow allowed between the layers. In this case, σ_h is larger in the layer characterized by larger values of λ_Y, λ_x, and λ_z, and by a smaller value of mean hydraulic conductivity. The introduction of a spatial trend into the model by allowing the components of $\underline{u}$ to differ from each other in a prescribed fashion, has been shown to have a strong influence on the results. Another interesting conclusion regards fluxes across the constant head boundaries. These will vary from one situation to another due to local heterogeneities, and so will the total flow rate across the system. The variability of the total flow rate is attributed by the authors to nonergodic effects arising from the bounded nature of the model.

The results of Smith and Freeze (1979b) confirm the conclusion of earlier researchers that when two-dimensional flow takes place in a stationary and isotropic medium under a uniform gradient, the effective hydraulic conductivity of this medium is given by the geometric mean. However, they demonstrated that the concept of an equivalent uniform porous medium is not appropriate for situations where the mean hydraulic gradient is nonuniform, or where the mean hydraulic conductivity varies with space.

A disadvantage of Monte Carlo simulation is that it often requires a large amount of computer time. An alternative approach would be to linearize the discrete stochastic differential equation governing the flow, as has been attempted by Tang and Pinder (1977) and Sagar (1978). To illustrate the concept of linearization, let us consider Sagar's approach, which relies on a numerical finite-element model of transient flow in a two-dimensional system. For simplicity, we will restrict our discussion to flow in an aquifer under steady-state conditions. Let us assume that the aquifer is subdivided into I triangular elements whose corners represent the nodal points of the finite-element grid. The transmissivity, T_i, in the ith element is uniform, but it can vary from one element to another. If h_n represents head at the nth node and Q_n is the rate of flow into or out of the aquifer in the vicinity of this node, then the finite-element equations can be written in matrix form as

$$\underline{\underline{A}}(\underline{T})\,\underline{h} = \underline{Q}, \tag{33}$$

where $\underline{T}$ is an I-dimensional vector consisting of all the T_i values in the elements; $\underline{h}$ and $\underline{Q}$ are N-dimensional vectors consisting of the h_n and Q_n values at the nodes, N being the number of nodes in the grid; and $\underline{\underline{A}}(\underline{T})$ is an N x N symmetric matrix of similar rank that is linear in $\underline{T}$. By premultiplying both sides of equation 33 by the inverse of $\underline{\underline{A}}(\underline{T})$, we can solve this equation for $\underline{h}$,

$$\underline{h} = Q/\underline{\underline{A}}\,\underline{T}\ . \tag{34}$$

Since $\underline{\underline{A}}(\underline{T})$ is linear in $\underline{T}$, this cannot be true about its inverse, and therefore $\underline{h}$ in equation 34 is a nonlinear function of $\underline{T}$.

In Sagar's (1978) model, $\underline{T}$ and $\underline{Q}$ are considered to be stochastic processes with prescribed first and second moments (that is, prescribed mean vectors and covariance matrices). The problem consists of determining the moments of $\underline{h}$ from those of $\underline{T}$ and $\underline{Q}$ (at constant head boundary nodes, h_n is known, and the corresponding equations can simply be eliminated from equation 34). In order to accomplish this, the right hand side of equation 34 is expanded in a Taylor series about the mean values of $\underline{T}$ and $\underline{Q}$, discarding all terms of order three or higher. In this manner, h becomes a linear function of $\underline{T}$, $\underline{Q}$, and the products and cross products of T_i and Q_n for all i = 1, 2, . . . , I and n = 1, 2, . . ., N, repectively. Since the first and second moments of $\underline{T}$ and $\underline{Q}$ are assumed to be known (including the covariances between T_i and Q_n for all i and n), $E[\underline{h}]$ can be explicitly evaluated by taking the expectation of the right hand side of the truncated equation. A similar Taylor expansion performed on the vector product $\underline{h}^T\underline{h}$ will lead to an explicit formula for the covariance matrix of $\underline{h}$, $\underline{\underline{V}}(\underline{h})$, whose typical term is defined as $V_{nm}(\underline{h}) = E[h_n h_m] - E[h_n]\,E[h_m]$. Sagar's expression for the mean value of $\underline{h}$ shows that $E[\underline{h}]$ differs from the hydraulic head computed from equation 34 by replacing $\underline{T}$ and $\underline{Q}$ on the right-hand side by their mean values. This is equivalent to saying that replacing $\underline{T}$ and $\underline{Q}$ in the model by deterministic quantities may lead to errors in the computation of $\underline{h}$.

Contrary to Monte Carlo simulation, linearization does not require any prior knowledge about the probability density functions associated with $\underline{T}$ and $\underline{Q}$. However, the reliability of the method is dependent to a considerable extent on the shape of these functions. According to Sagar (1978), who quotes Papoulis (1965), if the probability masses of T_i and Q_n are concentrated near their expected values and the functions are smooth in the vicinity of the mean, then retaining up to second order terms in the Taylor expansion will provide a good approximation for the first two moments of $\underline{h}$. Since formal proofs of stability and convergence are not available at the present time, a significant number of numerical experiments will have to be conducted before the reliability of the linearization method could be established for cases of practical interest, and particularly for the transient case. Unfortunately, Sagar did not have access to a computer during the development of his model and therefore he was able to work out only one very simple example of one-dimensional transient flow over a grid consisting of three elements and four nodes.

STOCHASTIC MODELS CONDITIONED UPON MEASURED VALUES OF AUTOCORRELATED PARAMETERS

All the stochastic models discussed thus far rely on consideration of an infinite (in the continuum case) or very large (in the discrete case) number of different possible realizations of the model parameters. However, in reality there is only one realization that corresponds to the true material properties of the medium. The above stochastic models are based on the assumption that the actual spatial distribution of the "true" parameters is completely unknown, but that there is enough information to characterize their spatial variability statistically by means of appropriate probability density functions or some of their moments. Thus, the different realizations generated by these models do not generally correspond to the "true" realization in the sense that parameter values at specific points in space are similar, but only in the sense that some of the global statistical properties of the true realization are preserved. For example, if these global properties include stationarity then, according to the former models, the probability that a predetermined parameter value will occur at a designated point in space will be the same as at all other points in the flow field. Even if the value of this parameter were actually measured at the designated point, the models would still allow it to vary from one realization to another in the same manner as the parameters at all other locations in the field, without direct regard to its measured value.

We must recognize that in order to determine the statistical parameters which enter into these stochastic models (for example, mean, variance, parameters of the autocovariance function), we must first measure the related material property at numerous different points in space. For example, the mean and autocovariance structure of transmissivities in a given aquifer cannot be determined without performing a sufficiently large number of field tests in this aquifer. The role of inverse methods in this regard will be discussed in the next section. However, once the material properties have been evaluated at specific locations, they could then be used not only to estimate the global statistical characteristics of the model parameters, but also to require that each realization of these parameters conform locally to the measured values. In other words, one should be able to reduce the uncertainty in the input as well as the output of the stochastic model by conditioning it upon measured values of the parameters at given points in the flow field.

A stochastic model capable of performing this task has been described recently by Delhomme (1979). The solution technique used by Delhomme is called conditional simulation. It is based on the geostatistical theory of Matheron (1971), known as kriging. This theory is sometimes referred to as the theory of regionalized variables. It relies on a knowledge of the mean trend (called drift in the kriging lingo) and autocovariance structure of a spatially-varying stochastic variable to estimate its values at the nodal points of a regular or irregular grid, given measured values of this variable at arbitrary points in space. Another purpose of kriging is to estimate spatial averages of the stochastic variable over specified subregions of its domain of definition. The method differs from other spatial interpolation and averaging techniques in that it provides not only estimates of point or average values, but also the variance of the corresponding errors of estimation.

In subsurface hydrology, kriging has so far been applied primarily to problems of horizontal flow in aquifers (Delhomme, 1978). The approach has been to view transmissivity data from pumping tests as point quantities representing measured values of a transmissivity field defined over the aquifer continuum. In most cases, the measurements are assumed to be exact. However, if the mean and variance of the measurement error is known, this can be formally accounted for in the model (for example, Delhomme, 1974). Usually, kriging is applied to the logarithm of the transmissivity, log T, which we shall represent by Z. If $Z(\underline{x})$ is known at N points $\underline{x}_n$, n = 1, 2, . . ., N, then the average value of $Z(\underline{x})$ over a finite subregion of the flow domain, Ω_i, is expressed as a weighted average of these data,

$$\hat{Z}(\Omega_i) = \sum_{n=1}^{N} \lambda_n^{(i)} Z(\underline{x}_n), \qquad (35)$$

where $\hat{Z}$ is the desired estimate and λ_n are the non-negative weights. The error of estimating the average in this fashion, $\varepsilon(\Omega_i)$, is defined as

$$\varepsilon(\Omega_i) = \hat{Z}(\Omega_i) - Z(\Omega_i) , \qquad (36)$$

where Ω_i is the area of the subregion and

$$Z(\Omega_i) = \frac{1}{\Omega_i} \int_{\Omega_i} Z(\underline{x})\, dx . \qquad (37)$$

The essence of kriging is to evaluate the weights, $\lambda_n^{(i)}$, in such a manner as to insure that the variance of ε, Var $[\varepsilon] = E[\varepsilon^2]$, will become minimum and its mean, $E[\varepsilon]$, will be zero. In other words, kriging strives to provide the best unbiased linear estimate of average log T over the subregion Ω_i. The corresponding value of transmissivity will represent the best estimate of the geometric mean of T over Ω_i, based on the available measurements.

In order to determine the optimum values of $\lambda_n^{(i)}$, it is customary (although not always necessary) to assume that $Z(\underline{x})$ is the realization of a weakly stationary and isotropic process with a constant mean (that is, without a drift). Then, according to equation 6, the autocovariance of $Z(\underline{x})$ corresponding to two points $\underline{x}$ and $\underline{x} + \underline{\zeta}$ becomes only a function of the scalar distance $\zeta \equiv |\underline{\zeta}|$ and can be expressed as $C_Z(\zeta)$. Furthermore, one can define the so-called variogram of $Z(\underline{x})$, $\gamma(\zeta)$, as

$$\gamma(\zeta) = C_Z(0) - C_Z(\zeta) . \qquad (38)$$

Since the mean is constant, it cancels out of the definition of the variogram. Thus, $\gamma(\zeta)$ can in principle be estimated without prior knowledge of the mean, merely by considering the amounts by which the measured data, $Z(\underline{x}_n)$, increase or decrease from one measurement point, $\underline{x}_n$, to another. In practice, the accuracy with which $\gamma(\zeta)$ is determined will depend to a considerable extent on the number of such data points and their spatial arrangement. It can be shown (for example, Matheron, 1971; Delhomme, 1978) that once $\gamma(\zeta)$ has been determined, the optimum weights for equation 35 can be evaluated by solving the following system of N simultaneous linear algebraic equations, known as kriging systems,

$$\sum_{m=1}^{N} \lambda_{nm} \lambda_m^{(i)} = \lambda_{ni} ; \quad n = 1, 2, \ldots, N, \qquad (39)$$

$$\sum_{n=1}^{N} \lambda_n^{(i)} = 1,$$

where $\gamma_{nm} = \gamma(|\underline{x}_n - \underline{x}_m|)$ and

$$\gamma_{ni} = \frac{1}{\Omega_i} \int_{\Omega_i} \gamma(|\underline{x}_n - \underline{x}|)\, \underline{dx} \,. \tag{40}$$

When the resulting values of $\lambda_n^{(i)}$ are substituted into equation 35, $\hat{Z}(\Omega_i)$ becomes the best unbaised linear estimate of $Z(\Omega_i)$. The variance of this estimate about $Z(\Omega_i)$ (that is, the variance of the estimation error) is given by

$$\mathrm{Var}[\varepsilon(\Omega_i)] = - \sum_{n=1}^{N} \sum_{m=1}^{N} \lambda_n^{(i)} \lambda_m^{(i)} \gamma_{nm} + 2 \sum_{n=1}^{N} \lambda_n^{(i)} \gamma_{ni} - \gamma_{ii} \,. \tag{41}$$

The covariance of the estimation errors corresponding to two different subregions, Ω_i and Ω_j, can be computed from

$$\mathrm{Cov}[\varepsilon(\Omega_i)\, \varepsilon(\Omega_j)] = - \sum_{n=1}^{N} \sum_{m=1}^{N} \lambda_n^{(i)} \lambda_m^{(j)} \gamma_{nm} + \sum_{n=1}^{N} \lambda_n^{(i)} \lambda_{nj} + \sum_{n=1}^{N} \lambda_n^{(j)} \gamma_{nj} - \gamma_{ij}, \tag{42}$$

where

$$\gamma_{ij} = \frac{1}{\Omega_i \Omega_j} \int_{\Omega_j} \int_{\Omega_j} \gamma(|\underline{x}_i - \underline{x}_j|)\, d\underline{x}_i\, d\underline{x}_j, \tag{43}$$

$\underline{x}_i$ and $\underline{x}_j$ being arbitrary points within X_i and Ω_j, respectively.

Recall that if the mean hydraulic gradient in a two-dimensional steady flow region is uniform, this region can be represented by an equivalent uniform material whose effective hydraulic conductivity (or, in our case, transmissivity) is given by the geometric mean, provided that T is lognormal. Thus, if Ω_o is made sufficiently small so that the mean hydraulic gradient within its confines does not vary significantly from one point to another, we should be justified in representing T over Ω_o by its geometric mean. The best estimate of this geometric mean, based only on measured values of T, is provided by equation 35.

Since $\hat{Z}(\Omega_i)$ is only an estimate of the true average of log T over Ω_i, it must not be confused with the latter. If the true average, $Z(\Omega_i)$, were known, it could be regarded as the "true" parameter to be assigned to Ω_i in a numerical model in which Ω_i represents a sufficiently small portion of the grid. If, instead, the model parameters are represented by their estimates, $\hat{Z}(\Omega_i)$, their spatial variability will be less pronounced than that of $Z(\Omega_i)$ (that is, the map of estimated average transmissivities will be smoother than that of the true averages). In order to study the effect of errors in estimating the model parameters by $\hat{Z}(\Omega_i)$, Delhomme (1979) used the method of conditional simulation. The principle of this method is to generate a large number of possible realizations of $\hat{Z}(\Omega_i)$ in a manner that would preserve both the autocorrelation structure of $Z(\underline{x})$ as represented by the variogram, $\gamma(\zeta)$, and the covariance structure of $\varepsilon(X_i)$ as represented by equations 41 and 42. Since equations 41 and 42 depend only on $\gamma(\zeta)$, the geometry of the system, and the location of the measurement points [but not on the measured values of $Z(\underline{x})$], preserving these factors will automatically insure that the covariance structure of $\varepsilon(\Omega_i)$ will remain unchanged. By preserving the structure of $\varepsilon(\Omega_i)$, $Z(\Omega_i)$ will be allowed to deviate from $\hat{Z}(\Omega_i)$ by large amounts where the variance of the estimation error is large, but only by small amounts where this variance is small. In this manner, the simulated realizations of $Z(\Omega_i)$ will be conditioned upon the measured values of $Z(\underline{x})$ at the N points $\underline{x}_n$. Each of these realizations will represent an equally likely version of reality, conditioned upon the knowledge of N specific values of $Z(\underline{x})$.

The conditional simulation procedure consists of two steps. In the first step, one synthesizes a large number of random realizations, $s(\underline{x})$, having the same variogram as $Z(\underline{x})$, but otherwise having arbitrary values everywhere in the field, including the measurement points, $\underline{x}_n$. Thus, $s(\underline{x})$ will not be conditioned upon the measured values, $Z(\underline{x}_n)$. The synthesis of $s(\underline{x})$ can be performed by various techniques such as, for example, spectral analysis (Mejia and Rodriguez-Iturbe, 1974b). Delhomme (1979) uses the so-called turning bands method developed by Matheron (1973) and his coworkers in France. The realizations generated by this method have a distribution which is approximately normal, as one usually requires for log T. In the second step, each realization, $s(\underline{x})$, is conditioned upon (that is, made consistent with) the measured values of $Z(\underline{x})$ at $\underline{x}_n$, n = 1, 2, . . ., N. This is done as follows: $Z(\Omega_i)$ can be expressed in the form $Z(\Omega_i) = \hat{Z}(\Omega_i) + [Z(\Omega_i) - Z(\Omega_i)]$. On the other hand, one can compute $s(\Omega_i)$ from the known values of $s(\underline{x})$, as well as obtain the kriging estimate $\hat{s}(\Omega_i)$ based on known $s(\underline{x})$ values at the N measurement points, $\underline{x}_n$. Thus, one can write $s(\Omega_i) = \hat{s}(\Omega_i) + [s(\Omega_i) - \hat{s}(\Omega_i)]$, which is analogous to the above expression for $Z(\Omega_i)$. Since the covariance structure of $[s(\Omega_i) - \hat{s}(\Omega_i)]$ is identical to that of $\varepsilon(\Omega_i)$, the former can be substituted for $[Z(\Omega_i) - \hat{Z}(\Omega_i)]$. In this manner, one can compute a conditional realization, $Z_s(\Omega_i)$, according to

$$Z\,(\Omega_i) = \hat{Z}(\Omega_i) + [s(\Omega_i) - s(\Omega_i)] \,. \tag{44}$$

As the mean of the term in brackets is zero, the mean of $Z_s(\Omega_i)$ is simply the kriging estimate,

$\hat{Z}(\Omega_i)$. Furthermore, the covariance structure of $Z_s(\Omega_i)$ is the same as that of $Z(\Omega_i)$. After a given realization of $Z_s(\Omega_i)$ has been generated with the aid of equation 44, the corresponding values of T are computed for all the subregions Ω_i, and the flow problem is solved by a suitable numerical method subject to given boundary conditions. Since this is done repeatedly for each realization, conditional simulation is seen to be a special form of Monte Carlo simulation.

Delhomme (1979) has performed conditional simulation on transmissivity data from up to 45 pumping tests conducted in the Bathonian aquifer in Normandy. The values of log T (T in m^2/s; logarithm to the base 10) range from -5.08 to -0.80, and their mean and variance are -2.45 and 0.98, respectively. There appears to be no drift, and the radius over which log T is positively correlated extends to over 10 km. According to Delhomme's experience, this radius can be less than 1 km in alluvial aquifers and 10 or even 20 km in limestone and chalk aquifers.

Delhomme considered two flow situations. In the first situation, steady flow takes place in a rectangular aquifer 10 km in length and 7.5 km in width. Three of the boundaries are impermeable, whereas one of the shorter boundaries is maintained at zero head. Flow out of the aquifer across that boundary is caused by a uniform recharge rate equal to 5 $\ell/s/km^2$. The aquifer is covered with a regular grid in which the nodes are positioned at intervals of 0.25 km parallel to the boundaries. On the basis of 50 conditional runs and 30 log T data clustered in two groups, Delhomme (1979) was able to verify that the mean transmissivity map is equal to the kriged map, and the map of variances obtained from the simulation is very similar to the map of estimation variances determined by kriging. The estimation variance is low in the two areas in which most of the log T data are clustered and large far from the sites of the pumping tests. For this reason, different conditional realizations of log T lead to different distributions of hydraulic head values. The fluctuations in head are small in comparison to those of log σ attesting to the strong smoothing (or filtering) effect of the governing stochastic differential equation. A similar filtering effect was noticed earlier by other workers. The standard deviation of hydraulic head, σ_h, increases with distance from the constant head boundary and stabilizes at a value of approximately 20 m.

The second flow situation concerns the Bathonian aquifer itself. The aquifer is bounded to the north by the sea, and steep hydraulic gradients in the southern part are due to local dewatering of mines. The numerical model consists of square cells 2 km^2 in size. Account is taken of withdrawal from wells, rainfall infiltration, the presence of streams, and leakage to or from an overlying alluvial aquifer and an underlying Bajocian aquifer, both of which are maintained at constant head. In this case, the analysis takes into account errors in the sample values of log T. After generating 50 different realizations of log T, Delhomme found that some of them led to a good reproduction of the observed hydraulic head distribution in the aquifer, whereas others did not. From this he concluded that the available log T data were consistent with the observed head, but were not sufficient to guarantee a good fit in each case. In the northern part of the aquifer, where a constant head boundary is present, the computed and observed heads agree fairly well for all the simulated log T values.

FURTHER CONDITIONING UPON HISTORICAL DATA RELATED TO FLOW REGIME: THE INVERSE METHOD

In comparing his results for flow in a rectangle with the analytical solution of Gelhar (1976), Delhomme (1979) found that both solutions led to a similar value of σ_h in the center of the flow field. This is somewhat discouraging because Delhomme's solution is conditioned upon 30 measured values of T, and one would therefore expect his σ_h to be less than that computed from the nonconditioned model of Gelhar. The phenomenon is explained by the fact that in Delhomme's solution, most of the pumping test sites are removed from the constant head boundary, and therefore the variance of the estimation error near this boundary is high. Since all the flow out of the aquifer takes place across the constant head boundary, the entire flow field is affected by the high uncertainty in log T and, consequently, σ_h is large. This shows that conditioning the solution on measured values of log T is not always sufficient to achieve a reduction in σ_h, but that the effect of such conditioning depends on the location of the measurement points with respect to the boundaries. In other words, the effect of conditioning depends not only on the transmissivity data, but also on the flow configuration in the system.

In the case of the Bathonian aquifer analyzed by Delhomme (1979), we saw that far from the sea coast, where head is kept constant, σ_h is relatively large, and agreement between observed and computed head values is generally poor (except for a few realizations of the log T field). On the other hand, the agreement closer to the coast is good even when the uncertainty in log T is large. This further emphasizes the fact that conditioning the solution on measured transmissivities may not be sufficient to significantly reduce the uncertainty in computed head values, and that the effect of such conditioning depends to a considerable extent on the prevailing flow regime.

Would it be possible to effect a further reduction in σ_h by taking into account the flow regime during the process of generating transmissivity inputs into a stochastic model of an aquifer? In other words, could the uncertainty in the hydraulic head values generated by such a stochastic model be reduced further by conditioning this model, not only on measured transmissivities, but also on the measurement of water levels in wells, and on some previous knowledge about flow rates into or out of the system (for example, pumping rates and recharge)? Delhomme (1979) seems to feel that the answer to this question is yes. He points out that a

> major limitation [of his approach] stems from the fact that all available information about the groundwater system has not yet been utilized.... Only transmissivity data have been taken into account.... The information given by a (e.g., steady state) distribution of hydraulic heads cannot be neglected when quantifying the uncertainty about transmissivity throughout the aquifer.... The results point out the need for additional investigation, using both [h] and log T data sets in a comprehensive stochastic approach of the uncertainty about groundwater flow.

Delhomme feels that the answer to this problem lies in the development of appropriate parameter estimation techniques, or inverse methods, which recognize the need to treat both h and log T as random variables.

An inverse method which is consistent with stochastic approach to groundwater modeling has been developed recently (Neuman and Yakowitz, 1979). Our method is based on the recognition that in many practical situations, the hydrologist will have much more information about water levels than about transmissivities, and measurements of n will be generally more accurate than those of T. For this reason, the kriged estimates of the geometric mean of T over subregions Ω_i of a given grid will be generally less reliable than similar estimates of point hydraulic head values, h_n, at the nodal points, n, of the same grid. Thus, if the kriged parameters are not able to reproduce the kriged hydraulic head map to within the known accuracy of this map, they must be modified until the level of uncertainty in the computed heads is reduced to that of the kriged heads. This inverse method (Neuman and Yakowitz, 1979) accomplishes this task through nonlinear programming, coupled with a statistical analysis of the residual errors arising from the inverse model.

Although the method is consistent with the kriging concept and can utilize kriged estimates of T and h as input parameters, kriging is not essential for its application provided that other estimates, together with some of their statistical properties, are available. When statistical properties of the estimates are not available, the inverse model can still be applied to realistic problems, but its output is then heuristic and must be regarded with a certain amount of caution.

This method relies on a finite element numerical model of steady-state flow in a two-dimensional system. For a horizontal aquifer, the finite element model is expressed by the matrix equation

$$\underline{\underline{A}}(\underline{T})\underline{h} = \underline{Q} \; , \tag{45}$$

where $\underline{T}$ is an I-dimensional vector representing the true geometric mean values of T over all I subregions, Ω_i, of the grid, each subregion consisting of one or more elements; $\underline{\underline{A}}(\underline{T})$ is an N x N symmetric matrix, called conductance or stiffness matrix, which is linear in $\underline{T}$, N being the number of nodes in the grid; $\underline{h}$ is an N-dimensional vector representing true head values at the nodes; and $\underline{Q}$ is an N-dimensional vector representing actual net rates of flow into or out of the aquifer through unique areas corresponding to all the individual nodes. Let us assume that $\underline{h}$ and $\underline{T}$ can be estimated from available field data with a method such as kriging. Then, we can express these known estimates, $\underline{h}^*$ and $\underline{T}^*$, as

$$\underline{h}^* = \underline{h} + \underline{\varepsilon} \tag{46}$$

and

$$\underline{T}^* = \underline{T} + \underline{\nu} \; , \tag{47}$$

where $\underline{\varepsilon}$ and $\underline{\nu}$ are the corresponding error of estimation. Similarly, let the estimate of $\underline{Q}$ be designated by $\underline{Q}^*$. The error of estimation associated with $\underline{Q}^*$ can often be neglected in comparison with other errors affecting the model (Neuman and Yakowitz, 1979). Assuming that this is true, we can substitute equation 46 and $\underline{Q}^*$ into equation 45 which leads to the stochastic finite element model

$$\underline{\underline{A}}(\underline{T})\,\underline{h}^* = \underline{Q}^* + \underline{\underline{A}}(\underline{T})\,\underline{\varepsilon} \; . \tag{48}$$

Premultiplying throughout by the inverse of $\underline{\underline{A}}(\underline{T})$, equation 48 becomes

$$\underline{h}^* = \underline{\underline{A}}(\underline{T})^{-1}\underline{Q}^* + \underline{\varepsilon} \; . \tag{49}$$

This can be viewed as a nonlinear regression model of h_n against Q_n in terms of the model parameters, T_i.

In order to estimate the parameters of this regression model, we assumed (Neuman and Yakowitz, 1979) that $\underline{\varepsilon}$ and $\underline{\nu}$ are uncorrelated with each other (which is true if $\underline{h}^*$ and $\underline{T}^*$ are evaluated independently, as in kriging), and their first two moments are given by

$$E[\underline{\varepsilon}] = 0, \tag{50}$$

$$E[\underline{\varepsilon}^T\underline{\varepsilon}] = \sigma_h^2 \underline{\underline{V}}_h, \tag{51}$$

$$E[\underline{\nu}] = 0, \tag{52}$$

$$E[\underline{\nu}^T\underline{\nu}] = \sigma_T^2 \underline{\underline{V}}_T. \tag{53}$$

Here $\underline{\underline{V}}_h$ and $\underline{\underline{V}}_T$ are known N x N and I x I symmetric positive-definitive matrices, respectively, and σ_h^2 and σ_T^2 are scalars, which may or may not be known. For example, if $\underline{T}^*$ is obtained by kriging based on a sufficiently large sample of point T values, then the right-hand side of equation 53 can be computed from a knowledge of Cov[log(T^*_i/T_i), log (T^*_j/T_j)], which, in turn, is given by equation 42. Note that $\varepsilon(\Omega_i)$ in equation 42 is equivalent to log T^*_i - log T_i.

A method for converting the statistics of logarithmc data to those of the arguments of the logarithm has been discussed by Matalas (1967) and Mejia and Rodriguez-Iturbe (1974a). However, if the number of sample points is small, or if these points are densely spaced in some parts of the aquifer but not in other parts, then the second moment of $\underline{\nu}$ cannot be computed with sufficient accuracy from the data. When this happens, the authors prefer to leave σ^2_T in equation 53 unspecified, treating it as an unknown parameter to be determined by the inverse procedure, thereby allowing for a certain degree of error in the magnitude (though not in the structure) of $E[\nu^T\nu]$. Similarly, if the available sample of water level data is also small, the method provides an option for treating σ^2_h in equation 51 as an unknown parameter.

The actual estimation of $\underline{T}$ is accomplished by minimizing the generalized least squares criterion

$$J(\hat{\underline{T}}) = J_h(\hat{\underline{T}}) + \lambda J_T(\hat{\underline{T}}) \tag{54}$$

with respect to $\hat{\underline{T}}$, where $\hat{\underline{T}}$ is an estimate of $\underline{T}$, λ is a non-negative coefficient, and J_h and J_T are defined as

$$J_h(\hat{\underline{T}}) = (\underline{h}^* - \hat{\underline{h}})^T \underline{\underline{V}}_h^{-1} (\underline{h}^* - \hat{\underline{h}}), \tag{55}$$

$$J_T(\hat{\underline{T}}) = (\underline{T}^* - \hat{\underline{T}})^T \underline{\underline{V}}_T^{-1} (\underline{T}^* - \hat{\underline{T}}), \tag{56}$$

$\hat{\underline{h}}$ being an estimate of $\underline{h}$ given by

$$\hat{\underline{h}} = \underline{\underline{A}}(\hat{\underline{T}})^{-1}\underline{Q}^* . \tag{57}$$

It can be shown (Schweppe, 1973, p. 96-106) that if $\underline{\underline{A}}(\underline{T})^{-1}$ were linear in $\underline{T}$ the optimum value of λ in equation 54, λ_{opt}, would be equal to σ^2_h/σ^2_T. The corresponding value of $\hat{\underline{T}}$, $\hat{\underline{T}}_{opt}$, would then represent the best unbiased linear estimate of $\underline{T}$ in that the error of estimation, $e_T = \hat{\underline{T}}_{opt} - \underline{T}$, would have minimum variance (as measured for example by the trace or determinant of $E[\underline{e}_T^T\underline{e}_T]$). Furthermore, the variance of $\underline{e}_T$ would never exceed that of the prior estimation (or kriging) error, $\underline{\nu}$, and would usually be smaller. In the nonlinear case actually dealt with here, the same will be true within a first order-approximation.

If σ^2_T or σ^2_h are considered to be unknown, the optimum value of λ in equation 54 will also remain unknown. In order to estimate λ_{opt} in this case, my co-worker and I (Neuman and Yakowitz, 1979) recommend evaluating a sequence of $\hat{\underline{T}}$ vectors by minimizing $J(\underline{T})$ in equation 54 for different values of λ. Each of these λ values will be associated with a different set of residual errors, $\eta_h = \hat{\underline{h}} - \underline{h}^*$ and $\eta_T = \hat{\underline{T}} - \underline{T}^*$. By analyzing the manner in which various statistics of these residual errors vary with λ, one should be able to define a narrow range of λ values within which λ_{opt} must be found. For example, the mean residual sum of squares,

$$s^2 = \frac{J(\hat{T})}{N}, \tag{58}$$

represents the best estimate of σ^2_h that one can derive from the inverse model. Thus, if σ^2_h can be estimated independently from the data (for example, by kriging), then the particular value of λ for which s^2 is closest to this former estimate of σ^2_h can be considered as an estimate of λ_{opt}. Other statistics that appear to be useful for the identification of λ_{opt} are the average value of η_h over all the nodes, the coefficient of variation of η_h, and the average autocorrelation between η_h values at all neighboring nodes (or average nearest-neighbor autocorrelation). These statistics will sometimes exhibit unusual behavior near λ_{opt}, which can be anticipated on theoretical grounds. When such behavior is apparent, the value of λ_{opt} can be estimated without necessarily having to rely on an independent estimate of σ^2_h.

Once λ_{opt} has been estimated in this fashion, the corresponding estimate of $\underline{T}$, $\hat{\underline{T}}_{opt}$, is adopted as the final solution of the inverse problem. If all the assumptions of the model are correct, the mean of the estimation error, $\underline{e}_T = \hat{\underline{T}}_{opt} - \underline{T}$, is zero, and $\hat{\underline{T}}_{opt}$ is unbiased (provided that $\underline{Q}^* \equiv \underline{Q}$). The covariance of $\underline{e}_T$ is given to a first order of approximation by

$$E[\underline{e}_T^T\underline{e}_T] = \sigma^2_h \, (\underline{\underline{S}}^T \underline{\underline{V}}_h^{-1} \underline{\underline{S}} \, \lambda_{opt} \, \underline{\underline{V}}_T^{-1})^{-1} , \tag{59}$$

where $\underline{\underline{S}}$ is an N x I sensitivity (or Jacobian) matrix, the typical term of which is defined as

$$S_{ni} = \left.\frac{\partial \hat{h}_n}{\partial \hat{T}_i}\right|_{\lambda=\lambda_{opt}} . \tag{60}$$

The error of estimating $\underline{h}$ by $\hat{\underline{h}}$, $\underline{e}_h = \hat{\underline{h}} - \underline{h}$, also has zero mean, and its covariance is given to a first order of approximation by

$$E[\underline{e}_h^T\underline{e}_h] = \underline{\underline{S}} \, E[\underline{e}_T^T\underline{e}_T] \, \underline{\underline{S}}^T . \tag{61}$$

An interesting feature of $E[\underline{e}_T^T\underline{e}_T]$ is that the the estimation errors, $\underline{e}_T$, often exhibit a high degree of positive autocorrelation parallel to the mean direction of flow, and small or even negative autocorrelation normal to this direction. My coworkers and I (Neuman and others, 1979) explain this phenomenon by the fact that the inverse problem is akin to the spatial Cauchy problem of classical mathematics, whose solution corresponds to our T. The solution of the Cauchy problem at any given point along a streamline shows a strong dependence on the solutions at all other points along that same streamline, but no dependence at all on solutions corresponding to points lying on other streamlines. This anisotropic nature of $\underline{e}_T$ reflects the fact that $\hat{\underline{T}}_{opt}$ is conditioned on the hydraulic head field which is itself anisotropic.

The minimization of $J(\hat{\underline{T}})$ in equation 54 with respect to $\hat{\underline{T}}$ for a given value of λ can be performed by a suitable nonlinear programming technique. My coworker and I (Neuman and Yakowitz, (1979) describe a highly efficient numerical method for minimizing $J(\hat{\underline{T}})$ which is valid when $\underline{e}$ in equation 46 is sufficiently small, and may sometimes be used to obtain rapid initial estimates of λ_{opt} when $\underline{\varepsilon}$ is large. We have applied our inverse method to a hypothetical situation in order to demonstrate its effectiveness in dealing with noisy $\underline{T}^*$ and $\underline{h}^*$ data. In another paper, application of the method to real data from the Cortaro Basin in southern Arizona has been described (Neuman and others 1979). In neither of these two examples has kriging been used to generate the input data into the inverse model, but an effort to do so is currently under way at the University of Arizona in Tucson. The transmissivity estimates generated by the inverse procedure for the Cortaro Basin compare reasonably well with other estimates obtained for the same aquifer by trial and error. Both sets of estimated transmissivities have been used with success to reproduce 25 years of recorded water level changes in the Cortaro Basin between 1940 and 1965 caused by temporal variations in recharge from streams and pumpage from wells. Although the estimates from the inverse method have a large variance, the ratios between these estimates in the direction of flow appear to be reasonably accurate because the corresponding estimation errors show a high degree of autocorrelation parallel to the stream lines.

Another inverse method which is consistent with the stochastic approach to groundwater modeling has been developed recently by de Marsily (1978). This method is based entirely on kriging. However, instead of expressing $\hat{Z}(\Omega_i)$ only in terms of measured transmissivity values as in equation 35, the author relies on an expanded form of this equation,

$$\hat{Z}(\Omega_i) = \sum_{k=1}^{K} \lambda_k^{(i)} Z(\underline{x}_k) + \sum_{m=1} \lambda_p^{(i)} Z(\underline{x}_p), \qquad (62)$$

where $\underline{x}_k$ are points at which T has actually been measured so that $Z(\underline{x}_k) = \log T(\underline{x}_k)$ is known for all $k = 1, 2, \ldots, K$, whereas $\underline{x}_p$ are so-called "pilot points" at which the values of $Z(\underline{x}_p)$ are unknown. The number, P, and location of the pilot points are determined subjectively by the modeler. Since the values of $\lambda_k^{(i)}$ and $\lambda_p^{(i)}$ depend only on the location of the points $\underline{x}_k$ and $\underline{x}_p$, they can be determined without knowing the values of $Z(\underline{x}_p)$, merely by solving a kriging system similar to that in equation 39. The inverse approach consists of estimating $Z(\underline{x}_p)$ so as to minimize the weighted least-squares criterion

$$J_h(\hat{\underline{Z}}) = (\underline{h}^* - \hat{\underline{h}})^T \underline{\underline{V}}_h^{-1} (\underline{h}^* - \hat{\underline{h}}), \qquad (63)$$

which is similar to equation 55 except that now the unknown parameters are not $\hat{\underline{T}}$, but $\hat{\underline{Z}} = (\hat{Z}_1, \hat{Z}_2, \ldots, \hat{Z}_p)$, where $\hat{Z}_p$ is the estimate of $Z_p \equiv Z(\underline{x}_p)$. Although the author's description of his method ends at this point, the reader should recognize that it is possible to compute the covariance matrix of the estimation errors, $(e_Z)_p = \hat{Z}_p - Z_p$, to a first order of approximation according to

$$E[\underline{e}_Z\underline{e}_Z^T] = \sigma_h^2(\underline{\underline{S}}_Z^T\underline{\underline{V}}_h^{-1}\underline{\underline{S}}_Z)^{-1}, \qquad (64)$$

where $\underline{\underline{S}}_Z$ is an N x P sensitivity (or Jacobian) matrix whose typical term is defined as

$$S_{np} = \frac{\partial \hat{h}_n}{\partial \hat{Z}_p}. \qquad (65)$$

The vector $\underline{e}_Z$ can then be treated in a manner not too different from that of measurement errors (see Delhomme, 1974) to compute the covariance matrix of the error involved in estimating $Z(\Omega_i)$ by $\hat{Z}(\Omega_i)$ according to equation 62.

Numerous other workers have developed inverse procedures that make lesser use of measured transmissivity data, or use these data only in an indirect manner. Some of these procedures have been reviewed (Neuman and Yakowitz, 1979), and the interested reader is referred to that paper for additional information.

In the past, parameter estimates obtained by inverse methods have been used only to provide deterministic predictions of aquifer behavior. However, it should be clear that these estimates are actually random variables and must therefore be used as input into stochastic aquifer models. In principle, most of the discrete stochastic models described in this paper could be used for this purpose. However, the actual use of outputs from inverse models as inputs for stochastic models remains an important and challenging task for future research.

FUTURE RESEARCH NEEDS

The stochastic models described in this paper are based on various assumptions that are not always satisfied in real-life situations. A considerable amount of research will be needed to establish the conditions under which these assumptions are valid. Most of these models have so far been applied only to steady-state problems, and are limited in their ability to deal with realistic problems involving complex boundary conditions in a computationally efficient manner. Thus, there is a need to develop efficient numerical schemes that solve the governing stochastic differential equations in a reliable manner for arbitrary boundary conditions. Furthermore, these schemes should be able to handle not only random parameters, but also random boundary conditions, initial conditions, and forcing functions (such as sink and source terms). One important prerequisite for practical applications is that the stochastic model be conditioned upon as much relevant information as may be available. This means that the numerical scheme must be compatible with statistically-based inverse methods, whose output would provide the input into the stochastic model. Thus, the development of reliable inverse methods which are consistent with the stochastic approach to groundwater modeling and can handle more than only steady-state data must be viewed as an important subject for future research. Once the development of stochastic models and inverse methods can be applied to realistic problems with confidence, research should continue toward their use in answering questions related to the design of data collection networks and the planning, design, and management of water resources systems in which groundwater plays a role. Other interesting areas of research include the transport of chemicals, flow through fractured rocks, and flow in nonisothermal systems such as geothermal reservoirs and aquifers used for the storage of thermal energy.

CONCLUSIONS

Based on all the work that has been summarized in this paper, we can draw the following conclusions:

1. The parameters of the equations governing subsurface fluid flow represent macroscopic material properties, which usually vary with space in a random manner. It is neither possible nor necessary to describe the spatial variability of these material properties by deterministic functions. The only feasible alternative is to characterize their spatial variability in statistical terms.
2. Various studies have implied that the probability distribution of hydraulic conductivities, transmissivities, and specific capacities is approximately lognormal, whereas that of porosities is approximately normal. Most of these studies were based on the assumption that measured values of points in space are statistically independent. We know today that, in reality, these values may show various degrees of spatial correlation (or autocorrelation) which must be taken into account in analyzing their statistical behavior. Furthermore, the spatial variability of one parameter (say hydraulic conductivity) will, in general, not be statistically independent of the variability of other parameters (such as porosity). It is not clear to what extent the autocorrelation of, and cross-correlation between, different parameters will affect the former conclusions about the distribution of various material properties.
3. In most cases, the material properties are measured over very small portions of the flow region. When this is true, the measured values can be viewed as point quantities of a macroscopic property defined over a continuum representing the flow field. This makes it possible to describe the flow process by partial differential equations whose parameters are stochastic variables defined over the same continuum.
4. When the governing stochastic differential equations are expressed in one- or two-dimensional form, they are suitable only for the description of spatially averaged quantities. For example, in working with two-dimensional aquifer models in the horizontal plane, the statistical information should be based on depth-averaged parameters such as transmissivities. In one-dimensional models, the parameters must be averaged over a plane perpendicular to the direction of flow. In general, space-averaged quantities will have a smaller variance than the related point quantities and will be correlated over larger distances. In dealing with point data such as those obtained from core samples, the appropriate form of analysis would be three-dimensional.
5. The stochastic differential equations that govern the flow in porous continua can be solved analytically for relatively simple problems. In more realistic situations where the flow is controlled by complex boundary conditions, the analytical approach must be supplemented by numerical methods.
6. In solving the governing stochastic differential equations by a numerical method, the flow domain is usually discretized into finite subregions, and the parameters are assumed to remain uniform (that is, constant) over each subregion. This is permissible provided that the uniform parameters are interpreted as effective values representing appropriate spatial averages of the related point quantities. However, the concept of an effective parameter will generally be valid under steady-state conditions only if each subregion is made small enough so that the mean hydraulic gradient does not vary significantly from one point to another over the subregion (that is, the ensemble average of the hydraulic gradient remains approximately uniform over the subregion). The hydraulic heads and gradients computed with such a model must be interpreted as mean values (that is, ensemble averages) of these quantities rather than their true point values. If the effective parameters are treated as stochastic variables, then these mean heads and gradients will also be stochastic.
7. In using a discrete model to analyze the effect of spatial variability on the actual (as opposed to mean) head values, the subregions must be small enough not to mask the autocorrelation structure of the material properties from which the model parameters are derived. This means that the stepwise variation of the model parameters must be such as to provide a reasonably good piecewise constant approximation for the autocovariance function of the true point macroscopic properties. Furthermore, the size of each subregion must never exceed the integral scale, which is the mean radius over which the point properties are correlated in any given direction in space. Obviously, this approach becomes impractical when the integral scale is much smaller than the dimensions of the flow region.
8. If hydraulic conductivites and transmissivities are assumed to be lognormal, their effective values in a two-dimensional region are given by the geometric mean. In one dimension, the effective values are given by the harmonic mean, which is generally smaller than the geometric mean, but becomes equal to it in the hypothetical case

where the material properties have zero variance. In three dimensions, the effective hydraulic conductivity is somewhat larger than the geometric mean, but approaches it rapidly as the variance of the conductivity decreases. All of this has so far been shown to hold true only for steady flow under a uniform mean hydraulic gradient.

9. The stochastic differential equations act as a smoothing mechanism which is capable of filtering out much of the noise present in the model parameters. This filtering effect is manifested by the fact that the fluctuations in computed hydraulic head values relative to the mean hydraulic head are invariably smaller than the relative fluctuations in the input parameters. Furthermore, the computed hydraulic head values are correlated over larger distances than the parameters, and their correlation structure is invariably anisotropic.

10. The fluctuations in hydraulic head computed by a stochastic model increase with the variance of the input parameters, and with the magnitude of the local mean hydraulic gradient. Another important factor affecting these fluctuations is the relationship between the radius over which the parameters are positively correlated and the size and nature of the flow field. In general, the larger the correlation radius relative to the dimensions of the flow field, the larger the variance of the computed hydraulic head will be. However, this variance will decrease sharply as one approaches Dirichlet-type boundaries, along which the hydraulic head is prescribed in a deterministic fashion.

11. In order to model the flow at a specific site, one must first obtain a statistical description of the material properties at this site. As a prerequisite for such a description, the material properties must actually be measured at a sufficiently large number of points in the field. Many of the stochastic models that have been described in the groundwater literature thus far have disregarded the fact that material properties will often be known at specific points. In these models, the variance of the parameters is assumed to be the same at every point in thc field, including points at which the material properties have actually been measured. A more realistic approach would be to condition the model upon these measurements and hopefully, in this fashion, affect a reduction in the variance of the computed hydraulic head values. A method for accomplishing this, called conditional simulation, has been described in a recent publication by Delhomme (1979).

12. In many practical situations, the hydrologist has more information about historical variations in the flow regime than about the material properties of the system. Furthermore, the information about hydraulic heads is usually more accurate than that about, say, transmissivities and storativities. Thus, instead of conditioning the stochastic model only on measured values of the parameters, it would make sense to take advantage of existing information about the flow regime and condition the model upon data describing this regime as well as the parameters. This can be accomplished by estimating the effective model parameters via an appropriate inverse method and then using these estimates, together with their statistics, as input into the stochastic model. At least two inverse methods that are fully consistent with the stochastic approach to groundwater modeling have been described in the literature. However, the task of using the output from these models as input into stochastic ground-water flow models has not yet been accomplished. This task is an important and challenging subject for future research.

13. The state of stochastic groundwater models is such that much additional research will be required before they will reach the degree of operational simplicity and theoretical reliability required to apply them in hydrologic practice. However, in the future, these models should become useful tools for providing more realistic predictions of aquifer behavior than the present-day deterministic models; for the design of more useful and economical data collection networks; and for more rational planning, design, and operation of water resource systems in which groundwater plays a role.

ACKNOWLDGMENTS

The author is grateful to Dr. R. Allan Freeze, Dr. Thomas Maddock III, Dr. Ghislain de Marsily, and Dr. Budhi Sagar for many useful comments regarding the original version of the manuscript.

REFERENCES CITED

Agterberg, F. P., 1974, Developments in geomathematics: New York, Elsevier, 596 p.

Bagnold, R. A., 1968, Deposition in the process of hydraulic transport: Sedimentology, v. 10, p. 45-56.

Bakr, A. A., 1976, Stochastic analysis of the effects of spatial variations in hydraulic conductivity on groundwater flow (Ph.D. dissertation): New Mexico Institute of Mining and Technology.

Bakr, A. A., and others, 1978, Stochastic analysis of spatial variability in subsurface flows, part I, comparison of one- and three-dimensional flows: Water Resources Research, v. 14, no. 2, p. 263-271.

Bartlett, M. S., 1975, The statistical analysis of spatial pattern: London, Chapman and Hall, 91 p.

Bear, J., 1972, Dynamics of fluids in porous media: New York, Elsevier, 764 p.

Bennion, D. W., and Griffiths, J. C., 1966, A stochastic model for predicting variations in reservoir rock properties: Trans AIME, no. 237, Part 2, p. 9-16.

Bibby, R., and Sunada, D. K., 1971, Statistical error analysis of a numerical model of confined groundwater flow, in Chiu, C. L., ed., Stochastic hydraulics, Proceedings, First International Symposium on Stochastic Hydraulics: p. 591-612.

Bras, R. L., and Rodriguez-Iturbe, I., 1976, Evaluation of mean square error involved in approximating the areal average of a rainfall event by a discrete summation: Water Resources Research, v. 12, no. 2, p. 181-184.

Bresler, E., and Dagan, G., 1979, Solute dispersion in unsaturated heterogeneous soil at field scale, part 2, applications: Soil Science Society of America Proceedings (in press).

Brust, K. J., van Bavel, C. H. M., and Stirk, G. B., 1968, Hydraulic properties of a clay-loam soil and the field measurement of water uptake by roots, part 3: Madison, Soil Science Society of America Proceedings v. 32, p. 322-326.

Bulnes, A. C., 1946, An application of statistical methods to core analysis data of dolomitic limestone: Trans. AIME, v. 165, p. 223-240.

Buyevich, Y. A., Leonov, A. I., and Safrai, V. M., 1969, Variations in filtration velocity due to random large scale fluctuations in porosity, part 2: Journal Fluid Mechanics, v. 37, p. 371-381.

Csallany, S., and Walton, W. C., 1963, Yields of shallow dolomite wells in northern Illinois: Illinois State Water Survey Report Investigations, no. 46.

Dagan, G., 1976, Comment on 'A stochastic-conceptual analysis of one-dimensional groundwater flow in nonuniform homogeneous media' by R. A. Freeze: Water Resources Research, v. 12, no. 3, p. 567.

___1979, Models of groundwater flow in porous formations of a statistically homogeneous random structure: Water Resources Research, v. 15, no. 1, p. 47-63.

Dagan, G., and Bresler, E., 1979, Solute dispersion in unsaturated heterogeneous soil at field scale, I, theory: Soil Science Society of America Proceedings (in press).

Davis, S. N., 1969, Porosity and permeability of natural materials, in De Wiest, R. J. M., ed., Flow through porous media: New York, Academic Press, p. 54-89.

Delhomme, J. P., 1974, La cartographie d'une grandeur physique a partir de données de differentes qualités, in Proceedings, Montpellier Meeting, International Association of Hydrogeologists: v. 10, no. 1, p. 185-194.

___1978, Kriging in the hydrosciences: Advances in Water Resources, v. 1, no. 5, p. 251-266.

___1979, Spatial variability and uncertainty in groundwater flow parameters: Water Resources Research (in press).

de Marsily, G., 1978, De l'identification des systemes hydrogeologiques, tome I, syntheses [Ph. D. dissertation]: Paris, Université P. et M. Curie.

Erdelyi, A., ed., 1954, Tables of integral transforms, in Bateman manuscript project: New York, McGraw-Hill, v. 1, 391 p.

Farengolts, Z. D., and Kolyada, M. N., 1969, Opyt primeneniya metodov matematicheskoy statistiki slya izutcheniya zakonov raspredeleniya gidrogeologicheskikh parametrov: Trudy Vsegingeo, v. 17, p. 76-112.

Freeze, R. A., 1975, A stochastic-conceptual analysis of one-dimensional groundwater flow in a non-uniform homogeneous media: Water Resources Research, v. 11, no. 5, p. 725-741.

Freeze, R. A., 1977, Probabilisitic one-dimensional consolidation: Journal of Geotechnical Engineering Division, American Society Civil Engineers, v. 103, (GT7), p. 725-742.

Gates, J. S., and Kisiel, C. C., 1974, Worth of additional data to a digital computer model of a groundwater basin: Water Resources Research, v. 10, no. 5, p. 1031-1038.

Gelhar, L. W., 1976, Effects of hydraulic conductivity variations on ground-water flows, in Proceedings, Second International IAHR Symposium on Stochastic Hydraulics: Lund, Sweden, International Association of Hydraulic Research.

Gelhar, L. W., and others, 1977, Comments on 'A stochastic-conceptual analysis of one-dimensional groundwater flow in a nonuniform homogeneous media' by R. A. Freeze: Water Resources Research, v. 13, no. 2, p. 477-479.

Gelhar, L. W., Gutjahr, A. L., and Naff, R. L., 1979, Stochastic analysis of macrodispersion in a stratified aquifer: Water Resources Research (in press).

Griffiths, J. C., 1967, The scientific method in analysis of sediments: New York, McGraw Hill.

Gutjahr, A. L., and others, 1978, Stochastic analysis of spatial variability in subsurface flows, part II, evaluation and application: Water Resources Research, v. 14, no. 5, p. 953-960.

Heller, J. P., 1972, Observations on diffusion and mixing in porous media, in Proceedings, Second International Symposium on Fundamentals of Transport Phenomena in Porous Media, IAHR and ISSS, v. 1: Guelph, University of Guelph, p. 1-26.

Ilyin, N. I., and others, 1971, Otsenka tochnosti opredeleniya vodopronitsayemosti gornykh porod: Moscow, Nedra.

Jetel, J., 1974, Complément régional de l'information sur les parametres pétrophysiques en vue de l'élaboration des modeles de systémes aquiféres, in Proceedings, Montpelier Meeting, International Association of Hydrogeologists.

Krasny, J., 1974, Les différences de la transmissivité statistiquement significatives dans les zones de l'infiltration et du drainage, in Proceedings, Montpelier Meeting, International Association of Hydrogeologists.

Krumbein, W. C., 1936, Application of the logarithmic moments to size frequency distributions of sediments: Journal Sedimentary Petrology, v. 6, no. 1, p. 35-47.

Law, J., 1944, A statistical approach to the interstitial heterogeneity of sand reservoirs: Trans AIME, v. 155, p. 202-222.

Lippmann, M. J., 1973, Two-dimensional stochastic model of a heterogeneous geologic system [Ph.D. dissertation]: Berkeley, University of California.

Lumley, J. L., and Panofsky, H. A., 1964, The structure of atmospheric turbulence: New York, John Wiley and Sons, Inc.

Martin, R., 1974, On spatial dependence, bias, and use of first spatial differences in regression analysis: Area, v. 6, p. 185-194.

Matalas, N. C., 1967, Mathematical assessment of synthetic hydrology: Water Resources Research, v. 3, no. 4, p. 937-945.

Matheron, G., 1967, Composition des permeabilitées en milieu poreux heterogene, méthode de

Schwydler et régles de ponderation: Revue de L'Institut Francais du Pétrole, p. 443-466.

___1971, The theory of regionalized variables and its applications: Fontainebleau, École des Mines.

___1973, The intrinsic random functions and their applications: Advances in Applied Prob., v. 5, p. 438-468.

McMillan, W. D., 1966, Theoretical analysis of groundwater basins operations: Berkeley, University of California, Hydraulics Laboratory Technical Report 6-25, 167 p.

Mejia, J. M., and Rodriguez-Iturbe, I., 1974a, Correlation links between normal and lognormal processes: Water Resources Research, v. 10, no. 4, p. 689-690.

___1974b, On the synthesis of random field sampling from the spectrum: an application to the generation of hydrologic spatial processes: Water Resources Research, v. 10, no. 4, p. 705-712.

Neuman, S.P., and Witherspoon, P. A., 1969, Theory of flow in a confined two-aquifer system: Water Resources Research, v. 5, no. 4, p. 803-816.

Neuman, S. P., and Yakowitz, S., 1979, A statistical approach to the inverse problem of aquifer hydrology, part I, theory: Water Resources Research (in press).

Neuman, S. P., Fogg, G. E., and Jacobson, E. A., 1979, A statistical approach to the inverse problem of aquifer hydrology, part II, case study: Water Resources Research (in press).

Nielsen, D. R., Biggar, J. W., and Erh, K. T., 1973, Spatial variability of field-measured soil-water properties: Hilgardia, v. 42, no. 7, p. 215-259.

Papoulis, A., 1965, Probability, random variables, and stochastic processes: New York, McGraw-Hill.

Rodriguez-Iturbe, I., and Mejia, J. M., 1974, The design of rainfall networks in time and space: Water Resources Research, v. 10, no. 4, p. 713-728.

Rousselot, D., 1976, Proposition pur une loi de distribution des permeabilités ou transmissivités: Rapport BRGM, Service Géologique Jura-Alpes.

Sagar, B., 1978, Galerkin finite element procedure for analyzing flow through random media: Water Resources Research, v. 14, no. 6, p. 1035-1044.

Schweppe, F., 1973, Uncertain dynamic systems: Englewood Cliffs, Prentice-Hall.

Seaber, P. R., and Hollyday, E. F., 1968, Statistical analysis of regional aquifers: Boulder, Geological Society of America, Special Paper no. 101, abstracts for 1966.

Smith, L., 1978, Stochastic analysis of steady state groundwater flow in a bounded domain [Ph.D. dissertation]: Vancouver, University of British Columbia.

Smith, L., and Freeze, R. A., 1979a, Stochastic analysis of steady state groundwater flow in a bounded domain, part 1, one-dimensional simulations: Water Resources Research (in press).

___1979b, Stochastic analysis of steady state groundwater flow in a bounded domain, part 2, two-dimensional simulations: Water Resources Research, (in press).

Tang, D. H., and Pinder, G. F., 1977, Simulation of groundwater flow and mass transport under uncertainty: Advances in Water Resources, v. 1, no. 1, p. 25-30.

Taylor, G. I., 1953, The dispersion of matter in a solvent flowing slowly through a tube, in Proceedings, Royal Society, Series A: London, Royal Society, v. 219, p. 189-203.

Walton, W.C., and Neill, I. C., 1963, Statistical analysis of specific capacity data for a dolomite aquifer: Journal Geophysical Research, v. 68, p. 8.

Warren, J. E., and Price, H. S., 1961, Flow in heterogeneous porous media: Society of Petroleum Engineers Journal, v. 1, p. 153-169.

Warren, J. E., and Skiba, F. F., 1964, Macroscopic dispersion: Transactions AIME, Society Petroleum Engineers, v. 231, p. 215-230.

Warren, J. E., Skiba, F. F., and Price, H. S., 1961, An evaluation of the significance of permeability measurements: Journal Petroleum Technology, v. 13, p. 739-744.

Warrick, A. W., Mullen, G. J., and Nielsen, D. R., 1977, Scaling field measured soil hydraulic properties using a similar media concept: Water Resources Research, v. 13, no. 2, p. 355-362.

Whittle, P., 1954, On stationary processes in the plane: Biometrika, v. 41, p. 434-449.

Willardson, L. S., and Hurst, R. L., 1965, Sample size estimates in permeability studies: Journal Irrigation Drainage Division, American Society Civil Engineers, v. 91 (IR1), p. 1-9.

Wu, T. H., Vyas, S. K., and Chang, N. Y., 1973, Probabilistic analysis of seepage: Journal Soil Mechanics Foundation Division, American Society Civil Engineers, v. 99 (SM4), p. 323-349.

Yakowitz, S., 1976, Model-free statistical methods for water table prediction: Water Resources Research, v. 12, no. 5, p. 836-844.

Manuscript Accepted by the Society October 14, 1980

Printed in U.S.A.

Geological Society of America
Special Paper 189
1982

Conceptual Aspects of Subsidence Due to Fluid Withdrawal

D. C. HELM
Lawrence Livermore Laboratory, University of California
Livermore, California 94550

ABSTRACT

A saturated reservoir or confined aquifer system responds to fluid withdrawal as a dynamic unit whose elemental volumes not only compress due to decrease in porosity but move horizontally and vertically in order for the system to remain contiguous.

Recent trends in the conceptual aspects of subsidence due to fluid withdrawal are toward developing a practical three-dimensional theory of reservoir movement. One approach to developing an adequate theory is to couple volume strain within a reservoir to calculated stress change. Transient displacement of reservoir material in a direction of interest is the sum of appropriate directional components of volume strain. It is not an integral sum of volume strain itself. To find directional components of strain in three dimensions requires so many parameters that, using this approach, a solution for matrix movement becomes intractable under most field conditions.

An alternative approach to developing a three-dimensional theory is proposed in this paper. It couples reservoir velocity fields in three dimensions to appropriate force fields. The generally neglected effect of the gravitational force field on transient movement of solids is included in the analysis. After calculating the transient velocity field of a saturated granular medium, it is conceptually possible to estimate the displacement of solids by integrating velocity over an appropriate interval of time.

INTRODUCTION

Subsidence is the "sinking or settlement of the land surface due to any of several processes. As commonly used, the term relates to the vertical downward movement of natural surfaces although small-scale horizontal components may be present. The term does not include landslides, which have large-scale horizontal components, or settlement of artificial fills" (Poland and others, 1972, p. 8).

Withdrawing fluid from depth at a discharging well necessarily takes fluid mass from storage within an underground porous formation. Although expansion of individual constituents (liquid, gas, and solids) contributes to the volume of discharge, the dominant mechanism for fluid production from a confined system is commonly found to be decrease in porosity. A decrease in thickness of buried sediments is one result of porosity loss. The process of decrease in thickness is called compaction by earth scientists and consolidation by soil engineers. Three-dimensional compression of a directly stressed formation (reservoir) from which fluid is being withdrawn affects skeletal movement in surrounding formations (including overburden) from which no fluid is being withdrawn. If and when the effects of matrix movement associated with fluid withdrawal migrate to the land surface, the term <u>subsidence</u> can be appropriately applied. Such subsidence is the surface expression of horizontal and vertical movements of solid material that are induced at depth.

The scientific study of subsidence due to fluid withdrawal has a brief, unfinished history. Fuller (1908, p. 33) was the first to theorize on the possible relation of fluid withdrawal to subsidence.

Early observations of the phenomenon as associated with ground-water withdrawal were reported in the 1930s (Rappleye, 1933; Tibbetts, 1933). Tolman's (1937) early textbook on ground-water hydrology contains a section (written in collaboration with J. F. Poland) on subsidence in the Santa Clara Valley, California. Tolman and Poland (1940) subsequently pointed out that subsidence in the Santa Clara Valley is caused not simply by declining artesian heads and the resulting compaction of permeable sands, but primarily by the non-recoverable compaction of slow-draining clay layers within the confined system. This marks the conceptual birth of an aquitard-drainage model that was developed quantitatively three decades later and will be described in the next section. In 1956, J. F. Poland established a research center in Sacramento, California (U.S. Geological Survey,

Water Resources Division), for the systematic study of subsidence and the related mechanics of stressed aquifer systems. A number of fine studies resulting from this effort were published in the U.S. Geological Survey Professional Paper Series 437 and 497 (Bull, 1964, 1972, 1975; Bull and Miller, 1975; Bull and Poland, 1975; Johnson and others, 1968; Lofgren, 1975; Lofgren and Klausing, 1969; Meade, 1964, 1967, 1968; Miller and others, 1971; Poland and others, 1975; Riley and McClelland, 1972).

Subsidence due to withdrawal of oil was recognized even before the 1930s (Minor, 1925; Pratt and Johnson, 1926). Quantitative investigation was initiated first by Nabor Carrillo, who analyzed the subsidence of Long Beach, California, caused by oil withdrawal (Carrillo, 1949), and of Mexico City, caused by ground-water withdrawal (Carrillo, 1948, 1969).

An outstanding survey by Poland and Davis (1969) of major areas in the world experiencing land subsidence due to withdrawal of fluids (water, oil, or gas) was published by the Geological Society of America in 1969. The same year witnessed the First International Symposium on Land Subsidence (Tison, 1970) in Tokyo, sponsored by UNESCO. A Second International Symposium on Land Subsidence (Johnson, 1977) was held seven years later in Anaheim, California. This meeting was swiftly followed in the United States by two specialty symposia on related subjects. One, on ground failure due to fluid withdrawal, was sponsored by the Geological Society of America (1978, p. 130f). The other was sponsored by the American Society of Civil Engineers (Saxena, 1979), and dealt with prediction and evaluation of subsidence. A third will be held by the Geological Society of America in 1980 to honor J.F. Poland (T.L. Holzer, written commun., 1980).

During the 1970s attention was drawn to the phenomenon of subsidence (Schiller, 1975) because of increased public interest in the environment. One response to this new concern is the U.S. Department of Energy's program (Lawrence Berkeley Laboratory, 1977) of encouraging universities, corporations, government agencies, and engineering consultants, under the general guidance of P. A. Witherspoon of Lawrence Berkeley Laboratory, to address the problems of land subsidence that may be associated with future geothermal production.

Table 1 lists major areas of land subsidence due to ground-water withdrawal. Remedial or protective measures that have been taken are also listed. Table 2 lists major areas of land subsidence that are associated with withdrawal of oil and gas. Field studies of subsidence are listed in Poland (1972) and Poland and Davis (1969). A publication entitled *A Guide to the Study of Subsidence Due to Ground-water Withdrawal* will soon be available from UNESCO.

The present report will review some important concepts that have been developed by earth scientists in their continuing effort to understand and to predict subsidence due to fluid withdrawal with quantitative accuracy. A survey of computational methods for use in a digital computer is beyond the scope of the present paper, but can be found in reports by the following four consulting firms: Systems Control, Inc. (Finnemore, 1971); R and D Associates (Grover and others, 1977); McClelland Engineers (Pollard and others, 1979); and Golder Associates (Miller and others, 1980).

Following a brief discussion of existing conceptual models, I shall make a fairly extensive statement on the relation of $\tilde{q}$ (specific discharge of fluid relative to solids) and $\tilde{v}_s$ (rate of movement of solids themselves), and their relation to appropriate force fields. For conditions of transient flow, the problem of magnitude and direction of $\tilde{v}_s$ is fundamental. Unless the relation between $\tilde{q}$ and $\tilde{v}_s$ in three dimensions is conceptually perceived and quantitatively investigated, little significant progress in calculating compaction or in predicting subsidence due to fluid withdrawal can be expected beyond the present state of the art. It will be seen that during transient flow the relation of $\tilde{q}$ and $\tilde{v}_s$ involves a generally neglected gravitational force--the submerged weight of solids. In terms of the two rates of movement, $\tilde{q}$ and $\tilde{v}_s$, the influence of the submerged weight of solids is great during early transient flow and gradually diminishes whenever a steady-state flow condition is approached. In terms of cumulative movement of the solid phase, however, the importance of the submerged weight of solids does not diminish. The present paper will end with a discussion of potential energy within the solid phase and of the role this energy plays in subsidence even when man is not withdrawing fluid. It is suggested that density contrast $\rho_s - \rho_w$ within a saturated and permeable elemental bulk volume contributes significantly to secondary movement of solids.

REVIEW OF CONCEPTUAL MODELS

In order to predict subsidence, one needs a conceptual model of skeletal compression or subsidence that is capable of being quantified. By necessity, such a model is a simplification of nature. Models that are reviewed in this section include the depth-porosity model, the half-space model, the viscoelastic reservoir model, and the aquitard-drainage model.

One of the most straightforward conceptual models (Schatz and others, 1978) uses downhole logging data from which a depth-porosity relation (Fig. 1) can be estimated. Most investigators (Athy, 1930; Hedberg, 1936; Dickinson, 1953; Magara, 1968, 1978) have plotted porosity data from shales or mudstones. Such data have been applied to the mechanics of overthrust faulting (Hubbert and Rubey, 1959; Rubey and Hubbert, 1959), to the natural generation of abnormal fluid pressures in response to a slowly compacting geologic system (Chapman, 1972a), and to primary migration of petroleum due to geologic compaction (Chapman, 1972b). In the one example that Schatz and others (1978) apply to estimating subsidence potential, they analyze porosity data from sands. Figure 1 can be treated as a type of in situ stress-strain curve at a site of interest. Under ideal circumstances, a porosity-dependent coefficient of volume compression can be derived.

TABLE 1. AREAS OF MAJOR LAND SUBSIDENCE DUE TO GROUND-WATER OVERDRAFT

Location	Depth range of compacting beds (ft)	Maximum subsidence (ft)	Area of subsidence (sq mi)	Time of principal occurrence	Remedial or protective measures taken
Osaka, Japan	30-1,300	10	75	1928-68	Reduced ground-water pumpage; built dikes; drainage pumping plants
Tokyo, Japan	30-1,000	14	75	1920-70+	Reduced ground-water pumpage; built dikes; drainage pumps; pumping stations
Mexico City, Mexico	30- 160	28	50	1938-70+	Reduced pumpage; imported water; built recharge wells
Taipei Basin, Taiwan	30- 800	4.3	50	1961-69+	Ground-water management code adopted; recharge planned
Shanghai Peoples Republic of China	150-1,000	8		1921-65	River water injected through wells
Central Arizona	300-1,800	7.5	250	1948-67	
Santa Clara Valley, California	180-1,000	13	250	1920-70	Built detention dams; increased local recharge; built dikes; imported water
San Joaquin Valley California	200-3,000	9.5 to 28	4,200 (>1 ft)	1935-70+	Built reservoirs and imported water to reduce ground-water pumpage
Lancaster area, California	200-1,000(?)	3	150	1955-67+	None
Las Vegas, Nevada	200-1,000	3	200	1935-63	Moved well field away from fine-grained deposits; imported Colorado River water
Houston-Galveston, Texas	200-2,000(?)	8	2,500 (>1 ft)	1943-73+	Plans for surface-water imports underway
Baton Rouge, Louisiana	150-2,000(?)	1	250	1934-65+	None

Modified from Poland, 1972.

TABLE 2. AREAS OF MAJOR LAND SUBSIDENCE DUE TO FLUID WITHDRAWAL FROM OIL AND GAS FIELDS

Location	Depth range of compacting beds (ft)	Maximum subsidence (ft)	Area of subsidence (sq mi)	Time of principal occurrence
Po Delta, Italy	300-1,300	6-10	300	1950-61
Niigata, Japan	300-3,000	8	250	1956-69+
Lake Maracaibo, Venezuela	2,000-4,500(?)	13(?)	80+	1930-60+
Inglewood, California	(?)	5.7	4	1918-62+
Wilmington, California	2,000-4,000+	29	30	1938-65
Goose Creek, Texas	600-4,500(?)	3+	4	1917-25

After Poland and Davis, 1969.

Note: Yerkes and Castle (1970, Table 1) list documented subsidence of 24 additional oil and gas fields in United States.

According to Schatz and others (1978), under actual field conditions it may give a depth-dependent order-of-magnitude estimate of an upper limit for ultimate nonrecoverable volume compression. Within a depth interval of interest, if volume strain is approximately isotropic, the vertical component of strain would be roughly one third the volume strain. Correspondingly, a coefficient for estimating vertical compression would be one third the calculated coefficient of volume compression. In addition, if the area πR^2 of the top of a vertically (or volumetrically) compressing reservoir of radius R (Fig. 2) is smaller than the area of land surface affected by subsidence, the average vertical component of subsidence at the land surface should be reduced proportionately. This mechanism will be discussed shortly. Finally, if bridging or arching effects (when they occur) are permanent on a regional scale, such a bridging mechanism would tend to attenuate ultimate surface subsidence.

In an area where little or no time-dependent field data are available, a simple ultimate-compression type of model is most appropriate. When time-dependent in situ stress and strain data become available, more sophisticated models such as an aquitard-drainage model that will be discussed later, become increasingly appropriate and precise in predictive power. The two most appealing features of a depth-porosity model are that it is simple and in situ. It can be used as an initial (or upper-bound) estimate of ultimate compression before developing a reservoir where good downhole data are available. Its compressibility values can also be used in conjunction with a half-space model that will be discussed later.

A depth-porosity model requires the buried sediments to be normally consolidated and normally pressured, although this assumption can in some cases be modified (Hubbert and Rubey, 1959; Chapman, 1972a). If different strata at a site were initially deposited with differing porosities, or if cementation or recrystallization processes have occurred in some formations and not in others, the method may pose difficulties in data interpretation. Such beds, if recognized, can be excluded from analysis. In any event, if a reasonable depth-porosity curve can be plotted for a site, important information becomes available to the investigator.

According to John Martin (oral commun., 1978) of Chevron Oil Company, Karl Terzaghi lectured at the Massachusetts Institute of Technology on the relative merits of in situ depth-porosity curves and laboratory time-consolidation curves for use in predicting subsidence at Long Beach, California. In the 1950s, Terzaghi opted for laboratory time-consolidation curves largely because the local depth-porosity curve gave what he thought were overly large predictions for ultimate subsidence. In retrospect, however, Terzaghi's predictions based on laboratory time-consolidation curves (Pacific Fire Rating Bureau, 1958, p. 5) were embarrassingly small (Christian and Hirschfeld, 1974).

A second conceptual model, which we shall call the half-space model (Fig. 2), was introduced to subsidence research by Carrillo (1949). The earth is represented as a homogeneous, isotropic, semi-infinite elastic (or poroelastic) medium. Land surface is represented as a flat upper boundary that is free to move. Neither calculated surface move-

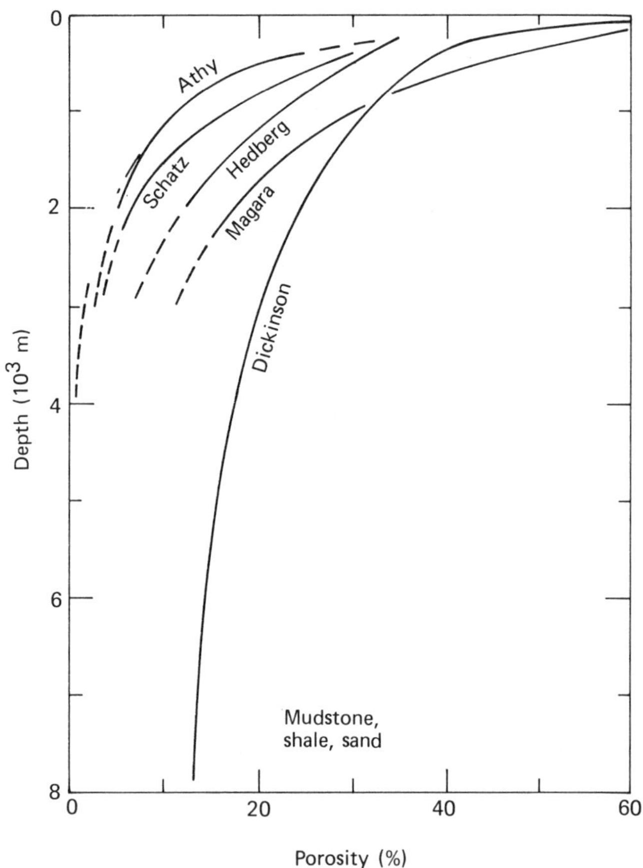

Figure 1. Depth-porosity model.

ment nor topographic relief affects the essential flatness of the idealized surface. Although a depressured zone at depth d below the land surface and with a center at radial distance r from an observer on the land surface is represented as a spherical tension center by some investigators (Carrillo, 1949; McCann and Wilts, 1951), and as a radially symmetric group of strain nuclei whose outer boundary is initially in the shape of a cylinder or disc by others (Geertsma, 1957, 1966, 1973a, 1973b; Finol and Farouq Ali, 1975), the two representations are conceptually similar. Gambolati (1972) has discussed the major conceptual distinctions between the tension-center representation (Carrillo, 1949; McCann and Wilts, 1951) and the strain-nucleus representation (Mindlin and Cheng, 1950; Sen, 1950; Geertsma, 1957). Briefly, a tension-center model tacitly assumes an infinitely compressible reservoir within an elastic half-space. The strain-nucleus model requires that the reservoir's compressibility be equal to the compressibility of the surrounding elastic half-space. Finite heterogeneity between reservoir rock and the surrounding half-space was introduced to this model by Gambolati (1972).

An important observation should be made here. The hydrodynamics of withdrawing fluid from storage within a reservoir and a resulting decrease in porosity would tend to give the reservoir a transient compressibility that is much larger than the surrounding nondraining half-space.

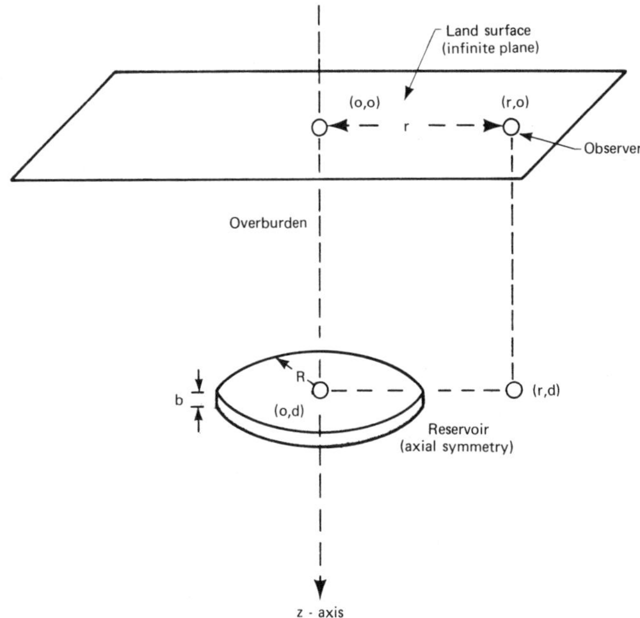

Figure 2. Half-space model.

Geertsma (1973a) reports that the strain-nucleus model is being modified to account for vertical heterogeneity of layered strata within the overburden. Geertsma points out that the presence of rigid basement rock beneath a depressured reservoir mathematically requires that the total volume of reservoir compaction appear at the land surface as subsidence. The areal distribution of this subsidence, however, is influenced by the ratio of depth d to radius R of a depressured reservoir. The ultimate effect of compaction on a deeply buried reservoir with small radius R will be spread over a large area at the land surface, and hence will minimize vertical subsidence at any overlying point. If the same reservoir were closer to the land surface, the volumetric effect at the land surface of an identical volume of reservoir compaction would in both cases be identical. However, because this volumetric effect would be spread over a smaller area at the land surface, it would in the latter case cause greater vertical subsidence at any specified point directly above the reservoir.

The half-space model is the first and most widely used method for estimating movement (vertical and horizontal) of material within the overburden. The ratio of depth d to radial extent R of a compacting reservoir may be critical in predicting movement of the land surface. To have developed a method for quantitatively estimating the influence and importance of this ratio is a significant step toward understanding the mechanics of subsidence.

The half-space model is based on equilibrium assumptions. The response of modeled elastic or poroelastic material to a specified pressure drop is mathematically required to be instantaneous throughout the modeled semi-infinite medium. Pressure drops within the reservoir itself can be allowed to change with time mathematically according to some preset formula or to a solution of a preset equation. Gambolati (1977) has coupled the equilibrium half-space model with a nonequilibrium equation for transient fluid flow within the modeled reservoir. Although the conceptual compatability of the two models (equilibrium and nonequilibrium) is not intuitively obvious, the coupling of them into a single solution scheme may be found in the future to be of practical benefit as a rough approximation.

Corapcioglu addressed the problem of modeling a generally observed time lag between stress change and deformational response within a reservoir itself (Corapcioglu, 1977; Brutsaert and Corapcioglu, 1976; Corapcioglu and Brutsaert, 1977). None of the previously discussed models can account for this observed phenomenon. Because the entire reservoir system is represented (Fig. 3) by Corapcioglu as a single undifferentiated porous material, the time lag of deformational response is mathematically accounted for by introducing a viscosity term directly into the compressibility coefficient.

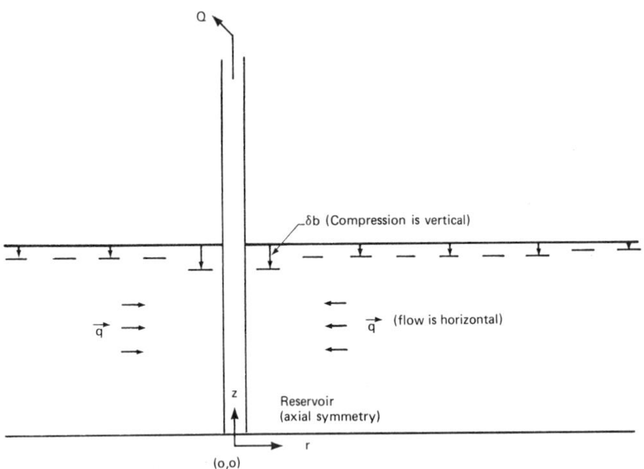

Figure 3. Viscoelastic reservoir model.

Corapcioglu's viscoelastic reservoir model computes both horizontal fluid flow and partly-delayed vertical compaction in response to a specified discharge rate Q. It requires, however, that no horizontal skeletal displacement be present within a depressured reservoir. On theoretical grounds (Verruijt, 1969) and according to empirical evidence (Yerkes and Castle, 1970; Wolff, 1970), horizontal movement may be of the same order of magnitude as vertical movement locally.

The viscoelastic reservoir model offers an alternative to the aquitard-drainage model, which will be discussed next. Both models are reservoir compaction models and ignore the behavior and influence of overburden material. For the special case of a shallow reservoir with large areal extent (such as a confined ground-water system pumped for irrigation), the d/R ratio is sufficiently small so that a reservoir compaction model can be used directly to calculate or predict subsidence. For this special case, ignoring the behavior and influence of overburden material is not critical.

P. W. Kasameyer (written commun., 1979) of Lawrence Livermore Laboratory has pointed out a problem in the way the viscoelastic model has been applied to field data. By simulating appropriate U.S. Geological Survey field data (Poland and others, 1975) that were gathered from several sites in California's San Joaquin Valley, Corapcioglu and Brutsaert (1977, Table 2) estimated values for the three viscoelastic coefficients used in their model. Two coefficients are for elastic properties and one is for viscous properties. From these tabulated values, Kasameyer found that time constants for the different sites range from one month to a maximum of one year. Such values cannot account for nonrecoverable compaction that continues over several years when previous maximum stress is not exceeded. This multiyear phenomenon is observed in the very field data simulated by the viscoelastic model. By such simulation, Corapcioglu and Brutsaert (1977) have inadvertently calculated several times more than the total ultimate strain possible according to their tabulated coefficient values. The problem might be avoided conceptually by adding a fourth viscoelastic coefficient that would allow an investigator to find an appropriately longer time constant. It may, however, also raise new practical problems, such as how many years of field data are required to estimate the value of a hypothetical fourth coefficient.

An alternative model, the aquitard-drainage model, has been used to simulate some of the identical field data from California that were simulated with the viscoelastic reservoir model discussed above. The resulting coefficient values (Helm, 1978b, Table 1) yield time constants that range from several years to several decades and even to centuries (Helm, 1978b, Table 2).

The aquitard-drainage model (Tolman and Poland, 1940; Riley, 1970; Helm, 1972, 1975, 1976) is essentially a nonlinear modification of Terzaghi's (1925) theory of one-dimensional consolidation. The reservoir (confined aquifer system) is represented (Fig. 4) as containing two basic types of porous material: (1) a group of fine-grained interbeds, each of which is completely surrounded by a connected system of (2) coarse-grained material. The fine-grained interbeds (aquitards) are considered much less permeable than the interconnected coarse-grained portion (aquifer) of a reservoir (confined aquifer system). Within a reservoir, such idealized slow-draining interbeds are conceptually distinct from caprock, a confining bed, or a semiconfining bed that serves as a reservoir's upper boundary and hydraulically separates one aquifer system from another. An aquitard does not serve as a hydraulic separator. The aquitard-drainage model conceptually attributes the observed time lag (of compaction response to stress change) to the vertical component of fluid flow from one idealized material (aquitard) to another (aquifer) within the system itself. The slow vertical drainage from highly compressible aquitards to less compressible aquifer material serves a similar function in this model that a viscous "dashpot" served in the previously described viscoelastic reservoir model.

The aquitard-drainage model is limited conceptually to the vertical component of behavior. When field measurements are consistent with this restriction, the model simulates observed field compaction

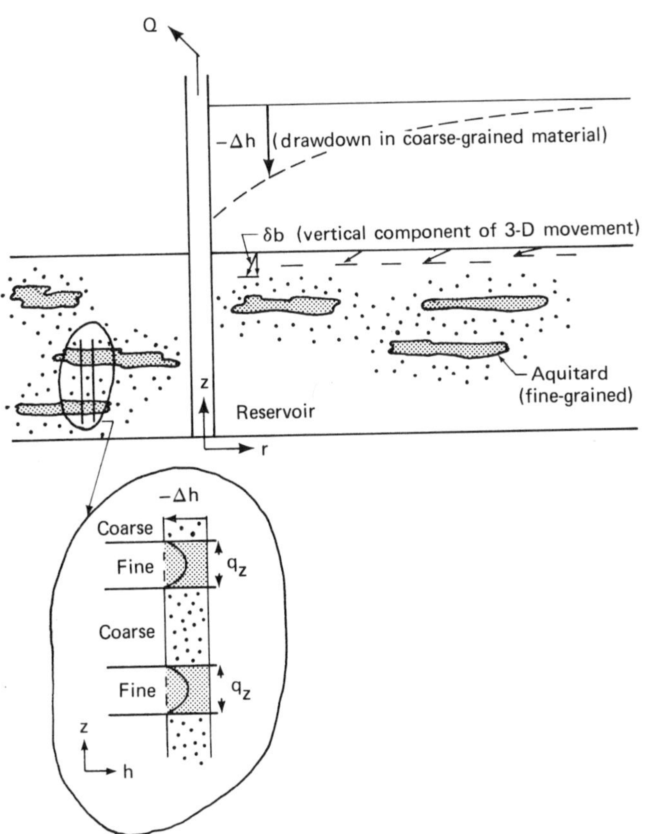

Figure 4. Aquitard-drainage model.

and expansion extremely well (Helm, 1977). A U.S. Geological Survey team under the guidance of J.F. Poland has developed a method of measuring vertical compaction and expansion of an aquifer system (Lofgren, 1970; Riley, 1970) as well as concurrent time-dependent changes of hydraulic head within coarse-grained members. As reported by Witherspoon and Freeze (1972, p. 23), the aquitard-drainage model was initially part of the U.S. Geological Survey's successful development of a field method (Tolman and Poland, 1940; Riley, 1970; Helm, 1972, 1975, 1976) for quantitatively estimating vertical components of confined aquifer-system parameters. These values are aggregate or average values for an entire system. In conjunction with appropriate field data, the model has been used (Helm, 1978b) to predict (1) residual nonrecoverable compaction within a system; (2) time-dependent in-situ preconsolidation stress (namely, a critical depth to water at which nonrecoverable compaction is stopped during the unloading phase and triggered during the reloading phase of a specified unloading-reloading cycle); and (3) a time-constant τ for a confined system at a site of interest. According to Terzaghi's theory of consolidation, τ is the length of time required for initially unstressed aquitards to reach 93% nonrecoverable compaction if water levels in adjacent aquifers (of a confined system) are instantaneously lowered a specified amount and then held constant.

Although the aquitard-drainage model considers only vertical components of behavior, it does not require the nonexistence of horizontal components. It could be coupled to a model of horizontal compression if three-dimensional analysis were of interest. However, the aquitard-drainage model has in actuality been coupled to a horizontal fluid flow model (Gambolati and Freeze, 1973) and a three-dimensional fluid flow model (Narasimhan and Witherspoon, 1977) without incorporating any theoretical model or computational capability for horizontal compression. Such partial coupling may contradict nature's processes. Trying to match field data in three dimensions to one such partially coupled model (Gambolati and Freeze, 1973) required uncoupling (Gambolati and others, 1974) before a successful match could be made, thus defeating the initial intent of the investigators.

The aquitard-drainage model has so far been applied to field data most successfully (Helm, 1976, 1977; Pollard and others, 1979) when arbitrary water-level fluctuations within permeable beds are specified as boundary conditions on aquitards at a site of interest, rather than when a regional discharge rate Q is specified.

One direction of current subsidence research is to combine three-dimensional fluid flow and three-dimensional matrix movement (which includes skeletal compression and expansion) into a more comprehensive theory (for example, Garg and others, 1975; Lewis and Schrefler, 1978). As with any new focus of scientific inquiry, the problem lies not in a paucity of suggested ideas, but in a superabundance. Sandhu (1979) has summarized the state of the art in major constitutive relations, from more classical rheological models (Biot, 1941) to theories of mixtures and of interacting continua (Green and Naghdi, 1965; Crochet and Naghi, 1966; Tabbador and Little, 1971). At present, theories of mixtures seem most readily applicable to multiphase flow of liquids and gases. The special role that solids play as one phase of multiphase flow within "porous" bulk material has begun to be recognized (Gray and O'Neill, 1976; Helm, 1979a). For saturated porous material, Sandhu (1979) perceptively points out the importance and some conceptual consequences of regarding solids as submerged (or "buoyant") within a fluid. This is an important shift in perspective that may be basic to effectively distinguishing between steady and nonsteady (transient) fluid flow. For steady fluid flow, the solid matrix serves merely as a stationary conduit through which a liquid moves. For transient flow, it may become necessary to treat solids as particles of a moving viscous fluid with special properties.

PURPOSE OF THE PRESENT PAPER

There are two conceptual paths an investigator might follow to develop a more comprehensive three-dimensional theory. One is to couple stress and strain in three dimensions; the other is to couple velocity and force fields in three dimensions. The first path is more commonly followed; the second will be followed in the present paper. It is appropriate here to contrast the two paths and to point out some of the advantages of the second.

The milestones along the stress/strain procedural

path are not entirely straightforward. First, one derives a governing equation for the transient distribution of stress. For illustrative purposes, such governing equations are derived in Appendix I in terms of the hydraulic head h, the fluid pressure p within saturated pores, and the sum of the normal components of effective stress σ'_{ii}. Such a governing equation can often be solved by specifying appropriate initial and boundary conditions. Using the concept of specific storage (or volumetric compressibility), one can also solve for volume strain. Let us assume we are clever enough not to require small deformation theory. At this point the stress/strain path reaches a fork in the road. In fact, a confusion of possible routes lies ahead. Using a preselected theory (rheological model) of three-dimensional stress/strain, one must find the directional components of strain. For field applications, this means that Poisson's ratio or Lame's constants or some equivalent parameters must be known a priori at the scale of interest. Integrating the directional components of strain from a boundary of zero displacement to a point $(\underset{\sim}{P},t)$ of interest, one eventually can find the cumulative displacement $\underset{\sim}{u}_{cum\ s}(\underset{\sim}{P},t)$ of skeletal material at the specified point of interest. To use Narasimhan's (oral commun., 1977) phrase, the basic problem is to determine the "shape factor" of three-dimensional strain if one wishes to calculate or predict the displacement field of skeletal material in three dimensions.

The alternative procedural path proposed in this paper is to relate the velocity field for solids $\underset{\sim}{v}_s$ in three dimensions to appropriate force fields. If a reasonable governing equation can be found, one can solve for $\underset{\sim}{v}_s(\underset{\sim}{P},t)$ from specified initial and boundary conditions. The cumulative displacement field $\underset{\sim}{u}_{cum\ s}(\underset{\sim}{P},t)$ is calculated simply by integrating $\underset{\sim}{v}_s$ over an appropriate interval of time.

The purpose of the present paper is to develop governing equations from which transient $\underset{\sim}{v}_s$ can theoretically be solved from specified initial and boundary conditions. After the cumulative displacement field $\underset{\sim}{u}_{cum\ s}$ is found directly by integrating $\underset{\sim}{v}_s$ over time, finite volume strain can be calculated uniquely from the divergence of $\underset{\sim}{u}_{cum\ s}$. Using these calculated transient distributions of volume strain, a transient distribution of fluid pressure can then be found from an appropriate stress/strain relation. The stress/strain path and the velocity/force-field path may not actually be two paths, but rather a single conceptual path being traversed in opposite directions.

CONCEPTUALIZATION OF SKELETAL MOVEMENT

In the remainder of this paper, I shall mathematically analyze skeletal movement during transient conditions of fluid flow, with special emphasis on the influence of gravitational forces. My discussion is restricted to behavior within a saturated isotropic reservoir. The isotropy requirement will be modified somewhat at the end of the paper. Although the source of some fluid at a discharging well is unquestionably due to expansion of individual constituents of a mixture--for example, solids and interstitial water--such expansion does not generally account for the total mass of fluid withdrawn. Only that part of the extracted fluid mass which comes from reservoir compression (porosity loss) is considered a potential cause of subsidence due to fluid withdrawal. For simplicity of analysis in this paper, skeletal structure of an idealized isotropic reservoir will be considered much more "compressible" in the sense of porosity loss than individual constituents of the saturated mixture.

It is known from Darcy's law (Hubbert, 1940) that fluid is pushed past solid material at a relative rate $\underset{\sim}{q}$ by a force per unit mass proportional to the negative gradient of hydraulic head $\underset{\sim}{\nabla} h$,

$$\underset{\sim}{q} = -K\underset{\sim}{\nabla} h, \tag{1}$$

where for isotropic material, hydraulic conductivity K can be considered a coefficient of proportionality. During transient conditions of fluid flow, the skeletal matrix itself deforms and moves. One possible method of solving for the cumulative displacement field $\underset{\sim}{u}_{cum\ s}$ of moving solids is to relate the velocity field $\underset{\sim}{v}_s$ of solids quantitatively to the transient gradient of hydraulic head $\underset{\sim}{\nabla} h$, and then to solve the straightforward relation

$$\underset{\sim}{u}_{cum\ s} = \int_0^t \underset{\sim}{v}_s\, dt. \tag{2}$$

The question to be addressed is the relation of $\underset{\sim}{v}_s$ to $\underset{\sim}{\nabla} h$ during conditions of transient flow. To analyze nonzero $\underset{\sim}{v}_s$, I shall show that the dynamics of interest is that part of a force diagram that has not reached a state of equilibrium.

Figure 5 schematically shows an equilibrium force diagram (Taylor, 1948, p 201f) that is associated with ultimate or steady-state fluid flow $\underset{\sim}{q}_{ult}$, which implies that enough time has elapsed so that

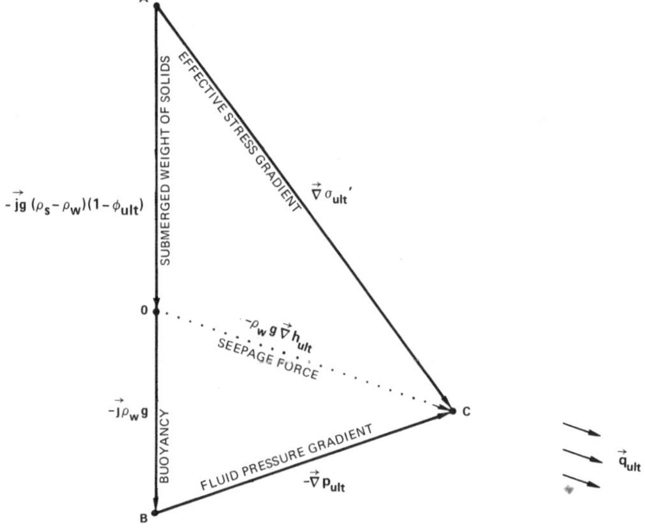

Figure 5. Force polygon (per unit bulk volume) of ultimate steady flow.

$\underset{\sim}{v}_s$ has been attenuated to zero. The lower triangle OBC in Figure 5 pertains to force intensities within the interstitial liquid. The upper triangle OAC pertains to force intensities within submerged solids. Force triangle AOC is not intended to show all body forces on the solids, but rather those forces that are affected by the presence and flow of water. Remember that, as represented in Figure 5, forces are in equilibrium and solids are not moving.

For saturated isotropic material, Darcy's law becomes

$$\underset{\sim}{q}_{ult} = -K\underset{\sim}{\nabla}h_{ult}, \tag{3}$$

where h_{ult} is the ultimate steady-state value of hydraulic head h. In Figure 5, vector $\underset{\sim}{AC}$ represents $\underset{\sim}{\nabla}\sigma_{ult}$, which is a resultant body force on the submerged granular skeleton per unit bulk volume; $\underset{\sim}{OC}$ represents $-\rho_w g\underset{\sim}{\nabla}h_{ult}$, which is a viscous drag or seepage force on skeletal material per unit bulk volume, where ρ_w is density of the interstitial fluid (water) and g is gravitational acceleration; and $\underset{\sim}{BC}$ represents $-\underset{\sim}{\nabla}p_{ult}$, which is a fluid-pressure force per unit bulk volume, or the resultant force intensity due to all fluid pressures during steady-state fluid flow. Gravitational body force intensity $\underset{\sim}{AB}$ in Figure 5 can be expressed by $-\underset{\sim}{j}g[\rho_w\phi_{ult} + \rho_s(1 - \phi_{ult})]$, which is the total weight per unit volume of solids and fluid within a saturated bulk volume. The final equilibrium porosity of an an elemental bulk volume is expressed by the term ϕ_{ult}, density of the solid constituent by ρ_s, and a unit vertical vector by $\underset{\sim}{j}$ (positive upward). Gravitational body force $\underset{\sim}{AO}$ in Figure 5 represents $-\underset{\sim}{j}g(\rho_s - \rho_w)(1 - \phi_{ult})$, which is the submerged (buoyant) weight of solids per unit bulk volume. Vector $\underset{\sim}{OB}$ represents $-\underset{\sim}{j}\rho_w g$, which is associated with the weight that a bulk volume would have if entirely occupied by water. The upward body force $\underset{\sim}{j}\rho_w g$, is called the buoyancy (Taylor, 1948, p. 202).

Hydrologists (Hubbert, 1940) have traditionally been interested in the relation of fluid pressure p and hydraulic head h. The general relation between p and h can be expressed as

$$\int_{p_o}^{p} \frac{1}{\rho_w} d\pi = g(h - z), \tag{4}$$

where π is a pressure variable of integration, p_o is an arbitrary reference pressure, and z (increasing upward) is the elevation of a point of interest. Standard force fields per unit volume that directly pertain to equation 4 for steady-state fluid flow are indicated in Figure 5 by triangle OBC.

It is important to emphasize a distinction between hydraulic head h of equation 4 and total energy Θ per unit mass in the liquid phase. By definition (Hubbert, 1940, p. 793), h includes pressure head and elevation head, as is indicated in equation 4, but excludes any component due to kinetic energy $v^2/2$. Total energy Θ includes a kinetic energy term. Only for the special case of negligibly small kinetic energy can gh be set equal to Θ (Hubbert, 1940, p. 802).

Soil engineers have traditionally been interested in the relation of fluid pressure p and effective stress σ', whose related force fields per unit volume are indicated for steady flow in Figure 5 by triangle ABC. Terzaghi's principle of effective stress can be expressed as

$$\sigma'_{ij} = \sigma_{ij} - \delta_{ij}p \quad (i,j = 1,2,3), \tag{5}$$

where σ_{ij} is a tensor representing total load and δ_{ij} is the Kronecker delta ($\delta_{ij} = 1$ for $i = j$; $\delta_{ij} = 0$ for $i \neq j$). In accordance with equation 5, mean normal effective stress $\sigma'_{ii}/3$ (an invariant) can be expressed by

$$\sigma'_{ii}/3 = \sigma_{ii}/3 - p \quad (i = 1,2,3), \tag{5a}$$

where double subscripts ii without parentheses indicate summation. The term σ' as used in this paper is appropriately restricted to denote the vertical normal component of the effective stress tensor (σ'_{ij} of equation 5), namely

$$\sigma' \equiv \sigma'_{(zz)} = \sigma'_{(zz)} - p. \tag{5b}$$

The term σ'_{ult} of Figure 5 is the ultinate value of transient σ' in equation 5b and in this paper is associated with conditions of steady flow.

A flux term for saturated bulk flow, $\underset{\sim}{q}_{bulk}$, can be specified in terms of velocity fields of the two constituents (water, $\underset{\sim}{v}_w$ and solids, $\underset{\sim}{v}_s$) by

$$\underset{\sim}{q}_{bulk} = \phi\underset{\sim}{v}_w + (1 - \phi)\underset{\sim}{v}_s, \tag{6}$$

where ϕ is porosity and $\underset{\sim}{v}_s$ is nonzero for transient flow. Relative velocity $\underset{\sim}{q}$ (specific discharge) of equation 1 can be defined either by

$$\underset{\sim}{q} = \underset{\sim}{q}_{bulk} - \underset{\sim}{v}_s \tag{7}$$

or by

$$\underset{\sim}{q} = \phi(\underset{\sim}{v}_w - \underset{\sim}{v}_s). \tag{8}$$

One can substitute either of the two foregoing equations into equation 6 to get the other. Equation 8 (Gersevanov, 1937; Verruijt, 1969) is the more commonly used.

Figure 6 shows schematically the vectorial relations of equations 6, 7, and 8. In Figure 6, vector $\underset{\sim}{O'E}$ represents $\underset{\sim}{q}_{bulk}$ of equation 6, vector $\underset{\sim}{O'(F)}$

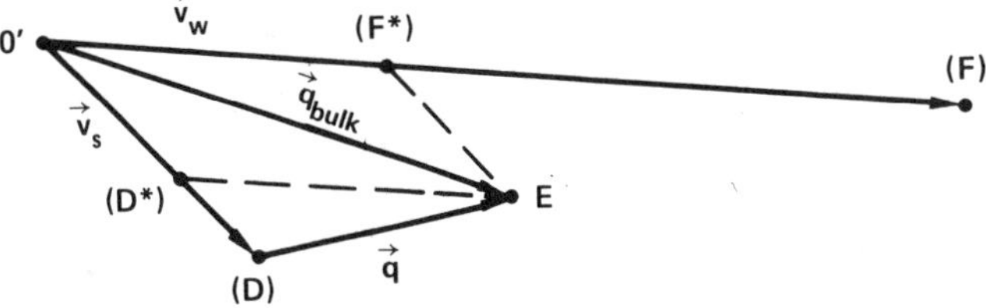

Figure 6. Velocity polygon of transient conditions of fluid flow through compressible isotropic and saturated porous material (with incompressible constituents).

represents $\underset{\sim}{v}_w$ vector $O'(D)$ represents $\underset{\sim}{v}_s$, and vector $(D)E$ represents q of equations 7 and 8. Furthermore, vector $O'(F^*)$ represents $\phi \underset{\sim}{v}_w$ of equation 6, and $O'(D^*)$ represents $(1 - \phi)\underset{\sim}{v}_s$ of equation 6. Parentheses in Figure 6 emphasize that points so designated are transient. Vectors associated with such points (Fig. 6) vary greatly with respect to time.

The bulk of this paper is an attempt to relate transient velocities to equilibrium and, more importantly, to nonequilibrium forces. Such relations are essential for associating the magnitude and direction of transient skeletal movement $\underset{\sim}{v}_s$ (Fig. 6) to existing steady-state theory (Fig. 5). Before making this attempt, however, we must lay some conceptual ground rules. Vector diagrams will be used to illustrate schematically the thrust of fundamental mathematical and conceptual arguments.

Skeletal movement $\underset{\sim}{v}_s$ is related to skeletal volume strain ε_{sk} by

$$d_P \varepsilon_{sk}/dt = \underset{\sim}{\nabla}_P \cdot \underset{\sim}{v}_s. \tag{9}$$

The subscript P specifies a point of interest where evaluation of any equation is to take place. If $\varepsilon_{sk(ii)}$ is the component of ε_{sk} in the ith direction, then skeletal volume strain is the invariant sum

$$\varepsilon_{sk} \equiv \varepsilon_{sk\ ii} \quad (i = 1,2,3), \tag{10}$$

where double subscripts without parentheses indicate summation.

An important step in combining velocity and force fields is to find a common reference frame. The vertical coordinate frame has already been tacitly specified by the z term in equation 4. Physical specifications, expressed as constitutive assumptions, inherently dictate the reference frame to be used. Triangle OBC in Figure 5 is based on equation 4 in that vector $\underset{\sim}{OB}$ represents the unit weight of water. By assuming the constitutive relation

$$\rho_w = \text{constant}, \tag{11a}$$

it follows that the unit weight of water $-\underset{\sim}{j}\rho_w g$ (Fig. 5) is a function of neither location nor time. Because the compressibility of water is generally greater than the compressibility of individual grains, equation 11a implies

$$\rho_s = \text{constant}. \tag{11b}$$

For saturated porous material, equations 11a and 11b specify incompressible constituents that, along with equation 4, tacitly require a nondeforming reference frame. If one wishes to introduce a deforming reference frame (for example, Cooper, 1966) the mathematical effect of such a frame on equations 4, 11a, and 11b must be spelled out. The constitutive equations and the definition of length z must be correspondingly altered. Unfortunately, mathematical hydrogeologists have tended to overlook this conceptual imperative. Because a nondeforming frame is used in the present paper, equations 4, 11a, and 11b can and will remain unaltered.

Specifying a reference frame, as was done above, is distinct from specifying a point of interest $\underset{\sim}{P}$ (Fig. 7), or an elemental volume of interest whose center of mass is located at $\underset{\sim}{P}$. For transient flow through saturated porous media, an elemental volume can be fixed in space ($\underset{\sim}{P} = \underset{\sim}{P}_o$), fixed in a solid continuum ($\underset{\sim}{P} = \underset{\sim}{P}_s$), or fixed in a liquid continuum ($\underset{\sim}{P} = \underset{\sim}{P}_w$). An elemental volume V_{P_o} fixed in space does not move; that is,

$$\underset{\sim}{v}_{P_o} = d\underset{\sim}{P}_o/dt = 0. \tag{12a}$$

Such a bulk volume does not deform, although porosity changes within it. Nor, during transient flow, does it contain the same solid or liquid material throughout time. In other words, matter (solid and liquid) flows through this abstract element V_{P_o}.

An elemental volume V_{P_s} fixed in the solid constituent contains the same solids throughout any time interval and moves with these solids at a rate

$$\underset{\sim}{v}_{P_s} \equiv d\underset{\sim}{P}_s/dt = \underset{\sim}{v}_s(\underset{\sim}{P}_s,t), \tag{12b}$$

where $\underset{\sim}{v}_s$ is a volume-average velocity term. Liquid, however, flows past grains and hence flows through bulk volume element V_{P_s}.

A third elemental volume V_{P_w} fixed in the liquid constituent contains the same liquid particles throughout any time interval, and moves with the liquid at a rate

$$\underset{\sim}{v}_{P_w} = d\underset{\sim}{P}_w/dt = \underset{\sim}{v}_w(\underset{\sim}{P}_w,t), \tag{12c}$$

where $\underset{\sim}{v}_w$ is a volume-average velocity term. Because liquid can flow past grains, the grains within (being passed by) bulk volume V_{P_w} need not remain the same.

Note (Fig. 7) that in the limit as $V_{P_s} \to 0$, the volume-average velocity term $\underset{\sim}{v}_{P_s}$ equals the flux rate of an abstract "particle" of a solid continuum. Similarly, for a small enough value of V_{P_w}, $\underset{\sim}{v}_{P_w}$ can be treated as the flux rate of an abstract "particle" of a liquid continuum.

The total derivative d_P/dt of a function $f[= f(\underset{\sim}{P},t)]$ is defined by

$$(d_P/dt)f = (\partial_P/\partial t + d\underset{\sim}{P}/dt \cdot \underset{\sim}{\nabla}_P)f \quad f = f(\underset{\sim}{P},t), \tag{13}$$

where $\partial_P/\partial t$ is a local derivative at $(\underset{\sim}{P}/\partial t)$. The subscript P once again emphasizes that one must that one must specify the location $(\underset{\sim}{P},t)$ where a transient function or equation is to be evaluated. According to convention (Mase, 1970, p. 111), the $d\underset{\sim}{P}/dt$ term of equation 13 specifies the rate of movement of a particle fixed within a specific continuum. Although equations 12b and 12c represent the movement of different materials, they are consistent with this standard interpretation of $d\underset{\sim}{P}/dt$. Equation 12a is very different; although it is not associated with any material continuum,

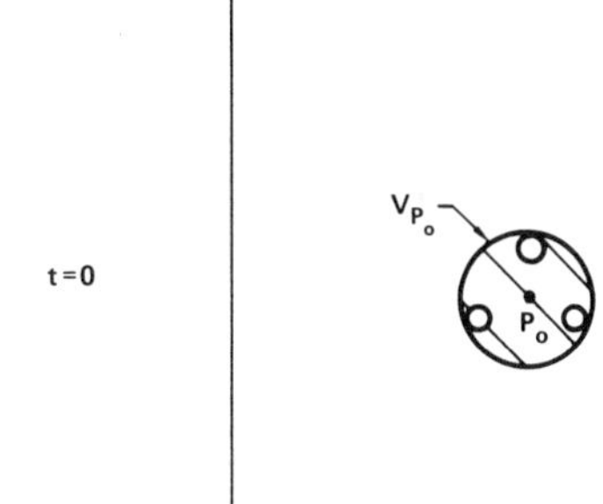

t>0

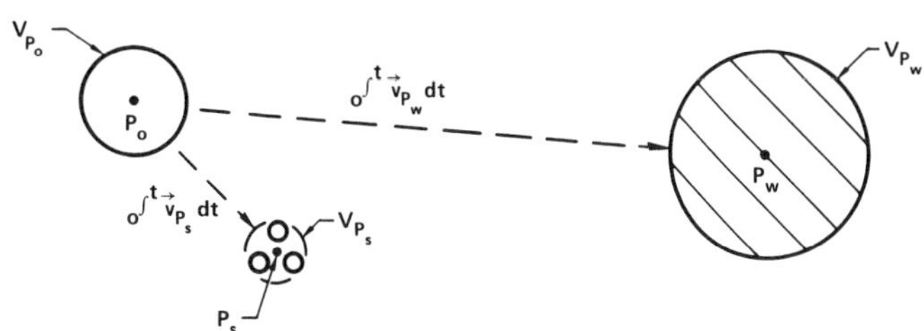

V_{P_w}

P_w

Figure 7. Schematic diagram of transient flow showing points $\underset{\sim}{P}_o$ (fixed in space), $\underset{\sim}{P}_s$ (fixed) in solid continuum), and $\underset{\sim}{P}_w$ (fixed in liquid continuum).

it is a physically valid and mathematically rigorous interpretation of $d\underset{\sim}{f}/dt$ of equation 13.

Substituting equations 12a, 12b, and 12c into equation 13 gives

$$d_{P_O}\underset{\sim}{u}_{cum\ s}/dt = \partial_{P_O}\underset{\sim}{u}_{cum\ s}/\partial t, \tag{13a}$$

$$d_{P_s}\underset{\sim}{u}_{cum\ s}/dt = \partial_{P_s}\underset{\sim}{u}_{cum\ s}/\partial t + (\underset{\sim}{v}_s(\underset{\sim}{P}_s,t)\cdot\underset{\sim}{\nabla}_{P_s})\underset{\sim}{u}_{cum\ s}, \tag{13b}$$

$$d_{P_w}\underset{\sim}{u}_{cum\ s}/dt = \partial_{P_w}\underset{\sim}{u}_{cum\ s}/\partial t + (\underset{\sim}{v}_w(\underset{\sim}{P}_w,t)\cdot\underset{\sim}{\nabla}_{P_s})\underset{\sim}{u}_{cum\ s}, \tag{13c}$$

where f in equation 13 is the cumulative displacement field of solids $\underset{\sim}{u}_{cum\ s}$. In accordance with Figure 7 and the present discussion, $\underset{\sim}{P}$ of equation 13 is $\underset{\sim}{P}_O$ in equation 13a, $\underset{\sim}{P}_s$ in 13b, and $\underset{\sim}{P}_w$ in 13c. Equation 13a has far-reaching conceptual consequences.

The skeletal strain rate of porous material at $\underset{\sim}{P}_s$ can be expressed by

$$\frac{d_{P_s}\varepsilon_{sk}}{dt} = \lim_{V_{P_s}\to 0}\frac{1}{V_{P_s}}\left[\frac{d_{P_s}V_{P_s}}{dt}\right] . \tag{14}$$

Equation 14 represents the strain rate of an idealized solid continuum, but not of a solid constituent (that is, not of individual solid grains).

Because bulk volume V_{P_O} is fixed in space, it does not change with respect to time in terms of a nondeforming frame. We can further specify that this is true regardless of the size of V_{P_O}, which means

$$\lim_{V_{P_O}\to 0}\frac{1}{V_{P_O}}\frac{d_{P_O}V_{P_O}}{dt} = 0 . \tag{15a}$$

Combining equations 9 and 14 for a bulk volume element V_{P_S} fixed in a solid continuum gives

$$\lim_{V_{P_s}\to 0}\frac{1}{V_{P_s}}\left[\frac{d_{P_s}V_{P_s}}{dt}\right] = \underset{\sim}{\nabla}_{P_s}\ \underset{\sim}{v}_s(P_s,t) , \tag{15b}$$

where the center of mass of V_{P_S} is located at $\underset{\sim}{P}_s$ (Fig. 7). Even though the following expression for the divergence of a liquid continuum need not be used in this paper, for completeness of development we write for a bulk volume element V_{Pw}, fixed in an idealized liquid continuum, the relation

$$\lim_{V_{P_w}\to 0} \frac{1}{V_{P_w}} d_{P_w} V_{P_w}/dt = \nabla_{P_w} \cdot v_w(P_w,t), \qquad (15c)$$

where the center of mass of V_{P_w} is located at P_w (Fig. 7). In general, equation 15c expresses divergence of a liquid continuum due to both porosity change and expansion of water itself. Because we assume that the constituent materials do not deform (see equations 11a and 11b), matrix deformation (equations 15b and 15c) is affected only by porosity change.

A general distinction between continuum mechanics and what I shall call reference-frame mechanics will now be introduced. Assumptions underlying standard continuum mechanics pertain directly to equations 15b and 15c, but do not pertain to equation 15a in the same way. A point P_o is specified to be fixed in abstract space in equation 15a. Continuum mechanics requires a "particle" (or point) of interest to be fixed in a specified continuum. This is done for a solid continuum by specifying P_s in equation 15b and for a liquid continuum by specifying P in equation 15c. If one chooses to apply continuum mechanics, solid grains within a finite volume element V_{P_s} are replaced by an abstract continuum that is assumed to be present everywhere. In addition, interstitial water within a finite volume element V_{P_s} can be replaced by another abstract continuum that is also assumed to be everywhere present. However, neither of these two abstract continua serves as a conceptual backdrop for equation 15a. Specification of an abstract space by constitutive relations 4, 11a, and 11b performs a similar role for equation 15a that specifying one continuum does for equation 15b and specifying another does for equation 15c. Regardless of one's reference frame, points P_o, P_s, and P_w must be kept conceptually distinct, as will be done in the following section. Otherwise, one may unwittingly introduce fundamental mathematical errors (see Appendix I). Conceptual ground rules have now been laid.

MASS BALANCE AND TRANSIENT FLOW

A general mass balance equation can now be expressed:

$$\frac{d_P}{dt} \int_{V_P} \rho^\alpha dV + \int_{S_P} \rho^\alpha (v - v_P) \cdot n dS = \int_{V_P} \dot{m} dV, \qquad (16)$$

where $\rho^\alpha(\alpha = s, w)$ is a partial density term that, for a two-constituent saturated mixture, can be expressed

$$\rho^s = (1 - \phi)\rho_s, \qquad (17)$$

$$\rho^w = \phi\rho_w, \qquad (18)$$

and n is a unit outward normal to the surface S that bounds bulk volume V_P whose center of mass is at P. Throughout the remainder of this section, we shall assume no rate of mass production $\dot{m}$ inside V_P. In other words, the right-hand side of equation 16 equals zero. If the bulk volume element is appropriately small such that ϕ can be treated as an average value over V_P, then equation 16 can be expanded for the solid and liquid phases

$$-V_P \frac{d_P}{dt}\phi + (1 - \phi_P)\frac{d_P}{dt}V_P + \int_{S_P} (1 - \phi)(v_s - v_P) \cdot n dS = 0, \qquad (19a)$$

$$V_P \frac{d_P}{dt}\phi + \phi_P \frac{d_P}{dt} V_P + \int_{S_P} \phi(v_w - v_P) \cdot n ds = 0, \qquad (19b)$$

where we have used equations 11a and 11b. Following Sith's (1971) method of development, we add equations 19a and 19b to yield

$$\frac{1}{V_P}\frac{d_P}{dt}V_P + \frac{1}{V_P}\int_{S_P} (q_{bulk} - v_P) \cdot n dS = 0, \qquad (20)$$

where we have divided through by V_P and used equation 6. By making V_P arbitrarily small, the divergence theorem can be applied to the surface integral in equation 20 so that

$$\lim_{V_P\to 0} \frac{1}{V_P}\frac{d_P}{dt}V_P + \nabla_P \cdot (q_{bulk} - v_P) = 0. \qquad (21)$$

For bulk volumes V_{P_o}, V_{P_s}, and V_{P_w},-which are associated respectively with point velocities V_{P_o} and v_{P_s}, equation 21 becomes

$$\nabla_{P_o} \cdot q_{bulk} = 0 \qquad (22a)$$

$$\nabla_{P_s} \cdot q_{bulk} = 0 \qquad (22b)$$

$$\nabla_{P_w} \cdot q_{bulk} = 0, \qquad (22c)$$

where we have used equation sets 12 and 15. Hence the equation

$$\nabla_P \cdot q_{bulk} = 0 \qquad (22)$$

is an expression of steady-state bulk flow and is valid everywhere within a system whose individual constituents are much less compressible than the skeletal structure (porosity) itself. If one wished to include expansion of individual constituents, appropriate terms would appear in the right-hand side of equation 22. The importance of equation 22 should be emphasized. It is applicable to bulk flow of a saturated ground-water system, even during transient conditions of fluid flow.

Most mathematical hydrogeologists (for example, Polubarinova-Kochina, 1952; Cooley, 1975) do not follow Smith's (1971) insightful development of adding equations 19a and 19b to get equation 20, which eventually leads to equation 22. In addition, physical events occurring at $\underset{\sim}{p}_o$, $\underset{\sim}{p}_s$, and $\underset{\sim}{p}_w$ have not generally been kept distinct. A critique of the more standard mathematical development is given in Appendix I. The significant role of gravitational forces during transient flow is also discussed in Appendix I.

DARCIAN EXPRESSION OF $\underset{\sim}{v}_s$

For solving the problem of skeletal deformation, we shall introduce a new approach for calculating the magnitude and direction of $\underset{\sim}{v}_s$. Combining equations 1 and 7, we find

$$\underset{\sim}{v}_s = \underset{\sim}{q}_{bulk} + K\underset{\sim}{\nabla}h. \tag{23}$$

Mechanical energy per unit mass is required within a fluid to move bulk material q_{bulk}. Using Helm's (1979a) postulate of viscous bulk movement, a Darcian type of expression can be applied to $\underset{\sim}{q}_{bulk}$, namely

$$\underset{\sim}{q}_{bulk} = -(K/g)\underset{\sim}{\nabla}_p\Theta. \tag{24}$$

The gradient of Θ in equation 24 is a fundamental impelling force on a mixture of saturated material. Detailed analysis of $\underset{\sim}{\nabla}_p\Theta$ is deferred to a discussion of secondary effects in Appendix II.

It is now possible to relate $\underset{\sim}{v}s$ to a force field per unit mass, namely

$$\underset{\sim}{v}_s = -K\underset{\sim}{\nabla}_p(\Theta/g - h), \tag{25}$$

where we have combined equations 23 and 24. Invoking equation 1 implies that equation 25 pertains to the component of skeletal movement that is directly related to flow of interstitial water past grains. Equation 25 thereby pertains to what will be called the primary component of skeletal movement in Appendix II. Correspondingly, the gradient $\underset{\sim}{\nabla}_p\Theta$ as used in equation 25 can be interpreted to equal $\underset{\sim}{\nabla}_p\Theta_w$ of equation 6** in Appendix II. Until sedimentary space is introduced in a later section of this paper, we limit our discussion to the primary component of skeletal movement.

For completeness of presentation, we write $\underset{\sim}{v}_w$ in terms of an appropriate force field. Combining equations 1, 8, and 25 for isotropic saturated material gives

$$\underset{\sim}{v}_w = -K\left(\underset{\sim}{\nabla}_p \frac{\Theta}{g} + \frac{1-\phi}{\phi}\underset{\sim}{\nabla}_p h\right). \tag{25a}$$

DIFFUSION EQUATION IN TERMS OF $\underset{\sim}{v}_s$

Although $\underset{\sim}{u}_{cum\ s}$ can be evaluated at any point $\underset{\sim}{p}$ of interest from $\underset{\sim}{v}_s$ in accordance with equation 2, an equation for evaluating $\underset{\sim}{v}_s$ itself has not yet been developed. Although at first glance equation 18** of Appendix II might appear to be such an equation, its calculation presupposes an even more fundamental equation for evaluating both transient and steady-state distributions of hydraulic head h. In light of serious questions raised in Appendix I regarding unqualified attempts to apply standard diffusion equations to the dynamics of skeletal displacement due to fluid withdrawal, we are faced with the challenging problem of how to find $\underset{\sim}{v}_s$.

For a radial distance far from a producing well, the rate of skeletal movement toward the well is a fairly straightforward problem to solve. A sample solution, though approximate, is given in Appendix IV. Nearer to a producing well, movement of solids radially inward is complicated by boundary constraints constraints that cause a zone of skeletal compression to develop within a confined system. Our remaining task is to derive a governing equation for $\underset{\sim}{v}_s$ that is applicable to skeletal movement both within a zone of skeletal compression near a producing well, and within a surrounding, more distant zone of skeletal extension.

A first step in such a derivation is to make a clear distinction between instantaneous and cumulative displacement fields.

Instantaneous Displacement of Skeletal Material

Previous attempts to estimate the cumulative displacement field $\underset{\sim}{u}_{cum\ s}$ of the skeletal matrix have made use of a displacement-strain relation such as

$$\underset{\sim}{u}_{cum\ s}(\underset{\sim}{p},t) \overset{(?)}{=} C\int_o^p \varepsilon_{sk}\,d\underset{\sim}{x}, \tag{26}$$

where

$$\underset{\sim}{u}_{cum\ s} = (u_{cum\ s\ 1}, u_{cum\ s\ 2}, u_{cum\ s\ 3}), \tag{26a}$$

$$\underset{\sim}{x} = (x_1, x_2, x_3). \tag{26b}$$

The coefficient C is a type of shape factor for cumulative volume strain ε_{sk} that makes equation 26 very complicated for three-dimensional analysis, even when a appropriate strain term can be calculated. Because volume strain is generally evaluated according to small deformation theory, it usually represents infinitesimal rather than cumulative strain. As a result, whenever small deformation theory is invoked, equation 26 is inappropriate. It should be replaced by a relation such as

$$\underset{\sim}{u}_s(\underset{\sim}{p},t) = C'\int_o^p \varepsilon'_{sk}\,d\underset{\sim}{x}, \tag{27}$$

where $\underset{\sim}{u}_s$ represents small or instantaneous displacement and e'_{sk} represents infinitesimal strain.

Even assuming C' can be found, solving equation 27 for $\underset{\sim}{u}_s$ does not directly give a value for $\underset{\sim}{u}_{cum\ s}$ of equation 26. To emphasize this distinction further, $\underset{\sim}{u}_s$ can be expressed in terms of $\underset{\sim}{v}_s$ by

$$\underset{\sim}{u}_s(\underset{\sim}{p},t) = \lim_{\delta t \to 0} \int_{t-\delta t}^{t+\delta t} \underset{\sim}{v}_s\,dt, \tag{28}$$

which contrasts to the more important $\underset{\sim}{u}_{cum\ s}$ $(\underset{\sim}{P},t)$ of equation 2.

Rate of skeletal displacement $\underset{\sim}{v}_s(P,t)$ is itself an instantaneous concept. It can, however, be expressed as a time derivative of either $\underset{\sim}{u}_{cum\ s}$ or $\underset{\sim}{u}_s$, namely

$$\underset{\sim}{v}_s(\underset{\sim}{P},t) = \lim_{dt\to 0} d_P\underset{\sim}{u}_{cum\ s}/dt, \tag{29a}$$

$$\underset{\sim}{v}_s(\underset{\sim}{P},t) = \lim_{dt\to 0} d_P\underset{\sim}{u}_s/dt. \tag{29b}$$

Although the rates at which $\underset{\sim}{u}_{cum\ s}$ and $\underset{\sim}{u}_s$ change with respect to time are identical at any specified point $\underset{\sim}{P}$ and instant t, the two displacement fields (cumulative and instantaneous) themselves are not equal. Note in passing that according to equation 29a, $\underset{\sim}{v}_s$ $(\underset{\sim}{P},t)$ can be legitimately applied to the left-hand sides of equations 13a, 13b, and 13c for specified points $(\underset{\sim}{P}_t,t)$, and $(\underset{\sim}{P}_w,t)$. Applying equation 29a to equation 13a will be a very important step in a subsequent section.

Basic Equations

Three new diffusion equations will be developed, one with $\underset{\sim}{u}_{cum\ s}$, another with $\underset{\sim}{v}_s$, and a third with $\tilde{q}$ as unknowns. In this section it is necessary to use concepts of externally applied stress σ_a and total gravitational stress σ_{gz} as developed in Appendix I. Equations numbered with an asterisk are found in Appendix I. In accordance with equations 4, 5b, 11*, and 12b*, we write

$$-\underset{\sim}{\nabla}_P h = \frac{1}{\rho_w g} \underset{\sim}{\nabla}_P\sigma' + \underset{\sim}{j}(G-1)(1-\phi) - \frac{1}{\rho_w g} \underset{\sim}{\nabla}_P\sigma_{(zz)\ a}, \tag{30}$$

where $\sigma_{(zz)\ a}$ is the vertical normal component of applied stress σ_a of equation 11*. The term σ' is defined by equation 5b as the vertical normal component of the effective stress tensor σ'_{ij}. For the sake of brevity, σ' will be referred to simply as effective stress. If $\underset{\sim}{\nabla}_P\sigma_{(zz)\ a}$ is assumed to be negligibly small, equation 30 simplifies to

$$\underset{\sim}{\nabla}_P\sigma' + \rho_w g\underset{\sim}{\nabla}_P h + \underset{\sim}{j}g(\rho_s - p_w)(1-\phi) = 0, \tag{30a}$$

which is shown schematically in Figure 8.

Figure 8 is a transient equivalent of Figure 5. It illustrates schematically the transient relation of effective force $\underset{\sim}{\nabla}\sigma'$ [vector $\widetilde{(H)(G)}$], the seepage seepage force $\underset{\sim}{\nabla}h$ [see vector $0\widetilde{(G)}$], and the submerged weight of solids [vector $\widetilde{(H)0}$]. Parentheses in Figure 8 indicate transient points. Equation 30a as written and Figure 8 as sketched are not intended to suggest that transient forces are in equilibrium. Equation 4 expresses the relation of observable hydraulic head h to transient fluid pressure p. The terms $\underset{\sim}{\nabla}\sigma'$ and $\underset{\sim}{\Delta}h$ as used in equation 30a and in and in Figure 8 indicate resultant forces during transient flow. They are not necessarily forces that

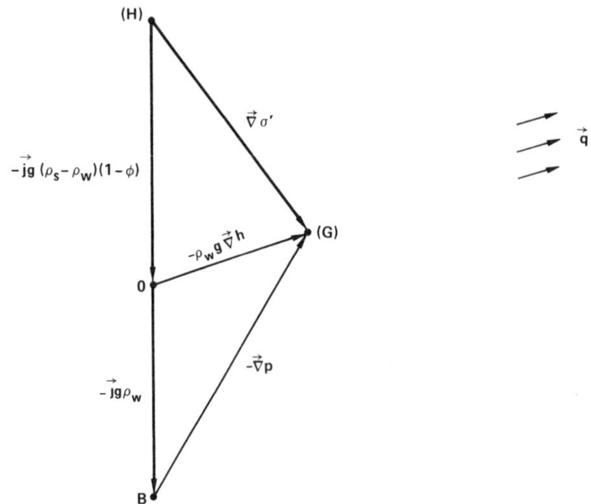

Figure 8. Force polygon of transient flow showing a nonequilibrium state of $\underset{\sim}{\nabla}\sigma'$.

impel a flow system toward an ultimate equilibrium state. Seepage force $\underset{\sim}{\nabla}h$ can best be interpreted as as an interactive force between water and solids. It represents a viscous resistance that solids exert on the flow of water.

An appropriate stress/strain relation is

$$(\rho_w g/S_{sk})\ \underset{\sim}{\nabla}_P\varepsilon_{sk} = -\underset{\sim}{\nabla}_P(\sigma' - \sigma'_{ref}), \tag{31}$$

where $\underset{\sim}{\nabla}_P\sigma'_{ref}$ represents effective force under an initial $(t < 0)$ reference condition of zero strain.

One stable initial reference condition is met when $\underset{\sim}{\nabla}_P\sigma'_{ref}$ is taken to equal the initial weight of submerged solids within a unit volume, namely

$$\underset{\sim}{\nabla}_P\sigma'_{ref} = -\underset{\sim}{j}g(\rho_s - \rho_w)(1-\phi_{initial}), \tag{32a}$$

which implies

$$\underset{\sim}{\nabla}_P p_{ref} = -\underset{\sim}{j}\rho_w g \tag{32b}$$

for an initial $(t<0)$ no-flow condition of

$$\underset{\sim}{\nabla}_P h_{ref} = 0 \tag{32c}$$

where p_{ref} and h_{ref} are reference values of fluid pressure and hydraulic head. It is important to keep in mind that the unstrained reference state given above for $t < 0$ is conceptually distinct from initial conditions at $t = 0^+$ that are analyzed and discussed in Appendix III. A condition of strain is essentially introduced to an elemental volume at $t = 0^+$.

In accordance with equation set 32, equation 31 becomes

$$(\rho_w g/S_{sk})\underset{\sim}{\nabla}_P(\underset{\sim}{\nabla}_P \cdot \underset{\sim}{u}_{cum\ s}) = \underset{\sim}{\nabla}\sigma' + \underset{\sim}{j}\rho_w g(G-1)(1-\phi_{initial}), \tag{33a}$$

$$(\rho_w g/S_{sk})\nabla_P(\nabla_P \cdot u_{cum\ s})$$

$$= \nabla_P p + j\rho_w g[1 - (G - 1)(\phi_{initial} - \phi)], \quad (33b)$$

$$(\rho_w g/S_{sk})\nabla_P(\nabla_P \cdot u_{cum\ s})$$

$$-\nabla_P h - j\rho_w g(G - 1)(\phi_{initial} - \phi), \quad (33c)$$

where we assume S_{sk} is associated with cumulative deformation and, similarly, that

$$\varepsilon_{sk} = -\nabla_P \cdot u_{cum\ s}. \quad (34)$$

We are now in a position to introduce some diffusion type of equations for granular movement. Combining equations 23 and 30 gives

$$v_s = q_{bulk} - (K/\rho_w g)\nabla_P \sigma' - jK(G - 1)(1 - \phi), \quad (35)$$

where $\nabla_P \sigma_{(zz)\ a}$ is assumed to be negligibly small. Substituting equation 33a into equation 35 gives

$$\frac{d_P u_{cum\ s}}{dt} - \frac{K}{S_{sk}} \nabla_P(\nabla_P \cdot u_{cum\ s})$$

$$= q_{bulk} - jK(G - 1)(\phi_{initial} - \phi), \quad (36)$$

where we have used equation 29b.

Taking the divergence of equation 36 gives

$$\frac{d_P \varepsilon_{sk}}{dt} - \nabla \cdot (\frac{K}{S_{sk}} \nabla_P \varepsilon_{sk})$$

$$= -K(G - 1)d_P(\phi_{initial} - \phi)/dz, \quad (37)$$

where we have used equations 22 and 34. Note that diffusion equation 37 is very similar to the more standardly derived diffusion equation 33*.

Volumetric S_{sk} is essentially defined by equation 31. Note in equation 31 that the gradient of volume strain is required to be a function of the gradient of a vertical normal component of incremental effective stress. Such a relation is commonly used to analyze compression tests in soils laboratories. This requirement is distinct from equations 6* and 30* in Appendix I; in these two equations, volume strain is assumed to be a function of incremental mean normal effective stress, which includes horizontal as well as vertical components. One class of materials whose properties allow S_{sk} of equation 31 to equal S_{sk}^{**} of equation 30* are those whose long-term rigidity is negligibly small (Jaeger and Cook, 1971, p. 105) There is no requirement a priori that S_{sk} and S_{sk}^{**} be equal.

Normal forces due to submerged weight of solids are accounted for in equations 37 and 33*. For sedimentary material whose long-term saturated rigidity is small, horizontal tectonic forces need not be small for equation 37 to be valid. By way of contrast, for equation 33* in Appendix I to be valid, the rigidity or competence of a saturated rock need not be small, but the divergence of all applied forces (including tectonic forces) must be small.

Taking the time derivative of equation 36 gives

$$\frac{d_P v_s}{dt} - \frac{K}{S_{sk}} \nabla_P (\nabla_P \cdot v_s)$$

$$= d_P q_{bulk}/dt + jK(G - 1)d_P\phi/dt, \quad (38)$$

where K and S_{sk} are assumed to be coefficients whose values are independent of time. A. M. Winslow (oral commun., 1980) has pointed out during his review of this paper that a similar, though distinct, diffusion equation can be expressed by

$$d_P q/dt - (K/S_{sk})\nabla_P(\nabla_P \cdot q) = -jK(G - 1)d_P\phi/dt, \quad (39)$$

which follows from an appropriate combination of equations 7, 22, and 38. In the next section, the rate of porosity change, which appears in the right-hand side of equations 38 and 39, will be expressed at $P = P_o$ as a function of either v_s or q.

At the present junction we can follow either of two possible strategies for finding $v_s(P,t)$. One is indirect; the other is direct. The first step of the indirect strategy is to develop a method for calculating $q(P,t)$ by use of governing equation 39. On the basis of equation 7, a value of $v_s(P,t)$ can be calculated indirectly by subtracting the resulting value of $q(P,t)$ from a corresponding value of $q_{bulk}(P,t)$, which can be found independently by solving governing equation 22. The direct strategy, on the other hand is simply to find a method for calculating $v_s(P,t)$ by solving governing equation 38. The remainder of this paper is a description of the direct strategy.

Equation 38 requires some simplification before it can be used as part of a practical solution scheme. In order to simplify equation 38, total derivatives will be expanded to reveal more fundamental parts, namely

$$\frac{\partial_P v_s}{\partial t} + (v_P \cdot \nabla_P)v_s - \frac{K}{S_{sk}} \nabla_P \left(\nabla_P \cdot v_s\right)$$

$$= \frac{\partial_P q_{bulk}}{\partial t} + (v_P \cdot \nabla_P)q_{bulk}$$

$$+ jK(G - 1) \left\{ \frac{\partial_P \phi}{\partial t} + (v_P \cdot \nabla_P)\phi \right\}, \quad (40)$$

where v_P is defined as dP/dt of equation 13. Note that, in general, v_s is itself expressed by a total derivative. For v_s in equation 40 we write

$$v_s = du_{cum\ s}/dt = \partial u_{cum\ s}/\partial t + (v_P \cdot \nabla_P)u_{cum\ s}, \quad (41)$$

which is in accordance with fundamental equations 13 and 29a. The distinction between $\underset{\sim}{v}_s$ and $\underset{\sim}{v}_P$ in equation 41 has far-reaching consequences (Helm, 1979b).

For constant $\underset{\sim}{Q}$ and incompressible constituents, it can be shown that the local derivative $\partial_P \underset{\sim}{q}_{bulk}/\partial t$ is everywhere valued at zero (see equation 22). When the points of interest are selected to be fixed in space (namely, $\underset{\sim}{P} = \underset{\sim}{P}_o$), each $\underset{\sim}{v}_P$ term in equations 40 and 41 equals $\underset{\sim}{v}_{Po}$ of equation 12a. For this special case, equation 41 reduces to equation 13a. Making this substitution simplifies equations 40 and 41 to

$$\partial_{P_o} \underset{\sim}{v}_s/\partial t - (K/S_{sk})\underset{\sim}{\nabla}_{P_o}(\underset{\sim}{\nabla}_{P_o} \cdot \underset{\sim}{v}_s)$$
$$= \underset{\sim}{j}K(G - 1)\partial_{P_o}\phi/\partial t, \tag{40a}$$

where

$$\underset{\sim}{v}_s = \underset{\sim}{v}_s(\underset{\sim}{P}_o, t) = \partial \underset{\sim}{u}_{cum\ s}/\partial t. \tag{41a}$$

For the sake of completeness, we express equations 40 and 41 at points $\underset{\sim}{P}_s$ and $\underset{\sim}{P}_w$ that move with the solid and liquid continua:

$$d_{P_s}\underset{\sim}{v}_s/dt - (K/S_{sk})\underset{\sim}{\nabla}_{P_s}(\underset{\sim}{\nabla}_{P_s} \cdot \underset{\sim}{v}_s)$$
$$= (\underset{\sim}{v}_s \cdot \underset{\sim}{\nabla}_{P_s})\underset{\sim}{q}_{bulk} + \underset{\sim}{j}K(G - 1)d_{P_s}\phi/dt, \tag{40b}$$

where

$$\underset{\sim}{v}_s = \underset{\sim}{v}_s(\underset{\sim}{P}_s, t) = \partial \underset{\sim}{u}_{cum\ s}/\partial t + (\underset{\sim}{v}_{P_s} \cdot \underset{\sim}{\nabla}_{P_s})\underset{\sim}{u}_{cum\ s}, \tag{41b}$$

and

$$d_{P_w}\underset{\sim}{v}_s/dt - (K/S_{sk})\underset{\sim}{\nabla}_{P_w}(\underset{\sim}{\nabla}_{P_w} \cdot \underset{\sim}{v}_s) = (\underset{\sim}{v}_w \cdot \underset{\sim}{\nabla}_{P_w})\underset{\sim}{q}_{bulk}$$
$$+ \underset{\sim}{j}K(G - 1)d_{P_w}\phi/dt, \tag{40c}$$

where

$$\underset{\sim}{v}_s = \underset{\sim}{v}_s(\underset{\sim}{P}_w, t) = \partial \underset{\sim}{u}_{cum\ s}/\partial t + (\underset{\sim}{v}_{P_w} \cdot \underset{\sim}{\nabla}_{P_w})\underset{\sim}{u}_{cum\ s}. \tag{41c}$$

Note that only in equations 40b and 41b do the velocity and displacement fields of solids, namely $\underset{\sim}{v}_s$ and $\underset{\sim}{u}_{cum\ s}$, coincide with the rate of movement and path of a group of solids whose identity remains unchanged. In equations 40a, 40c, 41a, and 41c, the specified velocity and cumulative displacement fields of solid material need not follow such a physically identifiable group of solids. This fundamental disassociation of vectors $\underset{\sim}{v}_s$ and $\underset{\sim}{u}_{cum\ s}$ from the movement and path of a continuing group of solids is true not only at a point $\underset{\sim}{P}_o$ fixed in space, but also at a point $\underset{\sim}{P}_w$ fixed in a distinct liquid continuum.

SKELETAL STRAIN RATE

The right-hand side of diffusion equation 40a can be transformed into a function of $\underset{\sim}{v}_s(\underset{\sim}{P}_o, t)$. To do this the concept of skeletal strain rate $d_P\varepsilon_{sk}/dt$ will be invoked. We define a rate of porous volume strain by

$$\frac{d_P\varepsilon_{sk}}{dt} = \lim_{V_P \to 0} \frac{1}{V_P}\frac{d_P V_v}{dt}. \tag{42a}$$

In accordance with equation 9, this implies

$$\underset{\sim}{\nabla}_P \cdot \underset{\sim}{v}_s = \lim_{V_P \to 0} \frac{1}{V_P}\frac{d_P V_v}{dt}. \tag{42b}$$

Bulk volume V_P equals the sum

$$V_P = V_v + V_s, \tag{43}$$

where V_v is the volume of voids and V_s is the volume of solids within an appropriately specified bulk V_P. Porosity ϕ is simply the ratio

$$\phi = V_v/V_P. \tag{44}$$

Recall that bulk volume V_{P_s} is defined such that the solid mass within a specified V_P remains constant. On the other hand, V_{P_w} requires that the mass of water within a saturated V_P remain constant. Finally, V_{P_o} simply requires that an idealized V_P retain a constant bulk volume.

For incompressible solids (see equation 11b), the volume of solids within V_{P_s} (Fig. 7) remains constant with respect to time, namely

$$d_{P_s}V_s/dt = 0.$$

Using this relation for V_{P_s}, the strain rate expressed by equation 42a equals, in view of equation 43,

$$\frac{d_{P_s}\varepsilon_{sk}}{dt} = \lim_{V_{P_s} \to 0} \frac{1}{V_{P_s}}\frac{d_{P_s}V_{P_s}}{dt}, \tag{45a}$$

which is equivalent to the divergence of an idealized solid continuum (see equation 15b). Equation 45a is identical to equation 14 and is associated with equation 2a* in Appendix I. Combining equation 14 or 45a with equation 2a* yields

$$\frac{d_{P_s}\varepsilon_{sk}}{dt} = \frac{1}{1 - \phi}\frac{d_{P_s}\phi}{dt}, \tag{45b}$$

which when combined with equation 9 gives

$$\underset{\sim}{\nabla}_{P_s} \cdot \underset{\sim}{v}_s = \frac{1}{1 - \phi}\frac{d_{P_s}\phi}{dt}, \tag{45c}$$

$$\nabla_{\underset{\sim}{P}_s} \cdot \underset{\sim}{q} = - \frac{1}{1 - \phi} \frac{d_{P_s} \phi}{dt}, \tag{45d}$$

where we have used equation 14* from Appendix I.

For completeness of presentation, we apply equation 42a to V_{P_w}, which gives

$$\frac{d_{P_w} \varepsilon_{sk}}{dt} = \lim_{V_{P_w} \to 0} \frac{1}{V_{P_w}} \frac{d_{P_w} V_v}{dt}. \tag{46}$$

Even under conditions of decreasing porosity (Fig. 7), equation 46 reduces to

$$d_{P_w} \varepsilon_{sk}/dt = 0 \tag{46a}$$

for the special case of saturated material with incompressible water (see equation 11a). Equation 46a essentially states that it is irrelevant to analyze skeletal strain rate at $\underset{\sim}{P} = \underset{\sim}{P}_w$. This is reasonable because in the limit as $V_{P_w} \to 0$, actual porous material has been conceptually replaced with an idealized liquid continuum. Substituting equation 9 into 46a gives the interesting result

$$\nabla_{\underset{\sim}{P}_w} \cdot \underset{\sim}{v}_s(\underset{\sim}{P}_w, t) = 0, \tag{46b}$$

$$\nabla_{\underset{\sim}{P}_w} \cdot \underset{\sim}{q}\,(\underset{\sim}{P}_w, t) = 0, \tag{46c}$$

where we have used equation 4* from Appendix I. Note in passing that in accordance with equation 46b, the second term in the left-hand side of equation 40c vanishes.

In contrast to equations 46b and 46c, the divergence of a liquid continuum is appropriately expressed by equation 15c. Recalling that V_v is constant within V_{P_w}, we find from multiplying the right-hand side of equation 15c by unity (in the form of V_v/V_v) that

$$\nabla_{\underset{\sim}{P}w} \cdot \underset{\sim}{v}_w(\underset{\sim}{P}_w, t) = - (1/\phi) d_{P_w} \phi/dt, \tag{47}$$

which is essentially identical to equation 3b*.

We are primarily interested in applying equation 42a to V_{P_o}, which gives

$$\frac{d_{P_o} \varepsilon_{sk}}{dt} = \lim_{V_{P_o} \to 0} \frac{d_{P_o}(V_v/V_{P_o})}{dt} = \frac{d_{P_o} \phi}{dt}, \tag{48a}$$

where bulk volume V_{P_o} is constant with respect to time. Combining equations 9 and 48a gives

$$\nabla_{\underset{\sim}{P}_o} \cdot \underset{\sim}{v}_s = d_{P_o} \phi/dt, \tag{48b}$$

$$\nabla_{\underset{\sim}{P}_o} \cdot \underset{\sim}{q} = - d_{P_o} \phi/dt, \tag{48c}$$

where we have used equation 4* from Appendix I. Recalling that, in accordance with equation 12a, $d\underset{\sim}{P}/dt$ of equation 13 is zero valued at $\underset{\sim}{P} = \underset{\sim}{P}_o$, the total rate of porosity change $d_{P_o}\phi/dt$, which appears in the right-hand side of equation set 48, equals the local rate of porosity change $\partial_{P_o}\phi/\partial t$, which appears in the right-hand side of equation 40a. Substituting equation 48b into the right-hand side of equation 40a and putting all terms into the left-hand side gives

$$\partial_{P_o} \underset{\sim}{v}_s/\partial t - \underset{\sim}{j} K(G - 1)(\nabla_{\underset{\sim}{P}_o} \cdot \underset{\sim}{v}_s)$$

$$- (K/S_{sk}) \nabla_{\underset{\sim}{P}_o} (\nabla_{\underset{\sim}{P}_o} \cdot \underset{\sim}{v}_s) = 0. \tag{49}$$

Equation 49 is a powerful equation. Interestingly enough, it is capable of further simplification. Before analyzing equation 49 by taking into consideration long-term "secondary" effects, the requirement of isotropy will be somewhat modified. In terms of cylindrical coordinates (r,θ,z), such a modification (described below) yields a set of two equations

$$\partial_{P_o} v_{s\,r}/\partial t - (K_{rr}/S_{sk}) \left\{ \partial_{P_o}(v_{s\,r}/r)/\partial r \right.$$

$$\left. + \partial^2_{P_o} v_{s\,r}/\partial r^2 + \partial^2_{P_o} v_{s\,z}/\partial z \partial r \right\} = 0, \tag{50a}$$

$$\partial_{P_o} v_{s\,z}/\partial t - K_{zz}(G - 1)(\nabla_{\underset{\sim}{P}_o} \cdot \underset{\sim}{v}_s)$$

$$- (K_{zz}/S_{sk}) \partial^2_{P_o} v_{s\,z}/\partial z^2 + \partial^2_{P_o} v_{s\,r}/\partial r \partial Z$$

$$+ \partial_{P_o}(v_{s\,r}/r)\partial z = 0. \tag{50b}$$

Equations 50a and 50b require the assumption of axially symmetric skeletal movement, namely

$$\underset{\sim}{v}_s = (v_{s\,r}, 0, v_{s\,z}). \tag{51}$$

The coefficients G, K_{rr}, K_{zz}, and S_{sk} are required to be constant with respect to time, and the reservoir is assumed to be homogeneous. Compressibility (specific storage) S_{sk} is inherently a coefficient for volume strain. What distinguishes equations 50a and 50b from equation 49 is that instead of requiring the reservoir material to be isotropic in three dimensions (namely, $K = K_{rr} = K_{\theta\theta} = K_{zz}$), it is merely assumed to be isotropic in a horizontal plane, namely

$$K_{rr} = K_{\theta\theta} \neq K_{zz}. \tag{52}$$

Such material is termed transversely isotropic. In addition, equations 50a and 50b require the principal directions of the hydraulic conductivity tensor to be vertical and horizontal and to be specified by K_{zz} and K_{rr}. For reasonable initial and boundary conditions on $v_{s\,r}$ in response to constant Q, equation 50a can be solved by standard methods (Helm, 1978a) assuming no shear.

The horizontal component of skeletal movement $v_{s\,r}$ is computationally the more straightforward one according to equation 50a. The second term in the left-hand side of equation 50b indicates that to estimate the vertical component $v_{s\,z}$, one must somehow be able to calculate the divergence of the rate of skeletal movement at P_o. Requiring no shear in equation 50b does not remove this term.

SEDIMENTARY SPACE

So far in this paper, we have derived equations in terms of an abstract space, which we can call "primary" space in accordance with Appendix II. Let us require primary space with its fixed points $\underset{\sim}{P}_o$ to move vertically relative to points $\underset{\sim}{P}_{sed}$, which are fixed in terms of a "sedimentary" space that is shown schematically in Figure 9. In sedimentary space, bulk volume V_{P_o} remains constant, equations 11a and 11b remain applicable, but point $\underset{\sim}{P}_o$, which is fixed in primary space (see equation 12a), moves in "sedimentary" space at a rate

$$d\underset{\sim}{P}_o/dt = -\underset{\sim}{j}K_{zz}(G - 1). \tag{53}$$

The right-hand side of equation 53 is suggested simply by the coefficient of the middle term in the left-hand side of equation 49. In other words, it is invoked here merely as a computational convenience for simplifying equation 49. It should be noted that the same result is obtained independently from a theoretical analysis of secondary movement (Appendix II).

Substituting equation 53 into equation 13 yields

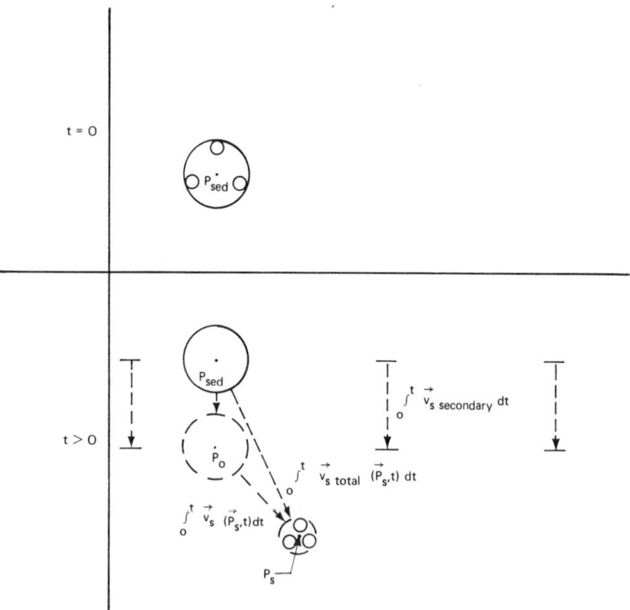

Figure 9. Schematic diagram of transient flow showing points $\underset{\sim}{P}_{sed}$ (fixed in sedimentary space), $\underset{\sim}{P}_o$ (fixed in primary space) and $\underset{\sim}{P}_s$ (fixed in solid continuum).

the relation in sedimentary space of

$$d_{P_o}\underset{\sim}{v}_s/dt = \partial_{P_o}\underset{\sim}{v}_s/\partial t - K_{zz}(G - 1)(\underset{\sim}{j} \cdot \underset{\sim}{\nabla}_{P_o})\underset{\sim}{v}_s. \tag{54}$$

For axially symmetric skeletal movement, equation 54 becomes

$$d_{P_o}v_{s\,r}/dt = \partial_{P_o}v_{s\,r}/\partial t - K_{zz}(G - 1)\partial_{P_o}v_{s\,r}/\partial z, \tag{55a}$$

$$d_{P_o}v_{s\,z}/dt = \partial_{P_o}v_{s\,z}/\partial t - K_{zz}(G - 1)\partial_{P_o}v_{s\,z}/\partial z, \tag{55b}$$

where material assumptions specified for equations 50a and 50b apply also to 55a and 55b. Combining equations 50a and 55a gives

$$d_{P_o}v_{s\,r}/dt + K_{zz}(G - 1)\partial_{P_o}v_{s\,r}/\partial z - (K_{rr}/S_{sk})\left\{\partial_{P_o}(v_{s\,r}/r)/\partial r + \partial^2_{P_o}(v_{s\,r}/r)/\partial r^2 + \partial^2_{P_o}v_{s\,z}/\partial z\,\partial r\right\} = 0. \tag{56a}$$

Combining equations 50b and 55b gives

$$d_{P_o}v_{s\,z}/dt - K_{zz}(G - 1)[v_{s\,r}/r \Big|_{P_o} + \partial_{P_o}v_{s\,r}/\partial r] - (K_{zz}/S_{sk})\left\{\partial^2_{P_o}v_{s\,z}/\partial z^2 + \partial^2_{P_o}v_{s\,r}/\partial r\partial z + \partial_{P_o}(v_{s\,r}/r)/\partial z\right\} = 0. \tag{56b}$$

Verruijt (1969, p. 343, 347) points out that following Jacob's (1946) classic theoretical development, mathematical hydrogeologists have, for computational convenience, required the horizontal displacement of solid material within an aquifer to vanish identically. Substituting this standard assumption into equations 56a and 56b gives

$$\partial^2_{P_o}v_{s\,z}/\partial z\partial r = 0, \tag{57a}$$

$$d_{P_o}v_{s\,z}/dt - (K_{zz}/S_{sk})\partial^2_{P_o}v_{s\,z}/\partial z^2 = 0. \tag{57b}$$

For both practical and fundamental physical reasons, Verruijt questions the vanishing of $v_{s\,r}$. By implication, this casts doubt on whether equations 57a and 57b are legitimate simplifications of equations 56a and 56b.

We shall simplify equations 56a and 56b by making a less restrictive assumption. Assuming a presupposed or physically determined value for shear allows equations 56a to yield a straightforward solution of $v_{s\,r}(\underset{\sim}{P}_o,t)$ for specified boundary and initial conditions. For the sake of illustration we shall assume no shear. Substituting this requirement into equations 56a and 56b gives

$$d_{P_o} v_{s\,r}/dt - (K_{rr}/S_{sk})[\partial_{P_o}(v_{s\,r}/r)/\partial r + \partial^2_{P_o} v_{s\,r}/\partial r^2] = 0, \tag{58a}$$

$$d_{P_o} v_{s\,z}/dt - (K_{zz}/S_{sk})\partial^2_{P_o} v_{s\,z}/\partial z^2 = K_{zz}(G-1)[v_{s\,r}/r_{P_o} + \partial_{P_o} v_{s\,r}/\partial r]. \tag{58b}$$

Substituting a straightforward solution of $v_{s\,r}(\underset{\sim}{P}_o,t)$ based on governing equation 58a into the the right-hand side of equation 58b allows one to solve in turn for $v_{s\,z}(\underset{\sim}{P}_o,t)$ in sedimentary space.

It is evident from examining equation sets 57 and 58 that assuming no shear within an aquifer is considerably less restrictive than assuming no horizontal displacement. Equations 58a and 58b are practical and powerful equations for calculating transient movement of skeletal material within a transversely isotropic aquifer system in response to constant discharge Q.

RESULTS

In this paper, I have suggested that instead of attempting to evaluate $\underset{\sim}{u}_{cum\,s}$ from an equation similar to equation 26, one can more fruitfully investigate the three-dimensional dynamics of $\underset{\sim}{v}_s$, which appears in the right-hand side of equation 2, namely

$$\underset{\sim}{u}_{cum\,s} = \int_o^t \underset{\sim}{v}_s dt. \tag{2}$$

By relating transient velocity fields to appropriate force fields per unit volume, a first step in this new conceptual direction has been taken.

Primary velocity fields $\underset{\sim}{q}$, $\underset{\sim}{q}_{bulk}$, $\underset{\sim}{v}_s$, and $\underset{\sim}{v}_w$ have all been related to a nonequilibrium dimensionless seepage force $\underset{\sim}{\nabla}_h$ and its ultimate equilibrium value $\underset{\sim}{\nabla}_{ult}$ for an isotropic reservoir through a conductivity coefficient K, namely

$$\underset{\sim}{q} = -K\underset{\sim}{\nabla}_P h, \tag{1}$$

$$\underset{\sim}{q}_{bulk} = -K\underset{\sim}{\nabla}_{P_o} h_{ult}, \tag{17**}$$

$$\underset{\sim}{v}_s = -K\underset{\sim}{\nabla}_{P_o}(h_{ult} - h), \tag{18**}$$

$$\underset{\sim}{v}_w = -K(\underset{\sim}{\nabla}_{P_o} h_{ult} + \frac{1-\phi}{\phi}\underset{\sim}{\nabla}_{P_o} h). \tag{18b**}$$

Equations numbered with two asterisks are found in Appendix II. Equation 1 is simply Darcy's law applied to nonsteady specific discharge (see equation 8). Equations 17**, 18**, and 18b** express bulk flow (see equation 6), velocity of solids, and velocity of water for the special case (see equations 12a** and 12b**) when nonzero steady-state specific discharge $\underset{\sim}{q}_{ult}$ is the ultimate response to producing a saturated isotropic reservoir at constant rate Q, and when only primary components of material movement are of interest.

It is important to note from Appendix II that during transient flow, the impelling force per unit mass for primary bulk flow q_{bulk} is postulated to be the gradient of total mechanical energy in the water phase $-\underset{\sim}{\nabla}_P\Theta_w$; that the impelling force per unit mass for primary specific discharge $\underset{\sim}{q}$ is the gradient of hydraulic head $-g\underset{\sim}{\nabla}h$; and finally that in accordance with these two postulates, the impelling force per unit mass for primary skeletal movement $\underset{\sim}{v}_s$ is the difference between the two. The term "primary" denotes a component of total behavoir that is directly attributable to a local viscous response to regional discharge Q. The flux terms mentioned in this paragraph are related to each other in accordance with

$$\underset{\sim}{q}_{bulk} = \underset{\sim}{q} + \underset{\sim}{v}_s. \tag{7}$$

It is postulated that even during transient flow, each velocity flux is related to the corresponding force field per unit mass through a coefficient of proportionality (K/g), where K is generally called hydraulic conductivity (see equations 1 and 6**).

Some equations for transient ground-water flow of the diffusion-type have been developed and discussed in this paper. Two of special interest are

$$\frac{d_P \underset{\sim}{u}_{cum\,s}}{dt} - \frac{K}{S_{sk}}\underset{\sim}{\nabla}_P(\underset{\sim}{\nabla}_P \cdot u_{cum\,s}) = \underset{\sim}{q}_{bulk} - \underset{\sim}{j}K(G-1)(\phi_{initial} - \phi), \tag{36}$$

and

$$\partial_{P_o}\underset{\sim}{v}_s/\partial t - \underset{\sim}{j}K(G-1)(\underset{\sim}{\nabla}_{P_o} \cdot \underset{\sim}{v}_s) - (K/S_{sk})\underset{\sim}{\nabla}_{P_o}(\underset{\sim}{\nabla}_{P_o} \cdot \underset{\sim}{v}_s) = 0. \tag{49}$$

Equation 49 requires the rate of regional discharge Q from a saturated aquifer system to be constant; equation 36 does not. For axially symmetric movement of transversely isotropic aquifer material, equation 49 simplifies to

$$d_{P_o} v_{s\,r}/dt - (K_{rr}/S_{sk})[\partial_{P_o}(v_{s\,r}/r)/\partial r + \partial^2_{P_o} v_{s\,r}/\partial r^2] = 0, \tag{58a}$$

$$d_{P_o} v_{s\,z}/dt - (K_{zz}/S_{sk})\partial^2_{P_o} v_{s\,z}/\partial z^2$$

$$= K_{zz}(G - 1)[v_{s\,r}/r\Big|_{P_o} + \partial_{P_o} v_{s\,r}/\partial r], \qquad (58b)$$

where we have assumed that any skeletal shear that occurs within a confined aquifer is negligibly small.

Several other equations of the diffusion type are developed in the text and in Appendix I. The difference between equations 37 and 33*, namely

$$\frac{d_P \varepsilon_{sk}}{dt} - \frac{K}{S_{sk}} \nabla_P^2 \varepsilon_{sk} = -K(G - 1)\, \frac{d_P(\phi_{initial} - \phi)}{dz}, \qquad (37)$$

$$\frac{d_P \varepsilon_{sk}}{dt} - \frac{K}{S_{sk}^{**}} \nabla_P^2 \varepsilon_{sk} = -\frac{K}{3}(G - 1)\, \frac{d_P(\phi_{initial} - \phi)}{dz}, \qquad (33^*)$$

is partly the interpretation of parameters S_{sk} and S_{sk}**. Equations numbered with one asterisk are found in Appendix I. Equation 37 requires in its derivation that the vertical normal component of applied force (per unit volume), namely $\underset{\sim}{\nabla}_P \sigma_{(zz)\,a}$, be negligibly small and that S_{sk} satisfy the relation

$$\underset{\sim}{\nabla}_P \varepsilon_{sk} = -\frac{S_{sk}}{\rho_w g} \underset{\sim}{\nabla}_P(\sigma'_{(zz)} - \sigma'_{(zz)ref}), \qquad (32)$$

where the subscript "ref" pertains to an initial unstrained state of stress. Equation 33* requires in its derivation that $\nabla^2\sigma'_{ii\,a}$ be negligibly small and that S_{sk}** satisfy the relation

$$\nabla_P^2 \varepsilon_{sk} = -\frac{S_{sk}^{**}}{3\rho_w g} \nabla_P^2(\sigma'_{ii} - \sigma'_{ii\,ref}), \qquad (30^*)$$

where repeated subscipts without parentheses indicate summation. It can be mentioned that equations 32 and 37 follow directly from selecting equation 5b rather than equation 5a to define a scalar type of quantity called "effective stress," σ', whose incremental change controls volumetric skeletal strain in the field. By way of contrast, equations 30* and 33* would follow if equation 5a had been selected as a definition of σ'. This second alternative definition is used in Appendix I.

After calculating $\underset{\sim}{v}_s$ directly from an expression such as equation 49 for specified initial and boundary conditions, it is conceptually possible to estimate $\underset{\sim}{u}_{cum\,s}$ by integrating $\underset{\sim}{v}_s$ over an appro- over an appropriate interval of time. Cumulative volume strain ε_{sk} is simply the divergence of $\underset{\sim}{u}_{cum\,s}$, from which the transient distribution of cumulative drawdown Δh can be calculated. Quantifying $\underset{\sim}{v}_s$ as suggested and developed in this paper not only yields a new theory for three-dimensional movement of porous material, it also leads directly to a new theory for transient ground-water flow within a confined aquifer system.

APPENDIX I: CRITIQUE OF STANDARD DIFFUSION EQUATIONS OF STRESS AND STRAIN

The question addressed in this Appendix is: To what extent can standard diffusion equations of stress and strain be used to quantify subsidence? The basic conceptual distinction between three locations, namely $\underset{\sim}{P}_o$, $\underset{\sim}{P}_s$, and $\underset{\sim}{P}_w$ as shown in Figure 7, will be maintained in this Appendix while deriving diffusion equations of stress and strain.

First, we divide equations 19a and 19b through by V_P and apply the divergence theorem to the surface integral after letting $V_P \to 0$. At $\underset{\sim}{P}_o$, $\underset{\sim}{P}_{sa}$, and $\underset{\sim}{P}_w$ the three sets of equations that follow from equations 19a and 19b are:

$$d_{P_o}(1 - \phi)/dt + \underset{\sim}{v}_s(\underset{\sim}{P}_o,t) \cdot \underset{\sim}{\nabla}_{P_o}(1 - \phi) + (1 - \phi)\underset{\sim}{\nabla}_{P_o} \cdot \underset{\sim}{v}_s = 0, \qquad (1a^*)$$

$$d_{P_o}\phi/dt + \underset{\sim}{v}_w(\underset{\sim}{P}_o,t) \cdot \underset{\sim}{\nabla}_{P_o}\phi + \phi\underset{\sim}{\nabla}_{P_o} \cdot \underset{\sim}{v}_w = 0, \qquad (1b^*)$$

$$\frac{1}{1 - d}\, d_{P_s}(1 - \phi)/dt + \lim_{V_{P_s} \to 0} \frac{1}{V_{P_s}}\, d_{P_s} V_{P_s}/dt = 0, \qquad (2a^*)$$

$$\frac{1}{\phi}\,(d_{P_s}\phi/dt + \underset{\sim}{\nabla}_{P_s} \cdot \underset{\sim}{q}) + \lim_{V_{P_s} \to 0} \frac{1}{V_{P_s}} \left(d_{P_s} V_{P_s}/dt\right) = 0, \qquad (2b^*)$$

$$\frac{1}{1 - \phi}\, d_{P_w}(1 - \phi)/dt - \frac{1}{1 - \phi}\underset{\sim}{\nabla}_P \cdot [\underset{\sim}{q}(1 - \phi)/\phi] + \lim_{V_{P_w} \to 0} \frac{1}{V_{P_w}}\, \frac{d_{P_w} V_{P_w}}{dt} = 0, \qquad (3a^*)$$

$$\frac{1}{\phi}\, d_{P_w}\phi/dt + \lim_{V_{P_w} \to 0} \frac{1}{V_{P_w}}\, d_{P_w} V_{P_w}/dt = 0. \qquad (3b^*)$$

A single asterisk denotes an equation that appears in Appendix I. Equations 1a*, 2a*, and 3a* pertain to mass balance of the solid phase. Equations 1b*, 2b*, and 3b* pertain to mass balance of the liquid (water) phase. Equations 2a* and 2b* pertain to continuity at points fixed within an abstract solid continuum. Equations 3a* and 3b* pertain to continuity at points fixed within an abstract liquid continuum. Equations 1a* and 1b* pertain to mass balance at points fixed within an abstract space defined by constitutive relations. Note that equation 13 must be applied to 1a* or 1b* in such a way that $d\underset{\sim}{P}/dt$ in equation 13 is identified only with $\underset{\sim}{v}_{P_o}$ of equation 12a. Unfortunately, standard theoretical developments cannot resist identifying $d\underset{\sim}{P}/dt$ in equation 13 with $\underset{\sim}{v}_{P_s}$ of equation 12b when equation 1a* is used, or with

$\underset{\sim}{v}_{P_w}$ of equation 12c when equation 1b* is used. The reader is referred to a more complete discussion (Helm, 1979b) of this conceptual error.

An even more significant conceptual problem will now be discussed. The standard equation for transient ground-water flow is generally expressed in terms of a single unknown. Hydraulic head h, fluid pressure p, the sum of normal components of effective stress σ'_{ii}, or skeletal volume strain ε_{sk} are frequent candidates. Transforming equation 22 into such an equation will illustrate the importance of gravitational forces that have been universally neglected by mathematical hydrogeologists. In accordance with equation 7, equation 22 can be written

$$\underset{\sim}{\nabla}_P \cdot \underset{\sim}{v}_s + \underset{\sim}{\nabla}_P \cdot \underset{\sim}{q} = 0. \tag{4*}$$

Substituting equations 1 and 9 into equation 4* gives

$$d_P\varepsilon_{sk}/dt - \underset{\sim}{\nabla}_P \cdot K\underset{\sim}{\nabla}_P h = 0. \tag{5*}$$

Equation 5* is a fundamental diffusion type of equation. Recalling equation 10, a volumetric compressibility term $S_{sk}{}^*$ can be introduced by the relation

$$d_P\varepsilon_{sk}/dt = -(S_{sk}{}^*/3\rho_w g)(d_P\sigma'_{ii}/dt - d_P\sigma'_{ii\ ref}/dt), \tag{6a*}$$

where σ'_{ii} /3 is mean normal effective stress and $\sigma'_{ii\ ref}$ pertains to an unstrained reference state of effective stress (t < 0). Both ε_{sk} and σ'_{ii} are invariant. Because unstrained $\sigma'_{ii\ ref}$ is inherently not a function of time, equation 6a* reduces directly to

$$d_P\varepsilon_{sk}/dt = -(S_{sk}{}^*/3\rho_w g)d_P\sigma'_{ii}/dt. \tag{6b*}$$

Substituting equation 6b* into equation 5* yields

$$(S_{sk}{}^*/3\rho_w g)d_P\sigma'_{ii}/dt + \underset{\sim}{\nabla}_P \cdot K\underset{\sim}{\nabla}_P h = 0. \tag{7*}$$

The present task is to express equation 7* in terms of only one unknown. Let us first transform mean normal effective stress $\sigma_{ii}'/3$ to hydraulic head h. Differentiating equations 4 and 5 with respect to time and combining the result leads to

$$(1/\rho_w g)d_P\sigma'_{ij}/dt = (1/\rho_w g)d_P\sigma_{ij}/dt + \delta_{ij}(d_P z/dt - d_P h/dt), \tag{8a*}$$

which implies

$$(1/3\rho_w g)d_P\sigma'_{ii}/dt = (1/3\rho_w g)d_P\sigma_{ii}/dt + d_P z/dt - d_P h/dt. \tag{8b*}$$

When substituted into equation 7*, equation 8b* leads to

$$d_P h/dt - \frac{K}{S_{sk}{}^*}\nabla_P{}^2 h = d_P z/dt + (1/3\rho_w g)d_P\sigma_{ii}/dt \tag{9*}$$

for isotropic homogeneous porous materials. The term $d_P z/dt$ in the right-hand side of equations 8b* and 9* expresses the vertical component of $d\underset{\sim}{P}/dt$ in in the right-hand side of equation 13. In accordance with equations 12a, 12b, and 12c, it is possible, at three conceptually distinct points of interest ($\underset{\sim}{P}_o$, $\underset{\sim}{P}_w$, and $\underset{\sim}{P}_w$), to specify $d_P z/dt$ as

$$d_{P_o} z/dt = 0, \tag{10a*}$$

$$d_{P_s} z/dt = v_{s\ z}(\underset{\sim}{P}_s,t), \tag{10b*}$$

$$d_{P_w} z/dt = v_{w\ z}(\underset{\sim}{P}_w,t), \tag{10c*}$$

where $v_{s\ z}$ is the vertical component of $\underset{\sim}{v}_s$ and $v_{w\ z}$ is the vertical component of $\underset{\sim}{v}_w$.

The term $d_P\ \sigma_{ii}/dt$ in the right-hand side of equation 9* deserves special attention. Total load σ_{ii} comprises two parts:

$$\sigma_{ii} = \sigma_a + \sigma_{gz}. \tag{11*}$$

The term σ_a in equation 11* represents the sum of the normal component of externally applied stress; σ_{gz} represents the weight of overlying material. The term σ_{gz} is a total gravitational stress whose influence on reservoir behavior cannot be ignored, namely

$$\sigma_{gz} = \int_{z_p}^{z_{wt}} \rho_w g\phi dz + \int_{z_p}^{z_{ls}} \rho_s g(1 - \phi)dz, \tag{12a*}$$

which implies

$$d_P\sigma_{gz}/dz = -\rho_w g[\phi + G(1 - \phi)], \tag{12b*}$$

where z_p is elevation of a point of interest, z_{wt} is elevation of the water table, z_{ls} is elevation of the land surface, and G is the specific gravity of solids. The term $d\sigma_{gz}/dz$ of equation 12b* is represented in Figure 8 by the magnitude of vector $\overparen{(H)B}$. For convenience, the weight of water in the unsaturated zone between z_{ls} and z_{wt} is considered in equation 12a* to be negligible. For the special case of no change in applied stress ($d_P\sigma_a/dt = 0$), which is assumed throughout this paper, equations 11* and 12a* can be combined to obtain the rate of change of total load

$$d_P\sigma_{ii}/dt = d_P\sigma_{gz}/dt = d_P/dt\int_{z_p}^{z_{wt}} \rho_w g d dz + d_P/dt\int_{z_p}^{z_{ls}} \rho_s g(1 - \phi)dz. \tag{13*}$$

Applying Leibnitz's rule for differentiation of integrals to equation 13*, assuming constant densities (see equations 11a and 11b), and substituting the result into equation 9* give

$$d_P h/dt - \frac{K}{S_{sk}^*} \nabla_P^2 h = d_P z/dt$$

$$+ \frac{1}{3} \left\{ \int_{z_P}^{z_{wt}} (d\phi/dt)dz + \phi d_P z/dt - \phi_{wt} d_{wt} z/dt \right.$$

$$+ G \int_{z_P}^{z_{ls}} [d(1 - \phi)/dt]dz + G(1 - \phi)d_P z/dt$$

$$\left. - G(1 - \phi_{ls})d_{ls} z/dt \right\} , \qquad (14^*)$$

where $d_{wt}z/dt$ is the rate of water-table fluctuation, $d_{ls}z/dt$ is the rate of vertical movement of of land surface (subsidence), and ϕ_{wt} and ϕ_{ls} are the values of porosity at the water table and land surface, respectively. Equation 14* can be simplified for the three conceptually distinct points (P_o, P_s, and P_w) by applying equations 10a*, 10b*, and 10c*

$$d_{P_o} h/dt - \frac{K}{S_{sk}^*} \nabla_{P_o}^2 h$$

$$= \frac{1}{3} \left\{ \int_{z_{P_o}}^{z_{wt}} (d\phi/dt)dz - \phi_{wt} d_{wt} z/dt \right.$$

$$\left. + G \int_{z_{P_o}}^{z_{ls}} [d(1 - \phi)/dt]dz - G(1 - \phi_{ls})d_{ls} z/dt \right\} , \qquad (14a^*)$$

$$d_{P_s} h/dt - \frac{K}{S_{sk}^*} \nabla_{P_s}^2 h = v_{s\,z}$$

$$+ \frac{1}{3} \left\{ \int_{z_{P_s}}^{z_{wt}} (d\phi/dt)dz + \phi v_{s\,z} - \phi_{wt} d_{wt} z/dt \right\} , \qquad (14b^*)$$

$$d_{P_w} h/dt - \frac{K}{S_{sk}^*} \nabla_{P_w}^2 h$$

$$= v_{w\,z} + \frac{G}{3} \left\{ \int_{z_{P_w}}^{z_{ls}} [d(1 - \phi)/dt]dz \right.$$

$$\left. + (1 - \phi)v_{w\,z} - (1 - \phi_{ls})d_{ls} z/dt \right\} , \qquad (14c^*)$$

where we have assumed no net change in solid mass between z_{P_s} and z_{ls} and no net change in liquid mass between z_{P_w} and z_{wt}, namely

$$d_P/dt \int_{z_{P_s}}^{z_{ls}} \rho^s dz = 0, \qquad (15a^*)$$

$$d_P/dt \int_{z_{P_w}}^{z_{wt}} \rho^w dz = 0. \qquad (15b^*)$$

Partial densities ρ^s and ρ^w are defined by equations 17 and 18. Equation 14* can be further simplified if a surface is assumed to exist at $z = b$ such that

$$\int_b^{z_{ls}} (d\rho^s/dt)dz = 0, \qquad (15c^*)$$

$$\int_b^{z_{wt}} (d\rho^w/dt)dz = 0. \qquad (15d^*)$$

The surface b can be considered the lowest upper boundary of a compressible reservoir. It represents a hydrologic boundary that separates draining material (reservoir) from nondraining material (overburden). This boundary may or may not coincide during early time with a lithologic surface (such as the base of a caprock or top of a semiconfining bed). There is possible upward migration of surface b. Separating the integrals in equations 14a*, 14b*, and 14c* into two parts above and below $z = b$ gives

$$d_{P_o} h/dt - \frac{K}{S_{sk}^*} \nabla_{P_o}^2 h$$

$$= -\frac{1}{3} \left\{ (G - 1) \int_{z_{P_o}}^{b} (d\phi/dt)dz + \phi_{wt} d_{wt} z/dt \right.$$

$$\left. + G(1 - \phi_{ls})d_{ls} z/dt \right\} , \qquad (16a^*)$$

$$d_{P_s} h/dt - \frac{K}{S_{sk}^*} \nabla_{P_s}^2 h = v_{s\,z}$$

$$+ \frac{1}{3} \left\{ \int_{z_{P_s}}^{b} (d\phi/dt)dz + \phi v_{s\,z} - \phi_{wt} d_{wt} z/dt \right\} , \qquad (16b^*)$$

$$d_{P_w}h/dt - \frac{K}{S_{sk}^*}\nabla_{P_w}^2 h = v_{w\,z} - \frac{G}{3}\left\{\int_{z_{P_w}}^{b}(d\phi/dt)dz - (1-\phi)v_{w\,z} + (1-\phi_{ls})d_{ls}z/dt\right\}, \quad (16c^*)$$

where we have used equations 15a* and 15b* for incompressible constituents of 11a and 11b.

The generally used equation for transient ground-water flow through a compressible isotropic homogeneous reservoir with incompressible constituents is

$$d_P h/dt - (K/S_s)\nabla_P^2 h = 0. \quad (17^*)$$

In light of equations 9*, 14*, and equation set 16*, which include the influence of gravitational forces, it is evident that the commonly-used equation 17* is an approximation at best. The vertical component of reservoir movement ($v_{s\,z}$ of equation 16b*) and the rate of land subsidence ($d_{ls}z/dt$ of equations 16a* and 16c*) have been shown to enter directly into the equation of transient ground-water flow itself through a universally-neglected gravitational term σ_{gz} of equation 12a*.

One might wonder how good an approximation equation 17* actually is. To get an idea of this for the sake of illustration, the first two terms in the right-hand side of equation 16b* will be expressed in terms of hydraulic head h. Combining equations 14, 2a*, and 5* gives

$$d_{P_s}\phi/dt = -(1-\phi)\underset{\sim}{\nabla}_{P_s}\cdot K\underset{\sim}{\nabla}_{P_s}h, \quad (18^*)$$

where $\underset{\sim}{P}_s$ represents an arbitrary point or "particle" fixed within a solid continuum. Substituting equation 18* into the first term in the right-hand side of equation 16b* gives

$$\int_{z_{P_s}}^{b}(d\phi/dt)dz = -K\int_{z_{P_s}}^{b}[(1-\phi)\nabla^2 h]dz. \quad (19^*)$$

Consider now the second term in the right-hand side of equation 16b*. The vertical component of $\underset{\sim}{v}_s$ can be expressed in terms of skeletal strain as

$$v_{s\,z}(\underset{\sim}{P}_s,t) = \int_{z_o}^{z_{P_s}}(d\varepsilon_{sk\,(zz)}/dt)dz, \quad (20^*)$$

where z_o is the elevation of an underlying surface where no vertical movement of matrix material occurs (such as rigid, impermeable basement rock) and $\varepsilon_{sk\,(zz)}$ is the vertical normal component of ε_{sk}. Note that z_o is fixed both in space and in the solid continuum. Substituting the vertical component of equation 5* into the right-hand side of equation 20* gives

$$v_{s\,z}(\underset{\sim}{P}_s,t) = K\int_{z_o}^{z_{P_s}}(d^2h/dz^2)dz = Kdh/dz\Big|_{z_o}^{z_{P_s}}. \quad (21^*)$$

If we require dh/dz to be zero valued at z_o, equation 21* reduces to

$$v_{s\,z}(\underset{\sim}{P}_s,t) = Kd_{P_s}h/dz. \quad (21a^*)$$

Substituting 19* and 21a* into the right-hand side of 16b* gives

$$d_{P_s}h/dt - \frac{K}{S_{sk}^*}\nabla_{P_s}^2 h = K(1+\frac{\phi}{3})d_{P_s}h/dz - \int_{z_{P_s}}^{b}[\frac{(1-\phi)}{3}\nabla^2 h]dz - \frac{\phi_{wt}}{3}d_{wt}z/dt. \quad (22^*)$$

If at a specified instant t one requires $\nabla^2 h$ to be uniformly distributed within a unit column of reservoir material between z_{P_s} and b, the integral in the right-hand side of equation 22* can be moved to the left-hand side to give

$$d_{P_s}h/dt - C_z\nabla_{P_s}^2 h = K(1+\frac{\phi}{3})d_{P_s}h/dz - \frac{\phi_{wt}}{3}d_{wt}z/dt, \quad (23^*)$$

where

$$C_z = \frac{K}{S_{sk}^*}\left\{1 - S_{sk}^*\int_{z_{P_s}}^{b}\frac{(1-\phi)}{3}dz\right\}. \quad (23a^*)$$

When the surface z = b is fixed in the solid constituent, C_z of equation 23a* is constant with respect to time for constant K/S_{sk}^*. Because of the lower limit of integration, however, C_z remains a function of depth.

In view of equation 23*, we can list some assumptions underlying the use of an equation such as

$$d_{P_s}h/dt - C_z\nabla_{P_s}^2 h = 0, \quad (23b^*)$$

which can be considered equivalent to equation 17* at $(\underset{\sim}{P}_s,t)$. These assumptions are:

1. The rate of water-table fluctuation is negligible throughout the time period of interest.
2. Specified points of interest $\underset{\sim}{P}_s$ are limited to locations where no vertical gradient of hydraulic head dh/dz develops during transient flow.
3. The vertical distribution of $\nabla^2 h$ is uniform throughout that portion of the reservoir (draining system) that overlies $\underset{\sim}{P}_s$. This overlying draining

system includes leaky caprock material (semiconfining bed).

4. The skeletal component of the coefficients of hydraulic diffusivity ($K/S_{SK}*$) should be replaced by a coefficient C_z of equation 23a*. For the simplest special case, C_z is a function of the initial elevation of a point P_s of interest.

If any of these assumptions is not valid for any portion of a flow system being analyzed, then equation 23b* should not be applied as a governing equation for that portion of the flow system. There would be even less justification in applying equation 17*.

To express transient ground-water flow in terms of fluid pressure p rather than in terms of hydraulic head h simplifies the mathematical expression only slightly. Equation 9* simplifies to

$$d_P p/dt - \frac{K}{S_{sk}*} \nabla_P^2 p = \frac{1}{3} d_P \sigma_{ii}/dt, \quad (24*)$$

and equations 16a*, 16b*, and 16c* simplify to

$$d_{P_o} p/dt - \frac{K}{S_{sk}*}\nabla_{P_o}^2 p = \frac{-\rho_w g}{3}\left\{ (G - 1)\int_{z_{P_o}}^{b} (d\phi/dt)dz + \phi_{wt} d_{wt} z/dt + G(1 - \phi_{ls}) d_{ls} z/dt \right\}, \quad (24a*)$$

$$d_{P_s} p/dt - \frac{K}{S_{sk}*}\nabla_{P_s}^2 p = \frac{\rho_w g}{3}\left\{ \int_{z_{P_s}}^{b} (d\phi/dt)dz + \phi v_{s\,z} - \phi_{wt} d_{wt} z/dt \right\}, \quad (24b*)$$

$$d_{P_w} p/dt - \frac{K}{S_{sk}*}\nabla_{P_w}^2 p = \frac{-\rho_s g}{3}\left\{ \int_{z_{P_w}}^{b} (d\phi/dt)dz - (1 - \phi) v_{w\,z} + (1 - \phi_{ls}) d_{ls} z/dt \right\}. \quad (24c*)$$

One might justifiably point out from equation 8b* that effects of change in total load (including not only gravitational force but also any changes in applied stress) are inherently accounted for in the $d\sigma'_{ii}/dt$ term in the left-hand side of equation 7*. Instead of expressing equation 7* in terms of hydraulic head as is done in equation 14*, one might consider expressing it in terms of the sum of the normal components of effective stress σ'_{ii}. This alternative will now be taken.

In accordance with equations 4 and 5, we find

$$\nabla_P^2 \sigma'_{ii} = \nabla_P^2 \sigma_{ii} - 3\rho_w g \nabla_P^2 h. \quad (25*)$$

In accordance with equation 11*, $\nabla_P^2 \sigma_{ii}$ comprises two parts,

$$\nabla_P^2 \sigma_{ii} = \nabla_P^2 \sigma_a + \nabla_P^2 \sigma_{gz}, \quad (26*)$$

which become

$$\nabla_P^2 \sigma_{ii} = \rho_w g (G - 1) d_P \phi/dz \quad (27*)$$

after equation 12b* is differentiated and $\nabla_P^2 \sigma_a$ is assumed to be zero-valued. Substituting equation 27* into equation 25* and eliminating $\nabla_P^2 h$ from the result and from equation 7* gives

$$(S_{sk}*/K) d_P \sigma'_{ii}/dt - \nabla_P^2 \sigma'_{ii} = \rho_w g (G - 1) d_P \phi/dz. \quad (28*)$$

Although the right-hand side of equation 28* is much simpler than the right-hand sides of equation sets 16* and 24*, it is evident that equation 28* does not reduce to the standard result

$$(S_{sk}*/K) d_P \sigma'_{ii}/dt - \nabla_P^2 \sigma'_{ii} = 0 \quad (29*)$$

except under unusual circumstances.

Equation 28* is not sensitive to distinctions between locations P_o, P_s, and P_w. Unfortunately, effective stress σ'_{ii} cannot be observed or measured directly. Transient effective stress is calculated (Lofgren, 1968) from observed drawdown, water table fluctuation, and subsidence. This practical fact returns us to equation set 16*.

For the sake of completeness, consider skeletal volume strain ε_{sk}. A volumetric compressibility term $S_{sk}**$ can be introduced through the relation

$$\nabla_P^2 \varepsilon_{sk} = -(S_{sk}**/3\rho_w g)(\nabla_P^2 \sigma'_{ii} - \nabla_P^2 \sigma'_{ii\,ref}). \quad (30*)$$

Note that both equations 6* and 30* relate volume strain ε_{sk} to an increment of mean normal effective stress.

Using equations 25*, 27*, and 30*, equation 5* becomes

$$(1/K) d_P \varepsilon_{sk}/dt - (1/S_{sk}**) \nabla_P^2 \varepsilon_{sk} = \frac{1}{3}\left\{ (G - 1) d_P \phi/dz - (1/\rho_w g) \nabla_P^2 \sigma'_{ii\,ref} \right\}. \quad (31*)$$

It is reasonable to make an assumption for the unstrained state that corresponds to equation 32a. Assuming thereby a relation such as

$$\nabla_P^2 \sigma'_{ii\,ref} = \rho_w g (G - 1) d\phi_{initial}/dz \quad (32*)$$

simplifies equation 31* to

$$d_P \varepsilon_{sk}/dt - (K/S_{sk}**) \nabla_P^2 \varepsilon_{sk} = -\frac{1}{3} K (G - 1) d_P (\phi_{initial} - \phi)/dz. \quad (33*)$$

Standard equations of time-consolidation theory that use skeletal volume strain ε_{sk} as the unknown (Mikasa, 1965; Gibson and others, 1967) require the right-hand side of equation 33* to be zero valued. The conceptual distinction between equations 33* and 37 is discussed in the main text.

Great pains have been taken in this Appendix to develop equation sets 16*, 24*, 28*, and 33*, yet we are no closer to being able to calculate the magnitude or direction of $\underset{\sim}{v}_s$ in three-dimensional space. This is precisely the point. Equation 17* has frequently been invoked as a first step in calculating reservoir movement $\underset{\sim}{v}_s$ and rate of subsidence $d_{1s}z/dt$ due to fluid withdrawal. Equation 17* should be applied to the problem of subsidence with extreme care because its unqualified use implies that reservoir movement $\underset{\sim}{v}_s$ and subsidence $d_{1s}z/dt$ do not occur.

APPENDIX II: PRIMARY AND SECONDARY MECHANISMS

Only one mechanism for skeletal movement is generally considered by hydrogeologists: movement associated with fluid withdrawal. Skeletal response to fluid withdrawal Q can be called primary compression or primary skeletal movement. When creep and other secondary effects are observed in the laboratory, additional forces are indicated. Even after the gradient of hydraulic head within a saturated soil specimen reaches zero, a specimen in a laboratory consolidometer slowly continues to compress, to decrease in porosity, and to have water squeezed from its skeletal frame. Secondary mechanisms, however small, may be significant to our attempts to understand natural subsidence. Eventually it will be necessary to distinguish a mechanism of natural ongoing geologic compaction in a sedimentary basin from the previously discussed mechanism of man-induced compaction due to an artificial increase in rate of fluid withdrawal.

In this Appendix, secondary effects are interpreted to represent a possible component of total skeletal movement (namely, $\underset{\sim}{v}_{s\ \text{secondary}}$) that may be induced by a gravitational force field per unit mass that is distinct from $\underset{\sim}{\nabla}_p\Theta_w$. The term $\underset{\sim}{\nabla}_p\Theta_w$ represents a force field per unit mass due to mechanical energy within the water phase. The following interpretation of $\underset{\sim}{v}_{s\ \text{secondary}}$ is not intended to include all secondary effects. For example, the term $-K(G-1)d_p(\phi_{\text{initial}}-\phi)/dz$ in right-hand side of equation 37 can legitimately be considered a secondary effect, because although the vertical gradient of porosity is a real and changing phenomenon within a stressed soil specimen, its presence in the governing equation is not accounted for quantitatively, even in the best of current soil mechanics theory (Mikasa, 1965; Gibson and others, 1967). Hence existing solutions for theoretical compression should be in error, at least to the extent that a porosity-gradient term influences a true solution. Such a term would represent a legitimate "secondary" effect because it requires an empirical discrepancy between behavior predicted from existing theory and the actual observed behavior in the laboratory. Other concepts have been used to explain empirical discrepancies such as material heterogeneity, coefficient nonlinearity, and rheological-model complexities. Although such efforts are worthwhile, it is of critical importance also to include within governing equations the influence of gravitational forces as expressed in the right-hand sides of equations 37 and 40a. Further discussion of gravitational forces can be found in Appendix I.

Impelling on a Saturated Mixture

A fundamental expression for mechanical energy is a generalization of Bernouli's theorem. The energy per unit mass of a single submerged constituent is simply

$$\Theta_\alpha^* = \int_0^{z_p} G_\alpha g\,dz + \int_{p_0}^{p} (1/\rho_w)\,d\pi + G_\alpha d\underset{\sim}{v}_\alpha/dt, \qquad (1^{**})$$

where π is a pressure variable of integration, G_α is the specific gravity of constituent α. For a unit mass of water ($\alpha = w$) and for a unit mass of solid material ($\alpha = s$), equation 1** becomes

$$\Theta_w^* = \int_0^{z_p} g\,dz + \int_{p_0}^{p} (1/\rho_w)\,d\pi + G_w d\underset{\sim}{v}_w/dt, \qquad (1a^{**})$$

$$\Theta_s^* = \int_0^{z_p} Gg\,dz + \int_{p_0}^{p} (1/\rho_w)\,d\pi + G\,d\underset{\sim}{v}_s/dt, \qquad (1b^{**})$$

where G is the specific gravity of solids. Equation 1a** was applied to steady flow through a porous medium by Hubbert (1940). The gradient of Θ_α^* can be considered an impelling force per unit mass.

It is important to distinguish between treating the gradient of Θ_α^* as an impelling force per unit unit mass of a single submerged constituent and treating it as an impelling force per unit mass of an interacting mixture. In the first case an elemental volume is so small that it contains only one constituent; in the second it contains a mixture. Whenever an asterisk is applied to the term Θ_α, namely R_α^*, a single submerged constituent is under consideration. A term Θ_α without an asterisk is used to indicate total mechanical energy within a multiconstituent mixture.

A force field per unit mass within a single submerged constituent can be expressed

$$\underset{\sim}{\nabla}_p\Theta_\alpha^* = \underset{\sim}{j}gG_\alpha + \underset{\sim}{\nabla}_p(p/\rho_w) + G_\alpha d\underset{\sim}{v}_\alpha/dt. \qquad (2^{**})$$

For $\alpha = w$ and $\alpha = s$, equation 2** reduces to

$$\underset{\sim}{\nabla}_p\Theta_w^* = \underset{\sim}{j}g + \underset{\sim}{\nabla}_p(p/\rho_w) + d\underset{\sim}{v}_w/dt, \qquad (2a^{**})$$

$$\underset{\sim}{\nabla}_p\Theta_s^* = \underset{\sim}{j}gG + \underset{\sim}{\nabla}_p(p/\rho_w) + G\,d\underset{\sim}{v}_s/dt. \qquad (2b^{**})$$

It is useful to define a volume-weighted total force per unit mass $\nabla_p\Theta$ mixture for a mixture of incompressible water and solids within a compressible bulk volume, namely

$$\nabla_p\Theta_{mixture} = \phi\nabla_p\Theta_w^* + (1 - \phi)\nabla_p\Theta_s^*. \qquad (3^{**})$$

For uniform ρ_w and G, equation 3** implies

$$\nabla_p\Theta_{mixture} = jg[\phi + G(1 - \phi)] + (1/\rho_w)\nabla_p p + d_p v/dt, \qquad (4^{**})$$

where

$$d_p v/dt \equiv \frac{1}{\rho_w}\sum_\alpha \rho^\alpha d_p v\alpha/dt. \qquad (4a^{**})$$

The partial density ρ^α for a two-constituent mixture is expressed by equations 17 and 18. An appropriate interpretation of the force that impels volume-weighted bulk flow q_{bulk} is volume-weighted $\nabla_p\Theta$ mixture of equation 3**.

It is instructive to distinguish between two bulk flows, namely $q_{bulk\ total}$ and $q_{bulk\ primary}$. The former is more nearly a true bulk flow. The latter, however, is more consistent with standard theory of fluid flow through porous media. To make the distinction more clearly, we write

$$q_{bulk\ total} = -(K/g)\nabla_p\Theta_{mixture}, \qquad (5^{**})$$

$$q_{bulk\ primary} = -(K/g)\nabla_p\Theta_w. \qquad (6^{**})$$

In the right-hand side of equation 6**, the term $\nabla_p\Theta_w$ is similar to, but conceptually distinct from $\nabla_p\Theta_w{}^*$ of equation 2a**. The latter term represents an impelling force on water (per unit mass) isolated from any other material; the former is an impelling force on water (per unit mass) as part of a mixture. The difference between $\nabla_p\Theta_w$ and $\nabla_p\Theta_w{}^*$ shows up in their respective inertial terms, namely

$$\nabla_p\Theta_w = jg + (1/\rho_w)\nabla_p p + d_p v/dt. \qquad (7^{**})$$

For uniform ρ_w, equation 7** contrasts with equation 2a** in that the inertial term in equation 7** pertains to two (or more) constituents and is defined by equation 4a**.

For the sake of completeness, we can similarly contrast $\nabla_p\Theta_s{}^*$ of equation 2b**, which pertains to a large block of submerged solid material, with $\nabla_p\Theta_s$ of the following equation, which for uniform ρ_w and G pertains to a solid constituent within an interacting saturated mixture,

$$\nabla_p\Theta_s = jgG + (1/\rho_w)\nabla_p p + d_p v/dt. \qquad (8^{**})$$

Figure 10 illustrates schematically how $\nabla_p\Theta_{mixture}$, $\nabla_p\Theta_w$, and $\nabla_p\Theta_s$ of equations 4**, 7**, and 8** are related.

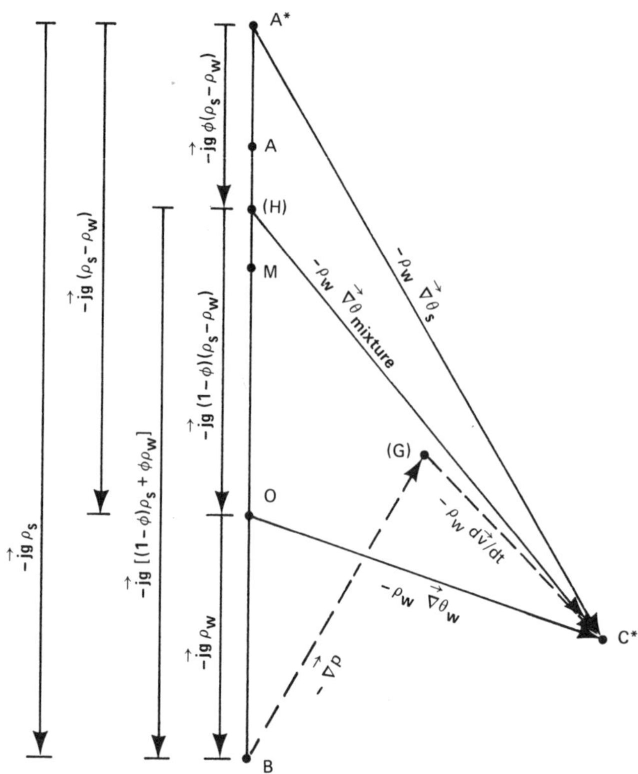

Figure 10. Force polygon showing a total impelling force per unit volume on a saturated granular mixture.

It is worthwhile to discuss the role of kinetic energy in the displacement of saturated solid material during transient fluid flow. The driving forces per unit mass, namely $-\nabla_p\Theta_{mixture}$ and $-\nabla_p\Theta_w$ in the right-hand sides of equations 5** and and 6**, can be expressed as a sum of two parts,

$$-\nabla_p\Theta_{mixture} = \nabla_p\sigma'/\rho_w - d_p v/dt, \qquad (9^{**})$$

$$-\nabla_p\Theta_w = -g\nabla_p h - d_p v/dt, \qquad (10^{**})$$

where we have used equation 4** and 7**. One can interpret $g\nabla_p h$ in equation 10** as a resistive force (per unit mass) on the water due to viscous drag exerted by solid particles. Similarly, in equation 9** one can interpret $-\nabla_p\sigma'/\rho_w$ as a force (per unit mass) arising from frictional resistance due to an effective viscosity of a saturated granular mixture.

For constant ρ_w (see equation 11a), Figure 11 schematically indicates force fields per unit volume in accordance with equation 10**. Figure 11 modifies Figure 8 by including vector (G)C*, which links transient vector O(G) to equilibrium vector OC*.

Primary Movement of Solids

We now take a fresh look at the theory of primary consolidation in soil mechanics and relate it to standard theories of transient ground-water flow.

Existence of nonzero steady-state or ultimate

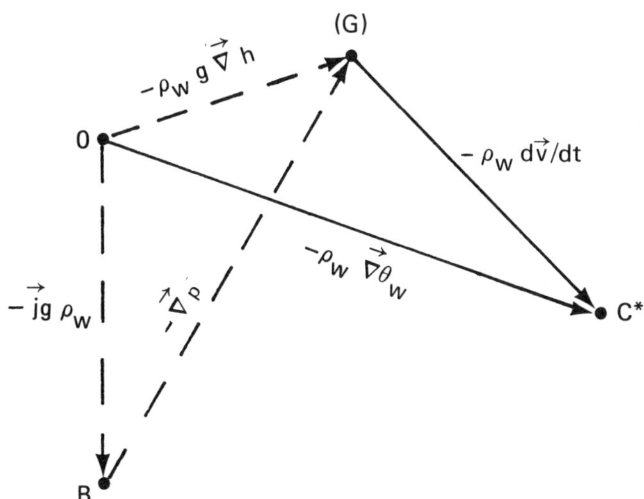

Figure 11. Force polygon of transient flow showing an inertial component of force.

flow, namely $\underset{\sim}{q}_{ult}$ of equation 2, is tacitly assumed in the discussion of Figure 5. The special case of applicability of nonzero $\underset{\sim}{q}_{ult}$ to transient flow as t approaches infinity is somewhat restrictive, but is no more restrictive than the theory of leaky aquifers in both its original (Jacob, 1946; Hantush and Jacob, 1955) and expanded forms (Hantush, 1960; Neuman and Witherspoon, 1969).

We now raise the following question: Under what conditions does point C of Figure 5 (when it exists) coincide with point C* of Figure 11? Consistent with assuming the ultimate existence of nonzero $\underset{\sim}{q}_{ult}$ (which is the underlying assumption of Figure 5) is a requirement that ultimately (t T B) the granular skeleton serves as a nonmoving conduit for fluid flow. These two conditions can be expressed by

$$\lim_{t\to\infty} \underset{\sim}{q} = \underset{\sim}{q}_{ult}, \qquad (11^{**})$$

$$\lim_{t\to\infty} \underset{\sim}{v}_s = 0. \qquad (12^{**})$$

Equation 11** is tacitly assumed by hydrogeologists when they analyze the hydraulics of flow to wells. Equation 12** is tacitly assumed by soil engineers when they mathematically analyze primary consolidation in the laboratory. Together with equation 3 in the text of this paper, requirements expressed by equation 11** or 12** lie at the very heart of what is called primary conditions of flow. Combining equations 7 and 12** gives

$$\lim_{t\to\infty} \underset{\sim}{q} = \lim_{\tau\to\infty} \underset{\sim}{q}_{bulk} \quad . \qquad (13^{**})$$

Similarly, we find from equations 11** and 13** that

$$\lim_{t\to\infty} \underset{\sim}{q}_{bulk} = \underset{\sim}{q}_{ult}. \qquad (14^{**})$$

On the basis of conservation of mass, it can be shown that if bulk acceleration $\partial_{p_o} \underset{\sim}{q}_{bulk}/\partial t$ of incompressible individual constituents is uniformly distributed (see equation 22), then whenever the rate of fluid withdrawal Q at a regional boundary remains constant, $\underset{\sim}{q}_{bulk}$ itself is everywhere ($\underset{\sim}{p} = \underset{\sim}{p}_o$) constant with respect to time. For constant Q, and for conditions that allow for nonzero $\underset{\sim}{q}_{ult}$, equation 14** simplifies to

$$\underset{\sim}{q}_{bulk}(\underset{\sim}{p}_o, t) = \underset{\sim}{q}_{bulk}(\underset{\sim}{p}_o) = \underset{\sim}{q}_{ult}(\underset{\sim}{p}_o). \qquad (15^{**})$$

Substituting equations 3 and 6** into equation 15** gives

$$(1/g)\underset{\sim}{\nabla}_{p_o} \Theta_w = \underset{\sim}{\nabla}_{p_o} h_{ult}. \qquad (16^{**})$$

Equation 16** implies that, in the neighborhood of $\underset{\sim}{p}_o$ fixed in space and for constant Q, the existence of nonzero $\underset{\sim}{q}_{ult}$ requires that point C in Figure 5 coincide with point C* in Figure 11.

Equation 16** requires that $\underset{\sim}{q}_{bulk}$ of equation 15** is $\underset{\sim}{q}_{bulk\ primary}$ of equation 6**. Substituting equation 15** into equation 3, we find

$$\underset{\sim}{q}_{bulk\ primary} = -K\underset{\sim}{\nabla}_{p_o} h_{ult}. \qquad (17^{**})$$

Recalling Darcy's law for transient flow (equation 1), equation 7 can be used to give a Darcian expression for the primary component of $\underset{\sim}{v}_s$,

$$\underset{\sim}{v}_{s\ primary} = -K\underset{\sim}{\nabla}_{p_o}(h_{ult} - h), \qquad (18^{**})$$

where we have used equation 17**. Equation 18** is a surprisingly simple equation for $\underset{\sim}{v}_s$ at ($\underset{\sim}{p}_o$,t). Figure 12 is essentially a schematic representation of equation 18**.

In passing, it is possible to relate $\underset{\sim}{v}_s$ to concept of excess pore-water pressure p_e (= p - p_{ult}). In accordance with equations 4 and 11a, an alternative expression for equation 18** is

$$\underset{\sim}{v}_{s\ primary}(\underset{\sim}{p}_o, t)$$

$$= -(K/\rho_w g)\underset{\sim}{\nabla}_{p_o}(p_{ult} - p) = (K/\rho_w g)\underset{\sim}{\nabla}_{p_o} p_e. \qquad (18a^{**})$$

Figure 12 helps us to visualize equation 18a**.

A misconception found in the literature should be clarified. For isotropic material, the impelling force on skeletal movement $\underset{\sim}{v}_s$ primary has been considered by some (Verruijt, 1970; Lofgren, 1971, 1978) to be a viscous drag, or seepage force, $-\rho_w g\underset{\sim}{\nabla}h$. In the present paper, it has been found found that what is significant for $\underset{\sim}{v}_s$ quantitatively is the difference between ultimate seepage force $-\rho_w g\underset{\sim}{\nabla}h_{ult}$ and transient seepage force $-\rho_w g\underset{\sim}{\nabla}h$. Equation 18** makes this clarification mathematically; vectors $\underset{\sim}{O(G)}$ and $\underset{\sim}{(G)C^*}$ in Figure 12 show it graphically.

For completeness of presentation we unite $\underset{\sim}{v}_w$ in in terms of an appropriate force field. Combining equations 1,8, and 18** for an isotropic saturated material gives,

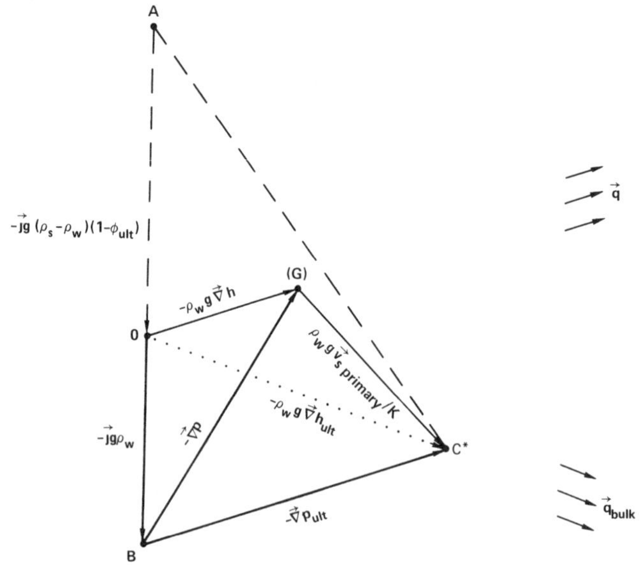

Figure 12. Force polygon of transient flow showing nonzero $\underset{\sim}{v}_{s\ primary}$ at $(\underset{\sim}{p}_o,t)$.

$$\underset{\sim}{v}_w = -K\left(\underset{\sim}{\nabla}_{p_o} h_{ult} + \frac{1-\phi}{\phi}\underset{\sim}{\nabla}_{p_o} h\right). \quad (18b^{**})$$

Before we conclude this section, we write the interesting theoretical relation,

$$\underset{\sim}{v}_{s\ primary}(\underset{\sim}{p}_o,t) = -(K/g)d_{p_o}\underset{\sim}{v}/dt, \quad (18c^{**})$$

in which we have combined 10**, 16** and 18**.

Secondary Movement of Solids

The problem of secondary effects in the laboratory can be expressed by distinguishing between two components of the total flow of water relative to solids, namely

$$\underset{\sim}{q}_{secondary} = \lim_{t\to\infty}(\underset{\sim}{q}\ total - \underset{\sim}{q}\ primary), \quad (19^{**})$$

where $\underset{\sim}{q}_{primary}$ satisfies Darcy's law:

$$\underset{\sim}{q}_{primary} = -K\underset{\sim}{\nabla}h. \quad (20^{**})$$

Equation 20** is essentially an interpretation of equation 1 and is justified by laboratory soil consolidation data. After the gradient of hydraulic head in the right-hand side of equation 20** reaches zero, a stressed soil specimen continues to compress in a consolidometer at a reduced rate and water continues to be squeezed out. In this Appendix, we shall investigate one of the possible mechanisms that contribute to $q_{secondary}$.

A key to the present investigation into secondary effects is the definition of $\underset{\sim}{q}_{bulk}$ found in equation 6 and $\underset{\sim}{q}$ found in equation 8. We generalize these definitions to be applicable to conditions of total flow and primary flow, namely

$$\underset{\sim}{q}_{bulk\ total} = \phi\underset{\sim}{v}_{w\ total} + (1-\phi)\underset{\sim}{v}_{s\ total}, \quad (21a^{**})$$

$$\underset{\sim}{q}_{bulk\ primary} = \phi\underset{\sim}{v}_{w\ primary} + (1-\phi)\underset{\sim}{v}_{s\ primary}, \quad (21b^{**})$$

and even to conditions of secondary flow

$$\underset{\sim}{q}_{bulk\ secondary} = \phi\underset{\sim}{v}_{w\ secondary} + (1-\phi)\underset{\sim}{v}_{s\ secondary}. \quad (21c^{**})$$

Similarly, we require

$$\underset{\sim}{q}_{total} = \phi(\underset{\sim}{v}_{w\ total} - \underset{\sim}{v}_{s\ total}), \quad (22a^{**})$$

$$\underset{\sim}{q}_{primary} = \phi(\underset{\sim}{v}_{w\ primary} - \underset{\sim}{v}_{s\ primary}), \quad (22b^{**})$$

$$\underset{\sim}{q}_{secondary} = \phi(\underset{\sim}{v}_{w\ secondary} - \underset{\sim}{v}_{s\ secondary}). \quad (22c^{**})$$

To estimate secondary effects, one can assume that primary effects (specified by equations 11** and 12b**) have ceased. Similar to equation 19** we write,

$$\underset{\sim}{q}_{bulk\ secondary} = \lim_{t\to\infty}(\underset{\sim}{q}_{bulk\ total} - \underset{\sim}{q}_{bulk\ primary}), \quad (23a^{**})$$

$$\underset{\sim}{v}_{s\ secondary} = \lim_{t\to\infty}(\underset{\sim}{v}_{s\ total} - \underset{\sim}{v}_{s\ primary}), \quad (23b^{**})$$

$$\underset{\sim}{v}_{w\ secondary} = \lim_{t\to\infty}(\underset{\sim}{v}_{w\ total} - \underset{\sim}{v}_{w\ primary}). \quad (23c^{**})$$

Recall that equation 12** can be written as

$$\lim_{t\to\infty}\underset{\sim}{v}_{s\ primary} = 0, \quad (12b^{**})$$

which when substituted into the right-hand side of equation 23b** gives

$$\underset{\sim}{v}_{s\ secondary} = \lim_{t\to\infty}\underset{\sim}{v}_{s\ total}. \quad (24^{**})$$

Because primary conditions of flow are directly associated with mechanical energy per unit mass within the water phase, it is reasonable to identify ultimate total flow of water with its ultimate primary flow, namely

$$\lim_{t\to\infty}\underset{\sim}{v}_{w\ total} = \lim_{t\to\infty}\underset{\sim}{v}_{w\ primary}. \quad (25^{**})$$

Equation 25** means that the flux of incompressible

water at $(\underset{\sim}{P},t)$ is assumed to be entirely in response to regional discharge Q (either natural or induced by man). We now have the essential postulates and equation for evaluating secondary movement of saturated porous material.

Substituting equation 25** into the right-hand side of equation 23c** gives

$$\underset{\sim}{v}_{w\ \text{secondary}} = 0. \qquad (26^{**})$$

Applying equations 5** and 6** to the right-hand side of equation 23a**, we find

$$\underset{\sim}{q}_{\text{bulk secondary}} = \lim_{t \to \infty} -(K/g)\underset{\sim}{\nabla}[(\sigma'/\rho_w) + gh], \qquad (27^{**})$$

where we have used equations 9** and 10**. Equation 27** reduces to

$$\underset{\sim}{q}_{\text{bulk secondary}} = \underset{\sim}{j}K(1 - \phi)(G - 1) \qquad (28^{**})$$

when we apply equation 30a. We now substitute equations 26** and 28** into equation 21c** to get

$$\underset{\sim}{v}_{s\ \text{secondary}} = -\underset{\sim}{j}K(G - 1). \qquad (29^{**})$$

Finally, combining equations 22c**, 26**, and 29** gives

$$\underset{\sim}{q}_{\text{secondary}} = \underset{\sim}{j}K\phi(G - 1). \qquad (30^{**})$$

Equations 26**, 28**, 29**, and 30** specify values of secondary movement for material within a saturated compressible porous formation.

Of special interest is equation 29**. A theoretical value for $\underset{\sim}{v}_s$ secondary has been found that agrees quantitatively with the value $d\underset{\sim}{P}_o/dt$ of equation 53 in the main body of this paper. Equation 53 expresses how a point $\underset{\sim}{P}_o$ fixed in primary space moves in terms of what we call sedimentary space. Equation 29** implies that a force other than interstitial fluid flow tends to cause skeletal collapse of nonrigid sedimentary material. This force is simply density contrast $(\rho_s - \rho_w)$ between solid particles and interstitial water. The reciprocal of K_{zz} can be considered a type of resistance to this collapse.

The reader may ask, How large is $\underset{\sim}{v}_s$ secondary for equation 29** (or $d\underset{\sim}{P}_o/dt$ of equation 53)? We shall estimate it in two ways. In the Gulf of Mexico Basin (Hardin and Hardin, 1961), maximum rates of deposition have risen from 2×10^{-4} m/yr for Eocene sediments to 8.1×10^{-4} m/yr for Miocene sediments and 1.2×10^{-3} m/yr for Pleistocene and Holocene sediments. These values can be considered reasonable approximations of subsidence rates of underlying formations under conditions of negligible fluid discharge Q. A second way to approximate $\underset{\sim}{v}_s$ secondary is by using the aquitard - drainage model that was discussed in an earlier section of this paper. Using that model, an effective value for K_{zz} of a compressible reservoir can be found by simulating observed rates of induced reservoir compaction. According to Helm (1978b, Table 1), values of K_{zz} estimated in this way at 15 sites in California range from 6×10^{-6} to 1.4×10^{-3} m/yr. Substituting these values of K_{zz} into equation 29** or 53 gives estimates of $\underset{\sim}{v}_s$ secondary that agree in general with those just estimated from reported rates of deposition of Gulf of Mexico sediments. Indirectly, this agreement suggests that for sedimentary material, a viscous or Darcian interpretation of movement of solids as proposed in this paper may be far from fortuitous.

APPENDIX III: THREE CANDIDATE DIRECTIONS OF $\underset{\sim}{v}_s$

Even when the direction of magnitude of $\underset{\sim}{q}_{bulk}$ (Fig. 6) are known and constant, the transient directions of $\underset{\sim}{q}$ and $\underset{\sim}{v}_s$ (or $\underset{\sim}{v}_s$ primary) are not are not necessarily known. This translates to mean that knowing an eventual steady-state value of $\underset{\sim}{\nabla}_{ult}h$ does not imply that the direction of transient transient $\underset{\sim}{\nabla}h$ is known. Hydrogeologists appear to be to be unaware of the existence of this conceptual problem. We shall now illustrate or restate the problem graphically: Even when vector $\widetilde{OC}$ in Figure 5 (or, correspondingly, vector $\widetilde{OC^*}$ in Fig. 12) is known and unchanging, the path of moving point (G) in Figure 12 is not necessarily known. In accordance with equation 12**, it is known for the special case of constant Q and nonzero $\underset{\sim}{q}_{ult}$ that moving point (G) in Figure 12 must in some manner approach point C* as t tends to infinity. For the sake of notational brevity, the term $\underset{\sim}{v}_s$ is used in in this Appendix to indicate the primary component of skeletal movement (see Appendix II).

Consider three paths that point (G) in Figure 12 might take toward point C*. The first possibility is merely an articulation of what has been tacitly assumed by most hydrogeologists. This can be defined by assuming either that the direction of specific discharge $\underset{\sim}{q}$ does not change during transient flow or that the direction of $\underset{\sim}{v}_s$ is identical to the direction of seepage seepage force $-\rho_w g\underset{\sim}{\nabla}h$. Either assumption has an identical geometric effect on the path of moving point (G); it must fall somewhere on a straight line joining two fixed points O and C* (Fig. 13B).

A second candidate path for point (G) to follow can be defined by assuming that the direction of $\underset{\sim}{v}_s$ is identical to the direction of effective force per unit volume $\underset{\sim}{\nabla}\sigma'$. In a graphic interpretation of this requirement, point (G) must fall somewhere on a straight line joining moving point (H) and fixed point C* (Fig. 14B).

A third candidate path requires that although the directions of $\underset{\sim}{v}_s \nabla\sigma'$ can be different from each other, neither direction changes when Q is constant (Fig. 17).

For ultimate steady flow, moving point (G) in Figures 13B, 14B, and 17 coincides with fixed point C* and moving point (H) coincides with fixed point A. The physical requirements represented by Figure 5 are thereby satisfied by all three candidate paths.

The physical implications of the first two candidate paths for point (G) will be examined first. The distinction between the first two paths is most clearly demonstrated by examining initial

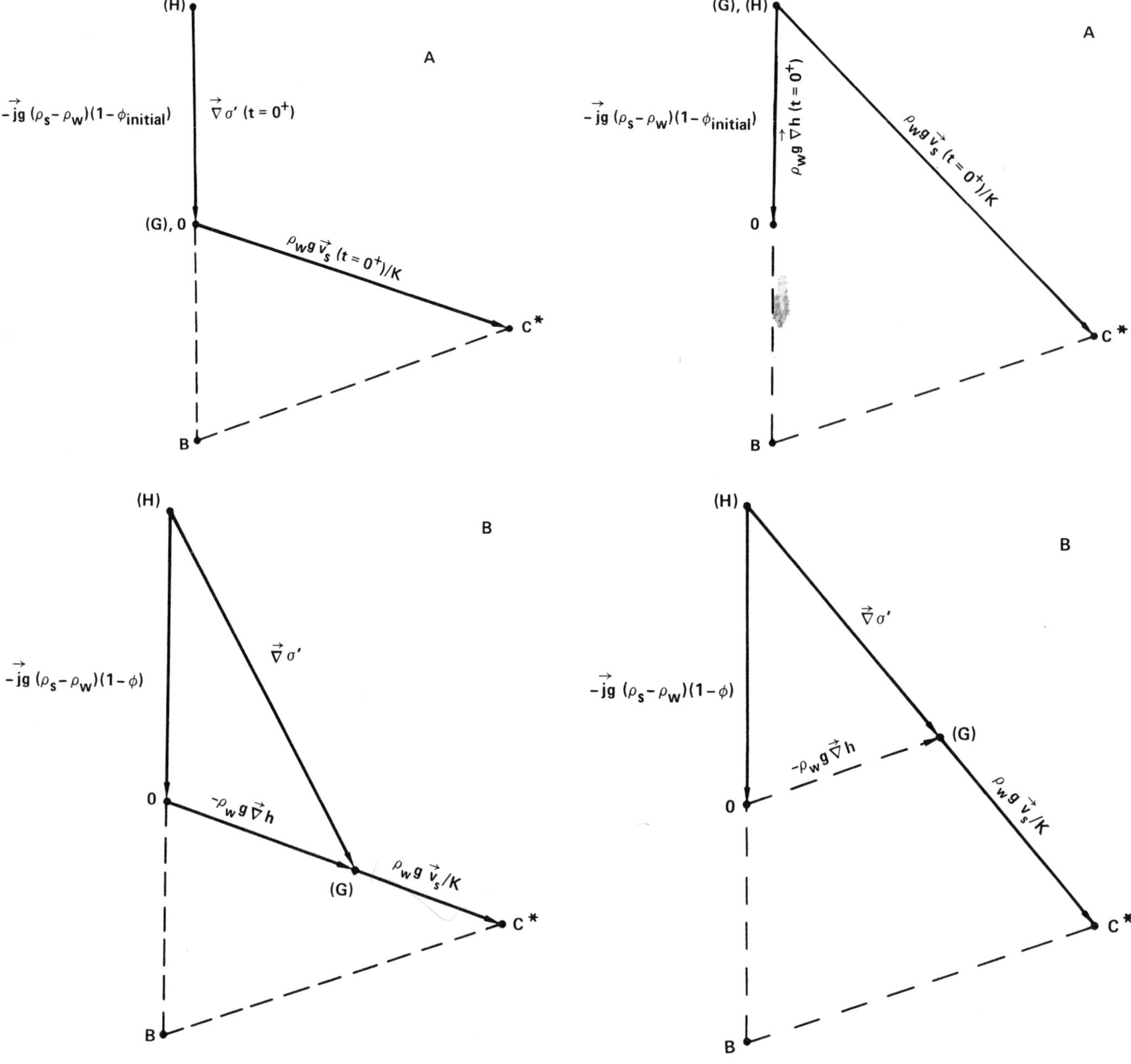

Figure 13. Force polygon of transient flow assuming $\underset{\sim}{v}_s$ is in the direction of seepage force per unit volume, $-\rho_s g\underset{\sim}{\nabla}h$: (A) Initial ($t = 0^+$) conditions; (B) Transient ($t > 0^+$) conditions.

Figure 14. Force polygon of transient flow assuming $\underset{\sim}{v}_s$ is in the direction of effective force per unit volume, $\underset{\sim}{\nabla}\sigma'$: (A) Initial ($t = 0^+$) conditions; (B) Transient ($t > 0^+$) conditions.

conditions. For simplicity, we require any horizontal component of effective force $\underset{\sim}{\nabla}\sigma'$ to be zero-valued at $t = 0^+$. Such an initial condition is consistent with the definition (in equation 5b) of σ' as being the vertical normal component of the effective stress tensor, σ'_{ij}. Geometrically, this means point (G) initially falls where an extension of line (G)C in Figures 13B and 14B intercepts the vertical (gravitational force) line joining fixed points A and B. For Figure 13B, this intercept point is fixed point 0; for Figure 14B, it is fixed point M, which is defined as the initial location of moving point (H). Figures 13A and 14A are the resulting force polygons for $t = 0^+$.

Some unexpected implications for velocity vectors (Fig. 6) can now be mentioned. Figure 15 is a velocity diagram for initial conditions of transient flow, assuming that granular movement $\underset{\sim}{v}_s$ is at all times ($t > 0^+$) in the direction of seepage force per unit volume $-\rho_w g\underset{\sim}{\nabla}h$, and that initially ($t = 0^+$) there is no horizontal component of $\underset{\sim}{\nabla}\sigma'$. This defines the first candidate path. Applying these requirements to the path of point (G), it can be demonstrated (Fig. 13B) that the direction of seepage force $-\rho_w g\underset{\sim}{\nabla}h$ cannot change during transient flow and that initially there is no horizontal component of seepage force $-\rho_w g\underset{\sim}{\nabla}h$. This means vector $\underset{\sim}{(G)O}$ in Figure 12 is initially zero valued (as shown in Fig. 13A) and correspondingly, vector (D)E in Figure 6 is initially zero-valued. This is shown in Figure 15. The implication of the initial conditions of the first candidate path can be expressed mathematically as

$$\nabla h(t = 0^+) = 0, \qquad (1^{***})$$

$$q(t = 0^+) = 0, \qquad (2^{***})$$

which in turn imply

$$v_s(t = 0^+) = v_w(t = 0^+) = q_{ult}, \qquad (3^{***})$$

where we have used equations 1, 7, 8 and 15. As shown graphically in Figure 13A, the initial gradient of the vertical normal components of effective stress must equal the initial submerged weight of solids per unit volume, namely

$$\nabla_{P_o} \sigma'(t = 0^+) = -jg(\rho_s - \rho_w)(1 - \phi_{initial}). \qquad (4^{***})$$

Remember that equations 1***, 2*** 3***, and 4*** are applicable only to one of the three candidate paths for moving point (G).

Figure 16 is a velocity diagram for initial conditions of transient flow, assuming that at all times ($t > o^+$) granular movement v_s is in the direction of effective force per unit volume $\nabla\sigma'$, and that initially ($t = o^+$) there is no horizontal component of $\nabla\sigma'$. This defines the second candidate path. Velocity vectors O'E, O'(D), and (D)E in Figure 6 are parallel to force vectors OC*, (G)C*, and O(G) in Figure 12. The same relation is true for corresponding vectors in Figures 16 and 14A. A relation exists between velocity vector (D)E in Figure 16 and force vector (G)O in Figure 14A such that the initial value of specific discharge q equals

$$q(P_o, 0^+) = jK(G - 1)(1 - \phi_{initial}). \qquad (5^{***})$$

Using equations 7, 8, and 15**, equation 5*** implies that the initial values of v_s and v_w can be written

$$v_s(P_o, 0^+)\ q_{ult} - jK(G - 1)(1 - \phi_{initial}), \qquad (6^{***})$$

$$v_w(P_o, 0^+) = q_{ult} + jK(G - 1)(1 - \phi_{initial})^2/\phi_{initial}. \qquad (7^{***})$$

Equations 5***, 6***, and 7*** pertain to primary components of flow (see Appendix II).

As pointed out earlier, points (H) and (G) initially coincide (Fig. 14A). More precisely, we write

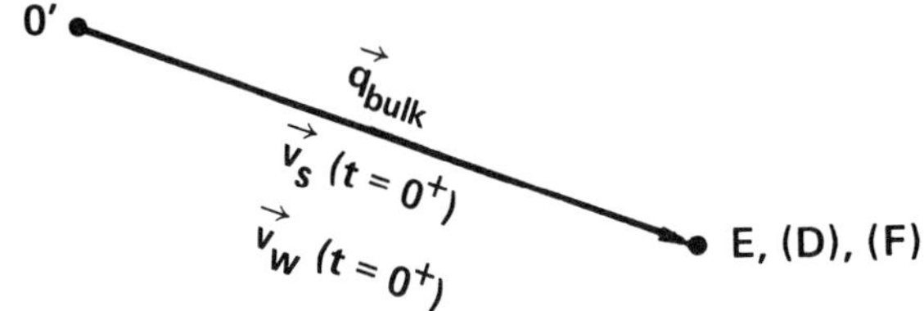

Figure 15. Velocity polygon of initial ($t = 0^+$) conditions in accordance with Figure 13A.

$$\nabla_{P_o} \sigma'(t = 0^+) = 0. \qquad (8^{***})$$

Equations 3***, 6***, and 7*** state in part that horizontal components of $v_s(P_o, 0^+)$ and $v_w(P_o, 0^+$ must equal the horizontal component of ultimate specific discharge q_{ult} at P_o. Equation 8*** states Equation 8*** states that for an instant at $t = 0^+$, the static buoyant weight of solids ceases to be felt as an effective force on any volume element V_{P_o} of interest. Equation 5*** states that specific such that for an initial instant a seepage force carries the weight of the solids. This initial dimensionless seepage force equals

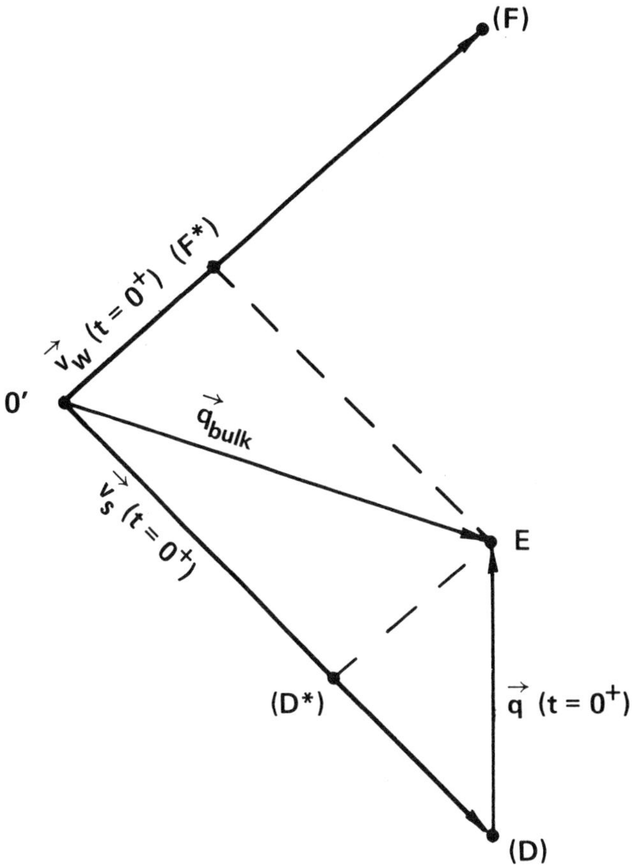

Figure 16. Velocity polygon of initial ($t = 0^+$) conditions in accordance with Figure 14A.

$$-\nabla h(\underset{\sim}{p}_o, 0^+)$$

$$= \underset{\sim}{j} dh(\underset{\sim}{p}_o, 0^+)/dz = \underset{\sim}{j}(G - 1)(1 - \phi_{initial}). \quad (9^{***})$$

Equation 9*** provides a possible mechanism that allows the granular frame to be weightless for an instant and thereby free to respond to a new stress state by moving and reassembling grains into a new bulk density. Eventually a new stable packing arrangement in equilibrium with the new stress state should be reached. This possible mechanism is important not only for reservoir compression in response to fluid withdrawal, but also for reservoir expansion in response to fluid injection. It may provide a key to examining the physics of transient fluid flow through nonrigid porous material.

It is necessary to clarify the meaning of $t = 0^+$ as used in equation 9***. In Appendix I, a surface b is defined as separating a reservoir system from overburden material. The reservoir serves as a source for fluid withdrawal Q. In contrast, the overburden is a nondraining system in that no fluid within it serves as a source for Q. In his modified theory of leaky aquifers, Hantush (1960) has in effect traced a theoretical upward migration of such a surface b through a semiconfining bed. Witherspoon and others (1967, p. 62ff) have done this in greater detail. Conceivably, a specified volume element at $\underset{\sim}{p}_o$ can lie within the overburden as defined above for a finite period of time, only to find that after surface b passes it, it becomes part of the reservoir as defined above. Initial time ($t = 0^+$) associated with Figures 14A and 16 and equation 9*** is actually triggered by the passing of a surface such as b. The concept of a moving hydrodynamic boundary b that migrates upward and perhaps outward radially from a producing well can be applied throughout an entire reservoir system. Its migration rate is undoubtedly affected, and perhaps locally arrested, by stratigraphic heterogeneity, facies changes, structural or bedrock surfaces of impermeability, and other material properties and boundary conditions at depth. The point of this discussion is simply that the initial instant $t = 0^+$ in 9*** is not necessarily identical for distinct points of interest within a reservoir.

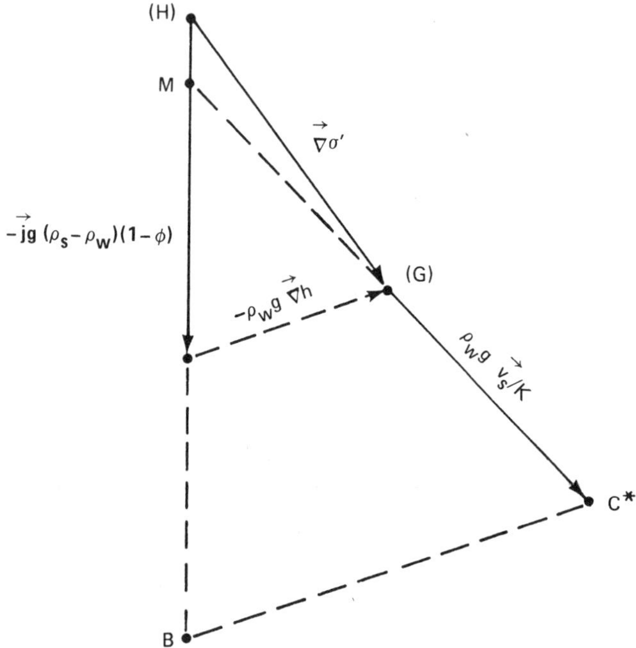

Figure 17. Force polygon of transient conditions ($t > 0^+$) of flow assuming the directions of $\underset{\sim}{v}_s$ and $\underset{\sim}{\nabla}\sigma'$ do not change for constant Q.

In contrast to equation 1***, equation 9*** requires either a preliminary rise in hydraulic head h beneath a point of interest $\underset{\sim}{p}_o$ or a preliminary decline in hydraulic head above a point of interest. Under some circumstances a preliminary rise in hydraulic head in response to pumping followed by a subsequent long-term decline has been observed in the field (Verruijt, 1969; Wolff, 1970). The preliminary rise in hydraulic head has previously been explained (Verruijt, 1969, 1971) in terms of Biot's (1941), Mandel's (1953), and Cryer's (1963) theoretical studies of three-dimensional stress-strain relations. The present Appendix has reached a similar conclusion by a different conceptual route involving exploration of the relation of force fields per unit volume (Fig. 5) for ultimate steady flow, and velocity fields (Fig. 6) for transient flow.

For the sake of illustration, consider a third candidate path where, similar to the first candidate path, the direction of $\underset{\sim}{v}_s$ is constant for constant Q. Instead of assuming the direction of $\underset{\sim}{\nabla}h$ remains constant, however, let us assume that the direction of $\underset{\sim}{\nabla}\sigma'$ remains constant. This assumption is plausible for a solid/liquid mixture (see equation 9** in Appendix II). For simplicity, we once more require any horizontal component of effective force $\underset{\sim}{\nabla}\sigma'$ to be zero-valued at $t = 0^+$. For this third and more plausible candidate path, point (G) falls somewhere on a straight line joining fixed points M and C*(Fig. 17). Ultimate conditions (Fig. 5) are once again satisfied. Of primary significance is that initial conditions are representable by Figures 14A and 16. Hence equations 5***, 6***, 7***, 8***, and 9*** hold true for the third candidate path. The discussion of equation 9*** is also applicable.

Because the dynamics of skeletal movement are not yet completely understood, it is premature to opt for one candidate path of point (G) over another. A fourth, as yet unsuspected, path may prove to be physically and mathematically necessary as well as sufficient. On the other hand, several paths may possibly be found to be equally likely. Boundary conditions (such as type of loading and locations of impermeable surfaces) will probably be found to be significant in determining how point (G) approaches point C* of Figure 11. The tacitly assumed dynamics of traditional hydrogeologists (namely, Figures 13 and 15) is only one, and certainly not the only, possibility.

In this Appendix, we have essentially posed the following conceptual problem. Under most field conditions the directions of $\underset{\sim}{v}_s$ and transient $\underset{\sim}{\nabla}h$ are not actually known. Even if they are tacitly or directly specified for convenience in order to solve a partial differential equation, we should take care that such specifications conform as closely as possible to physical behavior in the field. This represents an area for further field measurements and theoretical development.

APPENDIX IV: SAMPLE CALCULATION OF $\underset{\sim}{v}_s$ BEYOND THE RADIUS OF COMPRESSION

The importance of being able to calculate the value of $\underset{\sim}{v}_s$ has been discussed at length. We shall now give a sample calculation of $\underset{\sim}{v}_s$ for a special case. The term $\underset{\sim}{v}_s$ represents the primary component of skeletal movement (see Appendix II).

If a producing formation of thickness b is radially extensive, it is a fairly straightforward matter to estimate $\underset{\sim}{v}_s$ at a large distance r (> R) away from a well discharging at rate Q. During both steady-state and transient conditions of fluid flow through a confined system, $\underset{\sim}{q}_{bulk}$ can be estimated in accordance with equation 22 to approximate

$$\underset{\sim}{q}_{bulk} = -\underset{\sim}{i}Q/2\pi rb$$

for any r greater than the radius r_w of a discharging well and where $\underset{\sim}{i}$ (= $\underset{\sim}{r}/r$) is a unit radial vector (positive outward). It is convenient to assume a cylindrical reservoir (Fig. 2). For the sake of illustration and computational simplicity, assume that downward movement of incompressible solids is offset by upward movement of incompressible water due to porosity loss so that net bulk flow $\underset{\sim}{q}_{bulk}$ is radial. This assumption would need to be modified in a more rigorous analysis. Because most casings of producing wells preferentially pass liquids and stop solids from being removed from underground, a complex skeletal dynamic develops within a zone $r_w < r < R$, where R is the radius of a zone of compression (Fig. 2). An alternative way to conceptualize R is to perceive it as the radius of a type of zone constraint. Induced radial movement of solids $v_{s\ r}$ is constrained at $r = r_w$. The effect of this boundary constraint is felt throughout the system, $r_w < r < R$, causing a decrease in porosity. Beyond the radius of influence R, there is no decrease in porosity and hence no compression. At most there is only a change in shape of possibly moving pore volumes. For $r > R$, local increase in the tangential component of strain $\varepsilon_{\Theta\Theta}(= v_{s\ r}/r)$ due to inward radial movement is is offset by a decrease in the radial component of strain $\varepsilon_{rr}(= -dv_{s\ R}/dr)$ due to local elongation of pore volume along a radial line. The result is that no change in volume strain ε occurs beyond R even though skeletal material itself may be moving radially inward as part of a dynamic system. For incompressible constituents, there is no local movement of fluid past solids; $\underset{\sim}{v}_s = \underset{\sim}{v}_w$ for $r > R$. In accordance with equation 8, $\underset{\sim}{q} = 0$ for $r > R$. In accordance with equation 6, $\underset{\sim}{q}_{bulk} = \underset{\sim}{v}_s = \underset{\sim}{v}_w$ for $r > R$. By solving equation 22, the value of $\underset{\sim}{q}_{bulk}$ is known for all $r > r$. Hence for $r > R >> r_w$, because of the local numerical identity of $\underset{\sim}{q}_{bulk}$ with $\underset{\sim}{v}_s$, we find that

$$\underset{\sim}{v}_s = -\underset{\sim}{i}Q/2\pi rb \quad (r > R).$$

Assuming sample values of $Q = 5 \times 10^3\ m^3/d$ (= 917 gpm), $R = 2 \times 10^3$ m; and $b = 1 \times 10^2$ m, we find a maximum magnitude for $\underset{\sim}{v}_s$ near R to be approximately

$$\underset{\sim}{v}_s = 4 \times 10^{-3}\ m/d.$$

From the foregoing formula, the magnitude of $\underset{\sim}{v}_s$ generally decreases as r increases beyond R. Because of the boundary constraint at r_w, the magnitude of $\underset{\sim}{v}_s$ also tends to decrease with decreasing r within R. Evaluating $\underset{\sim}{v}_s$ within R is not a simple matter because it also decreases with time. In the main body of the paper, we have derived governing equations that permit $\underset{\sim}{v}_s$ to be calculated within R.

Obviously $\underset{\sim}{v}_s$ as calculated in this Appendix cannot continue at $r \geq R$ indefinitely at a constant rate or the reservoir would eventually collapse completely, which is physically impossible. The boundaries R or b should expand either smoothly or by quantum jumps until an "infinite" source of fluid is intercepted and steady-state fluid flow can be maintained. If instead of encountering such an infinite source, an impermeable boundary is encountered and the outward expansion of R and b is effectively arrested, Q will continue unchanged, but under conditions that allow drawdown Δh to increase dramatically. If such an impermeable boundary at depth (such as sloping bedrock, cemented facies change, or an impermeable fault plane) is also rigid in the sense that it cannot move horizontally, then one can expect a subsurface tension crack to be initiated.

ACKNOWLEDGMENTS

The 60th birthday celebration of Paul A. Witherspoon has provided the happy occasion for writing this paper. Professor Witherspoon imparts to his students not only knowledge, but respect for science, confidence, and disciplined optimism. I am fortunate and deeply grateful to have been one of his students.

Any author writing about subsidence due to fluid withdrawal owes a great debt to the pioneering work of the U.S. Geological Survey. My debt is greater than most because of my professional ties with Joseph F. Poland, Ben E. Lofgren, and Francis S. Riley of the Survey. Having learned subsidence research under the careful and gentle guidance of Joseph F. Poland is a singular honor.

In large part, this paper is a response to stimulating correspondence and conversation with J. Edward Smith of Phillips Petroleum Company. Alan M. Winslow, Paul W. Kasameyer, and John F. Schatz of Lawrence Livermore National Laboratory reviewed this paper with insight, patience, and humor. Finally, I should like to thank T. N. Narasimhan and J. A. Cheney of the University of California and S. K. Garg of Systems,

Science and Software for their challenging skepticism. Work was performed at the Lawrence Livermore Laboratory, Livermore, California, under the auspices of U.S. Department of Energy Contract No. W-7405-ENG-48.

REFERENCES CITED

Athy, L. F., 1930, Density, porosity, and compaction of sedimentary rocks: Bulletin of the American Association of Petroleum Geologists, v. 14, p. 1-24.

Biot, M. A., 1941, General theory of three-dimensional consolidation: Journal of Applied Physics, v. 12, no. 2, p. 155-164.

Brutsaert, W., and Corapcioglu, M. Y., 1976, Pumping of aquifer with viscoelastic properties: Journal, Hydraulic Division, American Society of Civil Engineers, v. 102(HY11), p. 1663-1675.

Bull, W. B., 1964, Alluvial fans and near-surface subsidence in western Fresno County, California: U.S. Geological Survey Professional Paper 437-A, 70 p.

___1972, Prehistoric near-surface subsidence cracks in western Fresno County, California: U.S. Geological Survey Professional Paper 437-C, 85 p.

___1975, Land subsidence due to ground-water withdrawal in the Los Banos-Kettleman City area, California, 2, Subsidence and compaction of deposits: U.S. Geological Survey Professional Paper 437-F, 90 p.

Bull, W. B., and Miller, R.E., 1975, Land subsidence due to ground-water withdrawal in the Los Banos-Kettleman City area, California, 1, Changes in the hydrologic environment conducive to subsidence: U.S. Geological Survey Professional Paper 437-E, 71 p.

Bull, W. B., and Poland, J. F., 1975, Land subsidence due to ground-water withdrawal in the Los Baños-Kettleman City area, California, 3, Interrelations of water-level change, change in aquifer-system thickness, and subsidence: U.S. Geological Survey Professional Paper 437-G, 62 p.

Carrillo, N., 1948, Influence of artesian wells in the sinking of Mexico City: Proceedings, Second International Conference Soil Mechanics and Foundation Engineering, Rotterdam, v. 7, p. 156-159.

___1949, Subsidence in the Long Beach-San Pedro area: Stanford, Stanford Research Institute, p. 67-69, 227-242.

___1969, The subsidence of Mexico City and Texcoco project: Mexico City, Sociedad Mexicana de Saelos, 328 p.

Chapman, R. E., 1972a, Clays with abnormal interstitial fluid pressures: Bulletin of the American Association of Petroleum Geologists, v. 56 no. 4, p. 790-795.

___1972b, Primary migration of petroleum from clay source rocks: Bulletin of the American Association of Petroleum Geologists, v. 56, no. 11, p. 2185-2191.

Christian, J. T., and Hirschfeld, R. C., 1974, Subsidence of Venice: predictive difficulties: Science, v. 185, no. 9, p. 1185.

Cooley, R. L., 1975, A review and synthesis of the Biot and Jacob-Cooper theories of ground-water motion: University of Nevada, Desert Research Institute, Technical Report H-W-25, 87 p.

Corapcioglu, M. Y., 1977, Mathematical modeling of leaky aquifers with rheological properties, in Johnson, A. I., ed., Land subsidence: International Association of Hydrologic Sciences Publication 121, p. 191-200.

Corapcioglu, M. Y., and Brutsaert, W., 1977, Viscoelastic aquifer model applied to subsidence due to pumping: Water Resources Research, v. 13, no. 3, p. 597-604.

Crochet, M. J., and Naghdi, P. M., 1966, On constitutive equations for flow of fluid through an elastic solid: International Journal of Engineering Science, v. 4, p. 381-401.

Cryer, C. W., 1963, A comparison of the three-dimensional consolidation theories of Biot and Terzaghi: Quarterly Journal of Mechanics and Applied Mathematics, v. 16, p. 401-412.

Dickinson, G., 1953, Geological aspects of abnormal reservoir pressures in Gulf Coast Louisiana: Bulletin American Association of Petroleum Geologists, v. 37, no. 2, p. 410-432.

Finnemore, E. J., 1971, A summary description of the present status of geothermal reservoir and subsidence models, in Atherton, R.W., ed, The analysis of subsidence associated with geothermal development, 2, Research report: Palo Alto, Systems Control Inc., Rept. 5139-2, p. 7/1-7/50.

Finol, A., and Farouq Ali, S. M., 1975, Numerical simulation of oil production with simultaneous ground subsidence: Journal of the Society of Petroleum Engineers, v. 15, no. 5, p. 411-422.

Fuller, M. L., 1908, Summary of the controlling factors of artesian flows: U.S. Geological Survey Bulletin 319, 44 p.

Gambolati, G. A., 1972, A three-dimensional model to compute land subsidence: Bulletin of the International Association of Hydrologic Sciences, v. 17, no. 2, p. 219-226.

___1977, Deviations from the Theis solution in aquifers undergoing three-dimensional consolidation: Water Resources Research, v. 13, no. 1, p. 62-68.

Gambolati, G. A., and Freeze, R. A., 1973, Mathematical simulation of the subsidence of Venice, 1, Theory: Water Resources Research, v. 9, no. 3, p. 721-733.

Gambolati, G. A., Gatto, P., and Freeze, R. A., 1974, Mathematical simulation of the subsidence of Venice, 2, Results: Water Resources Research, v. 10, no. 3, p. 563-577.

Garg, S. K., 1975, Simulation of fluid-rock interactions in a geothermal basin: La Jolla, Systems, Science and Software, Final Report No. SSS-R-76-2734, 63 p.

Geertsma, J., 1957, The effect of fluid pressure decline on volumetric changes of porous rocks: Transactions, American Society of Mechanical Engineers, v. 210, p. 331-340.

___1966, Problems of rock mechanics in petroleum production engineering: in Proceedings, First Congress of the International Society of Rock Mechanics, Lisbon, v. 1, p. 585-594.

___1973a, A basic theory of subsidence due to reservoir compaction: The homogeneous case: Verherdeligen Van Hel Konin Kligk Nederland Geologisch Mijnbouwkundig Genootschap, v. 28, p. 43-62.

___1973b, Land subsidence above compacting oil and gas reservoirs: Journal of Petroleum Technology, v. 25, p. 734-744.

Geological Society of America, 1978, Abstracts with programs, v. 10, no. 3, p. 93-160.

Gersevanov, N. M., 1937, The foundation of dynamics of soil (in Russian) (third edition): Leningrad, Stroiizdat.

Gibson, R. E., England, G. L., and Hussey, J. J. L., 1967, The theory of one-dimensional consolidation of saturated clays: 1, Finite nonlinear consolidation of thin homogeneous layers: Geotechnique, v. 17, no. 2, p. 261-273.

Gray, W. G., and O'Neil, K., 1976, On the general equations for flow through porous media and their reduction to Darcy's law: Water Resources Research, v. 12, no. 2, p. 148-154.

Green, A. E., and Naghdi, P. M., 1965, A dynamical theory of interacting continua: International Journal Engineering Science, v. 3, p. 231-240.

Grover, M. K., 1977, Evaluation of geothermal energy exploration and resource assessment--final report, v. 1: A review of geothermal subsidence modeling: Marina Del Rey, R and D Associates, Report TR-0400-001, 283 p.

Hantush, M. S., 1960, Modification of the theory of leaky aquifers: Journal of Geophysical Research, v. 65, no. 11, p. 3713-3725.

Hantush, M. S., and Jacob, C. E., 1955, Non-steady radial flow in an infinite leaky aquifer: American Geophysical Union Transactions, v. 36, no. 1, p. 95-100.

Hardin, F. R., and Hardin, G. C., Jr., 1961, Contemporaneous normal faults of the Gulf Coast and their relation to flexures: Bulletin American Association of Petroleum Geologists, v. 45, no. 2, p. 238-248.

Hedberg, H. D., 1936, Gravitational compaction of clays and shales: American Journal of Science, 5th Series, v. 31, no. 184, p. 241-287.

Helm, D.C., 1972, Simulation of aquitard compaction due to changes in stress (ab.): American Geophysical Union Transactions, v. 53, no. 11, p. 979.

___1975, One-dimensional simulation of aquifer system compaction near Pixley, California, 1, Constant parameters: Water Resources Research, v. 11, no. 3, p. 465-478.

___1976, One-dimensional simulation of aquifer system compaction near Pixley, California, 2, Stress-dependent parameters: Water Resources Research, v. 12, no. 3, p. 375-391.

___1977, Estimating parameters of compacting fine-grained interbeds within a confined aquifer system by a one-dimensional simulation of field observations, in Johnson, A. I., ed., Land Subsidence: International Association of Hydrologic Sciences Publication 121, p. 145-156.

___1978a, Radial movement of a confined aquifer towards a pumping well (ab.): Geological Society of America, Abstracts with Programs, v. 10, no. 3, p. 109.

___1978b, Field verification of a one-dimensional mathematical model for transient compaction and expansion of a confined aquifer system, in Verification of mathematical and physical models in hydraulic engineering: Proceedings, 26th Hydraulic Division Specialty Conference, College Park, Maryland, American Society of Civil Engineers, p. 189-196.

___1979a, A postulated relation between granular movement and Darcy's law for transient flow, in Saxena, S. K., ed., Evaluation and prediction of subsidence: American Society of Civil Engineers, p. 417-440.

___1979b, Comment on "Governing equations for geothermal reservoirs," by D. H. Brownell, S. K. Garg, and J. W. Pritchett: Water Resources Research, v. 15, no 3, p. 723-726.

Hubbert, M. K., 1940, The theory of ground-water motion: Journal of Geology, v. 48, no. 8, p. 785-944 (reprinted: New York, Hafner Publication Co., 1969).

Hubbert, M. K., and Rubey, W. W., 1959, Role of fluid pressure in mechanics of overthrust faulting: Geological Society of America Bulletin, v. 70, no. 2, p. 115-166.

Jacob, C. E., 1946, Radial flow in a leaky artesian aquifer: Transactions of the American Geophysical Union, v. 27, p. 198-205.

Jaeger, J. C., and Cook, N.G.W., 1971, Fundamentals of rock mechanics: London, Chapman and Hall Ltd., 515 p..

Johnson, A. I., ed., 1977, Land subsidence, Proceedings, 2nd International Symposium on Land Subsidence, Anaheim, California: Ann Arbor, International Association of Hydrologic Sciences, Publication no. 121, 669 p.

Johnson, A. I., Moston, R. P., and Morris, D. A., 1968, Physical and hydrologic properties of water-bearing deposits in subsiding areas in central California: U.S. Geological Survey Professional Paper 497-A, 71 p.

Lawrence Berkeley Laboratory, 1977, Geothermal subsidence research program plan: Berkeley, University of California, Publication no. LBL-5983, 111 p.

Lewis, R. W., and Schrefler, B., 1978, A fully coupled consolidation model of the subsidence of Venice: Water Resources Research, v. 14, no. 2, p. 223-230.

Lofgren, B. E., 1968, Analysis of stresses causing land subsidence: Geological Survey Research 1978, U.S. Geological Survey Professional Paper 600, p. B219-B225.

___1970, Field measurements of aquifer-system compaction, San Joaquin Valley, California, U.S.A., in Tison, L. J., ed., Land Subsidence: International Association of Scientific Hydrology Publication 88, p. 272-284.

___1971, Significant role of seepage stresses in compressible aquifer systems (ab.): American Geophysical Union Transactions, v. 52, no. 11, p. 832.

___1975, Land subsidence due to ground-water withdrawal, Arvin-Maricopa area, California: U.S. Geological Survey Professional Paper 437-D, 55 p.

___1978, Hydraulic stresses cause ground movement

and fissures, Picacho, Arizona: Geological Society of America, Abstracts with Programs, v. 10, no. 3, p. 113.

Lofgren, B. E., and Klausing, R. L., 1969, Land subsidence due to ground-water withdrawal, Tulare-Wasco area, California: U.S. Geological Survey Professional Paper 437-B, 101 p.

Magara, K, 1968, Compaction and migration of fluids in Miocene mudstone, Nagaoka Plain, Japan: Bulletin of the American Association of Petroleum Geology, v. 52, no. 12, p. 2466-2501.

___1978, Compaction and fluid migration: Amsterdam, Elsevier Scientific Publishing Co., 319 p.

Mandel, J., 1953, Consolidation des sols (etude mathematique): Geotechnique v. 3, no.7, p. 287-299.

Mase, G. E., 1970, Continuum mechanics: New York, McGraw-Hill Book Co. (Schaum's Outline Series), 221 p.

McCann, G. D., and Wilts, C. M., 1951, A mathematical analysis of the subsidence in the Long Beach-San Pedro area: Pasadena, California Institute of Technology, internal report, 117 p.

Meade, R. H., 1964, Removal of water and rearrangement of particles during the compaction of clayey sediments--review: U.S. Geological Survey Professional Paper 497-B, 23 p.

___1967, Petrology of sediments underlying areas of land subsidence in central California: U.S. Geological Survey Professional Paper 497-C, 83 p.

___1968, Compaction of sediments underlying areas of land subsidence in central California: U.S. Geological Survey Professional Paper 497-D, 39 p.

Mikasa, M., 1965, The consolidation of soft clay--A new consolidation theory and its applications: Civil Engineering in Japan, Japan Society of Civil Engineers, p. 21-26.

Miller, I., 1980, Simulation of geothermal subsidence: Berkeley, Lawrence Berkeley Laboratory, LBL 10794, 160 p.

Miller, R. E., Green, J. H., and Davis, G. H., 1971, Geology of the compacting deposits in the Los Banos-Kettleman City subsidence area, California: U.S. Geological Survey Professional Paper 497-E, 46 p.

Mindlin, R. D., and Cheng, D. H., 1950, Thermoelastic stress in the semi-infinite solid: Journal of Applied Physics, v. 21, p. 931.

Minor, H. E., 1925, Goose Creek oil field, Harris County, Texas: American Association of Petroleum Geologists Bulletin, v. 9, no. 2, p. 286-297.

Narasimhan, T. N., and Witherspoon, P. A., 1977, Numerical model for land subsidence in shallow ground-water systems: in Johnson, A. I., ed., Land Subsidence, International Association of Hydrologic Sciences Publication 121, p. 133-144.

Neuman, S. P., and Witherspoon, P. A., 1969, Theory of flow in a confined two aquifer system: Water Resources Research, v. 5, no. 4, p. 803-816.

Pacific Fire Rating Bureau, 1958, Subsidence in Long Beach-Terminal Island-Wilmington, California: Internal report, p. 38.

Poland, J. F., 1972, Subsidence and its control, in Cook, T. D., ed., Underground Waste Management and Environmental Implications: American Association of Petroleum Geologists, Memoir no. 18, p. 50-71.

Poland, J. F., and Davis, G. H., 1969, Land subsidence due to withdrawal of fluids, in Reviews in Engineering Geology, v. 2: Boulder, Geological Society of America, p. 187-269.

Poland, J. F., 1975, Land subsidence in the San Joaquin Valley, California, as of 1972: U.S. Geological Survey Professional Paper 437-H, 78 p.

Poland, J. F., Lofgren, B. E., and Riley, F. S., 1972, Glossary of selected terms useful in studies of the mechanics of aquifer systems and land subsidence due to fluid withdrawal: U.S. Geological Survey Water Supply Paper 2025, 9 p.

Pollard, W. S., Holcombe, R. F., and Marshall, A. F., 1979, Subsidence cause and effect, Harris-Galveston coastal subsidence district phase 1-A study: Houston, McClelland Engineers, Inc., 2 vols.

Polubarinova-Kochina, P. Ya., 1952, Theory of ground-water motion (translated by R. J. M. DeWiest): Princeton, Princeton University Press (1962).

Pratt, W. E., and Johnson, D. W., 1926, Local subsidence of the Goose Creek Field: Journal of Geology, v. 34, no. 7, p. 577-590.

Rappleye, H. S., 1933, Recent areal subsidence found in releveling: Engineering News-Record, v. 11, p. 848.

Riley, F. S., and McClelland, E. J., 1972, Application of the modified theory of leaky aquifers to a compressible multiple-aquifer system: U.S. Geological Survey Open-File Report, 96 p.

Rubey, W. W., and Hubbert, M. K., 1959, Role of fluid pressure in mechanics of over thrust faulting, 2, Over thrust belt in geosynclinal area of western Wyoming in light of fluid-pressure hypothesis: Geological Society of America Bulletin, v. 70, no. 2, p. 167-206.

Sandhu, R. S., 1979, Modeling land subsidence, in Saxena, S. K., ed., Evaluation and prediction of subsidence: American Society of Civil Engineers, p. 565-579.

Saxena, S. K., ed., 1979, Proceedings, International Conference on Evaluation and Prediction of Subsidence, Pensacola Beach, Florida: New York, American Society of Civil Engineers, 594 p.

Schatz, J. F., Kasameyer, P. W., and Cheney, J. A., 1978, A method of using in situ porosity measurements to place an upper bound on geothermal reservoir compaction: in Second Invitational Well Testing Symposium Proceedings: Berkeley, Lawrence Berkeley Laboratory, p. 90-94.

Schiller, R., 1975, The growing menace of our sinking lands: Reader's Digest, v. 107, no. 641, p. 124-127.

Sen, B., 1950, Note on the stresses produced by nuclei of thermoelastic strain in a semi-infinite elastic solid: Quarterly of Applied Mathematics, v. 8, no. 635.

Smith, J. E., 1971, The dynamics of shale compaction and evolution of pore-fluid pressure: International Association of Mathematical Geology Journal, v. 3, p. 239-263.

Tabbador, F., and Little, R. M., 1971, Interacting continuous medium composed of an elastic solid and an incompressible Newtonian fluid: International Journal Solids Structure, v. 7, p. 825-841.

Taylor, D. W., 1948, Fundamentals of soil mechanics: New York, John Wiley and Sons, 700 p.

Terzaghi, K., 1925, Principles of soil mechanics: IV, Settlement and consolidation of clay: Engineering News-Record, p. 874-878.

Tibbets, F. H., 1933, Areal subsidence: Engineering News-Record, v. 3, p. 204.

Tison, L. J., ed., 1970, Land subsidence, Proceedings of the Tokyo Symposium 1969: Louvian, Belgium, International Association of Scientific Hydrology, Publication no. 88 and 89 (2 vols), 661 p.

Tolman, C. F., 1937, Ground water: New York, McGraw-Hill Book Company, 593 p.

Tolman, C. F., and Poland, J. F., 1940, Ground-water, salt-water infiltration, and ground-surface recession in Santa Clara Valley, Santa Clara County, California: American Geophysical Union Transactions, 1, p. 23-34.

Verruijt, A., 1969, Elastic storage of aquifers: in DeWiest, R. J. M., ed., Flow through porous media: New York, Academic Press, p. 331-376.

___1970, Horizontal displacement in pumped aquifers (ab.): American Geophysical Union Transactions, v. 51, no. 4, p. 284.

___1971, Comment on Relationship between horizontal strain near a well and reverse water level fluctuations, Water Resources Research, v. 7, no. 5, p. 1369.

Witherspoon, P. A., and Freeze, R. A., 1972, The role of aquitards in multiple-aquifer systems, Penrose Conference of the Geological Society of America 1971: Geotimes, v. 17, no. 4, p. 22-24.

Witherspoon, P. A., 1967, Interpretation of aquifer gas storage conditions of water pumping tests: New York, American Gas Association, 273 p.

Wolff, R. G., 1970, Relationship between horizontal strain near a well and reverse water level fluctuations: Water Resources Research, v. 6, no. 6, p. 1721-1728.

Yerkes, R. F., and Castle, R. O., 1970, Surface deformation associated with oil and gas field operations in the United States: in Tison, L. J., ed., Land subsidence, v. 1: International Association of Scientific Hydrology, Publication 88, p. 55-56.

Manuscript Accepted by the Society October 14, 1980

Printed in U.S.A.

II. Measurement Aspects

Geological Society of America
Special Paper 189
1982

Physical Properties of Porous Geologic Materials

WILLIAM N. HOUSTON
Professor of Civil Engineering
University of California
Berkeley, California 94720

ADEL G. KASIM
Consulting Engineer
Hallenbeck McKay and Associates
Emeryville, California 94608

ABSTRACT

Porosity, permeability, and moduli are the properties most fundamental to assessing skeletal deformations and water flow rates in porous materials. For saturated materials, these properties are similar for both "soft" and "stiff" material skeletons, but greatly simplified models can be used when the skeleton is relatively soft. The effective stress is a key parameter for both soft and stiff materials when saturation is complete, but is much less powerful for partially saturated materials.

Ideally, the needed properties of porous materials should be determined from an integrated program of laboratory testing, field testing, and solution of the inverse problem using a numerical method capable of modeling the boundary conditions used in the field tests. Where such a large-scale program is impractical, properties can be estimated from extensive laboratory testing and from the typical values suggested herein.

INTRODUCTION

Since the 1940s, there has been a steadily increasing emphasis on the study and assessment of the behavior of porous materials. The two aspects that are of primary interest are 1) deformations and, in some cases, the stability of a mass of porous materials; 2) quantity and rate of fluid flow from and through the pores of the materials.

These two aspects of behavior have not been given equal attention by all disciplines of study. Geotechnical engineers, for example, are interested in both aspects but are primarily concerned with predicting the deformations produced by changes in the state of stress. Their interest in water flow derives not only from the economic aspects of flow quantities but also from the fact that water movements affect the state of stress and hence the resulting deformations.

Hydrogeologists, on the other hand, are primarily interested in assessing the amounts, rates, and patterns of water flow through the porous media. Although the emphasis and the scale of problems may differ from one discipline to another, the relationship between the two aspects of behavior of porous materials remains much the same. Consequently, analytical models developed and physical properties determined in one discipline may well be useful to work in other disciplines.

In this paper, we develop a more global approach to the study, assessment, and selection of the most fundamental physical properties of porous materials. It should then be possible to use physical property data more effectively regardless of which discipline provides them.

We first identify and define fundamental properties that are common to mathematical models that are usually employed to assess the deformations or the water flow. Whenever applicable, we discuss relationships among these properties. The second section of the paper focuses on each of the fundamental properties. Typical ranges of values are presented, together with direct and indirect methods of measurement. The remainder is devoted to summarizing and commenting on the effectiveness of the various methods used in assessing the physical parameters and to presenting conclusions as to the significance of some of the accomplishments to date in this field, current trends, and areas of potential improvements.

IDENTIFICATION OF FUNDAMENTAL PROPERTIES

The physical properties fundamental to the behavior of porous materials are those which relate a change in pressure in the system to the rate and amount of resulting deformation and fluid flow. The change in pressure (or stress) could be a result of:

1. Change in the external boundary stress, referred to hereafter as a "total" stress
2. Change in a body force, such as gravity or seepage force
3. Change in pore fluid pressure

Thus it is necessary to be more specific regarding the processes involved in order to identify fundamental properties. It will also be necessary to identify several properties as fundamental, rather than just one or two.

For the purpose of establishing terminology and a framework for discussion it is desirable to review briefly the concept of *effective stress* in porous materials.

Effective Stress

The principle of effective stress is the cornerstone on which soil mechanics is founded and is only slightly less important to the field of rock mechanics. Terzaghi (1925) first introduced the concept, which was developed by others, including Skempton (1960a, b). This development was reviewed and further explained by Mitchell (1976). Skempton (1960b) showed that equations 1 and 2 (below) validly express the "effective" stresses that control shearing resistance and volume change, respectively. These equations were checked experimentally and found to be applicable for saturated soil, rock, and concrete. The effective stress defined by equations 1 and 2 may differ in general from the intergranular pressure defined by Skempton (1960b).

$$\sigma' = \sigma - \left(1 - \frac{a_c \tan\psi}{\tan\phi'}\right) u, \tag{1}$$

where σ' = effective stress; that is, stress that uniquely controls shearing resistance,
σ = total or external stress,
u = pore water pressure, in units of stress,
a_c = average particle-to-particle or solid-to-solid contact area,
ψ = an intrinsic friction angle,
ϕ' = effective stress angle of shearing resistance;

$$\sigma' = \sigma - \left(1 - \frac{C_s}{C}\right) u, \tag{2}$$

where σ' = effective stress; that is, stress that uniquely controls volume change,
C = compressibility of the material "skeleton" or framework of particles,
C_s = compressibility of individual particles or of the "solid."

Compressibility is defined as the volumetric strain per unit of applied effective stress. For an isotropic linear elastic material it would be the reciprocal of the bulk modulus, K. In the case of the compressibility of the material skeleton, C, the volumetric strain is the volume change per unit bulk volume of the material. Compressibility is measured by applying an increment of pressure externally and measuring the resultant volumetric strain of the material skeleton after any pore fluid pressure has dissipated. That is, the pore fluid is allowed to move freely in or out until no gradient for flow exists. The compressibility, C, can be measured under "hydrostatic" loading conditions (Fig. 1a) or uniaxial compression conditions of saturated specimens (Fig. 1b), also called one-dimensional compression.

If the material were isotropic and elastic, the compressibility for case a in Figure 1 would be expected to be about 1.5 times the compressibility for case b. This difference is often neglected as a first approximation, however, and becomes relatively unimportant when ratios of compressibility are used.

Although equations 1 and 2 are very powerful, they are insufficient to answer the question of how much of an external load will be borne by the material skeleton and how much by the pore water of a saturated material when an increment in external stress is applied and the element is kept undrained. In addition, the relative values of the compressibility of water, C_w, and the material skeleton, C, are needed. Static equilibrium requires that the sum of the two load components must equal the externally applied load. However, the increment in effective stress, $\Delta\sigma'$, plus the increment in pore water pressure, Δu, is not necessarily equal to the increment in external (total) stress, $\Delta\sigma$, for low-porosity materials or materials for which C_s/C is significant.

The reason for nonequivalence is that the cross-sectional area on which the water pressure operates may differ significantly from the gross cross-sectional area in low-porosity materials. Relative values of stress may be determined using Figure 2.

In Figure 2, the material skeleton and the water phase are shown separated, even though the material is saturated. In general, the rigid membrane shown would not allow the increment in total stress to

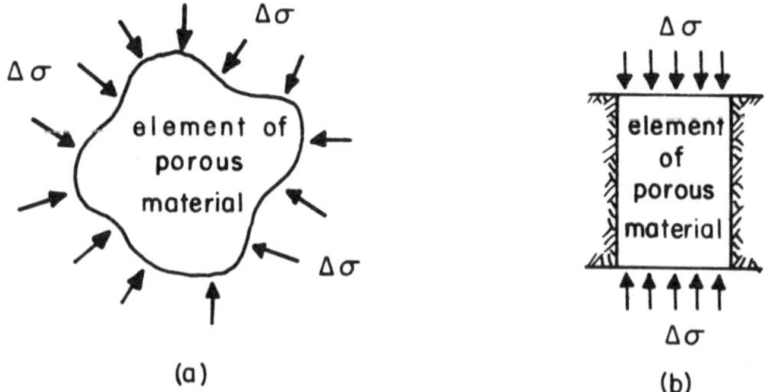

Figure 1. Boundary conditions for measuring compressibility.

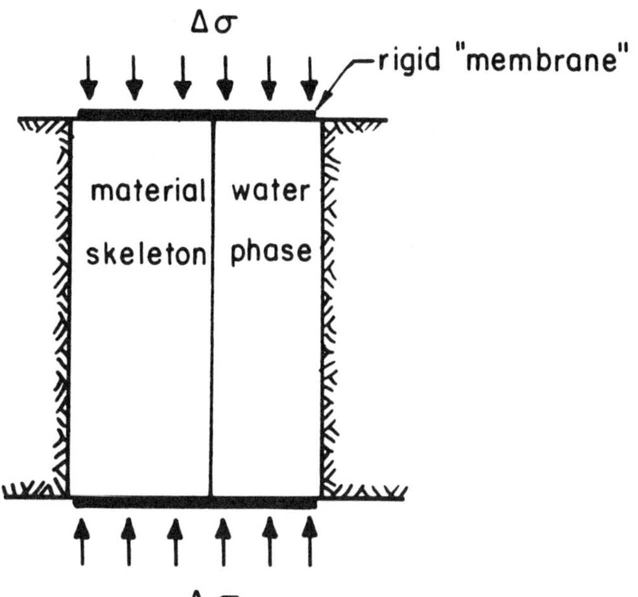

Figure 2. Schematic of saturated material with skeleton and water-phase separation.

be equally distributed to the water and the skeleton. In fact, we would expect the stiffer of the two to assume most of the load. Whatever the distribution of load (and stress) we would expect the volume change of the material skeleton to be identical to the volume change of the water because the element is undrained and required to remain saturated during loading. Thus,

$$C\Delta\sigma' = \frac{\Delta V_{skeleton}}{V}, \tag{3}$$

where $\Delta V_{skeleton}$ = change in bulk volume of skeleton,
V = bulk volume.

Then,

$$C_w\Delta u = \frac{\Delta V_w}{V_w} \tag{4a}$$

and

$$C_s\Delta u = \frac{\Delta V_{solids}}{V_{solids}}, \tag{4b}$$

where ΔV_w = change in volume of water,
V_w = volume of water = nV,
ΔV_{solids} = change in the volume of solid particles or solid phase,
V_{solids} = $(1 - n)V$,

n = porosity in decimal = volume of voids per unit bulk volume.

By equating ΔV_w to $(\Delta V_{skeleton} - \Delta V_{solids})$ from equations 3 and 4, equation 5 is obtained.

$$\Delta\sigma' = \frac{C_w\Delta u\, nV + C_s\Delta u\,(1 - n)V}{CV}$$

$$= \frac{nC_w\,\Delta u + (1 - n)C_s\,\Delta u}{C}. \tag{5}$$

Substitution of $\Delta\sigma$ from equation 5 into an incremental form of equation 2 yields:

$$\frac{\Delta u}{\Delta\sigma} = \frac{1}{1 + n\left(\frac{C_w}{C} - \frac{C_s}{C}\right)} = \text{"B"}. \tag{6}$$

The ratio $\Delta u/\Delta\sigma$ is Skempton's "B" parameter, which commonly is used in quantifying pore pressure changes in response to external load. From equations 5 and 6 the ratio $\Delta\sigma'/\Delta\sigma$ can also be computed readily:

$$\frac{\Delta\sigma'}{\Delta\sigma} = B\,n\,\frac{C_w}{C} - B(1 - n)\,\frac{C_s}{C}. \tag{7}$$

In the course of experimentally verifying equations 1 and 2, Skempton (1960b) gathered and developed considerable data on compressibility. Most of these data are shown in the first three columns of Table 1. It is immediately obvious that C_s/C is negligible for practical purposes for the dense sand and the materials listed below it. On the basis of data not shown in Table 1, Skempton also concluded that $a_c \tan\psi/\tan\phi'$ approaches zero for the same materials. With this simplification both equations 1 and 2 become:

$$\sigma' = \sigma - u \tag{2a}$$

or, in incremental form,

$$\Delta\sigma' = \Delta\sigma - \Delta u. \tag{2b}$$

Equations 2a and 2b correspond to the now-famous Terzaghi effective stress equation. Skempton was able to show that the reason this equation works so well for soils is the relatively high compressibility of the soil skeleton, C. In the case of rocks and concrete, however, $\tan\psi/\tan\phi'$ might range from 0.1 to 0.3, and a_c is not negligible. Likewise, C_s/C ranges from 0.08 to 0.46 for the rocks shown in Table 1. Thus, the approximation given by equation 2a may not be satisfactory for these materials, and the more general form, equation 2, is needed. Of course there are intermediate materials, sometimes called soft rocks, whose compressibility values must be determined before an applicable equation can be chosen.

The remainder of Table 1 is devoted to the "B" parameter, which quantifies the pore pressure response to an 'undrained' increment in external stress. Approximate values and ranges in values of porosity were estimated so that "B" could be computed from equation 6.

Table 1 shows that for the dense sand and other "soils" the B-value is essentially one.

TABLE 1. COMPRESSIBILITY AND "B" VALUES FOR SATURATED SOIL, ROCK, AND CONCRETE

Material	Compressibility* x 10^6 (m^2/kN)				Assumed porosity, n (%)‡	Skempton's B value
	C	C_s	C_s/C	$(C_w/C)^{\dagger}$		($\Delta\mu/\Delta\sigma$)
Quartzitic sandstone	0.059	0.027	0.46	8.3	5 - 10	0.718 - 0.56
Quincy granite (30-m deep)	0.076	0.019	0.25	6.4	1 - 3	0.942 - 0.844
Vermont marble	0.18	0.014	0.08	2.7	0.4 - 1	0.989 - 0.974
Concrete (approx.)	0.20	0.025	0.12	2.4	8 - 12	0.85 - 0.79
Dense sand	18	0.028	0.0015	0.027	35	0.991
Loose sand	92	0.028	0.0003	0.005	45	0.998
London clay (overconsolidated)	75	0.020	0.00025	0.007	50	0.997
Gosport clay (normally consolidated)	600	0.020	0.00003	0.0008	65	0.9995

* Compressibilities were measured at a value of $\sigma' \simeq 98$ kN/m^2.

† Compressibility of water = C_w was taken as 0.49×10^{-6} m^2/kN.

‡ Estimated porosity ranges here refer to pore space accessible to water.

$B = 1/[1 + n\,(C_w/C - C_s/C)] = (\Delta u_w/\Delta\sigma)$ undrained loading.

Likewise, the ratio $\Delta\sigma'/\Delta\sigma$ is essentially zero. This means that when an increment of external stress, $\Delta\sigma$, is applied to a saturated soil, the stress is at first assumed almost entirely by the pore water. If drainage subsequently occurs and the pore pressure generated by external loading is allowed to dissipate, $\Delta u \rightarrow 0$ as the pore pressure returns to its original value. If the exernal stress is held constant at its new value during this period, an increment in effective stress gradually increases to the value of $\Delta\sigma$ initially imposed: $\Delta\sigma' \rightarrow \Delta\sigma_{initial}$.

In the case of the concrete and rocks shown in Table 1, the pore pressure response is much more complex. The "B"-value can be significantly below one depending on the relative values of C_s, and C_w. No simple generalization can be made, and equations 2, 6, and 7 must be used to compute the response on a case-by-case basis.

The influence of the ratios C_s/C and C_w/C on the effective stress equation and on the pore pressure response makes the compressibility, C, fundamentally important. However, C is the reciprocal of a modulus, and further discussion of compressibility will be delayed until moduli are discussed. The value of C_w is constant for practical purposes, and the values of C_s shown in Table 1 are surprisingly similar.

It is begging the question to assume that the properties needed to employ commonly used solution techniques are in fact fundamental properties. Nevertheless, we will not resist the temptation to cite those properties essential to the most familiar methods for solving problems of deformation and flow.

Diffusion Equation

The basic equation governing pore fluid dissipation for one-dimensional flow was originally derived by Terzaghi (1925):

$$C_v \frac{\partial^2 u}{\partial z^2} = \frac{\partial u}{\partial t}, \tag{8}$$

where u = pore fluid pressure,
t = time,
z = direction of flow, taken here as vertical,
C_v = diffusivity coefficient or coefficient of consolidation for vertical flow in units of length squared per unit time.

Equation 8 describes the transient response in the time domain and is based on the assumption of time-independent boundary conditions. Terzaghi applied his derivation to the consolidation of saturated clays. Thus, equations 2a and 2b were assumed valid and the "B"-value was assumed to be 1. More explicitly, it was assumed that no change in external stress had occurred. Therefore, any change in pore pressure, Δu, was accompanied by an identical change in effective stress, $\Delta\sigma'$, which in turn produced a change in the volume of the material skeleton identical to the volume of fluid expelled or absorbed.

The deformations were assessed by employing C, the compressibility of the material skeleton, which Terzaghi called m_v. Darcy's law was assumed valid and the hydraulic conductivity, k, was used to govern the velocity of flow in response to a hydraulic gradient:

$$v_z = k_z \cdot i_z , \tag{9}$$

where z = direction of flow, taken as vertical,
v = velocity of flow, in units of length per unit time,
k = hydraulic conductivity, in units of length per unit time,
i = hydraulic gradient, unitless.

For transient flow, however, the rate of dissipation of pore pressure also depends on the coupled effect of water movement on the hydraulic gradient. Thus the coefficient of consolidation must and does incorporate both compressibility and hydraulic conductivity:

$$C_v = \frac{k}{m_v \gamma_w} , \tag{10}$$

where γ_w = unit weight of fluid.

It should be noted that m_v is a very nonlinear function of effective stress. The dependence of m_v on effective stress gradually lessens to a negligible quantity as the stiffness of the material increases to approach that of hard rock.

The hydraulic conductivity, k, of equation 10 should be distinguished from the absolute permeability, K. The absolute permeability is a property of the porous material only and it does not depend on the permeant. It depends only on the geometric size, configuration, orientation, and tortuosity of the flow path. The hydraulic conductivity depends on the properties of the permeant as well, as given by equation 11:

$$K = (\mu/\gamma_w)k \tag{11}$$

where K = absolute permeability in units of length squared,
μ = permeant viscosity in units of force × time per length squared,
γ_w = unit weight of permeant in units of force per length cubed,
k = hydraulic conductivity in units length per unit time.

Absolute permeability is no doubt a more fundamental property than hydraulic conductivity, but the latter is more commonly used, and most of the published data are given in terms of hydraulic conductivity for water at standard temperature and pressure.

The specific storage, S_s, is a porous material property very commonly employed by hydrogeologists in the solution of fluid-flow problems. It is defined as the volume of fluid absorbed or expelled (per unit bulk volume) in response to a unit change in pore fluid pressure. It is normally restricted to saturated materials. The definition can be construed to include partially saturated materials, but evaluation of S_s becomes exceedingly difficult for this case.

For materials with relatively soft skeletons such as the soils in Table 1 (where C_s/C and C_w/C are negligible), the specific storage, S_s, becomes:

$$S_s = C = m_v , \tag{12}$$

where S_s, C, and m_v have units of volumetric strain per unit change in "pressure." The units used for pressure for all terms in equation 12 may be either force per unit area or units of head.

The unit change in pressure (stress units) of equation 12 is either Δu or $\Delta\sigma'$ because they are equal for the conditions where equation 12 applies, if loading is undrained.

Narasimhan and Witherspoon (1977) presented this equation in slightly more rigorous form:

$$S_s = n\gamma_w\beta + \gamma_w \frac{a_v}{1 + e} , \tag{13}$$

where e = void ratio, n/(1 - n),
n = porosity,
a_v = $-(de/d\sigma')$,
σ' = effective stress,
$a_v/(1+e)$ = m_v,
β = C_{fluid} (C_w if fluid is water),
γ_w = unit weight of water.

The unit weight of water appears in each term because they express the unit change in pore water pressure in metres of water in their development. Equation 13 is more complete than 12 because the first term provides the component due to water compression, in response to a change in pore water pressure.

For materials for which C_s/C is not negligibly small, an incremental change in pore fluid pressure, Δu, is not necessarily accompanied by an equal change in effective stress, $\Delta\sigma'$, even if the corresponding change in total stress, $\Delta\sigma'$, is zero. Equation 2 shows the influence of C_s/C:

$$\Delta\sigma' = 0 - [1 - (C_s/C)\ [1 - (C_s/C)]\ \Delta u .$$

This means that if the definition of S_s is to be strictly adhered to for these materials, S_s must be measured directly by observing the volume of water expelled or absorbed in response to a unit change in pore fluid pressure, not to a unit change in effective stress.

Most of the data available from the literature relate to C, however. Thus, the most convenient approach would be to use equation 2c to compute $\Delta\sigma'$ for a unit change in Δu (assuming $\Delta\sigma$ is zero). Then the desired volume change of the material skeleton is simply $C\Delta\sigma'$: Volume change of skeleton per unit bulk volume, per unit change in $u = C\Delta\sigma' = C - C_s$.

For materials for which C_s is not negligible in comparison with C, the compressibility of water, C_w, will certainly not be negligible compared with C. If water compressibility and solid material compressibility are included then S_s for saturated materials becomes:

$$S_s = \underbrace{nC_w}_{\substack{\text{water}\\\text{compression}\\\text{term}}} + \underbrace{C - C_s}_{\substack{\text{skeleton}\\\text{term}}} + \underbrace{(1 - n)C_s}_{\substack{\text{compression}\\\text{solids}}}$$

$$= n(C_w - C_s) + m_v. \tag{14}$$

Equation 14 is, of course, essentially the same as equation 13 because C_w is typically about 25 times C_s.

In ground-water flow problems it is quite convenient to use the pore fluid pressure as the independent variable, and this practice will no doubt continue. The preceding discussions and derivations have shown, however, that when C_w/C is not negligible, a change in external pressure may not be fully reflected in the pore fluid pressure. This external pressure can cause changes in the effective stress and result in a different amount of fluid expulsion than would have been anticipated if only the pore fluid pressure were being monitored. Equation 2 could be used to evaluate the effects of the external pressure, $\Delta\sigma$, but it would probably be necessary to introduce one or more additional terms that quantify the volumetric change not accompanied by a change in pore fluid pressure.

In view of material heterogeneity, large depths for large-scale processes, and limited test data, C in equation 14 will rarely be assessed with the accuracy sufficient to justify the refinements associated with the third term, and in a great many cases the first term will be relatively unimportant.

Numerical Models

Terzaghi's (1925) closed-form solution to the diffusion equation has been used successfully in geotechnical engineering for homogeneous clays since the late 1920s. In the past few decades, however, finite-difference solutions have been widely employed to handle heterogeneous systems. One of the most highly developed numerical models of this type is the unified, integrated finite-difference model proposed by Narasimhan and Witherspoon (1977). This model uses a specific storage as given by equation 13 and includes, in addition, a term that accounts for the contribution of the partially saturated zone. Fluid movement is actually monitored in terms of mass transer so that mass balance can be assured. Among the advantages of the model are:

1. It handles saturated and unsaturated flow.
2. It guarantees mass balance.
3. It allows the user to track closely the physics of the process being modeled.
4. When one-dimensional compression occurs, it very accurately models the nonlinear dependence of compressibility on effective stress (using the conventional e-log p model).

When the strains are either one-dimensional (say vertical) or equal in all directions (hydrostatic compression), finite-difference models like the one described above can be expected to perform very well and exhibit the advantages noted. Difficulties can become significant when these boundary conditions are not approximately met, however. If shear loading (that is, unequally applied stresses in two or more directions) becomes significant, more general stress-strain data are needed.

The finite-element approach has the potential of accommodating general stress-strain relationships. Algorithms that incorporate a coupled approach to predicting deformation and fluid flow offer a very attractive alternative. Examples of such "solution packages" are those developed by Neuman (1973), Sandhu and Wilson (1969), and Chang and Duncan (1977).

The integrity of solutions obtained from such a powerful method as the finite-element one is strongly dependent on accurate assessment of the fundamental properties of the porous material. These properties can be divided into (1) moduli and (2) hydraulic conductivity values.

The moduli must represent the stress-strain relations, which are typically nonlinear and stress or strain path dependent. Tangent or secant values of moduli are usually used. Generally they are assumed to be some function of the state of stress or strain, but some type of "memory" of the stress or strain history is needed.

The general form in which permeability data are needed for such models is given by equation 15:

$$k_{ij} = \begin{bmatrix} K_{xx} & k_{xy} & k_{xy} \\ & k_{yy} & k_{yz} \\ \text{symm} & & k_{zz} \end{bmatrix}. \tag{15}$$

The first subscript refers to the direction of flow and the second subscript to the direction of gradient. The matrix is related to a reference material configuration.

Partially Saturated Materials

The discussions to this point have dealt almost exclusively with completely saturated soils. When the porous material is only partially saturated, substantial difficulties arise in evaluating both deformations and fluid-flow rates and amounts. Because of the great success attained in using the Terzaghi effective stress equation for saturated soils, attempts to apply the concept to partially saturated soils were to be expected. Several investigators, including Bishop (1960) and McMurdle and Day (1960), concluded that only part of the pore water pressure in unsaturated soil is convertible to equivalent mechanical stress. A procedure for evaluating an "equivalent" effective stress was suggested by Bishop and Blight (1963) and is given by equation 16:

$$\sigma_B' = (\sigma - u) + \chi(u_a - u_w), \quad (16)$$

where σ_B' = Bishop's "equivalent" effective stress,
σ = total stress,
u_a = pore air pressure,
u_w = pore water pressure,
(u_a-u_w) = soil "suction," also called Ψ,
χ = material parameter.

The parameter χ is usually evaluated as a function of the degree of saturation for a particular soil. However, recent experience of several investigators indicates a strong dependence of χ on several other factors. In particular, it appears that, even for a single value of degree of saturation, χ depends on whether the material is wetting or drying and whether the external stresses are primarily shear or compressional. Because of these difficulties, some investigators have chosen to uncouple the behavior of the water phase from that of the material skeleton. For unsaturated porous materials a set of constitutive relationships based on formulations suggested by Biot (1941), Jaeger and Cook (1976), and Fredlund and Morgenstern (1976) is given by equation 17. Each subequation contains consideration of the independent contribution of the Terzaghi effective stress, $\sigma_m' = \sigma_m - u_w$, and the water suction, $\Psi = (u_a - u_w)$, on both the volume change of the material skeleton, ε_v, and the change in the volumetric water content, θ. In essence the equations describe an uncoupled phase approach:

$$\varepsilon_v = \frac{\Delta\sigma_m'}{K} + \frac{\delta\Psi}{H} = \frac{\Delta V}{V} \text{ (for material skeleton)}, \quad (17a)$$

$$\theta = \frac{\Delta\Psi}{R} + \frac{\Delta\sigma_m'}{H_1} = \text{quantity of water absorbed or expelled per unit of bulk volume.} \quad (17b)$$

Here, K = a material modulus. (If the material were linear elastic, K would be the bulk modulus--here, a function of Young's modulus E and Poisson's ratio ν.) H = a material modulus, which quantifies the change in the volume of the material skeleton due to a change in Ψ. R and H_1 = similar moduli (or properties) quantifying separate effects of Ψ and σ' on the volume of water in the material; $\varepsilon_v = \Delta V_{skeleton}$ per unit bulk volume; and $\sigma_m' = \sigma_m - u_w = (\sigma_{11}' + \sigma_{22}' + \sigma_{33}')/3$.

The uncoupled phase equations 17a and 17b represent an approach that uses two independent stress-state variables to compute the deformation of the skeleton and amount of water absorbed or expelled. The first variable is the soil skeleton stress system, which may be expressed by a general full stress tensor, σ_{ij}'.

$$\sigma_{ij}' = \sigma_{ij} - u_w \delta_{ij}, \quad (17c)$$

where δ_{ij} = Kronecker delta = 0 for $i \neq j$, = 1 for $i = j$.

The second variable is the water-phase stress system, which may be expressed by an isotropic tensor, $\Psi\delta_{ij}$. Equations 17a and 17b could also be written with total stress, σ_m^*, as the stress state variable (Fredlund and Morgenstern, 1976):

$$\varepsilon_v = (\sigma^*_m/K) + (\Psi/H), \quad (17d)$$

$$\theta = (\Psi/R) + \sigma^*_m/H_1), \quad (17e)$$

where $\sigma_{ij}^* = \sigma_{ij} - u_a \delta_{ij}$,
u_a = pore air pressure,
$\sigma_m^* = (\sigma_{11}^* + \sigma_{22}^* + \sigma_{33}^*)/3$.

It must be noted, however, that definitions of moduli depend on the stress-state variable being used and consistency must be observed.

Use of Coupled vs. Uncoupled Approach

Equations 17a and 17b can also be written for saturated materials. If the Terzaghi effective stress, $\sigma'_T = \sigma - u_w$, is taken as the stress-state variable, the coefficients can be expressed in terms of now-familiar properties described earlier. Table 2 summarizes the form taken by these equations for saturated "soft" and "stiff" materials and presents the recommendations of the authors relative to the approach that is needed.

As stated earlier, in the case of saturated material with a soft skeleton, the equations degenerate to a single equation because $\varepsilon_v \simeq \theta$ and the coupled approach is quite satisfactory. In the case of saturated material with a stiff skeleton however, both C and C_s are needed for ε_v, and both C and C_w are needed for θ.

If σ'_{SK} from equation 2 were used in place of σ'_T, then only C could be needed to compute ε_v because the effect of C_s would be included in σ'_{SK}.

In the case of partially saturated materials, the use of an uncoupled approach becomes desirable for both soft and stiff skeletons, as suggested in Table 3. Fredlund and Morgenstern (1976) produced

TABLE 2. SATURATED MATERIALS

	Soft skeleton	Stiff skeleton
	Coupled approach satisfactory	Uncoupled approach needed
Skeletal deformations	$\varepsilon_V = (\Delta\sigma'_T/K) + (\Delta u_W/H)$ or	$\varepsilon_V = C\Delta\sigma'_T + C_S\Delta u_W$ C_S/C not negligible
	$\varepsilon_V = C\Delta\sigma'_T + C_S\Delta u_W \simeq 0$	σ'_T and u_W needed as i.v.
$\varepsilon_V = \Delta V_{skel}/V$	σ'_T needed as stress state	
	i.v.* but u_W can be used if $\Delta\sigma \simeq 0$	Note: coupled approach satisfactory if s'_{skem} is used
Amount of water inflow/outflow	Coupled approach satisfactory	Uncoupled approach needed
$\theta = (\Delta V_W/V)$	$\theta = (\Delta u_W/R) + (\Delta\sigma'_T/H,) \simeq 0$ or $\theta = n(C_W - C_S)\Delta u_W + C\Delta\sigma'_T$	$\theta = n(C_W - C_S)\Delta u_W + C\Delta s'_T$ $[n(C_W - C_S)/C]$ not negligible
	σ'_T needed as i.v. bt u_W can be used if $\Delta\sigma \simeq 0$	σ'_T and u_W needed as i.v. Note: u_W alone will work as i.v. if $\Delta\sigma \simeq 0$

* i.v. = independent variable.

ample experimental data to show that evaluation of the amount of water inflow and outflow requires both $\Delta\Psi$ and $\Delta\sigma_T'$ and that θ can be non-zero when $\Delta\Psi = 0$.

If ε_v only is sought it is possible to use Bishop's parameter of equation 16. The use of Bishop's effective stress from this equation is an attempt to lump the two stress systems into one, reducing the problem to the category of a coupled phase system. Evaluating tangent values of χ for all of the various loading and wetting/drying sequences of interest would involve an experimental effort comparable to that required to obtain tangent values for the various moduli of equation 17. It is believed that once the moduli of equation 17 have been evaluated, a more powerful tool is then available for prediction of response. It is on this basis that the recommendations of Table 3 rest.

Summary of Fundamental Properties

The preceding discussions have shown that for saturated soils the properties most fundamental to the solution of deformation and fluid-flow problems are moduli and permeability. The importance of the effective stress principle shows that compressibilities of the fluid and material skeleton are needed to determine if simplifying approximations can reasonably be made. However, the compressibility of the material skeleton, C, is simply the reciprocal of a particular modulus and could surely be computed from a comprehensive set of moduli needed for a finite-element analysis.

The specific storage, S_s, can be computed from the various compressibilities--C_w, C, and C_s. The coefficient of consolidation, C_v (or diffusivity coefficient), from equation 8 can be computed from the hydraulic conductivity and the compressibility of the material skeleton.

For partially saturated materials, an additional set of moduli are needed. If solution is attempted using a coupled material skeleton and water phase, then a parameter such as χ (equation 16) is needed. If the two phases are uncoupled then the additional moduli R, H, and H_1 are needed as indicated in equation 17.

The porosity, n, is a very important parameter, perhaps even a fundamental parameter because of its influence on specific storage (equation 15), the "B" value (equation 6), and permeability. For high porosity materials it can usually be estimated with sufficient accuracy without testing, but for low-porosity materials its assessment requires special attention.

The next section will be devoted to presentation of some typical values of these properties and an outline of some methods available for measuring them.

TABLE 3. PARTIALLY SATURATED MATERIALS

	Soft skeleton	Stiff skeleton
	Uncoupled approach needed	Uncoupled approach needed
Skeletal deformations	$\epsilon_v = (\Delta\sigma'_k/K) + (\Delta\Psi_w/H)$	$\epsilon_v = (\Delta\sigma'_T/K) + (\Delta\sigma/H)$
	$\sigma'_T = \sigma - u_w$ *and* Ψ	σ'_T Ψ needed as i.v.
$\epsilon_v = (\Delta V_{skel}/V)$	needed as i.v.* $\Psi = u_a - u_w$	but ϵ_v is relatively insensitive to Ψ
Amount of water inflow/outflow $\Theta = (\Delta V_w/V)$	Uncoupled approach needed $\Theta = (\Delta\Psi/R) + (\Delta\sigma'_T/H_1)$	Uncoupled approach needed $\Theta = (\Delta\Psi/R) + (\Delta\sigma'_T/H_1)$
	σ'_T *and* Ψ needed as i.v.	σ'_T and Ψ needed as i.v.

* i.v. = independent variable

The following properties are treated individually:

1. Permeability, which will be used instead of absolute permeability because the emphasis is on water as a pore fluid and most values are reported in terms of hydraulic conductivity
2. Porosity
3. Stress-strain moduli (One-dimensional compressibility values are cited as a special case.)
4. χ values and other moduli needed for dealing with partially saturated materials
5. Coefficient of consolidation, C_v (Although it can be calculated from compressibility and hydraulic conductivity, C_v is more stable than either of these properties. Also, values of C_v for several materials, soils in particular, are reported in the literature.)

PHYSICAL PROPERTIES: TYPICAL VALUES AND METHODS OF MEASUREMENT

Hydraulic Conductivity, k

Hydraulic conductivity is one of the most important properties of a porous material because it controls the rate of water movement through the material, and it is therefore fundamental to the solution of fluid-flow problems. Although both soils and rocks are porous, their permeability characteristics differ. Soils are usually regarded as consisting of discrete particles touching at isolated points, whereas materials such as sedimentary rocks are regarded as a solid skeleton traversed by a fine network of capillaries; in rocks of low porosity such as igneous rocks, the voids probably consist of grain-boundary cracks and joints.

Porosity does not in itself guarantee permeability because the voids must be interconnected to produce permeability. Presence of minute features in a porous medium such as shrinkage cracks or interbeds of silt or sand in clays may boost their permeabilities by several orders of magnitude. Furthermore, examination of the spectrum of typical values in Figure 3 shows that the hydraulic conductivity exhibits an extremely wide range for naturally occurring materials. This range shows why it is so difficult to estimate or measure k with precision.

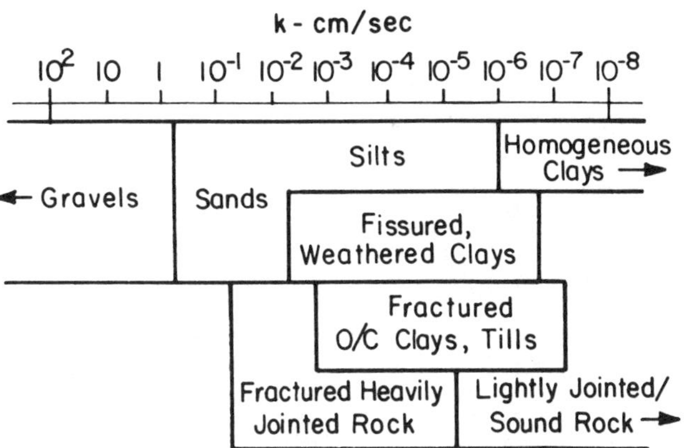

Figure 3. Approximate range of permeability (k) in soil and rock (from Milligan, 1975).

Methods of Determination

Darcy's law, equation 9, is the basis for measuring the hydraulic conductivity of porous materials. Because flow in porous materials is mostly laminar, experience shows that Darcy's law is satisfactorily valid. Reliable measurement of the permeability hinges on a reliable assessment of the relative roles of many factors, including

1. Heterogeneity of the medium
2. Discontinuities such as cracks and patterns, joints, fissures, and sand lenses
3. Fabric-related tortuosity

4. Changes in state of stress and its effect on the available void space for the flow

Various methods of measurement have been described in some detail by Milligan (1975) and Mitchell and others (1978). Some of the more important of these methods are discussed in the following sections.

The procedure common to all direct methods of permeability measurement is to impose a hydraulic gradient and measure the resulting flow for a finite volume of the medium.

Laboratory methods. The major advantage of conducting laboratory measurements of the permeability of a porous medium is that good control can usually be exercised over important factors such as the quantity of flow, size of the medium being tested, and material orientation with respect to direction of flow. The ability to measure the permeability for different orientations of the material with respect to the hydraulic gradient is indispensable in the effort to fully define the general matrix in equation 15, in order for porous materials to be incorporated in sophisticated methods of analysis.

Constant-head or falling-head permeability tests are suitable for relatively permeable materials such as gravel and sand and for some silts. For materials such as clay, which have relatively low k and high C, the consolidation test (ASTM D2435) is usually the most satisfactory method. A difficulty common to virtually all lab-test techniques is the uncertainty about the degree to which the sample being tested represents the porous medium in the field. It is difficult to obtain and test a lab sample large enough to include the discontinuities which may control the in situ permeability.

Field methods. Because of the limitations of lab measurements cited above, field methods are widely used, in spite of the greater expense of doing so. Field methods enable a better evaluation of the permeability of large masses of porous material because they include the effects of important macroscopic features such as sand lenses, fissures, and clay seams, which will most likely govern in situ values. Usually the permeability measured in the field is an average for the medium although it may be weighted more heavily for one direction. Field tests are either inflow (infiltration tests) or outflow (drawdown) tests. For both cases errors may be introduced in the measurements owing to disturbance of the material around the hole. The ratio of the values of permeability evaluated from both tests, K_{out}/k_{in}, may be as large as 500 in extreme cases (Milligan 1975). According to Milligan (1975), the best estimate of the in situ permeability, k, is obtained from

$$K = \sqrt{k_{out} \cdot k_{in}}. \qquad (18)$$

Three field methods are widely used:

1. Large-scale pumping test. This test generally provides the most accurate measure of in situ k. Water is pumped into or out of a screened or otherwise filtered well while water-level readings are made in several nearby observation wells. Wells may be drilled to penetrate the entire water-bearing formation or part of it. Both transient (nonequilibrium) or steady-state equilibrium tests may be conducted. For the steady-state test the Theim formula (Lange, 1967) is usually used. For the transient test the Theis (1935) formula is used to evaluate k.

2. Borehole test (USBR Method E-18). Because large-scale pumping tests are relatively costly, the borehole permeability test in one borehole is often used. It can furnish very useful permeability information, but it must be applied with care because it is not usually easy to check the results for accuracy. Typical conditions where misleading results could be obtained are illustrated by the two geologic profiles shown in Figure 4. To enhance the reliability of measurements several borehole tests should be conducted, but of course this drives the cost up toward that of a standard large-scale pumping test with moderation wells. In general the control of field-test parameters and reliability of data are better when an infiltration testing scheme is used.

3. Piezometer probe. If a hydraulic head difference is imposed between the water in a piezometer probe and the water in the nearby

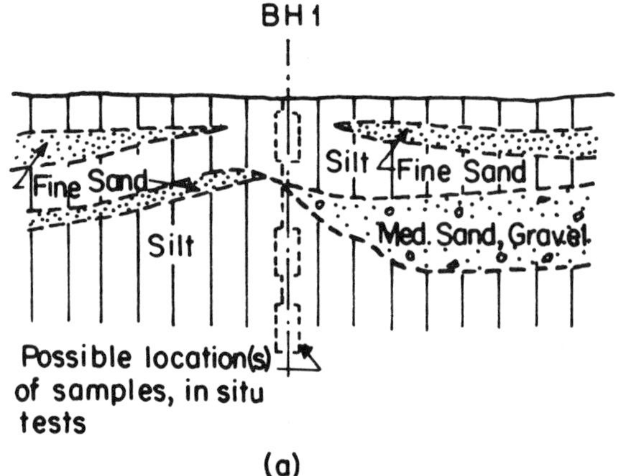

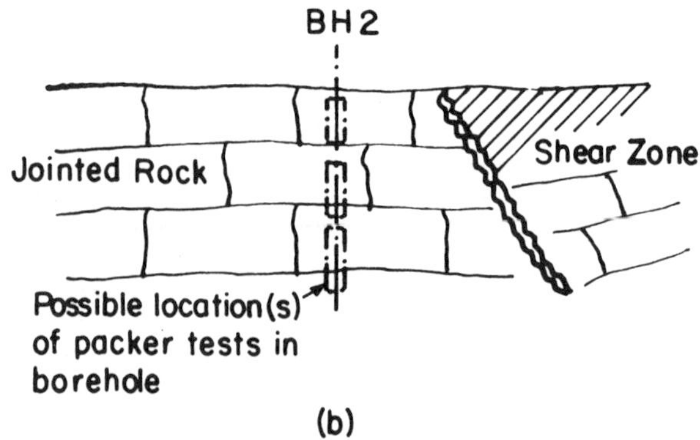

Figure 4. Limitations of sampling and in situ tests in a single borehole: (a) in soils; (b) in rock (from Milligan, 1975).

porous material, consolidation or swelling will occur. If the head difference and rate of water flow into or out of the medium are measured and the boundary conditions can be determined, the hydraulic conductivity can be calculated. Constant head or variable head tests may be conducted (Gibson, 1963). Some discrepancies between the hydraulic conductivity measured from full-scale pumping tests and that from piezometer probe tests have been reported, but in general, the accuracy of measuring the hydraulic conductivity using the piezometer probe test is better than that obtained from lab test data.

Empirical formulas. The permeability of a clean filter sand can be calculated from a number of formulas such as the following developed by Hazen (1911):

$$k_{cm/sec} = C\, D^2_{10}, \tag{19}$$

in which D_{10} is the mesh size in centimetres for which 10% of the material by weight is finer. The constant C is often assigned a value of 100. However, the presence of "fines" and the shape of the grain size distribution curve for the portion finer than D_{10} may significantly affect C.

When equation 19 is employed to estimate the permeability of layered porous materials the best average permeability is obtained by using an average D_{10} for the aquifer (Ahmad and others, 1975).

Theoretical Formulas. The formula developed by Poiseiulle for the permeability of a porous medium with irregular pores and laminar flow is given by equation 20 in functional form:

$$k = k_s \cdot G(e) \cdot F(S\%), \tag{20}$$

where k_s = a soil property,
K = hydraulic conductivity,
e = void ratio,
S = degree of saturation in percent
= volume of water/volume of voids × 100.

Chang and Duncan (1977) proposed the applicability of this formula to both clays and sands. They suggested that the value of k_s can be determined from laboratory tests duplicating field conditions as nearly as possible, and suggested equations 21 and 22 for G(e) and F(s):

$$G(e) = \frac{e^3}{(1 + e)} \Big/ \frac{e_o^3}{(1 + e_o)}, \tag{21}$$

where e_o is a reference void ratio at which k_s was determined. The void ratio is a measure of the amount of pore space per unit volume of solids,

$$F(s\%) = S^3\%, \tag{22}$$

where S% is the degree of saturation in percent.

Lambe and Whitman (1969) showed typical variations of the permeability with void ratio for a wide range of materials. When the material is only partially saturated, the coefficient of permeability will of course be smaller than when saturation is complete, as indicated by equations 20 and 22. The function F(S%), equation 22, is an approximate representation of experimental observations made by Mitchell and others (1965).

Summary and Recommendations Relative to k

The most reliable combination of methods for determinining k is to use full-scale pumping tests with observation wells together with laboratory tests and/or borehole permeability tests on selected sections of the boreholes (using packers to seal off other sections). The full-scale pumping tests produce an excellent average value of k for the zone tested but cannot be used alone to obtain k values for individual layers in heterogeneous materials. The laboratory tests on representative samples and/or the borehole tests cited can be used to provide at least a first approximation of the ratios of k values for the various layers and the ratio of horizontal to vertical permeability for each. Then iteration and adjustment by trial and error can be used to find the set of values most consistent with the field pumping test results.

Although a first approximation for k values can be obtained by assuming a rigid material skeleton for the foregoing computations, obviously a better solution can be obtained by incorporating any concurrent deformations of the material skeleton in a numerical model such as finite difference or finite element. This same general method (solving the inverse problem) will also be recommended in the next section as the optimum procedure for evaluating moduli.

Porosity, n

The porosity is important because of its effect on S_s (equation 13), its effect on the "B" value (equation 6), and its effect on moduli, hydraulic conductivity, and other properties in general. It can often be used as a fairly good indicator of the degree of weathering.

Typical values of porosity. Goodman (1979) has collected ranges of values of porosity for various soft to hard rocks; some of these values are shown in Table 4.

Goodman concluded that porosities for unweathered igneous rocks are typically 1% to 2% but increase up to 20% or more when weathering becomes severe. He also concluded that age and depth have remarkable effects on porosity. For instance, porosity could decrease to 0.33% with geologic aging. Rocks encountered at depths of several thousand feet might have porosities that are only half or one-third the porosities of the same type of rocks encountered at shallower depths.

Typical values for "soft" skeleton materials such as soils are markedly higher as shown in Table 4.

Methods of determination. Goodman (1979) has

TABLE 4. TYPICAL POROSITIES FOR POROUS MATERIALS

	Material	Range in n (%)	Commonly encountered values of n (%)
Soft skeleton	Sand	25 - 50	40 ± 8
	Silt	40 - 63	52 ± 8
	Clay	40 - 94	62 ± 12
Stiff* skeleton	Marble	0.3 - 1.1	0.7 ± 0.4
	Granite	0.1 - 20	2 ± 1
	Limestone	0.4 - 43	10 ± 20
	Sandstone	0.7 - 34	15 ± 79
	Shale	1.6 - 25	15 ± 7
	Tuff	14 - 40	25 ± 6
	Chalk	25 - 50	35 ± 7

* After Goodman, 1979 (originally from Clark, 1966; Brace and Riley, 1972).

described several methods for determining porosity in the laboratory. It is generally considered impractical to measure porosity in situ, although its value no doubt depends on the scale of the specimen tested in the laboratory. These methods can be considered generally satisfactory for soft to hard rocks that do not swell or shrink significantly when soaked or dried out. In brief the procedures are:

1. Soak specimen in water one or two days (a more thorough penetration of the water can be achieved by pressurizing it). Dry the specimen in an oven to constant weight. Calculate the n from the known water content and specific gravity. The specific gravity should be measured on a fragmented sample of the original porosity specimen.
2. Same as method 1 above except inject mercury instead of water.
3. Compute n directly from a very accurately measured dry density and specific gravity.

Summary and recommendations relative to n. The most reliable and convenient method of determining porosity of nonswelling rocks is probably the first method described above--soaking the rock in pressurized water, determining the water content by subsequent drying, and computing n from G_s measured on representative fragments of the same specimen.

When the porosity is relatively low it is difficult to avoid laboratory measurement errors that are significant. However, it should be noted that the greatest source of error will normally be hetereogeneity in the field. Large separations between discontinuities (compared with the laboratory specimen size) and variations in degree of weathering from point to point in the field will generally introduce more error into the porosity assessment than laboratory measurement errors on a single specimen. Hetereogeneity errors can be minimized only by increasing the number of individual measurements of porosity throughout the space of interest in the field.

Moduli

Moduli are usually thought of as being properties of the material skeleton. More generally, however, they could include the pore water as well. Numerically, the moduli are equal to changes in stress per unit strain. Thus, they can be used together with some function of stress--either on the material skeleton, the pore water, or both--to compute the strain in both the skeleton and the water phase. These data can then be used to assess the amount of water entering or leaving an element of material, assuming this is the primary objective.

The two distinct approaches discussed earlier are summarized and treated separately here.

Use of coupled phase equations. This type of formulation hinges on the implicit assumption that skeletal deformations and water inflow-outflow can be assessed from only one stress-state variable. That stress-state variable for saturated media could be

1. Water pressure, u_w
2. Suction, $\Psi = u_a - u_w$ = pore air pressure
3. Skempton's effective stress from equation 2 or 1, hereafter designated σ'_{SK}
4. Terzaghi's effective stress from equation 2a, hereafter designated $\sigma'_T = \sigma - w$

For skeletal deformations of partially saturated media Bishop's "equivalent" effective stress from equation 16

$$\sigma'_B = (\sigma - u_a) + \chi\ (u_a - u_w) \qquad (16)$$

can be used as the only stress state variable.

Use of uncoupled phase equations. This type of formulation is characterized by equation 17, and hinges on the presumption that two stress-state variables are needed to assess skeletal deformations and water inflow-outflow. One variable should describe the skeletal phase and the other variable

TABLE 5. MODULI, COMPRESSIBILITIES, AND POISSON'S RATIOS (ν) FOR VARIOUS ROCKS*

Description	E x 10^{-6}†	ν	C_{1-D} 10^8 (m^2/kN)	K 10^{-6} = $10^6/C_{3-D}$)	G x 10^{-6} ‡
Berea sandstone	18	0.38	2.8	26.0	6.7
Navajo sandstone	38	0.46	0.57	158	13
Tensleep sandstone	19	0.11	5.2	8	8.3
Hackensack siltstone	25	0.22	3.4	15	10.4
Monticello Dam greywacke	20	0.08	5.0	7.7	9
Solenhofen limestone	62	0.29	1.2	49	24
Bedford lime	27	0.29	2.7	22	10.7
Tavernalle limestone	54	0.30	1.3	45	21
Oneota dolomite	42	0.34	1.5	44	16
Lockport domomite	49	0.34	1.3	51	18
Flaming Gorge shale	5	0.25	15.6	3.5	2
Micaceous shale	11	0.29	7.0	8.5	4
Dworshak Dam gneiss	52	0.34	1.2	54	19
Quartz mica schist	20	0.31	3.6	17	7.6
Baraboo quartzite	85	0.11	1.14	36	38
Taconic marble	46.4	0.40	1.0	77	16
Cherokee marble	54.2	0.25	1.5	36	22
Nevada Test Site granite	71.6	0.22	1.2	43	29
Pikes Peak granite	68	0.18	1.35	36	29
Cedar City tonalite	185	0.17	5.0	9	8.6
Palisades diabase	79	0.28	1.0	60	31
Nevada Test Site basalt	33.7	0.32	2.0	31	13
John Day basalt	81	0.29	0.94	64	31
Nevada Test Site tuff	3.5	0.19	25.8	1.9	1.5

(After Goodman, 1979).

* Moduli are in kN/m^2 (1 kg/cm^2 = 98.1 kN/m^2).

† E.g., E = 18 x 10^6 kN/m^2 for Berea sandstone.

‡ $G = [E/2(1 + \nu)]$.

the water phase. Stress-state variables could be from the list above or from others suggested by Fredlund and Morgenstern (1976).

Typical Values of Moduli

Essentially all porous materials exhibit some nonlinearity in their stress-strain relations and nonlinearity is especially pronounced for soft skeleton materials such as soils. Thus, moduli depend upon stress state, direction of stress increment, stress, level, and strain history, temperature, time, and other factors. Nevertheless, some typical values will be presented for some common, simple loading conditions.

Goodman has compiled modulus data for a variety of rocks as shown in Table 5. These values are believed to apply to moderately low stress levels with monotonically increasing loads, but the E values reported include both recoverable and nonrecoverable deformations, mixed in unknown proportions.

Janbu (1963) has proposed that tangent values of constrained modulus M can be related to effective stress under one-directional compression by equation 23:

$$M = m\sigma_a\left(\frac{\sigma'}{\sigma_a}\right)^{1-a}, \tag{23}$$

where M = constraint or constrained tangent slope of the axial effective stress-axial strain under 1-D compression,
m = dimensionless modulus number; property of material,
σ' = effective axial stress,
σ_a = atmospheric pressure in same units as σ' and M,
a = dimensionless exponent; property of material.

Janbu examined data for a wide range of materials and found typical values and ranges in values as shown in Table 6. Note that for stiff skeleton materials where "a" is unity, the constraint modulus, M, is a constant, independent of stress level. This would be characteristic of an elastic material with weak or no coupling between the shear stresses and the volumetric strain. If it were also isotropic, M would be given by:

$$M = \frac{E\,(1 - \nu)}{(1 + \nu)(1 - 2\nu)}, \tag{24}$$

where E = Young's modulus and ν = Poisson's ratio.
For both stiff and soft skeleton materials, the compressibility under one-dimensional (1-D) compression is given by

$$C_{1-D} = 1/M. \tag{25}$$

Recall that the compressibility under three-dimensional or "isotropic" compression (C_{3-D}) is the reciprocal of the bulk modulus, K:

$$C_{3-D} = 1/K. \tag{26}$$

If the material were elastic and isotropic and exhibited negligible or no shear-stress-induced volumetric deformations, K would be given by

$$K = \frac{E}{3(1 - 2\nu)} \tag{27}$$

and the ratio of three-dimensional compressibility to one-dimensional compressibility would be:

$$\frac{C_{3-D}}{C_{1-D}} = \frac{3(1 - 2\nu)}{E} \cdot \frac{E\,(1 - \nu)}{(1 + \nu)(1 - 2\nu)}$$

$$= \frac{3(1 - \nu)}{(1 + \nu)}. \tag{28}$$

For $\nu = 0.33$ this ratio would be 1.5. As the properties of the material deviate from elastic, isotropic with uncoupled shear-volume behavior, the accuracy of equation 28 diminishes. In fact, for soft to medium clays the ratio of C_{3-D} to C_{1-D} is often very near 1.

For the case of soft skeleton materials the degree of nonlinearity is so pronounced that it is difficult to assign "typical" values of modulus. What is commonly done is to measure tangential values of moduli, which are assumed valid for only a small range in stress. In this way the material is characterized as "incrementally linear," and one of the most useful values thus quantified is the tangential value of the equivalent Young's modulus, E_t. It is common practice to consolidate a test specimen under hydrostatic stress conditions (that is, no shear stress) and then increase one of the principal stresses, holding the other two constant and equal. This is normally done by the triaxial test; the value of E_t therefrom starts from an initial value, E_i, and decreases to zero at the failure condition.

It should be noted that when E_t goes to zero, equation 27 predicts K going to zero and C_{3-D} going to infinity. Such is not the case, of course, and this observation simply illustrates the invalidity of Hook's Law (on which equation 27 is based) for these materials at near-failure stress conditions.

At very low stress levels, near the hydrostatic stress, some approximate ranges for E_i can be given for clays and sands. Equation 29 shows one of the most common empirical relationships used:

TABLE 6. TYPICAL VALUES AND RANGES IN VALUES FOR "A" AND "M" IN JANBU'S (1963) MODEL

	Material	m	a
Soft skeleton	Clay	15 ± 10	≃ 0
	Clean sand	150 ± 75	≃ 0.5
Stiff skeleton	Rocks	Wide range of values, which can be deduced from Table 5	≃ 1

$$E_i = (\sigma'/\sigma_a)^n, \quad (29)$$

where E_i = initial tangent modulus (an equivalent Young's modulus for drained loading),
K = dimensionless modulus number (material property),
σ_a = atmospheric pressure in units of σ' and E_i,
σ' = effective confining pressure under hydrostatic loading,
n = dimensionless exponent; material property.

As in the case of one-dimensional compression, the value of "n" for drained loading is in the neighborhood of 0.5 ± 0.1 for sands and approaches 1.0 for clays. Typical values of K range from 500 ± 200 for loose sand to 1,100 ± 400 for dense sand. For clays K is typically 300 ± 200.

Methods of Determination of Moduli

Measurements of the moduli are usually conducted according to their definitions from the equations in which they appear. The independent stress-state variables are varied, most commonly one at a time, and necessary measurements of deformations or water inflow-outflow are made.

Laboratory methods. A variety of laboratory tests are used to measure moduli. Tests differ primarily in the type of boundary conditions they impose. Thus the "best" tests depends on the characteristics of the porous material and the boundary conditions that will exist in the prototype.

Even though the test results may be used for very sophisticated constitutive laws, the most common laboratory tests for determining moduli are:

1. Triaxial compression tests with unloading cycles. For undrained tests pore pressures are measured and for drained tests volumetric strain is measured in addition to axial strain (Bishop and Henkel, 1962).

2. "Isotropic" or hydrostatic compreession tests with unloading cycle.

3. One-dimensional compression tests with unloading cycle (aASTM D-1557).

Some constitutive laws require data from triaxial extension tests, torsional shear tests, or simple shear tests to evaluate parameters that control the moduli. It is common, however, to attempt to formulate the constitutive laws in such a way that data from only the three tests above are needed ecause of the relative simplicity of the tests and widespread availability of the needed equipment.

Major sources of error in these laboratory moduli are:

1. Test specimens that are nonrepresentative because of hetereogeneity and scale effects.

2. Disturbance effects due to sampling; these generally decrease moduli, sometimes grossly. Often these disturbance effects can be erased by reconsolidation, but not always.

3. Failure to simulate prototype degree of drainage or rate of strain adequately.

4. Effects of boundaries such as end-plate friction on the triaxial test results. A procedure for removing these effects from the data has been developed by Kasim and Houston (1980, unpub. report).

The specific storage, S_s, is defined as the amount of water inflow-outflow per unit bulk volume per unit change in pore water pressure. Because its value can be calculated (equation 14), from known values of C and C_w and estimated values of C_s, it has not been treated separately as a fundamental property. However, it is not essential to calculate the value of S_s; it can be measured directly in the laboratory from a representative specimen.

The procedure is simply to change the pore water pressure by a known increment, say one unit, and concurrently hold the total stress constant (or impose whatever the concurrent change in the total stress that is expected in the prototype). Measure directly the amount of water inflow-outflow; this value, divided by the bulk volume, is S_s. This measured value includes all three terms of equation 14. It may be desirable for greater generality to separate out the first term, n C_w, which is constant and accounts for compressibility of water, and to relate the remainder of the measured S_s to the level of effective stress.

Field methods. Information on a variety of field methods of determining moduli has been collected and presented by Mitchell and others (1978). Among the most commonly used of these are the plate-load test and the Menard pressure-meter test. Plate-load tests are conducted at the ground surface or in relatively shallow test pits. The Menard pressure meter is used in a borehole but the depth is normally limited to a few tens of metres. In both tests the zone of material loaded is relatively small, although it is of course much larger than a laboratory test specimen.

Because the state of stress varies widely from point to point within the loaded zone for these tests, the most meaningful interpretation of the test data is obtained by analyzing the tests as mixed boundary-value problems.

By far the most effective method of applying stress to a large volume of material is a large-scale pumping test with observation wells. As suggested under the hydraulic conductivity section, the best estimates of both hydraulic conductivity and moduli are obtained from solving the large-scale pumping test as an inverse problem.

Indirect methods. Indirect methods of obtaining moduli include use of empirical correlations between moduli and index properties such as liquid limit, plasticity index, friction angle, relative density, and gradation parameters in the case of soft materials. Correlations between moduli and penetration resistance measured in situ are also used (Mitchell and others, 1978).

Geophysical methods are also widely used to

obtain sonic-wave-propagation velocities from which "low-strain" moduli can be computed. These methods are appropriate for both soft and stiff materials.

Summary and Recommendations Relative to Moduli

The most promising, and of course the most costly, approach is to combine both laboratory and field test results to find the best values of moduli. The large-scale pumping test with observation wells is recommended for this purpose. A numerical method such as finite difference or finite element is used to iterate a set of moduli (or moduli parameters) and hydraulic conductivity values that best fit the field observations. Laboratory tests on representative samples from the major material types can be used to obtain first estimates of the moduli and k values for each material type. These laboratory results can, at the very least, provide estimates of the ratios of values from material to material and indicate reasonable bounds for the values.

Ideally, the observation wells can be used to observe vertical and horizontal movements as well as water levels as a function of time. Vertical movements can be detected by installing casing segments that can move relatively and that have tabs or rims whose elevation can be measured. Inclinometer surveys can be used to determine the horizontal movements of the various points down in the observation wells. These measurements, together with measurement of horizontal and vertical movements of various points on the ground surface, can provide the data needed for back calculation to the best set of moduli and k values.

Nonuniqueness of the results is always a problem with such an approach, but the problem is minimized as the number of laboratory test specimens and number of observation wells is increased.

Values of χ and Other Moduli Needed for Partially Saturated Materials

As discussed in preceding sections, skeletal deformations of a partially saturated material can be evaluated by relating ε_v to the "equivalent" effective stress from equation 16. If, however, it is desirable to evaluate water inflow-outflow in addition to ε_v, then $dS/d\Psi$ must be evaluated over the range of interest, as was outlined by Narasimhan and Witherspoon (1977). The value of $dS/d\Psi$ represents the change in degree of saturation, S, per unit change in section, Ψ, and is a very nonlinear path-dependent function of Ψ.

It should be noted that when the change in external stress, $\Delta\sigma$, is zero, the function $dS/d\Psi$, together with $\Delta\Psi$, may be sufficient to evaluate both ε_v and θ. If, however, $\Delta\sigma$ as well as $\Delta\Psi$ is nonzero, an uncoupled approach as suggested by Fredlund and Morgenstern (1976) may be needed. This need is illustrated by the fact that a change in total stress, $\Delta\sigma$, may be accompanied by no change in suction, $\Delta\Psi = 0$, and the result will in general be nonzero ε_v and θ.

By either approach the scope of laboratory testing involved in quantifying the needed moduli is enormous, unless attention can be focused on a very limited set of initial and loading conditions. In general, field testing of partially saturated materials is impractical because of difficulties in evaluating or controlling the boundary conditions.

Partly because of the complexity and difficulty of testing partially saturated materials, published test data are very scarce. In fact, it is not reasonable to suggest typical ranges for the various moduli as was done for the other properties. Most of the meager quantity of data available for values of χ and $dS/d\Psi$ can be found in the following references: Barden (1965), Bishop (1960), Bishop and Blight (1963), Bishop and Donald (1961), and Jennings and Burland (1962).

The only generalization we offer for the value of χ is that it is essentially equal to the degree of saturation, S (as a decimal) when the degree of saturation is greater than 80% or 85%.

The use of uncoupled equations such as 17a and 17b is a recent development, and test data relating to these moduli are even more meager. Most of these data that are available are reported by Fredlund and Morgenstern (1976). A small amount of additional data was previously published by Burland (1965), Aitchison (1969), Matyas and Radhakrishna (1968), Barden and others (1969), and Brackley (1971).

Methods of Determination

The most commonly used laboratory tests for partially saturated materials are the triaxial test and the one-dimensional compression test, depending on the boundary conditions to be simulated.

If the triaxial test is used, for example, the specimen would be sealed in two rubber membranes surrounded by water in the triaxial cell, with a layer of silicon grease between the membranes. The total stress on the specimen, σ, can thus be made equal to the surrounding water pressure and the skeletal strain, ε_v, can be evaluated from the volume of water expelled from or taken into the cell.

The pore air pressure, u_a, can be applied or monitored through a relatively coarse porous stone at the top of the specimen. This stone would normally remain relatively dry. For measuring pore water pressure, u_w, a high air-entry-value stone, such as a ceramic stone, is needed at the base of the specimen. When the pore air pressure, u_a, is less than the air-entry value of the ceramic stone, air bubbles will not pass through the stone, but air can diffuse through the stone during long tests. In order to measure accurately the relatively small quantities of water expelled from or absorbed by the specimen, it may be necessary to flush and measure the quantity of diffused air as described by Fredlund (1975) and apply an appropriate correction.

The pore water pressure, u_w, can be measured with a pressure transducer or null device. The ceramic stone must, of course, be sealed to the lower platen so that water cannot pass around it. The water below the ceramic stone can go into tension, but only to about 80% or 90% of 1 atm, because cavitation will occur in the water.

However, u_a can be controlled and can be made relatively large--thus a wide range in the suction, $\Psi = u_a - U_w$, can be achieved.

Theoretically, the values of σ, u_a, and u_w can be independently controlled or measured. In practice, however, it is very difficult to measure u_a and it is a common practice to control it. In fact, u_a and σ are commonly controlled and u_w water inflow-outflow, θ, are measured or controlled. The value of u_w is commonly kept in the positive range.

Values of needed moduli can be computed from their definitions by measuring ε_v and θ in response to changes in either σ, u_a, or u_w (or functions or combinations of these stresses), while holding the remaining stresses constant as appropriate.

In the case of χ-value determinations, it is necessary to perform a companion set of tests on saturated (but otherwise identical) specimens for which a relation between ε_v and effective stress can be computed. Then the results from the two sets of tests can be combined to calculate the χ values, which will yield an "equivalent" effective stress as given by equation 16. The equivalent effective stress is that value which produces the values of ε_v obtained from the companion tests on saturated specimens.

Although the complexity of testing and test-data interpretation is correspondingly increased, it is possible to use the triaxial apparatus to apply stress increments involving shear stress as well. The observed values of ε_v can of course be divided into normal and deviatoric components and used in evaluating moduli accordingly.

Before embarking on a testing program of partially saturated materials we advise researchers to study carefully the experience gained and problems encountered by the various researchers cited in this section.

Coefficient of Consolidation, C_v

The coefficient of consolidation, C_v, can be computed from known values of k and m_v or C as given by equation 10:

$$C_v = \frac{k}{m_v \gamma_w} = \frac{k}{C\gamma_w}, \tag{10}$$

where $C = C_{1-D}$ or C_{3-D}, depending on the boundary conditions that control the ratio of horizontal to vertical strain in the prototype, although the subscript v is usually meant to imply only vertical strain.

C_v is treated separately here because it is often used as a material property for soft-skeleton materials, and because for these materials k and C generally decrease together as the effective stress increases; thus, C_v is somewhat more stable than either k or C.

Typical values of C_v. Table 7 shows typical ranges in C_v for "soft" and "stiff" skeleton materials. The range for rocks is rather wide because of the very pronounced effect of fracturing. When rocks become fractured, C increases but not nearly in proportion to k.

TABLE 7. RANGES IN C_v FOR "SOFT" AND "STIFF" SKELETON MATERIALS.

	Material	C_v (m^2/yr)
Soft skeleton	Clays	0.5 - 3
	Sands	10^2 - 0^3
Stiff skeleton	Rocks	10^2 - 10^9

Laboratory methods of determination. By far the most common procedure in the laboratory to use the consolidation test (ASTM D-2435). This test provides a basis for determining not only C_v but also k and m_v, including their variation with stress level.

In cases where k is extremely high and C is extremely low, it may be more practical to measure each independently in the lab and calculate C_v for use in computational models. When k is very high it is desirable to impose a small hydraulic gradient to prevent turbulent flow.

Field methods of determination. Mitchell and others (1978) have described several field tests that have been used to obtain C_v values with varying degrees of success. They concluded that field-measured values of C_v are generally higher than lab-measured values of C_v and that the differences generally arise from differences in flow direction, stress system, and sample disturbance.

The most reliable method is, of course, field observation of the rate and amount of ground-surface settlement. Values of C_v can then be back-calculated from conventional Terzaghi consolidation theory or from a numerical solution to the diffusion equation. If the layer dimensions are either large-scale or "full-scale" this procedure can require a very long time, perhaps many years.

In the case of San Francisco Bay mud, which has been loaded by numerous fills around the periphery of the bay, it has been possible to back-calculate values of C_v ranging from about 0.8 to 1.5 m^2/yr. This range is almost exactly the same range as has been obtained from hundreds of laboratory consolidation tests on samples. This close agreement is not observed for all materials, however.

Recommendations Relative to C_v

The primary problem with the use of laboratory tests for C_v is the question of representativeness of the samples as was discussed under the k and moduli sections. The recommendations given in those sections are considered appropriate here because C_v is calculated from k and C.

For the special case in which the layer of material of interest is relatively homogeneous and not fractured and when water flow and strain are both essentially vertical, direct laboratory consolidation tests for C_v are generally satisfactory.

CONCLUSIONS AND RECOMMENDATIONS

Porosity, permeability, and moduli are the main parameters which characterize the mechanical behavior of porous materials and are fundamental to the assessment of skeletal deformations and quantity and rate of water flow in porous materials.

For saturated materials these property parameters are not inherently different for "soft" and "stiff" skeleton materials, although the numerical values may differ radically. For soft saturated materials, some simplifying assumptions are justified which reduce the complexity drastically. Primarily these assumptions lead to the adoption of Terzaghi's effective stress to monitor both deformations and quantities of water flow. For stiff-skeleton saturated materials and for partially saturated materials, an uncoupled phase approach may be required for satisfactory accuracy in modeling, as we have suggested. The necessity of using a coupled or uncoupled phase approach depends mainly on whether the medium is fully saturated or partially saturated as well as on the type of response being sought. In an uncoupled approach the problem of assessing deformations and amount of water flow is tackled by employing two stress-state variables, namely, the water suction and an effective stress. In the coupled phase approach, one stress-state variable is employed.

In the case of saturated materials, both soft and stiff, we believe that assessing the "effective" stress is a key factor in predicting the skeleton deformations. For partially saturated materials the concept of an "effective" stress appears considerably less powerful.

We recommended that, when budget and time will allow, these properties be ascertained through a combination of laboratory and field testing, such as large-scale pumping tests. In general, redundancy in testing is desirable in order to explore the effects of various loading conditions and geologic details on the property-assessment process. The results of all these tests should be used together with a numerical method such as finite element or finite difference to solve the inverse problem. Iteration can be used to find that set of properties which is most consistent with all available test results.

Common to virtually all lab-test techniques is the difficulty and uncertainty about the degree to which the sample being tested represents the porous medium in the field. Heterogeneity of the medium in the field, sample disturbance, and failure of lab-testing schemes to duplicate field-loading conditions (or to allow lab data to be appropriately applied to field-loading conditions) may lead to errors in assessing the material response. In particular, sample disturbance significantly reduces moduli, and the presence of minute features may increase the permeability of a porous medium by several orders of magnitude. Errors in determining hydraulic conductivity and moduli would influence the integrity of models that directly employ these properties and models that use functions of these properties. Field tests, on the other hand, involve large masses with uncertainties about the prevailing boundary conditions.

If budget or time considerations do not permit field testing, and if the materials are relatively homogeneous within layers and free of fractures and fissures, then laboratory tests on representative samples as described herein may be used to estimate properties. If these conditions are not met, then the typical values suggested in the tables and references may be used to obtain very approximate estimates of the needed properties.

NOTATION

Arabic Symbols

a	Modulus exponent
a_c	Area of contacts of particles per unit cross-sectional area (L^2)
a_v	Coefficient of compressibility (L^2/F)
B	Skempton's pore-pressure parameter
C	Compressibility of skeleton = m_v (L^2/F) and parameter for an empirical permeability equation
C_c	Compression index
C_s	Compressibility of solid particles (L^2/F)
C_w	Compressibility of water (L^2/F)
C_{1-D}	C of skeleton for one-dimensional compression
C_{3-D}	C of skeleton for hydrostatic compression
C_v	Coefficient of consolidation (L^2/T)
D_{10}	Diameter for which 10% of the material by weight is finer
e	Void ratio
e_o	Reference void ratio
E	Tangent Young's modulus (F/L^2)
G	Shear modulus (F/L^2)
H	Modulus relating suction and skeleton deformation (F/L^2)
H_1	Modulus relating Terzaghi's effective stress and water phase
i	Hydraulic gradient
k	Hydraulic conductivity (L/T) or (Darcy's)
K	Bulk modulus, absolute permeability, modulus relating Terzaghi's effective-stress and skeletal-deformation modulus number
M	Constraint modulus
n	Porosity, exponent in modulus equation
R	Modulus relating suction and water phase
S	Degree of saturation
S_s	Coefficient of specific storage (1/L)
t	Time (T)
u	Pore water pressure (F/L^2)
u_a	Pore air pressure (F/L^2)
u_w	Pore water pressure (F/L^2)
v	Velocity of flow (L/T)
V	Bulk volume of a finite mass (L^3)
V_w	Volume of water

Greek Symbols

β	Compressibility (L^2/F)
γ_w	Unit weight of water (F/L^3)
δ_{ij}	Kronecker delta; = 0 for $i \neq j$; = 1 for $i = j$
Δ	Incremental change in a quantity
ε_v	Volumetric strain

Θ Volume of water absorbed or expelled per unit bulk volume
μ Viscosity of permeant (FT/L^2)
ν Poisson's ratio
σ Total stress (FT/L^2)
σ_a Atmospheric pressure (F/L^2)
σ_{ij} Total stress tensor (F/L^2)
σ_m Mean total stress; mean total stress minus pore air pressure (F/L^2)
σ' Effective confining pressue (F/L^2)
σ'_B Bishop's effective stress (F/L^2)
σ'_{ij} Terzaghi effective stress tensor (F/L^2)
σ'_{sk} Skempton's effective stress (F/L^2)
$\sigma_{T'}$ Terzaghi effective stress (F/L^2)

ϕ' Effective friction angle
χ Bishop's parameter relative to effect stress and pore water pressure
ψ Pressure head (L)

ACKNOWLEDGMENTS

We would like to acknowledge the cooperation and encouragement of P. A. Witherspoon and T. N. Narasimhan and to thank T. N. Narasimhan for his review of the manuscript. We are grateful to the Iraqi Ministry of High Education and Scientific Research for the scholarship that made it possible for the second author to pursue his studies at the University of California at Berkeley.

REFERENCES CITED

Ahmad, S., Lacroix, Y., and Steinbach, J., 1975, Pumping tests in an unconfined aquifer, in Proceedings, Conference on In-Situ Measurement of Soil Properties: Raleigh, North Carolina State University, v. 1, p. 1-21.

Aitchison, G.D., 1969, Soil suction in foundation design, in Proceedings, 7th International Conference on Soil Mechanics and Foundation Engineering: Mexico, v. 2, p. 1-8.

American Society For Testing and Materials, 1977, Annual book of ASTM standards, Soil and rock, building stones: Philadelphia, American Society for Testing and Materials, part 19, p. 369.

Barden, L., 1965, Consolidation of compacted and unsaturated clays: Geotechnique, v. 18, no. 3, p. 267-286.

Barden, L., Madedor, A. O., and Sides, G. R., 1969, Volume change characteristics of unsaturated clay: American Society of Civil Engineers, Journal of Soil Mechanics Foundation Division, v. 95, no. SM1, p. 33-51.

Biot, M. S., 1941, General theory of three-dimensional consolidation: Journal of Applied Physics, v. 12, p. 155-164.

Bishop, A. W., 1960, The principle of effective stress: Oslo, Norwegian Geotechnical Institute, publication 32, p. 1-5.

Bishop, A. W., and Blight, G. E., 1963, Some aspects of effective stress in saturated and unsaturated soils: Geotechnique, v. 13, p. 177-197.

Bishop, A. W., and Donald, I. B., 1961, The experimental study of partly saturated soils in the triaxial apparatus, in Proceedings, 5th International Conference on Soil Mechanics and Foundation Engineering: Paris, v. 1, p. 13-21.

Bishop, A. W., and Henkel, D. U., 1962, The measurement of soil properties in the triaxial test: London, Edward Arnold, second edition.

Brace, W. F., and Riley, D. K., 1972, Static uniaxial deformation of 15 rocks to 30 kb: International Journal of Rock Mechanics and Mining Science, v. 91, p. 271-288.

Brackley, I.J.A., 1971, Partial collapse in unsaturated expansive clay, in Proceedings, 5th Regional Conference on Soil Mechanics and Foundation Engineering: South Africa, p. 23-30.

Burland, J. B., 1965, Some aspects of the mechanical behavior of partly saturated soils, in Moisture equilibrium and moisture changes in soils beneath covered areas: Sydney, Australia, Butterworth and Company Ltd.

Chang, C. S., and Duncan, J. M., 1977, Analysis of consolidation of earth and rockfill dams: Berkeley, Institute of Transportation and Traffic Engineering, Department of Civil Engineering, University of California, report TE-77-3.

Clark, S. P., ed., 1966, Handbook of physical constants: Geological Society of America Memoir 97.

Fredlund, D. G., 1975, A diffused air volume indicator for unsaturated soils, Canadian Geotechnical Journal, v. 12, no. 4, p. 533-539.

Fredlund, D. G., and Morgenstern, N. R., 1976, Constitutive relations for volume change in unsaturated soils: Canadian Geotechnical Journal, v. 13, p. 261-276.

Gibson, R. E., 1963, An analysis of system flexibility and its effect on time lag in pore water pressure measurements: Geotechnique, v. 13, no. 1, p. 1-11.

Goodman, R. E., 1980, Introduction to rock mechanics: New York, John Wiley.

Hazen, A., 1911, Discussion of 'Dams on sand foundations' by A. C. Koenig: American Society of Civil Engineers Transactions, v. 73, p. 199.

Jaeger, J. C., and Cook, N.G.W., 1976, Fundamentals of rock mechanics: London, Chapman and Hall, Second edition.

Janbu, N., 1963, Soil compressibility as determined by odometer test and triaxial tests, in Proceedings, European Conference on Soil Mechanics and Foundation Engineering: v. 1, p. 19.

Jennings, J. E., and Burland, J. B., 1962, Limitations to the use of effective stresses in partly saturated soils: Geotechnique, v. 12, no. 2, p. 125-144.

Lamb, T. W., and Whitman, R. V., 1969, Soil mechanics: New York, John Wiley, 553 p.

Lang, S. M., 1967, Pumping tests for determining aquifer characteristics: American Society for Testing and Materials, STP 417, Permeability and capillarity of soils, p. 35-55.

McMurdie, J. L., and Day, P. R., 1960, Slow tests under soil moisture suction: Soil Science Society of America, Proceedings, v. 24,

p. 441-444.
Matyas, E. L., and Radhakrishna, H. S., 1968, Volume change characteristics of partially saturated soils: Geotechnique: v. 18, no. 4, p. 432-448.
Milligan, V., 1975, Field measurement of permeability in soil and rock, in Proceedings, Conference on In-Situ Measurement of Soil Properties: Raleigh, North Carolina State University, v. 2, p. 3-36.
Mitchell, J. K., 1976, Fundamentals of soil behavior: New York, John Wiley.
Mitchell, J. K., Hooper, D. R., and Campanella, R. J., 1965, Permeability of compacted clay: Journal of the Soil Mechanics and Foundations Division, American Society of Civil Engineers, v. 91, no. SM4, paper 4392, p. 41-65.
Mitchell, J. K., Guzikowski, F., and Villet, W.C.B., 1978, The measurement of soil properties in-situ: Present methods, their applicability and potential: Berkeley, Lawrence Berkeley Laboratory, University of California, LBL-6363.
Narasimhan, T. N., and Witherspoon, P. A., 1977, Numerical model for saturated-unsaturated flow in deformable porous media: Water Resources Research, v. 13, no. 3, p. 657-664.
Neuman, S. P., 1973, Saturated-unsaturated seepage by finite elements: Journal of the Hydrology Division, American Society of Civil Engineers, v. 99, no. HY12, p. 2233-2290.
Sandhu, R., and Wilson, E. L., 1969, Finite element analysis of land subsidence, in Land subsidence, Volume 2: International Association of Scientific Hydrologists, pub. 89, p. 393-400.
Skempton, A. W., 1960a, Significance of Terzaghi's concept of effective stress: From theory to practice in soil mechanics: New York, John Wiley, p. 4-16.
___1960b, Effective stress in soils, concrete and rocks, in Proceedings of Conference on Pore Pressure and Suction in Soils: London, Butterworths, p. 4-16.
Terzaghi, K., 1925, Principles of soil mechanics: A summary of experimental results of clay and sand: Engineering News Record, p. 3-98.
Theis, C. V., 1935, Relation between lowering of piezometric surface and duration of discharge of a well using groundwater storage: American Geophysical Union Transactions, p. 519-524.
U.S. Department of the Interior, Bureau of Reclamation, 1960, Earth manual: Washington, D.C., U.S. Department of the Interior, designation E-18, p. 541-546: designation E-19, p. 546-562.
Wong, K. S., and Duncan, J. M., 1974, Hyperbolic stress-strain parameters for nonlinear finite element analyses of stresses and movements in soil masses: Berkeley, Department of Civil Engineering, University of California, Geotechnical Engineering Report TE 74-3.

Manuscript Accepted by the Society October 14, 1980

Geological Society of America
Special Paper 189
1982

Assessing the Permeability Characteristics of Fractured Rock

J. E. GALE
Department of Earth Sciences
University of Waterloo
Waterloo, Ontario, Canada

ABSTRACT

In this paper the geometric and hydraulic characteristics of fractures and fractured rock masses are reviewed to assess the current state and future direction of fracture hydrology research. Laboratory data suggest that the parallel-plate analogy for flow through a single fracture is valid. Fracture-flux is a function of the cube of the fracture aperture. Flow through fractures is a function of normal stress, shear stress, and fracture surface characteristics such as roughness. Flow through fractured rock masses is determined by fracture orientation, spacing, fracture interconnection, and the stress field. These factors must be considered in assessing the directional permeabilities of fractured rocks. Consideration of these factors and the structural characteristics of fractured rock leads to the formulation of a conceptual framework for flow in fractured rock masses that is a form of coupled discrete-fractured porous media model. In this conceptual framework, the shear zones and fracture zones are described as discrete hydrogeologic features; individual fractures (joints) are defined as discrete features near zones of interest, such as underground excavations; and the fractured rock mass in general is described as a form of equivalent fractured porous medium. I propose that the properties of the equivalent continuous porous medium must be developed from discrete fracture properties, reflecting the dependence of fracture permeability on the stress tensor, fracture geometry, distribution of fracture apertures, and degree of fracture interconnection.

The sparse data base, lack of agreement on the appropriate conceptual models, and difficulty of conducting field studies in low-permeability fractured rocks require that calculations of total-flux and transit times in such rock masses be supported and confirmed by other lines of evidence, including geochemical and isotopic data.

INTRODUCTION

As the other papers in this symposium indicate and as the volume of literature referenced in these papers attests, considerable resources have been devoted to the study of fluid movement through porous media. In comparison, the field of fracture hydrogeology has been a distant and poorly-nurtured cousin. This difference in funding levels and attention from researchers has stemmed in part from the complexity of the problem and an apparent lack of economic incentives.

Most argillaceous and crystalline rock masses consist of rock blocks bounded by discrete fracture planes. The blocks may be of porous and permeable rock, such as sandstone, or may have low porosity and very low permeability, such as granite. It is the ratio of the permeability of the rock blocks to the permeability of the fractures that determines the significance of the fractures in a given flow system. In metamorphic and granitic rocks, it is generally accepted that the hydraulic conductivity of rock blocks is $<10^{-9}$cm/s and thus significant flows can occur only through the fracture system. Very little data are available on the effect of fractures on flow in shales and other low-permeability argillaceous rocks. In the more permeable argillaceous rocks, there has been almost no attempt at distinguishing the contributions to flow from the rock blocks and the fracture system. Ground-water studies in these rocks have assumed that the rock block-fracture system is a porous medium. Thus the very small data base in fracture hydrology is limited primarily to metamorphic and crystalline rocks. Hence, this paper will focus on these rock types. In addition, although extensive literature exists, no attempt will be made to discuss fluid movement through fractured carbonate rocks in which solution channels are the primary conduits.

To place a discussion of the current status and

future research directions of fracture hydrogeology in perspective, one should note the time framework of the main developments in the field of porous media hydrogeology. Thus, although the first experimental work on porous media flow was completed by Darcy in 1857, an adequate flow system theory was not introduced into the porous media literature until the early 1940s. The theory of modern well hydraulics was not well developed or widely used in the porous media field until the late 1940s and 1950s. In comparison, the first comprehensive set of experiments on flow through artificial fractures are the glass plate experiments of Lomize (1951). A review of the published literature suggests that a common agreement on the adequacy of existing conceptual and theoretical models for flow in fractured media is lacking.

This paper is not a comprehensive review of the field of fracture hydrogeology. Such reviews already exist (Wilson and Witherspoon, 1970; Gale, 1975; Witherspoon and Gale, 1977; Louis and others, 1977). Instead, I will review selected papers and contributions on the basic flow relationships and the factors that control flow in both single and multiple fracture systems. I will discuss these factors in the context of the basic geometric and hydraulic characteristics of fractured rock masses. This discussion will provide the background for assessing the methods of measuring the permeability of fractured rock masses, lead to the formulation of a conceptual framework for flow in fractured rocks, and serve as the basis for suggesting the direction of future research.

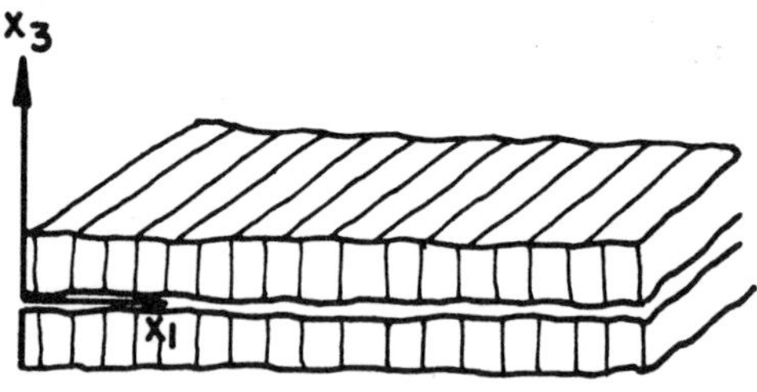

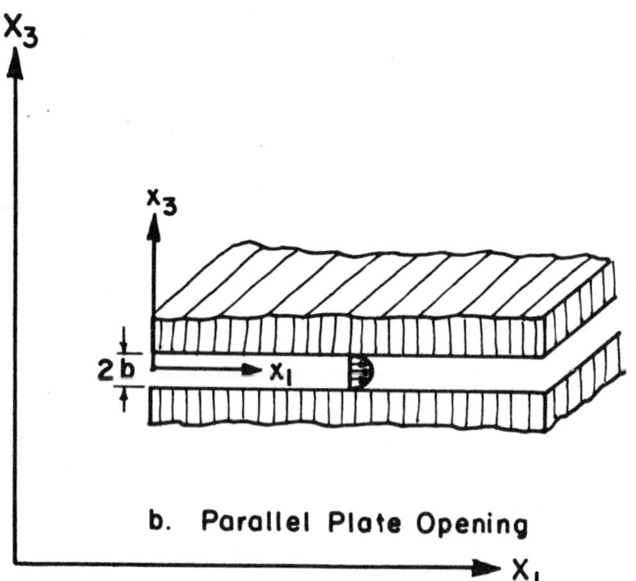

Figure 1. Flow in a single fracture.

FLOW THROUGH SINGLE FRACTURES

Basic Flow Relationships

Experimental work by Lomize (1951) and Huitt (1956) provided the impetus for the development of the basic equations governing fluid flow in fracture systems. It is assumed that fluid flow in a fracture is governed by Darcy's law where the permeability term, or the hydraulic conductivity (LT^{-1}), is derived by idealizing the fracture as a parallel plate opening (Fig. 1). Equations describing flow between parallel plates have been derived by a number of authors (Polubarinova-Kochina, 1962; Romm and Pozinenko, 1963; Snow, 1965; Gale, 1975; and several others). All of the derivations are based on the Navier-Stokes equation for single-phase, non-turbulent flow of a viscous incompressible fluid. In terms of hydraulic head, the parallel plate analogy for flow through a fracture is:

$$K_f = \frac{\rho g}{12\mu}(2b)^2, \qquad (1)$$

where ρ is the fluid density, g is gravity, μ is dynamic viscosity, 2b is the fracture aperture, and K_f is the hydraulic conductivity of a single fracture (LT^{-1}). For saturated flow in a single fracture, or a series of interconnected, continuous fractures, the constitutive and continuity equations are, respectively:

$$q = -K_f \frac{\partial H}{\partial \ell} \qquad (2)$$

and

$$\frac{\partial}{\partial \ell}\left(K_f \frac{\partial H}{\partial \ell}\right) = 0, \qquad (3)$$

where q is the fracture-flux [volume flow per unit time per unit length of cross-sectional fracture area (2b x 1)] and H is the hydraulic head. It can be shown from equation 1 that the flow through a unit length of fracture is a function of the aperture cubed. This cubic relationship has been confirmed by recent laboratory investigations of radial and linear flow in induced tension fractures (Iwai, 1976).

Figure 2 is a plot of fracture aperture versus fracture-flux under radial flow conditions and shows the close correspondence of the experimental data to the cubic law--represented by the solid lines. This functional, cubic-law relationship is thoroughly discussed by Witherspoon and others (1979a). In all of the experimental studies of flow through fractures it is accepted that the flow velocity of the fluid at the fracture wall is zero. If the aperture is large enough, this will fully

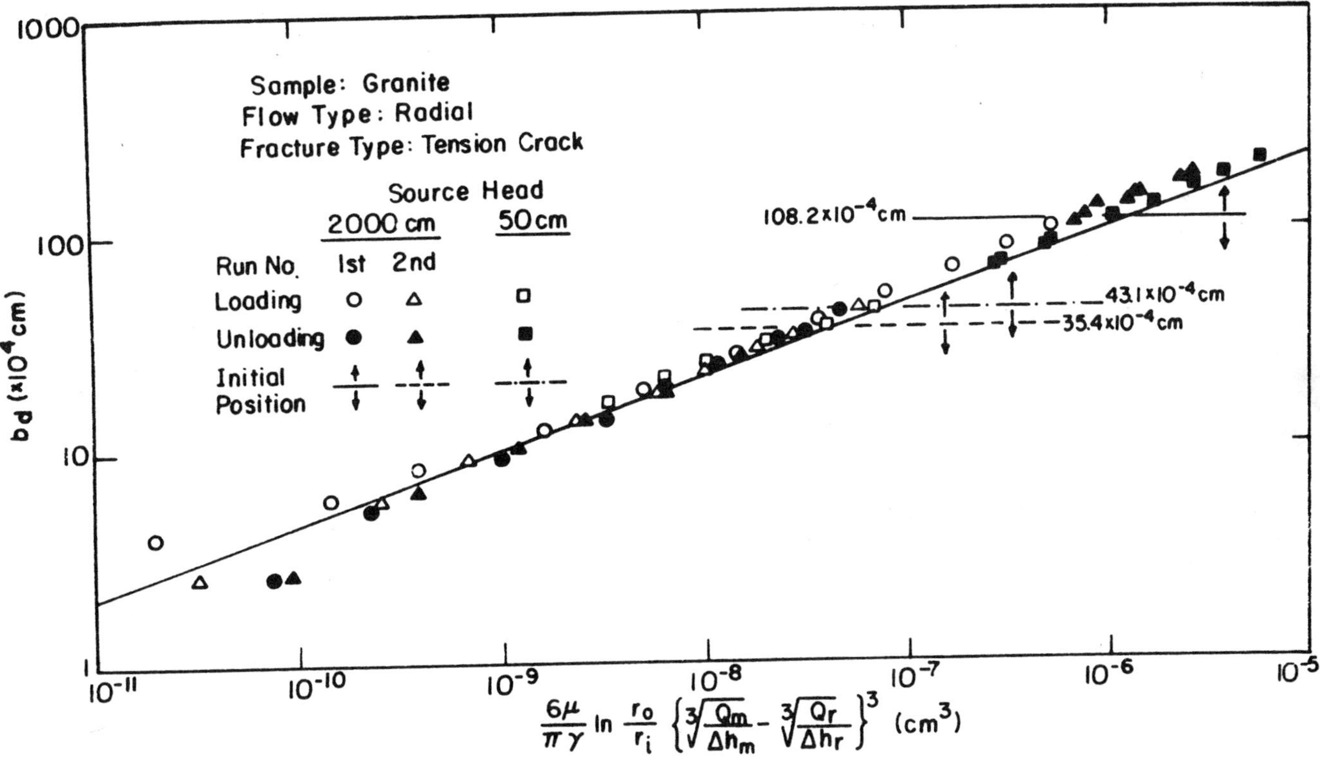

Figure 2. Comparison of experimental results for radial flow through tension fracture in granite with cubic law (after Iwai, 1976).

correspond to the real boundary conditions. But as pointed out by Romm (1966), if the fracture apertures are small, such as one would expect to find at depth, then there may be deviations from the classical law of hydrodynamics. These deviations may be associated with virtually stationary polymolecular layers of liquid on the fracture walls, the thickness of these layers being comparable in size to the crack opening (Romm, 1966). This phenomenon and the difficulties of measuring very low flow rates may explain some of the deviations from the cubic relationship present in Iwai's data for small apertures. As pointed out by Snow (1979, oral commun.), the surface chemistry effects may be small compared with the acute effect of tortuosity: as rock separation decreases, the effective aperture at inclined faces decreases faster, and path length exceeds the projected length. The relative importance of these effects varies with displacement in a manner that is not yet understood.

Factors Controlling Flow

Field and laboratory studies (Gale, 1975) have shown that pore pressure changes, producing changes in effective stress, can significantly change the fracture apertures, flow rates, and fluid pressure distributions in fractures. Because fracture-flux in a single fracture is a function of the aperture cubed (Fig. 2) and fractures are highly deformable (Fig. 3), it is reasonable to expect that flow through fractures will be highly stress-dependent.

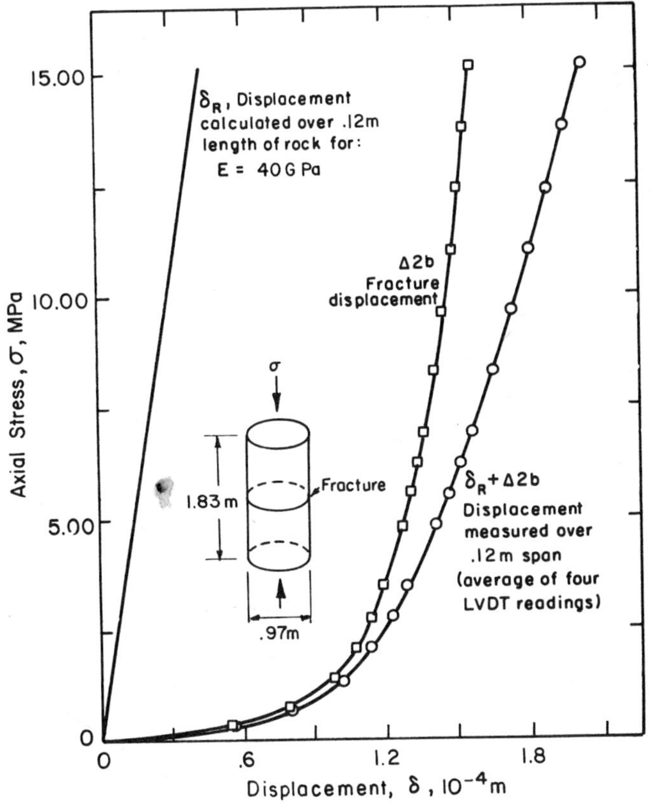

Figure 3. Fracture and rock deformation in sandblasted saw-cut surface (Gale, 1975).

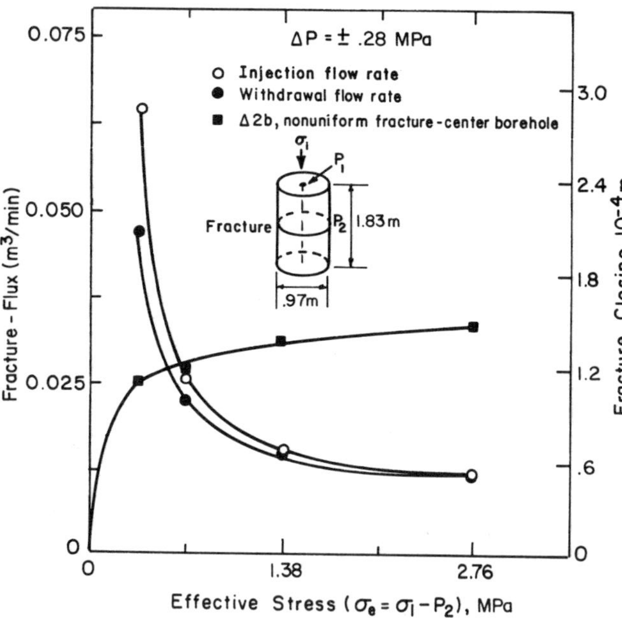

Figure 4. Changes in fracture-flux and fracture aperture as a function of effective stress in sandblasted saw-cut fracture surface (Gale, 1975).

Figures 4 and 5 show that there is a very rapid decrease in fracture flux with increase in effective stress. In addition, Figure 5 shows that both the initial and final fracture-flux values decrease with an increase in the number of loading and unloading cycles. This effect can be considered a form of hysteresis, resulting from seating of the fracture due to crushing of asperities in the fracture plane. The magnitude of this hysteresis decreases with increasing number of loading cycles, such that after three or four cycles the stress-permeability curves are very similar to each other.

Various authors (Snow, 1968a; Rayneau and Jouanna, 1971; Jones, 1975; Nelson, 1975; Louis and others, 1977; Witherspoon and others, 1977; Kranz and others, 1979) have proposed empirical models to describe the observed variation of fracture permeability with normal stress. Snow's model took the form of

$$K_h = K_o + [K_N(2b)^2/S](P - P_o), \quad (4)$$

where K_h is the permeability of horizontal fractures, K_o is the permeability at an initial pressure P_o, 2b is the fracture aperture, K_N is the normal stiffness of the fracture, and S is the fracture spacing. Because K_N is a nonlinear function of stress or pressure P, K_h becomes a nonlinear function of P. Jones (1975) proposed an empirical equation for the fracture permeability K of carbonate rocks

$$K = K_o[\log (P_h/P)]^3, \quad (5)$$

where K_o is a constant and P_h is the "healing" effective pressure, at which K = 0.

Equations presented by Louis and others,

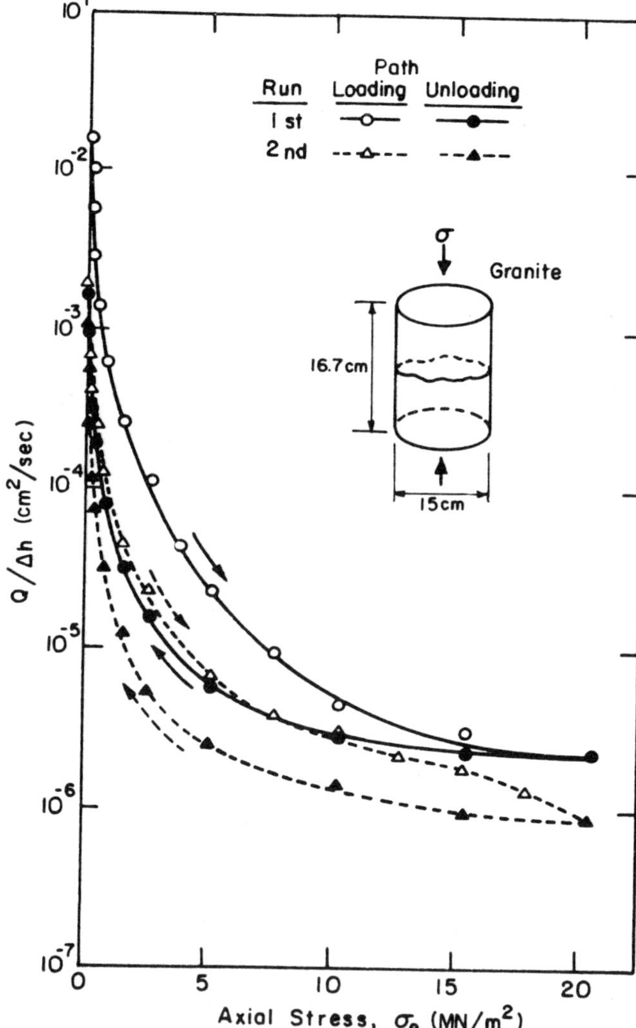

Figure 5. Effect of cyclic loading on permeability of tension fracture in granite with radial flow (after Iwai, 1976).

Witherspoon and others, and Kranz and others have the form of $K_f = \beta\ \sigma_e^{\alpha}$ where α is the slope of the line and β is the value of K_f, fracture hydraulic conductivity, at some prescribed value of σ_e, the effective normal stress.

Gangi (1978) concluded that most of the above models did not give sufficient insight into the physical mechanisms involved in the stress-fracture permeability relationships. He proposed a phenomenological model based on the bed-of-nails concept. Gangi's model is a welcome addition to the discussion. However, he assumes that the angles of the fracture surface relative to the common plane are less than 10°, resulting in the fracture and flow length differing by less than 2%, which unfortunately is not consistent with either the observed or measured roughness of fracture surfaces. It is not apparent to this writer how Gangi's model can account for the observed hysteresis in stress-fracture flow results or from the change in the overall stress-fracture permeability relationship with each loading cycle. Additional work on this subject will be required to establish generalized relationships that take into account the nonlinear dependence of normal stiffness on effective stress, the effects of weathering, roughness, rock type, grain size, and sample size.

Although a number of studies are currently under way, only limited test data on a few different rock types are available. Iwai's (1976) testing of induced tension fractures in three rock types--granite, marble, and basalt--provides the best data for comparing rock types. Iwai's data for the first loading cycle, shown as a semilog plot of the fracture-flux normalized with respect to the hydraulic head ($Q/\Delta h$), versus the axial stress (Fig. 6), show that the minimum fracture-flux values at the maximum axial stress of 20 MN/M^2 are similar for granite and marble, but at similar stress levels the fracture-flux values are significantly different for basalt. In Figure 6 we note that the initial fracture-flux values at the lowest stress conditions are also quite different for each rock type. Thus a direct

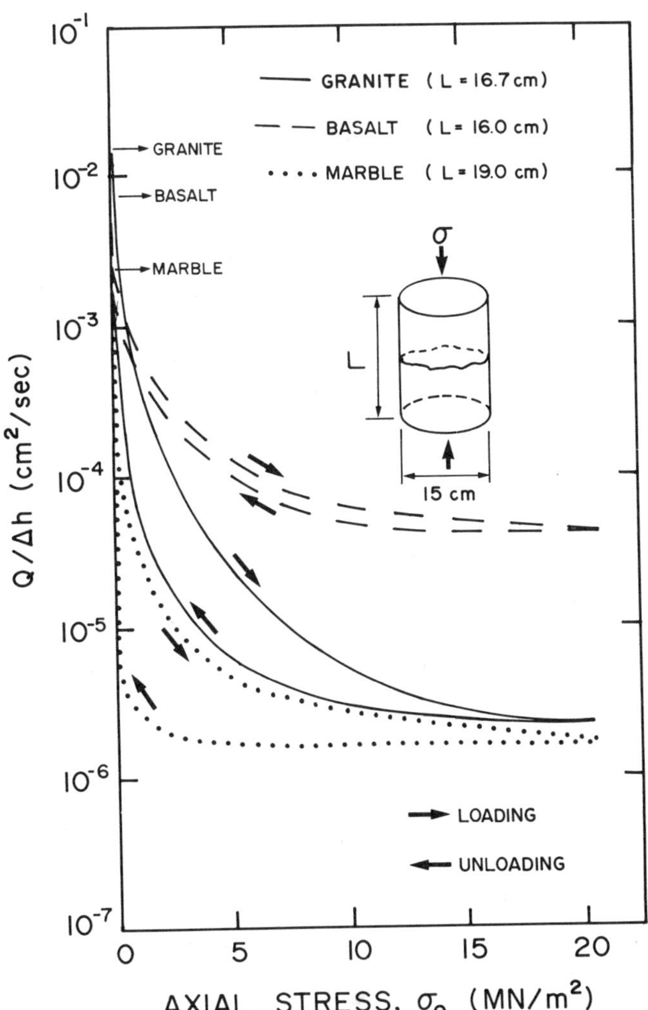

Figure 6. Fracture-flux versus stress data for three rock types (after Iwai, 1976).

TABLE 1. NORMAL STIFFNESS (K_N) OR STRESS PER UNIT DISPLACEMENT, AS A FUNCTION OF NORMAL STRESS, LOADING CYCLE, AND ROCK TYPE.*

	K_N secant modulus [(MN/M) x 10^{-4}]				
	(1)	(2)	(3)	(4)	(5)
Granite					
cycle 1	1.19	2.15	3.09	4.02	4.95
cycle 2	4.44	7.30	9.74	12.5	15.0
Marble					
cycle 1	4.17	7.55	10.9	14.1	17.1
cycle 2	6.25	11.5	16.4	21.3	26.2
Basalt					
cycle 1	3.45	5.75	7.89	9.95	11.9
cycle 2	3.64	5.99	7.94	10.2	12.3

(1) normal stress MN/M^2

* After Iwai, 1976

comparison of the fracture-flux versus stress curves in this form may not be completely valid. The approach taken by Brace (1978), in which the permeability is normalized with respect to some initial minimum stress, will make it easier to compare rock types over the entire fracture-flux range.

A comparison of Iwai's normal-force and normal-displacement (K_N) data (Table 1) suggests that there are significant differences between the K_N values at similar stress levels, but no trends are apparent in these data. From the theoretical discussion of the effects of stress on contact area (Gangi, 1978), one would expect that a high K_N value would indicate an increase in contact area and hence a corresponding decrease in fracture-flux. Thus on the first loading cycle the granite (Table 1) shows a K_N of 4.95 x 10^4 MN/M at an axial stress of 5 MN/M^2, while at the same axial stress on the second loading cycle the K_N value is 15.0 MN/M. In comparison, the K_N value for the marble in the second loading cycle is 26.2 MN/M, although the stress-permeability curves are somewhat similar (Fig. 6). On the other hand, the basalt sample shows very little hysteresis and has a K_N value of 12.3 MN/M, similar to that of the granite, but a significantly different stress-permeability curve than the granite. This difference probably resulted from the relatively uniform distribution of contact area in the basalt sample (Iwai, 1976). Obviously much more work needs to be done, with care being taken in preparing and testing both induced and natural fractures. It is only with the accumulation of basic empirical data that it will be possible to determine the connection, if any, between K_N values and the stress-permeability relationships.

Much of the laboratory testing in fracture hydrology, like mechanical property testing in rock mechanics, has been carried out on very small samples. Iwai (1976) has shown, both numerically

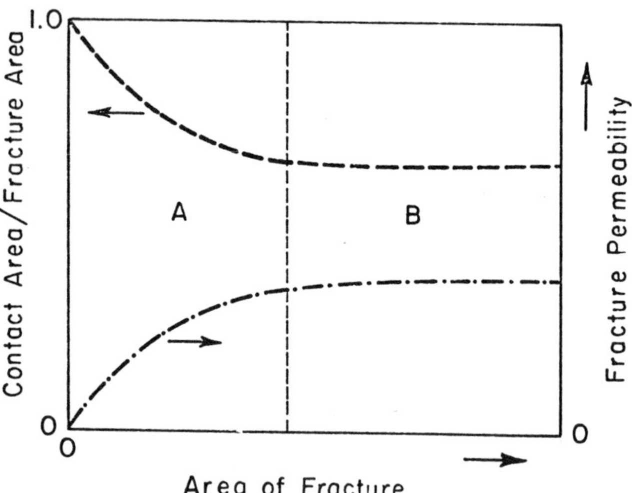

Figure 7. Effect of contact area on fracture permeability (Witherspoon and others, 1979).

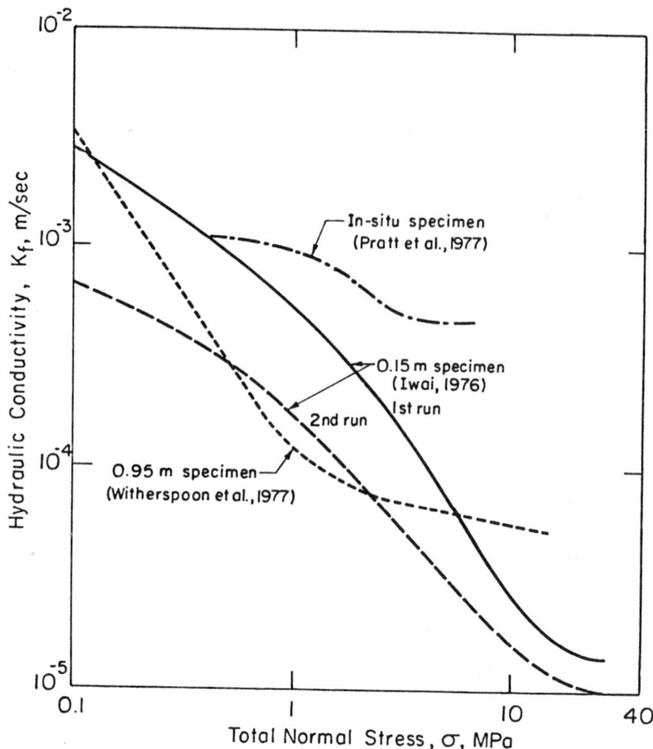

Figure 8. Variation in stress-fracture permeability relationship with fracture sample size (Witherspoon and others, 1979).

and experimentally, that fracture permeability is a function of contact area. Contact area can be changed by at least two methods: (1) increasing the stress acting across the fracture plane (Iwai, 1976), and (2) changing the size of the area of fracture surface being sampled. Questions have been raised (Witherspoon and others, 1979b) as to whether the stress-permeability relationships determined from laboratory tests on small samples, measuring hundreds of square centimetres in area, can be related to field conditions, where fracture surface areas measure on the order of tens of square metres. These questions are based on the conceptual model presented in Figure 7. Thus, as shown in Figure 7, as the area of the fracture surface approaches zero, ideally equal to the rock grain size, the ratio of the contact area to the fracture approaches 1, that is, a contact area of 100%. At some appropriate stress level this contact area would result in zero flow through the fracture, ignoring the micro-roughness in grain surfaces. Thus, given the normal distribution of contact points within the fracture plane, there would be a decrease in the ratio of the contact area to the fracture with increasing sample area, and a corresponding increase in the fracture permeability at the same stress level, as indicated by area A in Figure 7. At some given sample size (area B, Fig. 7) the sample of the fracture plane would contain an average number of contact points and thus similar samples of the fracture plane would give similar stress-permeability results. A comparison of existing stress-permeability data from different size samples suggests that such a scale effect may exist (Fig. 8). The existing data do not permit one to draw definite conclusions. Considering the potential importance of the scale effect, a systematic study should be undertaken to evaluate its importance and to determine the scaling factors required to correlate laboratory tests on small samples with field data. Considerable care must be taken to ensure that the test results are related to the surface characteristics of the fracture plane.

In addition to the studies referred to above, other researchers have studied the effects of normal stress on fractures (Sharp, 1970; Jouanna, 1972; Pratt and others, 1977). Additional studies on the effects of rock type, fracture type, fracture roughness, and sample size on the normal stress-fracture permeability relationship are currently under way at the University of Waterloo and at Lawrence Berkeley Laboratory, University of California. In contrast to the increasing number of studies on the effects of normal stress on fracture permeability, the effects of shear stress and shear displacement on fracture permeability have received very little attention from researchers. A preliminary experiment by Maini (1971) suggests that shear displacement has a major impact on the permeability of a fracture. Maini's results, when evaluated in the light of the mechanical behavior of fracture systems, suggest that more experiments with close control on the flow and stress-displacement boundary conditions are required in this area.

GEOMETRIC AND HYDRAULIC CHARACTERISTICS OF FRACTURED ROCK MASSES

Structural Framework

In any discussion of the hydrology of fractured crystalline and argillaceous rocks one must first consider the structural nature of the rock mass (Fig. 9). The main flow paths in fractured crystalline and argillaceous rocks are joints, fracture zones, and shear zones. Joints are discontinuous in

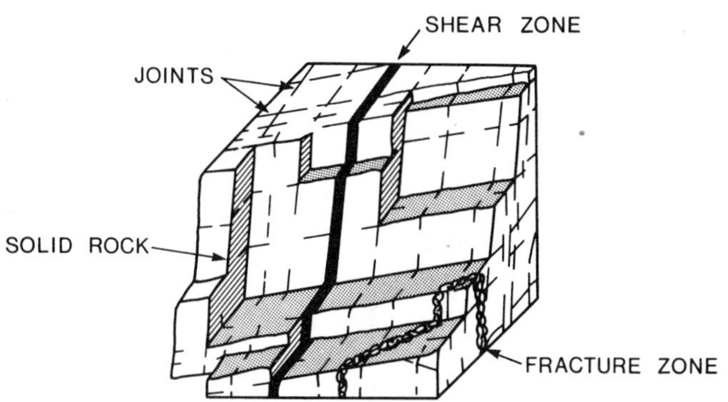

Figure 9. Conceptual model of the main flow paths in fractured rock.

their own planes, but when the ratio of joint length to joint spacing is large, they form continuous networks for flow. In rock masses dominated by joints, the hydraulic characteristics are to an unknown degree a result of the interconnection of the different joint sets. Fracture zones are defined as zones of closely-spaced and highly-interconnected discrete fractures that are generally not filled with clay or any other material. Fracture zones measure from less than a metre to tens of metres in width and are generally filled with broken and crushed rock. Depending on the rock type this material may be embedded in a clay matrix. Shear zones tend to be continuous throughout large parts of the rock mass, but may be either more or less conductive than the rock mass, depending on filling materials, cementation, age, and stress. Large-scale features can extend for tens of kilometres but their hydraulic properties can vary considerably over such distances. There are all transitions between these three classes of fractures.

As shown schematically in Figure 9, joints or fractures exist throughout the rock mass. Within a given rock mass a number of fracture sets can be defined, and each fracture set has a preferred orientation. This orientation can be referenced from the pole to the fracture plane and the distribution of poles can be plotted and contoured on stereonets. From this contouring the major fracture sets and the dispersion of orientations about the mean can be identified. From a fracture hydrology point of view, the variation of fracture orientations laterally and vertically must be determined. Several recent studies have shown that there is a general similarity between fracture orientations measured on the surface and those measured in the subsurface (Appleby, 1940; Kendorski and Mahtab, 1976; Raven and Gale, 1977). Figures 10 and 11 show the results of field measurements at a mine located in the Precambrian rocks of the Canadian Shield. In this case there is close similarity between the surface and subsurface fracture orientations.

From field observations, we know that the spacing of fractures within a given set is highly variable. Very few data exist on the nature of this variation in fracture spacing (except Snow, 1968b; Parsons, 1972) but it is a variable, like orientation, that can be measured on the surface, in underground openings, and in boreholes. Using a line-sampling technique, Snow (1967) has defined the spacing of fractures, Δ, belonging to a set having a dispersion of orientations, as the reciprocal of specific surface, S. The reciprocal of specific surface S can be computed from the angle Θ made by the normal to the fracture plane and the direction of a sampling line D. Thus for a number of fractures:

$$\Delta = 1/S = D/\sum_{i=1}^{n} 1/\cos\Theta_i. \qquad (6)$$

Because there are usually two or three sets present in a given rock mass it is obvious that the degree of fracturing or fracture intensity in a given rock mass is determined by both the orientation and spacing of the fractures in the different sets.

Porosity and Permeability: Field Observations

Most field measurements of permeability and estimates of porosity of fractured rocks have been made during site investigations. Such sites are generally located near major structural features. Other data have resulted from borehole tests during the domestic and industrial development of ground-water resources. These boreholes are generally restricted to depths of less than 150 m. In both cases the near-surface zone, where these measurements have been made, forms the most permeable zone within fractured crystalline and metamorphic rocks.

Field values that have been reported for the permeability of fractured crystalline rocks have been obtained using different testing procedures, and the results have been interpreted using widely varied theoretical models. Figure 12 is a compilation of porosity data that was summarized by Brace (1975) from field test results computed by Snow (1968b) and others. This figure also shows the range of crack and pore porosities of about 30 crystalline rocks determined by laboratory measurements.

These data suggest that fracture porosity decreases with depth. Crack or microcrack porosity, as determined from high pressure experiments, also decreases with depth, but at a much slower rate. Pore porosity is assumed to persist at the 10^{-3} level at all depths, at least down to the depths at which high-temperature effects take over. More recent measurements by Knapp (1975) gave total matrix porosities of 1 to 2% for a number of different granitic rocks. Knapp (1975) separated the total porosity, total void space in a rock, into (1) flow porosity, the void space within which fluid flow is the predominant aqueous component transport mechanism; (2) diffusion porosity, the void space within which ionic transport through the aqueous phase is the predominant transport mechanism; and (3) the residual porosity, which consists of the isolated voids that are neither connected to the flow porosity nor the diffusion porosity. Knapp's data suggest that for crystalline (granitic) rocks, the flow porosity is about 1%, the diffusion porosity is about 5%, and the residual porosity is about

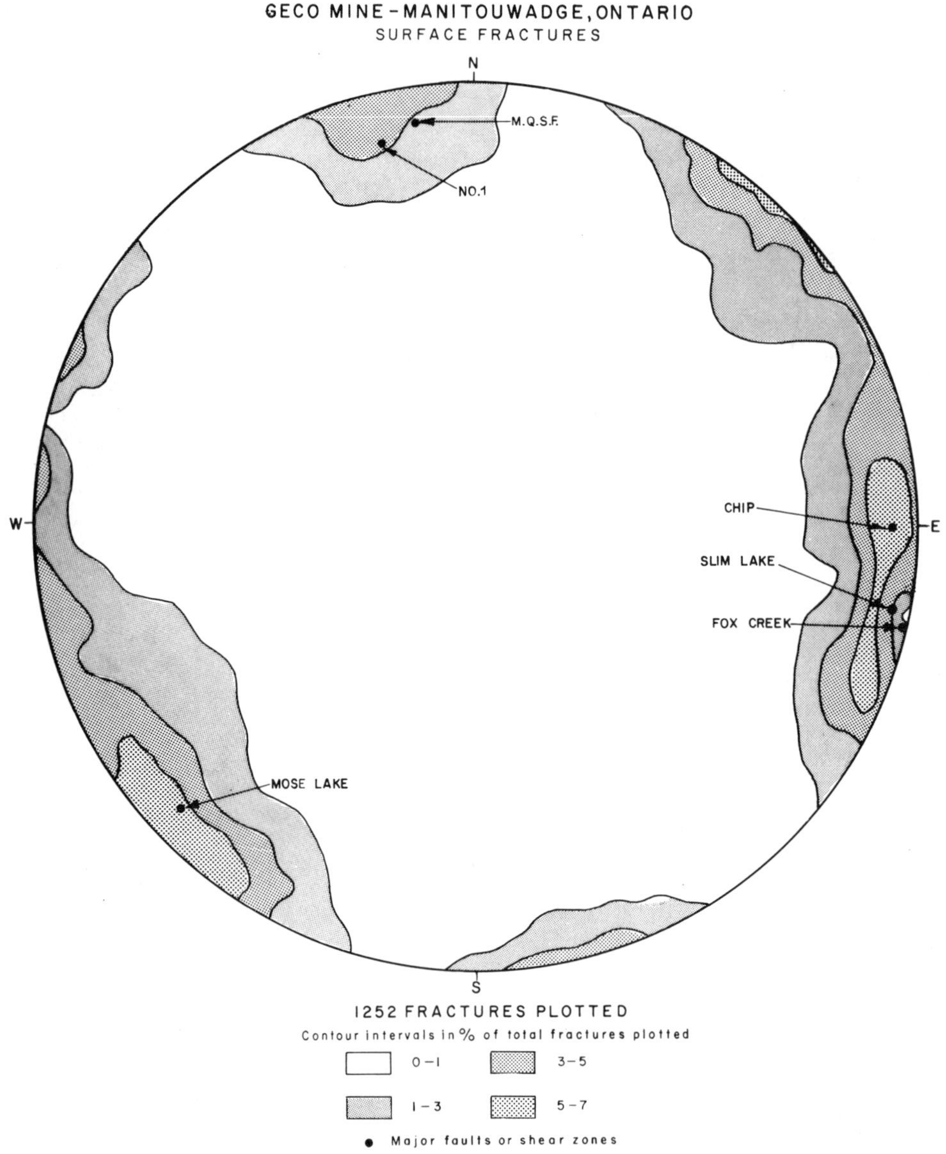

Figure 10. Contour plot of poles to fractures measured on surface outcrops (Raven and Gale, 1977).

94% of the total porosity. Thus the flow porosity of the intact granitic rocks is about 10^{-4}. Because the fracture porosities were computed from injection tests they represent the flow porosity of the fracture system. It is quite possible that the total fracture porosity is much higher than the computed flow fracture porosity.

The relative values of fracture porosity versus crack and pore porosity, and the ratio of flow to total porosity in the fractures, cracks, and pores, achieve some degree of importance if diffusion of radionuclides from the fractures into the rock blocks is assumed to be a significant mechanism of radionuclide attenuation (Cherry and Gale, 1979). These relative values achieve similar or greater degrees of importance in the use of isotopes to date ground waters if diffusion into the matrix or dead-end fractures under a concentration gradient is a significant mechanism.

Table 2 is a compilation of hydraulic conductivity values for different igneous and metamorphic rocks. In computing these values, no attempt was made to determine how many fractures intersected the borehole or borehole interval being tested, or what contribution each fracture made to the total permeability. The permeability was computed by assuming that the entire section of the borehole being tested was a permeable continuum. This method of computation conceals the distribution of fracture apertures and can result in unsuitable velocity values when one attempts to predict the transport of a contaminant in fractured media.

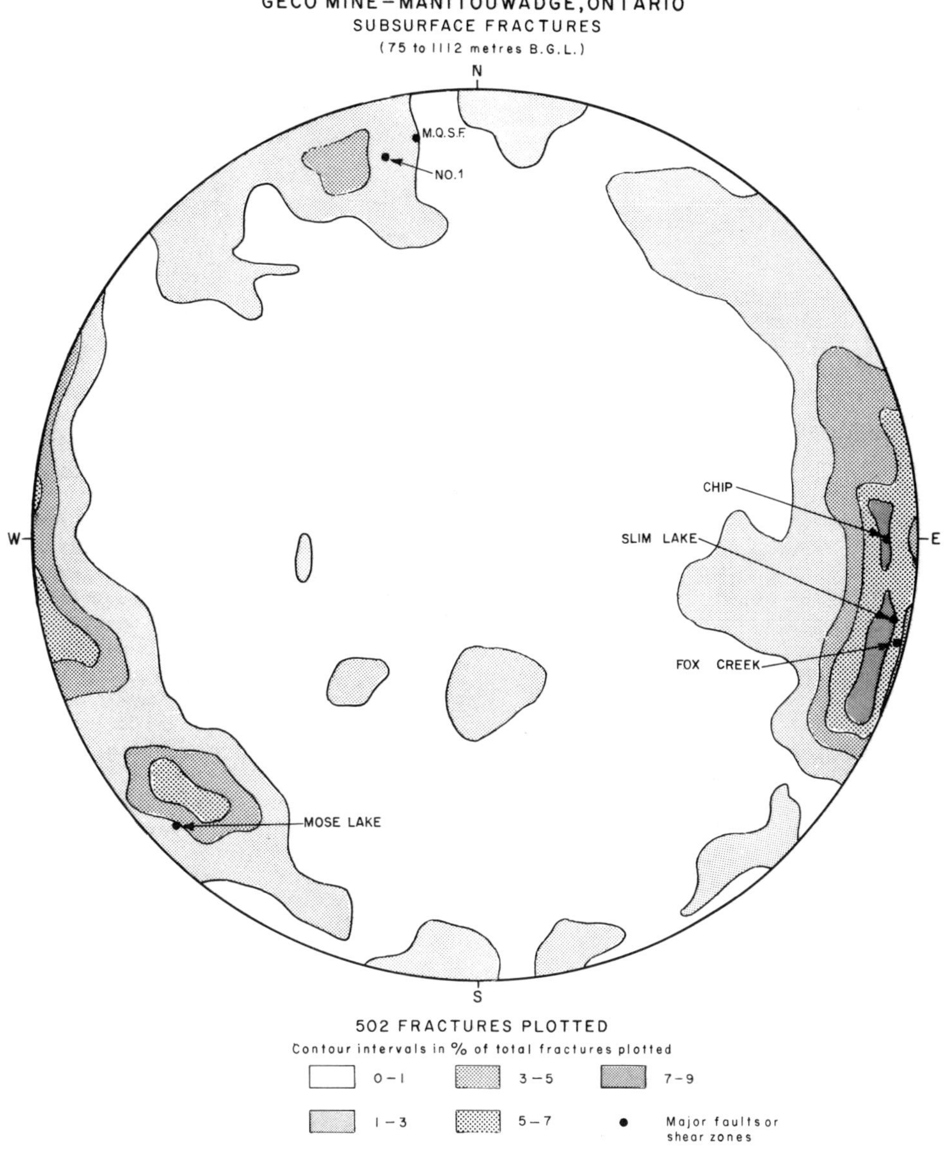

Figure 11. Contour plot of poles to subsurface fractures (Raven and Gale, 1977).

Measurement programs need to take into account the small-sample variability entailed in borehole methods.

A common characteristic of fractured rocks is that the permeability decreases with depth (Figs. 13 and 14). Both the well yield data (Fig. 13) tabulated by Davis and Turk (1964) and the injection test data (Fig. 14) tabulated by Snow (1968c) show a general logarithmic variation with depth.

Basic Flow Relationships For Multiple Fracture Systems

Romm and Pozinenko (1963), Snow (1965), Caldwell (1971), and others have shown that one can, using the parallel plate model, combine continuous fractures with different orientations, apertures, and spacing to produce a mathematical description of rock mass permeability in the form of a second-rank tensor. The basic approach used in calculating directional permeabilities of a fracture system is discussed below, and follows that of Snow (1965, 1969) and Bianchi and Snow (1969). Snow (1965) defined flow in a single conduit under a general field gradient as:

$$q_j = -\frac{(2b)^3}{12\mu} \rho g (\delta_{ij} - M_{ij}) I_i, \tag{7}$$

where q = fracture-flux vector per unit width of conduit $[L^3/T]$,

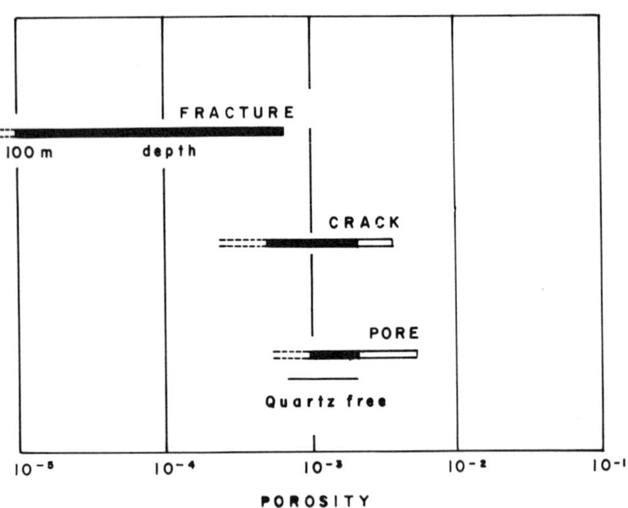

Figure 12. Variation in three types of porosity from data compiled by Brace (1975).

δ_{ij} = Kronecker delta,
M_{ij} = $n_i n_j$, a matrix formed by direction cosines of the normal to the conduit,
I_i = potential gradient vector.

Snow asserts that each fracture intersecting a sample line has its image at a distance L equal to the length of the sample line and in the direction of the sample line. Thus, the spacing between the fracture and its repetition is given as:

$$w = L|n_i\ D_i|, \tag{8}$$

where
w = spacing between fractures,
L = length of sample line,
n_i = direction cosines of normal to fracture plane,
D_i = direction cosines of sampling line.

The permeability of a medium containing a single fracture is then given by:

$$k_{ij} = \frac{2}{3L}\ \frac{b^3}{|n_i\ D_i|}\ (\delta_{ij} - M_{ij}),$$

where k_{ij} = intrinsic permeability tensor (L^2).

The permeability of the medium at a given sampling station is obtained by summing the contributions of the individual fractures. Where more than one sampling section is used, the average of all the stations is calculated (Bianchi and Snow, 1969). Having obtained the permeability tensor for a given rock mass, we must then determine the principal components of the tensor and their directions. This is equivalent to finding the eigenvalues and eigenvectors of the tensor and involves solving a cubic equation. An analytical solution to the eigenvalue problem exists, based on an analogy with the stress tensor (Westergard, 1964), and should be used to determine the accuracy of the method employed to calculate the principal components of the permeability tensor. The

TABLE 2. HYDRAULIC CONDUCTIVITY VALUES FOR DIFFERENT ROCK TYPES*

Rock type	Location	Hydraulic conductivity (cm/sec)	Remarks	Source
Basalt		$2.0 \times 10^{-9} - 5.0 \times 10^{-5}$		Richter (1968)
	Oakflat Dam, Calif.	4.0×10^{-5}		Snow (1968c)
	Hrazdan River, Yeravan, Aremenean SSR	$3.0 \times 10^{-5} - 5.0 \times 10^{-3}$	Varies with degree of folding & crushing	Ter Stepanian & Arakelian (1960)
	Sulky Site, Nevada N.T.S.	3.0×10^{-3}	Vesicular to dense basalt, computed from packer tests in holes penetrating full strata depth	Sherman and Banks (1970)
	Sulky Site, Nevada N.T.S.	5.6×10^{-3}	Vesiclar basalt, computed from packer test results	Lutton and Girucky (1966)
	Snake River, Idaho	1.0×10^{-1}		DeWeist (1969)
	Oahu, Hawaii	$1.0 \times 10^{-1} - 3.0 \times 10^{-1}$		DeWeist (1969)
	-	$1.0 \times 10^{-1} - 3.0 \times 10^{-1}$		Deweist (1969)
Chert	Marquette Iron Mining District, Michigan	1.84×10^{-10}		Stuart and others (1954)
Dolomite		9.66×10^{-7}		Murray (1960)
		$1.18 \times 10^{-3} - 1.91 \times 10^{-3}$	Field measurements	Stimpson (1976)
Gneiss		$1.4 \times 10^{-1} - 2.26 \times 10^{-1}$	Field measurements	Stimpson (1976)
Gabbro	Okanagan Highland, B.C.	4.28×10^{-7}	Depth 79.5 ft	Lawson (1968)
	Okanagan Highland, B.C.	5.25×10^{-8}	Depth 45.3 ft	Lawson (1968)
Granite	-	$4.91 \times 10^{-11} - 9.99 \times 10^{-11}$	Lab data (matrix)	Stimpson (1976)
			Field measurements	
Granodiorite		5.15×10^{-6}	22.5' depth	Stimpson (1976)
		2.30×10^{-7}	38.0' depth	Stimpson (1976)
		6.17×10^{-7}	46.5' depth	Stimpson (1976)
		9.31×10^{-7}	66.0' depth	Stimpson (1972)
		2.4×10^{-7}	96.0' depth	Stimpson (1976)

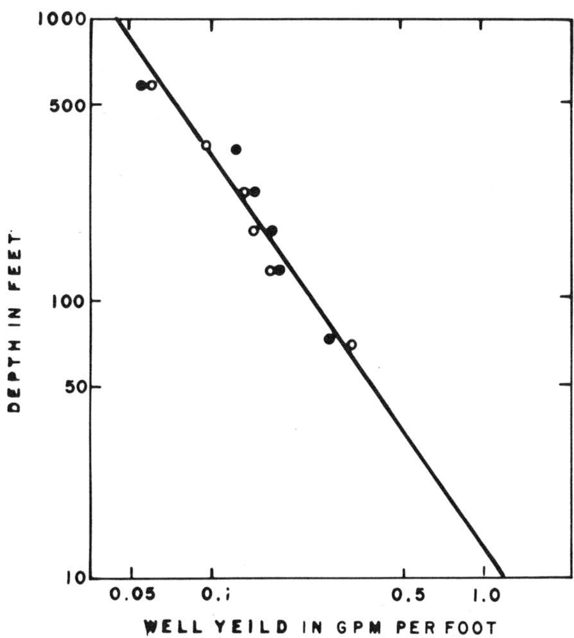

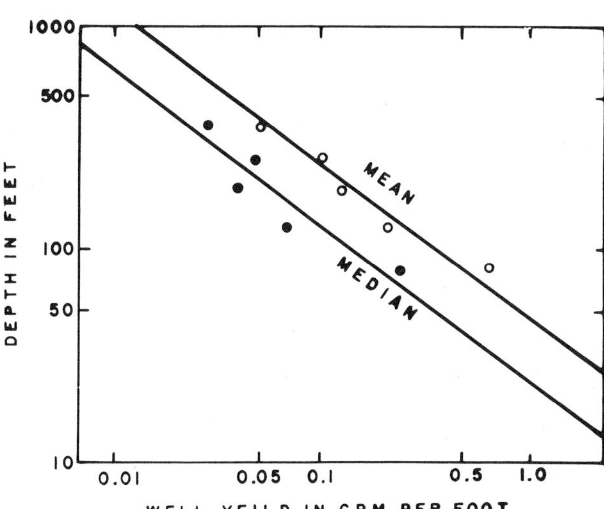

Figure 13. LEFT: Yields of wells in crystalline rocks of eastern United States. Open circles represent mean yields of granitic rocks based on a total record of 514 wells. Black dots represent mean yields of schist based on a total record of 1522 wells. RIGHT: Depth-yield relations for crystalline rocks of the Sierra Nevada, California, based on data from 239 wells. Most wells are in granodiorite or closely related rocks (after Davis and Turk, 1964).

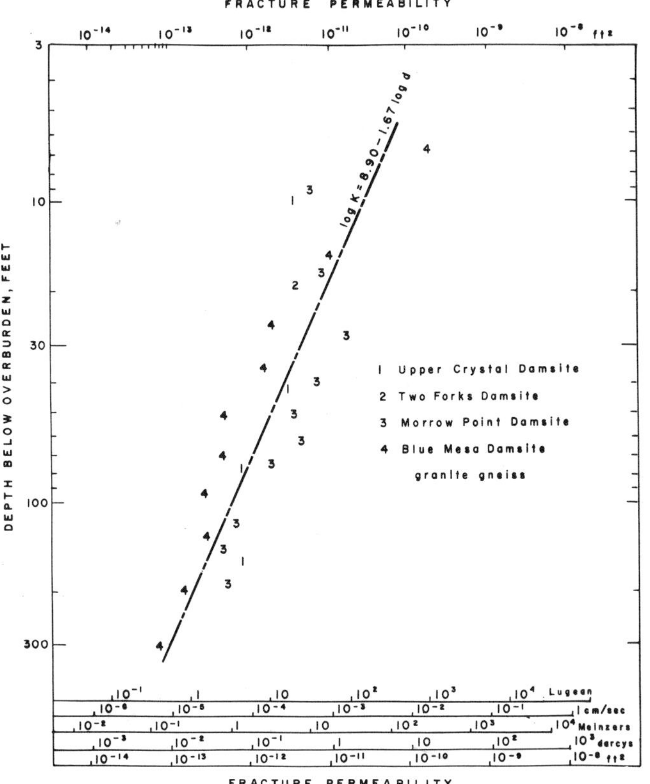

Figure 14. The mean permeability in each depth zone at four damsites versus the mean depth of midpoints of test sections. Values computed from injection tests (after Snow, 1968c).

calculated principal components of the permeability tensor represent the directional permeabilities of an equivalent anisotropic porous medium.

The work by Romm and Pozinenko (1963), Snow (1965, 1969), Caldwell (1971), and others represents a major advance in fracture hydrology. In addition to formulating the basic mathematical procedures, the main contribution of this work has been to show that the hydraulic characteristics of fractured rocks, subject to the considerations of scale (Rats and Chernyashov, 1967), can be described at the level of the individual conduits--that is, individual fractures or joints.

Many researchers have maintained that flow in fractured media can be treated as flow through porous media. Nearly all hydrogeologic analyses of ground-water flow in fractured rocks have proceeded on the assumption that fractured rock masses can be represented as an equivalent continuous porous medium. Figure 15 portrays the concept commonly used in representing a fractured medium as an equivalent porous medium. A parallel plate analogy is used for the fractures. In this example, the fracture-flux through apertures of 0.0034 cm spaced 1 m apart is equivalent to the flux through a porous medium with a hydraulic conductivity of 3.3 x 10^{-6} cm/s. In this development the fractures are assumed to be continuous in their own planes to some flow or pressure boundary. This, of course, is unrealistic. Fractures in most cases are continuous over distances that are probably on the order of three to four times the average fracture spacing. As shown schematically in Figure 15, it is the degree and nature of a fracture's interconnection with other fractures that determines to a great extent its contribution to the rock mass permeability.

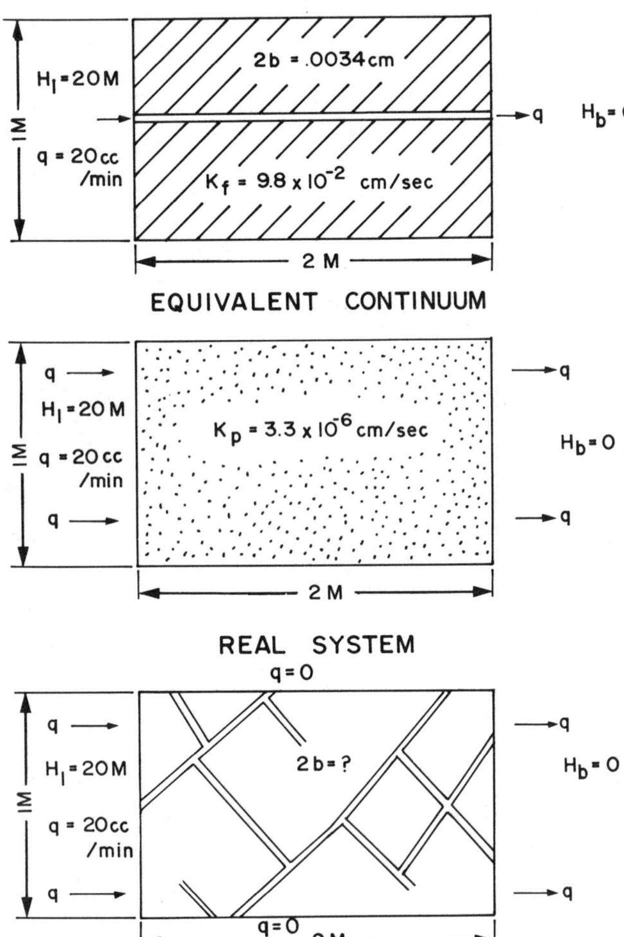

Figure 15. The equivalent continuum concept.

Furthermore, aperture is not constant along the length of a conduit.

Thus, if the equivalent continuum for a given fractured rock mass is not developed from the appropriate data base, large errors can be introduced in calculating the velocity field since the distribution of velocities is directly related to the distribution of the permeabilities of the individual conduits. This is clearly shown by extending the above comparison between the equivalent porous medium and a single fracture over a range of flow rates for flow into a borehole (Table 3). From these simple calculations it is obvious that, depending on the estimate of porosity, the assumption of an equivalent continuum presents a different view of hydraulic conductivities and velocity field than the assumption of continuous fractures. Undoubtedly the true hydraulic conductivities and flow velocities through a series of interconnected fractures lie somewhere between these two extreme values. Thus the effective apertures, calculated from flow rates and pressure gradients measured during packer injection tests, will be smaller than the real apertures of the fractures intersecting the borehole. This difference then will be due primarily to the degree of fracture interconnection and variations in fracture aperture.

Factors Controlling Rock Mass Permeability

It is apparent from the previous discussions and consideration of basic rock mass characteristics that many of the factors that control fluid flow through single fractures also control flow through multiple fracture systems. From equations 1 and 7 one can demonstrate that the distribution of fracture apertures will have a major effect on flow through a fractured rock mass. Direct measurements of fracture apertures in outcrops (Bianchi and Snow, 1969) suggest that the variation in fracture apertures can be approximated by a log-normal distribution. Thus, given the dependence of fracture-flux on the cube of the aperture, a few large fractures can completely dominate the flow system.

Rock masses are characterized by three-dimensional stress fields. Given the dependence of fracture permeability on effective normal stress, it is to be expected that fracture apertures will decrease in the direction of increasing stress. Thus the mean values for the aperture distribution model will decrease with increase in stress, which is normally associated with an increase in depth. Existing data suggest that, in general, there is a decrease in fracture permeability with depth.

Although the magnitude of the stress acting normal to a fracture plane has a major effect on the flow through that fracture, on the scale of a fractured rock mass the stress field is highly anisotropic. Because fractures have highly non-linear deformation characteristics, the deformation (change in aperture) per unit change in effective stress is highly dependent on the initial effective stress acting across the fracture plane. The initial effective stress is a function of the fluid pressure and the orientation of the fracture plane with respect to the principal stress components. Thus fractures at the same depth below the ground

TABLE 3. COMPARISON OF EQUIVALENT POROUS MEDIUM AND FRACTURE PERMEABILITIES

k_p (m/sec)	Q (cc/s)	Q (cc/min)	k_f (m/sec)	2b (M)
8.9×10^{-7}	10	600	8.8×10^{-3}	1.04×10^{-4}
8.9×10^{-8}	1	60	1.9×10^{-3}	4.82×10^{-5}
8.9×10^{-9}	0.1	6	4.1×10^{-4}	2.24×10^{-5}
8.9×10^{-10}	0.01	0.6	8.8×10^{-5}	1.04×10^{-5}
8.9×10^{-11}	0.001	0.06	1.9×10^{-5}	4.83×10^{-6}
8.9×10^{-12}	0.0001	0.006	4.1×10^{-6}	2.24×10^{-6}

Radial flow conditions: $r_w = 0.038$ m; $r_p = 10$ m; $H_b = 0$; $H_w = 10$ m; fracture spacing = 1 m.

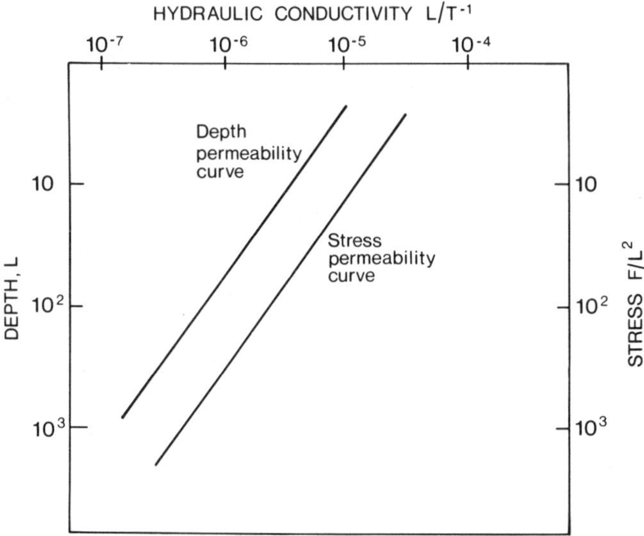

Figure 16. Comparison of stress permeability and depth permeability relationships.

surface but with different orientations may be subjected to different effective normal stresses and hence have different fracture permeabilities.

One factor that has not been included in previous analysis of fractured rocks is the degree of fracture interconnection. We know from field observations that fractures are discontinuous within their own planes. Given that a number of fracture sets, usually three or four, are present in most rock masses, and that orientations vary within sets and between sets, a fracture will usually intersect a number of other fractures. The number of fractures intersected will depend on the spacing of fractures within each set and the two-dimensional continuity of each fracture. The degree of fracture interconnection may well be a major factor controlling the movements of fluids through fractured rock masses. Figure 16 shows a schematic comparison of the stress-permeability relationship in a single fracture with the depth-permeability relationship (field data) for the same rock- fracture system. This may give some indication of the relative significance of stress versus fracture interconnection in controlling the flow of fluids through a fractured rock mass. Although it is assumed that stress and fracture interconnection may be the two most important variables controlling flow through fractured rocks, no quantitative data exist to verify this assumption.

As shown in Figure 9, shear zones and fracture zones represent significant structural features within fractured rock masses. As such, these features can generally be located and tested and their hydrogeologic characteristics defined. The effect of these major features on local and regional flow systems in fractured argillaceous and crystalline rocks is usually obscure. From spring discharges, we know these rocks are sometimes important conduits; elsewhere, steep gradients across faults indicate low transverse permeability. Figure 17, adapted from Stokes (1977), shows the interpreted results of a numerical model study of a regional ground-water flow system in crystalline rock from west-central Sweden. The flow system is short-circuited in numerous places by near-vertical shear or fracture zones. Between the shear and fracture zones, flow occurs in joints. It must be emphasized that the hydraulic characteristics of the shear and fracture zones were smeared out or averaged over large areas instead of being simulated at their true scale relative to the regional flow system. Also, the model results have not been confirmed by a rigorous comparison against field data.

METHOD OF MEASURING DIRECTIONAL PERMEABILITIES OF FRACTURED ROCKS

In discussing the different methods of measuring directional permeabilities of fractured rocks one must consider fractured argillaceous rocks that contain highly permeable porous layers which are isolated from argillaceous and crystalline rocks in which the fractures are the main conduits. In such cases current pump testing and analysis techniques permit determination of lateral permeabilities in the high-permeability zones and vertical permeabilities in the low-permeability zones (Witherspoon and others, 1967) using wells normal to the bedding. The principal permeabilities tend to be parallel and perpendicular to bedding due to the more permeable porous zones. Fracture-flow dominated rocks tend to be anisotropic and the principal axes of the permeability tensor need not be geometrically related to the lithologic boundaries. Because of the weakness of certain beds in shales, fractures tend to lie normal to bedding, which thus tends to make bedding a principal plane.

In most fractured argillaceous and crystalline rock masses, local flow is controlled by the fracture (joint) system and is usually anisotropic. Papadopulos (1967) and Hantush (1966) have suggested

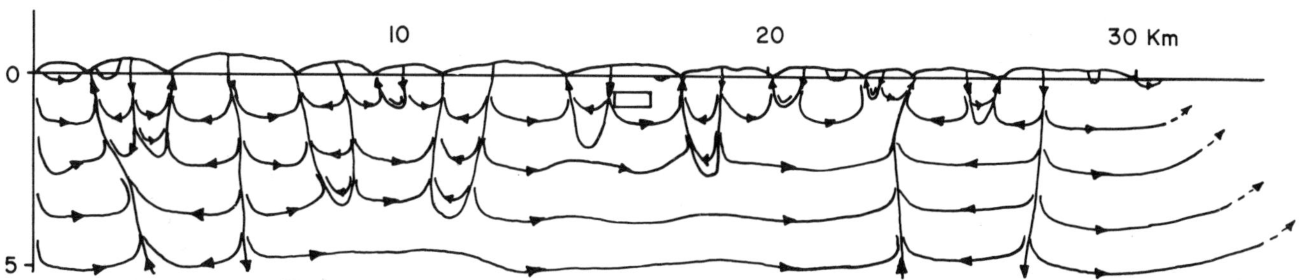

Figure 17. Results of numerical model study of regional flow (after Stokes, 1977).

observation well configurations that can be used to determine, from pump test data, the anisotropic permeability characteristics of fractured rocks. Both techniques are limited to testing the upper 100 m or so of a fractured rock mass by drilling normal to a known principal plane. Gringarten and Witherspoon (1972) proposed a single-well technique that relies on early time data to determine whether one is dealing with an equivalent vertical fracture or equivalent horizontal fracture intersecting the wellbore. The applicability of the pumping well with multiple observation wells used by Papadopulus (1967) and Hantush (1966) or the single pumping well techniques of Gringarten and Witherspoon (1972), has not been demonstrated, either empirically or theoretically, in fractured rock systems where the porosity and permeability decrease with well depth.

In the geotechnical field, where interest has focused primarily on the hydraulic characteristics of discrete zones rather than the overall rock-mass permeability, borehole packer injection tests have been preferred to pump-testing the entire well. The testing procedure usually consists of isolating the sections of interest in a borehole, using either single or multiple mechanical or inflatable packers, and performing either falling-head (Hvorslev, 1951) or constant-head injection tests (Glover, 1953; Louis, 1976). The lower limit of permeability measurement per unit length of borehole is usually controlled by the lower limit of the injection flow rate measurement system. This is usually of the order of a few cubic centimeters per minute. In some areas, such as in the nuclear waste disposal programs, it may be necessary to determine the hydraulic characteristics of a fracture or groups of fractures in the field. Figure 18 shows the results of packer tests in a borehole drilled from an excavation located 338 m below ground surface in fractured granite in Stripa, Sweden. The flow rates, in-situ pressures, and fracture distributions for this borehole suggest that at depths of 1000 m one must be able to measure injection rates over considerable ranges that can be as low as 10^{-2} to $10^{-3}cm^3$/minute. Recent work at the University of Waterloo has resulted in the development of a flow-rate measuring system (Gale and others, 1979) that permits one to accurately measure flow rates in this range. This method (Fig. 19) is based on a bubble technique used by the petroleum industry (Witherspoon, 1978, oral commun.) for measuring flow rates during laboratory tests. The method is very simple; an air bubble is injected into the line at point A, and its rate of travel is measured over a length of calibrated tubing, hence giving the required flow rate. The air bubble is bled off at point B, and the operation is repeated. If the air bubble system is used in conjunction with positive displacement tanks of different diameters (Fig. 19) a considerable range of flow rates can be measured.

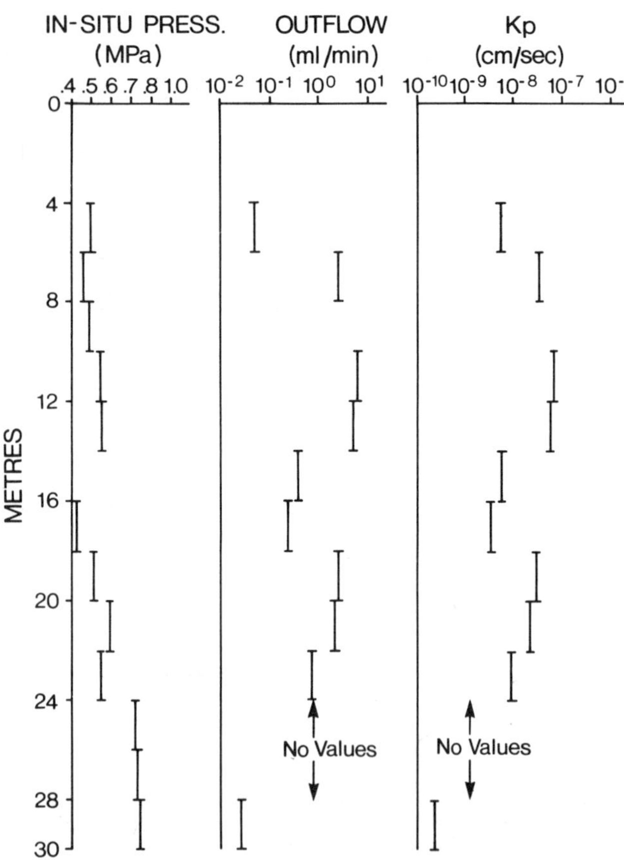

Figure 18. Results of packer outflow tests in the R4 borehole at the 338-m level, Stripa, Sweden (Gale and others, 1979).

Another testing technique that may be useful in measuring the permeability of discrete zones (1 to 2m or less in length) in low-permeability rocks is the pressure pulse technique. The pressure pulse technique was first used by Brace and others (1968) to measure the permeabilities of very small laboratory samples. This technique was subsequently applied to larger samples (Lin, 1977). Wang and others (1977) presented theory and type curves for interpreting borehole tests in fractured rocks. Laboratory experiments by Forster and Gale (1979) show that packer system compliance and temperature effects can seriously affect the test results.

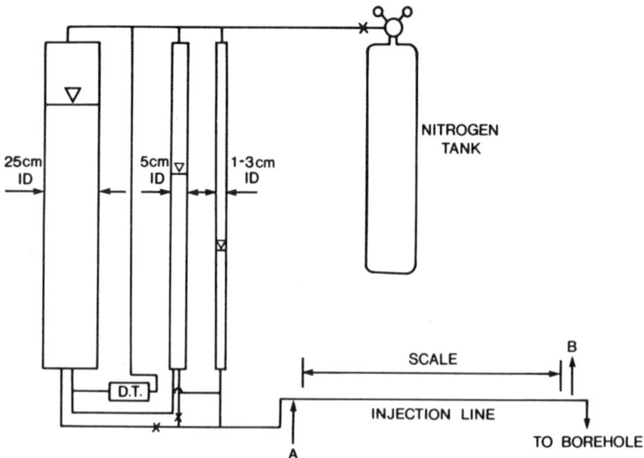

Figure 19. Schematic of flow rate measuring system (after Gale and others, 1979).

They suggest that additional work is required before the pressure pulse packer tests can be routinely applied in the field.

Thus a number of existing and developing borehole packer testing techniques will make it possible to determine the permeability of fractures in isolated sections of a borehole. A major step involved is developing the directional permeability characteristics of fractured rock masses from borehole packer measurements. Three possible approaches to this problem have been identified (Gale and Witherspoon, 1979). The first method, described by Louis and Pernot (1972), requires careful mapping of fracture orientations to calculate the directional permeability axes for the rock mass. A central borehole, with peripheral boreholes, is drilled parallel to two of the three principal permeability axes. Multiple packer injection tests are carried out in the central borehole and the fluid pressures are monitored in the peripheral boreholes. Using the measured flow rates and fluid pressures, the directional permeabilities of the rock mass can be determined using a graphical technique (Schneebeli, 1966) and the approaches outlined by Louis and Pernot (1972). Because fracture orientations do not necessarily coincide with the principal permeability axes and because the calculation of principal permeability directions requires one to assume an aperture distribution model, Louis and Pernot's approach is only applicable to symmetrical systems. In nonsymmetrical systems, in order to provide a reasonable degree of accuracy, nine peripheral boreholes are required (Hantush, 1966). This requirement limits the Louis and Pernot approach to near-surface studies or to the immediate vicinity of underground openings.

A second approach (Maini, 1971; Bourke and others, 1979) consists of drilling orthogonal boreholes oriented with respect to the fracture system and testing the boreholes with increasing packer spacing. It is anticipated that with increasing packer spacing the permeability should approach an average value. This approach has not been evaluated either theoretically or empirically. Carefully controlled field tests are needed in order to evaluate this method.

A third approach, based on earlier work by Snow (1966), is referred to as the "discrete data approach" and consists of using discrete fracture data to describe the rock mass in the form of a permeability tensor, that is, an anisotropic equivalent continuous porous medium. This approach requires careful mapping of the fracture system, the drilling of boreholes approximately perpendicular to the principal fracture sets, oriented core drilling, detailed logging of the drill core in order to describe the geometry and characteristics of the fracture planes, detailed injection testing to determine effective fracture apertures, and mathematical integration of these data into the form of a permeability tensor. This approach, which is currently being used in a field study at Stripa, Sweden (Gale and Witherspoon, 1979), is very time-consuming and requires processing a large volume of basic data. One advantage, from a conceptual point of view, is that the flow characteristics of the equivalent porous medium are developed at the level of the individual flow conduits, or single fractures.

In order to properly evaluate the various approaches and to build confidence in the more useful techniques, carefully designed tests will have to be carried out in a number of different rock masses representing different fracture systems and boundary conditions. It is important to be able to identify which of the approaches is most applicable to each of the fracture systems and permeability regimes present in the different fractured rock masses. No one method will be applicable to all of the fracture systems that will be encountered.

IMPLICATIONS AND CONCLUSIONS

The objective of describing factors controlling flow through single and multiple fractures and determining the permeability characteristics of fractured rock masses is to use these properties to develop realistic models of flow in fractured rock masses. Studies of local and regional flow systems and transport problems must begin with conceptual models that are based on the physics of flow in fractured rocks. Early theoretical work by Barenblatt and others (1960) on flow through fractured porous media was based on the two-porosity model; the porosity and permeability of the rock blocks were considered separately from the porosity and permeability of the fracture system. Flow between the rock blocks and fractures was coupled by an empirical relationship. This is an equivalent porous media approach, averaging the flow properties over a large scale. Other workers, especially in the petroleum industry, have taken a similar approach (Warren and Rook, 1963; Odeh, 1965; Kazemi, 1969). The most promising theoretical approach to non-steady fractured porous media flow is that of Duguid (1973). Duguid developed a more complete two-porosity formulation than that of Barenblatt and others, incorporating the elastic theory of Biot (1940) and the fractures as a statistical distribution. Duguid derived a theoretical relationship of flow between the fractures and the rock blocks, as opposed to Barenblatt's empirical relationship.

Other workers have attempted to include the major fractures as discrete features--a form of network analysis. The first theoretical and numerical models of flow through rigid fractures were the deterministic models by Wilson and Witherspoon (1970). Recent work with deterministic models has included coupling fluid forces and rock stresses, that is, the coupled stress-flow models (Noorishad and others, 1971; Gale, 1975), and developing a non-steady, coupled stress-flow, fractured porous media model (Ayatollahi, 1978). Deterministic models provide a useful tool for systematically evaluating how fluid forces and rock stresses interact. The effects of parameters such as fracture orientation, fracture apertures, fracture spacing, and fracture stiffness on flow rates and fluid pressure distributions have been studied (Gale, 1975). Deterministic models are limited in that computer storage does not permit one to include a very large number of the fractures that exist in a given rock mass. Also, it is impossible to specify, except in the immediate area of an underground opening, the exact location and geometry of each fracture.

Thus if one considers the structural framework and hydraulic characteristics of fractured rock masses, an appropriate theoretical and numerical model of flow through fractured rock masses must have a conceptual framework similar to that shown schematically in Figure 20. This conceptual model is a coupled discrete-fractured porous medium, where faults and shear zones are defined as individual features. Individual fractures are defined as discrete flow paths near the openings of an underground excavation, but are represented as an appropriately defined continuum at some distance from the opening. The properties of the continuum are developed in a manner that is consistent with the known geometric and hydraulic characteristics of fractured rocks. Thus the permeability tensor, K_{ij}, should be defined as a function of the space coordinates and should reflect (1) the true nature of the aperture distribution model; (2) the effects of the stress tensor, σ_{ij}, on fracture permeability; and (3) the geometry of the different fracture sets and the degree of fracture interconnection. Previous workers, in developing the permeability tensor from fracture aperture and geometry data, have erroneously assumed that the fractures are continuous within their own planes. This assumption may not be a major limiting factor in applying the discrete data approach. The effective aperture distribution model, which would be used in the permeability calculations, is determined from injection tests and hence the calculated effective apertures already reflect, in part, the fracture continuity or degree of fracture interconnection. Although there may be some uncertainty as to the need to define the degree of fracture interconnection for basic flow problems, there is a need to investigate this problem both theoretically and empirically, especially when one considers our poor understanding of the role of fractures in transporting contaminants. The most detailed work on contaminant transport in fractures is the parametric study of Castillo and others (1972). This work is limited in that it was based on the assumption that fractures are continuous within their own planes. A field study by Lundstrom and others (1978) using radionuclides gives some indication of velocities in a near-surface fracture system. However, the actual flow paths, although qualitatively described, appear to be poorly understood. Thus, there is still validity to Castillo's comment that "the classical concept of flow through a porous medium is generally inadequate to describe the flow behavior in jointed rock, and it becomes increasingly unsuitable for the analysis of dispersion (in fractured rocks)" (Castillo and others, 1972, p. 778). Advances in our understanding of dispersion and diffusion in fractured media will require additional conceptual and numerical models of flow in fractured media, yet to be developed and verified.

A number of areas that require additional research have been identified in the previous discussion. A major area of concern (especially

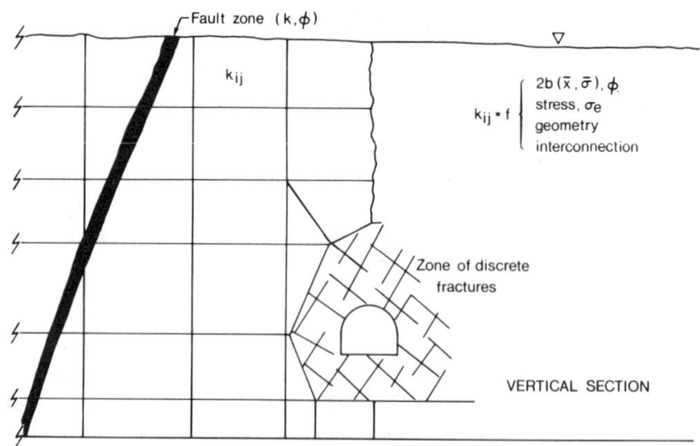

Figure 20. Conceptual model of flow through fractured rocks.

for researchers involved in studies of the disposal of high-level nuclear waste) is how the excavation, including shafts and boreholes, affects flow through the rock mass in their vicinity. These concerns will require additional studies on the factors that control the flow of fluids through a single fracture--normal stress, shear stress, fracture surface area, temperature, fracture geometry (roughness and waviness), and fracture-filling material. In all of these laboratory studies there is a need to coordinate testing procedures and techniques in order to permit comparison of data obtained at different laboratories. Additional research is needed to develop and improve borehole tests, especially the pressure pulse test, and to investigate the potential of rock mass tests. One promising rock mass test involves monitoring pore pressure changes due to earth and ocean tides. Given the low porosity, low permeability, and deformable nature of the individual fracture conduits, this could prove to be a fruitful but difficult area of research.

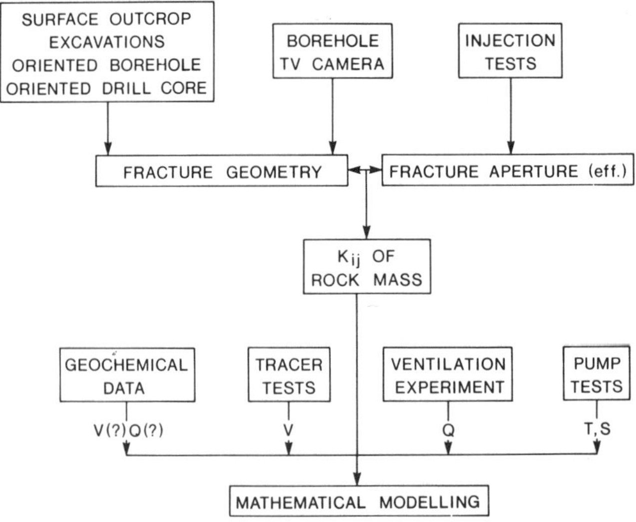

Figure 21. Approach to fracture hydrology studies at Stripa, Sweden (Gale and Witherspoon, 1979).

Very few studies have been made of deep flow systems in fractured crystalline rocks. This is an area that is currently receiving attention but still requires much additional work. In the past, most workers have focused on one line of research. Given the complexity of flow through fractured media and the demands of such problems as nuclear waste storage, future studies of fractured rock masses will have to be highly integrated. Figure 21 shows the approach being taken in a study of fractured granite at Stripa, Sweden. This study (Gale and Witherspoon, 1979) includes the discrete data approach, pump tests, a macropermeability experiment (ventilation experiment), and a detailed geochemical and isotopic study. Studies that attempt to integrate the physical geology, fracture porosity-permeability, geochemistry, and isotopic data are essential to the development of an understanding of how fluids move through fractured rock masses.

ACKNOWLEDGMENTS

This paper incorporates the work and ideas of many researchers, most of whom have been identified within the text, and I extend my sincere thanks to them. Although Paul Witherspoon is free to disagree with any of the ideas presented in this paper, he is its unofficial co-author. His special contributions to my efforts, and those of Dr. D. T. Snow, are gratefully acknowledged. John Cherry and Roger Macqueen critically reviewed an earlier version of the manuscript and substantial revision and reorganization resulted from this review. The final manuscript was reviewed by Dr. D. T. Snow and Dr. W. F. Brace and I extend my sincere thanks to them for their assistance. However, I am entirely responsible for any remaining errors or lack of organization. A special thanks is extended to Sheryl Sulliman for patiently typing and retyping this manuscript.

This work has been made possible by an operating grant provided by the National Research Council of Canada.

REFERENCES CITED

Appleby, A. N., 1940, Joint patterns in highly folded and crystalline rocks of the New Jersey Highlands: Geological Society of America Bulletin, v. 51.

Ayatollahi, M. S., 1978, Stress and flow in fractured porous media [Ph.D. dissertation]; Berkeley, University of California, p. 154.

Barenblatt, G., Zheltov, U. P., and Kochina, G. H., 1960, Basic concepts in the theory of seepage of homogeneous liquids in fissured rocks. Prikl. Matem. Mekh., v. 24, p. 852-864.

Bianchi, L., and Snow, D. T., 1969, Permeability of crystalline rock interpreted from measured orientations and apertures of fractures: Annals of Arid Zone, v. 8, no. 2, p. 231-245.

Biot, M. A., 1940, General theory of three-dimensional consolidation, Journal of Applied Physics, v. 12, p. 155-164.

Bourke, P. J., and others, 1979, Tests of porous permeable medium hypothesis for flow over long distances in fractured deep hard rock: Harwell, U. K., United Kingdom Atomic Energy Authority, AERE-R487.

Brace, F. W., 1975, Dilatancy related electrical resistivity changes in rocks: Pure and Applied Geophysics, Earthquake Prediction and Rock Mechanics, v. 113, p. 207-217.

Brace, F. W., 1978, A note on the permeability changes in geologic material due to stress. Pageoph, v. 116, p. 627-633.

Brace, F. W., Walsh, J. B., and Frangos, W. T., 1968, Permeability of granite under high pressure: Journal of Geophysical Research, v. 73, no. 6, p. 2225-2236.

Caldwell, M. A., 1971, The theoretical determination of the fluid potential distribution in jointed rocks [M.S. thesis]: University of Witwatersrand.

Castillo, E., Krizek, R. J., and Kardi, G. M., 1972, Comparison of dispersion characteristics in rock, in Proceedings, Second International Symposium on Fundamentals of Transport Phenomena in Porous Media: Guelph, Ontario, University of Guelph, Office of Continuing Education, v. 2, p. 778-797.

Cherry, J. A., and Gale, J. E., 1979, The Canadian Program for a high-level radioactive waste repository: A hydrogeological perspective: Geological Survey of Canada paper 79-10, p. 35-47.

Davis, S. N., and Turk, L. S., 1964, Optimum depth of wells in crystalline rocks: Ground Water v. 2, no. 2, p. 6-11.

De Weist, J. M., 1969, Flow through porous media: New York, Academic Press.

Duguid, J. O., 1973, Flow in fractured porous media [Ph.D. dissertation]: Princeton University, 111 pp.

Forster, C., and Gale, J. E., 1980, A laboratory assessment of the use of borehole pressure transients to measure the permeability of fractured rock masses: Berkeley, Lawrence Berkeley Laboratory, LBL-8674 (in press).

Gale, J. E., 1975, A numerical, field and laboratory study of flow in rocks with deformable fractures [Ph.D. dissertation]: Berkeley, University of California, 255 p. [Also Scientific Series #72, Inland Waters Directorate, Water Resources Branch, Environment Canada, Ottawa, Canada (1977).]

Gale, J. E., and Witherspoon, P. A., 1979, An approach to the fracture hydrology at Stripa, preliminary results: Berkeley, Lawrence Berkeley Laboratory, LBL-7079.

Gale, J. E., Forster, C. B., and Nadon, R., 1979, Permeability of fractured low-permeability rocks: Waterloo Research Institute, University of Waterloo.

Gangi, A. F., 1978, Variation of whole and fractured porous rock permeability with confining pressure: International Journal of Rock Mechanics, and Mining Sciences and Geomechanics, abstracts, v. 15, no. 5, p. 249-257.

Glover, R. E., 1953, Flow from a test hole located above groundwater level, in Zangar, C. N., ed.,

Theory and problems of water percolation: Engineering Monograph no. 8, U.S. Bureau of Reclamation, App. B, p. 69-71.

Gloyna, E. F., and Reynolds, T. D., 1961, Permeability measurements of rock salt: Journal of Geophysical Research, v. 66, no. 11, p. 3913-3921.

Gringarten, A. C., and Witherspoon, P. A., 1972, A method of analyzing pump test data from fractured aquifers, in Proceedings, Symposium on the Percolation of Fissured Rock: International Society of Rock Mechanics, T3-B:1-9.

Hantush, M. S., 1966, Wells in homogeneous anisotropic aquifers: Water Resources Research, v. 2, p. 273-279.

Huitt, J. L., 1956, Fluid flow in simulated fractures: Journal AICHE. v. 2, p. 259.

Hvorslev, M. J., 1951, Time lag and soil permeability in groundwater observations: Vicksburg, Miss. U. S. Army Corps of Engineers Waterways Experiment Station, Bulletin 36.

Iwai, K., 1976, Fundamental studies of fluid flow through a single fracture [Ph.D. dissertation]: Berkeley, University of California, 208 p.

Jones, F. O., 1975, A laboratory study of the effects of confining pressure on fracture flow and storage capacity in carbonate rocks: Journal Petroleum Technology, Jan. 1975, p. 21-27.

Jouanna, P., 1972, Effect des sollicitations mécaniques sur les ecoulements dans certains millieaux fissures [Ph.D. dissertation]: Université de Toulouse, 263 p.

Kazemi, H., 1969, Pressure transient analysis of naturally fractured reservoirs with uniform fracture distribution: Society of Petroleum Engineers Journal, v. 9, no. 4, p. 451-462.

Keller, G. V., 1960, Physical properties of tuffs of the Oak Springs Formation, Nevada: U.S. Geological Survey Professional Paper 400B.

Kendorski, F. S., and Mahtab, M., 1976, Fracture patterns and anisotropy of the San Manuel quartz monzonite: Association of Engineering Geologists Journal, v. 13, p. 23-52.

Knapp, R. B., 1975, An analysis of the porosities of fractured crystalline rocks [M.S. thesis]: University of Arizona, 90 p.

Kranz, R. L., and others, The permeability of whole and jointed Barre Granite: International Journal of Rock Mechanics, Mineral Science and Geomechanics Abstract, v. 16, p. 225-235.

Lawson, D. W., 1968, Groundwater flow systems in the crystalline rocks of the Okanagan Highland, British Columbia: Canadian Journal of Earth Sciences, v. 5, p. 813-823.

Lewis, D. C., Kriz, G. J., and Burgy, R. H., 1966, Tracer dilution sampling techniques to determine the hydraulic conductivity of fractured rock: Water Resources Research, v. 2, no. 3, p. 533-542.

Lin, W., 1977, Compressible fluid flow through rocks of variable permeability: Livermore, Lawrence Livermore Laboratory, UCRL-52304.

Lomize, G. M., 1951. Fluid flow in fractured rocks: Goseneroizdat, Moscow (in Russian).

Louis, C., 1976, Introduction a l'hydraulique des roches [Ph.D. dissertation]: Paris, Université Pierre et Marie Curie.

Louis, C., and Pernot, M., 1972, Three dimensional investigation of flow conditions of Grand Maison damsite, in Proceedings, Symposium on Percolation through Fissured Rock: International Society of Rock Mechanics, T4-F:1-16.

Louis, C., Dessene, J.-L., and Feuga, B., 1977, Interaction between water flow phenomena and the mechanical behavior of soil or rock masses, in Gudehns, G., ed., Finite elements in geomechanics: New York, John Wiley & Sons, 572 p.

Lundstrom, O., and others, 1978, In situ experiments on nuclide migration in fractured crystalline rocks. Stockholm, Sweden, KBS Technical Report No. 110.

Lutton, R. J., and Girucky, F. E., 1966, Geological and engineering properties project Sulky. PNE-720, Vicksburg, Miss., U.S. Waterways Experiment Station, 136 p.

Maini, Y. N. T., 1971, In-situ hydraulic parameters in jointed rock--their measurement and interpretation [Ph.D. dissertation]: Imperial College, University of London, 312 p.

Meyer, G., and Beall, R. M., 1958, Water resources of Carroll and Fredrick counties, Maryland: Department of Geology, Mines and Water Resources, Bulletin 22.

Murray, R. C., 1960, Origin of porosity in carbonate rocks: Journal Sedimentary Petrology, v. 30, no. 1, p. 59-84.

Nelson, R., 1975, Fracture permeability in porous reservoirs: experimental and field approach [Ph.D. dissertation]: College Station, Texas, Texas A & M University.

Noorishad, J., Witherspoon, P. A., and Brekke, T. L., 1971, A method for coupled stress and flow analysis of fractured rock masses: Berkeley, University of California, Geotechnical Engineering Publication 71-76, 128 p.

Odeh, A. S., 1965, Unsteady-state behavior of naturally fractured reservoirs: Society of Petroleum Engineers Journal, March 1965, p. 60-66.

Papadopulos, I. S., 1967, Nonsteady flow to a well in an infinite anisotropic aquifer, in Proceedings, Dubrovnik Symposium on Hydrology of Fractured Rocks: International Association of Scientific Hydrology, v. 1, no. 23, p. 21-31.

Parsons, M. L., 1972, Determination of hydrogeological properties of fissured rocks, in Proceedings, 24th Geological Congress: Ottawa, Canada, p. 89-99.

Polubarinova-Kochina, P. Ya., 1962, Theory of groundwater movement, Translated by R.J.M. De Wiest: Princeton, Princeton University Press.

Pratt, H. R., and others, 1977, Elastic and transport properties of an in-situ jointed granite: International Journal of Rock Mechanics and Mineral Science, v. 14, p. 35-45.

Rats, M. V., and Chernyashov, S. N., 1967, Statistical aspect of the problem of the permeability of the jointy rocks, in Proceedings, Dubrovnik Symposium on Hydrology of Fractured

Rocks, v. 2, p. 227-236.

Raven, K. G., and Gale, J. E., 1977, Project 740057: subsurface containment of solid radioactive waste: A study of the surface and subsurface structural and groundwater conditions at selected underground mines and excavations: Ottawa, Geological Survey of Canada, EMR-GSC-RW Internal Report #1-77.

Rayneau, C., and Jouanna, P., 1971, Influence de la temperature sur l'ecoulement en milieux fissures, in Symposium of the International Society for Rock Mechanics: International Society of Rock Mechanics, paper 2-5, 12 p.

Ribler, P., 1978. Determination of the water permeability of jointed rock: Vol. 5, Aachen 1978, Publication of the Institute for Foundation Engineering, Soil Mechanics, Rock Mechanics, and Water Ways Construction, RWTH (University) Aachen, Federal Republic of Germany.

Romm, E. S., 1966, Fluid flow in fractured rocks: Moscow, Nedra (in Russian).

Romm, E. S., and Pozinenko, B. V., 1963, Investigation of seepage in fractured rocks: Trudy VNIGRI, 214, (in Russian).

Schneebeli, G., 1966, Hydraulique souteraine: Paris, Eyrolles, 362 p.

Sharp, J. C., 1970, Fluid flow through fissured media [Ph.D. dissertation]: London, Imperial College of Science and Technology, 200 p.

Sherman, W. C., and Banks, D. C., 1970, Seepage characteristics of exposively produced waters in soil and rock: Vicksburg, Miss., U.S. Waterways Experiment/Technical/Report Station. NCO Technical Report No. 27.

Snow, D. T., 1965, A parallel plate model of fractured permeable media [Ph.D. dissertation]: Berkeley, University of California, 331 p.

___1966, Three-hole pressure test for anisotropic foundation permeability: Rock Mechanics and Engineering Geology, v. 4, no. 4, p. 298-316.

___1967, A discussion of Theme I, in Proceedings of the First International Congress on Rock Mechanics: International Society on Rock Mechanics, v. 3, p. 243-244.

___1968a, Fracture deformation and changes of permeability and storage upon changes of fluid pressure: Colorado School of Mines Quarterly, v. 63, no. 1, p. 201-244.

___1968b, Rock fracture spacings, openings and porosities: Journal Soil Mechanics and Foundation Division, Proceedings, ASCE, v. 94, (SMI), p. 73-91.

___1968c, Hydraulic characteristics of fractured metamorphic rocks of Front Range and implications to the Rocky Mountain Arsenal well: Colorado School of Mines Quarterly, v. 63, no. 1, p. 167-199.

___1969, Anisotropic permeability of fractured media: Water Resources Research, v. 5, no. 6, p. 1273-1289.

Stewart, J. W., 1964, Infiltration and permeability of weathered crystalline rocks, Georgia Nuclear Laboratory, Dawson County, Georgia: U.S. Geological Survey Bulletin 1133-D.

Stokes, J., 1977, Hydrologiskt och topografiskt betingade grundvatten rorelser: Stockholm, Just Kulturteknik, KTH, KBS Teknisk Rapport 45.

Stuart, W. T., Brown, E. A., and Rhodehamel, E. C., 1954, Groundwater investigations of the Marquette iron mining district: Michigan Geological Survey Division Technical Report 3.

Ter-Stephanian, G., and Arakelian, A., 1966, A case of deformation of lavas spread over Upper Pliocene landslides, in Proceedings of the First International Congress on Rock Mechanics, Lisbon: International Society of Rock Mechanics, v. 1.

Wang, J. S. Y., and others, 1977, Transient flow in tight fractures, in Proceedings, Invitational Well-Testing Symposium: Berkeley, Lawrence Berkeley Laboratory, LBL-7027, p. 103-116.

Warren, J. E., and Root, P. J., 1963, The behavior of naturally fractured reservoirs: Transactions, AIME, v. 228, p. 245-255.

Westergard, H. M., 1964, Theory of elasticity and plasticity: New York, Dover Publications Inc., 176 p.

Wilson, C. R., and Witherspoon, P. A., 1970, An investigation of laminar flow in fractured rocks: Geotechnical Report, 70-6, Berkeley, University of California, 178 p.

Witherspoon, P. A., 1979, Observations of a potential size effect in experimental determination of the hydraulic properties of fractures: Water Resources Research, v. 15, no. 5, p. 1142-1146.

Witherspoon, P. A., and Gale, J. E., 1977, Mechanical and hydraulic properties of rocks related to induced seismicity: Engineering Geology, v. 11, no. 1, p. 23-55.

Witherspoon, P. A., Amick, C. H., and Gale, J. E., 1977, Stress-flow behavior of a fault zone with fluid injection and withdrawal: Berkeley, University of California, Department of Materials Science and Mineral Engineering, Report #77-1.

Witherspoon, P. A., and others, 1967, Interpretation of aquifer gas storage conditions from water pumping tests: New York, American Gas Association Inc.

Witherspoon, P. A., and others, 1980, Validity of cubic law for fluid flow in a deformable rock fracture: Water Resources Research (in press).

Manuscript Accepted by the Society October 14, 1980

Printed in U.S.A.

Geological Society of America
Special Paper 189
1982

Porous Rock–Fluid Systems at Elevated Temperatures and Pressures

W. H. SOMERTON
Department of Mechanical Engineering and Lawrence Berkeley Laboratory, University of California, Berkeley, California 94720

ABSTRACT

This paper is a review and analysis of the effects of elevated temperatures and pressures on the physical properties of porous rock-fluid systems. It is well known that the physical properties of rock vary considerably depending on the magnitude and history of the stresses, pore fluid pressures, and temperatures to which the rocks are subjected as well as the type and amount of fluid saturation. The review shows that the available data are limited and sometimes contradictory. Although general trends have been established, these need to be quantified for use in subsurface performance calculations.

It has been concluded that three major problems need to be solved before appropriate values for the properties and behavior of rocks in their subsurface environment may be assigned. The first of these is how to simulate subsurface stress and temperature conditions in laboratory measurements. The second is the need to correlate the physical properties required in reservoir-performance analysis and log interpretation with some simpler properties of the rock-fluid system. All desired properties used in the correlation should be measured concurrently on the same test specimen and under identical test conditions. The third need is to develop models that will make it possible to predict physical properties and behavior of rocks from easily determined characteristics of the rock-fluid system. Examples of successful modeling of the thermal properties and behavior of rocks are presented in the paper.

INTRODUCTION

The physical properties of most rocks vary considerably depending on the stress, pore fluid pressure, temperature, and the type and amount of fluid saturation. Thus, measurements made at surface conditions of pressure and temperature (as in routine core analyses) may be substantially different in value from those measured at simulated reservoir conditions. Calculations on storage capacity and flow behavior that are based on routine core-analysis data may be very misleading. Interpretation of well-log data may also be in serious error if the effects, particularly of temperature on sonic velocities and formation resistivity factors, are not properly accounted for.

Fatt was one of the first to study the effects of subsurface stresses on the physical properties of rocks. He and his co-workers evaluated the effects of stress on such properties as porosity, permeability, formation resistivity factors, compressibilities, and elastic moduli (Fatt, 1953, 1957, 1958; Gray and others, 1963; Mann and Fatt, 1960). Dobrynin (1962), using Fatt's data and some of his own, attempted to quantify the effects of stress on rock properties and has continued to make contributions in this area (Dobrynin, 1977). Chierici and others (1967) made extensive studies of the effects of stress on rock properties and correlated their findings with rock type and clay content. Many other investigators have contributed to the knowledge in this field.

Despite the large amount of work that has been done in this area, relatively little use has been made of the accumulated data. A. Timur (1979, oral commun.) has pointed out that serious errors in log interpretation and subsurface analyses have been and are still being made. This seems to be because the available data are limited and are restricted to specific rock-fluid systems that may bear little relationship to actual reservoir systems. Consequently, industry is beginning to measure desired properties on actual reservoir samples at simulated reservoir conditions. (Getty Oil Company's University Research Program lists several topics in this area.) This in itself may offer only a limited solution to the problem, owing to the heterogeneous nature of most reservoir formations.

To compound the above difficulties, the effects of elevated temperatures on the physical properties of rocks have only recently been recognized (Von Gonten and Choudhary, 1969; Somerton and others, 1974; Casse, 1974; Weinbrandt and others, 1975). With deeper drilling of wells and development of geothermal resources, the effects of temperature can no longer be ignored.

This paper reviews the state of present technology in relation to the effects of stress and temperature on the properties and behavior of rock-fluid systems. This work is limited to porous reservoir-type rocks at the moderate pressures and temperatures encountered in fluid-bearing reservoirs. Also discussed are the yet-to-be-solved problems and some ideas that may lead to their solution.

PORE PROPERTIES OF ROCK

Drilling to increased depths to tap new oil resources, rapid development of thermal oil-recovery methods, and active interest in geothermal power development have made it more important to study the effects of high temperatures and pressures on the storage capacity and flow behavior of rocks. A great deal of research has been done on the effects of various stress states on these properties, but comparatively little work has been done on the effects of elevated temperatures and stress combined.

The effects of temperature and stress on rock properties may vary according to when they are measured. Measurements taken before and after heating and applying stress may show quite different results from those taken before and during these processes. For example, a sandstone that is heated under stress and then cooled commonly displays increased permeability due to internal damage. However, when a sandstone remains heated under stress, its permability is generally less than when measured at room temperature.

The storage capacity of a rock is determined primarily by its effective porosity--that is, fraction of the bulk volume of the rock is occupied by the interconnected pore spaces and depends on the temperature, pore pressure, and volume factor of the stored fluid. The effective porosity is probably also a function of temperature and pore pressure and may also be a function of the nature of the contained fluid if reactive clays are present in the rock. Clay swelling may reduce the effective porosity.

Although data on the effects of temperature and stress on porosity are sparse, there are such data on related properties. For example, the effects of temperature and stress on bulk and pore compressibilities have been investigated and results are discussed below.

The fluid flow capacity of rocks is generally defined by the Darcy law coefficient, permeability. Although permeability is certainly related to the amount of pore space in the rock, the size and continuity of the interconnected pore spaces are much more important. Permeability is very sensitive to temperature, stress, and reactive fluids that may cause clay swelling. Closure of a few of the small interconnections between pore spaces can cause drastic reduction in permeability, yet the reduction in porosity may be negligible unless the closures completely isolate groups of pores. Because of the many connections between pores and because even dead-end pores contribute to effective porosity, there is only a small statistical probability that closure of a few pore interconnections will have much effect on porosity.

As in the case of porosity, there are other properties that are closely related to permeability, such as the formation electrical resistivity factor. Formation resistivity factor and permeability may be interrelated by the tortuosity, as expressed by the Kozeny-Carmen equation. This relationship is not necessarily unique in that other factors may be involved.

There are a number of other considerations that we must address. In gases flowing at low pressures, the "slip," or Klinkenberg effect, must be considered. In dilute aqueous solutions flowing through quartz-rich rocks, streaming potentials may arise which could affect the fluid-flow capacity. Swelling of clay minerals and migration of finely divided particles into the pore spaces may have large effects on permeability. And finally, we must consider the presence of even a small amount of a second fluid phase in interpreting the observed effects of temperature and stress on fluid-flow capacity.

Storage Capacity of Rocks

Although a moderate amount of information has appeared in the literature on the effect of stress on porosity, no direct work has been reported on the effect of temperature on porosity. The most notable work on the effect of stress on the porosity of sandstones has been done by Fatt (1957), Wyble (1958), and Wilhelmi and Somerton (1967). The stress conditions used by each investigator were different. Fatt applied hydrostatic loading, Wyble varied radial loading with the ends constrained, and Wilhelmi used triaxial loading. Work by Chierici and others (1967) has shown that change in porosity is only a

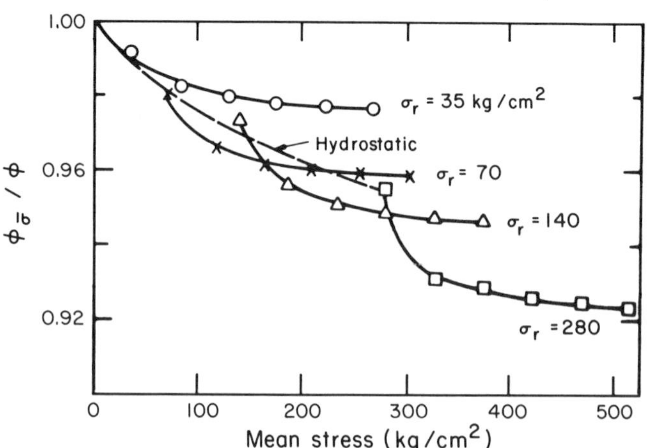

Figure 1. Effect of triaxial loading on porosity of Berea sandstone (data from Wilhelmi and Somerton, 1967).

function of the hydrostatic component of the stress tensor. Thus, if stress is evaluated at its mean effective value ($\bar{\sigma} - p_p$), change in porosity should be independent of the individual stress values. A replotting of Wilhelmi's data in terms of mean $\bar{\sigma} = (\sigma_v + 2\sigma_h)/3$, where $p_p = 0$ (Fig. 1), appears to contradict this. Except for the first triaxial loading point at each confining pressure, the effect of triaxial loading on porosity is small compared with the effect of hydrostatic loading for equivalent effective stress levels.

The effect of stress on porosity for most sandstones is small, 5% to 10% decrease from the unstressed value in the moderate stress range investigated in the above work. Some typical results from Fatt's (1957) data are shown in Figure 2. Because the decrease in porosity with increase in stress is generally exponential, Wyble (1958) has proposed an equation of the following form:

$$\phi_{\bar{\sigma}}/\phi = (1.00 - \underline{b})\ 10^{-\alpha\bar{\sigma}} + \underline{b}, \tag{1}$$

where $\phi_{\bar{\sigma}}$ = porosity at stress $\bar{\sigma}$,

ϕ = porosity at unstressed conditions,

$\underline{b}$ = the $\phi_{\bar{\sigma}}/\phi$ value to which the curve is asymptotic, and

α = the exponential coefficient.

Values of $\underline{b}$ and α are characteristic of a given rock but are also found to be dependent on the direction in which the stress is applied relative to the bedding planes. The latter point is a clue that anisotropy may be a factor in the effect of the applied-stress direction on reduction in porosity.

At this writing, no published data have been found that directly show the effect of temperature on porosity. Von Gonten and Choudhary (1969) have shown the effect of pressure and temperature on pore-volume compressibility, and my co-workers and I (Somerton and others, 1974) have shown the effect of temperature on bulk and rock-solids compressibility. From these data, the effect of temperature on porosity may be estimated. In both cases, the applied stress was hydrostatic. Figure 3 shows the fractional change in pore volume for Berea sandstone at two temperatures--20 and 200 °C. Figure 4 shows the fractional change in bulk volume

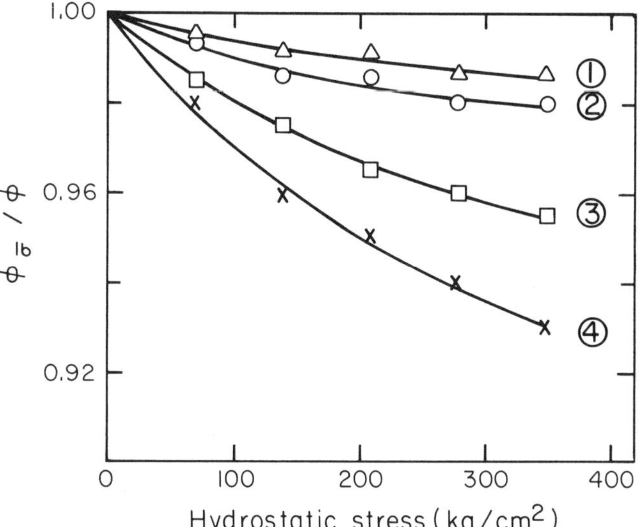

Figure 2. Effect of hydrostatic loading on porosity of various sandstones: 1, Lyons; 2, Bradford; 3, Allison; 4, Sespe (data from Fatt, 1957).

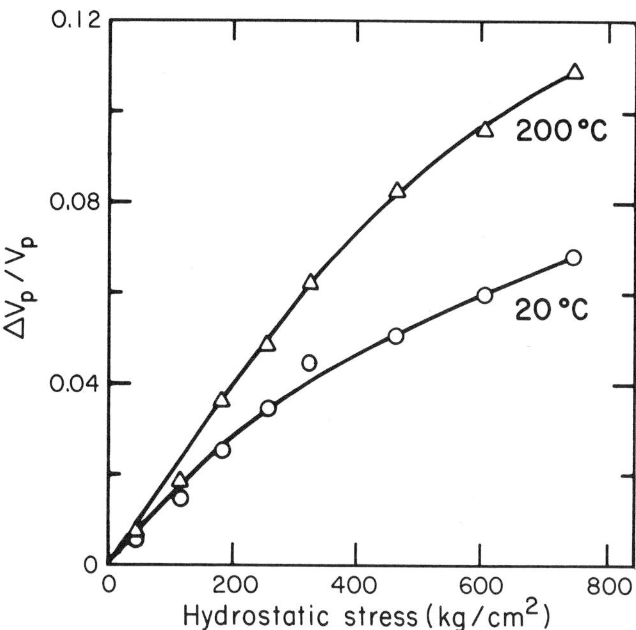

Figure 3. Change in pore volume of Berea sandstone with temperature and stress (data from Von Gonten and Choudhary, 1969).

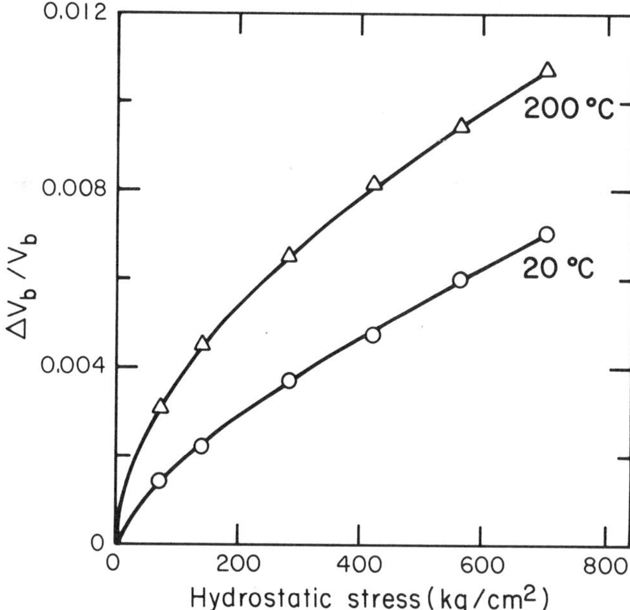

Figure 4. Change in bulk volume with temperature and stress for Berea sandstone (data from Somerton and others, 1974).

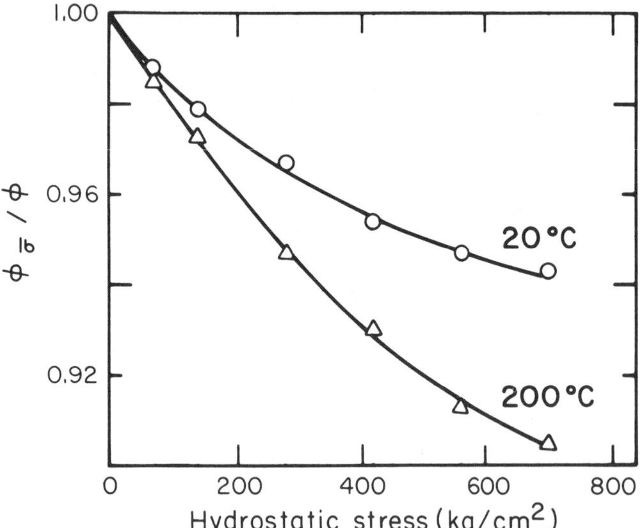

Figure 5. Calculated change in porosity of Berea sandstone with temperature and stress.

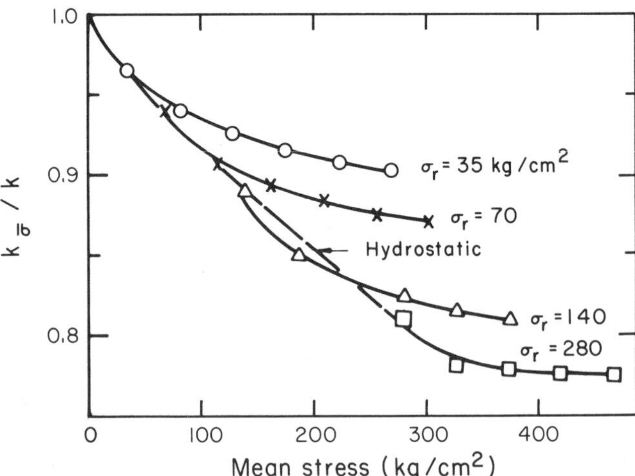

Figure 7. Effect of triaxial stress on permeability of Berea sandstone (data from Wilhelmi, 1967).

for Berea sandstone at the same two temperatures. Figure 5 shows the calculated values of change in porosity of Berea sandstone with stress at the two temperature levels. The greater reduction in porosity with stress at higher temperature is substantial for this one example. We cannot generalize this result, however, since the data were for different specimens of the same sandstone type run by different investigators on different test equipment.

Fluid-Flow Capacity of Rocks

Permeability of porous rocks is much more sensitive than porosity to changes in stress and temperature. The reduction of permeability is roughly five times the reduction in porosity at moderate stress levels (Fig. 6). According to Chierici and others (1967), the deviative component of the stress tensor has no effect on the permeability of clean sandstones. Their experimental data seem to have confirmed this within the limits of experimental accuracy. This result is somewhat unexpected, except that it is significant to note that in their triaxial loading tests, the maximum stress was applied in the same direction as the fluid flow. Wilhelmi and Somerton's (1967) results (Fig. 7), on the other hand, showed a marked difference between hydrostatic and triaxial loading under flow conditions similar to those of Chierici. Results of Gray and others (1963), when replotted in terms of mean stress as shown in Figure 8, again indicated a indicated a difference between triaxial and hydrostatic loading. In this case the maximum stress was applied in the direction perpendicular to the fluid flow. Although the data are scattered and the amount of permeability reduction seems low for Berea sandstone, the reduction in permeability for triaxial loading was again less than that for hydrostatic loading at the same mean effective stress.

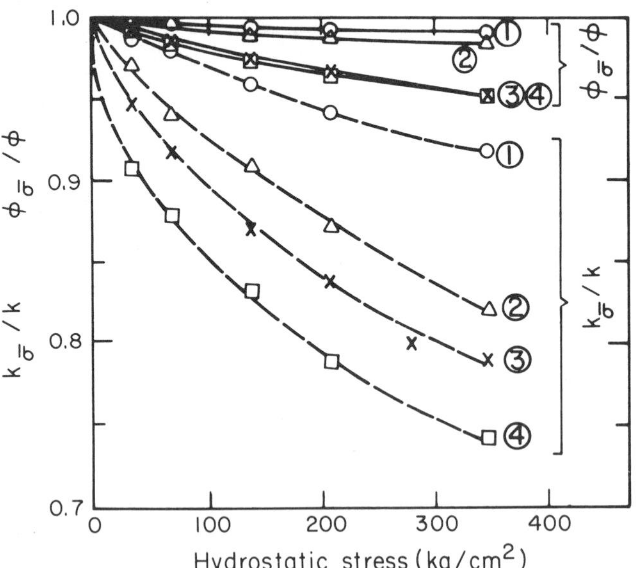

Figure 6. Comparison of effect of stress on porosity and permeability of various sandstones: 1 and 2, basal Tuscaloosa; 3, southern California coast; 4, Los Angeles basin (data from Fatt, 1953).

The first significant report on the effect of temperature on the permeability of natural rocks was made by Afinogenov (1969). He showed a substantial reduction in permeability (as much as a hundred fold) with increase in temperature from 21 to 94 °C. The validity of these findings may be questionable since the tests were run during a period of rapid temperature increase, and nonequilibrium test conditions prevailed. Afinogenov attributed the large change in permeability to some physicochemical interactions of the flowing fluid with the mineral surfaces of the rocks and in part to changes in the pore structure with temperature and stress.

Work on the effect of temperature on fluid-flow properties of rocks has been done by several investigators at Stanford. Weinbrandt and others (1975), while studying the effect of temperature on relative permeability, also noted a substantial decrease in

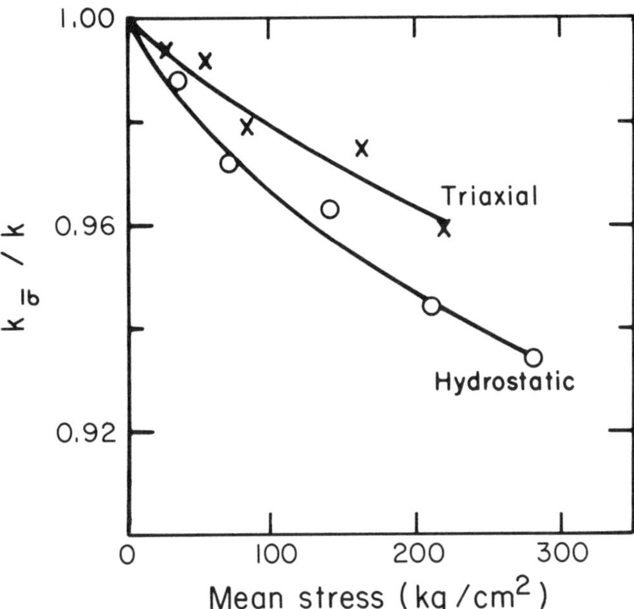

Figure 8. Effect of stress conditions on permeability of Berea sandstone (data from Gray and others, 1963).

absolute permeability. The permeability of Boise sandstone was observed to decrease to less than half its room-temperature value upon increasing temperature to 80 °C. Casse (1974) continued this work with tests on three outcrop sandstones (Bandera, Berea, and Boise), using several flowing fluids. Figure 9 shows the results of tests on Berea sandstone with the use of distilled water and white

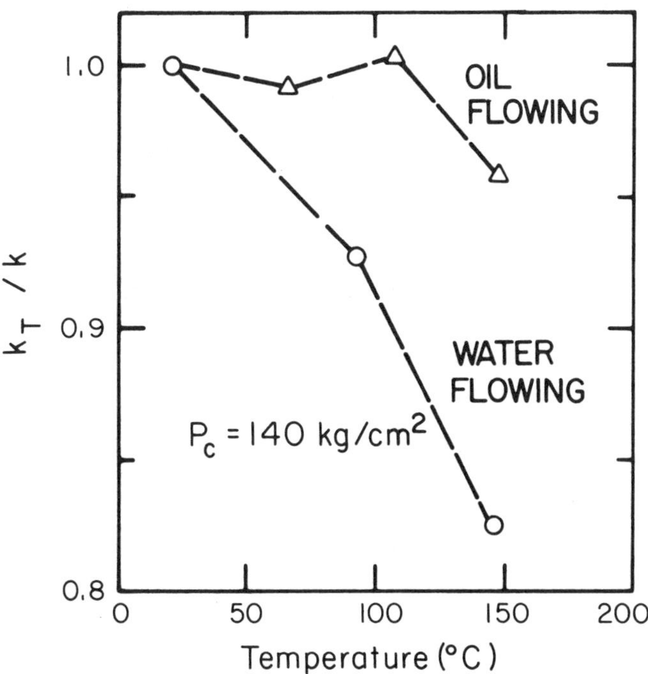

Figure 9. Effect of temperature on permeability of Berea sandstone to oil and water (data from Casse, 1974).

mineral oil as the flowing fluids. Whereas distilled water flow caused a 17% reduction in permeability at 146 °C, mineral oil flow showed only a 4% decrease--about the same order of magnitude as the range of accuracy of the flow tests. Temperature had no effect on absolute permeability to gas when data were corrected for the Klinkenberg effect. However, the slip factor (slope of the Klinkenberg plot) increased in a linear manner with temperature.

The combined effects of temperature and stress on permeability of Berea sandstone to distilled water are shown in Figure 10, which is based on Casse's (1974) data. Increasing the temperature greatly increases the effect of stress on reduction of permeability. This might give credence to the importance of structural changes, except that temperature has little effect on the permeability to oil, even at high confining stress levels.

Stanford investigators have generally concluded that reduction in absolute permeability, when flowing distilled water, is primarily due to physicochemical effects associated with the mineral grain surfaces. However, these researchers have not dismissed the probable contribution of structural changes in permeability reduction with temperature. Structural changes in porous rock accompanied by change in temperature might be analyzed by considering the Kozeny equation as given by Pirson (1958, p. 101):

$$k = \frac{\phi^3}{2\tau S_v^2(1-\phi)^2}, \tag{2}$$

where k = absolute permeability,
ϕ = fractional porosity,
τ = tortuosity, and
S_v = specific surface area.

The decrease in porosity with temperature is small and does not explain the large observed changes in permeability. For example, with the use of data from

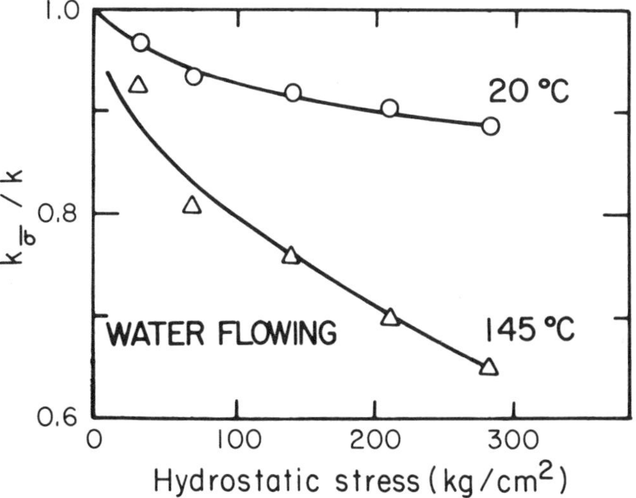

Figure 10. Change in permeability to water of Berea sandstone with temperature and stress (data from Casse, 1974).

Figure 5 for change in porosity of Berea sandstone from 20 to 200 °C at 280 kg/cm^2 confining stress (less than 2% porosity decrease), the decrease in permeability estimated from equation 2 would be 2.2%. This compares with Casse's observed reduction in permeability of 27% over a small temperature range (20 to 145 °C) as given in Figure 10. Thermal expansion data reported by Somerton and Selim (1961) showed that permeability would be reduced by less than 1% by this factor. The major reduction in permeability would probably be caused by changes in the tortuosity, if structural change were the cause of permeability reduction.

Tortuosity can be most easily calculated from formation resistivity factor (F) and porosity determinations according to the relation by Pirson (1958, p. 109):

$$\tau = \phi F. \tag{3}$$

Equation 2 then becomes:

$$k = \frac{\phi^2}{2FS_v^2(1-\phi)^2} \,. \tag{4}$$

A reciprocal relationship between permeability and formation resistivity factor would thus be expected. Using Fatt's (1953, 1957) data for basal Tuscaloosa Sandstone, values of $\tau_{\bar{\sigma}}/\tau$ were calculated from equations 2 and 3 and are shown plotted in Figure 11 as a function of hydrostatic stress. The specific surface area (S_v) was assumed not to change with stress. (Recent communication with A. Timur, Chevron Oil Field Research, has indicated that this assumption may not be valid and that further analysis is needed.) Rather surprising, the change in tortuosity based on formation resistivity factor was about twice the change based on permeability. Insufficient data were available to make similar comparisons for the effect of temperature on tortuosity. However, in Figure 12 the change of permeability with temperature at two constant stress values (Casse data) and the change of resistivity factor at two constant but different stress values (Sanyal and others, 1972; Brannon and Von Gonten, 1973) are compared. Again, the change in resisitivity factor is much greater than the change in permeability with temperature.

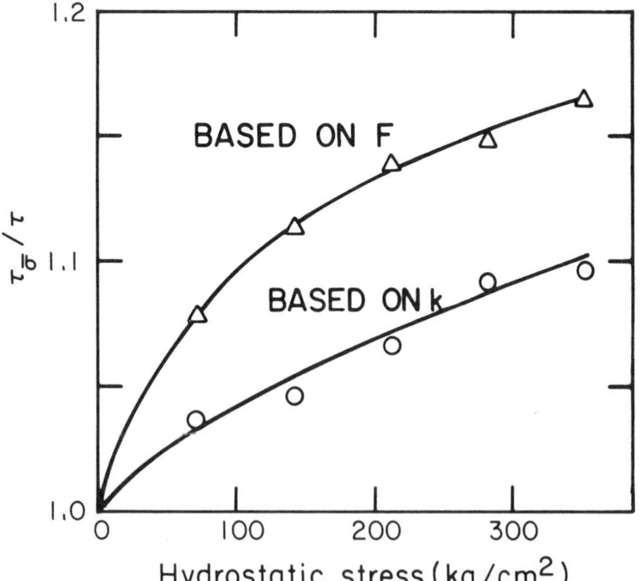

Figure 11. Change in tortuosity of basal Tuscaloosa Sandstone with stress (data from Fatt, 1953, 1957).

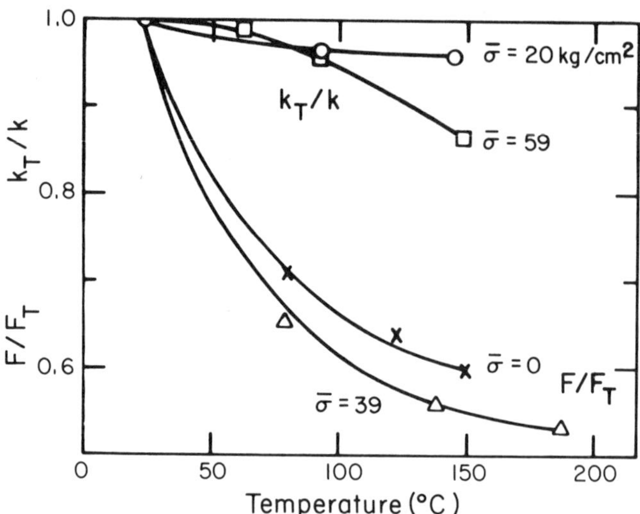

Figure 12. Change in permeability and resisitivity factor of Berea sandstone with temperature (data from Brannan and Von Gonten, 1973).

From the above analysis it would appear that structural changes could explain reduction in permeability with increased temperature except for the fact that such changes are of small magnitude when oil or gases are the flowing fluids (Casse, 1974). Permanent changes in permeability are known to be caused by clay reactions that lead to migration and plugging by clay particles. Mungan (1965) has done extensive work on this subject and has found, for example, that for Berea sandstone (which contains the clay minerals illite and kaolinite) a sudden change from brine to distilled water as the flowing fluid can cause permeability reduction of from 180 to less than 1 md. He also noted that rapid heating of a Berea core to 400 °F (~204 °C) caused comparable reduction in permeability. That this reduction was due to clay mineral migration and plugging was confirmed by the presence of clay particles in the effluent. A number of acidic and alkaline solutions can also cause reduction in permeability by clay plugging. These reductions are more or less permanent and can hardly explain the reduction in permeability with temperature noted by Casse (1974). Distilled water was used throughout the tests, and change in temperature was probably fairly gradual, minimizing the effect of thermal shock. In addition, the reduction in permeability was not permanent, but rather, the cores returned nearly to their original permeability values on the cooling cycle.

Reduction in permeability to flow of low-salinity brine or distilled water in quartz-rich sandstone has been suggested as being due to a reverse electrokinetic effect. Chilingar and others (1970) have reported that the fluid-flow capacity of a sand pack or sandstone can be increased many times by application of an electric potential to the flowing system. A so-called streaming potential is known to develop in systems in which liquids of low ionic concentration are flowing along silica surfaces. This potential decreases as the ionic concentration of the liquid increases. Thus, for freshwater flow, a maximum effect on flow capacity would be expected, whereas for high-salinity brine, little or no electrokinetic effect would be expected. Oil or gas flow would not be expected to have any effect on permeability by this cause.

To link the electrokinetic effect to the reduction of water permeability with temperature, the electrokinetic effect itself must be sensitive to temperature. Pirson (1958, p. 110) has given the Perrin streaming potential equation in the following form:

$$E_s = \frac{D\zeta R_w \nabla P}{4\pi\mu} \times 10^{-7}, \qquad (5)$$

where E_s = streaming potential,

D = dielectric constant of flowing fluid,

R_w = resistivity of flowing fluid,

ΔP = pressure differential,

μ = fluid viscosity, and

ζ = zeta potential.

The dielectric constant for water decreases with increased temperature by a factor of about two in the temperature range of the current investigations. The electrical resistivity of water also decreases with increased temperature. The resistivity of a low-salinity brine may decrease by a factor of five, but this is offset by an equivalent reduction in brine viscosity in this temperature range. The effect of temperature on zeta potential has not been clearly established, although, according to D. W. Fuerstenau (1976, oral commun.), the effect is small. The net effect would appear to be a decrease in streaming potential with increased temperature. Thus, this phenomenon could not explain the effect of temperature on the reduction of permeability.

Figure 13 shows the results of some of our recent permeability studies. This test was run on Berea sandstone with the use of distilled water as the flowing fluid until the system was stabilized at a test temperature of about 195 °C. After test data were taken at this condition, KCl was added to the distilled water in several steps. When stable conditions were obtained at each step, including constant resistivity of the effluent, permeability and resistivity data were again recorded. After

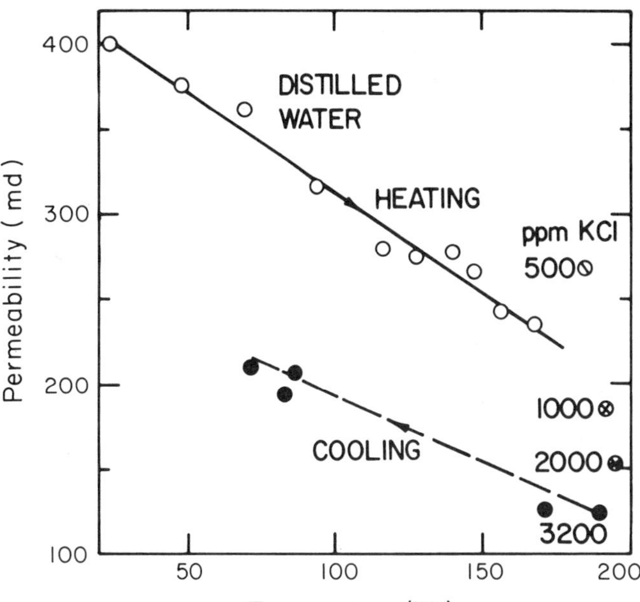

Figure 13. Preliminary results of permeability measurements of Berea sandstone.

a KCl concentration of 3,200 ppm was reached, the heaters were turned off and the system allowed to cool. A few data points were taken on the cooling cycle. All tests were run at a constant confining pressure of 87.5 kg/cm^2 and pore pressure of 17.5 kg/cm^2.

The permeability reduction with increased temperature was somewhat greater than that reported by Casse (1974) but was almost identical to that reported by Weinbrandt and others (1975) for Berea sandstone. The data points taken during the heating cycle showed some scatter owing to minor fluctuations in the back pressure regulator. Upon the first addition of KCl, the permeability increased. Further additions of KCl, however, caused progressive decreases in permeability. After the final addition of KCl and stabilization of the test conditions, the system was cooled and the permeability increased.

Because of the high resistivities of the core saturated with distilled water, no formation resistivity factors were measured during the heating cycle. Upon addition of KCl, the resisitivity dropped to a reasonable level and formation factors were measured at the four KCl concentrations as shown in Table 1.

Table 1. MEASURED FORMATION RESISITIVITY FACTORS

KCl concentration (ppm)	Temperature (°C)	F
500	191	17.1
1,000	190	17.5
2,000	196	17.7
3,200	193	17.9

The formation resistivity factor at room temperature for a group of Berea cores with a KCl concentration of 5,000 ppm ranged between 13.1 and 13.8. This increase of about 33% is less than that reported by Brannon and Von Gonten (1973) for the same temperature range.

Although the above results cannot be interpreted quantitatively, they are qualitatively significant. The reduction in permeability of Berea sandstone with temperature has been confirmed. Although it is possible that the rather sudden change in KCl concentration of the flowing fluid caused some plugging of the core, probably due to clay migration, the recovery of a portion of the permeability upon cooling confirms a significant temperature effect. If the permeability reduction were due to a reverse electrokinetic effect, an increase in the salinity of the flowing fluid should have resulted in an increase in the permeability. Although an increase in permeability was noted on the first addition of KCl, further increases in KCl concentration caused the permeability to decrease rather substantially. If the decrease in permeability was due to clay migration and plugging, results of this test are then conclusive. Further tests in which the salinity change is very gradual (Mungen, 1965) may help to separate these effects.

Although probably more a function of the matrix characteristics than pore properties, formation resistivity factor has been included in this section because of its relation to permeability. Recent work by Pivinskii and Weed (1976) has shown that formation resisitivity factors are strongly time dependent when saturated with low-salinity waters (tap water). It may take as long as 4,500 h for the value to stabilize, owing to the kinetics of ion transport between rock and saturating solution. This is probably not a factor in determining formation resistivity factor when normal brine concentrations are used.

CONCEPT OF EFFECTIVE STRESS

The terms *stress* and *pressure* have been used interchangeably here. Technically this is not correct, and reference should instead be made to confining or overburden stress and to pore fluid pressure. Effective stress is given as

$$\sigma_{eff} = \sigma_{conf} - N \times P_p, \tag{6}$$

where N is an empirical constant expressing the fraction of the pore fluid pressure (P_p) which effectively counteracts the confining stress (σ_{conf}). Chierici and others (1967) have contended that N is a strong function of clay or shale content of sandstones, its value decreasing with increase in clay content.

One should note that if in the above relationship for effective stress the value of N is less than 1.0 (N is often taken as 0.85), change in pore fluid pressure would have less effect on effective stress than would change in overburden or confining stress. Recent work by Zoback and Byerlee (1975) showed that at least in the case of pore properties (porosity and permeability), change in pore pressure has a substantially greater effect than change in confining stress. To explain this effect, these investigators proposed a model consisting of two parts--the solid framework of the rock and a softer lining with grain-cementing material. Changes in pore fluid pressure would thus have a greater effect on the softer pore lining and cementing material than on the harder solid framework of the rock. The "effective stress," as far as pore properties are concerned, would thus require the value of N to be greater than unity.

We must conclude from the above that to relate laboratory-measured data directly to subsurface reservoir performance, we need to know more about the response of rocks to different subsurface stress conditions. For pore properties, the major effect is change in pore-fluid pressure. For matrix-dominated properties, the effective stress as expressed by equation 6 is probably the appropriate representation of the effect of stress on these properties.

Some additional problems arise in relating laboratory-test stress conditions to subsurface stress behavior. Uncertainty as to the state of stress in the subsurface compounds this difficulty. A triaxial state of stress may exist in which, in the absence of lateral tectonic forces, the horizontal stress is substantially less than the vertical stress as shown by equation 7:

$$\sigma_h = \frac{\nu}{1 - \nu} \cdot \sigma_v, \tag{7}$$

where ν is Poisson's ratio. In deep, high-temperature reservoirs, on the other hand, the "plastic" nature of the reservoir rocks leads one to the assumption that the stress is hydrostatic (lithostatic); that is, that stress is equal in all directions. The work of Wilhelmi and Somerton (1967) has shown that these two stress states cannot be equated by applying the mean-stress concept [$\bar{\sigma} = (\sigma_v + 2\sigma_h)/3$] to the triaxial case.

Another problem in applying laboratory-test data to subsurface analyses is the question of loading cycles. Most porous rocks show some "permanent" or unrecoverable deformation on the first loading cycles. The usual procedure with competent rocks is to stress cycle the rock until the physical property being measured has stabilized. This may involve four or five loading cycles within the stress range of the measurements. Similarly, heating cycles may also be applied. Some question arises about whether this procedure really "restores" the rock to its in situ condition. Mattax and others (1975) argued to the contrary at least for the cases of consolidated and friable sands. They concluded that first-loading-cycle compressibility data obtained in the laboratory compared favorably with values calculated from acoustic and pressure-pulse field tests.

MATRIX-DOMINATED PROPERTIES

A clear distinction between pore- and matrix-dominated properties cannot be made because they are obviously mutually dependent. However, for purposes of the present discussion, matrix-dominated properties will include elastic-wave velocities, strength and deformation characteristics (including bulk and pore compressibilities), and thermal properties. These properties are all dependent upon such characteristics as mineral composition, degree of compaction and cementation, grain size and size distribution, and porosity. All of these properties are also dependent on environmental conditions of stress, temperature, and fluid saturation.

Elastic-Wave Velocities and Deformation Properties

The effects of temperature, stress, and fluid saturation on compressional-wave (P) and shear-wave (S) velocities of a typical outcrop sandstone (Berea) are shown in Figure 14. The effect of stress is generally represented by an equation of the form

$$V_{p,s} = A_{p,s}(\sigma)^{1/x_{p,s}}, \tag{8}$$

where subscripts p and s refer to compressional and shear, respectively. According to Palen (1978), for shear velocities, the values of constant A vary widely for different rock types, ranging from 500 for softer rocks to 3,000 for hard rocks. Constants for compressional-wave velocities are approximately 50% higher. The exponent x ranges from a value of 6 for unconsolidated sands to well over 40 for hard rocks.

The effects of temperature and fluid saturations on wave velocities in rocks have not yet been quantified but are the subjects of current investigations.

Measured compressional and shear velocities for a Cerro Prieto sandstone are shown in Figures 15 and 16, respectively. This sandstone is much more friable than Berea. The effects of effective stress, temperature, and fluid saturation are shown in the figures. The pore-fluid pressure was maintained high enough only to keep the saturating brine (6,000 ppm KCl) in liquid phase.

The value of the exponent x in equation 8 averages about 9.0 for this soft sandstone in a dry condition, and is much closer to the theoretical value of 6.0 for unconsolidated sands than would be expected for harder sandstones. For the liquid-saturated core, the value of x increases to about 20, indicating a much lesser effect of stress in this case. The value of constant A varies depending on temperature, fluid saturation, and whether it is for shear or compressional velocity. No analytical

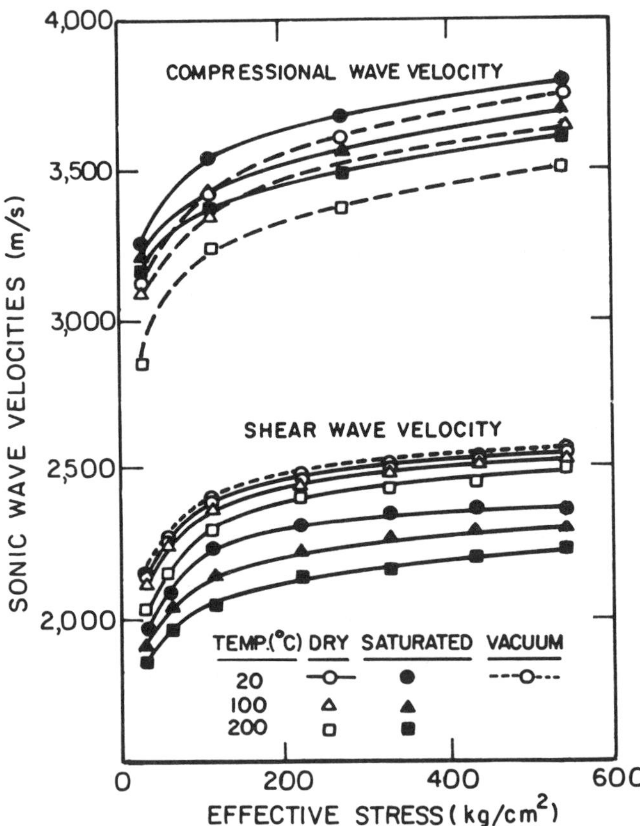

Figure 14. Effects of temperature and stress on sonic velocities for Berea sandstone (data from Palen, 1978).

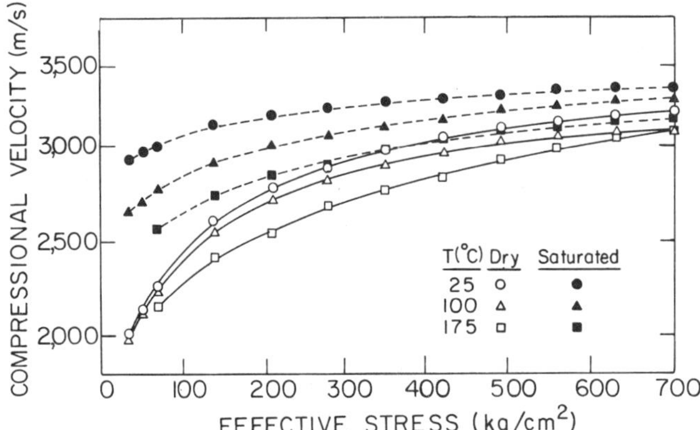

Figure 15. Compressional-wave velocities for Cerro Prieto sandstone (data from Palen, 1978).

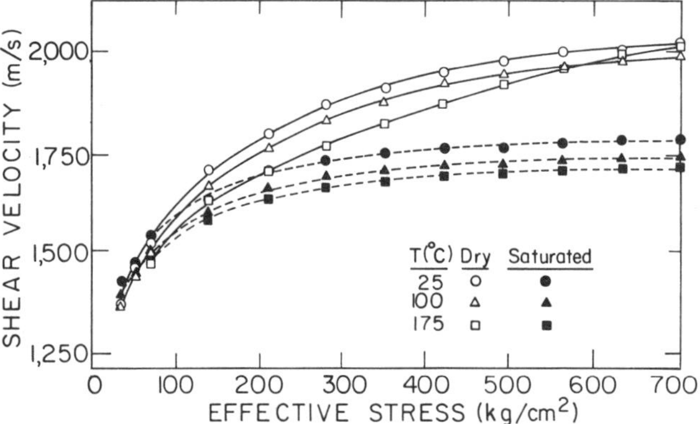

Figure 16. Shear-wave velocities for Cerro Prieto sandstone (data from Palen, 1978).

expression has yet been developed relating the value of A to the above parameters.

On the basis of well-known relationships between compressional- and shear-wave velocities and dynamic-elastic modulus, Palen (1978) has calculated Young's modulus, bulk modulus, and Poisson's ratio as plotted in Figures 17, 18, and 19, respectively. Bulk modulus should provide useful data for subsidence calculations. Young's modulus and Poisson's ratio data may be useful for well-bore stability studies as, for example, in the mechanical-properties log (Tixier and others, 1975).

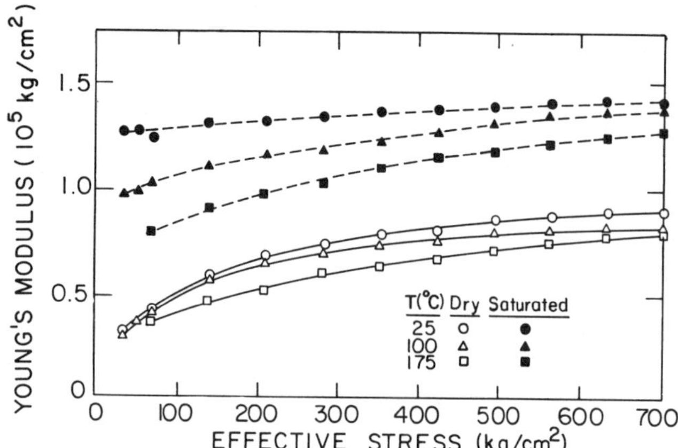

Figure 17. Dynamic Young's modulus for Cerro Prieto sandstone (data from Palen, 1978).

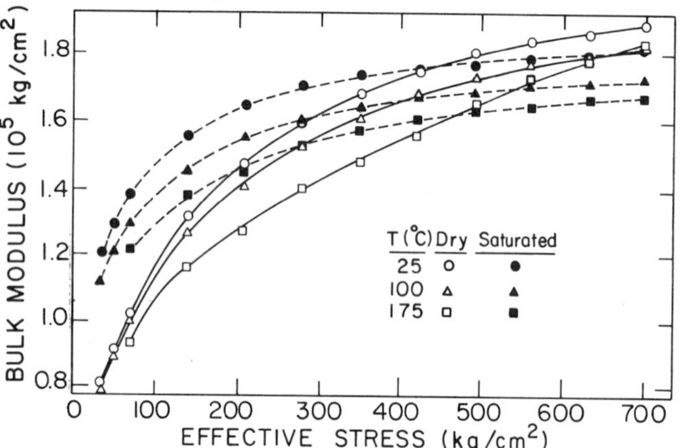

Figure 18. Dynamic bulk modulus for Cerro Prieto sandstone (data from Palen, 1978).

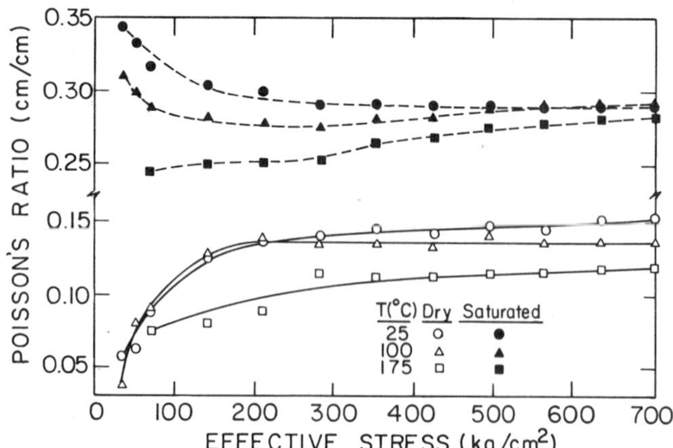

Figure 19. Dynamic Poisson's ratio for Cerro Prieto sandstone (data from Palen, 1978).

An interesting feature of Figure 19 is the two-fold difference in Poisson's ratio for dry and liquid-saturated rocks. This offers the possibility of detecting liquid/gas interfaces in subsurface formations where both shear-and compressional-wave velocities are measured in boreholes.

The effects of temperature on P-wave velocity need to be dealt with in the interpretation of sonic or transit time logs. The observed decrease in velocity with increased temperature would result in a prediction of increased porosities using the time-average equation. Because it was shown earlier that porosity decreases with increased temperature, large errors could result in sonic-log interpretation in high-temperature formations. Mobarak (1971) has estimated conservatively that porosity could be in error on the high side as much as 30%.

Compressibility

Figure 20 shows the effects of stress and temperature on the bulk compressibility of Berea sandstone. The values calculated at 20 °C are from earlier work (Somerton and others, 1974), which showed that a correction factor could be applied to compressibilities calculated from sonic data as

$$C_{bd}/C_{bs} = 5.0\ \sigma^{1/6}, \tag{9}$$

where C_{bd} = compressibilities calculated from sonic data,

C_{bs} = compressibilities from static data, and

σ = effective stress.

The quantity C_{bd} is calculated from sonic data using the following expression for dynamic-bulk modulus:

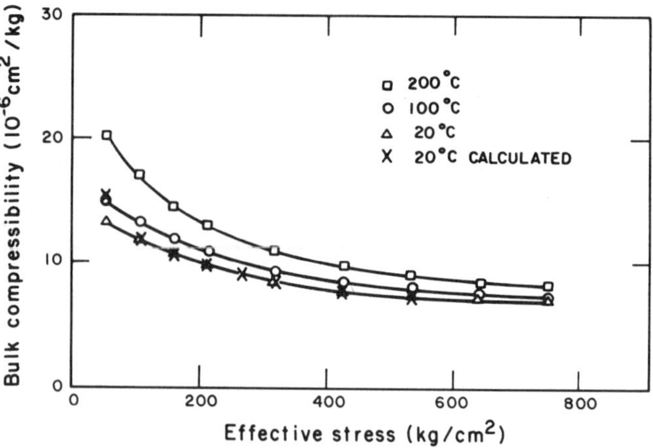

Figure 20. Effects of temperature and stress on bulk compressibility of Berea sandstone (data from Palen, 1978).

$$K = 1/C_{bd} = \rho[V_\rho^2 - (4/3)V_s^2], \quad (10)$$

where V_p = compressional velocity,
V_s = shear velocity, and
ρ = bulk density.

Except for the lowest stress value, corrected dynamic-bulk-compressibility values shown in Figure 20 agree quite well with measured static values.

Compressibility values have particular applications in reservoir performance studies and in surface subsidence calculations. Pore-compressibility values are used in estimating reservoir-storage volume changes that result from reduction in pore fluid pressure. Very few data are available on the effect of temperature on pore compressibility, but a comparison of values calculated from the expression

$$C_p = C_b/\phi, \quad (11)$$

with values reported by Von Gonten and Choudhary (1969), showed excellent agreement (Somerton and others, 1974). The effect of increased temperature is to magnify the amount of decrease in reservoir volume with decrease in pore-fluid pressure.

Bulk and matrix compressibilities were used by Geertsma (1973) in predicting surface subsidence upon reduction of pore-fluid pressure in subsurface reservoirs. Geertsma's analysis is based on relating the compacting subsurface reservoir to the nucleus-of-strain concept, which gives expressions similar to McCann and Wilts' (1951) spherical-tension model. Required in the analysis is a quantity called the "uniaxial compaction coefficient," defined by:

$$C_m = 1/3\ [(1 + \nu)/(1 - \nu)]\ (1 - \beta)\ C_b, \ldots \quad (12)$$

where ν = Poisson's ratio,
$1 - \beta$ = Biot constant ($= 1 - C_r/C_b$),
C_r = rock-matrix compressibility, and
C_b = bulk-volume compressibility.

The difficult quantity to evaluate in equation 12 is Poisson's ratio. Geertsma suggested the use of estimated values between 0.25 to 0.35 for Poisson's ratio, although measured values are observed over a much larger range and are dependent on stress, temperature, and fluid saturation. The effect of increased temperature would be to increase the magnitude of the compaction coefficient and thus to increase the amount of surface subsidence upon decreasing pore fluid pressure as fluids are withdrawn from the subsurface reservoir. The large reduction in Poisson's ratio from liquid to vapor saturation condition would indicate a substantial reduction in the compaction coefficient and thus less surface subsidence in vapor- or gas-saturated subsurface formations.

Thermal Properties

The most important thermal property needed in subsurface applications is thermal conductivity. In transient heating or cooling calculations, values of thermal diffusivity may be required, but if thermal conductivity is known, diffusivity may be easily calculated (Somerton, 1958). This latter calculation requires knowledge about specific heat, which may be estimated from a mineral or oxide analysis. In transient heating or cooling, an understanding is needed of thermal expansion and mineral-phase changes which consume or liberate heat (Somerton and Selim, 1961).

Thermal conductivity of rocks is primarily a function of conductivities of the rock solids and saturating fluids, porosity, and mineral grain size. Environmental conditions of temperature, stress, and fluid saturation have important effects on thermal conductivity.

Because thermal properties of rocks are difficult to measure, require elaborate equipment and taking a great deal of time, much effort has gone into development of models for predicting thermal properties from simple properties of the system (see Ozbek, 1976, for review of these models).

One of the best models for predicting thermal conductivity is that of Krupiczka (1960); it has the general form

$$\lambda/\lambda_f = (\lambda_s/\lambda_f)^x, \quad (13)$$

where λ = effective conductivity of the rock-fluid system,
λ_f = conductivity of saturating fluid,
λ_s = conductivity of rock solids,
$x = A + B\phi + C \log (\lambda_s/\lambda_f)$,
ϕ = fractional porosity, and
A, B, C = constants (0.280, -0.757, and -0.057, respectively).

Equation 13 predicts conductivity better for low values of the ratio λ_s/λ_f, which would be used for liquid-saturated rocks. Martínez-Baez (1979) obtained better agreement with experimental values by modifying the exponent x to:

$$x = A + B \log \phi + C \log \delta_m + D \log (\lambda_s/\lambda_f) + E \log (\delta_{90}/\delta_{10}), \quad (14)$$

where δ_m = median grain size in millimetres,
δ_{90} = grain size of 90 percentile larger than, and
δ_{10} = grain size of 10 percentile larger than.

For the particular group of samples he was working with, the constants A, B, C, D, and E were 0.417, -0.160, 0.077, -0.160, and 0.459, respectively. Although the fit of calculated to experimental values was excellent for liquid-saturated samples (correlation coefficient = 0.951), the modification still did not give satisfactory results for air-(vapor-) saturated samples. Ghaffari (1978) has been developing a new model that provides for three-dimensional heat flow and contact resistance; this model shows promise of giving more general agreement for any fluid saturant.

My co-workers and I (Somerton and others, 1974) have shown that the following expression gives an excellent representation of the effect of partial-liquid, partial gas (vapor) saturation:

$$\lambda_{S_\ell} = \lambda_{(S_\ell = 1.0)} - 0.363\lambda_s(1.0 - S_\ell^{1/2}), \quad (15)$$

where λ_{S_ℓ} = thermal conductivity of partial liquid saturated rock,
$\lambda_{(S_\ell = 1.0)}$ = thermal conductivity of rock fully saturated with liquid (may be calculated from equation 13),
λ_s = conductivity of rock solid, and
S_ℓ = fractional-liquid saturation.

The effect of temperature on thermal conductivity may be predicted by applying a modification of the Tikhomirov (1968) equation (Somerton, 1975) as follows:

$$\lambda_T = \lambda_B - 1.28 \times 10^{-3} (T - T_B)(\lambda_B - 3.3)$$

$$\cdot [\lambda_B(T \times 10^{-3})^{-0.06\lambda_B} + 3.0]\lambda_B - 0.94, \quad (16)$$

where λ_B = thermal conductivity at base temperature 20 °C (mcal/s-cm- °K),
T = temperature (°K), and
T_B = base temperature (°K).

Figure 21 shows the agreement between calculated and experimentally determined effects of temperature on thermal conductivity. For lower thermal conductivity values (<6.0 mcal/s.cm. °K) the following simplified form of equation 16 may be used:

$$\lambda_T = \lambda_B - 0.3 \times 10^{-2}(T - T_B)(\lambda_B - 3.3). \quad (17)$$

In cases where the saturating liquid might boil, an additional factor must be added to account for the large amount of heat transferred by the

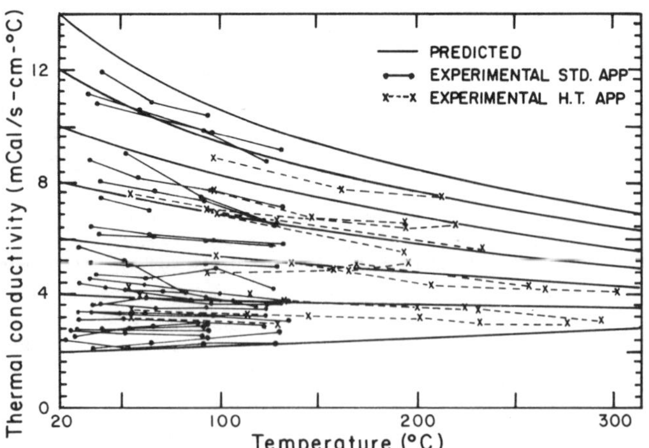

Figure 21. Effect of temperature on thermal conductivity of rocks.

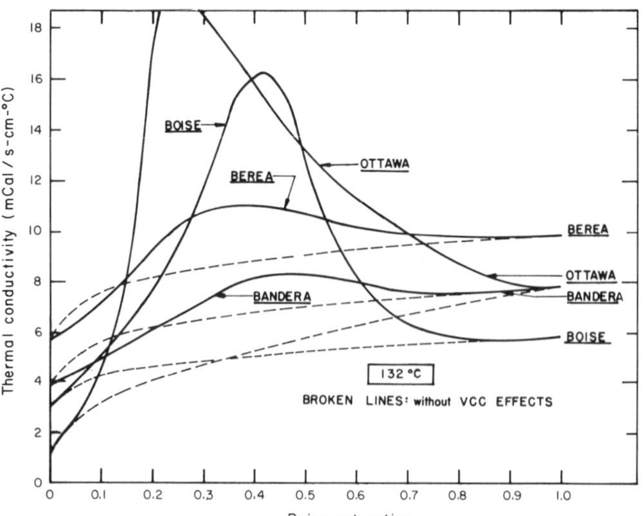

Figure 22. Comparison of VCC effect in several sandstones.

vaporization-condensation-capillary (VCC) effect:

$$\lambda_{S_\ell a} = \lambda_{S_\ell} + \lambda_{VCC}(S_\ell), \quad (18)$$

where $\lambda_{S_\ell a}$ = apparent thermal conductivity,
λ_{S_ℓ} = true thermal conductivity, and
$\lambda_{VCC}(S_\ell)$ = VCC contribution to thermal conductivity.

This effect is clearly demonstrated in Figure 22. The broken lines are the conductivities as functions of liquid saturation that would be expected without the VCC effect. The solid lines are the actual measured values of apparent thermal conductivity.

The VCC heat-transfer reaction is similar in principle to the well-known "heat-pipe" phenomenon (Sun and Tien, 1972). At the heat-input end of the pipe, a liquid is vaporized and the vapor travels to the output end of the system where it is condensed and gives up its latent heat of vaporization. The condensed liquid is then returned to the input end by capillary forces acting in an annular wick along the inner surface of the pipe. In the thermal conductivity apparatus, a temperature gradient of about 10 to 15 °C is imposed across the 3.2-cm length of the 5.1-cm-diameter cylindrical sample. Pore pressure at the midpoint of the sample is maintained at vapor pressure corresponding to the midpoint temperature. Thus, at the higher-temperature end of the sample, the liquid saturant is vaporized. Under the existing vapor-pressure gradient, the vapor travels to the lower-temperature end of the sample and condensation occurs. The condensed liquid travels by capillary action through the finer pore spaces to the low-liquid-saturation region at the higher-temperature end of the sample. Liquid flow occurs under a capillary pressure gradient hindered or aided by the gravity gradient, depending on the direction of the temperature gradient.

The continuous cycle of fluid motion within the cell causes the transfer of heat in the form of latent heat of vaporization as well as by conduction. The magnitude of heat transfer is determined by the mass fluid flux and the magnitude of the heat of vaporization of the flowing liquid. Gomaa (1973) has analyzed the VCC phenomenon in porous media and has concluded that the magnitude of the effect is dependent upon absolute permeability; heat of vaporization of the flowing liquid; porosity; temperature gradient; viscosities of fluid phases; and liquid saturation as well as upon properties of the fluid-rock system that are saturation dependent, including relative permeability and capillary pressure.

The major importance of absolute permeability on the VCC effect is clearly shown by comparing the magnitudes of the ratio of the maximum value of the VCC effect with the conductivity without VCC ($\lambda_{s\ell}$) for several sands, as given in Table 2.

TABLE 2. MAGNITUDE OF VCC EFFECT

Sample	$\lambda_{VCC}/\lambda_{s_\ell}$	(md)$^\kappa$
Bandera	0.23	38
Berea	0.26	190
Boise	2.15	2,500
Ottawa	3.47	6,500

These data indicate that the maximum magnitude of the VCC effect for these sandstones, with the exception of the Berea, is approximately proportional to the square root of the absolute permeability.

Other analyses have shown that the VCC effect is directly proportional to the heat of vaporization of the saturating fluid. The effect of liquid saturation is complex, although it is clear that there is a close relation to relative permeability of the rock-fluid system. Since multiphase fluid flow cannot occur below critical vapor saturation or below residual or "irreducible" liquid saturation, no VCC effects can occur beyond these limits.

The importance of the VCC effect in analysis of reservoir performance where fluid-phase changes occur is currently under investigation by Su (1978).

OTHER PROPERTIES

Other properties of rock-fluid systems are stress and temperature dependent. Fatt (1953) showed that although absolute permeability may be strongly affected by overburden stress, relative permeability does not appear to change with stress. Weinbrandt and others (1975) have shown, however, that relative permeability does change with temperature. This is attributed primarily to the observation that irreducible water saturation increases and residual oil saturation decreases with increased temperature, causing a corresponding shift of the relative permeability curves. Thus, for a given saturation value, the relative permeability to oil increases and the water-oil relative permeability ratio (k_w/k_o) decreases with increased temperature. Lo and Mungen (1973) found similar effects for capillary pressure curves; that is, irreducible water saturation increased markedly with increased temperature for several sandstones. There was also an indication that residual oil saturations decreased with increased temperature. Interestingly, they observed that capillary pressure curves for limestones showed no temperature sensitivity.

FUTURE ADVANCES

The foregoing review of the effects of temperature and stress on the physical properties of rock-fluid systems is certainly not comprehensive. Rather, I wished to highlight some of the important problems in order to provide guidance for future work. There are three major problems that need to be solved before we can assign appropriate values for the properties and behavior of rocks in their subsurface environment.

The first problem, that of evaluating subsurface environmental conditions and then simulating these conditions in the laboratory, has been discussed above. The state of stress in the subsurface--triaxial or hydrostatic (lithostatic)--generally is not known, and we must develop techniques to evaluate it. The effect of fluid pressure is not fully understood. We need to evaluate the role of pore pressure as it affects pore properties in particular. The effect of temperature on rock properties has only been considered important in recent years. Rock-fluid reactions appear to be more important at higher temperatures than at lower temperatures, and this matter needs further study. It is possible that as we learn more about the behavior of rocks, we may be able to evaluate subsurface conditions from our logging data, knowing the effect subsurface conditions may have on the measured rock properties.

Second, efforts are under way in at least three research centers to develop techniques and apparatus which will make it possible to determine several different rock properties all within one test on the same test specimen. With cooperation and funding from Lawrence Berkeley Laboratory, the U.S. Department of Energy, several industrial concerns, and the Department of Mechanical Engineering, at the University of California, Berkeley, we are constructing an apparatus that will measure P- and S-wave velocities, bulk and pore compressibilities, permeability, formation resistivity factor, thermal conductivity, and thermal diffusivity all on the same test specimen and at the same temperature and stress conditions. Results of this work will be used in establishing correlations of the different properties with the characteristics of the rock-fluid system. Similar work is being done by Terra

Tek in Salt Lake City and by the University of Texas at Austin. At least three major oil company research centers are activating or expanding efforts in this area.

The third need is for models with which we can predict properties from simpler, more readily measurable characteristics of the rock-fluid system. Other models will be needed to predict the behavior under changing environmental conditions. Perhaps the greatest progress in this direction has been made in the case of thermal properties as discussed earlier. These models will be extremely valuable for use in subsurface-reservoir simulation studies. The final goal of this work should be to develop a universal model which will describe all of the physical properties and behavior of rock-fluid systems.

ACKNOWLEDGMENT

This work was supported by the Director, Office of Energy Research, Office of Basic Energy Sciences, Division of Engineering, Mathematics, and Geosciences of the U.S. Department of Energy under Contract No. W-7405-ENG-48.

REFERENCES CITED

Afinogenov, Y. A., 1969, How the liquid permeability of rocks is affected by pressure and temperature: Novosivirsk, Fizika Tekhnicheskie Problemy Ruzrabotki Poleznykh Iskopoemykh, no. 6, p. 34-42.

Brannan, G. O., and Von Gonten, W. D., 1973, The effect of temperature on the formation resistivity factor of porous media, in Transactions, 14th Symposium, Society of Professional Well Log Analysts: Houston, Society of Professional Well Log Analysts.

Casse, F. J., 1974, The effect of temperature and confining pressure on fluid flow properties of consolidated rocks [Ph.D. thesis]: Stanford, Stanford University.

Chierici, G. L. and others, 1968, The effect of overburden pressure on some petrophysical characteristics of sandstone reservoir rocks, in Proceedings, World Petroleum Congress, 7th Mexico City: Amsterdam, Elsevier.

Chilingar, G. V., El Nassir, A., and Stevens, R. G., 1970, Effect of direct electrical current on permeability of sandstone cores: Journal of Petroleum Technology, v. 22, no. 7, p. 830-36.

Dobrynin, V. M., 1962, Effect of overburden pressure on some physical properties of sandstones: American Institute of Mechanical Engineers Petroleum Transactions, v. 225, no. 2, p. 360 II-360.

___1977, Behavior of reservoir rocks at high pressure and temperature; Berkeley, University of California, Petroleum Engineering Research Seminar Series.

Fatt, I., 1953, The effect of overburden pressure on relative permeability: Journal of Petroleum Technology, v. 5, no. 10, p. 15-17.

___1957, Effect of overburden and reservoir pressure on electric logging formation factor: American Association of Petroleum Geologists Bulletin, v. 41, no. 11, p. 2456-66.

___1958, Pore volume compressibilities of sandstone reservoir rocks: American Institute of Mechanical Engineers Transactions, v. 213, p. 326.

Geertsma, J., 1973, Land subsidence above compacting oil and gas reservoirs: Journal of Petroleum Technology, v. 25, no. 6, p. 734.

Ghaffari, A., 1978, Thermal conductivity of porous systems: Berkeley, University of California, Petroleum Engineering Laboratories, Research Progress Report.

Gomaa, E. E., 1973, Thermal behavior of partially liquid saturated porous media, [Ph.D. thesis]: Berkeley, University of California.

Gray, D. H., Fatt, I., and Bergamini, G., 1963, The effect of stress on permeability of sandstone cores: Society of Petroleum Engineers Journal, v. 3, no. 2, p. 95-100.

Krupicska, R., 1960, Analysis of thermal conductivity in granular materials: International Chemical Engingeering, v. 6, no. 1, p. 71.

Lo, H. Y., and Mungen, N., 1973, Effect of temperature on water-oil relative permeabilities in oil-wet and water-wet systems: Dallas Society of Petroleum Engineers Preprint 4505.

Mann, R. L., and Fatt, I., 1960, Effect of pore fluids on the elastic properties of sandstone: Geophysics, v. 25, no. 2, p. 443.

Martínez-Baez, L. F., 1979, Experimental investigation of thermal conductivity of core samples from the Cerro Prieto geothermal field and an improved prediction method, in Proceedings, First Symposium on the Cerro Prieto Geothermal Field, Baja California, Mexico, September 20-22, 1978: Berkeley, Lawrence Berkeley Laboratory, LBL-7098, p. 342-351.

Mattax, C. C., McKinley, R. M., and Clothier, A. T., 1975, Core analysis of unconsolidated and friable sands: Journal of Petroleum Technology, v. 27, no. 12, p. 1423.

McCann, G. D., and Wilts, C. H., 1951, A mathematical analysis of subsidence in the Long Beach-San Pedro area: California Institute of Technology Report.

Mobarak, S. M., 1971, The effect of temperature on wave velocities in rocks [M.S. thesis]: Berkeley, University of California.

Mungan, N. E., 1965, Permeability reduction through changes in pH and salinity: Journal of Petroleum Technology, v. 27, no. 12, p. 1449.

Ozbek, H., 1976, Thermal conductivity of multi-fluid saturated porous media [Ph.D. thesis]: Berkeley, University of California.

Palen, W. A., 1978, The effects of temperature, pressure, and saturation on sonic velocities and dynamic elastic moduli of reservoir sandstones [M.S. thesis]: Berkeley, University of California, p.

Pirson, S. J., 1958, Oil reservoir engineering, (second edition): New York, McGraw-Hill.

Piwinskii, A. J., and Weed, H. C., 1976, A study of rock-solution interaction and its effect on Archie's Law: Institute for Electrical and Electronics Engineers Transactions on Geoscience Electronics, v. GE-14, no. 4, p. 221.

Sanyal, S. K., Marsden, S. S., and Ramey, H. J., Jr., 1972, The effect of temperature on electrical resistivity of porous media, in Transactions, Society of Professional Well Log Analysts: Houston Society of Professional Well Log Analysts.

Sinnokrat, A. A., Ramey, H. J., and Marsden, S. S., 1971, Effect of temperature level upon capillary pressure curves: Society of Petroleum Engineers Journal, v. 11, no. 1, p. 13.

Somerton, W. H., 1958, Some thermal characteristics of porous rocks: Journal of Petroleum Technology, v. 10, no. 5.

___1975, Thermal properties of partially liquid saturated rocks at elevated temperatures and pressures, in API Research Project 155 Final Report: Berkeley, University of California.

Somerton, W. H., and Selim, M. A., 1961, Additional thermal data for porous rocks--Thermal expansion and heat of reactions: Society of Petroleum Engineers Journal, v. 1, no. 4, p. 249-53.

Somerton, W. H., El-Shaarani, A. H., and Mobarak, S. M., 1974, High-temperature behavior of rocks associated with geothermal type reservoirs: Society of Petroleum Engineers Preprint 4897.

Su, H. J., 1978, Heat transfer in porous media with phase changes, in Research project report: Berkeley, University of California, Petroleum Engineering Laboratories.

Sun, K. H., and Tien, C. L., 1972, Simple conduction model for theoretical steady-state heat pipe performance: American Institute of Aeronautics and Astronautics Journal, v. 10, no. 8, p. 1051.

Tikhomirov, V. M., 1968, Conductivity of rocks and their relationship with density, saturation, and temperature: Neftianoe Khoziarstro (in Russian), v. 46, no. 4, p. 36.

Tixier, M. P., Loveless, G. W., and Anderson, R. A., 1975, Estimation of formation strength from the mechanical properties log: Journal of Petroleum Technology, p. 283.

Von Gonten, W. D., and Choudhary, B. K., 1969, The effect of pressure and temperature on pore volume compressibility: Society of Petroleum Engineers Preprint 2526.

Weinbrandt, R. M., Ramey, H. J., Jr., and Casse, F. J., 1975, Effect of temperature on relative and absolute permeabilities of sandstones: Society of Petroleum Engineers Journal, v. 15, no. 5, p. 376.

Wilhelmi, B., and Somerton, W. H., 1967, Simultaneous measurement of pore and elastic properties of rocks under triaxial stress and conditions: Society of Petroleum Engineers Journal, v. 7, no. 3, p. 283-94.

Wyble, D. O., 1958, Effect of applied pressure on the conductivity, porosity and permeability of sandstones: American Institute of Mechanical Engineers Petroleum Transactions, v. 213, p. 430-32.

Zoback, M. D., and Byerlee, J. D., 1975, Permeability and effective stress: American Association of Petroleum Geologists Bulletin, v. 59, p. 154.

Manuscript Accepted by the Society October 14, 1980

Printed in U.S.A.

Geological Society of America
Special Paper 189
1982

Instrumentation for Well Tests

R. C. SCHROEDER
Earth Sciences Division
Lawrence Berkeley Laboratory
University of California
Berkeley, California 94720

ABSTRACT

In this paper, the types and uses of well tests in the disciplines of hydrology, petroleum engineering, and geothermal engineering are reviewed. Geologic and economic factors that influence the design of well tests are summarized along with the difficulties of designing tests for each category of applications. The types of data needed for hydrothermal tests are discussed in detail, and the problems commonly encountered when taking measurements under various operating conditions are reviewed.

A historical review of well-testing instruments includes information on some of the latest equipment advances in the field. The ideal instrument array for well testing would measure pressure, temperature, flow rate, enthalpy, and concentration simultaneously and at several locations. The accuracy and durability limits of commercially available equipment are evaluated against this ideal according to the properties the equipment measures. Finally, the instruments that come closest to meeting the standards of well log analysts are identified.

INTRODUCTION

Well tests are designed and carried out to determine the properties of ground-water aquifers, petroleum and natural gas deposits, and (more recently) aquifers with abnormally high temperatures, known as geothermal reservoirs. The measurements obtained from well tests allow us to make quantitative estimates of the condition of the well and of the formation properties and hydrological characteristics of the aquifer or rock formation in which the well is completed. Well tests are also a necessity in the relatively new fields of underground storage of energy and radioactive wastes. Paul A. Witherspoon's significant and pioneering contributions in each of these Earth-science disciplines are indicative of the breadth of his interest and his scientific and engineering virtuosity; it is fitting that this volume honors his accomplishments.

Three types of wells are used in well tests: (1) production wells, from which fluid is withdrawn; (2) injection wells, used for fluid disposal and underground storage; and (3) observation wells, used neither for injection nor withdrawal but for observation of flowing wells. During short-term production or injection well tests, information on the condition of the well and estimates of formation properties are sought. When production or injection wells are flowed for extended periods of time, observations can be made in static nonflowing wells to provide estimates of the average formation properties, hydrological continuity, inhomogeneities, and other complicating conditions of the flowing wells.

Well testing is one of the primary functions in three disciplines--hydrology, petroleum engineering, and geothermal engineering. However, there are significant differences in the design and execution of well tests in each of these disciplines. The primary differences are related to the characteristics of the geological formations in which ground water, hydrocarbons, and geothermal fluids are usually found, and in which hot fluids or radioactive wastes are stored. In general, ground-water aquifers are shallow deposits with strata that may be unconsolidated, and usually have relatively high transmissivity and storage. Hydrocarbon deposits are found in a huge variety of geological formations. They range from shallow oil deposits (most of which are now depleted) to deep, low-permeability deposits of natural gas (for example, the Devonian shale formations). In the case of the over pressured zones near the Gulf of Mexico, deep aquifers contain both hot, saline water and large amounts of natural gas (methane). The hydrocarbon deposits, in particular oil, are usually found in thin layers of moderate- to high-permeability sedimentary rocks. In the case of geothermal reservoirs, both sedimentary deposits and volcanic formations are known in which high temperatures are present. Like petroleum reservoirs, geothermal reservoirs are found in a wide variety of rock types, geological settings, and types of fluids.

Owing to the differences in the geological characteristics, a well test designed to determine formation parameters in a shallow, unconfined ground-water aquifer may differ considerably from a test in a thin, confined oil reservoir or a very thick, deep, unconfined geothermal reservoir. Similarly, special problems arise when hot fluids are to be stored underground or when a formation is to be evaluated for use as a waste repository. The choice of well-test instrumentation often depends heavily upon the purpose of the test, the expected formation parameters, and the geological nature of the resource to be investigated.

The geological differences are not the only factors that influence the design of well tests. The economics of the resource also determine, in part, the methods and equipment used to obtain the required information. Water is a cheap commodity and petroleum a more expensive one. Ground-water well tests are generally designed and executed with a minimum of cost as one of the objectives; for oil and gas well tests, less consideration is given to cost and more flexibility is allowed in obtaining the necessary information. As a result, the instrumentation developed in ground-water hydrology is often simpler and cheaper than are the types of equipment used in oil fields. In the case of geothermal reservoirs, the special problems associated with the high temperatures make well testing not only more difficult technically but also more costly.

The geological differences associated with the different types of well tests result in special problems in the three different cases. In ground-water well tests the special problems include multilayered systems, unconfined-aquifer behavior, and in some cases, saturated-unsaturated flow or free-surface problems. In the case of petroleum reservoirs, the well-test problems often are related to water or gas drive in confined systems. In the case of underground energy or waste storage, the problems relate to the evaluation of particular formation properties. For energy storage the formation must be porous and have high permeability. For waste storage the conditions must be just the opposite.

Hydrothermal resource temperatures can range from slightly above ambient to close to the critical point of water. The low-temperature hydrothermal resources are usually found in or near shallow ground-water systems. High-temperature hydrothermal resources are generally characterized by some form of cap rock, usually a shale, clay, or metamorphic layer that tends to insulate the fluid deposit and hold in the heat. In the case of the low-temperature geothermal resources, the well-testing problems are similar to those associated with ground-water aquifers. In the case of high-temperature geothermal resources, the problems are more closely related to the temperature limitations of the equipment. In addition to the geothermal rock systems, which are saturated with fluid, there is current great interest in the possibility of extracting energy from hot dry rock. Here the well-testing problem is one of determining the characteristics of an artificial hydraulic fracture when the temperature of the rock and fluid can be higher than the critical temperature of water.

Since natural geothermal reservoirs are usually found in regions of high tectonic activity, the rocks (whether sedimentary or igneous) commonly are also highly fractured with complicated stratigraphy and intruded volcanics. This poses problems for the analyst and affects well-test design and instrumentation because interpretation of fracture parameters usually requires high-precision, early-time data. The complicated stratigraphy and intruded volcanics make tests in wells with large open intervals particularly problematical, as the effects of partial penetration and multilayered flow are difficult to analyze even if efforts are made to isolate the effects from other layers. Recent studies have pointed out that reliance upon only a single instrument or measurement in a multi-layered open interval may not provide sufficient information concerning formation parameters regardless of the type of analysis (Dogru and Seinfeld, 1979).

High temperatures make precision measurements technically difficult because available electronic instrumentation cannot for a sufficient amount of time withstand temperatures above about 150 °C. Although there are instruments commercially available that can be used to make geothermal measurements at temperatures as great as 400 °C, they do not provide satisfactory precision and control for the test engineer. The primary problems for continuous readout measurements are the cable insulation above 250 °C and the downhole electronic components at temperatures above 150 °C. Work is currently in progress on these high-temperature instrumentation problems at the Sandia Laboratory in Albuquerque (Veneruso, 1978) and in industrial laboratories (Lamers, 1979).

The fluids in geothermal reservoirs are characterized by both the temperatures and the rock types in which the resource is found. Both liquid and two-phase resources are known and are being exploited. The water, whether liquid or two-phase, often has large mass fractions of CO_2 present in the reservoir. This is particularly true of geothermal reservoirs before large-scale withdrawal of fluids. In addition to CO_2, many gases such as H_2S and NH_3 can be found in small amounts. The presence of large amounts of dissolved noncondensible gases can result in two-phase wellbore flow. Wellbore storage effects complicate the analysis, and corrosive gases such as H_2S and NH_3 are hazardous and destructive to many tool materials.

REQUIRED MEASUREMENTS

The purpose of well tests, as noted above, is to obtain data that can be used to evaluate either the well, a formation, or a resource. Geothermal resource assessment includes evaluating the total power available and the environmental impact of extracting that power. The power availability is dependent upon the resource size and the degree of difficulty in bringing the energy to the surface. The primary environmental effects of energy storage, waste storage, and resource utilization are the hydrothermal impacts on existing neighboring aquifers and the ground-surface instabilities that can accompany fluid withdrawal or injection. Subsidence measurements do not ordinarily fall into the category of well tests, and have not been included in the remainder of this report. Only hydrothermal measurements and their associated instruments are discussed. Table 1 summarizes the types of measurements that are usually needed for hydrothermal tests.

The accuracy requirements of various well-test measurements may vary with the type of measurement and the eventual use of the data. Well-testing instruments have three primary components: (1) a reaction element (transducer); (2) a signal conditioner (transmission system); and (3) a recorder (display). The accuracy of the measurement depends on the accuracy of all the components and can never exceed the minimal accuracy of any individual component. For example, if a transducer has infinite resolution and the data are recorded to four decimal places, the measurement has a resolution of four decimal places--not the infinite resolution of the transducer. There are several basic accuracy concepts, not all of which apply to the well-test measurements listed in Table 1. Table 2 summarizes these principles of measurement accuracy.

TABLE 1. TYPES OF MEASUREMENTS REQUIRED FOR HYDROTHERMAL WELL TESTS

Measurement	Comments
Time	Used to determine transient measurements of all kinds
Depth	Used for all well-test measurements and for profiles (logs) as a function of depth
Pressure	The primary quantity in flow tests of wells from which the hydrological parameters of the rocks and the flow characteristics of the reservoir (aquifer) can be determined
Water level	Similar to pressure measurements, but not always possible to measure; also used to define the aquifer free surface
Temperature	Indicates the location of the heat in a hydrothermal system, and provides data on the thermal production and injection characteristics of hydrothermal systems
Flow rate	A required quantity in any flowing well test; also provides a measure of the power associated with the flow of thermal waters
Quality	The mass fraction of steam-to-water in a two-phase mixture; this quantity is necessary in some cases to determine the total flow and to determine the fluid enthalpy.
Enthalpy	When referred to a common reference temperature enthalpy can give the power associated with a given flow from a thermal well; it is also required when determining the flow rate in certain cases
Concentration	Required in all cases of water quality and brine handling in surface equipment; essential in all cases of tracer studies and waste migration, both natural and artificial

In geothermal wells, particularly at high temperatures, the required measurements cannot always be made with the precision that the analyst would like. In some cases, this is due to the limitations of the existing high-temperature instruments. However, there are additional problems in hot wells, such as thermal expansion (stretching) of the cable or wire line in deep well measurements, which are not easily calibrated and are seldom corrected for.

In general, well-test requirements for precise time measurements are not severe and are within the capability of any commercial quartz clock. However, most equipment does not include a free choice of the time interval for sampling or recording. Accurate measurements of well depth are often difficult to achieve in the field, but this is primarily a problem with technique rather than a fundamental instrumentation deficiency. For example, if a wire line or cable is run into a well at too high a rate, the cable counter can slip or lose accuracy. In addition, there are the thermal considerations referred to above. Although a tool can usually be returned to its original location with satisfactory accuracy by marking the wire, the depth is seldom known with high precision. Fortunately, this is not usually a serious problem in well testing.

In ground-water aquifer well tests, only the pressure (or water level) and flow rate usually need

TABLE 2. PRINCIPLES OF MEASUREMENT ACCURACY

Concept	Definition
Accuracy	The degree of conformity of a measurement to an accepted standard
Precision	The recording (or display) accuracy of an instrument
Linearity	The relationship between the signal input and output
Stability	The inherent accuracy of an instrument during prolonged continuous use
Resolution (sensitivity)	The accuracy to which an instrument can discriminate
Dependence	The instrument's relationship to its environmental conditions
Response	The instrument's reaction to a signal

to be determined. Since quite often a downhole pump is required, it is generally necessary to make the pressure measurement downhole, below the pump, while measuring discharge at the surface. However, for well tests of hydrothermal systems, information can be desirable for temperature, flow rate, and water quality in addition to downhole pressure. Information on the characteristics of layers of rock or an individual fracture may be required. For these tests the transient downhole flow rate and temperature are helpful in evaluating the data. Downhole flow measurements are also of value when water flashes or when gas comes out of solution above the sandface in the wellbore or at the surface When a pump is in place in a well, a downhole flow-rate measurement is not usually possible. In addition to these considerations, measurements of concentration of certain components generally are of great importance. The presence of water in oil wells and the quantities of gas or dissolved solids in geothermal wells usually are of interest. In energy storage projects, the dissolved solids and tracer materials are important. In waste storage projects, the migration of tracers or waste products is the key issue. In all cases measurements of small concentrations of gases, chemicals, or isotopes are important when tracer studies are made to determine the anisotropy, inhomogeneity, or dispersion in an underground system. In general, all of the measurements indicated in Table 1 are required at a particular point in a well for the duration of a test, but in many cases not all of the measurements can be made, owing to costs or associated complications.

HISTORICAL REVIEW OF WELL-TEST INSTRUMENTATION

The term "instrumentation" is used so casually that we forget that most simple instruments were first developed less than 250 years ago. Although instruments as we know them did not exist in ancient times, measurements have been made with the use of ingenious (although crude) devices throughout recorded history. For example, water-level measurements were recorded as long ago as 3000 B.C. At Roda, Egypt, records of the water level of the Nile are available as far back as A.D. 641 (Biswas, 1970). (Figure 1 shows a reconstruction of the Roda nilometer with gauging marks on a central column.) However, it was not until 1831 that Palmer recorded a patent for a mechanical float-operated water-level gauge to monitor the level of the River Thames (Casperlein, 1974).

Although water levels were measured from stream gauging with no apparent problem, the same was not true of discharge measurements. Initially the flow-rate measurements were primarily of concern with regard to the quantity of water in streams and rivers. However, the discharge from artesian springs and in man-made aqueducts soon was of great interest to scientists and hydraulic engineers. There is no known record of the correct stream-flow relationship prior to Hero of Alexandria. (His lifetime is not well known but was in the interval 150 B.C. to A.D. 250, according to Biswas.) Unfortunately, his formula

Q = Area x Velocity of flow = Volumetric discharge

died with him, and even the marvelous Roman waterways and aqueducts were built and maintained with the erroneous idea that stream flow was dependent only on the cross-sectional flow area (Rouse and Ince, 1952). An amusing example was given by Biswas (1970) of a Roman water commissioner who tried to balance his account books by calculating the discharge into and out of a particular aqueduct. Finding it impossible to balance the flow, he concluded that someone was drawing off water from the system surreptitiously. The problem may have been due to the incorrect flow relationship he was using, since Herschel concluded 14 centuries later that there probably was no discrepancy when the correct flow calculation was made.

The value of experiments by Archimedes, Hero, and other ancient Greeks, and later of Leonardo da Vinci, cannot be denied (Reti, 1972). The collections of da Vinci's notes and drawings show that he had invented a downhole bellows-type diaphragm well pump, a vertical-axis flow meter, and a vane-type anemometer for use as a flow meter. However, because of his secrecy, these ideas were not adopted during his time.

The birth of the experimental method has generally been attributed to Galileo Galilei (D'Abro,

Figure 1. Reconstruction of the Roda nilometer (Source: Popper, 1951, as quoted by Biswas, 1970).

1939). The construction of instruments such as the thermometer (Foley, 1933) and telescope (Lodge, 1893) during Galileo's lifetime signaled the beginning of instrumentation as we know it.

Following Galileo, the experiments of Descartes, Pascal, Boyle, Newton, Joule, and others not only led to the fundamental principles used in modern instruments but resulted in many of the first models of the instruments themselves--the manometer gauge is an example. Newton derived the first approximation for the flow of fluid through a sharp-edged orifice, and it was the Italian, Poleni, who improved the approximation used by Newton, and who also first derived an expression for flow over a sharp edge having a free surface. Hence, the basic weir equation

$$Q = (2/3)\ Cb\sqrt{2gj^{3/2}}$$

is sometimes named after him (Rouse and Ince, 1952). A French contemporary of Poleni's was Henri de Pitot, after whom the pitot tube is named. During this same period Robert Hooke designed a sophisticated horizontal-axis flow meter (Frazier, 1974).

Although the invention of the thermometer is attributed to Galileo, it was not until the nineteenth century that the "caloric" theory was put to rest and the foundations of thermodynamics were constructed (Edser, 1910). The work of Mayer, Joule, and Von Helmholz (Lindsay, 1976) between 1840 and 1850 resulted in the final understanding of the relationship between mechanical work and heat energy. About 1850 the "new" understanding of the energy concept resulted in fundamental studies by Clausius (1850), Thomson (1851), and others. The basic principle of thermal calorimetric measurements to determine the heat content of a solid or liquid had already been carried out (Joule, 1843, 1845). In the latter reference, Joule reported on the famous porous plug measurements.

Precision instrumentation for measuring pressure, temperature, and water level accompanied the technological developments of the twentieth century.

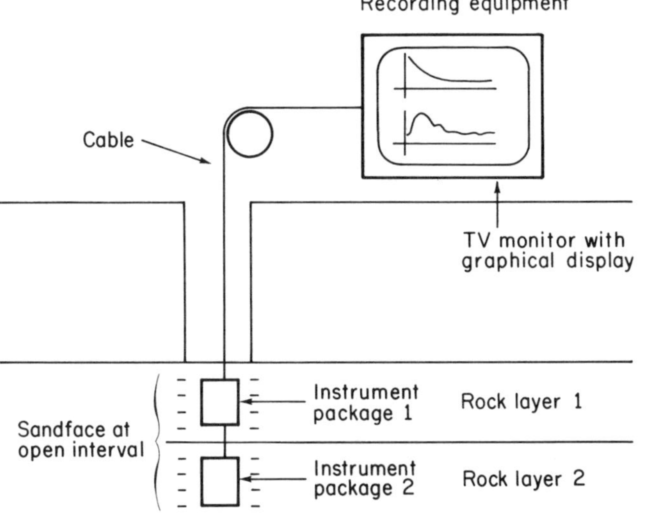

Figure 2. Idealized diagram of downhole instrumentation for well tests.

Although the inventions and their refinements are too numerous to mention, there is one device which deserves special attention, as it can truly be said to have ushered in a new era in instrumentation. That device is the computer. The machines that initiated this era began with Babbage in 1833, although the mathematicians Leibnitz and Pascal constructed special-purpose calculating machines in the seventeenth century, as did several others before Babbage. In 1946, International Business Machines Company (IBM) introduced the first commercial electronic calculator, and in 1950 Remington Rand delivered the first computer (the UNIVAC) to the U.S. Census Bureau (Reid-Green, 1978). A parallel and related development in instrumentation was the design of digital circuits and readouts. The commercial appearance of the computer was preceded by machines that used mechanical relays (IBM-Mark I) at Harvard University, and the first electronic machine (the ENIAC), which was invented in 1943.

IDEAL INSTRUMENTATION FOR WELL TESTING

Rather than review in detail the equipment commercially available for well-test measurements, I review the features that would be ideally desirable for different types of measurements. Along the way I indicate the limitations of instruments that currently are used, and identify the commercial instruments that come closest to meeting the needs of well-test analysts.*

In the section on well-test measurements, I concluded that, ideally, measurements of pressure, temperature, flow-rate, enthalpy, and concentration should be made continuously and simultaneously at a given point in a well. In addition, it would often be of interest to have simultaneous measurements at more than one point. Needless to say, no instrument or group of instruments is presently available to accomplish this task. Figure 2 shows a diagram of the ideal system. The downhole instrument packages include several transducers, and there could be several packages to the system. The signals are multiplexed and sent to the surface where they are recorded and displayed in graphic form for evaluation by the test engineer. With this real-time data, the on-site investigator can make decisions regarding the well test by using a computer that is time shared for the recording, display, and analysis requirements. Clearly the recording should have priority.

In the past, such a system could be built only at great cost. But since 1977 a new computer development has brought the price of such a system within reason. That new development is the rapid improvement of microcomputers. Many of these are available with nearly all of the features and computing power required to accomplish the tasks outlined above. These computers are based on the new circuitry and memory technology that has produced a complete

*Reference to a company or product name does not imply approval or recommendation of the product by the University of California or the U.S. Department of Energy to the exclusion of others that may be suitable.

central processing unit on a single semiconductor chip (Toong, 1977). The entire microcomputer can be assembled on a single circuit board. This new area of computer technology is in its infancy, and proponents currently are struggling with standardization of the data input and output configurations.

In the case of ground-water well measurements, ordinarily only the pressure, flow rate, and concentration are required even in the most sophisticated experiments. The most important two measurements are pressure and flow rate, with temperature equally important in hydrothermal tests. Equipment that has been used to make simultaneous pressure, temperature, and turbine-flow measurements at temperatures in excess of 275 °C is available from the Geophysical Research Corporation and the Kuster Corporation. Because all three instruments are placed in a well using wire-line suspension and are mechanical-clock driven, there is no control over the instruments during the well test. Hence, the investigator is uncertain of the outcome until the tools are removed from the well. In addition, the record of the measurement consists of a line scribed on a brass sheet and does not provide high precision. A downhole sampler is also available from Kuster, and with modification the entire string could (in principle) be used to make simultaneous measurements at a given location in the well. In practice, the tool string would be so long that it would be difficult to insert into the well, and one would not be sampling (measuring) the conditions at approximately a single point. In addition, the practical problems of synchronizing the clocks and ensuring that all clocks operate properly is an extremely difficult task at all temperatures. Hence, simultaneous measurements are seldom carried out using these tools.

One of the major developments in high-precision pressure measurements was introduced by the Hewlett-Packard Corporation in the early 1970s. Their device uses a pair of quartz-crystal resonators and provides the most accurate pressure measurement currently available for field use. A system is now available from the Gearhart-Owens Company that includes the Hewlett-Packard quartz gauge coupled with a temperature tool and surface electronics. This system provides the test engineer with a temperature-compensated record of the downhole pressure. The correct pressure and temperature are automatically recorded in the appropriate units at the surface. This tool (which is suspended from a single conductor cable) can also be modified to carry a turbine flow meter and could (in principle) provide simultaneous measurements with high precision. Unfortunately, although the Gearhart-Owens tool has been extensively used in oil and gas measurements and for low-temperature (<150 °C) geothermal measurements, it cannot be used at high temperatures due to the failure of the downhole electronics components. For applications at temperatures less than 150 °C, this equipment provides all of the necessary requirements for precision simultaneous pressure-temperature measurements.

Pressure, temperature, and turbine flow measurements can be made at extremely high temperatures with the use of available equipment. But the accuracy, reliability, and control of the instruments are poor. On the other hand, very accurate pressure and temperature measurements are available from equipment with armored cables, but these measurements are limited to relatively low temperatures. There is, however, a wide range of devices which are, in a sense, intermediate in accuracy and durability at high temperatures. Tables 3 and 4 list the existing available types of instrumentation and their present adequacy.

Pressure and temperature measurements at high temperatures with moderate precision and response are currently limited by the lack of economically priced high-temperature multiconductor cable. Transducers are at present available that can withstand temperatures in excess of 300 °C, but these devices require as many as seven conductors for simultaneous measurements. Lamers (1979) has summarized the accuracies and relative merits of various temperature and pressure transducers. The geothermal well-test group at Lawrence Berkeley Laboratory has recently built several downhole instruments that combine a quartz-crystal transducer (Paroscientific) with a standard resistive temperature device (RTD) temperature probe. This tool gives high-precision simultaneous pressure and temperature capabilities at temperatures up to about 150 °C (limited by the quartz transducer). The LBL group has also completed tools that provide simultaneous temperature, pressure, and flow measurements up to 250 °C (the approximate temperature limit of commercially available cable). A review of alternate high-temperature pressure transducers has been given by Bernstein (1977).

Instruments that measure heat content (enthalpy) include calorimeters, Gardon gauges, and heat-transfer probes (Jones, 1977). None of the methods is considered to be high precision and none has been adapted for use in well tests, although attempts have been made to measure enthalpy using calorimetry at the wellhead (Lamers, 1979). A technique of determining the enthalpy of two-phase water-steam

TABLE 3. INSTRUMENT METHODS USED FOR THE VARIOUS TYPES OF MEASUREMENTS

Measurement	Instrument method
Time	Quartz, electrical, mechanical
Depth	Counters, stepping motors
Pressure	Strain gauges, variable reluctance, piezoelectric, Bourdon tubes
Water level	Bubblers, capacitance, conductance, float
Temperature	Resistance (RTD), thermistor, thermocouple, bimetallic
Flow rate	Magnetic, acoustic, turbine, differential pressure, weirs and flumes
Quality	Separators, acoustic, laser doppler
Enthalpy	Calorimetry, critical discharge
Concentration	Chemical analysis, laser scatter, hot wire, and film anemometry

mixtures has been developed in New Zealand (James, 1962). However, the James (or critical discharge) method requires simultaneous measurements of the critical pressure and the liquid flow rate. Because liquid flow rate is difficult to measure for two-phase flows, the James technique is not very accurate. The measurements of the critical discharge pressure and liquid flow rate are often used to obtain simultaneous measurements of both total flow and enthalpy. In general, instrumentation for accurate two-phase flow measurements remains one of the primary problems in high-temperature well tests.

An interesting method of measuring two-phase (non-interacting) flows in vertical pipes has been developed by the Sunstrand Corporation (Alford and others, 1978). In that approach, a quartz pressure gauge is used as the sensing instrument. The fluctuations in pressure in the liquid phase are random and can be used to derive the velocity of the liquid in the pipe (Bird and others, 1960).

In general, the simplest devices for single-phase flow-rate measurements are differential-pressure orifice plates, or water levels in weir or flume boxes (World Meteorological Organization, 1971). More recent developments include magnetic flow meters and acoustic transducers. Both have drawbacks. The magnetic meters require large amounts of power and are degraded by corrosive or abrasive fluids. The acoustic meters require particles or voids (bubbles) to be present in the fluid and thus cannot measure clean fluid flows unless gas or particles are introduced to the flow. These latter meters are also expensive. Neither of these techniques has been used downhole for evaluation of flow from layers.

Measurements of concentrations of particular constituents of the well-test fluids are generally made by sampling the fluid either at the wellhead or downhole, followed by chemical analysis. If a well is pumped, the wellhead sample may be representative of downhole conditions. But if the fluid is two-phase or expands or cools appreciably, the wellhead sample may not be representative of the reservoir conditions. On the other hand, no general procedures for analyzing samples at the downhole conditions have been developed. Some methods are being studied and include hot-wire techniques, scattering methods (Jones, 1977), and radioactive detector calibrations for isotope-tracer measurements. None of these types of instrumentation has been used downhole in hot wells. In addition, there are several methods for measuring the quality of two-phase flow that depend on density measurements to identify the two phases. Laser, ultrasonic, and acoustic techniques are currently under development.

TABLE 4. EVALUATION OF COMMERCIALLY AVAILABLE INSTRUMENTATION FOR WELL TESTING

Instrumentation	Evaluation
Time	Adequate (wellhead), marginal downhole at high temperature
Depth	Usually adequate
Pressure	Adequate below 150 °C for high precision: adequate at low precision above 150 °C: limited by cables at moderate precision to less than 250 °C
Water level	Adequate, although commercial equipment must be modified above 100 °C
Temperature	Adequate, although limited in accuracy by cables at high temperature (<250 °C) and subject to drift over long periods at high temperature
Flow rate	Adequate for single phase: adequate at high cost for two-phase (wellhead separators): marginal for downhole at high temperatures, not available for downhole two-phase flow
Quality	Low accuracy capabilities (wellhead)
Enthalpy	Adequate at high cost (wellhead)
Concentration	Needs more emphasis (wellhead)

SUMMARY

The purpose of well testing and the consequent need for well-test instrumentation is to provide data on well conditions and to determine formation parameters and hydrological characteristics of the geologic system. The choice of instrumentation and test requirements defining size, accuracy, and reliability depend upon several considerations. One is the type of well test--production, interference, or other type. In addition to the actual test design, the particular application determines the instrumentation needs. For example, a simple water-level indicator may suffice for an interference test in a water well, whereas a high-precision, high-temperature device may be necessary for a hydrothermal test. The details of the tests also play a role in choosing the instrumentation requirements. The presence of several layers of different rock, the presence of fractures, the expected temperature and pressure variations, and the length of time required for the test all play a role in determining the choice of well-test instrumentation.

Several measurement quantities were summarized in Table 1 as the most important for well testing. Obviously, not all of the quantities need to be measured for every well test, and in some cases the necessary instrumentation is either not available or not adequate to make the measurement. Because the instruments for a measurement are composed (in general) of transducers, transmission systems, and recording equipment, their availability and adequacy are dependent upon all three components. For low-temperature applications, instrumentation for well tests is available, and only precision-related instrumentation problems exist. However, for high-temperature tests only low-accuracy instrumentation systems are available. Although some components are available for high-temperature applications, total measuring systems are not available. This is primarily due to the need for high-temperature multiconductor cable at a reasonable cost. The lack of high-temperature active circuits is also a problem, and both of these deficiencies are essentially a materials problem. The final problem in high-temperature instrumentation is not related to materials but rather to technique. Currently, two-phase flows can be measured to sufficient accuracy, but only with the use of expensive separators with separate flow instrumentation for the individual phases.

ACKNOWLEDGMENT

This work was supported by the Assistant Secretary for Conservation and Renewable Energy, Office of Renewable Technology, Division of Geothermal and Hydropower Technologies of the U.S. Department of Energy under Contract No. W-7405-ENG-48.

REFERENCES CITED

Alford, B. M., Patterson, M., and Womack, F., 1978, Development and field evaluation of the production surveillance monitor: Journal of Petroleum Technology, v. 30, p. 160-166.

Bernstein, L., 1977, Measurement of unsteady pressures, forces and acceleration, in Measurement of unsteady fluid dynamic phenomena: Washington, Hemisphere Publishing Corp.

Bird, R. B., Stewart, W. E., and Lightfoot, E. N., 1960, Transport pheonmena: New York, John Wiley and Sons.

Biswas, A. K., 1970, History of hydrology: Amsterdam, North-Holland Publishing.

Casperlein, A., 1974, Historical development of hydrometry, in Three centuries of scientific hydrology: UNESCO.

Clausius, R., 1850, On the motive power of heat, and on the laws which can be deduced from it for the theory of heat, in Mendoza, E., ed., 1960, Reflections on the motive power of fire and other papers: New York, Dover Publications.

D'Abro, A., 1939, The rise of the new physics: New York, Van Nostrand, v. 1.

Dogru, A. H., and Seinfeld, J. H., 1979, Design of well tests to determine the properties of stratified reservoirs, in Proceedings, Fifth Society of Petroleum Engineers Symposium on Reservoir Simulation: Dallas, Society of Petroleum Engineers.

Edser, E., 1910, Heat: New York, MacMillan.

Foley, A., 1933, College Physics: London, P. Blakiston's Son and Co.

Frazier, A., 1974, Water current meters: Washington, D.C., Smithsonian Institution Press.

James, R., 1962, Steam-water critical flow through pipes, in Proceedings, Institute of Mechanical Engineers: v. 176, no. 26.

Jones, T. V., 1977, Heat transfer, skin friction, total temperature, and concentration measurements, in Richards, B.E., ed., Measurement of unsteady fluid dynamic phenomena, New York, McGraw-Hill, p. 63-102.

Lamers, M. D., 1979, Measurement requirements and methods for geothermal reservoir systems parameters--An appraisal: Berkeley, Lawrence Berkeley Laboratory, LBL-9090, GREMP-6.

Lindsay, R. B., ed. 1976, Applications of energy, nineteenth century: Stroudsburg, Pennsylvania, Dowden, Hutchinson and Ross, Benchmark Papers on Energy, v. 2.

Lodge, O., 1893, Pioneers of science: New York, MacMillian.

Popper, W., 1951, the Cairo nilometer: Berkeley, University of California Press, Publications in Semitic Philology, v. 12.

Reid-Green, K. S., 1978, A short history of computing: Byte, v. 3, no. 7, p. 84-94.

Reti, L., ed., 1974, The unknown Leonardo: New York, McGraw-Hill.

Rouse, H., and Ince, S., 1952, History of hydraulics: New York, Dover.

Thomson, W. (Lord Kelvin), 1851, on the dynamical theory of heat, with numerical results deduced from Mr. Joule's equivalent of a thermal unit and M. Regnault's observations on steam, in Lindsay, R. B., ed., 1976, Applications of energy, nineteenth century: Stroudsburg, Pennsylvania, Dowden, Hutchinson, and Ross, Benchmark Papers on Energy, v. 2, p. 66-85.

Toong, H.M.D., 1977, Microprocessors: Scientific American, v. 237, no. 3, p. 146-161.

World Meteorological Organization, Working Group, 1971, The use of weirs and flumes in steam gauging: Geneva, Technical Note no. 117.

Veneruso, A. F., 1978, High temperature instrumentation, in Proceedings, Invitational Well Testing Symposium: Berkeley, Lawrence Berkeley Laboratory, LBL-7027, p. 46-52.

Manuscript Accepted by the Society October 14, 1980

Geological Society of America
Special Paper 189
1982

Advances in Borehole Geophysics for Hydrology

PHILIP H. NELSON
Earth Sciences Division
Lawrence Berkeley Laboratory
University of California
Berkeley, California 94720

ABSTRACT

Borehole geophysical methods provide vital subsurface information on rock properties, fluid movement, and the condition of engineered borehole structures. Within the first category, salient advances include the continuing improvement of the borehole televiewer, refinement of the electrical conductivity dipmeter for fracture characterization, and the development of a gigahertz-frequency electromagnetic propagation tool for water saturation measurements. The exploration of the rock mass between boreholes remains a challenging problem with high potential; promising methods are now incorporating high-density spatial sampling and sophisticated data processing.

Flow-rate measurement methods appear adequate for all but low-flow situations. At low rates the tagging method seems the most attractive. The current exploitation of neutron-activation techniques for tagging means that the wellbore fluid itself is tagged, thereby eliminating the mixing of an alien fluid into the wellbore. Another method uses the acoustic noise generated by flow through constrictions in and behind casing to detect and locate flaws in the production system.

With the advent of field-recorded digital data, the interpretation of logs from sedimentary sequences is now reaching a sophisticated level with the aid of computer processing and the application of statistical methods. Lagging behind are interpretive schemes for the low-porosity, fracture-controlled igneous and metamorphic rocks encountered in geothermal reservoirs and in potential waste-storage sites. Progress is being made on the general problem of fracture detection by use of electrical and acoustical techniques, but the reliable definition of permeability continues to be an elusive goal.

INTRODUCTION

As evidenced by the diversity of topics addressed in this volume, the scope of the hydrological disciplines is rapidly increasing. Concurrently, the problems encountered in making borehole measurements are growing more complex. A dazzling variety of physical-measurement techniques have been adapted to the borehole environment while new techniques and new applications are continually making their debut. Because of the diversity of implementation, it often appears that the only constraint on the subject of "borehole geophysics" is that of gathering information from a probe suspended at the end of a long cable within a small cylindrical cavity in rock. Hence, a review of the subject is impossible without clear delineation of the topics of interest.

The terms "borehole logging" or "well logging" are often used interchangeably with "borehole geophysics." A log is a continuous record of some physical property. In this review, I emphasize continuous or semi-continuous modes of data acquisition and presentation, but not to the exclusion of other, single-station measurements that can best be executed from a wire-line system. I have included any measurement which fits this loose description, but have biased the review to include subjects of primary interest to the hydrologist, with an emphasis on developments of the last few years. I have omitted those measurements commonly undertaken by the hydrologist and reservoir engineer, such as the direct measurement of pressure and flow by conventional well-test methods.

This review relies heavily upon the available literature. One of the most important contributions is the monograph by Keys and MacCary (1971), an excellent introduction and manual which covers the fundamentals and applications of the basic tools applicable to ground-water work. This monograph is valuable to a reviewer of the field because it serves as a reference of extant methodology; topics covered in it, which are routinely available, are not reviewed here.

Many of the methods used for hydrological investigations are the same as for oil and gas field investigations where the techniques were first developed, although the probes used in the smaller diameter holes commonly drilled for hydrological investigations are necessarily simplified versions of the probes used in the petroleum field environment. Also adaptable from the well-logging industry are many of the interpretive techniques developed for use in sedimentary rocks. Hence this review relies heavily upon the literature developed by the commercial well-logging industry. Most of the references are to papers and presentations accomplished since 1977, available as preprints and publications of the Society of Petroleum Engineers and the Society of Professional Well Log Analysts.

This review is organized according to the purpose of the measurement, rather than the measurement techniques. The categories chosen are rock properties, fluid properties and flow, engineering needs, and data interpretation. Specific tools are highlighted in each of these categories, either because of recent or promising developments. I conclude with a few opinions on developments concerning the detection of fractures and the measurement of permeability.

ROCK PROPERTIES

Borehole logging is usually performed to obtain quantitative porosity estimates and information on lithology, structure, and rock permeability as well as to establish other parameters relevant to the definition of the hydrological system. For purposes of discussion, probes used to obtain these physical parameters are classified into three groups of varying radii of investigation.

View of the Borehole Wall

The depth of borehole-wall investigations ranges from the few angstroms obtained in an optical view of the surface to the few millimetres of penetration achieved by acoustic televiewer probes.The goal of these studies is to obtain a representation of the borehole surface adequate to depict rugosity and the planar features intersected by the borehole such as fractures, dikes, contacts, structural discontinuities, and bedding planes.

A view of the borehole wall is of little interest in homogeneous, porous media, but in fractured media it has great utility. Nonlogging methods used to acquire information on fractures include coring and the impression packer, both limited in practice because of cost and recoverability in the case of core, and because of cost, resolution, and reliability for packer methods. Logging methods that have been developed to obtain the "borehole view" include the following:

Optical methods. These use a wide range of camera types and, more recently, the videorecorder. Here equipment cost, operating time, and data reduction present formidable barriers to widespread use. Can recording formats be improved and standardized? Can analysis be automated in an economical fashion? Will fiber optics or laser imaging be useful in hostile borehole environments? Optical methods warrant continued development because of their inherent resolution and the relative ease of interpreting a visual image. Obviously, optical methods are restricted by opaque borehole fluids.

Mechanical methods. These employ the unsung workhorse of borehole instruments, the caliper, which comes in a variety of designs. There are 1-, 3-, 4-, 6-, or 40-arm versions available for various hole sizes and resolutions. For fracture detection in hard rock, the mechanical caliper appears to warrant improvement. There is no reason why we cannot have a borehole micrometer with many transducers around the borehole circumference supplying multiple measurements of diametric fluctuations of a fraction of a millimetre. Such multiple transducer tools have already been developed for downhole engineering applications. Nonmechanical methods of calipering, such as optical and acoustic, offer the potential for great improvement in both vertical and diametrical resolution.

Acoustic imaging methods. The borehole televiewer is continuing to receive the development effort it deserves for hydrologic investigations in fractured rock (Keys, 1978). It is an important tool because of its good resolution and useful output format on film and because it can be run in mud-filled boreholes. Besides providing the view of the borewall, the sensor output can be modified to produce a four-component, high-resolution acoustic-caliper log (Keys, 1979). Given favorable operating conditions, the results can be quite impressive. (Figure 1 shows examples of televiewer logs.) On the other hand, under adverse conditions as reported by Smith and others (1978), the image can be difficult or impossible to interpret.

Electrical methods. The four-arm dipmeter is a sophisticated package of (1) micro-focused conductivity pads applied against the borehole wall to sense electrical conductivity on the scale of centimetres; (2) borehole-orientation-and inclination-sensing devices; and (3) telemetry. With the application of computer processing, the system is quite useful in establishing the true strike and dip of sedimentary features in sedimentary formations. Beck and others (1977) have given examples of dipmeter applications in fractured carbonates, one of which is shown in Figure 2. Conductivity increases whenever one of the four pads crosses a fracture zone. All curves deflect equally at the same depth where a horizontal fracture is detected, as shown near the bottom of the figure. When one or two pads contact a vertical fracture the overlaid curves separate, as indicated by the hatched zones. In this example, hole size is quite uniform and the tool orientation is stable, as indicated by the caliper and azimuth traces of the log. This and other examples demonstrate the effectiveness of the dipmeter in

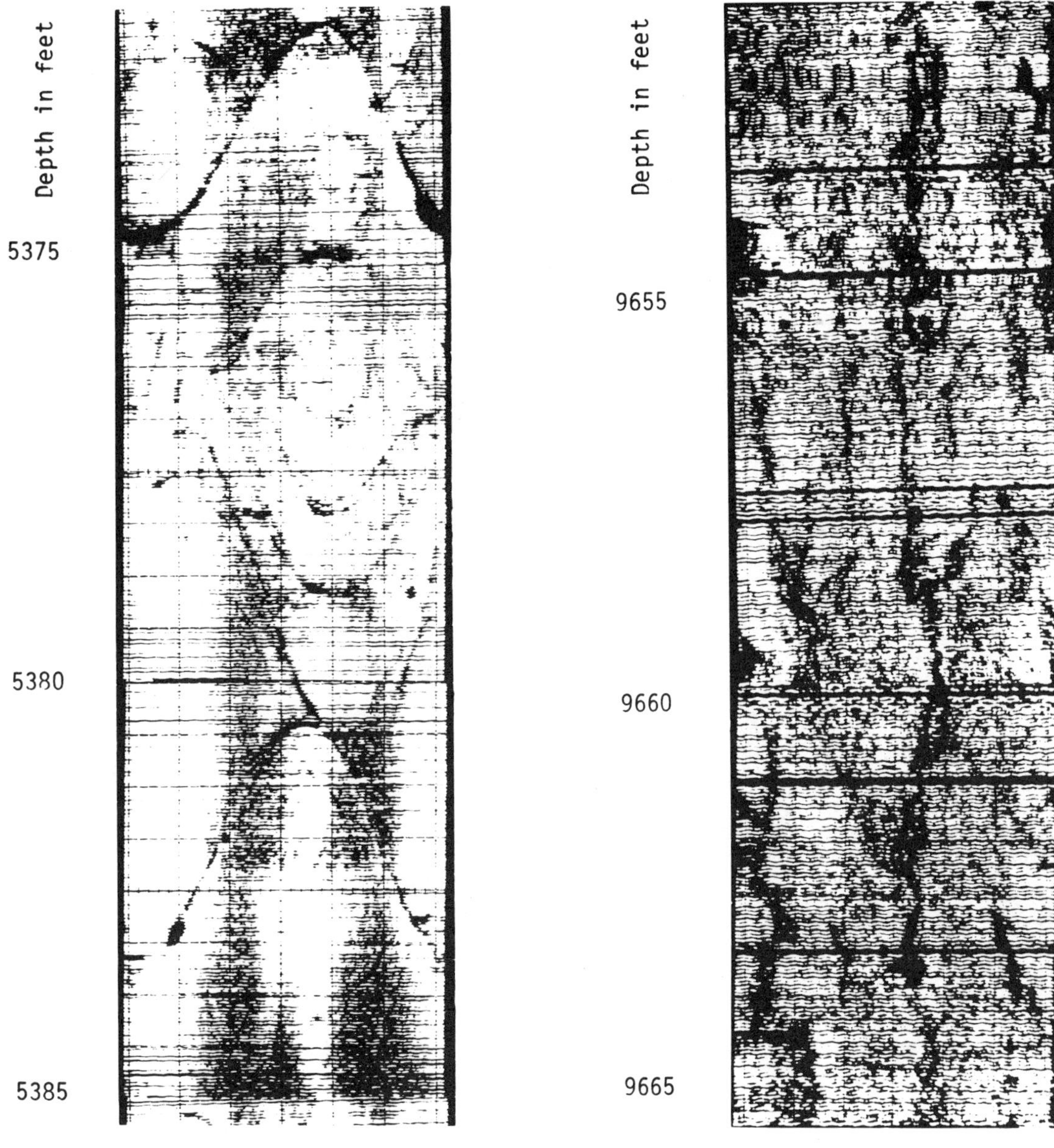

Figure 1. Acoustic televiewer log illustrating the modes of fracturing. Horizontal scale is greatly exaggerated with respect to the vertical. The sinusoidal signatures on Log A are caused by fractures intersecting the wellbore at fairly high angle. Log B shows vertical fractures induced by hydraulic fracturing. (Source: Keys, 1976)

delineating fractures and planar features. Other sensors, such as electromagnetic induction and infrared, might also be miniaturized and deployed in the dipmeter configuration.

Bulk Properties within Metres of the Borehole

To measure bulk properties of the rock from a few centimetres to a few metres distant from the borehole, we use the now-conventional "open hole" logging devices such as natural gamma, neutron, gamma-gamma, acoustic velocity, and electrical tools. Most of these have been developed and refined by service companies within the petroleum field, and the literature on their applications is extensive, almost overwhelming. The previously cited monograph on borehole geophysics by Keys and MacCary (1971) deals largely with this category of borehole measurements. Hence these tools, designed to measure porosity and the type of saturating fluid and to provide information on lithology and bulk properties, are the most mature and most

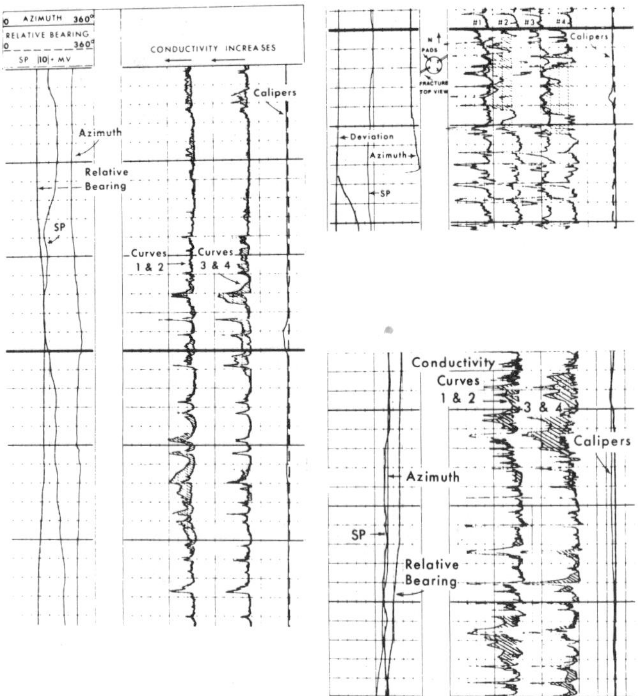

Figure 2. Four electrical conductivity curves in fractured carbonates recorded on a dipmeter sonde. The individual conductivity curves are measured on small arrays on pads pressed against the borewall spaced 90 degrees aport. The traces on the left column give the orientation of the first pad. (Source: Beck and others, 1977).

widely applied of the several categories discussed in this paper. Since they are now so routinely applied and so well described in the literature, further discussion of the common porosity tools is omitted here. Bear in mind, however, that most of the radioactive tools can operate in the presence of a well screen or casing regardless of the type of fluid that fills the hole.

One new tool of potential interest to hydrologists is an electromagnetic propagation tool (Calvert and others, 1977), which provides a direct measurement of water content based on the dependence of high-frequency (1.1 GHz) propagation velocities upon the dielectric constant of water. The tool is configured in a geometry similar to acoustic velocity probes (see Fig. 3), measuring the transit times and attenuation of continuous-wave signals between a pair of receivers. An example of the resulting log is shown in Figure 4. The reciprocal of the velocity, travel time, is plotted with values increasing to the left in the center track. Like the dipmeter, the electromagnetic propagation tool differs from conventional open-hole devices in its higher resolution, resulting in a record with a spiky character. According to the theory, the measurement reflects the total water present in the rock. Corrections for water temperature and matrix effects are required. Freedman and Vogiatzis (1979) discussed the method further, computing the correction factors required to account for the mud cake and for finite antennas.

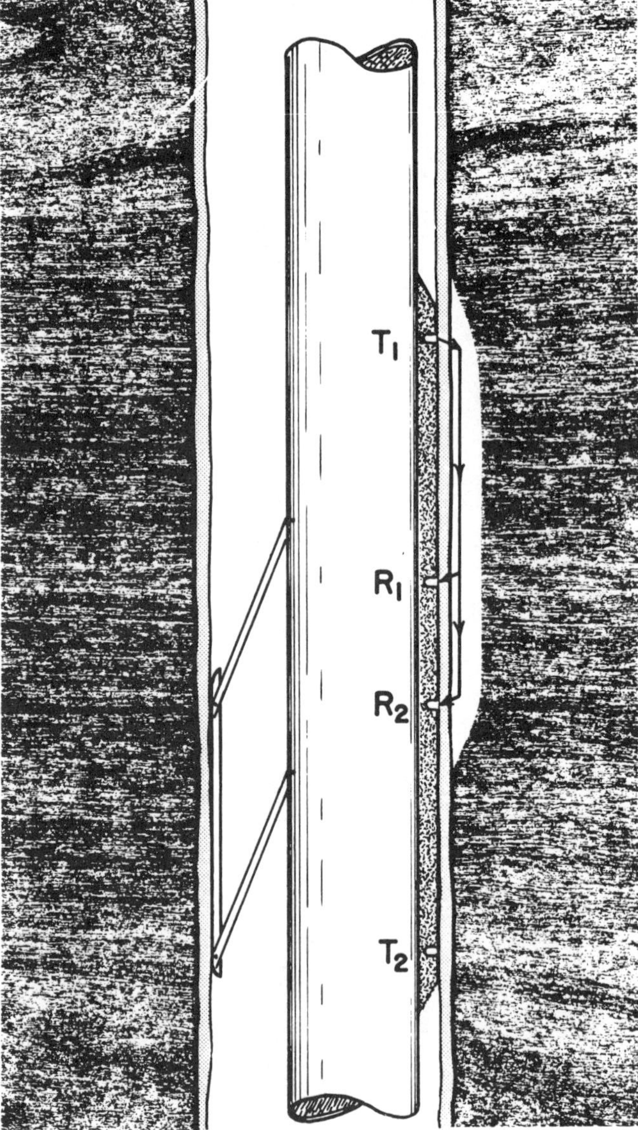

Figure 3. Schematic of electromagnetic propagation tool. Travel time of high-frequency pulse is measured between receivers R_1 and R_2. Transmitters T_1 and T_2 are fired sequentially. (After Calvert and others, 1977).

The electromagnetic propagation method has potential for application to other hydrologic problems. In conjunction with other porosity tools, such as the gamma-gamma, it should be straightforward to compute water saturation above the water table. Unless the water is highly saline, the changes of conductivity and dielectric permittivity with salinity cancel each other, so the tool is relatively independent of water quality. A potential application of this tool is the detection of fractures in rocks of low porosity, as the vertical resolution is quite good (receiver spacing is 4 cm) and can be improved even further.

Another electrical method, seldom used outside the mineral exploration industry, is the induced-polarization or complex-resistivity method.

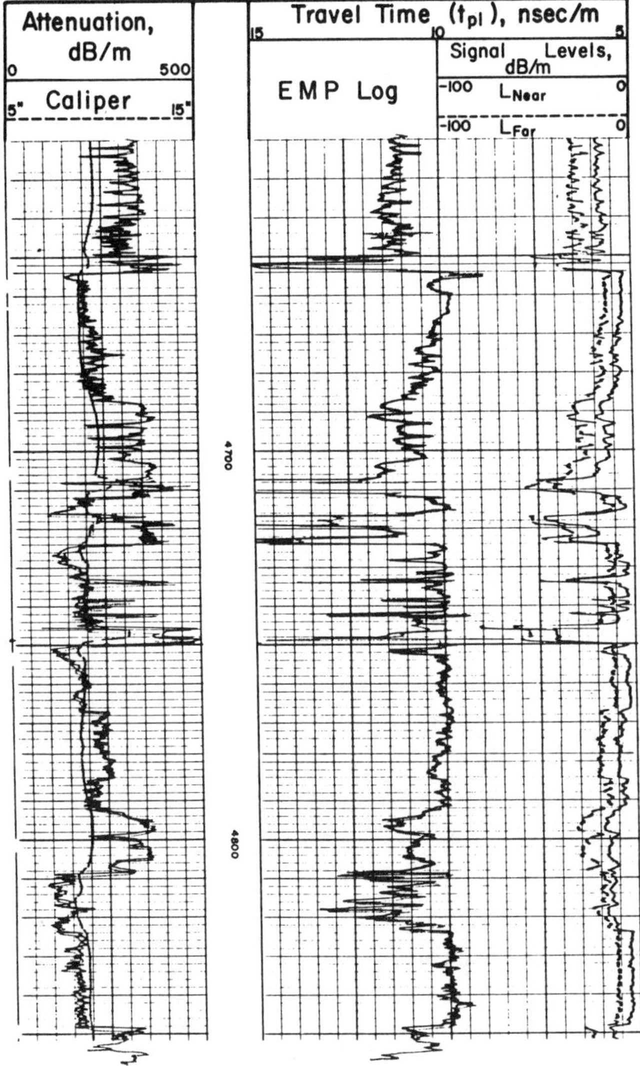

Figure 4. Example of electromagnetic pulse log. Travel time (center track) increases to the left as water content increases. (After Calvert and others, 1977).

These terms refer to the measurement of the out-of-phase (quadrature) component of the conventional resistivity. Snyder and others (1977) reviewed the literature in the field and the applications to the borehole environment. The literature suggests that the out-of-phase component responds proportionally to the clay content in sediments, as reflected by the cation-exchange capacity of core samples. The metallic mineral content also causes an induced-polarization response, and for this reason the method is widely used in prospecting for sulfide mineralization. Hence another application here would occur when sulfide minerals can be used for lithologic correlation among several boreholes.

Rock Properties between Boreholes
Separated by as much as 100 M

Measurement tools in this category are the least commonly applied, the most poorly developed, and the most expensive. On the other hand, cross-hole methods probably hold the greatest potential for the development of useful new techniques. Currently the impetus to develop cross-hole tools comes from civil engineers, who need them for site investigations for dams and nuclear power plants. By and large, the approach is to work with acoustic probes in two coplanar boreholes, although some investigators are also applying electromagnetic propagation methods. Except for very specific applications, the following four ingredients are expected to be crucial to the continued development of viable cross-hole methods:

1. Accurate surveys of borehole locations. These are necessary because the conversion of our measurement (transit time) to an interpreted parameter (velocity variations) requires knowledge of the inter-hole distance.

2. A high redundancy of spatial data points, to be gathered either by borehole arrays or by sequential logs. A transmitter in one borehole is held at one elevation while the second borehole is logged; then the transmitter is moved to a new elevation and the sequence is repeated. These procedures are necessary to adequately sample the rock mass and are illustrated schematically in Figure 5.

3. Redundancy in physical parameters. We will not be content in sampling just a single parameter such as P-wave arrival time in cross-hole applications, because interpretation is aided

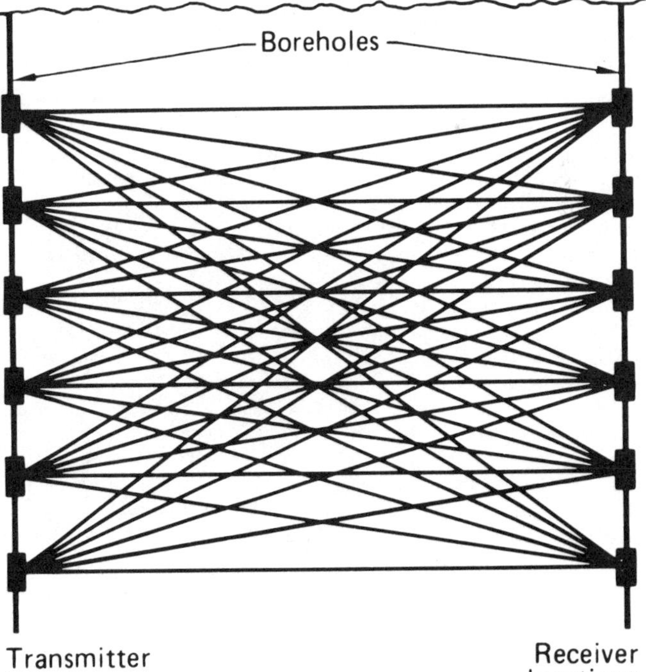

Figure 5. Schematic illustrating how cross-hole probing techniques sample subsurface volume when multiple positions are occupied by transmitter and receiver probes. The area of greatest coverage is in the central portions, with lesser coverage around the periphery. (After Lytle and others, 1978).

tremendously when the second and third parameters are supplied. Examples are the use of shear-wave data to supplement compressional-wave data, electromagnetic attenuation to supplement velocity data, and the use of wide-band equipment to examine changes in attenuation and velocity with varying frequency.

4. Data processing. Obviously, the redundancy of physical spatial sampling requires a great deal of processing to reduce the data to an interpretable format. Fortunately, with modern electronics both the developmental and the production systems are feasible. For example, Aron and others (1978) described a four-receiver system that used all four signals simultaneously in the correlation process to determine shear-wave arrivals. Although the processing methods are being applied to an in-hole probe, similar techniques could be readily applied to cross-hole data. The lessons of seismic-reflection profiling for petroleum are readily available to us for application in both data analysis and physical properties. Inevitably we are headed toward complex systems when we address the problem of measuring properties between holes. With good understanding and design, these complexities can be incorporated into the processing, with resulting simplicity for the user.

To demonstrate these four points, Figure 6 shows the results of a cross-borehole electromagnetic survey performed by Lytle and others (1978). Their technique is to propagate continuous-wave, high-frequency (50 MHz) electromagnetic energy between stations in adjacent boreholes, using the geometry and high sampling density illustrated in Figure 5. Received power and phase-shift measurements are converted to cross-sectional representation of attenuation and velocity in the rock mass between the two holes. The conversion is accomplished with iterative algorithms now widely used in medical tomographic techniques where ray-path optics are assumed.

Figure 6 shows the attenuation measured between

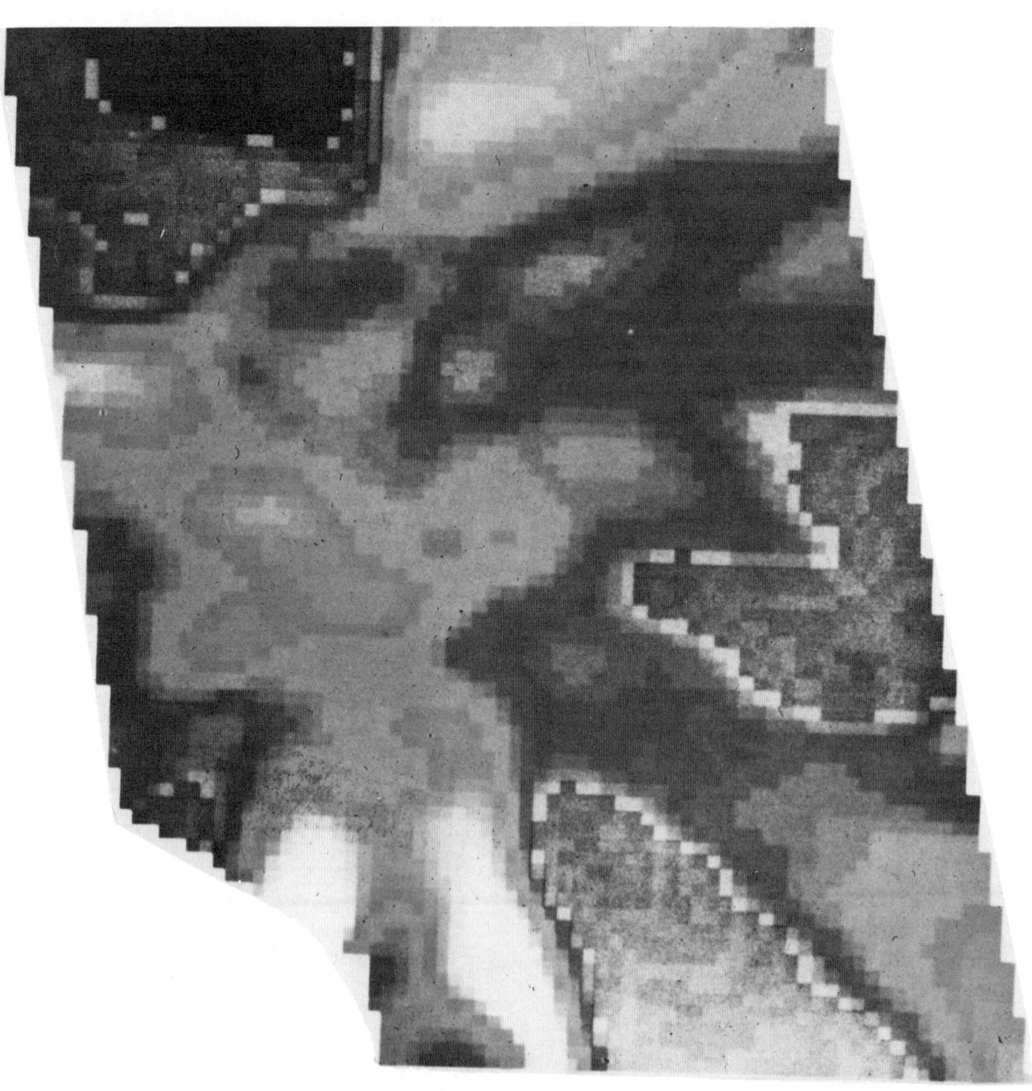

Figure 6. Plot of electromagnetic attenuation between boreholes. Two of the four slant boreholes bound the figure at the left and right sides, two others lie within the data plane. (After Lytle and others, 1978).

four coplanar boreholes. The original figure presented the attenuation data on a color scale, thereby presenting a more detailed picture than is shown in the black-and-white reproduction in Figure 6. The high attenuation zones reportedly correlate well with high fracture density observed in core.

This example touches on the four points made earlier. First, accurate position is necessary, or distortion results, as mentioned by Lytle and others. Second, some 3,000 data points were taken, providing data on overlapping ray paths. Third, both attenuation and phase shift were recorded. In this case the inverted attenuation contained the useful information in terms of fracturing. Fourth, the data collection, processing, and graphic display procedures obviously require extensive computer processing.

FLUID PROPERTIES AND MOVEMENT

Evidence of fluid movement and indicators of water quality are often sought in logging programs. Because the total dissolved solids are linearly related to the specific conductance of water, this quantity or its reciprocal fluid resistivity is often sought in borehole work. If the water in the borehole column is representative of the water in rock (uncontaminated by drilling fluid), then a specific fluid-conductivity probe can be used. Another approach is to infer this value for the water in an aquifer with the use of resistivity logs or self-potential measurements. In either case, compensation for temperature effects is necessary before converting to a water-quality parameter such as total dissolved solids.

Fluid movement is a critical measurement with many practical ramifications. For example, a flow profile or injectivity profile of a borehole is quite handy when planning or interpreting more detailed hydrological tests aimed at defining permeability.

The simplest but crudest log for estimating fluid flow is a temperature log. Even if it is simply run after drilling, useful information can be obtained by observing thermal anomalies induced in the rock by loss of drilling fluid in permeable zones. With more control on the temperature and the quantity of water injected, and if a number of temperature logs are run sequentially after injection, even more detail on the flow regime can be inferred. A new application of temperature logs is to detect fluid flow in channels behind casing (Cooke, 1978) by measuring the temperature of the casing from a rotating temperature sensor pressed against it. The casing temperature is compared with the fluid temperature and presented as a difference at high resolution. Small cooling or heating anomalies are then indicative of fluid movement behind the casing. This application illustrates the extreme sensitivity of temperature measurements, as significant perturbations are of the level of 0.005 °C.

Detailed modeling of the thermal history is required to relate the temperature information to quantitative flow estimates, however, and a more practical route is to measure the flow profile directly with borehole flow-measurement probes.

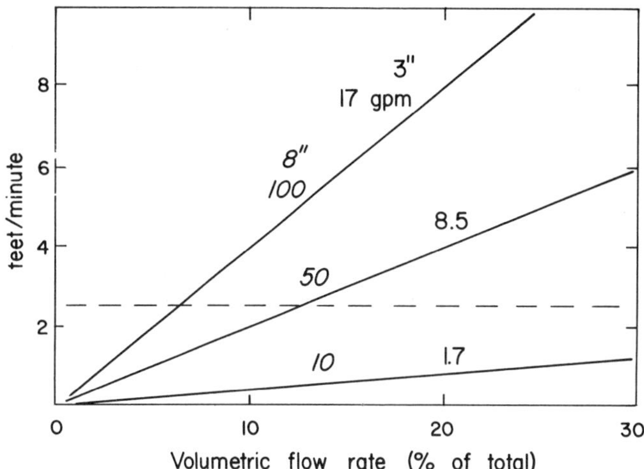

Figure 7. Required resolution of a flow meter to detect 10% and 20% changes of total flow in 76-mm (3-in.) and 203-mm (8-in.) diameter boreholes. For example a probe with a resolution of 20 mm/s (3 ft/min) operating in a 203-mm (8-in.) borehole, flowing at 5 x 10^{-3} m^3/s (80 gpm) can resolve losses of about 13% of total flow or about 0.6 x 10^{-3} m^3/s (8 gpm). The horizontal dashed line at 13 mm/s (2.5 gpm) represents the resolution of a continuous spinner flowmeter under good operating conditions.

Two types of probes are readily available on a commercial service basis: the radioactive trace ejector and the borehole spinner. The spinner converts flow rate to electric pulses by sensing the rotation of an impeller exposed to the water column. Its lower limit of resolution is on the order of 1 m/min as indicated in Figure 7. What does this mean in terms of flow rate? This must be answered in terms of the acceptable resolution for a given total flow rate in a specific hole size. If it is known that one major flow zone controls the loss of fluid during injection, then a resolution of 50% of total flow may be acceptable if the concern is simply to locate the single zone. However, usually the concern will be to resolve as much of the fluid loss as possible. Figure 7 shows the minimum detectable flow rates required of a flowmeter if 10% or 20% resolution of total flow is required. The change of horizontal scale given for the 8-in and 3-in boreholes reflects the dependence of the average flow velocity upon the inverse square of the diameter so that, for a given velocity resolution, much lower volumetric flows can be detected in a small-diameter bore than in a large one. In many hydrologic applications, more resolution is desirable than is currently available, both to extend the method to lower total flow rates and to enhance the resolution at moderate flow rates. The spatial resolution of zones of fluid loss depends directly upon the flow resolution.

A spinner flow tool has advantages over the tracer method in that continuous data are obtained and the data are generally more easily acquired and simpler to reduce. But a spinner is only one of numerous types of flowmeters that might be configured to the borehole measurement. Various flow sensors for process control have been developed and applied to metering flow in pipes. Such sensors are well described in the literature (Dondell and Wendt, 1974). Adapting these sensors to the borehole involves "turning the geometry inside-out" and meeting the requirements of temperature, pressure, and ruggedness. With enough incentive, it would seem that a significant improvement over the present-day spinner can be achieved with continuous metering--that is, without resorting to tagging methods. An advantage of the borehole configuration is that the measurement can be made with the tool in motion, resulting in a high velocity of fluid relative to the sensing element. Hence high resolution need not be achieved at minimum flow, although other problems are introduced by movement of the sensor.

The radioactive tracer method belongs to the class of tagging, or discrete, methods in which the time of travel of a tagged volume of water is measured. With the tracer technique, the passage of radioactivity-tagged water past a pair of gamma-ray detectors is timed. Knowing the travel time and distance, and correcting for the borehole diameter, the volumetric flow rate can be estimated. With a number of measurements at different depths, a flow profile of the well is produced. The examples shown in Figure 8 were run at different times and at different flow conditions, demarcating three principal flow zones in the well. Note the corresponding temperature anomalies.

The tagging geometry can also be applied by ejecting a saline slug from the tool into the column and timing its passage past a pair of fluid-conductivity cells. A third tagging method is by heating a portion of the water in the flow and observing the temperature change downstream. A heat-pulse flowmeter has been developed (Dudgeon and others, 1975) that achieves a stated resolution of 1 mm/s (0.2 ft/min) in a 38-mm-diameter probe.

Quite recently, neutron-activation methods have been employed in tagging techniques for flow measurement. The advantage with activation methods is that the tagged element is manufactured in situ rather than introduced, so there is no contrast in physical or chemical properties between the tagged and the untagged fluids. The nuclear magnetic-resonance flowmeter can be considered a member of this class, although it has not been adapted to boreholes. Neutron activation of isotopes in the water is another method, and two recent applications to the borehole environment may be cited. Scott and others (1978) have developed a single-hole method to measure horizontal flow velocity in a reservoir. The method requires saline formation water, as it relies upon activating ^{24}Na (half-life of 15 hr) and observing the gamma-ray activity produced during its decay. The authors cited field cases where horizontal velocities of 0.1 to 1.5 m/d were measured. A second neutron-activation method is presented by Arnold and Paap (1979), activating ^{16}O nuclei in the water to ^{16}N (half-life of 7 s) and observing the resulting gamma-ray flux at detectors downstream from the neutron source. The technique has been successfully applied to the quantitative estimation of flow rate in channels behind casing.

ENGINEERING NEEDS

In the two preceding sections I have concentrated on physical measurements of rock and fluid in open boreholes. I assumed that the in-situ features were of prime interest; that is, the purpose of logging is to obtain information on the natural state of the geologic or hydrologic system. However, in many cases the borehole has been purposely modified with engineered structures, and logs are run either to inspect those structures or to define what changes have occurred in the natural system as a result of installing and operating the downhole equipment. The latter category is referred to here as logging for engineering needs. In the petroleum service industry, the distinction is made clear by referring to the former as open-hole logging and the latter as cased-hole or production logging.

Operational problems and their cures are so commonplace that they seldom receive attention in the literature. Nevertheless, there is a broad class of problems where quick and effective

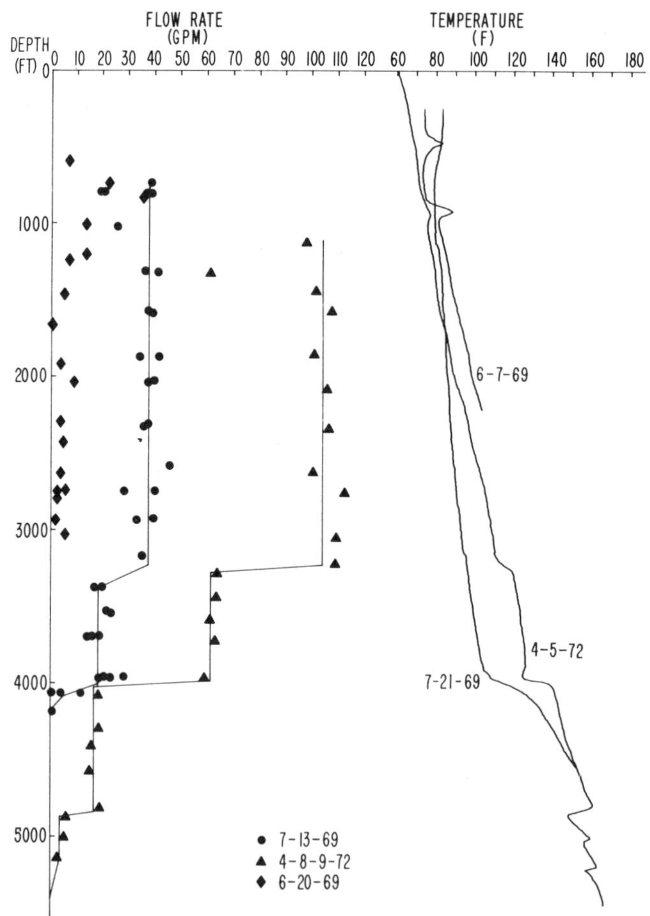

Figure 8. Temperature logs and flow profiles based on radioactive tracer logs. Symbols represent individual measurements of velocity. (After Apps and others, 1979.)

diagnostic action has direct economic impact. These problems usually do not require a "high technology" solution, although improvements in measurement technology often have immediate application to prosaic problems. Typically, borehole probes are required to diagnose malfunctions of downhole equipment, locate cemented zones, locate the free point of stuck pipe, determine the quality of cement bonds, or measure the depth to borehole washouts. Other requirements include locating the zones of fluid loss or entry, determining leaks behind casing, evaluating changes in reservoir conditions, guiding hydraulic fracturing or well-stimulation operations, or simply determining the depth to water.

Usually, the diagnosis of borehole-engineering problems requires no more than the interpretation of a few well-chosen logs. Oliver (1977) described three cases where various combinations of temperature, electrical, caliper, gamma-ray, casing-collar-locator, flowmeter, and television logs were used to diagnose problems that had developed in water-supply boreholes. One of these cases occurred at a pumping station built before 1900 in a fissured limestone. The water yield from a submersible pump in a large-diameter shaft fell

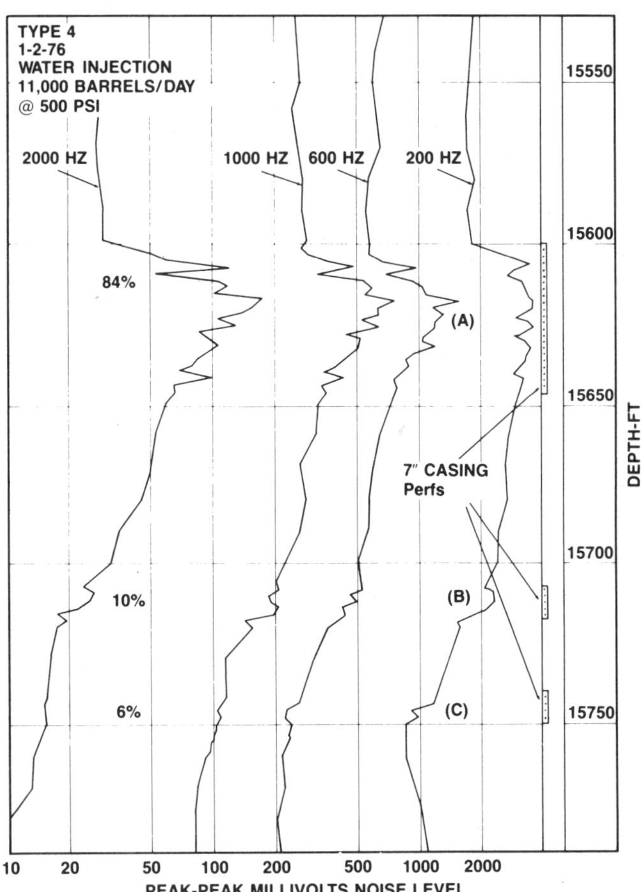

Figure 9. Noise log with various high-pass filter settings. Logs were recorded with the well under injection at 11,000 bbl/d of water. Three sets of perforations in the casing are shown. (After Robinson, 1976.)

suddenly and the pump began to deliver large quantities of air. Logs from an adjacent borehole, where surging noises were audible, showed that water was entering the boreholes at a defective portion of the casing where the diameter was reduced. The water entering the borehole at this point then became aerated during a 7-m cascade to the water surface. The temperature, caliper, flowmeter, and television logs were particularly useful in diagnosing the problem.

A specialty tool often useful in detecting leaks under flow conditions is the noise meter, which consists of an acoustic transducer such as a piezoelectric crystal sensitive to acoustic energy in the 200- to 2,000-Hz frequency range. The tool simply records the amplitude and frequency character of acoustic noise generated by fluid movement in the wellbore. The device is especially useful for determining gas flow through constrictions under high differential pressures, but can also be used to detect anomalous water flow. Robinson (1976) described several field examples, including the one reproduced in Figure 9, where three sets of perforations through the casing are shown. According to Robinson, the noise log was run to determine the quantity of water going into each set of perforations when water is being injected into the well, and also to get an indication of flow behind the casing between the perforated zones. The four traces correspond to different high-pass filter settings, the 1,000-Hz trace recording noise at frequencies above 1,000 Hz. Qualitatively, it is clear that most of the flow in the well occurs in the uppermost perforated zone. Quantitative estimates of the relative losses were obtained from the 1,000-Hz data, with a consideration of the number of perforations in each zone (Fig. 9). Additionally, the noise levels immediately below the uppermost perforations were greater than the noise levels above the perforations, which is interpreted as an indication of flow behind the casing for some distance below the perforated zone.

DATA INTERPRETATION

Most log analysts would underscore this recommendation: the more logs the better. This is especially true for those who are looking at new areas, new problems, or an entire hydrologic system using data from a number of wells. Keys and MacCary (1971 p. 15) state it more clearly:

> As a general rule, the more types of geophysical logs that are available for a single well and the more wells that are logged within a given geohydrologic environment, the greater the benefits that can be expected from logging. The synergistic character of logs is due to the fact that each type of log actually measures a different parameter, and when several are analyzed together, each will tend to support or contradict conclusions drawn from the others.

The corollary to this deserves mention: the time and expertise required to examine the logs, with access to analytical determinations of physical properties of core or cuttings, should be allocated as necessary.

Analysis of calibrated well logs in sedimentary sequences has become increasingly standardized and automated over the past decade, a trend accelerated by digital data acquisition in the field. Graphic techniques for determining lithology and geologic trends are well described in the literature. The technique of plotting one tool response against a second can be used to infer the character of porosity, to spot tool malfunctions and miscalibrations, and to categorize lithology. MacCary (1978) reviewed these analytical approaches, then applied them to two holes cored in a carbonate aquifer in Texas. Detailed information on lithology and porosity is the chief outcomes of such analysis, and on a spot basis can be performed quickly by hand to examine trends in unfamiliar environments. Where the lithologies can be handled by routine analysis, commercial services provide continuous computed stratigraphic analyses presented in a log-type format.

However, interpretation schemes are far less developed for igneous and metamorphic rock sequences than for sediments, although the former are now receiving more attention under the impetus of geothermal development (Glenn and Hulen, 1979; Keys, 1979; Sanyal and others, 1979). The various techniques employed by different authors include histograms and cross-plots to distinguish different lithologies, direct comparison of logs with core, and the development of new plotting techniques to account for the heavy mineral content. Sanyal and others (1978) have presented a discussion of the problems and possible avenues of approach in geothermal exploration environments. High temperature is still a barrier to the routine acquisition of data in geothermal areas, but even if all field problems were resolved, interpretation schemes must be modified. The different lithologies require that the tool response be examined carefully, as the neutron tool, for example, responds in unexpected ways in the presence of elements with high-absorption cross sections. Geothermal reservoirs are characteristically fracture controlled, whereas the permeability of petroleum reservoirs is usually controlled by the intergranular porosity. And finally, objectives of geothermal exploration, such as definition of high formation temperature and rock thermal properties, are new and are not yet routinely treated.

Borehole logging is being applied to the problems associated with the disposal and storage of radioactive waste--both low-level waste, which is currently being stored near the surface, and high-level waste, for which deep underground repositories are planned. In both situations, good understanding of the hydrologic regime is critical; for the near surface sites this requires the interpretation of neutron, gamma-gamma, gamma-ray, and other logs to characterize the porosity and permeability in soils and poorly consolidated sediments.

Another useful technique is to acquire gamma-ray spectra in the borehole to examine the migration of radioactive species away from the storage site. Keys and others (1979) have presented borehole spectra obtained from shallow-land burial sites, along with a table of the more prominent gamma-emitting isotopes likely to be present in waste material. Borehole gamma spectrometry has several advantages over water sampling. The measurement is representative of the species present in a particular lithologic unit rather than of just the unit from which a sample can be obtained. Also, the borehole spectrometer can operate in an encased hole.

Salt domes and bedded salt layers are prime geological candidates for deep disposal of radioactive waste. In these formations, the location of boundaries, pre-existing cavities, and boreholes, and the location of water-bearing zones will be of importance in site exploration and design. Unterberger (1978) has described the application of radar reflection methods that make use of antenna arrays mounted on small, wheeled vehicles that can be moved about in underground salt mines. Reflection data are interpreted as geologic features having a spatial resolution dependent on the frequency of the reflected wave and the constancy of the velocity of propagation.

In hard rock, also a potential medium for underground waste storage, the permeability is controlled by fractures. The definition of fractures and the ability to discriminate among different fracture types and fillings will require new approaches and an increased equipment sensitivity to thin fractures. My co-workers and I (Nelson and others,1979) have described the preliminary results of a logging suite obtained at a granitic site in Sweden. A portion of the suite is shown in Figure 10. The goal of this work was to characterize the rock permeability before performing detailed hydrologic testing. We selected zones where, on the basis of temperature, neutron, gamma-ray, sonic, and electrical responses, we expected the permeability to be anomalous. Of these techniques, the sonic travel time and sonic waveform were the most useful.

NEW APPROACHES AND UNSOLVED PROBLEMS

A primary use of well logs is to correlate depositional or formational characteristics between boreholes or even across a sedimentary basin or an intrusive complex. Thorough regional studies are a multidisciplinary effort in which logs are a useful, often vital, component. The first step beyond visual examination of the logs involves machine correlations of log features with allowance for the thinning, thickening, or complete disappearance of geologic features. A further level of refinement was explored by Vincent and others (1979) in applying pattern-recognition techniques to dipmeter logs. In their work, a log was first decomposed into individual features or elements such as peaks, troughs, and spikes. Depth bounds and a set of parameters describing the feature were assigned to each element. Correlation among logs was based on a comparison of the parameter sets for individual features. Correlations were adopted successively on multiple passes rather than sequentially with depth. This approach allowed them to correlate first those features which shared a higher level of similarity, thereby constraining the correlation of lower-level

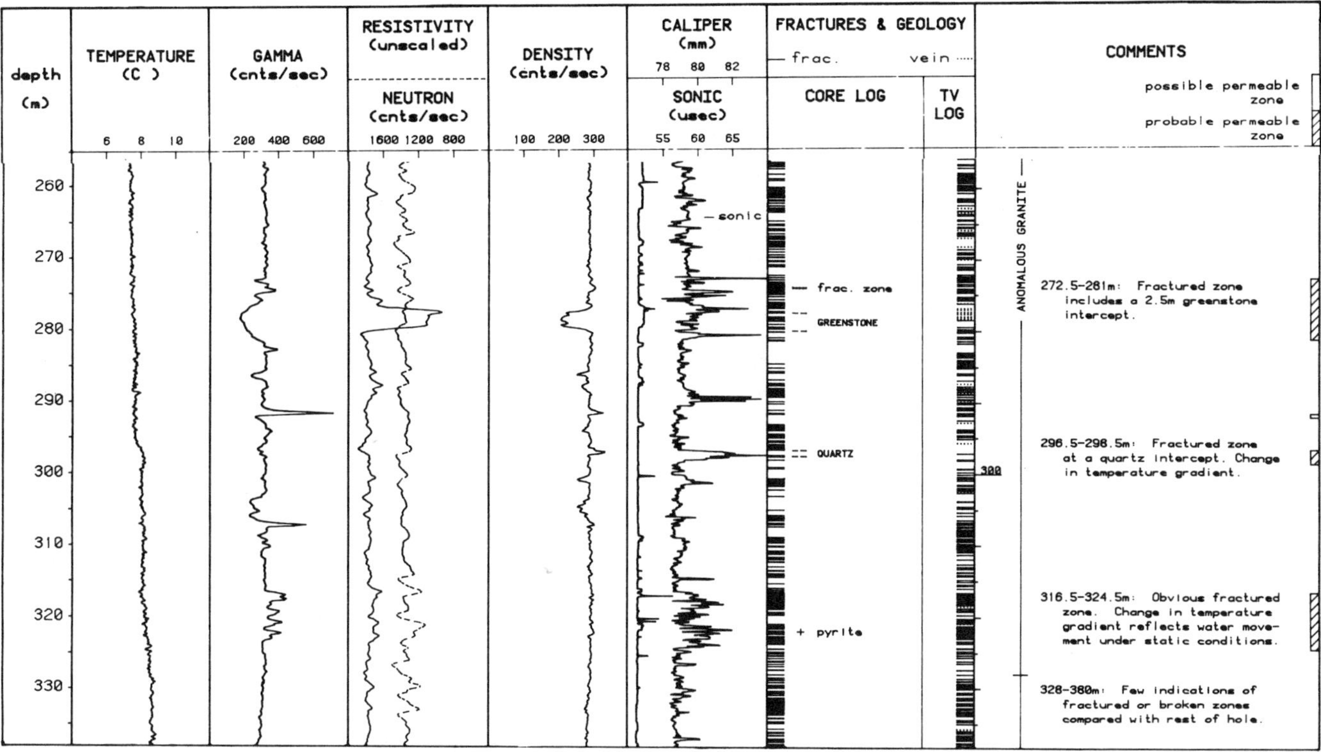

Figure 10. Suite of logs from an experimental test site for radioactive waste storage in granite (Nelson and others, 1979).

features by disallowing correlations that crossed over one already established. The method appears to approximate more closely the approach that can be carried out by a human analyst than do conventional machine procedures. The authors note the potential extension of the method for well-to-well correlations.

Investigations in igneous and metamorphic terranes can benefit from several methods not widely employed in sedimentary studies. Results from the magnetic-susceptibility and induced-polarization probes in base-metals exploration have been discussed by Glenn and Nelson (1979). The susceptibility log is dominated by the presence of magnetite, a common accessory mineral controlled by both lithology and hydrothermal alteration patterns. The induced polarization, already discussed in a previous section, is controlled by sulfide mineralization, clay minerals, and zeolites. In special situations, highly anomalous polarization responses can be produced by other geologic materials such as disseminated carbon.

A third method gaining acceptance in sedimentary studies is use of the differential (or spectral gamma-ray) tool, which identifies and quantifies the abundance of the three naturally occurring radiogenic elements--potassium, uranium, and thorium. Hassan and others (1976) reviewed the mineralogical associations of these three elements and the implications of their presence in various sedimentary settings. West and Laughlin (1976) evaluated a spectral gamma-ray log obtained in crystalline basement rocks.

As already mentioned, progress has been made in delineating natural fractures with the use of acoustical and electrical techniques. Further improvements can be expected soon. But some studies are ignoring the basic nature of fractures--the fact that they are planar, or nearly so. Attempts to extract fracture information from combinations of existing logs by using empirical techniques may be successful in specific areas or rock types, but it is doubtful that this approach will result in a generally applicable method. Instead, we must take advantage of the physical features of fractures to come up with a generally valid methodology. We are dealing with features that are planar on a local scale (metres) and that are partially or completely filled with material different from the matrix.

Reliable definition of permeability continues to be an elusive goal. As is the case for fracturing, the indirect approach to determining permeability will not be generally applicable. This means that empirical relationships between the porosity and permeability or, say, between the irreducible water saturation and permeability, may meet with success in specific rock types or geologic settings, but that appreciable reworking of empirical data will be required to apply the method in a different setting. Hence it seems better to work on methods that will directly measure the quantity needed, which is the rate of movement through rock under a known pressure gradient. Successful attempts have been made along these lines with such techniques as the nuclear magnetic resonance method, which establishes a "free fluid index" estimate of the water not bound to surfaces.

However, we need to measure the water present in the interconnected pore space and the properties of that interconnected pore space. The best practical approach will be to combine logging measurements with injection programs. The simplest approach here is to use the flow profile, discussed in a previous section; certainly, improved flow-logging tools warrant attention from this standpoint. We might also consider other phenomena that could be measured during flow. One of these is the streaming potential generated by the flow of water in porous media. Haines (1978) presented an application of streaming potential to map the loss of water from a reservoir by using repeated measurements at different heads (reservoir level). In another field of endeavor, the attenuation of acoustic energy at ultrasonic frequencies was related to pore-fluid-attenuation mechanisms (Winkler and Nur, 1978). One model ascribed the attenuation to fluid movement between cracks that occurs as an acoustic wave passes. If so, there is promise that a borehole measurement of ultrasonic wave attenuation might be linked to the permeability.

ACKNOWLEDGMENT

This work was supported by the Assistant Secretary for Nuclear Energy, Office of Waste Isolation of the U.S. Department of Energy under Contract No. W-7405-ENG-48. Funding for this project is administered by the Office of Nuclear Waste Isolation at Battelle Memorial Institute.

REFERENCES CITED

Apps, J. A., and others, 1979, Geohydrological studies for nuclear waste isolation at the Hanford Reservation: Berkeley, Lawrence Berkeley Laboratory, LBL-8764, v. II.

Arnold, D. M., and Paap, H. J., 1979, Quantitative monitoring of water flow behind and in wellbore casing: Journal of Petroleum Technology, v. 31, no 1.

Aron, J., Murray, J., and Seeman, B., 1978, Formation compressional and shear interval-transit-time logging by means of long spacings and digital techniques: Society of Petroleum Engineers Paper 7446.

Beck, J., Schultz, A., and Fitzgerald, D., 1977, Reservoir evaluation of fractured Cretaceous carbonates in south Texas: Society of Professional Well Log Analysts 18th Annual Symposium Transactions, p. M1-M25.

Calvert, T. J., Rau, R. N., and Wells, L. E., 1977, Electromagnetic propagation: A new dimension in logging: Society of Petroleum Engineers Paper 6542.

Cooke, C. E., Jr., 1978, Radial differential temperature (RDT) logging--A new tool for detecting and treating flow behind casing: Society of Petroleum Engineers Paper 7558.

Dondell, R. B., and Wendt, R. E., Jr., ed., 1974, Flow measuring devices, in Flow--Its measurement and control in science and industry: Pittsburgh, Instrument Society of America, v. 1, part 2.

Dudgeon, C. R., Greer, M. J., and Smedmor, W. J., 1975, Heat-pulse flowmeter for boreholes: Marlow, Bucks, Great Britain, Water Research Center, Medmenham Laboratory Technical Report TR4.

Dycke, J. H., Keys, W. S., and Meneley, W. A., 1972, Application of geophysical logging to ground-water studies in southern Saskatchewan: Canadian Journal of Earth Sciences, v. 9, no. 1.

Freedman, R., and Vogiatzis, J. P., 1979, Theory of microwave dielectric constant logging using the electromagnetic propagation method: Geophysics, v. 44, no. 5, p. 969-986.

Glenn, W. E., and Hulen, J. B., 1979, A study of well logs from Roosevelt Hot Springs KGRA, Utah, in Society of Professional Well Log Analysts 20th Annual Logging Symposium Transactions: p. ZZ-1.

Glenn, W. E., and Nelson, P. H., 1979, Borehole logging techniques applied to base metal ore deposits, in Proceedings, Exploration 77 Symposium: Ottawa, Geological Survey of Canada.

Haines, B. M., 1978, The detection of water leakage from dams using streaming potentials, in Society of Professional Well Log Analysts Nineteenth Annual Logging Symposium, p. K1-K14.

Hassan, M., Hossin, A., and Combaz, A., 1976, Fundamentals of the differential gamma ray log, in Society of Professional Well Log Analysts Seventeenth Annual Logging Symposium Transactions: p. H1-H18.

Keys, W. S., 1976, Borehole geophysics in geothermal areas--problems and progress, in Second Workshop, Geothermal Reservoir Engineering: Stanford, Stanford University.

___1978, Borehole geophysics delineates fractures in geothermal wells, in Abstracts 48th Annual Meeting, Society of Exploration Geophysicists, San Francisco: paper GL-6.

___1979, Borehole geophysics in igneous and metamorphic rocks, in Society of Professional Well Log Analysts Twentieth Annual Logging Symposium Transactions: v. II, p. 001-0026.

Keys, W. S., and MacCary, L. M., 1971, Application of borehole geophysics to water-resources investigations, in Techniques of water-resources investigations: U. S. Geological Survey, book 2, chap. 1.

Keys, W. S., Eggers, D. E., and Taylor, T. A., 1979, Borehole geophysics as applied to the management of radioactive waste--site selection and monitoring, in Carter, M. W., and others, eds., Management of low-level radioactive waste: New York, Pergamon Press, v. 2.

Lytle, R., and others, 1978. Electromagnetic cross-borehole survey of a site proposed for an urban transit station: Livermore, Lawrence Livermore Laboratory, UCRL-52484.

MacCary, L. M., 1978, Interpretation of well logs in a carbonate aquifer: U.S. Geological Survey, Water Resources Investigations, paper 78-88.

Nelson, P. H., and others, 1979, Preliminary report on geophysical and mechanical borehole measurement at Stripa: Berkeley, Lawrence Berkeley Laboratory, LBL-8280.

Oliver, D., 1977, The use of wireline logging to investigate problems in water supply boreholes: Marlow, Bucks, Great Britain, Water Research Centre, Medmenham Laboratory, Report ILR 693.

Robinson, W. S., 1976, Recent application of the noise log, in Society of Professional Well Log Analysts Seventeenth Annual Logging Symposium, pp. Y1-Y25.

Sanyal, S. K., Wells, L. E., and Bickham, R. E., 1978, Geothermal well log interpretation mid-term report: Los Alamos, Los Alamos Scientific Laboratory.

Sanyal, S. K., Juprasert, S., and Jusbache, M., 1979, An evaluation of a rhyolite-volcanic ash sequence from well logs, in SPWLA Twentieth Annual Logging Symposium Transactions: p. TT1-TT14.

Scott, H. D., Paap, H. J., and Arnold, D. M., 1978, A logging method for determining horizontal velocity of water in oilfield formations: Dallas, Society of Petroleum Engineers, paper no. 7433.

Smith, M. B., and others, 1978, The azimuth of deep, penetrating fractures in the Wattenberg field: Journal Petroleum Technology, v. 30, no. 2.

Snyder, D. D., Merkel, R. H., and Williams, J. T., 1977. Complex formation resistivity--the forgotten half of the resistivity log, in Society of Professional Well Log Analysts Eighteenth Annual Logging Symposium Transactions: p. Z1-Z39.

Unterberger, R. R., Radar progagation in rock salt: Geophysical Prospecting, v. 26, no. 2, p. 312-328.

Vincent, P. H., Gartner, J. E., and Attali, G., 1979. An approach to detailed dip determination using correlation by pattern recognition. Journal of Petroleum Technology, v. 31, no. 2, p. 232-240.

West, F. G., and Laughlin, A. W., 1976, Spectral gamma logging in crystalline basement rocks: Geology, v. 4, p. 617-618.

Winkler, K., and Nur, A., 1978. Pore fluids and seismic attenuation in rocks: Proceedings, 48th Annual Meeting of Society of Exploration Geophysicists, San Francisco.

Manuscript Accepted by the Society October 14, 1980

Printed in U.S.A.

III. Mathematical Aspects

Geological Society of America
Special Paper 189
1982

Analytical Solutions in Subsurface Fluid Flow

IRAJ JAVANDEL[1]
Shiraz University
Shiraz, Iran

ABSTRACT

The most commonly used analytical methods for solving ground-water problems are reviewed here. A simple example follows each method, illustrating its application. Literature cited indicates typical cases in which the method has been applied. Details of the methods and their rigorous analyses are not discussed here. The particular property of each method that is of considerable importance in solving the ground-water problems is emphasized.

INTRODUCTION

In 1898, Slichter showed that steady-state motion of ground water through porous media could be characterized by the Laplace equation. As a result, the analogy between heat conduction in solids and ground-water motion through porous media, in the steady-state case, was established before the start of the twentieth century. Considering that quite a large number of heat conduction problems had already been solved by that time, this analogy naturally opened the door to a treasure of solutions for steady-state ground-water problems.

Some 37 years later, in his classic paper, Theis (1935, p. 519) wrote:

> Therefore, the mathematical theory of heat conduction developed by Fourier and subsequent writers is largely applicable to hydraulic theory. This analogy has been recognized, at least since the work of Slichter, but apparently no attempt has been made to introduce the function of time into the mathematics of groundwater hydrology.

He was indeed correct in the strict sense of the phrase "groundwater hydrology." However, if we talk about a more general scope of work, that is, flow of fluid through porous media, the works of Hurst (1934) and Muskat (1934) should be mentioned.

[1] Present address: Earth Sciences Division, Lawrence Berkeley Laboratory, University of California, Berkeley, California 94720.

Using the continuity equation and considering the compressibility of the fluid only, Hurst (1934) seems to have been the first to develop a parabolic partial differential equation for the nonsteady flow of fluid through porous media. Two months later, Muskat (1934), being aware of the work of Hurst before submitting his own work, published a more general paper in the same journal.

The concept of the compressibility of the elastic artesian aquifer and its contribution to the amount of water released from the aquifer had been pointed out by Meinzer (1928). Nevertheless, this physical concept was not introduced in the works of Hurst and Muskat. Later on, Theis (1935), recognizing the analogy between ground-water flow and heat conduction in a nonsteady condition, but without giving consideration to the governing partial differential equation, simply introduced the line-sink solution, which had been brought to his attention by Lubin from the book of Carslaw (1921). It is interesting to mention that the concept of line source in non-steady-state heat conduction had been introduced by Lord Kelvin before 1880 (see Thomson, 1878), more than half a century before the work of Theis.

Jacob (1940) later developed the diffusivity equation based on the assumption of the aquifer being elastic artesian with the release of the stored water immediately following the decline of pressure. The concept of delayed yield, that is, the existence of a time-lag between the reduction of pressure within the aquifer and the release of part of the water, was first mentioned by Jacob (1940). Later, this concept was introduced mathematically into the governing equation by Boulton (1954).

ANALYTICAL METHODS OF SOLUTION

In this section, I review the most commonly used analytical methods for solving ground-water problems. A simple example will illustrate the application of each method and literature will be cited to indicate typical cases in which the method has been applied. Details of the methods and their rigorous analysis are beyond the scope of this paper and are not given here. Ignoring the effect of delayed yield, in most cases the problem is to find a solution for the following partial differential equation, subject to certain initial and boundary conditions:

$$\text{div} \cdot \left([K] \text{ grad } (h)\right) = S_s \frac{\partial h}{\partial t}, \quad (1)$$

where h is the hydraulic head, [K] is a 3 x 3 symmetric matrix representing the permeability, S_s is the specific storage, and t is the time.

Because [K] is a symmetric matrix, one can find three coordinate axes such as x, y, and z to write equation 1 in the following form

$$\frac{\partial}{\partial x}\left(K_x \frac{\partial h}{\partial x}\right) + \frac{\partial}{\partial y}\left(K_y \frac{\partial h}{\partial y}\right) + \frac{\partial}{\partial z}\left(K_z \frac{\partial h}{\partial z}\right) = S_s \frac{\partial h}{\partial t}. \quad (2)$$

In general, K_x, K_y, and K_z are functions of x, y, z, and h. When K_x, K_y, and K_z are functions of h, such as in the case of flow of fluid through unsaturated media, then a nonlinear partial differential equation (PDE) is obtained.

A homogeneous but anisotropic medium corresponds to K_x, K_y and K_z remaining constant in space. When the medium remains saturated, one may be able to change the problem into a homogeneous, isotropic problem, usually with some sort of space transformation that may be presented by a linear homogeneous PDE.

$$\frac{\partial^2 h}{\partial x^2} + \frac{\partial^2 h}{\partial y^2} + \frac{\partial^2 h}{\partial z^2} = \frac{S_s}{K}\frac{\partial h}{\partial t}. \quad (3)$$

It should be noted that the existence of any source or sink has been ignored in the development of equation 3. Such a phenomenon should be considered as a boundary condition. If, however, one were to incorporate the effect of sinks or sources within such a differential equation, the result would be a linear nonhomogeneous PDE.

$$\frac{\partial^2 h}{\partial x^2} + \frac{\partial^2 h}{\partial y^2} + \frac{\partial^2 h}{\partial x^2} = \frac{S_s}{K}\frac{\partial h}{\partial t} + A(x, y, z). \quad (4)$$

Equation 4 is analogous to the governing equation for the conduction of heat in solids, as well as that for the slowing down of neutrons, diffusion, and some electrical problems.

Because some of these fields are much older than the fields of hydrology or of flow through porous media, much more work has been done on them. Therefore, equation 4 has been solved with many different boundary conditions before the time this equation was introduced into the field of flow through porous media. As a result, during a period of time, many of the works from the heat-conduction field have been directly or indirectly adapted to problems of flow through porous media.

Separation of Variables

The method of separation of variables, which was introduced by Fourier, is convenient for the solution of linear homogeneous problems. Here, one attempts to convert the given PDE into several ordinary differential equations. The assumption is made that the solution can be presented as a product of two or more functions, each being a function of one variable only. Substitution of the assumed solution into the PDE usually leads to two or more simple ordinary differential equations which can be readily solved.

In the process of determining the integration constants, one usually encounters eigenvalues and characteristic equations. As a result, an infinite number of possible solutions will appear. At this stage, using the linear property of the PDE and the properties of the orthogonal functions, one can usually apply the initial condition to reach the final solutions, which appear in the form of summations. This method can be applied to both steady and nonsteady problems. A simple example illustrates the method.

Example 1. Find the solution of the homogeneous equation

$$\frac{\partial^2 h}{\partial x^2} = \frac{1}{a}\frac{\partial h}{\partial t} \quad (5)$$

that satisfies the initial condition

$$h(x, 0) = f(x) \quad (6)$$

and the boundary conditions

$$h(0, t) = 0; \; h(L, t) = 0. \quad (7)$$

We shall consider solutions of the form

$$h(x, t) = X(x)T(t). \quad (8)$$

Substituting the assumed solution into equation 5 and dividing both sides of equality by XT, we obtain

$$\frac{1T'}{aT} = \frac{X''}{X} = -\lambda^2. \quad (9)$$

Since (1/a)(T'/T) is a function only of t and X"/X is a function only of x, the only way equality could hold is that they both be equal to the same constant $(-\lambda^2)$.

Hence it follows that

$$X'' + \lambda^2 X = 0, \quad (10)$$

$$T' + a\lambda^2 X = 0. \quad (11)$$

The solution to equation 10 subject to boundary conditions 7 is

$$X_n(x) = \sin(\lambda_n x), \quad (12)$$

where

$$\lambda_n = \frac{n\pi}{L} \qquad (n = 1, 2, 3, \ldots) \tag{13}$$

are called eigenvalues.

The solution to equation 11 that corresponds to these values of λ_n is

$$T_n(t) = C_n e^{-a\lambda_n^2 t}. \tag{14}$$

Therefore, there are an infinite number of solutions of the form

$$h_n(x, t) = X_n(x)T_n(t)$$

$$= C_n e^{-a\lambda_n^2 t} \sin \frac{n\pi}{L}x . \tag{15}$$

Because equation 5 is linear, the sum of all the above relations is also a solution. Hence

$$h(x, t) = \sum_{n=1}^{\infty} C_n e^{-a\left(\frac{n}{L}\right)t} \sin \frac{n}{L} x. \tag{16}$$

To fulfill the initial condition (equation 6) we have

$$f(x) = h(x, 0) = \sum_{n=1}^{\infty} C_n \sin \frac{n\pi}{L}x. \tag{17}$$

In some cases, such as this example, C_n happens to be a Fourier sine coefficient of the function f(x), which can be easily calculated from

$$C_n = \frac{2}{L}\int_0^L f(y) \sin \left(\frac{n\pi}{L}y\right)dy. \tag{18}$$

Nonhomogeneous Problems

Note that partial differential equation 5 and its boundary conditions (7) are both linear and homogeneous. Both of these restrictions are necessary in order to be able to use the method of separation of variables. Linear, homogeneous equations allow partial solutions to be added together. Linearity and homogeneity of the boundary conditions are required for the eigenfunctions to be orthogonal.

Nonhomogeneous problems, however, can be solved through application of the following procedures, which change the problem into some simpler problems that can be handled by the method of separation of variables.

Simplification of the General Problem

1. *Boundary conditions independent of time.* Formation of such a problem may be presented as:

$$\nabla^2 h - \frac{S_s}{K}\frac{\partial h}{\partial t} = A(x, y, z), \tag{19}$$

$$h = f(x, y, z) \text{ at } t = 0, \tag{20}$$

$$h = \phi(x, y, z) \text{ at the boundary.} \tag{21}$$

Here, the differential equation and the boundary conditions are nonhomogeneous. Simplifying the problem will reduce the following two problems, of which one is steady and the other nonsteady but homogeneous. Let

$$h = h_1 + h_2, \tag{22}$$

such that

$$\begin{cases} \nabla^2 h_1 = A(x, y, z) & (23) \\ h_1 = \phi(x, y, z) \text{ at the boundary} & (24) \end{cases}$$

and

$$\begin{cases} \nabla^2 h_2 - \frac{S_s}{K}\frac{\partial h_2}{\partial t} = 0 & (25) \\ h_2 = f(x, y, z) - {}_1h \quad \text{at } t = 0 & (26) \\ h_2 = 0 \quad \text{at the boundary.} & (27) \end{cases}$$

2. *Boundary conditions as prescribed functions of time.* Such a problem may be formulated as:

$$\nabla^2 h - \frac{S_s}{K}\frac{\partial h}{\partial t} = A(x, y, z), \tag{28}$$

$$h = f(x, y, z) \quad \text{at } t = 0, \tag{29}$$

$$h = \phi(x, y, z, t) \quad \text{at the boundary.} \tag{30}$$

This problem may be reduced to the following simpler ones. Let

$$h = h_1 + h_2, \tag{31}$$

where

$$\begin{cases} \nabla^2 h_1 - \frac{S_s}{K}\frac{\partial h_1}{\partial t} = A(x, y, z), & (32) \\ h_1 = f(x,y,z) \quad \text{at } t = 0, & (33) \\ h_1 = 0 \quad \text{at the boundary,} & (34) \end{cases}$$

and

$$\begin{cases} \nabla^2 h_2 - \frac{S_s}{K}\frac{\partial h_2}{\partial t} = 0, & (35) \\ h_2 = 0 \quad \text{at } t = 0, & (36) \\ h_2 = \phi(x, y, z, t) \quad \text{at the boundary.} & (37) \end{cases}$$

The first problem may be solved by the procedure mentioned in part 1, the second one by the help of Duhamel's theorem.

3. *Problems whose solutions can be expressed as a product of solutions of similar problems.* This is a special form of the method of separation of variables which can only be applied when the following conditions hold: (a) the initial condition can be expressed as a product of two or more functions, each of which is only a function of a single variable, and (b) boundary conditions are of either constant-head or no-flow type.

Let $h_1(r, t)$ be the solution to the following problem:

$$\frac{1}{r}\frac{\partial}{\partial r}\left(r\frac{\partial h_1}{\partial r}\right) = \frac{S_s}{K}\frac{\partial h_1}{\partial t} \quad \text{for } a < r < b, \tag{38}$$

$$h_1 = H_1(r) \quad \text{at } t = 0, \tag{39}$$

$$\alpha_1 \frac{\partial h_1}{\partial r} - \beta_1 h_1 = 0 \quad \text{at } r = a, \tag{40}$$

$$\alpha_1' \frac{\partial h_1}{\partial r} + \beta_1' h_1 = 0 \quad \text{at } r = b. \tag{41}$$

Also let $h_2(z, t)$ be the solution to this problem:

$$\frac{\partial^2 h_2}{\partial z^2} = \frac{S_s}{K}\frac{\partial h_2}{\partial t} \quad z_1 < z < z_2, \tag{42}$$

$$h_2 = H_2(z) \quad \text{at } t = 0. \tag{43}$$

$$\alpha_2 \frac{\partial h_2}{\partial z} - \beta_2 h_2 = 0 \quad \text{at } z = z_1, \tag{44}$$

$$\alpha_2' \frac{\partial h_2}{\partial z} - \beta_2' h_2 = 0 \quad \text{at } z = z_2. \tag{45}$$

Then

$$h = h_1(r)\, h_2(z) \tag{46}$$

is the solution to the following problem:

$$\frac{1}{r}\frac{\partial}{\partial r}\left(r\frac{\partial h}{\partial r}\right) + \frac{\partial^2 h}{\partial z^2} = \frac{S_s}{K}\frac{\partial h}{\partial t}, \tag{47}$$

$$h = H_1(r)\, H_2(z) \quad \text{at } t = 0, \tag{48}$$

$$\alpha_1 \frac{\partial h}{\partial r} - \beta_1 h = 0 \quad \text{at } r = a, \tag{49}$$

$$\alpha_1' \frac{\partial h}{\partial r} + \beta_1' h = 0 \quad \text{at } r = b, \tag{50}$$

$$\alpha_2 \frac{\partial h}{\partial r} - \beta_2 h = 0 \quad \text{at } z = z_1, \tag{51}$$

$$\alpha_2' \frac{\partial h}{\partial z} + \beta_2' h = 0 \quad \text{at } z = z_2. \tag{52}$$

The method of separation of variables can be used both for steady- and nonsteady-state ground-water problems. It can be used either alone or in combination with other methods, which we shall see later.

One of the earliest applications of this method to ground water problems is found in the work of Jacob (1946). In his paper, Jacob first established a partial differential equation for radial flow in an elastic artesian aquifer with a vertical leakage proportional to the drawdown. Then he looked for a solution for nonsteady state as a combination of two functions: one for steady-state with boundary conditions, and the other for nonsteady state with initial conditions and homogeneous boundary conditions. He then applied the method of separation of variables by using properties of the orthogonal functions to find the solution for the nonsteady part of the problem.

Once again, if one wants to look into the more general field of flow through porous media, the works of Hurst (1934) and Muskat (1934) should also be considered. After developing the partial differential equation for flow of compressible fluid through porous media, Hurst then solved the equation for two different kinds of radial flow toward a well in a limited reservoir, one for constant well pressure and the other for constant production rate. In both problems he used the basic method of separation of variables, but the initial steps toward finding the elementary solutions were omitted.

Later, Muskat (1934), using the method of separation of variables together with Duhamel's theorem, gave the solution for three radial-flow problems where boundary conditions were time-dependent.

The method of separation of variables has also been used in more complex problems of multilayer aquifers. Among others, the works of Katz (1960) and Freeze (1966) may be mentioned. Using this method, Katz has solved the problem of fluid flow in a stratified, bounded reservoir, produced with constant fluid pressure at the well. Freeze has also used the method to handle the problem of two-dimensional steady-state regional ground-water flow through multilayered aquifers.

Integral Transform

Another powerful method for solving partial differential equations is the method of integral transform. When a function $K(\alpha, x)$ is a known function of the two variables α and x and the integral

$$F(\alpha) = \alpha \int_a^b f(x)\, K(\alpha, x)\, dx \tag{53}$$

is convergent, then the function $F(\alpha)$ defined by the above integral is called the integral transform of the function $f(x)$ by the kernel $K(\alpha, x)$. It can be easily shown that such an operation is linear. The limits of integration are usually the same as the domain of definition of the variable x.

Depending on the form of the kernel $K(\alpha, x)$ and the nature of the limits of integration (finite or infinite), several different kinds of integral transforms are obtained.

The use of an integral transform will often reduce a partial differential equation with n independent variables into one with n-1 independent variables, thus reducing the complexity of the problem under discussion. In some instances, successive operations of this type will ultimately reduce the problem to a boundary-value problem with an ordinary differential equation, which is often much easier to solve.

Laplace Transform

Although the development of the transformation and the corresponding operational calculus was begun by Laplace (1749-1827) and Cauchy (1789-1857) (Cooper, 1952), most of the theory of the Laplace transform that is used today has come from the combined work of Heaviside (1890), Bromwich (1916), and Carson (Carslaw and Jaeger, 1959).

In this case, the limits of integration are zero and infinity, and the kernel is $K(p, t) = e^{-pt}$. Thus, the Laplace transform of a function of f(t) defined for $t > 0$ may be written as

$$F(p) = \int_0^\infty e^{-pt} f(t)\, dt. \tag{54}$$

One of the fundamental properties of this transformation is

$$\int_0^\infty e^{-pt} \left(\frac{\partial f}{\partial t}\right) dt = pF(p) - f(0), \tag{55}$$

which is the result of integration by parts. Although application of the Laplace transform simplifies many mathematical problems, its particular importance in solving initial-boundary value problems should be emphasized. When dealing with a parabolic differential equation such as a diffusivity equation, using the Laplace transform with respect to time and its property given in equation 55 not only eliminates the dependency of the function f(r, z, t) on t but also immediately uses the initial condition f(0). At this stage the problem, which deals only with the space variables, may be solved either by separation of variables or by successive application of other suitable kinds of integral transforms.

Once the simplified problem is solved, one faces the task of finding a function whose transform is the result of the simplified problem. This process is called Laplace inversion. There are at least four methods of finding the inverse function.

1. Tables. The easiest way is to look into available tables of the Laplace transformation. This is possible for relatively simple functions.

2. Partial fraction. In this method, if the function is in the form of a complicated fraction which is not available in an inversion table, then it is split into simpler fractions.

3. Application of Mellin's inversion. The formula (Carslaw and Jaeger, 1941) is given here:

$$f(t) = \frac{1}{2\pi i} \int_{\gamma - i\infty}^{\gamma + i\infty} e^{\lambda t} F(\lambda)\, d\lambda. \tag{56}$$

4. Numerical inversion. If the transform function happens to be very complex and does not lend itself easily to any of the above methods, one can always find the value of the inverse function at discrete values of time by means of several available numerical methods (Javandel, 1976).

Example 2. Find the solution to the following equation:

$$\frac{\partial^2 h}{\partial x^2} = \frac{1}{\alpha}\frac{\partial h}{\partial t} \qquad \text{for } x > 0, \tag{57}$$

subject to:

$$h = 0 \qquad \text{at } t = 0, \tag{58}$$

$$h = H \qquad \text{at } x = 0, \tag{59}$$

$$h = 0 \qquad \text{as } x \to \infty. \tag{60}$$

In order to use the Laplace transform with respect to time, one can multiply both sides of equations 57 and 59 by e^{-pt} and integrate them between $t = 0$ to ∞; if $\bar{h}$ is the transform function, one obtains:

$$\frac{d^2\bar{h}}{dx^2} = \frac{p}{\alpha}\bar{h} - h(0), \tag{61}$$

$$\bar{h} = \frac{H}{p} \qquad \text{at } x = 0. \tag{62}$$

The solution to the ordinary differential equation 61, subject to conditions 58, 60, and 62, is then:

$$\bar{h} = \frac{H}{p}\, e^{-x\sqrt{p/\alpha}}. \tag{63}$$

Using a table of Laplace transforms, one can find

$$h = H \operatorname{erfc}\left(\frac{x}{2\sqrt{\alpha t}}\right). \tag{64}$$

One of the earliest applications of this method to the problems of flow through porous media was in the work of van Everdingen and Hurst (1949). In their paper, van Everdingen and Hurst used the method of the Laplace transform to solve the problem of radial flow to a well in a reservoir. They obtained a solution for "constant terminal rate" and "constant terminal pressure" for both finite and infinite reservoirs.

In the 1950s, Hantush introduced the method of Laplace transform into the field of ground-water flow. In 1956 Hantush presented a paper at a regional meeting held at Stanford University which was later published (Hantush, 1959). In that paper, he employed the Laplace transform to find the transient drawdown distribution in a leaky aquifer drained by a flowing well. Since then, the Laplace transform has been used in the analytical solution of almost all transient ground-water problems.

Hankel Transform

The Hankel transform of zero order of a function h(r) is defined as

$$\bar{h}(\lambda) = \int_0^\infty rJ_0(rg)h(r)\, dr, \tag{65}$$

where $J_0(r\lambda)$ is the Bessel function of the first kind and zero order.

In this case also, the limits of integration are zero and infinity, but the kernel is $K(\lambda, r) = rJ_0(r\lambda)$. Here, the inversion formula is

$$h(r) = \int_0^\infty \lambda J_0(r\lambda)\, \bar{h}(\lambda)\, d\lambda. \tag{66}$$

One of the important properties of this transformation, which is very helpful for the solution of axisymmetric-flow problems in aquifers with infinite radial extent, is

$$\int_0^\infty rJ_0(r\lambda)\left(\frac{\partial^2 h}{\partial r^2} + \frac{1}{r}\frac{\partial h}{\partial r}\right) dr = -\lambda^2\bar{h}(\lambda). \tag{67}$$

This is true provided that both $rhJ_1(\lambda r)$ and $r\, \partial h/\partial r\, J_0(\lambda r)$ vanish when r goes to zero and infinity.

It is clear that the use of the above transformation in axisymmetric flow problems where r is defined between zero and infinity will temporarily eliminate the variable r, and this normally will simplify the problem to a great extent. Relaxation of the restriction on the inner boundary, without introducing any complexity, will take care of the problems in which the flux is assigned along the inside boundary, because

$$\int_0^\infty rJ_0(r\lambda)\ \frac{\partial^2 h}{\partial r^2} + \frac{1}{r}\frac{\partial h}{\partial r}\ dr$$

$$= -\lambda^2 \bar{h}(\lambda) - \lim_{r\to 0} r\frac{\partial h}{\partial r}. \tag{68}$$

Example 3. Find the solution to the following differential equation:

$$\frac{\partial^2 s}{\partial r^2} + \frac{1}{r}\frac{\partial s}{\partial r} = \frac{1}{\alpha}\frac{\partial s}{\partial t}, \tag{69}$$

subject to

$$s(r, t) = 0 \qquad \text{at } t = 0, \tag{70}$$

$$\lim_{r\to\infty} s(r, t) = 0, \tag{71}$$

$$\lim_{r\to 0}\left(r\frac{\partial s}{\partial r}\right) = -\frac{Q}{2\pi T}. \tag{72}$$

Multiply both sides of equation 69 by $rJ_0(r\lambda)$ and integrate with respect to r. Application of equation 68 gives

$$-\lambda^2\bar{s} + \frac{Q}{2\pi T} = \frac{1}{\alpha}\frac{d\bar{s}}{dt}. \tag{73}$$

The solution to this ordinary differential equation, considering the initial condition, is

$$\bar{s} = \frac{Q}{2\pi T\lambda^2} - \frac{Q}{2\pi T\lambda^2}e^{-(\alpha t)\lambda^2}. \tag{74}$$

The inverse Hankel transform of equation 74 is

$$s = \frac{Q}{2\pi T}\int_0^\infty J_0(\lambda r)\lambda\left(\frac{1}{\lambda^2} - \frac{\exp(-\alpha t\lambda^2)}{\lambda^2}\right) d\lambda. \tag{75}$$

Differentiating with respect to t and using the following formula (from tables),

$$\int_0^\infty \lambda J_0(\lambda r)e^{-\lambda^2 b}\, d\lambda = \frac{1}{2b}\exp\left(-\frac{r^2}{4b}\right), \tag{76}$$

gives

$$\frac{\partial s}{\partial t} = \frac{Q}{4\pi T}\frac{\exp\left(-\frac{r^2}{4\alpha t}\right)}{t}. \tag{77}$$

Hence

$$s = \frac{Q}{4\pi T}\int_{\frac{r^2}{4\alpha t}}^\infty \frac{e^{-y}}{y}\, dy. \tag{78}$$

Hantush (1960) was perhaps the first to use the Hankel transform technique for solving ground-water flow problems. In his paper, Hantush applied a combination of the Laplace and Hankel transforms to find the equation of drawdown and the rate of leakage in an aquifer confined by semipervious elastic strata. Since then, the combination of the Laplace and Hankel transforms has been frequently used to find a solution to nonsteady-state ground-water problems with axisymmetric flow patterns.

Finite Integral Transforms

In the integral transform techniques mentioned above, the domain of definition of the particular variable of interest was infinite. That kind of transformation, as was noted before, is suitable for handling variables such as time with the domain of definition zero to infinity and any space variable defined in that range.

Finite integral transforms, first suggested by Doetsh (1935), may be suitable to problems in which the field of variation of an independent variable is finite. Here we shall present those finite integral transforms that have proved to be instrumental in solving ground-water problems.

Finite Fourier Transforms

Finite cosine transform. Given a function $f(x, y, z, t)$ defined in the interval $0 \leqslant z \leqslant b$, its finite cosine transform with respect to z may be given as

$$f_c(n) = \int_0^b f(z) \cos \frac{n\pi z}{b}\, dz \qquad n = 0, 1, 2, \ldots \tag{79}$$

The inverse formula is

$$f(z) = \frac{1}{b} f_c(0) + \frac{2}{b} \sum_{n=1}^{\infty} f_c(n) \cos \frac{n\pi z}{b}. \tag{80}$$

The important operational property of this transformation is

$$\int_0^b \frac{\partial^2 f}{\partial z^2} \cos\left(\frac{n\pi z}{b}\right) dz$$

$$= -\left(\frac{n\pi}{b}\right)^2 f_c(n) + (-1)^n \frac{\partial f(b)}{\partial z} - \frac{\partial f(0)}{\partial z}. \tag{81}$$

Owing to the form of equation 81, the finite cosine transform is particularly useful for problems in which the gradients normal to two parallel boundaries are given.

Finite sine transform. The finite sine transform and its inverse formula with respect to a variable z of a function $f(x, y, z, t)$ defined in the interval $0 \leqslant z \leqslant a$ may be given as

$$f_s(n) = \int_0^a f(z) \sin\left(\frac{n\pi z}{a}\right) dz \qquad n = 1, 2, 3, \ldots, \tag{82}$$

$$f(z) = \frac{2}{a} \sum_{n=1}^{\infty} f_s(n) \sin\left(\frac{n\pi z}{a}\right). \tag{83}$$

The important property of this transformation may be written as

$$\int_0^a \frac{\partial^2 f}{\partial z^2} \sin \frac{n\pi z}{a}\, dz = -\left(\frac{n\pi}{a}\right)^2 f_s(n)$$

$$+ \left(\frac{n\pi}{a}\right) [f(0) - (-1)^n f(a)], \tag{84}$$

thus making this transformation quite useful when the value of the function itself is given on two parallel boundaries.

Example 4. Find the solution to the following equation:

$$\frac{\partial^2 s}{\partial r^2} + \frac{1}{r}\frac{\partial s}{\partial r} + \frac{\partial^2 s}{\partial z^2} = 0, \tag{85}$$

subject to

$$\frac{\partial s}{\partial z} = 0 \qquad \text{at } z = 0 \text{ and } z = b, \tag{86}$$

$$s = 0 \qquad \text{at } r = r_e, \tag{87}$$

$$\lim_{r \to 0} \left(r \frac{\partial s}{\partial r}\right) = \frac{-Q}{2\pi K \ell} \qquad 0 < z < \ell$$
$$= 0 \qquad \ell < z < b. \tag{88}$$

One may note that the above formulation introduces steady flow to a partially penetrating well.

Application of the cosine transform with respect to z and use of the boundary condition in equation 86 yields

$$\frac{d^2 s_c}{dr^2} + \frac{1}{r}\frac{ds_c}{dr} - \left(\frac{n\pi}{b}\right)^2 s_c = 0, \tag{89}$$

$$s_c = 0 \qquad \text{at } r = r_e, \tag{90}$$

$$\lim_{r \to 0} r\left(\frac{ds_c}{dr}\right) = \frac{-Qb}{2\pi^2 K \ell n} \sin\left(\frac{n\pi \ell}{b}\right). \tag{91}$$

A solution to equation 89 may be written as

$$s_c = AI_0\left(\frac{n\pi r}{b}\right) + BK_0\left(\frac{n\pi r}{b}\right). \tag{92}$$

Finding constants A and B through the use of equations 90 and 91 and substituting,

$$S_c = \frac{-Qb}{2\pi^2 K \ell n} \sin \frac{n\pi \ell}{b}\left[I_0 \frac{n\pi r}{b} \frac{K_0(n\pi r_e/b)}{I_0(n\pi r_e/b)} - K_p \frac{n\pi r}{b}\right]. \tag{93}$$

One can now find the solution by finding the inverse of equation 93:

$$s(r, z) = \frac{Q}{2\pi Kb}\left[\ln\frac{r_e}{r} + \sum_{n=1}^{\infty}\frac{2b}{n\pi\ell}\sin\frac{n\pi\ell}{b}\cos\frac{n\pi z}{b}\right.$$

$$\left.\cdot\left\{K_0\left(\frac{n\pi r}{b}\right) - I_0\left(\frac{n\pi r}{b}\right)\frac{K_0(n\pi r_e/b)}{I_0(n\pi r_e/b)}\right\}\right]. \qquad (94)$$

Finite Hankel Transform

Sneddon (1946) introduced the finite integral transform with the kernel of the Bessel function. Depending on the range of variation of the variable to be excluded, one of the following transformations may be adopted.

The finite Hankel transform of a function f(x), when x varies between 0 and a, is defined by

$$\bar{f}(y_i) = \int_0^a xf(x)\, J_0(xy_i)dx, \qquad (95)$$

where y_i is a root of the transcendental equation

$$J_0(ay_i) = 0. \qquad (96)$$

Note that in general the kernel of the integral in equation 95 can be $xJ_n(xy_i)$.

The inverse function f(x) is

$$f(x) = \frac{2}{a^2}\sum_i \bar{f}(y_i)\frac{J_0(xy_i)}{[J_1(ay_i)]^2}. \qquad (97)$$

One of the important properties of this transformation is

$$\int_0^a x\left(\frac{\partial^2 f}{\partial x^2} + \frac{1}{x}\frac{\partial f}{\partial x}\right)J_0(xy_i)dx$$

$$= ay_i f(a)\, J_1(ay_i) - y_i^2\,\bar{f}(y_i). \qquad (98)$$

Thus, application of this transformation to the problems concerned with radial flow in a finite reservoir, where the radial coordinate varies between 0 and a, will simplify them to a great extent.

If the variable x varies between b and a such that $0 < b \leqslant x \leqslant a$, then

$$\bar{f}(y_i) = \int_b^a xf(x)\left[J_0(xy_i)Y_0(ay_i) - Y_0(xy_i)J_0(ay_i)\right]dx, \qquad (99)$$

where y_i is a root of transcendental equation

$$J_0(y_i b)Y_0(y_i a) - J_0(y_i a)Y_0(y_i b) = 0, \qquad (100)$$

and the inversion formula is

$$f(x) = \sum_i \frac{2y_i^2 J_0^2(y_i b)\bar{f}(y_i)}{J_0^2(ay_i) - J_0^2(by_i)}$$

$$\cdot\,[J_0(xy_i)Y_0(ay_i) - J_0(ay_i)Y_0(xy_i)]. \qquad (101)$$

The summations of equations 97 and 101 extend over all positive roots of the respective cases. In this case the important property is

$$\int_b^a x\left(\frac{\partial^2 f}{\partial r^2} + \frac{1}{r}\frac{\partial f}{\partial r}\right)[J_0(xy_i)Y_0(ay_i) - J_0(ay_i)Y_0(xy_i)]dx$$

$$= \frac{J_0(ay_i)}{J_0(by_i)}f(b) - f(a) - y_i^2\,\bar{f}(y_i), \qquad (102)$$

which, therefore, makes this transformation useful for radial-flow problems in bounded reservoirs having a finite inner boundary.

Example 5. Find the solution to the following equation:

$$\frac{\partial^2 s}{\partial r^2} + \frac{1}{r}\frac{\partial s}{\partial r} = \frac{1}{\alpha}\frac{\partial s}{\partial t}, \qquad (103)$$

subject to:

$$s = 0 \quad \text{at } t = 0, \qquad (104)$$

$$s = 0 \quad \text{at } r = a, \qquad (105)$$

$$s = 1 \quad \text{at } r = b. \qquad (106)$$

Use of the finite Hankel transform given by equation 99 and application of boundary conditions (equations 105 and 106) yields

$$\frac{J_0(ay_i)}{J_0(by_i)} - y_i^2\,\bar{s} = \frac{1}{\alpha}\frac{d\bar{s}}{dt}. \qquad (107)$$

Equation 107 is a linear ordinary differential equation whose solution, considering the initial condition-equation 104, is

$$\bar{s}(y_i) = \frac{J_0(ay_i)}{J_0(by_i)\cdot y_i^2}(1 - e^{-\alpha y_i^2 t}). \qquad (108)$$

Use of the inverse formula given by equation 101 gives

$$s = \sum_i \frac{2J_0(by_i)\, J_0(ay_i)}{[J_0^2(ay_i) - J_0^2(by_i)]}$$

$$\cdot\,(1 - e^{-\alpha y_i^2 t})[J_0(ry_i)Y_0(ay_i) - J_0(ay_i)Y_0(ry_i)]. \qquad (109)$$

method of Green's functions to find the steady-state solution to leaky systems where the aquifers are in the form of infinite half-planes, infinite quadrants, infinite strips, and rectangles.

Nonsteady State

For nonsteady-state problems, the Green's function is considered as the head at the point (x,y,z) at the time t due to an instantaneous point source of unit strength generated at the point (x',y',z') at the time τ, the head being initially zero all over the domain of definition and the surface being kept either at no-flow condition or at h = 0.

If such a function can be found, then the head at any point (x',y',z') at the time t, due to the initial distribution f(x,y,z) and the appropriate boundary condition is given by

$$h(x',y',z',t) = \iiint (G)_{\tau=0} f(x,y,z)\,dxdydz - \alpha \int_0^t \left[\iint \phi(x,y,z,t) \frac{\partial G}{\partial n}\, ds \right] dt \quad (120)$$

when the value of head on the boundary s is given by $\phi(x,y,z,t)$, and

$$h(x',y',z',t) = \iiint (G)_{\tau=0} f(x,y,z)\,dxdydz + \alpha \int_0^t \left[\iint G \frac{\partial \phi}{\partial n}\, ds \right] dt \quad (121)$$

when the value of flux is given on the boundary s. In equations 120 and 121, α is the diffusivity.

For further discussion about nonsteady Green's function as well as its application to ground-water problems see DeWiest (1969) and Gringarten (1973).

Other Methods

Other methods such as Boltzmann's transformation, the application of point sources, and the image method have long played an important role in solving ground-water problems.

The idea of the instantaneous point source of heat and its systematic application was introduced by Lord Kelvin (Thomson, 1878). In the field of ground water, the idea may be expressed in the following form: drawdown at any point (x,y,z) and a time t, in an infinite porous medium, with hydraulic diffusivity α, due to an instantaneous release of a finite quantity of fluid, $Q = CS_s$ at the point (x',y',z') and time t = 0 is given by

$$s = \frac{C}{8(\pi\alpha t)^{3/2}} \cdot \exp\left\{ - \frac{(x - x')^2 + (y - y')^2 + (z - z')^2}{4\alpha t} \right\}. \quad (122)$$

Equation 122 satisfies the diffusivity equation and is the starting point for the solution of many ground-water problems. Equation 122 may be integrated with respect to z' or x' to yield the drawdown due to a vertical or horizontal instantaneous line sink. Further integration with respect to time leads to an expression giving the drawdown due to a continuous line sink.

Since in most practical cases the radius of the pumping well is very small relative to other dimensions of the problem, the well is commonly assumed to be a line sink. This assumption introduces some errors in calculating the drawdown. However, it has been shown (Mueller and Witherspoon, 1965) that, except for small values of time and for distances very close to the well (approximately 20 times the radius of the well), the magnitude of the error is quite negligible.

Example 8. Let us consider the problem of flow to a fully penetrating well in a horizontal, confined aquifer of uniform thickness. Adapting the idea from the heat-conduction theory, Theis (1935) introduced the infinite line-sink solution to express drawdown in an aquifer with the above properties.

If we choose a set of coordinate axes (r, z), such that the z-axis coincides with the axis of the well, then drawdown due to an instantaneous line sink, which is obtained by integrating equation 122 with respect to z', between $-\infty$ and $+\infty$, is given by

$$s = \frac{c}{4\pi\alpha t} \exp\left\{ - \frac{r^2}{4\alpha t} \right\}. \quad (123)$$

The reason for considering an infinite line sink is to eliminate the dependency of the drawdown on z, which indirectly provides the condition for confinement of the aquifer. Substitution of t in equation 123 by t − t' and its integration with respect to t' between 0 and t gives the drawdown due to a continuous line sink. The rate of discharge per unit length of this line is $q = CS_s$.

$$s = \frac{C}{4\pi\alpha} \int_{\frac{r^2}{4\alpha t}}^{\infty} \frac{e^{-u}}{u}\, du = \frac{q}{4\pi K} \int_{\frac{r^2}{4\alpha t}}^{\infty} \frac{e^{-u}}{u}\, du = \frac{Q}{4\pi T} \int_{\frac{r^2}{4\alpha t}}^{\infty} \frac{e^{-u}}{u}\, du, \quad (124)$$

where $Q = q\ell$ is the total rate of discharge of the well, and K and T are permeability and transmissibility of the aquifer, respectively.

Appropriate application of the image method on equation 122 gives the effect of a point sink in a semi-infinite or strip-type medium. Consecutive integration of the result with respect to z and time could lead to the more complex solution of partial penetration (Hantush, 1957). Stallman (1952) has shown how the image method could be used to account for the no-flow or constant-head boundary in a bounded aquifer.

CONCLUSIONS

The most commonly used analytical techniques for solving ground-water problems have been reviewed. The following comments may be drawn from the previous discussion:

1. The method of separation of variables may be adopted for both steady- and nonsteady-state problems. This method is restricted only to linear and homogeneous boundary-value problems. Nonhomogeneous linear problems may, however, be converted to simpler ones, some of them solvable by the method of separation of variables.

2. The Laplace transform is particularly useful for handling time-dependent problems. This method can also be used for steady-state problems provided the variable to be excluded is defined for all positive values. The major difficulty with this method is the inversion, which is sometimes quite complicated.

3. The infinite Hankel transform is very useful for solving radial flow problems when the range of definition of the radial variable is from zero to infinity.

4. If the domain of definition of an independent variable is finite, then one of the finite integral transforms could simplify the problem. The finite Hankel transform may be used for radial coordinates, and sine or cosine transforms for cartesian coordinates.

5. Conformal mapping is quite a useful method for handling steady two-dimensional seepage problems. A special type called the hodograph method is particularly useful when dealing with problems involving seepage and free surfaces at their boundaries.

6. The Green's function method is useful for solving steady- and nonsteady-state problems. As Muskat (1946) has noted, the disadvantage of this method is that it is difficult to find the Green's function for relatively complex problems.

Although it is sometimes believed that analytical methods are no longer suitable for solving complex ground-water problems, this kind of judgment is rather unfair considering that, in general, engineers have always had a time lag in applying the newly developed methods in mathematics.

REFERENCES CITED

Aravin, V. I., and Numerov, S. N., 1965, Theory of fluid flow in undeformable porous media: Translated from Russian: Jerusalem, Israel Program for Scientific Translation.

Boulton, N. S., 1954, Unsteady radial flow to a pumped well allowing for delayed yield from storage: International Association of Scientific Hydrology Pub. 37, p. 472-477.

Carslaw, H. S., 1921, Introduction to the mathematical theory of the conduction of heat in solids: New York MacMillan.

Carslaw and Jaeger, 1941, Operational methods in applied mathematics: London, Oxford University Press.

___1959, Conduction of heat in solids: London, Oxford University Press.

Cooper, J. L. B., 1952, Heaviside and the operational calculus: Math Gazette, v. 36, p. 5-19.

DeWiest, R., 1969, Flow through porous media: New York, Academic Press.

Doetsch, G., 1935, Integration von Differential-gleichungen ver mittels der endlichen Fourier Transformation: Mathematische Annalen, v. 62, p. 52-68.

Freeze, R. A., 1966, Theoretical analysis of regional groundwater flow [Ph.D. dissertation]: Berkeley, University of California.

Gringarten, A. C., 1973, The use of source and Green's functions in solving unsteady flow problems in reservoirs: Society of Petroleum Engineers Journal, p. 285-296.

Hamel, G., 1934, Zeitschrift fur Angewandte mathematik und mechanik: v. 14, p. 129.

Hantush, M. S., 1957, Nonsteady flow to a well partially penetrating an infinite leaky aquifer: Iraqi Scientific Society Proceedings, v. 1, p. 10.

___1959, Nonsteady flow to flowing wells in leaky aquifers: Journal of Geophysical Research, v. 64, no. 8, p. 1043-1051.

___1960, Modification of the theory of leaky aquifers: Journal of Geophysical Research, v. 65, no. 11, p. 3713-3725.

___1967, Flow to wells in aquifers separated by a semipervious layer: Journal of Geophysical Research, v. 72, no. 6, p. 1709-1719.

Hantush, M. S. and Jacob, C. E., 1954, Plane potential flow of groundwater with linear leakage: American Geophysical Union Transactions, v. 35, no. 6, p. 917-936.

Harr, M. E., 1962, Groundwater and seepage: New York, McGraw-Hill.

Hurst, W., 1934, Unsteady flow of fluids in oil reservoirs: Physics Journal of Applied Physics, v. 5, p. 20-30.

Jacob, C. E., 1940, On the flow of water in an elastic artesian aquifer: American Geophysical Union Transactions, v. 21, no. 2, p. 574-586.

___1946, Radial flow in leaky artesian aquifer: American Geophysical Union Transactions, v. 27, no. 2, p. 198-205.

Javandel, I., 1976, Application of numerical inversion of Laplace transformation for solving groundwater problems, in Proceedings, International Seminar on Regional Groundwater Hydrology and Modeling, Venice, Italy: International Business Machines, p. 273-401.

Katz, M. L., 1960, Fluid flow and heat transfer in stratified systems [Ph.D. thesis]: University of Michigan.

Meinzer, O. E., 1928, Compressibility and elasticity of artesian aquifers: Economic Geology, v. 35, p. 263-291.

Mueller, T. D., and Witherspoon, P. A., 1965, Pressure interference effects within reservoirs and aquifers: American Institute of Mining and Metallurgical Engineers Transactions, v. 234, p. 471.

Muskat, M., 1934, The flow of compressible fluids through porous media and some problems in heat conduction: Physics (Journal of Applied Physics), v. 5, p. 71-94.

___1946, Flow of homogeneous fluids through porous media: Ann Arbor, Edwards Inc.

Polubarinova-Kochina, 1962, Theory of groundwater movement: Translated from Russian by J.M.R. deWiest, Princeton University Press, New Jersey.

Russell, D. G., and Prats, M., 1962, Performannce of layered reservoirs with cross flow-single-compressible-fluid case: Transactions, Society of Petroleum Engineers, American Institute of Mining and Metallurgical Engineers, v. 225, p. 53-67.

Slichter, C. S., 1898, Theoretical investigation of the motion of groundwater: U.S. Geological Survey 19th Annual Report, Part 2, p. 295-384.

Sneddon, I. N., 1946, Finite Handel transforms: Philosophical Magazine v. 37, p. 17-25.

Stallman, R. W., 1952, Nonequilibrium type curves modified for two well systems: U.S. Geological Survey Groundwater Note 3.

Theis, C. F., 1935, The relation between the lowering of the Piezometric surface and the rate and duration of discharge of a well using groundwater storage: American Geophysical Union Transactions, p. 519-524.

Tranter, C. J., 1966, Integral transforms in mathematical physics: New York, John Wiley.

Thomson, W., 1978, Heat: Encyclopedia Brittanica, v. 11, p. 554.

van Everdingen, A. F., and Hurst, W., 1949, The application of the Laplace transformation to flow problems in reservoirs: American Institute of Mining and Metallurgical Engineers Transactions, v. 186, p. 305.

Manuscript Accepted by the Society October 14, 1980

Printed in U.S.A.

Geological Society of America
Special Paper 189
1982

Flow-Test Evaluation of Fractured Reservoirs

A. C. GRINGARTEN
FLOPETROL, BP 592
77005, Melun Cedex, France

ABSTRACT

This paper summarizes available methods for interpreting flow tests in fissured formations. The methods are taken from the geotechnical, ground-water, and petroleum literature. These are usually based on "equivalent reservoir" solutions, which attempt to relate the actual fissured reservoir behavior to that of a known theoretical model, homogeneous or heterogeneous, of lower complexity. The double-porosity concept, used by many authors for ground-water and petroleum engineering problems, is critically reviewed and the relationship between the various available solutions is established. The methods presented in the paper are illustrated with well-test data from aquifers and oil reservoirs of various characteristics.

INTRODUCTION

The movement of underground fluids is of interest in many different engineering fields, and has been the subject of numerous studies since the 1950s. Flow-analysis procedures, however, are well established only for porous fluid-bearing reservoirs that are considered reasonably homogeneous.

The problem of fluid flow in heterogeneous formations is still the subject of much debate. It is generally agreed that conventional methods primarily developed for homogeneous reservoirs are inadequate, and that new, specific approaches are required in order to provide a convincing explanation for some commonly observed flow peculiarities.

In particular, fractured formations have received much attention, mainly in the literature on engineering geology, in connection with slope stability and seepage problems of large dams and tunnels; in ground-water hydrology, in relation to pollution-oriented problems such as subsurface waste disposal and salt-water intrusion; and in petroleum engineering, because of the development of hydraulic well fracturing and the recognition of naturally fissured reservoirs as important producers.

This paper discusses various approaches for analyzing the flow of fluid in fissured formations in connection with ground-water and petroleum engineering problems. Solutions used in geotechnical engineering are only briefly mentioned, and the reader is referred to specific publications on that subject for more details (Wilson and Witherspoon, 1970; Parsons, 1972).

This paper illustrates some of the analysis methods with results of a number of tests made in fissured reservoirs of different rock characteristics. A few of the tests were especially designed for fracture-flow analysis.

GENERAL CONCEPTS OF WELL-TEST INTERPRETATION

Inverse versus Direct Problem

The principles governing well-test analysis are more easily understood when one considers well-test interpretation as a special pattern-recognition problem.

In a well test, a known signal I (for instance, the constant withdrawal of reservoir fluid) is applied to an unknown system S (the well + reservoir). The response of that system, O (the change in reservoir pressure), is measured during the test. Schematically, I → S → O.

The purpose of well-test interpretation is to *identify* the system S, knowing only the input I and output O signals and possibly some other reservoir characteristics, such as initial or boundary conditions and shape of drainage area.

This type of problem is known in mathematics as the *inverse problem:* S = O/I. It can be solved directly by dividing the Fourier transform of the output signal by that of the input signal, obtaining the transfer function of the system. This method is useful when one is only interested in the system response to different input signals, but it does not allow any physical description of the actual reservoir (black box approach).

A more descriptive approach involves the search of a well-defined *theoretical reservoir,* whose response to the same input signal is as close as possible to that of the actual reservoir. The response of the theoretical reservoir is computed for specific initial and boundary conditions, which must correspond to those of the actual one, when they are known. This is called the *direct problem:* O = I x S.

Interpretation thus relies on models, whose characteristics are assumed to represent the characteristics of the actual reservoir.

If the wrong model is selected, then the parameters calculated for the actual reservoir will not be correct. On the other hand, the solution of the inverse problem is usually not unique: it may be possible to find several reservoir configurations that would yield similar responses to a given input signal. However, the number of alternative solutions decreases as the number and the range of output signal measurements increase.

Theoretical Models

Theoretical models are always constructed in the same manner. As sketched in Figure 1, they include a basic model, inner boundary conditions, and outer boundary conditions.

The basic model is of infinite lateral extent and may be homogeneous or heterogeneous (double-porosity, multilayered, or composite). It comprises initial conditions, usually with uniform pressure.

The basic model is of practical interest only in interference tests, to analyze the pressure response of observation wells away from the active well. To be useful for practical applications at the active well, the basic model must be associated with inner boundary conditions, most commonly wellbore storage and skin effect, fractures, and partial penetration.

Finally, outer boundary conditions may have to be added, depending upon the actual situation. Possibilities include no-flow boundaries (sealing faults) and constant pressure (recharge).

Most models presented in the oil literature assume impermeable upper and lower boundaries, so that flow eventually becomes radial in the formation. Until recently, these models did not take into account inner boundary conditions at the active well, and were valid only for interpreting *long-term* well-test data. They were not adequate for analyzing early-time data (data affected by wellbore conditions) that were mistakenly considered unreliable. Inner boundary conditions are also absent in most models used in ground-water hydrology because they are mainly used to analyze interference tests. Outer boundary conditions are well documented in both the oil and ground-water literature.

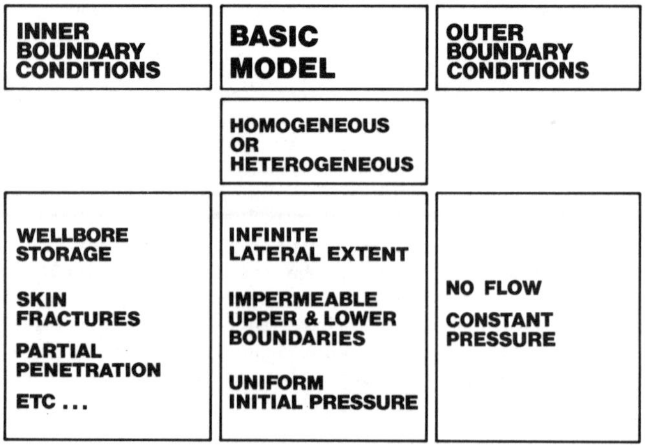

INNER BOUNDARY CONDITIONS	BASIC MODEL	OUTER BOUNDARY CONDITIONS
	HOMOGENEOUS OR HETEROGENEOUS	
WELLBORE STORAGE SKIN FRACTURES PARTIAL PENETRATION ETC ...	INFINITE LATERAL EXTENT IMPERMEABLE UPPER & LOWER BOUNDARIES UNIFORM INITIAL PRESSURE	NO FLOW CONSTANT PRESSURE

Figure 1. Schematic of a theoretical model.

Model Identification

To be appropriate for the system under study, the theoretical model selected for analysis should include the various dominant features of the real system.

In most practical cases, inner boundaries, the basic model, and outer boundaries dominate at different time periods in the test, and exhibit special behaviors that are easily recognizable. In fact, the effect of inner and outer boundary conditions on the pressure behavior of a model is independent of the nature of the basic model itself (homogeneous or heterogeneous). The reverse is also true but to a more limited extent.

The distinctive impact of the various components of a theoretical model is best seen on a log-log plot, with the change in pressure during the test, Δp, plotted as a function of the time, Δt, elapsed since the start of the test.

For example, wellbore storage yields a log-log straight line of slope unity[1] at early times (Δp is proportional to Δt; Ramey, 1970) and, conversely, a unit slope log-log straight line passing through early time data is usually indicative of wellbore storage. On the other hand, because Δp is proportional to Δt the same data points must also be located on a straight line passing through the origin when Δp is plotted versus Δt in Cartesian coordinates. Such a plot, *specific to a given flow regime,* is called a *specialized plot.*

In the same way, a high-conductivity fracture communicating with the wellbore yields a log-log straight line with half unit slope[2] (Δp is proportional to $\sqrt{\Delta t}$; Gringarten and others, 1975), whereas a low-conductivity fracture yields a log-log straight line with a one-fourth slope (Cinco and Samaniego, 1978b). Radial flow (Δp proportional to $\log \Delta t$; Van Everdingen and Hurst, 1949), spherical flow (Δp proportional to $1/\sqrt{\Delta t}$; Moran and Finklea, 1962) and pseudo-steady-state flow (Δp, linear function of Δt; Jones, 1956) also exhibit distinctive log-log shapes.

The specialized plots associated with these various flow regimes are, respectively:

Δp vs $\sqrt{\Delta t}$ (high-conductivity fracture),
Δp vs $\sqrt[4]{\Delta t}$ (low-conductivity fracture),
Δp vs $\log \Delta t$ (radial flow),
Δp vs $1/\sqrt{\Delta t}$ (spherical flow),
Δp vs Δt (pseudo-steady-state flow, late time data),

and should show a straight line through the points of interest.

The log-log plot is called a *diagnostic plot:* it allows identification of dominating flow regimes, which yield straight lines on specialized plots.

[1]Log-log slope of unity = 1 log cycle (drawdown) for 1 log cycle (time).

[2]Log-log slope of 0.5 = 1 log cycle (drawdown) for 2 log cycles (time).

The log-log behavior of a complete model (basic model + inner boundary conditions + outer boundary conditions) is simply obtained as the *superposition* on the same log-log graph of the individual log-log behaviors of each component of the model. The resulting log-log curve is called a *type-curve* and characterizes the model.

Type-Curve Analysis

Type-curves are usually graphed as a dimensionless pressure versus a dimensionless time, with each curve being characterized by a dimensionless number that depends upon the specific reservoir model. Dimensionless parameters are defined as the real parameter times a coefficient that includes reservoir characteristics. When the appropriate model is being used, real and theoretical pressure versus time curves are identical in shape but displaced with respect to each other, with the displacement factors for both pressure and time axes being proportional to certain reservoir parameters.

Plotting real data as log-log pressure versus time curves thus provides *quantitative* as well as *qualitative* information on the reservoir.

Qualitative information--recognizing dominating flow regimes from their characteristic shapes--helps in selecting the most appropriate theoretical reservoir model, and to break down the test data into segments that can be analyzed by specialized methods. Specialized methods (such as Jacob's, 1950) use the parameters of the straight line on the specialized plots, and usually provide more accurate results than quantitative log-log analysis.

Quantitative log-log analysis implies matching the log-log plot of the test data against a type-curve from a theoretical model that includes the various features identified on the actual data. For a given theoretical model, however, not all type curves are equivalent. Depending upon the choice of dimensionless pressure and time parameters, one type curve may be easier to use within a specific data range, and different graphs of the same type-curve data are common in the literature. As a general rule, type-curve matching is easier when all the theoretical curves on the type-curve graph merge into one curve where the actual well data are the most numerous.

Reservoir Information Obtained from a Test

Because inner boundaries, the basic model, and outer boundaries dominate the test at different times, the information obtained from a test depends on the period in the test that is being analyzed. In other words, different ranges of test data yield reservoir parameters characterizing the reservoir on different scales. Basically, the earlier the test data, the more detailed the reservoir information that can be extracted.

This point is illustrated in Figure 2, which represents the pressure behavior of a well during a typical constant rate drawdown test. Three time periods can be identified.

Period 1 corresponds to late-time data, where boundary effects are predominant. This period was the first to be investigated by well testing in the

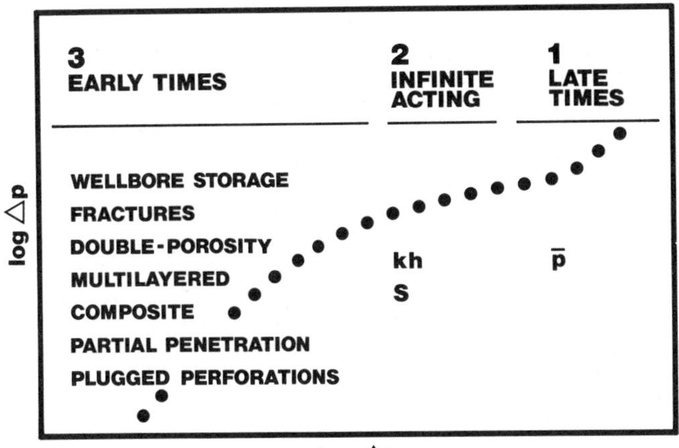

Figure 2. Information obtained from well-test data.

1920s and 1930s (and is still the main target of geotechnical testing). Production wells were shut in at regular intervals and downhole pressure-point measurements were taken to obtain the reservoir average pressure. A material balance (zero-dimensional) model was then used to estimate the reserves. Because all closed systems exhibit the same pseudo steady-state behavior, no other information can be extracted from these data.

Researchers then realized that the validity of spot pressure measurements was dependent upon the duration of the shut-in period; the less permeable the formation, the longer the shut-in period necessary to reach average pressure. Transient testing was thus introduced and was well developed in the 1950s and 1960s. This corresponds to Period 2 on Figure 2. Data from Period 2 are analyzed to obtain the permeability-thickness product, kh, of the formation, and the skin, S. However, these parameters only represent a gross reservoir behavior, and cannot be used to describe the system in greater detail.

For instance, any horizontal reservoir of infinite lateral extent with impermeable upper and lower boundaries will eventually exhibit radial flow behavior during the infinite acting Period 2, and the same kh value could represent a homogeneous, multilayered, or fissured reservoir.

In the same way, a positive skin could indicate either a damaged well or an undamaged well with partial penetration; and a negative skin characterizes a stimulated well that could be acidized, hydraulically fractured, or that intersects a natural fracture.

This kind of detailed information is obtained only from early-time data (Period 3 on Fig. 2). Period 3 has been the subject of many studies in the oil literature since the late 1960s, and is the usual target of the so-called modern well-test analysis.

THEORETICAL MODELS FOR FISSURED RESERVOIRS

Two approaches have been used for analyzing flow problems in fissured rocks. One is based on representing the fracture formation as a *discontinuous medium*, and considers individual fracture behavior and the *exact geometry* of the fissured system; the other assumes *continuity* and *statistically homogeneous* rock and fluid properties. The choice of either approach depends essentially upon the relative scale of the phenomenon being studied, and on the availability of the necessary input parameters.

Deterministic models--which attempt to duplicate the actual geometry of the fissure/porous block system by specifying individual fissure locations, orientations, and aperture variations--are usually limited to relatively small-scale, geotechnical projects. The continuum approach is more realistic in ground-water and petroleum engineering problems.

Deterministic Models

Deterministic models rely upon a knowledge of the flow characteristics of individual fracture segments and take into account the discontinuous character of the fissured rock masses, their heterogeneity in a given field, and the completely arbitrary orientation of the fissure network. Different methods have been suggested to deal with the nature of the fracturation, starting from the simplest case (a rock mass with a single system of conductive fractures) to the complex case of arbitrarily oriented sets of fractures (Louis, 1975).

Most models are based on parallel plate flow theory and the concept of directional hydraulic conductivity. It has been shown that the steady-state flow of water in a fissure follows the potential theory when one uses a velocity potential, $k_f\Psi$, which is related to the hydraulic potential, $\Psi = p + \rho gz$. This property also extends to a system of plane parallel fissures if the hydraulic conductivity, k_f, of the single fissure is replaced by the directional hydraulic conductivity, K, of the system of fissures (Louis, 1969).

The problem in space is thus broken down into a series of two-dimensional problems for each fissure or system of parallel fissures. Knowing the directional hydraulic conductivities at the scale of the lattice of the model used, it is then possible to find the hydraulic potential distribution by means of any usual method of solution.

For example, Louis (1969) used a two-dimensional steady-state analytical model, whereas Sharp (1970) employed a line-element numerical technique for two-dimensional problems, and a planar-element numerical technique for three-dimensional problems. In the same way, Wilson (1970) developed a line-element model for purely parallel plate flow between impermeable blocks, and a triangular-element model for combined fissure and block flow problems.

Statistical Models

When fracture density is large and many measurements of fracture properties are available, mean values can be determined for aperture, frequency, and orientation in any set of fractures, and a permeability tensor can be calculated, assuming that the medium can be considered continuous. This approach was used by several authors with deterministic models such as the ones described above (Serafim and del Campo, 1965; Serafim, 1968; Romm, 1966; Snow, 1965, 1968).

All these models have been used for solving geotechnical problems under steady-state flow conditions. Another type of statistical model, applicable to transient flow, has been used by various authors in connection with regional ground-water and petroleum engineering problems. In these models, porous blocks and fissures cannot be located exactly and are assumed to be uniformly distributed throughout the formation. The reservoir is thus treated as an "equivalent" system, either homogeneous or heterogeneous, as dictated by the general behavior of the actual test data.

An "equivalent" is a system whose calculated behavior is equivalent to the observed behavior of the real system. The word "equivalent," however, only applies to the basic theoretical model. Boundary conditions, whether inner or outer, should be the actual ones, when they are known.

In the following, I examine various theoretical models that have been used in the literature for interpreting the behavior of fissured reservoirs. They are introduced in a logical order, starting in each case with the basic model, and adding boundary conditions as appropriate. Homogeneous models are described first because they are simpler, well-documented, and readily available. Heterogeneous models are introduced next. Among these, the double-porosity model, especially designed for interpreting fissured systems, is analyzed in great detail.

Homogeneous Model. Usual homogeneous models have often been extended to analyze tests in fissured formations simply by including anisotropy.

The anisotropic effect of a fractured medium is easy to visualize in the case of a vertical fracture system. Experiments conducted by Huskey and Crawford (1967) indicate that permeability should increase when the fractures are parallel to the direction of flow, whereas there should be no change in permeability due to fractures at right angles to the direction of flow. As a result, the maximum permeability should be representative of the fracture system, while the minimum permeability will reflect the effects of the matrix.

The same problem was investigated by Prats (1972) and Asfari and Witherspoon (1973). Prats was able to develop an analytic expression relating the permeability anisotropy to fracture geometry for infinitely conductive fractures in offset systems of fractures. Prats' model and results are presented in Figures 3 and 4, respectively. The ordinate in Figure 4 gives the ratio of the apparent permeability of the fractured formation normal to the fracture plane (k_m) to that parallel to the fracture planes (k_f), that is, the ratio of the matrix block permeability to the fissure permeability. In his model, the fissures are not necessarily connected, and flow to the well occurs through the blocks and through the fissures intersecting the well.

Asfari and Witherspoon (1973) used a finite-element numerical model to study the case of finite and infinite conductivity fractures in regular and offset systems of fractures. Results for infinite conductivity fractures in offset systems were in agreement with Prats' findings. The effect of regular vertical fracture systems is shown in Figure 5 for infinitely conductive fractures and

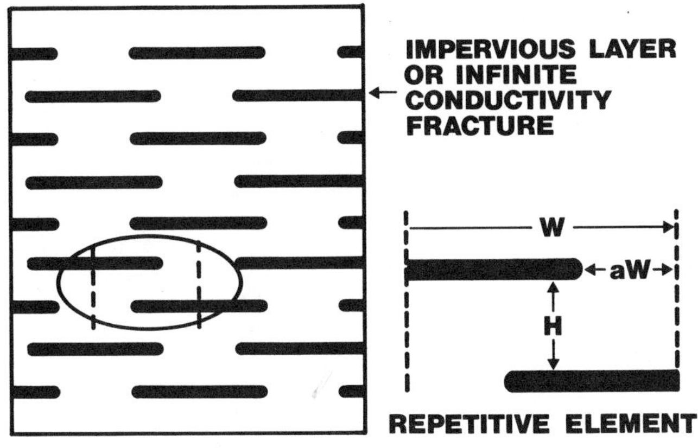

Figure 3. Schematic of an offset infinite-conductivity fracture pattern (Prats, 1972).

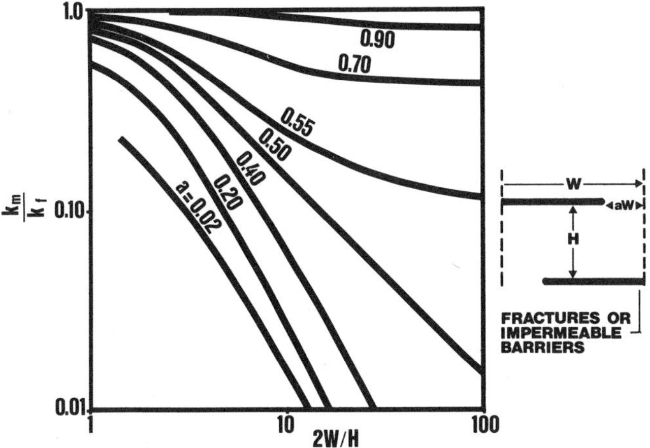

Figure 4. Permeability anisotropy in an offset infinite-conductivity fracture pattern (Prats, 1972).

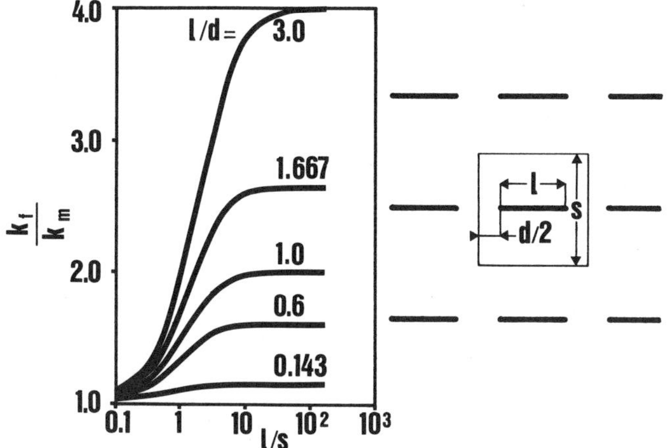

Figure 5. Permeability anisotropy in a regular infinite-conductivity fracture pattern (Asfari and Witherspoon, 1973).

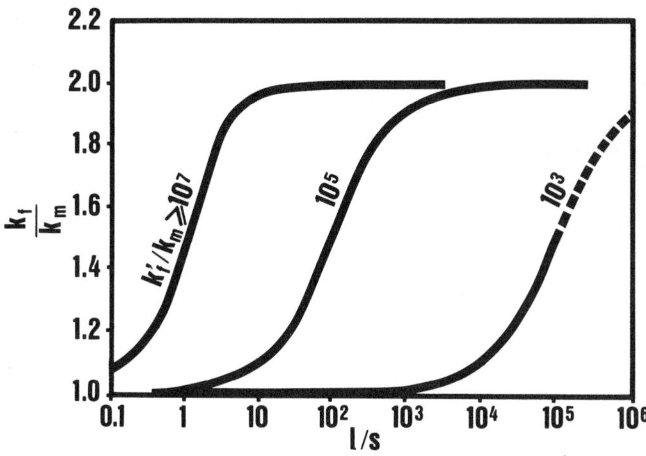

Figure 6. Effect of fracture conductivity on permeability anisotropy [(l/d) = l] (Asfari and Witherspoon, 1973).

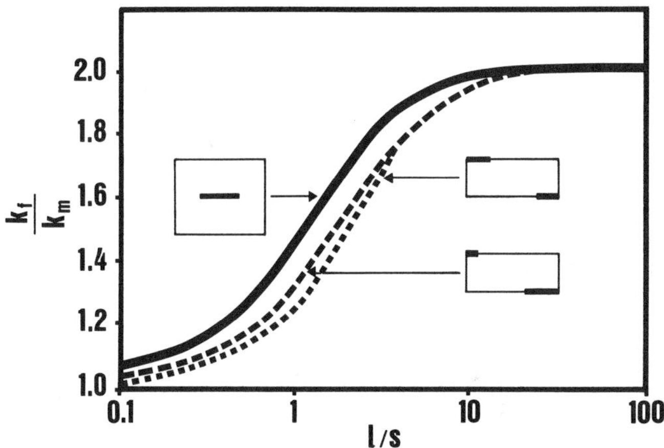

Figure 7. Effect of various patterns on permeability anisotropy [(l/d) = l] (Asfari and Witherspoon, 1973).

in Figure 6 for finite conductivity fractures. The influence of the fracture pattern is indicated in Figure 7.

The equivalence between naturally fractured porous rock and anisotropic homogeneous media was further documented by Parsons (1966) and Kiraly (1971). Parsons indicated that the property of equivalence would hold only for fracture distributions that are not too dispersed.

Basic Model. The anisotropy concept was used by Elkins and Skov (1960) to analyze pressure transients observed during the development of the Spraberry field, a low-permeability oil reservoir (1 md or 10^{-8} m/sec) cut by an extensive system of vertical fractures. The authors employed an equation developed by Collins (1961) for the pressure drawdown at the location of one well, due to constant production at another well in an extensive anisotropic reservoir of uniform thickness (Fig. 8). The method, however, required a knowledge of the principal axes of perm-

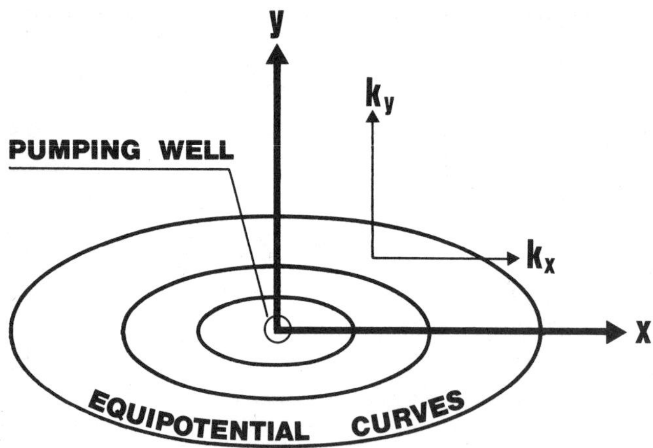

Figure 8. Equivalent homogeneous permeability tensor for fractured reservoirs with anisotropy in the x and y directions.

eability, which in this case were along and at right angles to the major fracture trend. A more general equation for the drawdown was later derived by Papadopulos (1965) with respect to an arbitrary system of axes. Papadopulos also developed a method for finding the hydraulic properties of the system from pump test data and the directions of maximum and minimum permeabilities.

A similar method of analysis was proposed by Saad (1967) for investigating reservoirs with natural horizontal fractures. The technique suggested would allow computation of the horizontal and vertical permeabilities, provided that the pumping well partially penetrates the aquifer (Hantush, 1957).

However, constructed from the basic model alone, these methods are only adequate for interference tests and cannot be used to analyze tests where data are taken only at the production well (except for late-time data, which are not characteristic).

Inner Boundary Conditions--Fractured Well. When the system is highly fractured, the pumping well is likely to intersect at least one of the fractures, and this affects the unsteady-flow behavior of the reservoir in a characteristic way. For such a situation, a method of interpretation was proposed by Gringarten and Witherspoon (1972), based on mathematical models originally developed by Gringarten and Ramey (1974) and Gringarten and others (1974) for the special situation of hydraulically-fractured reservoirs.

These models described the pressure behavior of a well that intersects a single horizontal fracture (Gringarten and Ramey, 1974) or a single vertical fracture (Gringarten and others, 1974) while the well produces at constant rate from an otherwise homogeneous porous medium. Two types of solutions were obtained. One solution was based on the assumption of a uniform head along the fracture (a fracture of infinite conductivity), and was found to match field data from hydraulically-fractured wells. The other solution assumed that fluid enters the fracture at the same rate per unit area (a fracture with uniform flux) and matched the behavior of a well that intersects a natural fracture. This work was extended to finite conductivity hydraulic fractures by Cinco and others (1978). Inspection of a number of pump test results indicates that these solutions should be useful in evaluating data from fractured formations. One characteristic of these solutions is that, for highly conductive fractures, a log-log plot of fluid-level drawdowns versus producing time yields a straight line with a slope of 0.5 at early times, indicating linear flow from the reservoir matrix into the fracture. At long times, the drawdown behavior is the same as that indicated by the exponential integral solution plus a constant that depends on the point where measurements are being made.

Vertical Fracture. The solution for a well intersecting a single-plane vertical fracture in a radially infinite homogeneous porous medium (Gringarten and others, 1974), has been modified by Gringarten and Witherspoon (1972) to account for anisotropy. A plane view of the two-dimensional model is sketched in Figure 9. The permeability k_f in the x direction parallel to the fracture can be interpreted as representing the increased permeability due to the existence of a major trend of vertical fractures in the x direction. The matrix permeability is k_m. The pumping well is located along the vertical fracture axis.

The pressure drawdown for a uniform flux fracture, obtained by means of the Green's function and product solution method (Gringarten and Ramey, 1973), can be expressed as

$$p_{fD} = \frac{2\pi\sqrt{k_f k_m}\,h}{q\mu}(p_i - p_f)$$

$$= \frac{\sqrt{\pi}}{4}\int_0^{t_D} \left[\operatorname{erf}\frac{1-\frac{x}{x_f}}{2\sqrt{\tau}} + \operatorname{erf}\frac{1+\frac{x}{x_f}}{2\sqrt{\tau}}\right] \cdot \exp\left[-\frac{1}{4\tau}\left(\frac{y}{x_f}\right)^2 \frac{k_f}{k_m}\right]\frac{d\tau}{\sqrt{\tau}} \quad (1)$$

$$t_D = \frac{k_f t}{\phi_m \mu (c_t)_m x_f^2}\,. \quad (2)$$

Here x_f represents the fracture half-length, and the subscript m refers to matrix properties. At long times, p_{fD} becomes

$$p_{fD} = \frac{1}{2}\ln(2.25t_D') + \pi\left\{1 - \frac{1}{4}\frac{x}{x_f}\ln\frac{k_f y^2 + (x+x_f)^2 k_m}{k_f y^2 + (x-x_f)^2 k_m} - \frac{y\sqrt{k_f k_m}\,h}{2x_f}\arctan\frac{y x_f\sqrt{k_f k_m}}{k_f y^2 + (x^2 - x_f^2)k_m}\right\}, \quad (3)$$

where

$$t'_D = \frac{t}{\phi_m \mu (c_f)_m}$$

$$\cdot \frac{k_f k_m}{\sqrt{[k_f y^2 + (x + x_f)^2 k_m][k_f y^2 + (x - x_f)^2 k_m]}} . \quad (4)$$

Equations 1 and 3 are the basis for analyzing pump test data.

The dimensionless drawdown at the pumping well ($x = y = 0$) is obtained from equation 1 as

$$p_{fD} = \pi \sqrt{t_D} \operatorname{erf}\left(\frac{1}{2\sqrt{t_D}}\right) - \frac{1}{2} \operatorname{Ei} \frac{1}{4t_D} . \quad (5)$$

A log-log plot of p_{fD} versus t_D from equation 5 is shown in Figure 10. This plot is characterized by an initial half-unit slope straight line, and can be used as a type curve in analyzing drawdown data that exhibit the same characteristics. Customarily, drawdowns observed at the pumping well are plotted versus time on logarithmic paper of the same scale and then matched with the type curve. From the dual coordinates of an arbitrary matching point (Δp, t) and (p_{fD}, t_D), one obtains the relations

$$\sqrt{k_f k_m} = \frac{q\mu}{4\pi h} \frac{p_{fD}}{p_i - p_f} , \quad (6)$$

$$\frac{k_f}{\phi_m \mu (C_t)_m x_f^2} = \frac{t_D}{t} . \quad (7)$$

The first equation yields the geometric mean of the maximum and minimum permeabilities. The second equation can be solved for k_f/x_f^2 if a value for storage is available.

As indicated by equation 3 for large values of time, the analysis can also be performed by means of the "straight line" method. The time at which the semi-log straight line begins is indicated by an arrow in Figure 10.

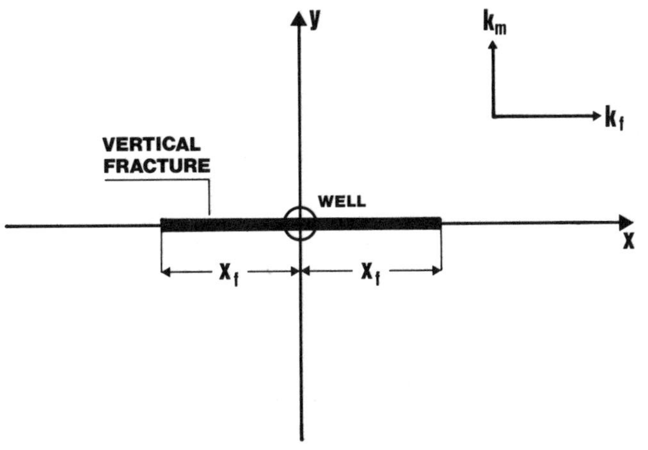

Figure 9. Vertical fracture model.

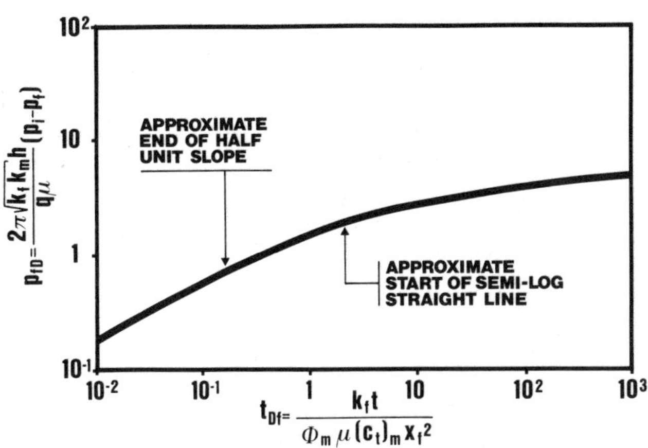

Figure 10. Type curve for vertical-fracture model with drawdowns measured at pumping well (Gringarten and others, 1974).

Since the hydraulic properties of the system cannot be individually determined from the above analysis, it is apparent that observations at other points in the system are needed. Drawdown measurements at a minimum of two observation wells are required to provide the necessary data.

The transient response at observation wells differs from that at the pumping well (see Fig. 11). No initial half-unit slope straight line is observed on log-log coordinates, except where the observation well intersects the same vertical fracture as the pumping well (Fig. 11, top left). The observed pressure response depends upon the location of the observation well and the transmissibility ratio, which makes the "type curve matching" method difficult to use. However, far enough from the pumping well the behavior of the system becomes identical to that of a line source well in an anisotropic porous medium. This is apparent for $r_D > 5$ in Figure 11, and can be verified from equation 3. The second term on the right-hand side of equation 3 vanishes as

$$r_D = \frac{1}{x_i} \sqrt{x^2 + y^2 \frac{k_f}{k_m}}$$

increases or d_f decreases. The first term in equation 3 becomes identical to the long-time approximation for the drawdown created by a line source well in an anisotropic aquifer, written with respect to the principal axes of transmissibility (Collins, 1961; Papadopulos, 1965). This is realized within 1% when

$$x^2 + y^2 \frac{k_f}{k_m} \geqslant 25 x_f^2 . \quad (8)$$

In such a case, methods described by Papadopulos (1965) can be used for analysis. Type-curves for a vertically fractured well in a closed rectangle are also available (Gringarten, 1978).

Horizontal Fracture. A sketch of a horizontal fracture model is given in Figure 12. The pumping

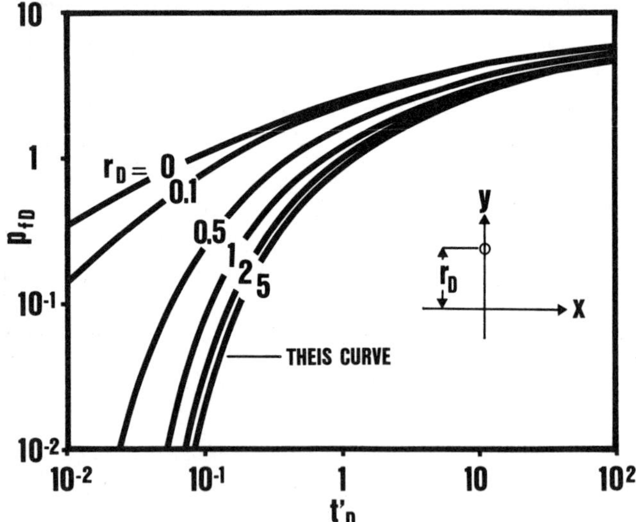

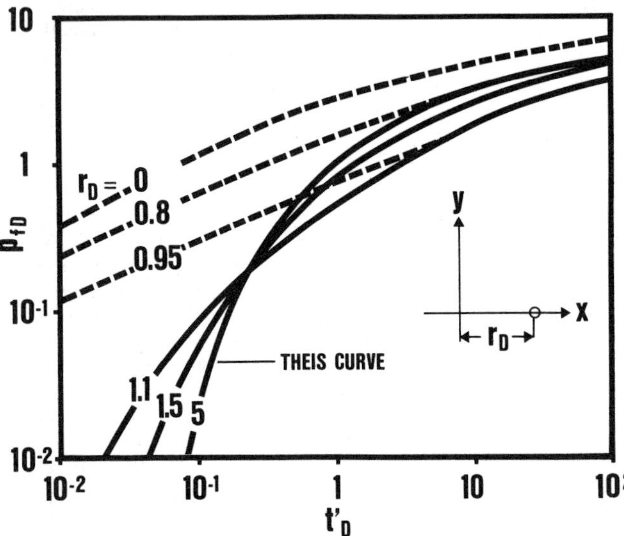

$$p_{fD} = \frac{2\pi\sqrt{k_f\, k_m}\, h}{q\mu}\,(p_i - p_f)$$

$$r_D = \frac{1}{x_f}\sqrt{x^2 + y^2\,\frac{k_f}{k_m}}$$

$$t'_D = \frac{t}{\Phi_m\,\mu\,(c_t)_m}\;\frac{k_f\, k_m}{\sqrt{\left[k_f y^2 + (x+x_f)^2 k_m\right]\left[k_f y^2 + (x-x_f)^2 k_m\right]}}$$

Figure 11. Type curves for vertical-fracture model with drawdowns measured at observation wells (Gringarten and Witherspoon, 1972).

well is located along the axis of the plane horizontal fracture. The permeability k_f in the radial direction is parallel to the fracture and can be interpreted as representing the increased permeability due to the existence of a horizontal fracture system. The permeability k_m in the vertical direction represents the matrix permeability.

An analytical solution for this case, as obtained by Gringarten and Ramey (1974), can be written as:

$$p_{fD} = \frac{1}{H_D}\int_0^{t_D} \frac{\exp\left[-\frac{1}{4\tau}\left(\frac{r}{r_f}\right)^2\right]}{\tau}$$

$$\cdot \int_0^1 I_D\left(\frac{r\nu}{2r_f\tau}\right)\exp\left(-\frac{\nu^2}{4\tau}\right)\nu\, d\nu$$

$$\cdot \left[1 + 2\sum_{n=1}^{\infty}\exp\left(-\frac{n^2\pi^2\tau}{H_D^2}\right)\cos n\pi\,\frac{Z_f}{H}\,\cos n\pi\,\frac{Z}{H}\right] d\tau, \qquad (9)$$

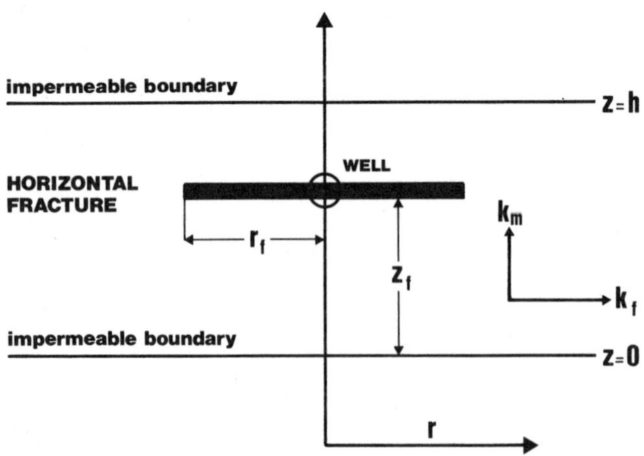

Figure 12. Horizontal fracture model.

where I_o is the Bessel function of the first kind of zero order and

$$t_D = \frac{k_f t}{\phi_m \mu (c_t)_m x_f^2}, \tag{10}$$

$$p_{fD} = \frac{2\pi\sqrt{k_f k_m}\, r_f}{q\mu}(p_i - p_f), \tag{11}$$

$$H_D = \frac{h}{r_f}\sqrt{\frac{k_f}{k_m}}, \tag{12}$$

where r_f is the radius of the horizontal fracture.

As in the vertical fracture case, type curve matching and the semilog straight-line method (when applicable) can be used for analysis. Type curves for analyzing drawdowns at the pumping well ($r = 0$, $z = z_f$), when the fracture is at the center of the formation are presented in Figure 13. A log-log plot of p_{fD} versus t_D is given for various values of dimensionless thickness, H_D. It should be noted that the definitions of p_{fD} and t_D are not the same as in the case of the vertical fracture model. Again, at early times, a straight line with a half-unit slope is obtained. Aside from this similarity, the shapes of the horizontal fracture curves in Figure 13 are sufficiently different from that of the vertical fracture curve in Figure 10 (except for $1 < H_D < 3$) to make it possible to distinguish between the two types in a pumping test. The horizontal fracture curves corresponding to $1 < H_D < 3$ are very similar to the vertical fracture curve on Figure 10, and caution should be exercised in interpreting field data for these special cases.

For high H_D values, the latter part of the horizontal fracture curves has a shape similar to that of the double-porosity curves, or the unconfined aquifer ones. Different possible interpretations would therefore have to be considered, especially in shallow aquifers.

By matching drawdown data with the type curves on Figure 13, one can obtain three equations:

$$\sqrt{k_f k_m}\, r_f = \frac{q\mu}{2\pi h}\,\frac{p_{fD}}{p_i - p_f}, \tag{13}$$

$$\frac{k_f}{\phi_m \mu (C_t)_m r_f^2} = \frac{t_D}{t}, \tag{14}$$

$$h_D = \frac{h}{r_f}\sqrt{\frac{k_f}{k_m}}, \tag{15}$$

from which k_f, k_m, and r_f can be calculated if a value for $\phi_m(c_t)_m$ is available or can be estimated.

In some cases, however, h is so large that the data plot follows the type curve $H_D = \infty$ over the entire test period. In this event, it is not possible to determine k_f, k_m, and r_f from data measured only in the pumping well. The radial flow

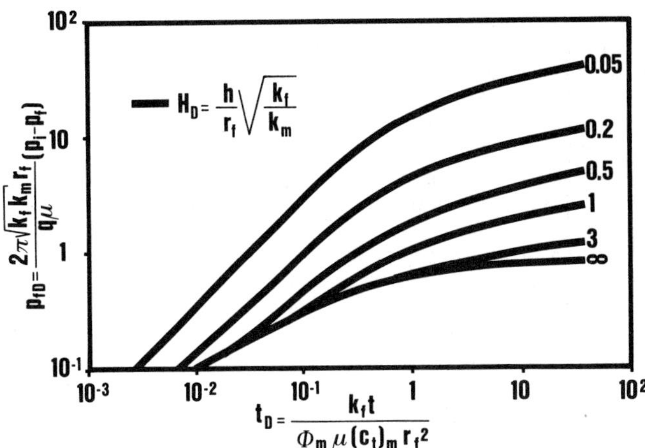

Figure 13. Type curves for horizontal-fracture model with drawdowns measured at pumping well (Gringarten and Witherspoon, 1972).

straight-line methods will not help because the time at which the semi-log straight line begins is given by

$$t = \max\left[\frac{\phi_m \mu (c_t)_m h^2}{2k_m}\ ;\ 12.5\,\frac{\phi_m \mu (c_t)_m \left(r^2 + r_f^2\right)}{k_f}\right] \tag{16}$$

and may be very large if h is large.

As in the vertical fracture case, field data from the pumping well can be matched to the exponential integral solution (basic homogeneous model; Theis, 1935) if early-time data are neglected. It is obvious that erroneous results should then be expected.

Drawdown behavior at observation wells in a horizontally-fractured system depends upon the distance to the pumping well (Gringarten and Ramey, 1974). If the observation well intersects the fracture, an initial half-unit slope straight line appears on a log-log plot. If the distance to the pumping well is greater than the fracture radius, this is no longer so, and there exists a critical distance from the pumping well equal to

$$r_i = r_f + 2h\sqrt{\frac{k_f}{k_m}}, \tag{17}$$

beyond which the drawdown is the same as that created by a line-source well in a homogeneous reservoir with isotropic radial permeability, k_f. Drawdown data from such observation wells can be analyzed with the Theis solution to provide values of k_f and $\phi_m(c_t)_m$.

The method presented here provides a way of analyzing fractured aquifers as an "equivalent" anisotropic, homogeneous porous medium, with a single fracture of much higher permeability intersecting the pumping well. It should be applied to those field cases where a log-log plot of drawdown versus pumping time, as observed at the pumping well, reveals an initial half-unit slope straight line. By matching the data plot with the type curves presented in this chapter, it should be possible to determine whether the fracture trend is essentially horizontal, vertical, or neither. In those cases where a match can be obtained, the hydraulic properties of the system can be computed as outlined above.

These models should also be applicable to non-porous systems such as fractured crystalline rock for which an initial straight line of half-unit slope on log-log paper is apparent from field data measured in the pumping well. In such cases, however, the specific storage $\phi_m(c_t)_m$ used in the formula would represent the storage factor of an equivalent porous medium. If possible, this value should be confirmed by analyzing data that have been measured at observation wells.

Wellbore Storage and Skin. In some tests, the initial half-unit slope straight line is not apparent, and a unit slope straight line is obtained instead. This indicates the existence of a large storage volume connected to the pumping well, and corresponds to a fracture of large dimensions rather than a planar fracture as used in the models. Solutions corresponding to this situation exist for vertically fractured wells (Ramey and Gringarten, 1975; Fig. 14) and can be easily obtained for the horizontal-fracture case by means of a numerical method (Cinco and Samaniego, written commun., 1978).

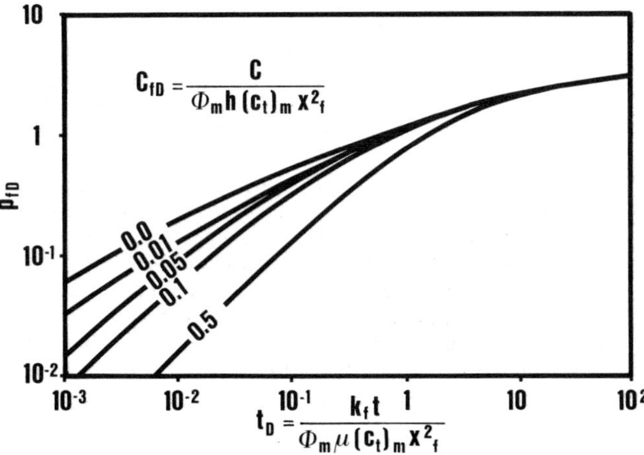

Figure 14. Type curve for an infinite-conductivity vertical fracture with wellbore storage (Ramey and Gringarten, 1975).

In the case of a vertical fracture of infinite conductivity, it has been shown by Gringarten and others (1979) that the addition of wellbore storage yields a type curve very similar to that for a non-fractured well with wellbore storage and a negative infinitesimal skin. These authors presented a set of type curves where both models are included (Fig. 15), without obvious transition from one model to the other. The type curves are plotted as a dimensionless drawdown,

$$p_D = \frac{2\pi kh}{q\mu} \Delta p \tag{18}$$

versus a dimensionless time

$$\frac{t_D}{C_D} = \frac{2\pi kh}{\mu C} \Delta t, \tag{19}$$

where k is the reservoir permeability and C the wellbore storage constant. C_D is the dimensionless wellbore storage constant, equal to:

$$C_D = \frac{C}{2\pi\phi c_t h r_w^2}. \tag{20}$$

Each curve is labeled in terms of C_De^{2S}, where S represents an infinitesimal skin effect.

Damaged wells usually correspond to C_De^{2S} greater than 10^3, undamaged wells to C_De^{2S} values between 10^3 and 5, and acidized wells to C_De^{2S} between 5 and 0.5. These limits are approximate and may vary slightly. It was proved, however, that C_De^{2S} values less than 0.5 could not be obtained with an infinitesimal skin, and had to imply fractures in communication with the well. On the other hand, C_De^{2S} could be greater than 0.5 with fractures.

Analysis of a test with the type curves of Figure 15 yields the reservoir transmissivity, the skin factor, and the wellbore storage constant. It was learned from experience that the quantitative value of these last two parameters could suggest the existence of a fissured system: in such a case, the skin is usually highly negative (less than -5), and the wellbore storage constant is 40 to 100 times higher than what would be expected from the wellbore volume. This situation may be visualized as a limiting case of a double-porosity behavior which will be discussed in the next chapter.

Heterogeneous Model. In many practical cases, the behavior of a fissured system is so obviously different from that of an homogeneous one that a special model is required for analysis. Attempts have thus been made to find a suitable equivalent heterogeneous reservoir. Possibilities include double-porosity, multilayer, and composite models.

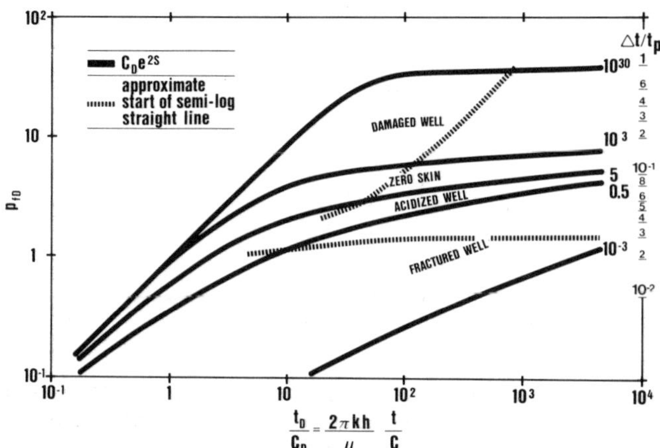

Figure 15. Wellbore storage and skin type curves for a well with wellbore storage and skin in a homogeneous reservoir (after Gringarten and others, 1979).

Double-porosity Model. The concept of a double-porosity medium as representative of the behavior of a fissured aquifer was first introduced by Barenblatt and others (1960). Their model assumed the existence of two regions of different porosities and permeabilities within the formation. The first porous medium consisted of relatively wide pores of the first order (fissures and blocks). The blocks in themselves were porous, consisting of grains separated by fine pores of the second order. The blocks fed liquid to the fissures and acted as a uniformly distributed source.

Variations of this model have been investigated by a number of authors in the field of petroleum engineering. The investigations led to various analytical expressions for the reservoir transient pressure behavior (Warren and Root, 1963; Odeh, 1965; Kazemi and others, 1969; de Swaan, 1976; Najurieta, 1976). In general, the workers have not been in agreement, perhaps because different results are possible for fractured systems depending upon the reservoir characteristics encountered. However, a careful inspection of the published analytical solutions indicates that these are essentially identical. Apparent differences come only from the definition of the various parameters used in the derivation.

To warrant this point, the double-porosity theory of Barenblatt and others is presented in detail hereafter. As before, the basic model is considered first, then inner and outer boundary conditions. Contributions of the various authors to the basic theory are stressed where appropriate and equivalences or differences among the various available solutions are pointed out.

Basic Model. One basic assumption in the model of Barenblatt and others is that any infinitesimal reservoir volume contains a large number of fissures and a large number of blocks. As a consequence, each point in space is associated with two pressures, namely, (1) the average liquid pressure p_f in the fissures in the vicinity of the point, and (2) the average liquid pressure p_m in the blocks in the vicinity of that same point.

Let V_f be the ratio of the total volume of the fissures to the bulk volume, and V_m that of the total volume of the blocks to the bulk volume.[1] If ϕ_f is the fissure porosity (ratio of pore volume in the fissures to the total volume of the fissures[2]) and ϕ_m the block porosity (ratio of pore volume in the blocks to the total volume of the blocks), the average reservoir porosity ϕ is given by

$$\phi = \phi_f V_f + \phi_m V_m \tag{21}$$

with

$$V_f + V_m = 1. \tag{22}$$

Assuming Darcy's law is applicable to the flow of fluids in both the fissures and the blocks, a mass balance on an elementary volume of fissure yields

$$\frac{\partial(\phi_f V_f \rho_f)}{\partial t} + \mathrm{div}\left(-\frac{k_f}{\mu}\rho_f \nabla p_f\right) - \rho_f q = 0, \tag{23}$$

where ρ_f is the density of the liquid in the fissures, μ its viscosity, k_f the fissure permeability, and q the volume of liquid flowing from the blocks into the fissures per *unit bulk volume* per unit time.

In the same way, a mass balance in the blocks yields

$$\frac{\partial(\phi_m V_m \rho_m)}{\partial t} + \mathrm{div}\left(-\frac{k_m}{\mu}\rho_m \nabla_m\right) + \rho_m q = 0, \tag{24}$$

where the parameters with subscript m have the same meaning for the blocks as the corresponding ones with subscript f have for the fissures.

In the case of constant overburden pressure, $\phi_f V_f$ depends on the fluid pressure in the fissures and can be assumed to be independent of the fluid pressure in the blocks. Hence

$$d(\phi_f V_f) = \phi_f V_f (c_r)_f dp_f, \tag{25}$$

where $(c_r)_f$ is the rock compressibility in the fissure system. Combining equations 3 and 25 with that for a slightly compressible fluid, finally yields

$$\frac{k_f}{\mu}\nabla^2 p_f = \phi_f V_f\left[(c_r)_f + (c)_f\right]\frac{\partial p_f}{\partial t} - q$$

$$= \phi_f V_f (c_t)_f \frac{\partial p_f}{\partial t} - q, \tag{26}$$

where $(c)_f$ and $(c_t)_f$ represent the fluid and the total compressibilities in the fissures, respectively.

In the same manner, equation 24 leads to

$$\frac{k_m}{\mu}\nabla^2 p_m = \phi_m V_m (c_t)_m \frac{\partial p_m}{\partial t} + q. \tag{27}$$

Equations 26 and 27 represent the general equations for a double-porosity medium.

In fractured systems, it is further assumed that the permeability of the blocks is low, so that the term on the left-hand side of equation 7 is negligible compared with the term on the right-hand side. Equations 6 and 7 then become, respectively,

$$\nabla^2 p_f = \frac{\phi_f V_f \mu (c_t)_f}{k_f}\frac{\partial p_f}{\partial t} + \frac{\phi_m V_m \mu (c_t)_m}{k_f}\frac{\partial p_m}{\partial t}$$

$$= \frac{1}{\eta_f}\frac{\partial p_f}{\partial t} + \frac{k_m}{k_f}\frac{1}{\eta_f}\frac{\partial p_m}{\partial t}, \tag{28}$$

where η is the hydraulic diffusivity and

[1]Table 1 gives the correspondence between the symbols used in this paper and those employed by previous authors.

[2]If fissures do not contain infill, $\phi_f = 1$.

TABLE 1. CORRESPONDENCE AMONG SYMBOLS USED BY VARIOUS AUTHORS FOR DOUBLE POROSITY MODEL.

	Barenblatt et al. (1960)	This paper	Warren and Root (1963) Kazemi et al. (1969)	This paper	Odeh (1965)	This paper	Kazemi (1969)	This paper	de Swaan (1976) Najurieta (1976)	This paper
Fissure pressure	p_1	p_f	p_2	p_f	p_f	p_f	$\rho(o)\int_0^{p_f} \frac{dp}{\rho(p)} + gz$	p_f	p_f	p_f
Block pressure	p_2	p_m	p_1	p_m	p_m	p_m	$\rho(o)\int_0^{p_m} \frac{dp}{\rho(p)} + gz$	p_m	p_m	p_m
Fissure pore volume to bulk volume ratio	m_1	$\phi_f V_f$	ϕ_2	$\phi_f V_f$	$\phi_f \beta$	$\phi_f V_f$	$\phi_f \frac{\delta}{h}$	$\phi_f V_f$		
Block pore volume to bulk volume ratio	m_2	$\phi_m V_m$	ϕ_1	$\phi_m V_m$	$\phi_m (1 - \beta)$	$\phi_m V_m$	$\phi_m \frac{h - \delta}{h}$	$\phi_m V_m$		
Fissure porosity	1	ϕ_f	1	ϕ_f	ϕ_f	ϕ_f	ϕ_f	ϕ_f	ϕ_f	ϕ_f
Block porosity			ϕ_m	ϕ_m	ϕ_m	ϕ_m	ϕ_m	ϕ_m	ϕ_m	ϕ_m
Fissure permeability	k_1	k_f	$\overline{k}_2 = \sqrt{k_{2x} k_{2y}}$	k_f	k_f	k_f/V_f	k_f	k_f/V_f	k_f	k_f/V_f
Block permeability	0		k_1	k_m	k_m	k_m/V_m	k_m	k_m/V_m	k_m	k_m/V_m
Fissure total compressibility	0		c_2	$(c_t)_f$	c_f	$(c_t)_f$	c_f	c_f	c_f	c_f
Block shape parameter	β_o/m_o	$(c_t)_f$	c_1	$(c_t)_m$	c_m	$(c_t)_m$	c_m	c_m	c_m	c_m
Interporosity parameter	α	αk_m	α	αk_m	α	α				
Formation thickness	h	h	h	h	h	h	$h - \delta$	h		
Fracture thickness							δ	h_f	h_f	$h_f = V_f h$
Parameters characteristic of fissured formation	$\frac{r}{\sqrt{\eta}}$	$r\sqrt{\frac{\alpha k_m}{k_f}}$	λ	$\lambda = a r_w^2 \frac{k_m}{k_f}$						
			ω	$\omega = \frac{\phi_f V_f (c_t)_f}{\phi_f V_f (c_t)_f + \phi_m V_m (c_t)_m}$						

$$\frac{\partial p_m}{\partial t} = \frac{-q}{\phi_m V_m (c_t)_m} . \tag{29}$$

Pseudo-steady-state Block to Fissure Flow: Zero-Fissure Compressibility. In their solution, Barenblatt and others (1960) considered that the fissure compressibility was negligible in equation 26 (due to the low volume of the fractures), and that the flow of fluid from blocks to fissures, represented by q, was occurring *under pseudo-steady-state conditions*

$$q = \frac{\alpha}{\mu} k_m (p_m - p_f) . \tag{30}$$

Hence

$$\frac{\partial p_m}{\partial t} = \frac{-\alpha k_m}{\phi_m V_m \mu (C_t)_m} (p_m - p_f) = - \alpha \eta_m (p_m - p_f) . \tag{31}$$

In equation 30, α is a parameter characteristic of the fissured rock and has the dimension of a reciprocal area. The pseudo-steady-state assumption has been used by other authors (Warren and Root, 1963; Odeh, 1965; Kazemi and others, 1969). The assumption has been justified by the fact that any transient effect was likely to be short, which could be inferred from published formulae describing the initiation of pseudo-steady-state flow (Muskat, 1946). However, this is not always correct, as will be discussed later.

Barenblatt and others rearranged equations 28, 29, and 30 to obtain

$$\phi_m V_m (c_t)_m \frac{\partial p_f}{\partial t} + \operatorname{div} \left[- \frac{k_f}{\mu} \operatorname{grad} p_f - \frac{k_f}{\alpha} \frac{\phi_m V_m (c_t)_m}{k_m} \frac{\partial}{\partial t} \operatorname{grad} p_f \right] = 0 , \tag{32}$$

which has a form similar to that of the usual diffusivity equation, with a fluid velocity equal to the term between brackets.

Equation 31 was then solved for the transient linear flow into a trench for the transient radial flow into a line source well. For the latter case, the usual reservoir initial and boundary conditions were assumed, namely, $p_f = p_i$ for $t \leq 0$, all r and $p_f = p_i$ for $r \to \infty$, all t, but the boundary condition at the well was written as

$$q = -\frac{2\pi k_f h}{\mu}\left[\left(r\frac{\partial p_f}{\partial r}\right) + \frac{1}{\alpha\eta_m}\frac{\partial}{\partial t}\left(r\frac{\partial p_f}{\partial r}\right)\right]_{r=r_w}, \quad r_w \to 0, \tag{33}$$

where h is the total formation thickness.

After integration, and with the condition

$$\left(r\frac{\partial p}{\partial r}\right)_{r\to 0} = 0 \quad \text{for } t = 0$$

(which indicates that the initial flow comes entirely from the blocks), the well boundary condition was obtained as

$$\left(r\frac{\partial p_f}{\partial r}\right)_{r=r_w} = -\frac{q\mu}{2\pi k_f h}\left[1 - \exp(-\alpha\eta_m t)\right], \quad r_w \to 0. \tag{34}$$

Substituting

$$p_{fD} = \frac{\sqrt{2\pi k_f h}}{q\mu}(p_i - p_f)$$

into equations 31, 32, and 34 and taking the Laplace transform, finally yields the solution

$$\bar{p}_{fD} = \frac{1}{s}\frac{1}{1 + \dfrac{r^2 s}{k_f/\phi_m V_m \mu (c_t)_m}\left(\dfrac{k_f}{\alpha k_m r^2}\right)} \cdot k_0\left(\frac{r\sqrt{s}}{\sqrt{k_f/\phi_m V_m \mu (c_t)_m}} \cdot \frac{1}{\sqrt{1 + \dfrac{r^2 s}{k_f/\phi_m V_m \mu (c_t)_m}\left(\dfrac{k_f}{\alpha k_m r^2}\right)}}\right), \tag{35}$$

where s is the Laplace transform parameter and K_0 the modified Bessel function of the second kind of zero order. At long times, equation 35 becomes

$$\bar{p}_{fD} = \frac{1}{s}k_0\left(\frac{r\sqrt{s}}{\sqrt{k_f/\phi_m V_m \mu (c_t)_m}}\right). \tag{36}$$

Because the fissure volume is considered to be small, V_m can be taken as unity, and inversion of equation 16 gives the well-known exponential integral function

$$p_{fD} = -\frac{1}{2}\,\mathrm{Ei}\left(-\frac{\phi_m \mu (c_t)_m r^2}{4k_f t}\right)$$
$$= \frac{1}{2}\left(\ln\frac{k_f t}{\phi_m \mu (c_t)_m r^2} + 0.80907\right). \tag{37}$$

The complete solution, obtained by inverting equation 35 is shown on Figure 16, where p_{fD} is plotted versus $k_f t/[\phi_m V_m \mu (c_t)_m r^2]$ as a function of

$$\lambda\left(\frac{r}{r_w}\right)^2 = \alpha r^2\frac{k_m}{k_f},$$

where

$$\lambda = \alpha r_w^2\frac{k_m}{k_f}.$$

The solution of Barenblatt and others, equation 35, was later modified to be read in terms of the block pressure p_m and has appeared as such in subsequent Russian literature on the subject (Romm, 1966). The differential equation for the flow in porous blocks is actually identical to equation 31 (or equation 28, except for the symbol p_m replacing p_f). The well-boundary condition, however, is written as

$$q = \frac{2\eta k_f h}{\mu}\left(r\frac{\partial p_m}{\partial r}\right)_{r=r_w}, \quad r_w \to 0 \tag{38}$$

on the assumption that the fluid produced at the well is that released from the porous blocks.

On the other hand, if movement of fluid to the well occurs only in the fracture system (Warren and Root, 1963), the well-boundary condition for the fissures becomes

$$q = \frac{2\pi k_f h}{\mu}\left(r\frac{\partial p_f}{\partial r}\right)_{r=r_w}. \tag{39}$$

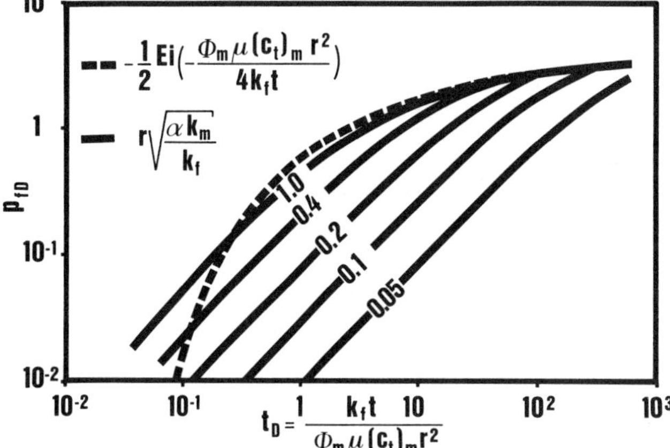

Figure 16. Zero-fissure compressibility-block pressure-type curves (from Barenblatt and others, 1960.

Because of the relation between p_f and p_m given by equation 31, the well-boundary condition for the block pressure is given by equation 33 with p_m replacing p_f.

As a consequence, *the same differential equation* (equation 31, or equation 28 in the general case) *leads to two different pressure solutions, depending upon which type of well-boundary condition is used* (equation 33 or 39). Relating one solution to the fissure pressure and the other to the block pressure is only a matter of choice. Actually, the same solutions have been presented as representative of the block pressure in some publications (Streltsova, 1976) and of the fissure response in subsequent ones (Streltsova-Adams, 1978a).

Finite Fissure Compressibility. The general equation for a double-porosity fractured system, equation 28, has been solved by Warren and Root (1963), Odeh (1965), and Kazemi and others (1969), in the case of a well producing at a constant rate q in an infinite radial reservoir. Warren and Root's and Odeh's solutions were analytical, whereas that of Kazemi and others was numerical. These authors used the assumption given by Barenblatt and others of pseudo-steady-state flow from blocks to fissures (equation 31).

Warren and Root and Kazemi and others used the well-boundary condition defined by equation 39, whereas Odeh (1965) wrote

$$q = \frac{2\pi k_f V_f h}{\mu}\left(r\frac{\partial p_f}{\partial r}\right)_{r=r_w}. \qquad (40)$$

This well-boundary condition, equation 39, has also been written (de Swaan, 1976; Najurieta, 1976) as

$$q = \frac{2\pi k'_f h_f}{\mu}\left(r\frac{\partial p_f}{\partial r}\right)_{r=r_w}. \qquad (41)$$

Equations 40 and 41 are, of course, equivalent to equation 39, and are simply obtained by using $k'_f V_f$ instead of k_f. This takes into account the area of flow available to the fractures at the wellbore ($h_f = V_f h$ is the cumulative fracture thickness). In the same way, $k'_m V_m$ is used in place of k_m. Substituting $k'_f V_f$ and $k'_m V_m$ for k_f and k_m in equations 28 and 31 yields (Odeh, 1965; de Swaan, 1976)

$$\nabla^2 p_f = \frac{\phi_f \mu (c_t)_f}{k'_f}\frac{\partial p_f}{\partial t} + \frac{k'_m V_m}{k'_f V_f}\frac{\phi_m \mu (c_m)_t}{k'_m}\frac{\partial p_m}{\partial t}$$

$$= \frac{1}{\eta_f}\frac{\partial p_f}{\partial t} + \frac{k'_m V_m}{k'_f V_f}\frac{1}{\eta_m}\frac{\partial p_m}{\partial t} \qquad (42)$$

and

$$\frac{\partial p_m}{\partial t} = -\frac{a k'_m}{\phi_m \mu (c_m)_t}(p_m - p_f) = \alpha\eta_m(p_m - p_f). \qquad (43)$$

Although equations 42 and 43 are equivalent to equations 28 and 31, respectively, *they imply a different type of system, where two media exist, but are not necessarily uniformly distributed.* This would be the case, for instance, of a two-layered reservoir. For a double-porosity fractured system, as defined by Barenblatt and others, I prefer using equation 39 by analogy with the corresponding homogeneous porous medium equation. There, the total reservoir thickness is used, although the thickness available to flow from the pores is only ϕh, where ϕ is the formation porosity.

Substituting

$$p_{fD} = \frac{2\pi k_f h}{q\mu}(p_i - p_f)$$

into equations 28, 31, 32, and 39, and taking the Laplace transform yields the transformed differential equation

$$\nabla^2 \bar{p}_{fD} = \frac{s}{\eta(s)}\bar{p}_{fD} \qquad (44)$$

where

$$\frac{1}{\eta(s)} = \frac{1}{\eta_f} + \frac{k_m}{k_f}\frac{\alpha}{\alpha + s/\eta_m}\frac{1}{\eta_m}. \qquad (45)$$

The solution of equation 18 is

$$\bar{p}_{fD} = \frac{1}{r_w}\frac{\sqrt{\eta(s)}}{s\sqrt{s}}\frac{K_0\left(r\sqrt{\frac{s}{\eta(s)}}\right)}{K_1\left(r_w\sqrt{\frac{s}{\eta(s)}}\right)} \qquad (46)$$

for a finite radius well (k_1 is the modified Bessel functions of the second kind of first order) and

$$\bar{p}_{fD} = \frac{1}{s}K_0\left(r\sqrt{\frac{s}{\eta(s)}}\right) \qquad (47)$$

for a line source well ($r_w \to 0$).

Equation 47 can also be written:

$$\bar{p}_{fD} = \frac{1}{s}K_0\left[\frac{r\sqrt{s}}{\sqrt{\eta_f}}\sqrt{1 + \frac{1}{\frac{\phi_f V_f (c_t)_f}{\phi_m V_m (c_t)_m} + \left(\frac{r^2 s}{\eta_f}\right)\left(\frac{k_f}{\alpha k_m r^2}\right)}}\right] \qquad (48)$$

which indicates clearly that p_{fD} is a function of three dimensionless variables, namely:

1. a dimensionless time $k_f t/[\phi_f V_f \mu (c_t)_f r^2]$ based on the fissure diffusivity,
2. the dimensionless group $\lambda(r/r_w)^2 = \alpha r^2(k_m/k_f)$ representing interporosity flow, which was already included in Barenblatt and others' solution (equation 35), and
3. a new parameter

$$\frac{\phi_f V_f (c_t)_f}{\phi_m V_m (c_t)_m} \quad \text{or} \quad \omega = \frac{\phi_f V_f (c_t)_f}{\phi_f V_f (c_t)_f + \phi_m V_m (c_t)_m}.$$

Equations 46 and 47 (or 48) are similar to the

corresponding equations for a finite radius and for a line-source well in a homogeneous infinite reservoir, respectively (van Everdingen and Hurst, 1949), except for the fact that $\eta(s)$ is a function of s. The inverted forms of equations 46 and 47 are therefore readily available for values of s such that $\eta(s)$ is a constant, and, in particular, at long and early time values.

1. *At early times* $(s \to \infty)$ equation 45 reduces to

$$\frac{1}{\eta(s)} = \frac{1}{\eta_f} \tag{49}$$

and the solution for the line-source well, obtained by inverting equation 47, is

$$p_{fD} = \frac{2\pi k_f h}{q\mu}(p_i - p_f) = -\frac{1}{2}\,\mathrm{Ei}\left(-\frac{\phi_f V_f (c_t)_f r^2}{4k_f t}\right). \tag{50}$$

Flow due to fluid compressibility thus comes only from the fractures, and the pressure response is the homogeneous reservoir equation for the fracture system. Flow from the matrix blocks is essentially zero.

2. *At long times* $(s \to 0)$ equation 45 yields

$$\frac{1}{\eta(s)} = \frac{1}{\eta_f} + \frac{k_m}{k_f}\frac{1}{\eta_m} = \frac{\phi_f V_f (c_t)_f + \phi_m V_m (c_t)_m}{k_f}\,\mu, \tag{51}$$

and inversion of equation 47 gives

$$p_{fD} = \frac{2\pi k_f h}{q\mu}(p_i - p_f)$$

$$= \frac{1}{2}\,\mathrm{Ei}\left\{\frac{\phi_f V_f (c_t)_f + \phi_m V_m (c_t)_m \quad \mu r^2}{k_f t}\right\}$$

or (52)

$$p_{fD} = \frac{1}{2}\left\{\ln\frac{k_f t}{\phi_f V_f (c_t)_f + \phi_m V_m (c_t)_m \quad \mu r^2} + 0.80907\right\}.$$

At long times, the fractured reservoir behavior is therefore equivalent to that of a homogeneous porous medium with a permeability equal to the fissure permeability, and a porosity-compressibility product equal to the arithmetic sum of that for each medium.

It is interesting to notice that the long-time equivalent hydraulic diffusivity given by equation 51 is independent of α and only depends upon the shape of the matrix blocks. Two particular distributions are of interest. If the blocks and fissures can be represented as infinite horizontal slabs with thicknesses h_m and h_f, respectively (Fig. 17), then

$$V_f = \frac{h_f}{h_m + h_f} \quad \text{and} \quad V_m = \frac{h_m}{h_m + h_f},$$

and equation 51 becomes

$$\frac{1}{\eta} = \frac{1}{\eta_f} + \frac{k'_m}{k'_f}\frac{h_m}{h_f}\frac{1}{\eta_m}, \tag{53}$$

where k'_m and k'_f are as defined in equation 22. In the same way, spherical matrix blocks of radius r_m (Fig. 18) give

$$V_f = \frac{2\pi r_m^2 h_f}{2\pi r_m^2(2r_m + h_f)} \quad \text{and} \quad V_m = \frac{\frac{4}{3}\pi r_m^3}{2\pi r_m^2(2r_m + h_f)},$$

and equation 51 yields

$$\frac{1}{\eta} = \frac{1}{\eta_f} + \frac{k'_m}{k'_f}\frac{2}{3}\frac{r_m}{h_f}\frac{1}{\eta_m}. \tag{54}$$

3. *At intermediate times*, there is a transition from fracture flow to flow from fractures and matrix blocks, during which the drawdown remains constant. The drawdown value at which the transition occurs

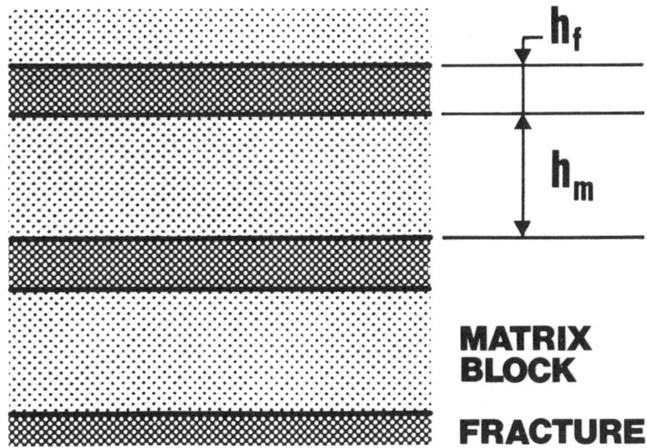

Figure 17. Schematic of a fissured reservoir with horizontal fissures and blocks.

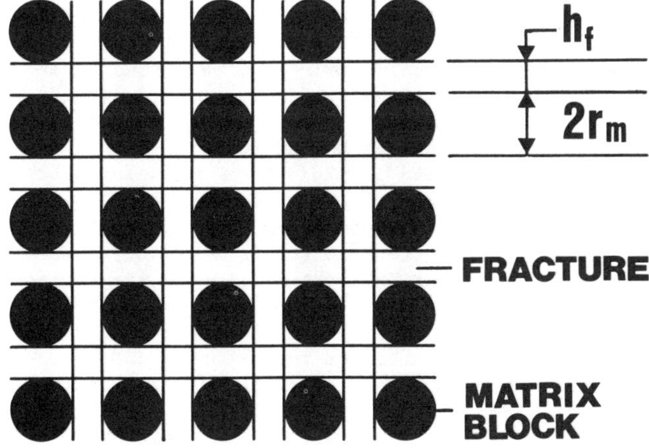

Figure 18. Schematic of a fissured reservoir with spherical matrix blocks.

is independent of early- and long-time behaviors, and is only a function of λ; it is equal to

$$(p_{fD})_{transition} = K_0\left(\frac{r}{rw}\sqrt{\lambda}\right), \tag{55a}$$

(Streltsova-Adams, 1978b), which in practical cases reduces to

$$(p_{fD})_{transition} = \ln \frac{2r_w}{\gamma r\sqrt{\lambda}}, \tag{55b}$$

(Bourdet and Gringarten, 1980).

4. *In the general case,* equation 47 can be inverted numerically, or by means of the classical inversion formula. In fact, the exact inversion of equation 47 (or 48) is available in a paper on unconfined aquifers by Boulton (1963).

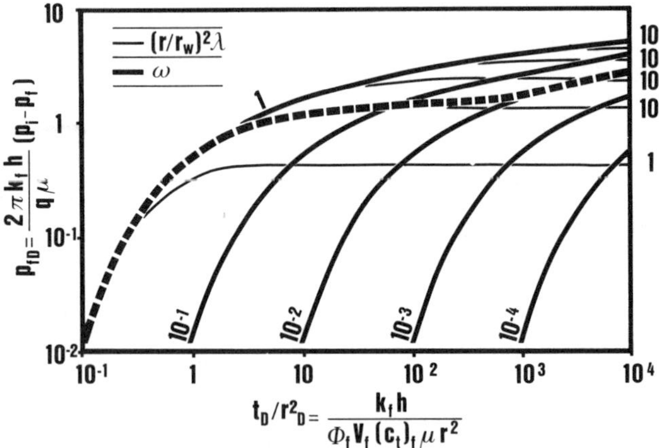

Figure 19. Log-log type curves for the fissure pressure in a double-porosity fractured reservoir (Bourdet and Gringarten, 1980).

Values of p_{fD} obtained by Bourdet and Gringarten (1980) by means of a numerical Laplace inverter as in Mavor and Cinco (1979), are presented in log-log coordinates on Figure 19 as a function of $t/n_f r_w^2$ for various values of ω and $(r/r_w)^2\lambda$.

As mentioned before, for a given value of ω, two asymptotic solutions described by equations 50 and 52, respectively, limit the family of curves; their horizontal displacement corresponds to the "delay yield" from the various blocks equal to ω.

The same function is plotted on a semilog graph in Figure 20, where all the curves are now located between parallel straight lines with slopes equal to 1.151, which correspond to the log approximation of the Ei functions in equations 50 and 52, respectively.

It must be pointed out that the type curves of Figure 19 are identical to those corresponding to the time-drawdown curves of an unconfined aquifer (Boulton, 1963). There is therefore a possibility of misinterpretation of tests in aquifers exhibiting such behavior.

An approximate solution to equation 47 was given by Warren and Root (1963) and Odeh (1965) as

$$p_{fD} = \frac{1}{2}\left\{\ln \frac{k_f t}{[\phi_f V_f (c_t)_f + \phi_m V_m (c_t)_m]\mu r^2} + 0.80907 - \mathrm{Ei}\left[\frac{\alpha k_m t}{\phi_m V_m \mu (c_t)_m}\right] + \mathrm{Ei}\left[-\frac{\alpha k_m t}{\phi_m V_m \mu (c_t)_m}\,\frac{\phi_f V_f (c_t)_f + \phi_m V_m (c_t)_m}{\phi_f V_f (c_t)_f}\right]\right\}. \tag{56}$$

Equation 56 becomes identical to equation 52 at long times, and reduces to

$$p_{fD} = \frac{1}{2}\left(\ln \frac{k_f t}{\phi_f V_f \mu (c_t)_f r^2} + 0.80907\right) \tag{57}$$

at intermediate times.

Warren and Root (1963) included anisotropy for the fracture permeability, and also considered the case of a finite radial reservoir. Taking anisotropy into account results in using $\sqrt{k_{xf}k_{yf}}$ in place of k_f and in subtracting

$$\ln\left[(k_{xy}/k_{yf} + 1)/2\sqrt{k_{xf}/k_{yf}}\right]$$

from p_{fD} in equation 56.

For the isotropic case, Warren and Root presented a semilog plot of p_{fD} versus

$$\frac{k_f t}{[\phi_f V_f (c_t)_f + \phi_m V_m (c_t)_m]\mu r_w^2}$$

for various values of λ and of

$$\omega = \frac{\phi_f V_f (c_t)_f}{\phi_f V_f (c_t)_f + \phi_m V_m (c_t)_m}$$

and stressed the existence of the *two semilog*

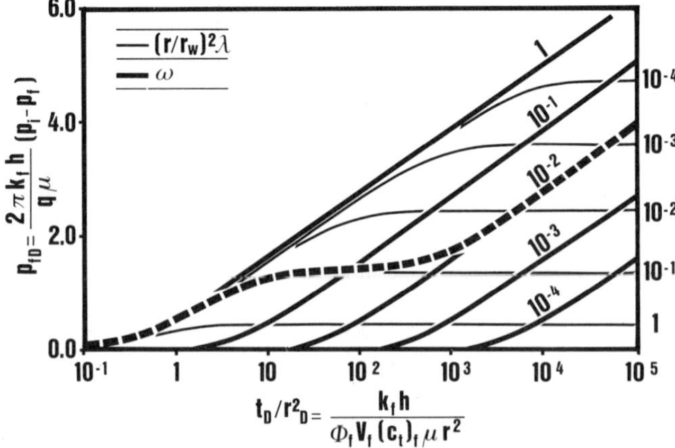

Figure 20. Semi-log type curves for the fissure pressure in a double-porosity fractured reservoir (Bourdet and Gringarten, 1980).

straight lines corresponding to equations 57 and 52, as a characteristic of double porosity fractured reservoir behavior. Their graph was identical to Figure 20 (within the precision of Warren and Root's approximation) but concerned mainly large dimensionless time values. Warren and Root's results were later confirmed by Kazemi and others' (1969) numerical solution.

On the other hand, Odeh (1965) only considered the long-time approximation of equation 47 (equation 52), and therefore did not recognize the existence of the first straight line.

The explanation of this discrepancy between Warren and Root's and Odeh's conclusions can be found in Figure 19; a semilog straight line develops only if the match of the corresponding data points on a Theis curve extends beyond the start time of the semilog straight line. If this occurs during both fissure flow and fissure-plus-block flow, two straight lines will be seen on a semilog graph, as indicated by Warren and Root (1963). Such a situation would only occur at small $\lambda(r/r_w)^2$ values (less than 10^{-2}).

On the other hand, at higher values of $\lambda(r/r_w)^2$, transition from fissure flow to fissure-plus-block flow occurs before the start of the semilog approximation on the fissure Theis curve, and only one semilog straight line, representing fissure plus block flow will eventually appear, as observed by Odeh (1965).

In any case, as stressed in the Introduction, the use of the basic model is limited to interference tests. When feasible, type-curve matching of drawdown data with the type curve of Figure 19 yields k_fh, ω, λ, and both fissure and matrix storativities. Then λ can be used to compute the size of the blocks if a value for k_m is available from core analysis. Semilog analysis provides k_fh in the usual way, when the last (fissure plus block) straight line is present but the two semilog straight lines are required to obtain ω, which is then given by

$$\omega = 10^{-\delta p/m},$$

where δp is the vertical distance between the semilog straight lines, and m the semilog straight line slope (Warren and Root, 1963). Here, λ is usually not accessible.

Transient Block to Fissure Flow. Transient flow from blocks to fissures was considered by de Swaan (1976). His model is identical to that of Warren and Root, but he used equations 42 and 41, with $k'_f = k_f/V_f$ and $k'_m = k_m/V_m m$ instead of equations 28 and 39, respectively.

The source term corresponding to equation 29 represents, in this case, the flow from blocks to fissures, *per unit volume of fracture*, per unit time. It was expressed by de Swaan as a convolution integral

$$\left(\frac{q}{V_f}\right) = \frac{-2}{A_m h_f} \int_0^t \frac{\partial \Delta p_f(\tau)}{\partial \tau} q_m(t-\tau)d\tau. \qquad (58)$$

Here $(1/2\ A_m h_f)$ is the volume of fissure associated with a block, under the assumption that the fracture thickness h_f is small enough so that the flow from every matrix block is evenly and instantaneously distributed in the associated fracture volume. A_m is the surface area of a matrix block, and q_m the flow across the block surface due to a unit-step pressure drop of the fissure pressure

$$q_m = \frac{k'_m}{\mu} A_m(-\nabla \Delta p_m) \text{at the block surface.} \qquad (59)$$

De Swaan computed q_m for an infinite reservoir with horizontal fractures and matrix blocks in the shape of infinite slabs, and for cubical or parallelepiped matrix blocks whose behavior was assumed to be similar to that of spherical blocks of equal volume.

Laplace transform of the resulting dimensionless fissure pressure function is still given by equation 46 for a finite radius well, and by equation 47 for a line-source well with

$$\frac{1}{\eta(s)} = \frac{1}{\eta_f} + \frac{2k'_m}{k'_f h_f} \frac{1}{\sqrt{s\eta_m}} \tanh\left(\frac{h_m}{2}\sqrt{\frac{s}{\eta_m}}\right)$$
$$= \frac{1}{\eta_f} + \frac{2k_m}{k_f h_m} \frac{1}{\sqrt{s\eta_m}} \tanh\left(\frac{h_m}{2}\sqrt{\frac{s}{\eta_m}}\right) \qquad (60)$$

for the horizontal slab block model, and

$$\frac{1}{\eta(s)} = \frac{1}{\eta_f} + \frac{2k'_m}{k'_f h_f} \frac{1}{r_m s}\left[r_m\sqrt{\frac{s}{\eta_m}} \coth\left(r_m\sqrt{\frac{s}{\eta_m}}\right) - 1\right]$$
$$= \frac{1}{\eta_f} + \frac{3k_m}{k_f r_m} \frac{1}{r_m s}\left[r_m\sqrt{\frac{s}{\eta_m}} \coth\left(r_m\sqrt{\frac{s}{\eta_m}}\right) - 1\right] \qquad (61)$$

for spherical blocks of radius r_m.

Both equations 60 and 65 reduce to equation 49 at early times (fissure flow, $s \to \infty$); and to equation 51 at long times (fissure plus block flow, $s \to 0$), when k_f and k_m are used. It can actually be shown that equations 60 and 61 become equivalent to equation 45 at long times, with $\alpha = 12/h_m^2$ and $\alpha = 15/r_m^2 m$ respectively.

At intermediate times, however, the pressure behavior of a double porosity system with the transient flow assumption is very different from that with the pseudo-steady-state assumption. This is apparent on Figure 21, where equation 47 was plotted after numerical inversion for both pseudo-steady-state and transient block to fissure flow conditions, at the same ω and λ values. No stabilization exists in the transition curves with the transient block to fissure flow assumption. These transition curves are functions of $(r/r_w)^2(\lambda/\omega)$ times a coefficient that depends upon the shape of the blocks (Fig. 22; Bourdet and Gringarten, 1980).

Inner Boundary Conditions. The effect of wellbore storage and skin was investigated by Mavor and Cinco (1979) who obtained the pressure in the fissure system at the well and at any point in the reservoir.

With the nomenclature used in the present paper, Mavor's solution at the well can be written in the Laplace domain as

$$\bar{p}_{fD} = \frac{1}{s\left\{\frac{s\mu C}{2\pi k_f h} + \frac{1}{S + \frac{K_0\left(r_w\sqrt{\frac{s}{\eta(s)}}\right)}{r_w\sqrt{\frac{s}{\eta(s)}}\,K_1\left(r_w\sqrt{\frac{s}{\eta(s)}}\right)}}\right\}}, \quad (63)$$

where C is the wellbore storage constant and S the infinitesimal skin. Equation 63 was derived under the assumption of pseudo-steady-state flow from blocks to fissures, with $\eta(s)$ given by equation 45. It is also valid, however, for transient block-to-fissure flow, with $\eta(s)$ given by either equation 60 or equation 61. Equation 63 reduces to equation 46 when C = S = 0, and to equation 47 if, in addition, $r_w \to 0$. As found in the basic model case, equation 63 is similar to the corresponding equation for a homogeneous reservoir (Agarwal and others, 1970).

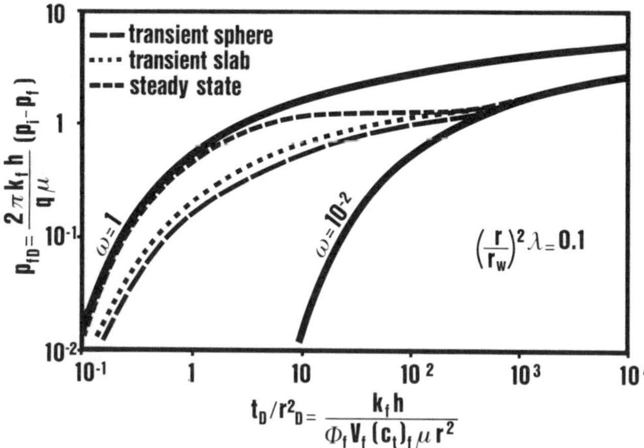

Figure 21. A comparison between solutions with pseudo-steady-state and transient flow from blocks to fissures: ω=0.01, λ(r²/rw) = 0.1 (Bourdet and Gringarten, 1980).

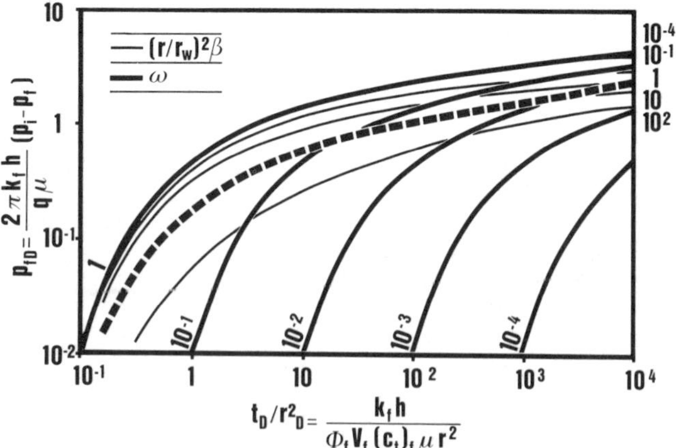

Figure 22. Log-log type curves for the fissure pressure in a double-porosity fractured reservoir with transient flow from blocks to fissures (Bourdet and Gringarten, 1980).

Bourdet and Gringarten (1980) showed that, in the presence of skin, equation 63 could be approximated in all practical cases by

$$\bar{p}_{fD} = \frac{1}{s\left[\frac{s\mu C}{2\pi k_f h} + \frac{1}{\ln\left(2/\gamma\sqrt{\frac{s}{\eta(s)}}\,r_w^2 e^{-2S}\right)}\right]}, \quad (64)$$

where γ = 1.78 is the exponential of Euler's constant.

1. *At early times*, equation 46 reduces to equation 49 and equation 64 takes the form

$$\bar{p}_{fD} = \frac{1}{s\left\{\frac{s\mu C}{2\pi k_f h} + \frac{1}{\ln\left(2/\gamma\sqrt{\frac{s\mu C}{2\pi k_f h}\frac{1}{(c_D)_f e^{2S}}}\right)}\right\}}, \quad (65)$$

which corresponds to wellbore storage and skin behavior in the fissure system alone. $(c_D)_f$ is the dimensionless wellbore storage constant defined as in equation 20, but with respect to the fissure system parameters,

$$(c_D)_f = \frac{c}{2\pi\phi_f V_f (c_t)_f h r_w^2}. \quad (66)$$

2. *At late times*, equation 46 reduces to equation 51, and equation 64 becomes

$$\bar{p}_{fD} = \frac{1}{s\left\{\frac{s\mu C}{2\pi k_f h} + \frac{1}{\ln\left(2/\gamma\sqrt{\frac{s\mu C}{2\pi k_f h}\frac{1}{(c_D)_{f+m} e^{2S}}}\right)}\right\}} \quad (67)$$

representing wellbore storage and skin behavior in the total (fissure plus block) system, with

$$(c_D)_{f+m} = \frac{c}{2\pi[\phi_f V_f (c_t) + \phi_m V_m (c_t)_m] h r_w^2}. \quad (68)$$

Equations 65 and 67 define, in fact, a single family of curves, plotted as p_D versus t_D/c_D (equation 19) in Figure 15: pressure data measured at the active well during constant rate drawdown in a double-porosity reservoir follow first one of the $c_D e^{2S}$ curves of Figure 15, and then shift to another curve corresponding to a lower $c_D e^{2S}$ value. The first $c_D e^{2S}$ curve represents flow in the fissures, whereas the second $c_D e^{2S}$ curve represents flow in the total system. Thus ω is simply obtained as the ratio of the last $c_D e^{2S}$ value (equation 68) to the first $c_D e^{2S}$ value (equation 67).

3. *At intermediate times*, $[s r_w/\eta(s)]$ is equal to λ when the pseudo steady-state assumption of equation 30 is used. Transition curves from fissure flow to total system flow thus *only depend on* λe^{-2S} and are obtained from equation 64 as

$$p_{fD} = \ln \frac{2}{\gamma \sqrt{\lambda^{-2S}}} \left[1 - \exp - \left\{ \frac{2\pi k_f ht/\mu c}{\ln \dfrac{2}{\gamma \sqrt{\lambda e^{-2S}}}} \right\} \right] . \quad (69)$$

The problem of a well with wellbore storage and skin in a double-porosity reservoir can thus be broken down into two parts.

1. The problem of a well with wellbore storage and skin in a homogeneous reservoir, whose solution is represented by the type curves in Figure 15.

2. The effect of the double-porosity reservoir, represented by the λe^{-2S} curves from equation 69.

The corresponding drawdown type curves are simply obtained by superposing these last curves onto those of Figure 15 (Fig. 23, Bourdet and Gringarten, 1980).

The type curves of Figure 23 are, in fact, more general than indicated by equation 63. Like those of Figure 15, they also include the case of an infinite-conductivity vertical fracture with wellbore storage ($c_De^{2S} < 0.5$). As a result, type-curve analysis yields information on the quality of the well (damaged, undamaged, acidized, or fractured) the usual wellbore storage and skin parameters (S, C) and parameters of the double-porosity reservoir (k_fh, ω, and λ).

Restrictions on the use of semilog analysis methods are the same as in the basic model case: two parallel semilog straight lines are present on drawdown data only when the match on the first and last c_De^{2S} curves extends beyond the start of the semilog approximation (Figure 23). This is not very common in practice, and usually only the last semilog straight line is available. The shape of the Horner (1951) semilog plot, however, is always very typical.

Bourdet and Gringarten (1980) also derived a set of type curves for transient flow conditions from blocks to fissures, with $\eta(s)$ given by either equation 60 or equation 61. The difference between these and the type curves of Figure 23 is in the transition curves, which are c_De^{2S} curves displaced downwards along a 45° direction, and labeled in terms of ($c_De^{2S}/\lambda e^{-2S}$). Contrary to the type curves of Figure 23, however, these last type curves appear to find little practical application.

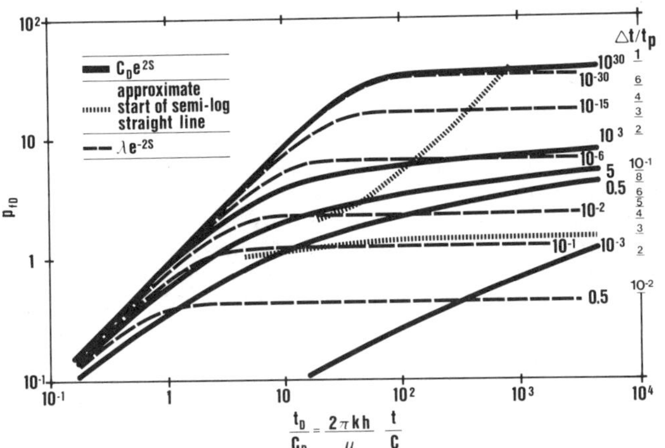

Figure 23. Log-log type curves for a well with wellbore storage and skin in a double-porosity fractured reservoir (Bourdet, D. and Gringarten, A. C., 1980).

Outer Boundary Conditions. As mentioned before, the case of a line-source well in a finite radial reservoir was studied by Warren and Root (1963). The same problem, but with wellbore storage and skin, was investigated by Mavor and Cinco (1979).

A solution for fissure flow in a fractured unconfined aquifer was actually obtained by Boulton and Streltsova (1978). They found that double-porosity effects were predominant at early times and unconfined aquifer effects at late times. The initial portion of each time drawdown curve represents the fissure elastic response and is approximated by equation 50 at very early times. As time increases, the delayed elastic response of the porous blocks contributes to the fissure flow, whose response approaches the exponential integral function equation 52. The effect of the water table change as time increases finally supersedes the aquifer elastic response, and the final segments of the type curves merge with the late-time pressure function for unconfined aquifers.

Multilayered Model. Many authors (such as Kazemi, 1969, or Boulton and Streltsova, 1977a, b) have represented fissured reservoirs as a series of uniformly spaced horizontal matrix and fissure layers, where flow is allowed in both radial and vertical directions. Because of the symmetry, only one layer of fissure and one layer of block is actually included in the model. Governing equations are 26 and 27, written in terms of k'_f and k'_m, with a source term given by either equation 30 or 58. The well-boundary dary condition is expressed by equation 41.

As mentioned before, such a two-layered system is equivalent to a double-porosity medium if k_f and k_m are substituted for k'_f and k'_m, respectively. In other words, if one considers a two-layered system, where the fissure layer has a permeability k'_f and a thickness h_f, and the block layer, a permeability k'_m and a thickness h_m, the corresponding double-porosity solution can be obtained from the two-layered solution by simply replacing

$$k'_f \text{ by } k_f = k'_f V_f = k'_f \frac{h_f}{h_f + h_m}$$

and

$$k'_m \text{ by } k_m = k'_m V_m = k'_m \frac{h_m}{h_f + h_m} .$$

The permeability-thickness product, obtained from analysis, however, remains unchanged:

$$k_f h_f = k'_f (V_f h) = k_f h.$$

Composite Model. An example of the use of such a model for analyzing pressure drawdowns, measured at the producing well, was given by Adams and others

(1968). The authors noticed a peculiar behavior of a low-permeability, fractured dolomite gas reservoir that could be interpreted as an increase in permeability at some distance from the well. Unlike the previous double-porosity model, the well could communicate with the fracture system via the relatively tight matrix. The fracture reservoir was thus analyzed by means of an equivalent "composite" reservoir; that is, the system consists of two concentric, homogeneous and isotropic annular regions of distinctly different permeability (Hurst, 1960).

FIELD EXAMPLES

Although many studies have dealt with the problem of flow in fissured formations, convincing field examples are scarce in the literature, especially concerning the validity of the double-porosity approach.

Warren and Root (1965) presented build-up data for two wells that appear to exhibit "two parallel semilog straight lines" (Fig. 24). In the absence of log-log, however, it is not possible to ascertain that these two straight lines are indeed the ones predicted by the double-porosity fissure medium theory. As pointed out by Bourdet and Gringarten (1980), the first straight line may simply be due to wellbore storage effects.

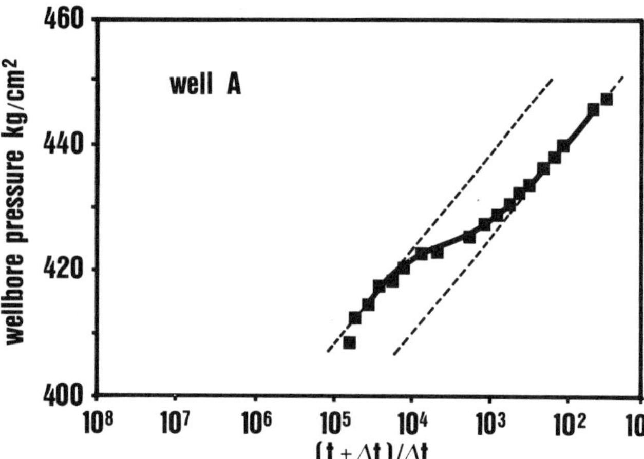

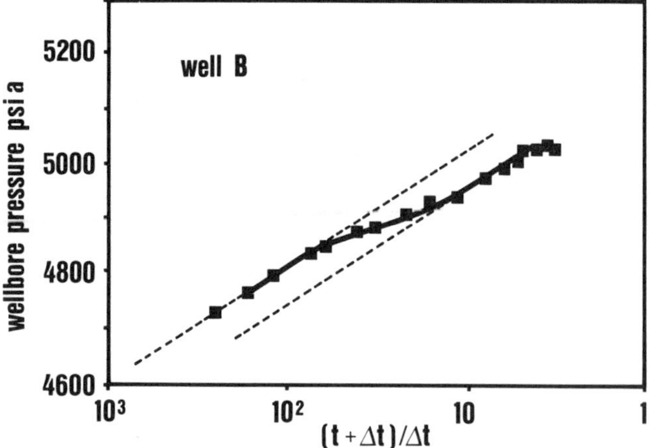

Figure 24. Example of parallel semi-log straight lines (Warren and Root, 1965).

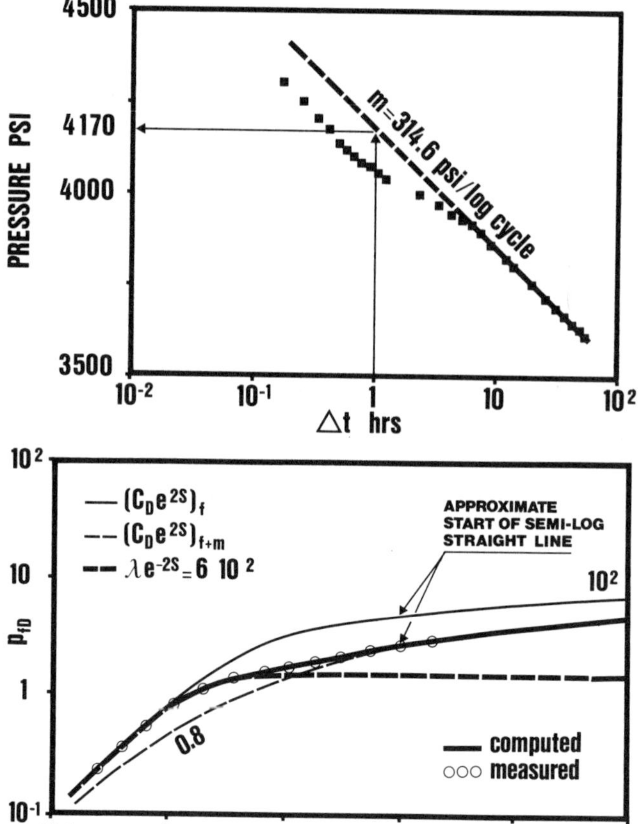

Figure 25. Well-test data from a fractured oil reservoir (Bourdet and Gringarten, 1980).

Another example of the "two semilog straight line behavior," from Borevsky and others (1973), has been reported by Streltsova-Adams (1978a). On the other hand, Odeh (1965) gave an example of a test where the flow behavior of the fractured reservoir was that of a homogeneous system and speculated that, in most cases, the early semilog straight line would be superseded by wellbore storage effects. Examples similar to Odeh's are presented by Bourdet and Gringarten (1980) (Fig. 25), and Gringarten and others (1980). In fact, I encountered many such examples, whereas tests with two parallel semilog straight lines have been very rare. As already discussed, the "two parallel semilog straight line" behavior is only possible for very small e^{-2S} values, which are only reached in damaged wells or in fissured reservoirs with large blocks. The general behavior of the test data, however, is usually different from that of a homogeneous system.

Other commonly reported pecularities of the flow behavior of fractured systems, especially in ground-water hydrology, include evidence of fluid communication between widely separated wells, resulting in almost instantaneous and similar response to fluid withdrawal; changes in slope on semilog drawdown plots; and drawdowns away from the pumping well greater than in its vicinity. Examples of such behaviors, published in the oil and ground-water literatures, have been gathered by Streltsova-Adams (1978a).

In order to evaluate the various methods of interpretation summarized in the present paper, I conducted a series of pumping tests at the Bureau de Recherches Géologiques et Minières (BRGM, Orléans, France), in fissured aquifers of various characteristics.

One common feature observed on all of the pumping tests was the very short duration of what is referred to in the models as the "early time period." These early time periods last only a few minutes at most, compared with a few hours or tens of hours in oil reservoirs. As a consequence, very precise and fast data acquisition systems had to be used in order to record data from the very beginning of the test. (For this reason, most well test data already available were not adequate for analysis.) The data acquisition systems we used included strip-chart recorders, x-y plotters, and tape recorders. Pressure measurements were made with electronic transducers, or bubbler-type equipment. Flow rates at the pumping wells were also continuously monitored and controlled with a constant head tank such as the one described by Gale (1975).

Another problem was the large variety of flow behaviors that followed the early time periods, because of the widely differing aquifer geometries and boundary conditions encountered. These included infinite reservoir thickness, leakage, recharge, unconfinement, and impermeable upper and lower boundaries. As a result, many fractured-well models had to be derived for applying the "equivalent homogeneous reservoir" concept, in order to match well test results for the whole test period. This, however, was easily achieved with the Green's function and product solution method (Gringarten and Ramey, 1973).

Figure 26 presents the results of a test on a limestone karst aquifer, with two perpendicular systems of vertical fractures, and stratification subhorizontal joints (Auzet and others, 1975; Auzet and Gringarten, 1976). Drawdown and build-up data are included for the pumping well and three observation wells. Although the three observation wells are at about the same distance from the pumping well, their responses are very different. Furthermore, for all of the wells, build-up data are not identical to the drawdown data but appear to have been shifted toward increasing time values for the pumping well and toward decreasing time values for the observation wells. This shift may be caused by wellbore storage effects in the pumping and observation wells.

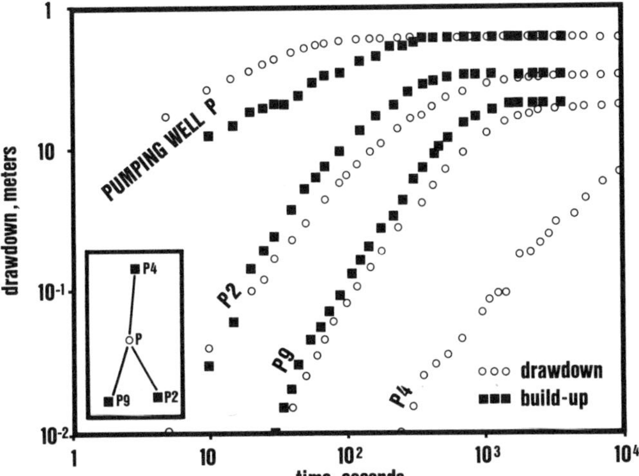

Figure 26. Well-test data from a limestone karst aquifer (Auzet and Gringarten, 1976).

At long times (after 1h of pumping), the pressures in all of the wells become stabilized, which can be attributed to a constant head boundary (aquifer open to the aquifer above).

At early times, a log-log half-unit slope straight line is noticeable on drawdown and build-up pumping well data, but not on observation well data, although two of these (P 2 and P 9) are known to be connected to the pumping well by a horizontal fracture. In order to check this point, a second test was performed with packers positioned in P 2 and P 9 above the main horizontal fractures for eliminating wellbore storage effects (Pitard, 1976). The results are shown in Figure 27, where an early-time log-log half-unit slope straight line is apparent on P 2 and P 9, as predicted by the horizontal fracture model theory (Gringarten and Ramey, 1973).

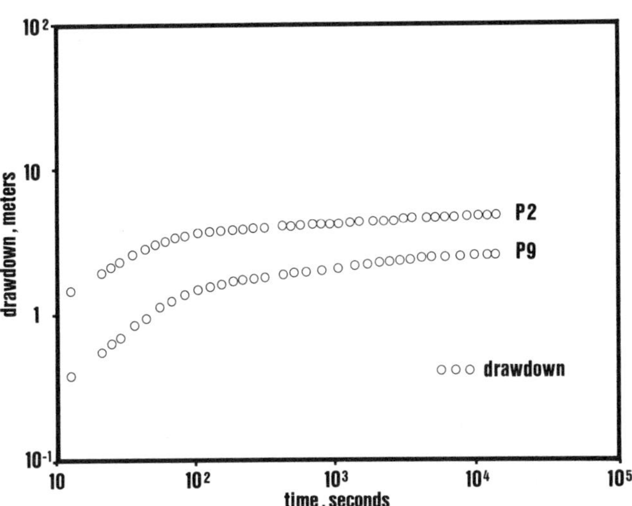

Figure 27. Well-test data from a limestone karst aquifer with packers in observation wells (Pitard, 1976).

The model used to interpret the tests shown on Figures 26 and 27 was that of a single horizontal fracture in a homogeneous, horizontally anisotropic aquifer of infinite lateral and vertical extent, with a constant-pressure upper boundary. This model yields an average permeability value equal to $\rho g/\mu \sqrt{k_{xf}k_{yf}} = 10^{-6}$ m/s, which seems reasonable (Pitard, 1976).

Figure 28 presents the results of a pump test in a fractured confined chalk aquifer, where blocks have a cubic shape with a side length of the order of 0.1 m (Gringarten and Bertrand, 1978). On Figure 28, three pressure measurements could be made within the first second of the test. The early-time flow period lasts only for 2 min. and is entirely dominated by wellbore storage. Buildup data appear

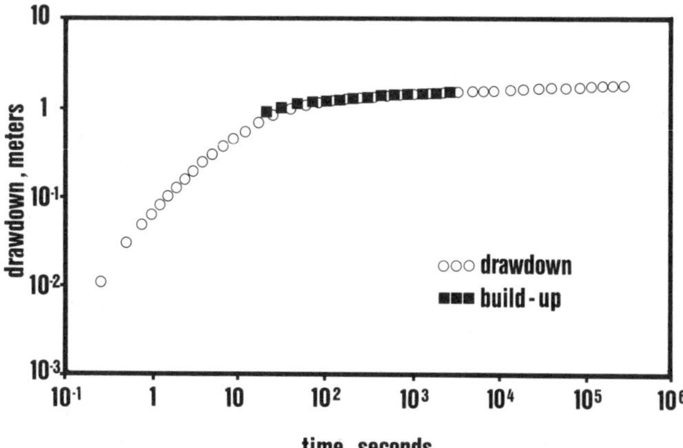

Figure 28. Well-test data from a chalk aquifer (Gringarten and Bertrand, 1978). Log-log plot.

to be identical to the drawdown data, although most of the early time buildup data are missing because of the inability of the bubbler to follow rapid recovery pressures.

The wellbore storage obtained from the drawdown data in Figure 28 is equal to twice that computed from the actual well radius, and can be attributed to the presence of fractures intersecting the wellbore. These should extend to some distance into the formation, as can be inferred from a Cartesian plot of drawdown pressure versus the square root of producing time. A straight line is obtained for producing times up to 16s, which indicates linear flow from the matrix into the fractures communicating with the well.

The data in Figure 28 were analyzed with a model consisting of a single horizontal fracture with storage at the midpoint of an infinite reservoir with impermeable upper and lower boundaries (Fig. 13). With $H_D = 20$, this yields a transmissivity $(\rho g/\mu)k_f h = 1.8 \times 10^{-2}$ m^2/s, compared to a semilog analysis result of 1.5×10^{-2} m^2/s (Fig. 29). Assuming an aquifer storage constant equal to $S = \phi_m c_m h \rho g = 10^{-2}$, one obtains the matrix permeability $(\rho g/\mu)\ k_m = 5 \times 10^{-6}$ m/s, and an "equivalent" horizontal fracture radius equal to $r_f = 11.4$ m. Alternatively, interpretation could be made with the double-porosity type curve of Figure 23.

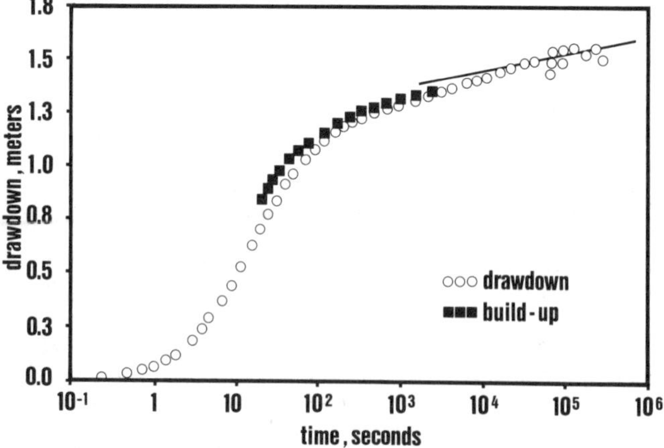

Figure 29. Well-test data from a chalk aquifer (Gringarten and Bertrans, 1978). Semi-log plot.

The results of a pumping test (test #2) in a porphyritic granitic rock with three main directions of fractures (horizontal, vertical, and at a 45° angle) are shown in Figure 30 for the pumping well and one observation well 14 m from the pumping well. One striking feature of the plots is the lack of symmetry between drawdown and buildup data in each well, and the similarity of the pressure behavior of both wells, except at early times (less than 2 min.) where wellbore storage effects are dominant at the pumping well. The early-time behavior of the pumping well actually represents a case of changing wellbore storage in which the storage coefficient changes from a value that corresponds to actual well radius to a value that is three times as large.

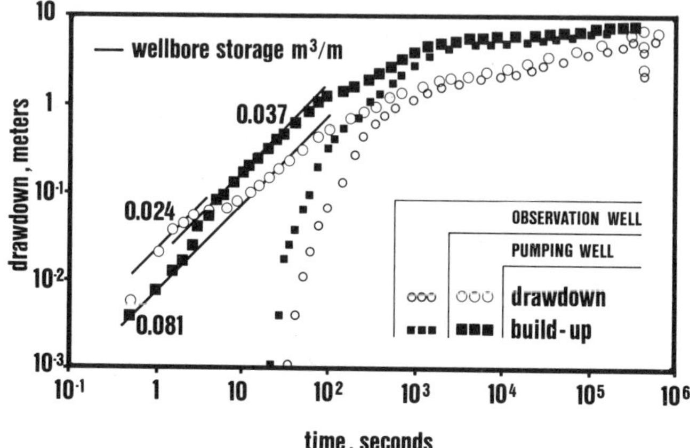

Figure 30. Well-test data from a granitic rock. Log-log plot.

It was speculated that this changing wellbore storage was due to fractures intersecting the pumping well and participating in the flow during either the early time period (as a storage effect) or the transient flow period, thus possibly changing the flow configuration. This interpretation is supported by the results from a previous test on the same wells, performed just after air drilling with a rock bit, and before well clean-up (test # 1 on Fig. 31). Because the fractures intersecting the pumping well were plugged by cuttings, the pressure behavior of the pumping well is similar to that corresponding

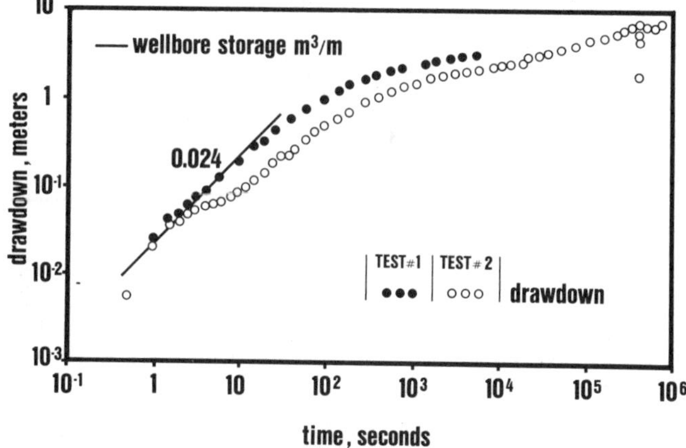

Figure 31. Well-test data from a granitic rock, before and after clean-up. Log-log plot.

to the usual homogeneous radial flow, with a wellbore storage effect due to the actual well radius. After clean-up (test #2), the flow configuration changed into the fissure-type flow already mentioned. However, the clean-up had no effect on the observation well.

The test of Figure 30 was interpreted with a model consisting of a vertical fracture with wellbore storage in a closed square (Ramey and Gringarten 1975), for the drawdown data, and with a model consisting of a horizontal fracture for the buildup data (Fig. 32). Results of the analysis appear consistent, with a transmissivity $[(\rho g/\mu)k_f h]$ equal to 6×10^{-5} m^2/s from the drawdown and 8×10^{-5} m^2/s from the buildup, a vertical fracture half-length of 13 m, and a horizontal fracture radius of 11 m. Here again, interpretation is possible with the double-porosity model.

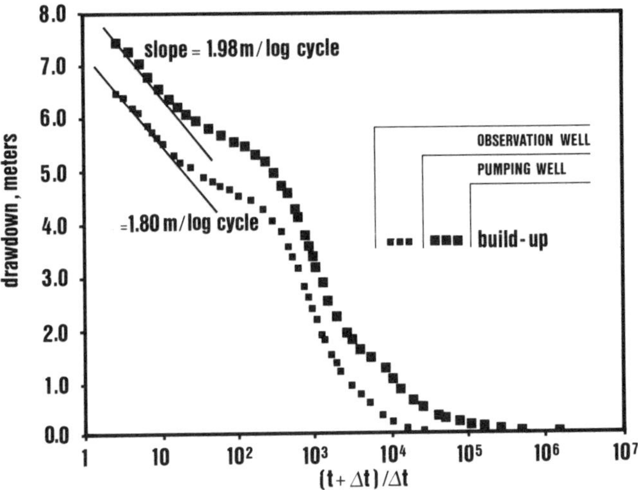

Figure 32. Well-test data from a granitic rock. Semi-log Horner plot.

Figure 33 shows the results of two pumping tests performed on a granitic rock at Lawrence Berkeley Laboratory. As in the previous example, there is a lack of symmetry between draw-down and buildup data corresponding to the first test; wellbore storage changes during the early part of the drawdown (first 10 min), decreasing from a storage value about three times that corresponding to the actual wellbore radius, to the actual well radius storage value. The storage coefficient remains constant during the buildup, and corresponds to the actual well radius.

The pressure behavior during the drawdown appears similar to homogeneous transient radial flow, whereas during buildup it corresponds to fracture flow. As in the previous example, the change in pressure behavior during the test could be caused by fractures participating or not participating in the flow.

As a check, a second test was performed with a lower pumping rate, so that the liquid level in the well would remain within the top layer at all times (test #2 in Fig. 33). This time, drawdown and buildup data are identical, and have a shape similar to that of the buildup in test 1, indicating a fracture-type flow.

The test was interpreted with the double-porosity type curves of Bourdet and Gringarten (1980) for the buildup of test 1, and the drawdown and the buildup of test 2. Results are consistent, giving a transmissivity equal to 6×10^{-6} m^2/s with $\lambda = 4 \times 10^{-3}$ and $\omega = 0.04$ (Fig. 34).

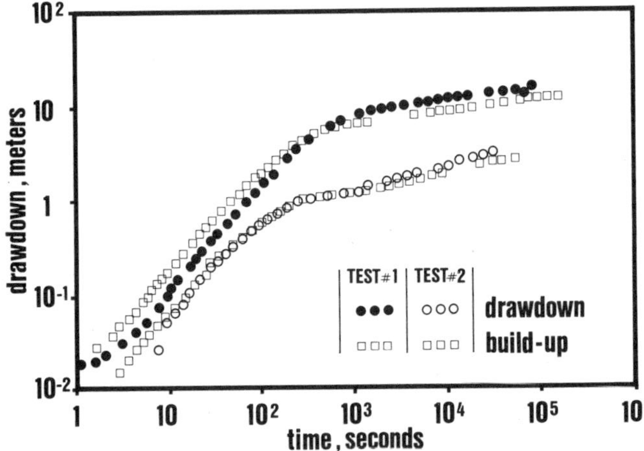

Figure 33. Well-test data from a granitic rock. Log-log plot.

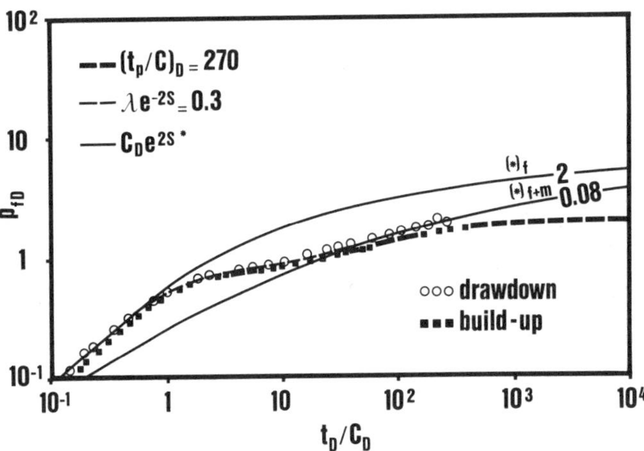

Figure 34. Well-test data from a granitic rock. Log-log match with double-porosity type curve of Figure 23.

CONCLUSIONS

From the numerous models that have been proposed in the past, and in view of the pump test data presented in this paper, it is clear that flow in fractured formations is a very complex problem and that no method of analysis is universal.

What has been proposed so far can be organized into three main types of solutions:

1. The deterministic approach, mainly used in geotechnical engineering for small-scale problems, which is based on an accurate and detailed description of individual fracture systems.

2. The double-porosity medium approach, which assumes a uniform distribution of matrix blocks and fissures throughout the reservoir, and yields governing differential equations that are the

exact equivalent of the diffusivity equation in homogeneous porous systems.

3. The "equivalent homogeneous reservoir" approach, which considers only the main trends of the pressure behavior of the fissured reservoir, and tries to relate them to a known model of lower complexity.

The distinction among the three types of solutions is actually very artificial, and future theoretical development in the study of the flow behavior of fissured systems will have to consider a combination of these types. A detailed description of the wellbore environment will be necessary to fully understand the well-pressure behavior. Along that line, new methods, such as the one proposed by Jouanna and Fras (1979), based on the signal theory and frequency analysis of harmonic pumping tests, appear very promising.

The double-porosity model, which now incorporates wellbore storage effects, skin, and fractures communicating with the producing well (Fig. 23) has been found to be particularly applicable by the author in many cases where other methods were unsatisfactory or yielded results of questionable physical significance. The value of this model, however, should be validated by checking analysis results with available information on block size and fissure volume.

The area where research is most needed remains that of pressure measurement. From the field examples presented in this paper, it is clear that very important information is available from very early time data, which has not been available in the past because of the lack of accuracy in these time ranges. This is similar to what has been experienced in the study of porous reservoirs, where early time data have been disregarded for a long time as unreliable, until it was realized that they could be related to wellbore or near-wellbore effects such as wellbore storage, skin effect, and hydraulic or natural fractures.

Well test data from many different fractured formations, obtained under carefully controlled conditions as described in this paper, would help our understanding of fissured reservoir flow behavior. Although many different results are expected, a few specific patterns will probably emerge, which could be used to classify fissured reservoirs. Development of adequate theoretical models would then be greatly simplified.

NOMENCLATURE

A_m	Area of matrix block
C	Wellbore storage coefficient
(c)	Fluid compressibility
(c_r)	Rock compressibility
(c_t)	Total compressibility
g	Gravity
h	Total reservoir thickness
h_f	Cumulative fracture thickness, or thickness of fissure layer
h_m	Thickness of block layer
k	Permeability
p	Pressure
q	Flow rate from blocks to fissures, or discharge rate at production well
r	Distance to production well
r_f	Horizontal fracture radius
r_m	Radius of spherical matrix blocks
r_w	Radius of production well
s	Laplace transform parameter
S	Infinistesimal skin
t	Production time
V	Ratio of total volume of one porous system to bulk volume
z, y, z	Coordinates
x_f	Vertical fracture half length
α	Block shape parameter
η	Hydraulic diffusivity
λ	Interporosity flow parameter
μ	Fluid viscosity
ρ	Fluid density
ϕ	Porosity
ω	Ratio of fissure to total system (blocks plux fissures) storativities

SUBSCRIPTS

D	Dimensionless
f	Fractures
i	Initial
m	Matrix blocks

APPENDIX

The formulae used in this paper are written in the Darcy system of units employed in the petroleum engineering literature. By definition, a permeability of 1 darcy (D) corresponds to a flow of 1 $cm^3/s/cm^2$ of a liquid with a viscosity of 1 centipoise, for a pressure gradient of 1 atm/cm.

The permeability k is a property of the porous medium alone, and is independent of the flowing fluid. It has the dimension of an area (1 darcy = 0.987×10^{-8} cm^2).

The units in the Darcy system of the parameters used in the formulae are as follows:

c	Compressibility	vol/vol/atm
ϕ	Porosity	fraction
h	Thickness	cm
k	Permeability	darcy
μ	Viscosity	centipoise (cp)
p	Pressure	atm
q	Flow rate at reservoir level	cc/sec
r	Radius	cm
t	Time	sec

In ground-water hydrology, the permeability, or hydraulic conductivity K includes the fluid properties μ and ρ : $K = (k/\mu\ \rho g$ and 1 darcy (at 20°C) is equal to $9.8\ 10^{-6}$ m/s. The transmissibility is $T = Kh = (kh/\mu\ \rho g$.

In the same manner, the aquifer storage, equal to the volume of water that a unit decline in head releases in a vertical prism of aquifer with unit cross-sectional area, is given by $S = \phi ch\rho g$.

The drawdown is related to the pressure as: $s = (p_i - p)/\rho g$.

REFERENCES CITED

Agarwal, R. G., Al-Hussainy, R., and Ramey, H. J., Jr., 1970, An investigation of wellbore storage and skin effect in unsteady liquid flow: I, Analytical treatment: Society of Petroleum Engineers Journal, v. 10, no. 3, p. 279-290.

Adams, A. R., Ramey, H. J., Jr., and Burgess, R. J., 1968, Gas well testing in a fractured carbonate reservoir: Journal of Petroleum Technology, v. 20, no. 10, p. 1187.

Asfari, A., and Witherspoon, P. A., 1973, Numerical simulation of naturally fractured reservoirs, presented at 3rd Numerical Simulation of Reservoir Performance Symposium of SPE of AIME, Houston, Tex., Jan. 10-12, 1973: Dallas, Society of Petroleum Engineers, SPE- 4290.

Auzet, J. P., Drogue, C., and Gringarten, A. c., 1975, Analyse d'essais de pompage dans les massifs calcaires fissurés. Interprétation du début du régime transitoire, 3ème Réunion Annuelle des Sciences de la Terre, Montpellier, France (abstract): Montpellier, France, Université des Sciences et Techniques du Languedoc.

Auzet, J. P., and Gringarten, A. C., 1976, Interprétation de pompages d'essais dans un massif calcaire fissuré. Etude de la nappe karstique du Lez (Hérault): Orleans, France, Bureau de Recherches Géologiques et Mineìres, Rapport 76 SGN 276 GTH.

Barenblatt, G. E., Zheltov, I. P., and Kochina, I. N., 1960, Basic concepts in the theory of homogeneous liquids in fissured rocks. Journal of Applied Mathematics and Mechanics (USSR) v. 24, no. 5, p. 1286-1303.

Borevsky, B. V., Samsonov, B. G., and Jazvin, L. S., 1973, Metodika opredelenia parametrov vodonosnich gorizontov po dannivn otkacheck (Aquifer parameters determination from pumping test analysis): Moscow, Nedra.

Boulton, N. S., 1963, Analysis of data from non-equilibrium pumping tests allowing for delayed yield from storage: Proceedings of the Institute of Civil Engineering, v. 26, p. 469-482.

Boulton, N. S., and Streltsova, T. D., 1977a, Unsteady flow to a pumped well in a fissured water-bearing formation: Journal Hydrology, v. 35, p. 257-269.

___1977b, Unsteady flow into a pumped well in a two-layered water-bearing formation: Journal of Hydrology, v. 35, p. 245-256.

___1978, Unsteady flow to a pumped well in an unconfined fissured aquifer: Journal of Hydrology, v. 37, p. 349.

Bourdet, D., and Gringarten, A. C., 1980, Determination of fissure volume and black size in fractured reservoirs by type curve analysis, presented at the SPE-AIME 55th Annual Technical Conference and Exhibition, Dallas, Texas, Sept. 21-24: Dallas, Society of Petroleum Engineers, SPE-9293.

Cinco, H., and Samaniego, F., 1978, Transient pressure analysis for fractured wells, presented at the SPE-AIME 53rd Annual Fall Technical Conference and Exhibition, Houston, Texas, Oct. 1-3: Dallas, Society of Petroleum Engineers, SPE-7490.

Cinco, H., Samaniego, F., and Dominguez, N., 1978, Transient pressure behavior for a well with a finite conductivity vertical fracture: Society of Petroleum Engineers Journal, v. 18, no. 4, p. 253.

Collins, R. E., 1961, Flow of fluids through porous material: New York, Reinhold Publishing Corp., p. 115.

Elkins, L. F., and Skov, A. M., 1960, Determination of fracture orientation from pressure interference: Transactions of AIME, v. 219, p. 301.

Gale, J., 1975, A numerical, field and laboratory study of flow in rocks with deformable fractures [Ph.D. Dissertation]: Berkeley, University of California.

Gringarten, A. C., 1978, Reservoir limit testing for fractured wells, presented at the SPE-AIME 53rd Annual Fall Technical Conference and Exhibition, Houston, Texas, Oct. 1-3: Dallas, Society of Petroleum Engineers, SPE-1452.

Gringarten, A. C., and Bertrans, L., 1978, Détermination des caractéristiques hydrauliques des aquifères fissurées par pompages d'essai en régime transitoire - Application aux nappes de la craie: Presented at Colloque Regional Hydrogéologie de la Craie du Bassin de Paris, Rouen, France, May 25-26, Bureau de Recherches Géologiques et Minières, Rapport 78 SGN-669 GEG.

Gringarten, A. C., and Ramey, H. J., Jr., 1973, The use of source and Green's functions in solving unsteady flow problems in reservoirs: Society of Petroleum Engineers Journal, v. 13, no. 5, p. 285.

___1974, Unsteady-state pressure distributions created by a well with a single horizontal fracture, partial penetration, or restricted entry: Society of Petroleum Engineers Journal, v. 14, no. 4, p. 413.

Gringarten, A. C., Ramey, H. J., Jr., and Raghavan, R., 1974, Unsteady-state pressure distributions created by a single infinite conductivity vertical fracture: Society of Petroleum Engineers Journal, v. 14, no. 4., p. 347.

___1975, Applied pressure analysis for fractured wells: Journal of Petroleum Technology, v. 27, no. 7, p. 887.

Gringarten, A. C., and Witherspoon, P. A., 1972, A method of analyzing pumping test data from fractured aquifers, in Proceedings, Symposium on Percolation in Fissured Rock: Stuttgart, International Society of Rock Mechanics, v. 3, p. B 1 - B 9.

Gringarten, A. C., and others, 1979, A comparison between different skin and wellbore storage type curves for early-time transient analysis, presented at the SPE-AIME 54th Annual Technical Conference and Exhibition, Las Vegas, Nevada, Sept. 23-26: Dallas, Society of Petroleum Engineers, SPE-8205.

Gringarten, A. C., and others, 1980, Horner type-curve analysis: presented at the SPE-AIME 55th Annual Technical Conference and Exhibition, Dallas, Texas, Sept, 21-24: Dallas, Society of Petroleum Engineers, SPE-9291.

Hantush, M. S., 1957, Non-steady flow to a well

partially penetrating an infinite leaky aquifer: Proceedings of the Iraqui Science Society, no. 1.

Horner, D. R., 1951, Pressure build-up in wells, in Proceedings, 3rd World Petroleum Congress: Leiden, E. J. Brill, v. II, p. 503.

Hurst, W., 1960, Interference between oil fields: Transactions of AIME, v. 219, p. 175-192.

Huskey, W. L., and Crawford, P. B., 1967, Performance of petroleum reservoirs containing vertical fractures in the matrix: Society of Petroleum Engineers Journal, v. 19, no. 6, p. 221.

Jacob, C. E., 1950, Flow of Groundwater, in Rouse, H., ed., Engineering hydraulics: New York, John Wiley and Sons, Ch. 5.

Jones, P., 1956, Reservoir limits tests: Oil & Gas Journal, v. 54, no. 59, p. 184.

Jouanna, P., and Fras, G., 1979, Introduction á la reconnaissance, dans l'espace, des fréquences, de milieux fissurées par essais d'eau transitoires. Notion de signatures spectrales et application au cas d'horizons fissurés reconnus par pompage harmonique. Unicité des solutions: Comptes rendus de l'Académie des Sciences de séance du 15 janvier 1979, Paris, v. 288, no. 2.

Kazemi, H., 1969, Pressure transient analysis of naturally fractured reservoirs with uniform fracture distribution: Society of Petroleum Engineers Journal, v. 9, no. 11, p. 451.

Kazemi, H., Seth, M. S., and Thomas, G. W., 1969, The interpretation of interference tests in naturally fractured reservoirs with uniform fracture distribution: Society of Petroleum Engineers Journal, v. 4, no. 4, p. 463-472.

Kiraly, L., 1971, Groundwater flow in heterogeneous anisotropic fractured media: A simple two-dimensional electric analog: Journal of Hydrology, v. 12, no. 3, p. 225.

Louis, C., 1969, A study of groundwater flow in jointed rock and its influence on the stability of rock masses: Imperial College, University of London, Rock Mechanics Research Report no. 10.

Louis, C., 1975, Introduction à l'Hydraulique des Roches, [Doctorate of Sciences thesis]: Paris, Paris VI University.

Mavor, M. J., and Cinco, H., 1979, Transient pressure behavior of naturally fractured reservoirs, presented at the 1979 California Regional Meeting of the Society of Petroleum Engineers of AIME, Ventura, California, April 18-20. Dallas, Society of Petroleum Engineers, SPE-7977.

Moran, J. H., and Finklea, E. E., 1962, Theoretical analysis of pressure phenomena asociated with the wireline formation tester: Journal of Petroleum Technology, v. 14, no. 8, p. 889.

Muskat, M., 1946, The flow of homogeneous fluids through porous media: Ann Arbor, Michigan, J. W. Edward, Inc.

Najurieta, H. L., 1976, A theory for the pressure transient analysis in naturally fractured reservoirs: Instituts Mexicano del Petroleo.

Odeh, A. S., 1965, Unsteady-state behavior of naturally fractured reservoirs: Society of Petroleum Engineers Journal, v. 17, no. 3, p. 245.

Papadopulos, S. S., 1965, Nonsteady flow to a well in an infinite anisotropic aquifer, in Proceedings, Dubrovnik Symposium on Hydrology of Fractured Rocks, v. 1, p. 21-31.

Parsons, M. L., 1972, Determination of hydrogeological properties of fissured rocks, in Proceedings, 24th Geological Congress, Montreal, Canada: Section II, Hydrogeology, p. 89-99.

Parsons, R. W., 1966, Permeability of idealized fractured rock: Society of Petroleum Engineers Journal, v. 6, no. 2, p. 126.

Pitard, J., 1976, Contribution à l'interprétation des essais par pompage, dans les roches fissurés [Thèse de Doctorat de Spécialité (Géologie Appliquée]: Montpellier, France, Université des Sciences et Techniques du Languedoc.

Prats, M., 1972, The influence of oriented arrays of thin impermeable shale lenses or of highly conductive natural fractures on apparent permeability anisotropy: Journal of Petroleum Technology, v. 24, no. 10, p. 1219-1221.

Ramey, J. J., Jr., 1970, Short-time well test data interpretation in the presence of skin effect and wellbore storage: Journal of Petroleum Technology, v. 22, no. 1, p. 97.

Ramey, H. J., Jr., and Gringarten, A. C., 1976, Effect of high-volume vertical fractures on geothermal steam well behavior, in Proceedings, Second U. N. Symposium on the Development and Use of Geothermal Resources, San Francisco, May 20-29: Washington, D. C., U.S. Government Printing Office (Lawrence Berkeley Laboratory), v. 3, p. 1759-1762.

Romm, E. S., 1966, Filtratsionnie ssvoistsa teschinovatich porod (Flow phenomena in fractured rocks): Moscow, Nedra (in Russian).

Saad, K. F., 1967, Determination of the vertical and horizontal permeabilities of fractured water bearing formation: Bulletin IASH, v. 3, p. 22.

Serafim, L. J., 1968, Influence of interstitial water on the behavior of rock masses, in Stagg, K. G., and Zienkiewics, O. C., eds., Rock mechanics in engineering practice: London, Willen & Sons, p. 55-97.

Serafim, L. J., and del Campo, A., 1965, Interstitial pressures on rock foundations of dams: Journal of Soil Mechanics and Foundations Division ASCE, v. 91, no. 5, p. 65.

Sharp, J. C., 1970, Fluid flow though fissured media [Ph.D. dissertation]: London, Imperial College of Science and Technology.

Snow, D. T., 1965, A parallel plate model of fractured permeable media [Ph.D. dissertation]: Berkeley, University of California.

Snow, D. T., 1968, Anisotropic permeability of fractured rock, in De Wiest, R., ed., Hydrology and flow through porous media (Muskat Volume): New York, Academic Press.

Strelstova, T. D., 1976, Hydrodynamics of groundwater flow in a fractured formation: Water Resources Research, v. 12, no. 3, p. 405-414.

Strelstova-Adams, T. D., 1978a, Well hydraulics in heterogeneous aquifer formations in Chow, V. T., ed., Advances in Hydrosciences: v. 11, p. 357-423.

___1978b, Fluid flow in naturally fractured reservoirs, in Proceedings, Second Invitational Well Testing Symposium, October 25-27, Berkeley, California: Berkeley, Lawrence Berkeley

Laboratory, LBL-8883.
de Swaan, A. O., 1976, Analytic solutions for determining naturally fractured reservoir properties by well testing: Society of Petroleum Engineers Journal, v. 16, no. 3, p. 117.
Theis, C. V., 1935, The relation between the lowering of the piezometric surface and the rate and duration of discharge of a well using groundwater storage: Transactions, American Geophysical Union, p. 519-524.
van Everdingen, A. F., and Hurst, W., 1949, The application of the Laplace transformation to flow problems in reservoirs: Transactions, AIME, v. 186, p. 305.
Warren, J. E., and Root, P. J., 1963, Behavior of naturally fractured reservoirs: Society of Petroleum Engineers Journal, v. 3, no. 3, p. 245.
Warren, J. E., and Root, P. J., 1965, Discussion on "Unsteady-State Behavior of Naturally Fractured Reservoirs": Society of Petroleum Engineers Journal, v. 5, no. 2, p. 64.
Wilson, C. R., 1970, An investigation of laminar flow in fractured porous rocks [Ph.D. dissertation]: Berkeley, University of California.
Wilson, C. R., and Witherspoon, P. A., 1970, An investigation of laminar flow in fractured porous rocks, Berkeley, University of California, Department of Civil Engineering, Pub. 70-6.

Manuscript Accepted by the Society October 14, 1980

Printed in U.S.A.

Geological Society of America
Special Paper 189
1982

Well-Loss Function and the Skin Effect: A Review

HENRY J. RAMEY, JR.
Department of Petroleum Engineering
Stanford University
Stanford, California 94305

ABSTRACT

The well-loss function, CQ^2, and the effective wellbore radius constitute two very important ideas in Jacob's classic study (1947). Both ideas were accorded extreme importance in the petroleum engineering field. Swift and Kiel published results of their award-winning study on non-Darcy flow in 1962, and the skin-effect concept (related to effective wellbore radius) was presented independently by both Hurst and van Everdingen in 1953.

Jacob's well-loss function has been generalized to the form CQ^n in the current literature, although this form seems to work well only over the range of flow rates used in testing. The petroleum engineering approach suggests that the form should be $C(S_T + DQ)Q$, where S_T is the skin effect, a constant, and D is the non-Darcy flow coefficient, often a constant also. The new form contains CQ^n as a special case for a limited range of flow; however, n values would lie between 1 and 2. Field experience in water-well testing indicates n values from 2.5 to 2.7. Fetkovich (1973) presented information on oil and gas flow indicating that transmissivity depends upon pressure (thus also upon flow rate) for solution gas-drive oil production. There appears to be an analogy possible for high-rate production of gas-saturated water.

INTRODUCTION

In a classic study published in 1947, C. E. Jacob introduced the important concepts of "well loss" and the "effective radius" of a well. Jacob specifically defined "well loss" to include the head loss caused by friction as water flows across the well screen upward inside the casing to the pump intake. He said that this required an additional term t for laminar flow of water through the porous medium to the well screen:

$$H_w = BQ + CQ^2. \tag{1}$$

The first term on the right in equation 1 represents the head loss due to laminar flow through the porous medium. The second term indicates the approximate head loss due to the frictional pressure drop when water flows through the screen and up the casing to the pump intake. Jacob made it clear that the flow within the well could represent frictional losses not exactly proportional to the square of the flow rate. It is well known that frictional effects on flow in pipe lead to a head loss proportional to the flow rate for laminar flow, and proportional to the square of the flow rate for fully turbulent flow. For intermediate flow regimes, the power of the flow-rate term could lie between the values of unity and 2. In an answer to a prepared discussion by Carl Rohwer, C. W. Jacob responded that high-precision measurements might warrant the assumption that the drawdown obeys the law

$$H_w = BQ + CQ^n \quad n < 2. \tag{2}$$

Jacob indicated that the exponent n in equation 2 should be less than 2. This would be a reasonable expectation if the main cause of the second term on the right in equation 2 were frictional resistances to flow through the well screen and within the well casing to the "well intake." The exact meaning of Jacob's phrase "well intake" is not clear for a surface pump with a tubing string. He infers that in this case the well loss should include frictional losses in the tubing. However, measurement of a pumping liquid level in the annulus would actually yield the head at the tubing shoe. It appears that the proper interpretation of "well intake" is the entrance of a tubing string connected to the pump.

Several misconceptions of Jacob's work appear to have arisen over the years. Hantush (1964) credits Rorabaugh (1953) with originating equation 2, although it appears that Jacob actually proposed this form in his 1947 publication discussion. Also, Rorabaugh cites Jacob (1947) for suggesting the assumption that flow near the well face in the formation would be turbulent. Rorabaugh gives the impression that Jacob intended that the well-loss function should include head losses within the porous medium and near the sand face caused by high-velocity flow. If the velocity were high enough, results would violate Darcy's law for laminar flow. High-velocity flow in porous media was not mentioned in Jacob's paper, and turbulent flow was indicated only for flow within the well screen and well casing. However, Rorabaugh cited discussions with Jacob, R. W. Stallman, N. J. Lusczynski, and W. F. Guyton for several years before his remarkable 1953 study of high-velocity (nonlaminar) flow near the sandface of the well.

Both Jacob and Rorabaugh presented step drawdown testing methods that could be used to estimate the coefficients B and C in equations 1 and 2 and coefficient n in equation 2. Jacob (1947) also made the very important suggestion that a step-drawdown test could be used to obtain the effective wellbore radius. Jacob defined the effective wellbore radius as the distance, measured radially from the axis of the well, at which the theoretical drawdown based on the logarithmic head distribution equals the actual drawdown just outside the screen. From this and other discussions, Jacob clearly proposed that the effect of a change in permeability in the formation immediately outside the well screen be evaluated.

Jacob made another important observation concerning transient flow into a well with a well loss and an effective well radius. Equations 1 and 2 are written for steady-state flow. Jacob extended their application to the case of transient flow in an infinitely large system. He considered the effect of the well-loss function for head changes measured at the pumping well, and described interference and pressure-recovery effects for transient phenomena. Jacob also presented field-test data to illustrate his concepts. Truly his work remains one of the remarkable papers in the development of both ground-water and petroleum technology on well-test analysis.

SKIN EFFECT AND WELLBORE STORAGE

One of the key contributions to the petroleum engineering literature concerning wellbore effects was that of van Everdingen and Hurst (1949). They presented dimensionless response functions for a finite-radius well (see also Hantush, 1964) and introduced the wellbore storage concept later used by Cooper and others (1967).

Although both were landmark studies, they were not necessarily the first papers in this field. Jacob frequently referred to earlier work by Theis (1935), and the work of van Everdingen and Hurst had roots in earlier work by Moore and others (1933), Hurst (1934), Muskat (1937a), and Carslaw and Jaeger (1947).

Although a discussion of all these references is not necessary here, one example is worthwhile. Moore and others (1933) used sonic liquid-level determination to find the amount of fluid withdrawn from the annulus between the casing and tubing. With this, they corrected the surface production rate to a sand-face production rate, which is a sophisticated use of the concept of wellbore storage. The 1949 publication by van Everdingen and Hurst presented a log-log type-curve for the case of a well with wellbore storage (see also Chatas, 1953) such as that later used by Ramey (1964) and Cooper and others (1967).

In addition to solutions for wells with a finite radius, an important contribution of van Everdingen and Hurst (1949) was the addition of the concept of wellbore storage, which they called "annulus unloading." They stated that although a constant well-production rate could be established at the surface, the initial fluid withdrawal came from depletion of the fluids stored within the wellbore. That is, the flow through the sand face would start at a very low rate and increase toward the constant surface-production rate over some time interval. In 1950, Miller and others examined the effect of wellbore storage on pressure recovery after production ceased. They selected the term "afterflow" because they observed that closing a valve of a well at the surface would not necessarily stop flow through the sand face. Flow into the wellbore would continue until the well reached a static pressure. Both annulus unloading and afterflow caused distinct changes in the shape of the early portion of a pressure drawdown or build-up.

In 1953, van Everdingen and Hurst, in two separate papers, presented the concept of the "infinitesimal skin effect," which has had a great impact on pressure-transient analysis in the petroleum engineering field. Both authors stated that attempts to match field-performance data with their 1949 solutions were a problem. Although graphs of producing head versus the logarithm of time did yield semilog straight lines (as indicated by Jacob in 1947), the level of the semilog straight lines differed from the analytic solutions. The analytic solution of 1949 had taken the form

$$\frac{2\pi T H_w \rho}{Q} = P_D(t_D). \qquad (3)$$

For constant-rate production in an infinitely large medium, the dimensionless function on the right in equation 3 is

$$P_D(t_D) \simeq \frac{1}{2}\ln(4t_D) - 0.5772$$

$$= \frac{1}{2}\ln(t_D + 0.80907) \quad t_D > 5, \qquad (4)$$

where the dimensionless time t_D is

$$t_D = \frac{kt}{\phi \mu c_t r_w^2} = \frac{Tt}{S r_w^2}. \qquad (5)$$

Equation 4 suggests that a graph of drawdown versus the logarithm of time should produce a straight line whose slope is inversely proportional to the transmissivity.

In 1953, van Everdingen and Hurst defined a new problem. They proposed that the main reason for a mismatch between their previous analytic solutions and field data was the presence of an "infinitesimal skin" on the producing well face. This "skin effect," sometimes called the "skin factor," was caused by a reduction in the sand-face permeability caused by mud filtrate and mud-particle invasion during the completion of the well. Their concept of the skin effect was reciprocally related to the heat-transfer film coefficient in convective heat transport in pipes. However, both van Everdingen and Hurst coupled the skin-effect concept with wellbore storage, or annulus unloading. Each produced an analytic solution for a line-source well. This was a special case of a more general solution for a finite-radius well with a heat-transfer film coefficient produced by Jaeger (1940, 1941). The problem considered by van Everdingen and Hurst may be represented by modifying equation 3:

$$\frac{2\pi T H_w \rho}{Q} = P_D(s, C_D, t_D), \tag{6}$$

where the new dimensionless group C_D is

$$C_D = \frac{C_1}{2\pi h \phi C_t r_w^2}. \tag{7}$$

For a liquid-level change, the dimensionless coefficient C_D is related to the dimensionless group of Cooper and others (1967), α, by the relationship

$$\alpha = \frac{1}{2C_D}. \tag{8}$$

Both van Everdingen and Hurst presented Mellin inversion integral forms for the function defined by equations 6 and 7 but did not evaluate the integrals. It appears that the first evaluations of the finite skin-effect cases were presented by Carslaw and Jaeger (1947) and Agarwal and others (1970).

One major contribution of the studies by van Everdingen and Hurst was a long-time approximating form for equation 6:

$$\frac{2\pi T H_w \rho}{Q} = P_D(t_D) + s \tag{9}$$

NON-DARCY FLOW

From 1953 through the early 1960s, a number of important papers concerning the well-loss function or the skin effect appeared in both the ground-water hydrology and petroleum engineering literatures. Many of the publications in the petroleum engineering literature dealt with measurements of the well condition that were essentially flow efficiencies (ratios of the actual to an ideal specific capacity). Muskat (1937a) used the term "flow efficiency" in connection with the concepts of partial penetration and perforations to express well productivity as a fraction of the productivity if the entire sand face had been open to flow. The main problem with flow-efficiency measurements of well condition was that they were generally time dependent for transient-flow conditions. On the other hand, the skin effect defined by van Everdingen and Hurst was essentially constant and independent of the duration of the test. (The skin effect could change as the well condition changed. An example would be increasing damage with continued production due to fines migration or precipitation. The skin effect could also be changed by well treatment, such as acidization.) As a result, the van Everdingen and Hurst skin effect became the dominant well measurement by the early 1960s, although the flow efficiency continued to be used to discuss well condition and treatment results.

In 1961, R. V. Smith concluded, on the basis of field tests on natural gas wells, that the van Everdingen and Hurst skin effect was a function of the flow rate during the test. Smith's field data indicated that the skin effect was a linear function of the flow rate:

$$s = s_T + DQ. \tag{10}$$

Smith's finding was derived empirically. In the next year, a theoretical foundation for Smith's observation was published independently by Tek and others (1962) and by Swift and Kiel (1962). The reason was non-Darcy flow in the porous medium near the sand face. That is, the reason for the existence of the DQ term on the right in equation 10 was believed to be a flow velocity high enough to violate Darcy's law. The Forchheimer (1901) equation was necessary to represent flow through the formation near the well:

$$-\frac{ds}{dr} = \frac{\mu}{k}\bar{V}_r + \beta\rho\bar{V}_r^2. \tag{11}$$

All of the petroleum engineering equations (3 to 6) assume that the producing head, H_w, is measured in the wellbore immediately opposite the producing formation midpoint. As such, flowing friction within the liner screen or slots is included, but flowing friction in the wellbore to the tubing inlet is not. This is not likely to be an important factor in most cases. The main point is that the second term on the right in equation 11 was conceived as a result of high-velocity flow in the formation adjacent to the well. Katz and others (1959) had measured "turbulence" coefficients, β, for gas flow in dry sandstones.

In the studies by Swift and Kiel (1962) and Tek and others (1962) the terms "non-Darcy" and "turbulent" flow were used. In 1973, Fetkovich pointed out that certain phase-equilibrium and relative-permeability mechanisms could cause apparent non-Darcy flow effects at flow velocities below the minimum needed for departure from laminar flow. These effects were pressure-level dependent, but appeared rate sensitive because of the usual coupling between drawdown and flow rate. One example of this is gas coming out of solution in oil at the bubble point. Another is condensation of liquid from a gas at the dew point. Both phenomena cause a reduction in relative and effective permeability and thus appear as rate-dependent skin effects. Because the two-phase fluid banks would increase in size with time, the effects could also appear to be time dependent.

The need for the second term on the right in either equation 1 or equation 11 is most certainly a result of a number of factors, some of which depend on high-velocity flow. It is known that many wellbore factors may combine to magnify the skin effects. Bilhartz and Ramey (1977) showed that partial penetration could increase non-Darcy flow effects by causing the total flow to pass through a limited opening at the sand face.

In the petroleum literature, many investigators of partial penetration and other completion configurations use the term "pseudo-skin effects." This is intended to differentiate between skin effects caused by formation damage in drilling and completion and apparent skin effects caused by blanking off part of the sand face in the completion design (slotted liners, partial penetration). In 1965, I proposed that the total apparent skin effect should be the sum of the skin effect, the pseudo-skin effects, and the flow-rate-dependent term, DQ (Ramey, 1965). Let us now return to the studies by Swift and Kiel (1962) and Tek and others (1962).

Katz and co-workers had presented many studies of high-velocity (nonlaminar) flow in porous media before 1962. However, before the study by Smith (1961), it had not been recognized that these effects would be evident in the behavior of wells. Tek and co-workers and Swift and Kiel explained that the main reason for the Smith observation was high-velocity flow near the well sand face. Swift and Kiel presented a test-analysis technique strongly reminiscent of that used by Jacob and Rorabaugh. However, the Swift and Kiel study was based mainly on the premise of high-velocity flow near the sand face, and represented an original contribution that was eventually awarded the Ferguson Medal of the Society of Petroleum Engineers. In 1965, I observed that the true skin effect, s_T, and the non-Darcy flow coefficient, D, in equation 10 could be determined by ordinary pressure-transient testing (Ramey, 1965).

The skin effect may be determined from equations 4 and 9. The true skin effect, s_T, and the high-velocity flow coefficient, D, may be determined from equation 10 by graphing two or more determinations of the skin effect, S, versus the flow rate, Q. This graphing should be done, of course, on Cartesian coordinates.

Combining equations 9 and 10 gives

$$\frac{2\pi T H_w \rho}{Q} = P_D(t_D) + s_T + DQ. \tag{12}$$

Solving equation 12 for the drawdown yields

$$H_w = \frac{Q}{2\pi T\rho}\left[P_D(t_D) + s_T + DQ\right] \tag{13}$$

or

$$H_w = \left[\frac{P_D(t_D)}{2\pi T\rho} + \frac{s_T}{2\pi T\rho}\right] Q + \frac{D}{2\pi T\rho}Q^2. \tag{14}$$

In the event the flow system goes to steady state, the term $P_D(t_D)$ becomes constant and equation 14 is then similar to Jacob's (1947) postulate given by equation 1. It can be shown to be *exactly* the same if the skin effect, s_T, is coupled with the wellbore radius to form the effective well radius:

$$r_w' = r_w e^{-s_T}. \tag{15}$$

(see Matthews and Russell, 1967, p. 21). Replacement of $P_D(t_D)$ in equation 14 by equation 4 yields

$$H_w = \frac{Q}{4\pi T\rho}\left[\ln(4t_D) - 0.5772 + 2s_T\right] + \frac{D}{2\pi T\rho}Q^2, \tag{16}$$

and from equation 15

$$2s_T = \ln(r_w/r_w')^2. \tag{17}$$

Substituting equation 17 in equation 16 yields:

$$H_w = \frac{D}{4\pi T\rho}\left[\ln(4t_D)\left(\frac{v_w^2}{v_w'^2}\right) - 0.5772\right] + \frac{D}{2\pi T\rho}Q^2. \tag{18}$$

This is the same as Jacob's equation 13b (1947) if his dimensionless time is based on an effective well radius and his constant C is

$$C = \frac{D}{2\pi T\rho}. \tag{19}$$

The main result is that use of the skin-effect rather than an effective well radius suggests that equation 1 in this study should have been written

$$H_w = BQ + \frac{1}{2\pi T\rho}(s_T + DQ)Q. \tag{20}$$

The Exponent n

Field experience often indicates that performance matches an equation of the form

$$H_w = BQ + CQ^n, \tag{21}$$

rather than of equations 1 or 20. Hantush (1964) observed that although equation 21 did seem to match performance with test flow-rate ranges quite well, extrapolation to other rates on the basis of equation 21 frequently gave poor results. Equation 21 may be a special case of equation 20. If equations 21 and 20 are equated, then

$$\frac{1}{2\pi T\rho}(s_T + DQ)Q = CQ^n, \tag{22}$$

or

$$\frac{s_T Q}{2\pi T\rho} + \frac{DQ^2}{2\pi T\rho} = CQ^n. \tag{23}$$

Equation 23 can be presented graphically as shown in

Figure 1. The two terms on the left in equation 23 are shown by the two solid lines of slope unity and 2, respectively. The sum of the two solid lines represents CQ^n and is shown by the dashed line.

The dashed line is a curve whose slope ranges between unity and 2. Short portions of the curve appear straight. Consider that a well test is conducted at four flow rates shown by the heavy dots on Figure 1. An expanded version is shown in Figure 2. A straight line may be passed through the dots as shown, and the value n determined from the slope. Although this line would represent behavior within the range of the test flow rates, it would be a poor representation of actual behavior (dashed line) for either lower or higher flow rates. It would be far better to represent skin effect by the coefficients in equation 20 rather than by those in equation 21.

The sort of analysis used in Figure 1 was first applied to gas well deliverability curves by Carter and others (1963), who showed it was possible to relate n to the skin effect and to the non-Darcy flow coefficient.

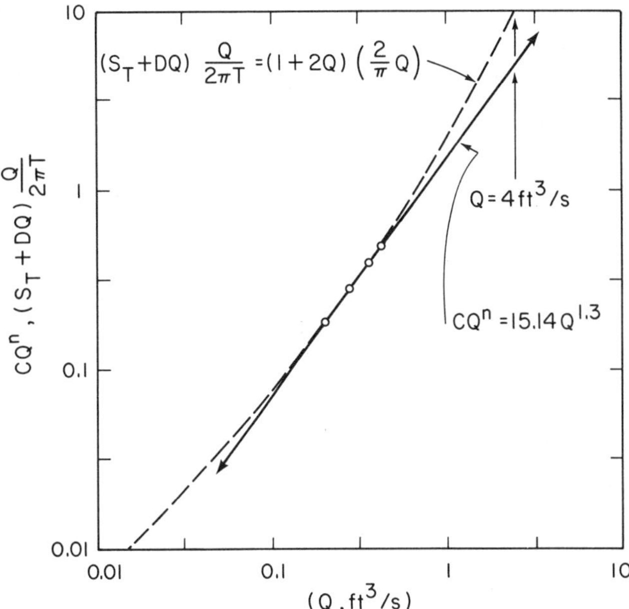

Figure 2. Non-Darcy flow and skin effect and well-loss function versus flow rate.

Water Well Test Results

The preceding theory applies for many gas- and oil-well test analyses and seems convincing except for one fact. Field results from water-well testing commonly indicate an n value near 2.5 (Rorabaugh, 1953). Fetkovich (1973) has discussed a similar problem for solution gas-drive oil flow. The transmissivity for depletion-solution gas-drive systems has been found to be a function of pressure, which involves considering relative permeability and pressure-dependent physical properties of oil. This mechanism is not directly applicable to water flow. Dissolved gas evolving from solution could provide a similar mechanism. Stress-sensitive transmissivity near the producing well face could also provide a similar effect. It appears worthwhile to pursue the reason for the strange values of n observed in water-well tests to date.

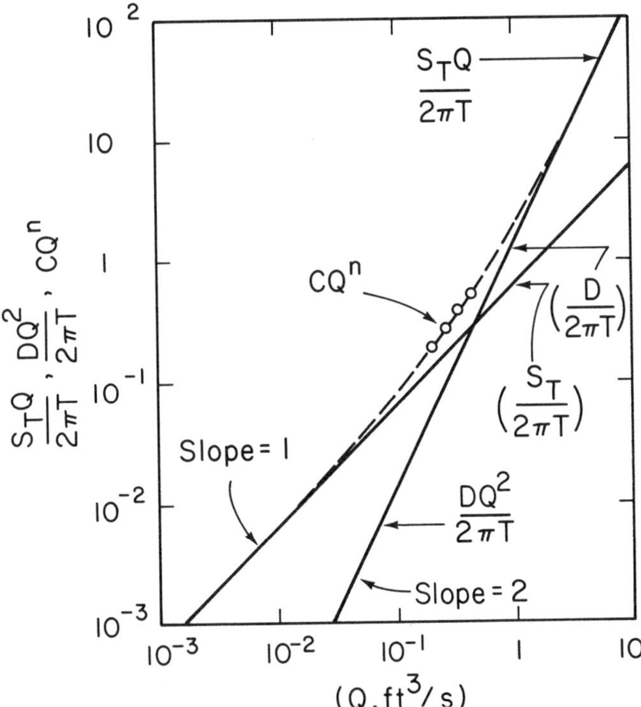

Figure 1. Graphic presentation of equation 23: skin effect and non-Darcy flow components and well-loss function versus flow rate.

CONCLUSIONS

The main purpose of this study was to clarify the relationship between the well-loss function and the concepts of skin effect and non-Darcy flow as used in petroleum engineering. This investigation indicated that an equality may be reached between these two concepts. A restatement of equation 1 in terms of the skin effect rather than an effective wellbore radius led to equation 20. Equation 20 reasonably matches the high-rate performance of gas wells. However, recent studies of oil flow suggest that relative permeability effects caused by gas evolution could lead to higher-order apparent rate effects, as observed in many water well tests. This appears to be a worthwhile area for further study.

The classic concepts of effective well radius (skin effect) and non-Darcy flow first appeared in the most sophisticated form in the ground-water literature; wellbore storage was first developed in the petroleum literature. These three ideas--the skin effect or effective well radius, non-Darcy flow near the well, and wellbore storage--are important keys to understanding the behavior of transient pressures measured in a wellbore.

Since 1950, the major thrusts of technology development in the parallel literatures on water-well and gas- and oil-well testing have been in different directions. In the petroleum literature, ideas concerning the well-loss function (skin effect and non-Darcy flow) and wellbore storage were developed; in ground-water work, interference testing was a major concern. The reasons for these differences may be clear in retrospect. Water wells were shallow and inexpensive compared with oil wells. Drilling an observation well for an interference test was practical, especially in shallow formations of unknown porosity that frequently "leaked" water from adjacent formations (Jacob, 1946). On the other hand, oil formations were deep and frequently consolidated, the fluids were very compressible, drilling was too expensive for observation wells, and extended periods of shut-in were not feasible because of the high value of petroleum.

We have come full cycle in a technological sense. Oil technicians are avidly interested in interference testing because of the development of new, high-sensitivity devices for measuring pressure. Interference tests can be run in short periods of time despite the high compressibility of petroleum fluids. Likewise, hydrologists are now interested in well performance. Few wells are still inexpensive and the real value of water is becoming more apparent. The future promises closer communication between these two technologies, with obvious benefits to the public.

NOMENCLATURE

B = coefficient (equation 1)

C = coefficient (equation 1)

C_1 = wellbore storage coefficient (volume/pressure)

C_D = dimensionless wellbore storage (equation 7)

c_t = total system effective isothermal compressibility

D = non-Darcy coefficient (units of reciprocal flow rate)

H_w = producing head (ft)

h = formation thickness

n = exponent of Q (see equation 2)

$p_D(t_D)$ = dimensionless pressure (see equation 3)

Q = flow rate

r_w = well radius (actual)

S = storativity ($\phi h c_t$)

s = infinitesimal skin effect

T = transmissivity ($T = kh/\mu$)

t = time

t_D = dimensionless time (see equation 5)

α = wellbore storage parameter (equation 8)

μ = viscosity (cp)

ϕ = fractional porosity

ρ = specific weight of fluid

REFERENCES CITED

Agarwal, R. G., Al-Hussainy, R., and Ramey, H. J., Jr., 1970, An investigation of wellbore storage and skin effect in unsteady liquid flow, I, Analytical treatment: Society of Petroleum Engineers Journal, v. 10, no. 3, p. 279.

Billhartz, H. L., and Ramey, H. J., Jr., 1977, The combined effects of storage, skin, and partial penetration on well test analysis: Dallas, American Institute of Mechanical Engineers, Society of Petroleum Engineers SPE-6754.

Carslaw, H. S., and Jaeger, J. C., 1947, Conduction of heat in solids (First edition) Oxford University Press, p. 16, 284.

Carter, R. D., Miller, S. C., and Riley, H. C., 1963, Determination of stabilized gas well performance from short flow tests: Journal of Petroleum Technology, v. 15, no. 6, p. 651.

Chatas, A. T., 1953, A practical treatment of nonsteady-flow problems in reservoir systems: Petroleum Engineering, v. 25, no. 5, 6, 8.

Cooper, H. H., Jr., Bredehoeft, J. D., and Papadopulos, I. S., 1967, Response of a finite-diameter well to an instantaneous charge of water: Water Resources Research, v. 3, no. 1, p. 263-269.

Fetkovich, M. J., 1973, The isochronal testing of oil wells: Dallas, American Institute of Mechanical Engineers, Society of Petroleum Engineers, SPE-4529.

Forchheimer, Ph., 1901, Wasserbewegung durch Boden: Zeits V. deutsch Ing., 45, p. 1782.

Gladfelter, R. E., Tracy, G. W., and Wilsey, L. W., 1955, Selecting wells which will respond to production-simulation treatment: Drilling and Production Practice, American Petroleum Institute.

Hantush, M., 1964, Hydraulics of wells, in Ven Te, C., ed., Advances in hydrosciences, Volume 1: New York, Academic Press.

Hurst, W., 1934, Unsteady flow of fluids in oil reservoirs: Physics, Petroleum Engineering, v. 5, p. 20-30.

___1953, Establishment of the skin effect and its impediment to fluid flow in a wellbore: Petroleum Engineering, v. 25, p. B-6.

Jacob, C. E., 1946, Radial flow in a leaky artesian aquifer: American Geophysical Union Transactions, v. 27, p. 198-208.

___1947, Drawdown test to determine effective radius of artesian wells: American Society of Civil Engineers Transactions, v. 112, p. 1047-1070.

Jaeger, J. C., 1940, Radial heat flow in circular cylinders with a general boundary condition, II: Journal of the Royal Society, N.W. Wales, v. 74, p. 342.

___1941, Radial heat flow in circular cylinders with a general boundary condition, II: Journal

of the Royal Society, N.W. Wales, v. 75, p. 130.
Katz, D. L., and others, 1959, Handbook of natural gas engineering: New York, McGraw-Hill.
Matthews, C. S., and Russell, D. G., 1967, Pressure build-up and flow tests in wells: Dallas, Society of Petroleum Engineering Monograph series, v.1, p. 4.
Miller, C. C., Dyes, A. B., and Hutchinson, C. A., Jr., 1950, Estimation of permeability and reservoir pressure from bottom-hole pressure build-up characteristics: American Institute of Mechanical Engineers Transactions, v. 189, p. 91-104.
Moore, T. V., Schilthuis, R. J., and Hurst, W., 1933, The determination of permeability from field data: American Petroleum Institute Proceedings API Bulletin 211, p. 4.
Muskat, M., 1937a, The flow of homogeneous fluids through porous media: New York, McGraw-Hill, 763 p.
____1937b, Use of data on the build-up of bottomhole pressures: American Institute of Mechanical Engineers Transactions, v. 123, p. 44-48.
Ramey, H. J., Jr., 1965, Non-Darcy flow and wellbore storage effects in pressure build-up and drawdown of gas wells: Journal of Petroleum Technology, v. 17, no. 2, p. 223-233.
Rorabaugh, M. I., 1953, Graphical and theoretical analysis of step-drawdown test of artesian well: American Society Civil Engineers Proceedings, v. 79, sec. 362, 23 p.
Smith, R. V., 1961, Unsteady-state gas flow into gas wells: Journal of Petroleum Technology, November, p. 1151-1159.
Swift, G. W., and Kiel, O. G., 1962, The prediction of gas-well performance including the effect of non-Darcy flow: Journal of Petroleum Technology, v. 14, no. 7, 791-798.
Tek, M. R., Coats, K. H., and Katz, D. L., 1962, The effect of turbulence on flow of natural gas through porous reservoirs: Journal of Petroleum Technology, v. 24, no. 7.
Theis, C. V., 1935, The relation between the lowering of the piezometric surface and the rate and duration of discharge of a well using ground-water storage: American Geophysical Union Transactions, p. 519-524.
van Everdingen, A. F., 1953, The skin effect and its influence on the productive capacity of a well: American Institute of Mining Engineers Transactions, v. 198, p. 171-176.
van Everdingen, A. F., and Hurst, W., 1949, The application of the Laplace transformation to flow problems in reservoirs: American Institute of Mining Engineers Transactions, v. 186, p. 305-324.

Manuscript Accepted by the Society October 14, 1980

Printed in U.S.A.

Geological Society of America
Special Paper 189
1982

Numerical Modeling in Hydrogeology

T. N. NARASIMHAN
Earth Sciences Division
Lawrence Berkeley Laboratory
University of California
Berkeley, California 94720

ABSTRACT

The equation governing isothermal flow of water in deformable media is one of mass conservation. For an appropriately small volume element satisfying certain averaging criteria, the conservation equation helps convert the rate of accumulation of mass to an equivalent, average time-derivative of the potential function. The goal of numerical modeling is to evaluate the rate of mass accumulation in the volume element by integration in space and time. In order to carry out the integration, the flow region has to be partitioned and spatial gradients of potential evaluated. Depending on the procedure by which spatial partitioning is achieved and on the manner in which potential gradients are evaluated, different numerical schemes such as the integral finite difference method and the finite element method arise. These methods can lead to a set of either explicit or implicit discretized equations. In setting up the equations, time-dependent coefficients may be handled in either a quasi-linear or an iterative fashion. The final set of implicit equations, giving rise to a sparse coefficient matrix, may be handled through direct solvers or through iterative solvers. Numerical models may be validated with the help of analytic solutions to partial differential equations or with experimental data. Based on set-theoretic concepts, it is reasoned that numerical models possess far greater generality than merely providing analytic solutions to the partial differential equation. To derive the real benefit of numerical models, therefore, techniques should be developed to validate numerical solutions based on their own axiomatic foundations, rather than relying upon analytic solutions. Although our current computing abilities transcend our ability to generate sophisticated field data, there is considerable scope for further research on numerical methods. Among new areas of research one could include: (a) incorporating the algebra of probabilistic distributions into existing deterministic models, and (b) developing new techniques to validate numerical models in their own right.

INTRODUCTION

A strong desire, or even obsession, to predict the future is a basic human trait. Man desires to predict the behavior of the environment around him in order that he may manage it efficiently for his own benefit. A very important component of man's surroundings is the hydrogeologic environment. As a vital source of dwindling fresh water supplies in many parts of the world, the importance of this environment is becoming increasingly obvious. Furthermore, as man perturbs this environment, he causes manifold effects related to earth deformations, instabilities, and contamination, in addition to resource depletion. The ability to reliably predict the evolution of hydrogeologic systems is thus of paramount importance. Earth scientists seek to achieve hydrogeologic predictions through the use of mathematical models.

The development and use of mathematical models of natural systems for predictive purposes consists of two steps. First, one sets up conceptual equations that relate observable physical quantities in mutual cause-effect relationships. Second, these conceptual equations are appropriately translated into computational equations, which are quantitatively evaluated through various techniques. In general, there exist two broad classes of mathematical techniques which enable the implementation of the computational equations. These are: (1) analytic solutions, and (2) numerical methods. The purpose of this paper is to review the current state of our knowledge in relation to the use of numerical methods in the field of hydrogeology. In scope, we shall restrict our attention principally to the flow of water in isothermal, deformable ground water systems. We will neglect consideration of heat or solute transport in these systems.

ANALYTICAL AND NUMERICAL METHODS

Before distinguishing between analytical and numerical methods, it is necessary for us to first set down the governing equations which relate the physical parameters to each other. It is now well recognized that the governing equation describing the transient flow of ground water in subsurface systems is an equation of mass conservation. Specifically, the equation of mass conservation is written for a macroscopic subdomain of the flow region which is large enough so that physical parameters can be statistically defined over the subdomain in some meaningful fashion. Yet, the subdomain is small enough so that the average property of interest varies smoothly over it. Consider such an arbitrary subdomain shown in Figure 1. Any average property associated with this subdomain is represented at the interior nodal point, designated for convenience as ℓ. The rate of accumulation of mass in this volume element through fluid crossing its boundaries as well as due to any sources or sinks that may exist is given by the integral expression shown in equation 1.

$$\rho_\ell G_\ell - \int_\Gamma \rho_\Gamma \vec{q} \bullet \vec{n}_\Gamma d\Gamma = M_{c,\ell} \frac{\partial \phi_\ell}{\partial t}, \quad (1)$$

in which

ρ_ℓ is the average density of water in ℓ,

G_ℓ is the rate at which water is generated from ℓ expressed as vol/time,

ρ_Γ is the density of water at $d\Gamma$,

$\vec{q}$ is the mean Darcy velocity over $d\Gamma$,

$\vec{n}$ is the outer normal to $d\Gamma$,

$d\Gamma$ is a surface segment,

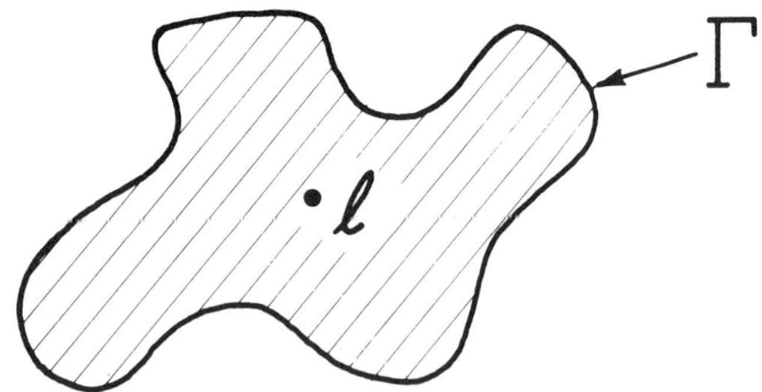

Figure 1. A subdomain or volume element of the flow region whose average properties are associated with a representative nodal point ℓ, interior to the subdomain.

Γ is the surface bounding volume element,

$M_{c,\ell}$ is the fluid mass capacity of ℓ, defined as the mass of water required to change the average fluid potential over ℓ by unity and includes expansion of water, deformation of matrix, and desaturation of the voids (Narasimhan and Witherspoon, 1977a),

ϕ_ℓ is average fluid potential over ℓ expressed in equivalent head of water and $\phi_\ell = z_\ell + \psi_\ell$, where z_ℓ is elevation above datum, and

ψ_ℓ is pressure head.

Equation 1 is subject to appropriate boundary conditions and initial conditions. Note that at this level equation 1 is a conceptual equation. Now we wish to consider how this conceptual equation can be transformed into a computational equation. The simplest way to do this is to consider the flow region as a whole and partition the flow region into appropriately small subdomains such as ℓ, satisfying the condition that each subdomain is sufficiently large so that average quantities will have statistical meaning over it, and yet small enough so that the average property varies smoothly over it. If, indeed, such a partition is carried out, then in order to compute the time derivatives on the right-hand side of equation 1, one simply has to evaluate the surface integral shown on the left-hand side of equation 1. With the help of this surface integral and the known source terms as well as boundary conditions, one can then compute the required time derivative. The basic problem is thus one of carrying out integrations over finite subdomains or finite subsets of the flow region. However, successful evaluation of integrals over a large number of arbitrarily shaped subdomains requires a very sophisticated ability to do fast computing. Such an ability has become possible only in the past two decades with the advent of fast computing devices. Suppose for a moment that such a computing ability is not available. Then how can one go about practically implementing the conceptual equation given in equation 1? Is there any other way by which equation 1 can be made tractable? The answer is indeed yes, and such a requirement or desire leads us to the consideration of a differential equation. In order to derive the differential equation from equation 1 we merely normalize both sides of the equation with reference to the bulk volume of the element ℓ and assume that in the limit, the element ℓ vanishes to an infinitely small volume. In this case, the surface integral on the left hand side of equation 1 leads to the well-known definition of divergence and one gets as a result

$$\rho_\ell g_\ell - \rho_\ell \operatorname{div} \vec{q} = m_{c,\ell} \frac{\partial \phi_\ell}{\partial t}, \quad (2)$$

where g_ℓ is volumetric water generation rate per unit volume of subdomain ℓ, div $\vec{q}$ is the divergence of the Darcy velocity, defined by

$$\text{div}\ \vec{q} = \lim V \to 0 \quad \frac{1}{V} \int_{\Gamma} \vec{q} \bullet \vec{n}\ d\Gamma ,$$

and $m_{c,\ell}$ is the specific fluid mass capacity of ℓ and is defined as $M_{c,\ell}$ per unit volume. It so happens that if the divergence term in equation 2 is evaluated for any particular coordinate system and the volume element has a regular shape with sides normal to the principal coordinate axes, then the divergence term leads to the spatial derivatives of the Darcy velocity or, equivalently, to the second spatial derivatives of potentials. Such an equation for a Cartesian system in three dimensions is given in equation 3:

$$\text{div}\ \vec{q} = \frac{\partial q}{\partial x_1} + \frac{\partial q}{\partial x_2} + \frac{\partial q}{\partial x_3}$$

$$= -K_{x_1} \frac{\partial^2 \phi}{\partial x_1^2} + K_{x_2} \frac{\partial^2 \phi}{\partial x_2^2} + K_{x_3} \frac{\partial^2 \phi}{\partial x_3^2} , \qquad (3)$$

where K_{x_1}, K_{x_2}, and K_{x_3} are the principal hydraulic conductivities oriented parallel to the axes x_1, x_2 and x_3. In view of equation 3, equation 2 now becomes

$$\rho_\ell g_\ell + \rho_\ell \left[K_{x_1} \frac{\partial^2 \phi}{\partial x_1^2} + K_{x_2} \frac{\partial^2 \phi}{\partial x_2^2} + K_{x_3} \frac{\partial^2 \phi}{\partial x_3^2} \right]$$

$$= m_{c,\ell} \frac{\partial \phi_\ell}{\partial t} , \qquad (4)$$

where $\phi_\ell = \phi_\ell(x_1, x_2, x_3, t)$. It should be noted here that the partial differential equation 4 has helped us to pose the fundamental problem of integration given in equation 1 as an equivalent problem involving spatial and temporal gradients of the potential function.

In order to mathematically solve equation 4, one merely has to find an analytical expression for the potential function ϕ in terms of x_1, x_2, x_3 and t such that the function satisfies both sides of the equation. Depending on the complexity or simplicity of the problem, the required analytical expression can be generally found in a systematic fashion with the help of a number of mathematical techniques developed over the past two centuries or more. Analytical solutions to differential equations so obtained have provided a tremendous amount of insight into many hydrogeological problems. Yet it is well recognized that these solutions become extremely cumbersome for most realistic field problems. For this reason many workers have turned to the use of numerical models, relying heavily on the availability of fast computing devices. Despite the fact that numerical solutions have to be obtained for each specific problem individually, the numerical models have the advantage of much greater generality and flexibility than the analytic solution.

NUMERICAL METHODS

Fundamental Tasks

In light of what has been said above, we will treat the numerical problem basically as one of integration. In particular, the elementary component of integration is the evaluation of the flux over a small, but finite interval of time, across a single discrete surface segment $d\Gamma$ of the chosen subdomain. This recognition directly leads to the identification of the two following fundamental tasks of numerical modeling.

Task 1: Partition the flow domain into subsets ℓ, (ℓ = 1,2,3, ..., L) and geometrically define its bulk volume as well as its bounding surface. A partition may be defined as "a finite collection of disjoint sets" (James and James, 1976, p. 282). Two sets are disjoint if there is no object which belongs to each of the sets.

Task 2: Evaluate fluxes across each identified surface segment $d\Gamma$ over a chosen, discrete time-interval. However, since fluxes are related to the spatial gradient of potential, the second fundamental task is to evaluate potential gradients across (or even along) each surface segment $d\Gamma$.

The differences between the various numerical methods used in hydrogeology can be discussed systematically in terms of the two aforementioned tasks. In fact, recognition of these differences provides us with a base for unifying the different methods and rationally evaluating their relative strengths and weaknesses. We therefore now move on to consider in detail the techniques that have been developed to accomplish these two tasks.

Techniques for Achieving Spatial Partition

From a computational point of view, spatial partitioning consists in subdividing the flow region into non-overlapping subdomains and generating certain geometrical as well as topological information. These include 1) the volume of the subdomain, to enable computing its fluid mass capacity; 2) the area of each discrete surface segment bounding the subdomain; 3) the spatial location of a nodal point interior to the subdomain at which the averaged, intensive properties of the subdomain are represented; 4) an identification index for the subdomain; 5) the identifying index of each neighboring subdomain with which the given subdomain communicates across a particular surface segment; and 6) the distance between the nodal points of communicating subdomains to enable the evaluation of spatial gradients. These data could be generated and expressed either explicitly, in a straightforward manner or implicitly incorporated into equivalent computational operations. (For a set-theoretic view of partitioning, see Narasimhan, 1978.)

Explicit Partitioning. The explicit partitioning approach has been used by many workers (MacNeal, 1952; Tyson and Weber, 1964; Edwards, 1972) in what has since been termed the integrated finite difference method (IFDM) by Narasimhan and Witherspoon (1976). In this method (for example,

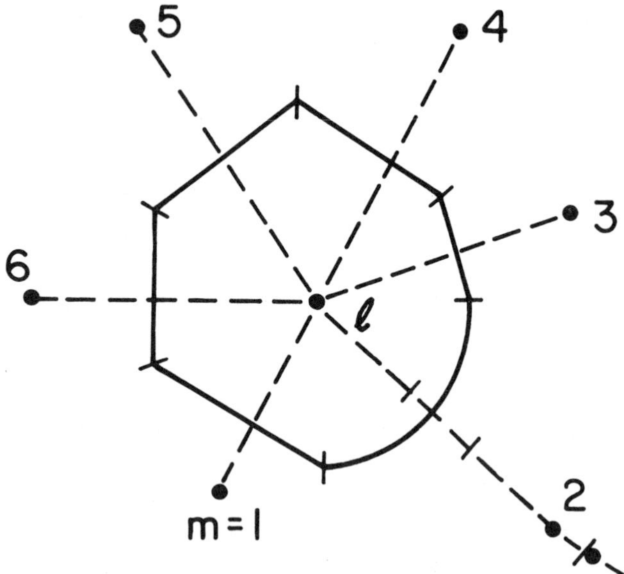

Figure 2. An explicitly defined subdomain in the Integrated Finite Difference Method.

Fig. 2) a volume element of arbitrary shape is bounded by many discrete surface segments. In three dimensions, the surface segments are planar or curved surfaces, whereas in two dimensions they are straight or curved lines. It is assumed that the element is appropriately small to assure smooth variation of averaged quantities over it. It is also assumed that the surface segments are small and are normal to the line joining the two nodal points on either side so that the gradients evaluated between the nodal points represent the "average" gradient over the surface segment. Conversely, if, for a given problem, the distribution of potentials is prescribed at many different spatial points, then these "point" data are assumed to denote the average values over a small subdomain around the point of interest. Thus, the spatial point where potential is known becomes a representative nodal point for a finite subdomain surrounding it which satisfies the smoothness and averaging criteria already mentioned.

Once the partitioning has been carried out, the geometric data, namely, volumes, surface areas, and nodal-point distances are measured as accurately as needed and used as the required input information for the computational mass conservation problem.

If instead of letting the volume elements (in this paper we shall treat the term volume element to be synonymous with subdomain) assume general polyhedral or polygonal shapes, one restricts the subdomains to rectangles, parallelepipeds, and cylindrical or spherical shells with the bounding surfaces perpendicular to the principal-coordinate axes (Fig. 3), then the resulting partition is the same as that which has been used in the classical finite difference method (FDM) of solving partial differential equations. The only difference between the numerical equations of the IFDM and the FDM in this case is that the latter expresses mass conservation per unit volume of each subdomain while the former expresses the same for the subdomain itself. Thus, for all practical purposes, the FDM can be treated as a special case of the IFDM. One should also point out here that it is customary to consider (for example, Zinkiewicz, 1971) the FDM as a technique that discretizes and approximates the differential equation at each point of interest. However, if one recognizes that the FDM is but a special case of the IFDM, then through the trivial operation of multiplying by the bulk volume of the subdomain, the FDM could also be viewed as an integral scheme.

For the volume element shown in Figure 2, we can now write the equation of mass conservation as

$$\rho_\ell G_\ell + \sum_m \rho_{\ell m} \vec{q} \bullet \vec{n}_{\ell m} \Delta\Gamma_{\ell m} = M_{c,\ell} \frac{\Delta\phi_\ell}{\Delta t}, \tag{5}$$

in which G_ℓ is the volume of fluid generated from element ℓ, ℓ and m denote the neighboring volume elements that communicate across a common interface, and ρ_ℓ and $\rho_{\ell m}$ are average fluid densities over ℓ and at the interface between ℓ and m. The geometric quantities $\Delta\Gamma_{\ell m}$, the area of the surface segment separating ℓ and m as well as V_ℓ, and the volume of ℓ which is needed for computing $M_{c,\ell}$ are assumed to be known because of the explicit partitioning.

As a special case (the IFDM), if the line joining nodal points ℓ and m is perpendicular to $\Delta\Gamma_{\ell m}$ and if the materials in ℓ and m are isotropic (or if the line ℓm coincides with a principal direction of anisotropy), then one could substitute the following expression for the dot product $\vec{q} \bullet \vec{n}$ in equation 4:

$$\vec{q} \bullet \vec{n}_{\ell m} = - K_{\ell,m} \frac{(\phi_m - \phi_\ell)}{d_{\ell m} + d_{m\ell}}, \tag{6}$$

where $K_{\ell m}$ is the mean hydraulic conductivity at the interface ℓm, and $d_{\ell m}$ and $d_{m\ell}$ are distances to the interface ℓm from nodal points ℓ and m, respectively. Note that K is related to absolute permeability or intrinsic permeability by the relation $K = k\rho g/\mu$ where k is intrinsic permeability, g is acceleration due to gravity, and μ is coefficient of viscosity.

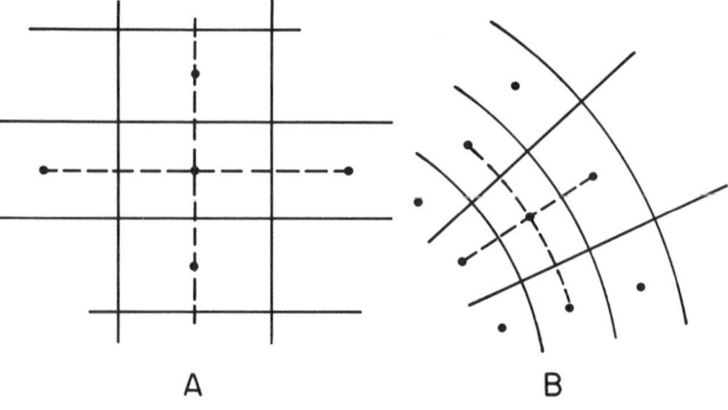

Figure 3. An explicitly defined subdomain in the FDM scheme. (A) Two-dimensional, Cartesian coordinates. (B) Polar coordinates.

Implicit Partitioning. Instead of proceeding from a well-defined volume element to a representative nodal point in its interior, one could start from a nodal point at a given spatial location and treat the data at this point as representative average over some finite subdomain surrounding it. While one may not find it necessary to explicity define this subdomain, one may implicitly give consideration to it in the computational procedure. We will consider the actual mechanism of this implicit procedure in a later section that deals with the Galerkin method. First we shall consider its conceptual basis.

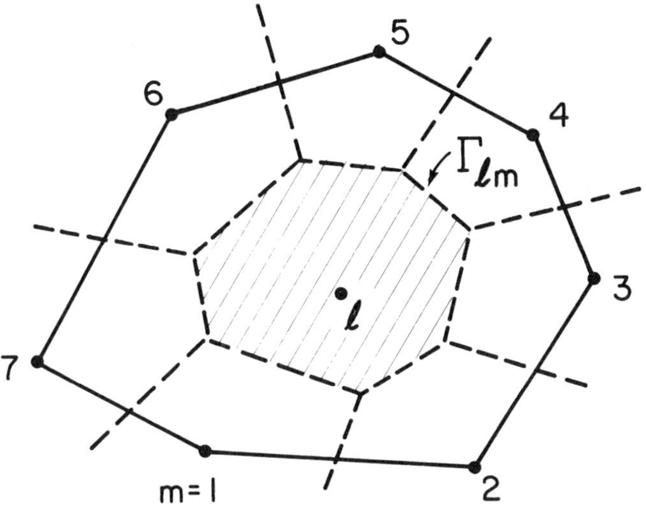

Figure 4. An implicitly defined subdomain of nodal point ℓ.

In Figure 4 the point ℓ denotes the location at which the average values of the ill-defined hatched subdomain are represented. Note that this subdomain forms a fractional part of the larger subdomain defined by the solid line connecting nodal points 1 through 7, which communicate with ℓ.

The conceptual equation of mass conservation requires that we evaluate the rate of fluid accumulation over the hatched region V_ℓ by summing up the fluxes over $\Gamma_{\ell m}$. Mathematically, however, we can achieve the same goal by dividing the hatched region V_ℓ into many smaller subdomains and summing up the rate of fluid accumulation in each. That is, one replaces the surface integration by an equivalent volume integration. This simple equivalence is mathematically stated in the form of the well-known divergence theorem (Sokolnikoff and Redheffer, 1966) which states,

$$\int_V \operatorname{div} \vec{q}\, dV = \int_\Gamma \vec{q} \bullet \vec{n}\, d\Gamma \tag{7}$$

Now, since the subdomain of ℓ is smaller than the subdomain formed by the surface connecting the neighboring nodes, we can, in principle, compute the rate of accumulation of fluid in V_ℓ through a process of volume integration, not over V_ℓ itself (since it is not clearly defined) but through a process of "weighted" volume integration over the larger region bounded by the solid line in Figure 4. That is, if we evaluate the rate of fluid accumulation over a small element dV somewhere in the larger region, then a fractional part of it, W_ℓ, will be associated with the subdomain V_ℓ. The magnitude of W_ℓ ($0 \leq W_\ell \leq 1$) will depend in general on the proximity of dV to the nodal point ℓ. In practice, this approach is attractive because the weighted volume integration can be effectively achieved purely in terms of the coordinates of the nodal points. Note that specifying the nodal point coordinates is generally less cumbersome than generating the three types of geometric quantities required in the explicit partition.

It must be emphasized here that, conceptually speaking, the weighted volume integration over V_ℓ, when carried out rigorously, is equivalent to the surface integral over $\Gamma_{\ell m}$. In fact, for simple cases, one could identify the exact nature of the subdomain V_ℓ satisfying the condition that the surface integral exactly equals the weighted volume integral (Narasimhan, 1978). Thus, the operation of volume integration can be ultimately justified in terms of the appropriate surface integral.

Techniques for Evaluating Spatial Gradients

As shown in equation 5, fluid flux across a surface segment is proportional to the potential gradient across the interface. In the numerical approach these gradients are evaluated in terms of the average potentials associated with the representative nodal points. Two basic techniques (that of the finite differences and that of the finite elements) are commonly used to evaluate gradients numerically.

The Finite Difference Approach. This oldest and simplest method consists in calculating the gradient of potential between two points by the relation

$$\vec{\nabla}\phi_{\ell m} = \frac{\phi_m - \phi_\ell}{D_{\ell m}}, \tag{8}$$

where $\vec{\nabla}\phi_{\ell m}$ is the gradient in ϕ between points ℓ and m and $D_{\ell m}$ is the distance between the points. This technique is used in the IFDM as well as the FDM.

Despite its simplicity, this technique has a disadvantage. That is, for measuring the gradient in a given direction two nodal points are required in that direction. However, if the material of interest is anisotropic and the surface segment of interest is not normal to a principal axis of the permeability tensor (Fig. 5) then the normal flux across the segment depends on the gradient normal to the surface as well as parallel to the surface. Thus, in general, for a surface segment of three dimensions,

$$\vec{q}_1 = K_{11}\frac{\Delta\phi}{\Delta x_1} + K_{12}\frac{\Delta\phi}{\Delta x_2} + K_{13}\frac{\Delta\phi}{\Delta x_3}, \tag{9}$$

where K_{11}, K_{12} and K_{13} are the components of the permeability tensor and x_1, x_2, x_3 are local coordinate axes with x_1 normal to the interface in question (Fig. 5). The finite difference gradient provides only the first of the three terms on the right hand side of equation 8 and hence is inadequate for evaluating q_1, except in the special case when $K_{12} = K_{13} = 0$, which occurs when x_1 coincides with one of the principal directions of anisotropy. A more general gradient evaluation technique is obviously required to handle equation 8.

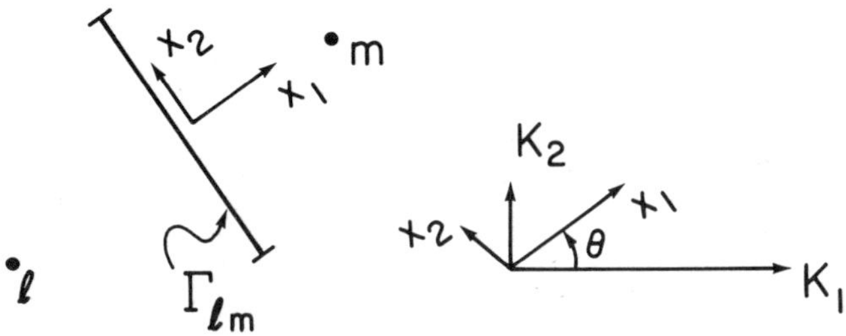

Figure 5. Schematic of a surface segment oriented obliquely to the principal directions of anisotropy.

The Finite-Element Approach. The finite-element approach takes a more general view of gradients. That is, this method first mathematically defines the potential surface over any small region of interest and then evaluates the gradient in any given direction by evaluating the derivative in that direction. Thus, the first step in the finite element method is to set up the equation for the smooth surface, which connects the known potentials at the nearby nodal points. The minimum number of points required to set up such an equation is (n + 1) where n is the dimensionality of the subregion. Thus, a minimum of three noncolinear points are needed in two dimensions and four noncolinear points in three dimensions.

Consider for example, a simple triangular region in two dimensions shown in Figure 6. If we assume

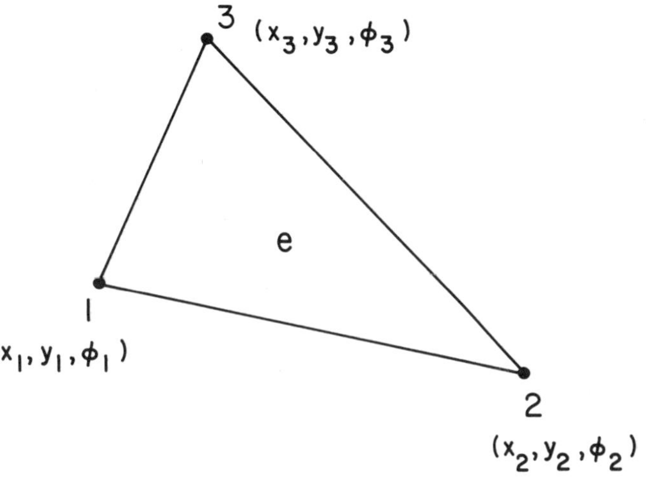

Figure 6. A triangular finite element.

that the surface connecting ϕ_1, ϕ_2 and ϕ_3 over the triangular region is a plane, then ϕ at any point within the region can be expressed by the simple relation

$$\phi = N_1\phi_1 + N_2\phi_3 + N_3\phi_3, \tag{10}$$

where $0 \leq N_1, N_2, N_3 \leq 1$ are functions of the coordinates of the point of interest (x_1, x_2) as well as the coordinates of the corners of the triangle, and the ϕs are only functions of time. Since, for the triangle, the Ns are non-negative numbers and $N_1 + N_2 + N_3 = 1$, they are also weighting functions (also known as Barycentric functions), whose magnitudes depend on spatial location within the triangle. Inasmuch as ϕ within the triangle depends only on the values at the corner, the discrete triangle is called a "finite element" (designated e in Fig. 6). Note particularly that the finite element is an artifice designed principally for the purpose of evaluating gradients.

If we now want to evaluate the gradient of ϕ anywhere within e in any direction x_i, then by simple differentiation we obtain

$$\frac{\partial\phi}{\partial x_i} \approx \frac{\partial}{\partial x_i}(N_1\phi_1 + N_2\phi_2 + N_3\phi_3). \tag{11}$$

However, since only N_1, N_2, and N_3 are spatially dependent, and ϕ depends only on time at each point, it follows that

$$\frac{\partial\phi}{\partial x_i} \approx \phi_1\frac{\partial N_1}{\partial x_i} + \phi_2\frac{\partial N_2}{\partial x_i} + \phi_3\frac{\partial N_3}{\partial x_i}\ . \tag{12}$$

Also, since N_1, N_2, and N_3 depend on the shape of e, which is chosen to be a triangle, the derivatives on the right hand side of equation 12 are known a priori in terms of the nodal point coordinates and the coordinates of the point of interest.

The analog of the triangle in three dimensions is the tetrahedron, which has four vertices. The assumption of linear variation in equation 9 implies that the functions N_1, N_2 and N_3 involve only the first powers of x_1 and x_2. It is possible, however, to pick other shapes besides a simple triangle and let the potential surface be curved surface over the subdomain. Some two-dimensional examples of higher order elements are given in Figure 7. In the case of the examples in Figure 7, ϕ is expressed by

$$\phi = N_i\phi_i,\ i = 1,2,3,\ \ldots,\ L\ , \tag{13}$$

where L = 9 for Case B in Figure 7. Unlike the linear case in equation 10, the N_i's in equation 13 involves higher powers of x_1 and x_2, and hence the elements are known as "higher" order elements. In Case A, Figure 7, ϕ varies according to a quadratic relation on each side of the triangle, while ϕ varies according to a cubic relation

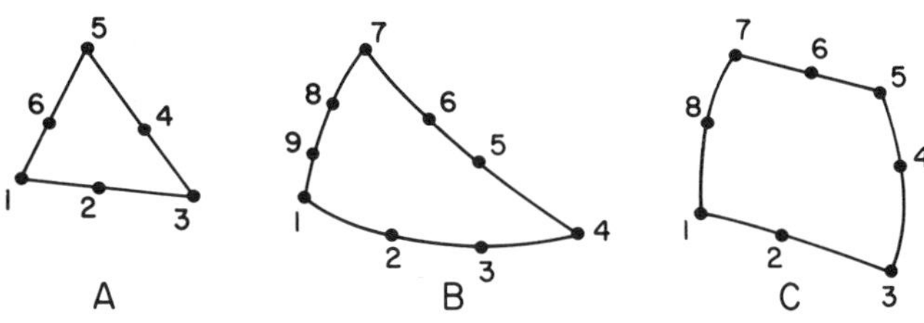

Figure 7. Some two-dimensional examples of higher-order elements.

along the sides of the curvilinear triangle in Figure 7B. These are therefore called quadratic and cubic elements, respectively.

That gradients could be evaluated in a more general fashion than the finite difference approach was known before the 1950s. For example, Tasny-Tschiassny (1949) used such a technique for a mesh of triangles. However, this method did not gain widespread recognition until the mid 1960s when structural engineers (such as Zienkiewicz and others, 1966) applied the technique to the solution of deformation problems. Since deformation problems invariably involve shear terms (analogous to non-zero terms in the permeability tensor), they found it very useful to be able to get away from the limitation of the finite difference technique of evaluating gradients. From structural engineering literature it was formally introduced into the groundwater literature by Javandel and Witherspoon (1968). However, the technique had been used earlier to solve steady heat-conduction-type electromagnetic problems by Winslow (1966).

Discretized Equations Using Mathematical Operators

The discretized equations for the IFDM and its special variant, the FDM, were developed in a straightforward manner in equations 5 and 6. In both these cases the subdomains of mass conservation are explicitly defined, while gradients of potential at the interface were evaluated using finite differences. As opposed to the IFDM and the FDM, the finite element method, as it is used widely now, combines the implicit method of partitioning the flow region through weighted volume integration in conjunction with the finite element concept of evaluating gradients. The generation of the discretized finite element equations, however, is generally carried out through techniques involving mathematical operations rather than with the help of the subdomain concepts (set theoretic concepts) introduced in the earlier sections. We will now consider the use of the techniques for setting up the finite-element equations.

For transient ground-water problems (and the analogous heat-conduction problem) the earliest operator technique to be used was that involving variational principles (Gurtin, 1964; Javandel and Witherspoon, 1968). Although it exists for the linear, steady-state heat conduction problems, a true variational principle does not exist for the transient case. To overcome this difficulty, Gurtin (1964) first applied the Laplace transform to the partial differential equation to eliminate time dependence and then defined a variational principle in the Laplace domain. Gurtin's variational principle for the conduction problem thus involves convolution integrals. It was soon pointed out by Finlayson and Scriven (1967) that insofar as one is concerned with the development of discretized finite element equations, there is no advantage to be gained in looking for a variational functional which may or may not exist. Instead, it is far more efficient to develop the equation through a procedure such as the Galerkin method or the method of weighted residual, which basically involve the weighted volume integration approach already discussed under implicit partitioning. We will therefore cease further consideration of the variational approach and proceed to consider in some detail the weighted-integral finite-element scheme.

Consider, for simplicity, a point ℓ in the flow region which is connected to its neighbors by straight lines, giving rise to a system of triangular elements (Fig. 8). Note that each triangular element is shared by three nodal points. Therefore, the mass of fluid contained in each element should be shared by the subdomains associated with each corner point. Moreover, since there is no bias for any of the corner nodes, each triangle is to be shared equally by its corner points. Thus we associate 1/3 of the area of each triangle, e_1, e_2, ..., e_6, with nodal point ℓ by drawing the medians of each triangle as shown in Figure 8. This equal division of each triangle among the three corners can also be accomplished by integrating the weighting function over each triangle (Narasimhan, 1978). In other words, it

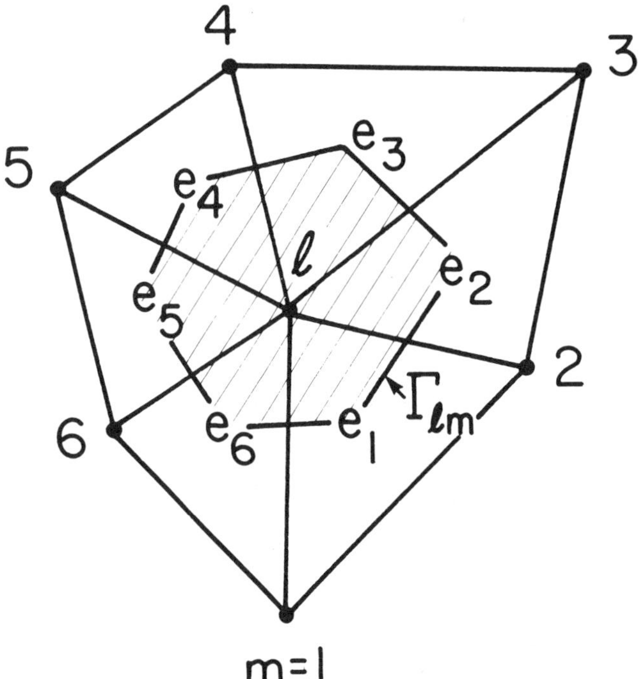

Figure 8. Subdomain of nodal point ℓ formed by assembling portions of contributing finite elements e_1, e_2, ... , e_6.

can be shown that

$$\int_{V_e} N_\ell \, dV = 1/3 \, V_e.$$

We now wish to evaluate the rate of accumulation of mass over the hatched region by a process of weighted integrations.

Writing the conservation equation for the hatched region in Figure 8, we get

$$\sum_{\substack{e_m \\ m=1}}^{6} \int_{\Gamma_{\ell m}} \rho_\Gamma K_\Gamma \vec{\nabla}\phi \bullet \vec{n}_\Gamma d\Gamma$$

$$= \left(\frac{1}{3} \sum_{\substack{e_m \\ m=1}}^{6} V_{e_m} m_{c,e_m} \right) \frac{\partial \phi_\ell}{\partial t}, \tag{14}$$

where V_{e_m} is the "volume" of the element e_m and m_{c,e_m} is the specific fluid mass capacity of the element e_m. Applying the divergence theorem to the left-hand side to convert the surface integral to a volume integral, and recognizing that

$$\int_{V_{e_m}} N_\ell dV = 1/3 \, V_{e_m} \, ,$$

we can write equation 14 as

$$\sum_{\substack{e_m \\ m=1}}^{6} \int_{V_\ell \cap e_m} \vec{\nabla} \bullet \rho K \vec{\nabla}\phi dV = \frac{1}{3} \sum_{\substack{e_m \\ m=1}}^{6} V_{e_m} m_{c,e_m} \frac{\partial \phi_\ell}{\partial t}, \tag{15}$$

where $V_\ell \cap e_m$ is the intersection of the sets V_ℓ and e_m.

In equation 15, the domain of integration is the hatched region in Figure 8. We now wish to extend the domain of integration to include the whole of the triangular elements and introduce, at the same time, a weighting function. Thus, let

$$\sum_{\substack{e_m \\ m=1}}^{6} \int_{V_\ell \cap e_m} \vec{\nabla} \bullet \rho K \vec{\nabla}\phi dV \equiv \sum_{\substack{e_m \\ m=1}}^{6} \int_{V_{e_m}} w_\ell \vec{\nabla} \bullet \rho K \vec{\nabla}\phi dV, \tag{16}$$

where w_ℓ is the weighting function with reference to subdomain ℓ, V_{e_m} is the region of element e_m, and

$$\sum_{\substack{e_m \\ m=1}}^{6} \int_{V_\ell \cap e_m} m_{c,e_m} \frac{\partial \phi_\ell}{\partial t} dV \equiv \sum_{\substack{e_m \\ m=1}}^{6} \int_{V_{e_m}} w_\ell \, m_{c,e_m} \frac{\partial \phi_\ell}{\partial t} dV. \tag{17}$$

Then, in view of equations 16 and 17, equation 14 becomes

$$\sum_{\substack{e_m \\ m=1}}^{6} \int_{V_{e_m}} w_\ell \left[\vec{\nabla} \bullet \rho K \vec{\nabla}\phi dV - m_{c,e_m} \frac{\partial \phi_\ell}{\partial t} \right] dV = 0. \tag{18}$$

Note that in equation 18, the terms within the parentheses represent the partial differential equation of groundwater flow, in which the time derivative is treated as the volume averaged time derivative at the nodal point ℓ. In other words, we could have started from the partial differential equation for a volume element centered at ℓ, multiplied by the weighting function w_ℓ (for example, see Finlayson, 1972; Pinder and Frind, 1972; Neuman, 1973) integrated over all the elements having ℓ as a corner point, and arrived at equation 18.

We now recognize that over each finite element triangle, ϕ is given by equation 10 in terms of the weighting functions, N_i, i = 1,2,3, ..., . We substitute this approximation into equation 18 and obtain

$$\sum_{\substack{e_m \\ m=1}}^{6} \int_{V_{e_m}} w_\ell \vec{\nabla} \bullet \rho K \vec{\nabla} N_i \phi_i dV =$$

$$\sum_{\substack{e_m \\ m=1}}^{6} \int_{V_{e_m}} w_\ell m_{c,e_m} dV \frac{\partial \phi_\ell}{\partial t}, \tag{19}$$

in which the repeated subscript i denotes summation over i = 1, 2, 3. Note that the approximation given by equation 10 was introduced expressly for the purpose of measuring spatial gradients. Therefore, we substitute equation 10 only in the spatial derivative and not in the time derivative.

We now ask the critical question, what is the correct weighting function w_ℓ for the transient diffusion problem? Recall from the right-hand side of equation 14, and from discussions immediately preceding it, that nodal point ℓ commands 1/3 of each triangle. Hence we should have the condition

$$\int_{V_{e_m}} w_\ell dV,$$

which yields 1/3 V_{e_m}. We already know, however, that the weighting function N_ℓ does in fact satisfy this condition. So we choose $w_\ell = N_\ell$ and get the well-known "Galerkin" formulation

$$\sum_{\substack{e_m \\ m=1}}^{6} \int_{V_{e_m}} N_\ell \left[\vec{\nabla} \bullet \rho K \vec{\nabla} N_i \phi_i - m_{c,e_m} \frac{\partial \phi_\ell}{\partial t} \right] dV = R_\ell \approx 0, \qquad (20)$$

where R_ℓ is the residual error associated with subdomain ℓ arising due to errors inherent in the approximate relation in equation 11. N_ℓ is a spatial partition function required by set-theoretic considerations (Narasimhan, 1978). However, from a mathematical point of view (Finlayson, 1972), N_ℓ is treated as a function that weights residuals over the region of integration. The set theoretic interpretation gives a conceptual credibility to the choice of N_ℓ as the most suitable weighting function, but the notion that N_ℓ is a function that weights residuals does not provide a logical basis for considering the Galerkin weighting function as the most appropriate one for the transient ground-water flow problem.

Consider now the left-hand side of equation 20. It can be shown that $\nabla N_i \phi_i \equiv 0$ at any point within a triangular element since the assumed linear variation of potential satisfies the Laplace equation over the triangular finite element. Hence, the first volume integral on the left hand side of equation 20 cannot be evaluated in its present form. To make the integration meaningful, we apply Green's First Identity to it (Sokolnikoff and Redheffer, 1966) and replace it with an equivalent volume integral and a surface integral. Thus,

$$\sum_{\substack{e_m \\ m=1}}^{6} \int_{V_{e_m}} N_\ell \vec{\nabla} \bullet \rho K \vec{\nabla} N_i \phi_i dV =$$

$$- \sum_{\substack{e_m \\ m=1}}^{6} \int_{V_{e_m}} \vec{\nabla} N_\ell \bullet \rho K \vec{\nabla} N_i \phi_i dV$$

$$+ \sum_{\substack{e_m \\ m=1}}^{6} \int_{\Gamma_{e_m}} N_\ell \, \rho K \vec{\nabla} N_i \phi_i \bullet n_\Omega d\Gamma. \qquad (21)$$

Again, the nature of N_ℓ is such that the surface integral on the right-hand side can be non-zero only if the point ℓ lies on an outer boundary of the flow region. Therefore, in the weighted-volume integral, finite-element method, the final discretized mass-conservation equation for a subdomain ℓ (neglecting sources) reduces to

$$\sum_{\substack{e_m \\ m=1}}^{6} \int_{V_{e_m}} \vec{\nabla} N_\ell \bullet \rho K \vec{\nabla} N_i \phi_i dV$$

$$+ \left(\sum_{\substack{e_m \\ m=1}}^{6} \int_{V_{e_m}} N_\ell m_{c,e_m} \, dV \right) \frac{\partial \phi_\ell}{\partial t}$$

$$= \sum_{\substack{e_m \\ m=1}}^{6} \int_{\Gamma_{e_m}} N_\ell \rho K \vec{\nabla} N_i \phi_i \bullet \vec{n}_\Gamma d\Gamma. \qquad (22)$$

Recognizing that the right-hand side incorporates known boundary conditions and that

$$\int_{V_e} N_\ell \, dV = 1/3 \; V_e,$$

we can rewrite equation 21 as

$$\sum_{\substack{e_m \\ m=1}}^{6} \int_{V_{e_m}} \vec{\nabla} N_\ell \bullet \rho K \vec{\nabla} N_i d_i dV$$

$$+ \left(\frac{1}{3} \sum_{\substack{e_m \\ m=1}}^{6} m_{c,e_m} V_{e_m} \right) \frac{\Delta \phi_\ell}{\Delta t} = Q_\ell, \qquad (23)$$

where Q_ℓ incorporates known boundary conditions and sources, if any.

The volume integral in equation 23 reduces to a simple algebraic expression in the case of a simple triangular finite element (Felippa, 1966; Narasimhan and others, 1978). Thus for a given triangle with corners designated ℓ, 2, 3,

$$\int_{V_{e_m}} \nabla N_\ell \bullet \rho K \nabla N_i \phi_i dV$$

$$= \frac{\rho K}{4\Delta_{e_m}} [b_\ell (b_\ell \phi_\ell + b_2 \phi_2 + b_3 \phi_3)$$

$$+ c_\ell (c_\ell \phi_\ell + c_2 \phi_2 + c_3 \phi_3)], \qquad (24)$$

where Δ_{e_m} is the volume of element e_m and

$$\begin{aligned} b_\ell &= y_2 - y_3 & c_\ell &= x_3 - x_2 \\ b_2 &= y_3 - y_\ell & c_2 &= x_\ell - x_3 \\ b_3 &= y_\ell - y_2 & c_3 &= x_2 - x_\ell \end{aligned} \qquad (25)$$

in which x, y are local coordinate axes and the subscripts 2,3 denote the other two nodal points of the triangular element besides ℓ.

The algebraic expressions in equation 24 can be easily extended to include nonisotropic material properties (for example Neuman, 1975; Narasimhan and others, 1978a).

Now, returning to equations 18 and 19, recall that in the differential equation within the parenthesis we have substituted $\phi = N_i\phi_i$ only in the spatial derivative term. We treated $(\partial\phi/\partial t) = (\partial\phi_\ell/\partial t)$ as an average over the subdomain commanded by point ℓ. We chose to do this on the basis of the conceptual foundations of the integral equation of mass conservation in which ϕ_ℓ is the average value of ϕ over subdomain ℓ bounded by the surface Γ. Instead, suppose one were to ignore the aforesaid reasoning and treat the partial differential equation in a purely mathematical sense. Then one could substitute the relation $\phi = N_i\phi_i$ in both the terms in equation 19, giving rise to

$$\sum_{\substack{e_m \\ m=1}}^{6} \int_{V_{e_m}} w_\ell \nabla \bullet \rho K \nabla N_i \phi_i dV = \sum_{\substack{e_m \\ m=1}}^{6} \int_{V_{e_m}} N_\ell N_i m_{c,e_m} \frac{\partial \phi_i}{\partial t} dV, \quad (26)$$

where the index i denotes the corner nodal points of the element e_m. If we compare equations 20 and 26 we will immediately see that while the integrals involving the spatial gradient of ϕ in both cases are exactly equivalent, the term involving the time derivative in equation 26 takes on a more complex form than in equation 20. Here again, it has been shown (for example Felippa, 1966) that

$$\int_{V_e} N_{i_d} N_i d_v = 1/6\, V_e \text{ for } i = \ell$$

and $1/12V_e$ for $i \neq \ell$. Thus, the right-hand side of equation 26 involves the time derivative of all the neighboring nodal points of ℓ while in equation 20 the corresponding term involves only the time derivative of ℓ. Whereas equation 20 leads to a "lumped mass matrix," equation 26 leads to what has been called a "consistent" or "distributed mass matrix."

The weighted volume integral in equation 23 reduces to a simple algebraic expression only for the simplest elements (triangles or tetrahedra). In fact it can be shown (Narasimhan, 1975) that the algebraic expression is exactly equal to the fluxes across the surfaces $\Gamma_{\ell m}$ of the hatched subdomains in Figure 8. However, when one resorts to the use of higher-order approximations, then the volume integral does not lead to a simple algebraic expression. Instead, one has to subdivide each finite element into smaller subdomains (dV's) and carry out the volume integration numerically, using techniques such as Gaussian quadrature (Zienkiewicz, 1971). In this case, however, this writer has not been able to demonstrate for himself (Narasimhan, 1975) that the volume integral is fully equivalent to the primitive surface integral. For this reason this writer has some skepticism about the appropriateness of using higher-order approximations for transient problems. In other words, while higher-order approximations may provide increased accuracy in computing spatial gradients, the increased accuracy may be offset by errors introduced due to the non-equivalence of volume and surface integrals. It must be emphasized that such errors do not exist in steady-state problems, for under steady-state conditions the surface integral and the volume integral of any and every subdomain over the flow region is zero. Therefore, higher-order approximations should be expected to yield improved accuracy for steady-state problems.

In the ground-water literature (for example Pinder and Frind, 1972) it is customary to use higher-order approximations in conjunction with elements with curved sides. In these cases, the weighted volume integration procedure helps approximate gradients as well as takes into account the curved nature of the boundaries. For this, the elements are first transformed into a rectangle, parallelepiped, or any such simple shape through curvilinear coordinate transformation, and the volume integrals evaluated in the transformed domain. These are the so-called isoparametric elements (Zienkiewicz, 1971).

Summary of Discretized Equations

On the basis of what has been discussed so far, we can identify the three following types of discretized numerical equations. These equations pertain to a volume element ℓ of the flow region over which mass is conserved.

Explicit partition, finite difference gradients:

$$\rho_\ell G_\ell + \sum_m \rho_{\ell m} K_{\ell m} \frac{\phi_m - \phi_\ell}{D_{\ell m}} \Delta\Gamma_{\ell m} + \sum_b \rho_{\ell b} K_{\ell b} \frac{\phi_b - \phi_\ell}{D_b} \Delta\Gamma_{\ell b} = M_{c,\ell} \frac{\Delta\phi_\ell}{\Delta t}, \quad (27)$$

in which m includes all volume elements interior to the flow region communicating with ℓ and b denotes boundary elements incorporating the boundary conditions.

Explicit partition, finite element gradients, lumped mass matrix:

$$\rho_\ell G_\ell - \sum_{\substack{e_m \\ m=1}}^{6} \int_{V_{e_m}} \vec{\nabla} N_\ell \bullet \rho K \vec{\nabla} N_i \phi_i dV + \sum_{\substack{e_m \\ m=1}}^{6} \int_{\Gamma_{e_m}} N_\ell \rho K \vec{\nabla} N_i \phi_i \bullet \vec{n}_\Gamma d\Gamma$$

$$= \left(\sum_{\substack{e_m \\ m=1}}^{6} \int_{V_e} N_\ell \, m_{c,e_m} \, dV \right) \frac{\nabla \phi_\ell}{\nabla t} , \qquad (28)$$

where m is number of neighbors to which ℓ is connected.

Implicit partition, finite element gradients, distributed mass matrix:

$$\rho_\ell G_\ell - \sum_{\substack{e_m \\ m=1}}^{6} \int_{V_{e_m}} \vec{\nabla} N_\ell \bullet \rho K \vec{\nabla} N_i \phi_i dV$$

$$+ \sum_{e_m}^{6} \int_{\Gamma_{e_m}} N_\ell \rho K \vec{\nabla} N_i \phi_i \bullet \vec{n}_\Gamma d\Gamma$$

$$= \sum_{\substack{e_m \\ m=1}}^{6} \int_{V_{e_m}} N_\ell N_i \, m_{c,e_m} \frac{\Delta \phi_\ell}{\Delta t} dV. \qquad (29)$$

To account for anisotropy, the argument of the volume integral on the left-hand side of equations 28 and 29 may be replaced by

$$\frac{\partial N_\ell}{\partial x_i} \rho K_{ij} \frac{\partial \phi_i}{\partial x_j} .$$

Note that one equation such as 27, 28, or 29 is to be set up for every volume element in the flow region for which time derivatives of potential are to be computed. Moreover, if we define the mean permeability, $K_{\ell m}$, to be zero when element m does not communicate with element ℓ, or if we define N_ℓ to be zero over all finite elements that do not have ℓ as a vertex, then equations 26 through 28 can be conveniently written in matrix form, as follows:

$$A_{\ell m} \phi_m = Q_\ell \qquad (30)$$

or

$$[A] \{\phi\} = \{Q\}. \qquad (31)$$

In equations 30 and 31 all the known sources, boundary conditions, and initial conditions are incorporated in the term Q on the right-hand side. The coefficient $A_{\ell m}$, which incorporates permeabilities, the weighting function, surface areas, and related terms as appropriate, includes the "conductance" between the subdomain ℓ and m. Hence [A] is called the conductance matrix. However, since the finite element method originated in the structural engineering literature, [A] is sometimes called the "stiffness" matrix.

Reflections on Nomenclature

At this juncture it is pertinent to reflect upon certain terminologies that have become established in the literature on numerical simulation. The terms finite differences and finite elements are now widely used. Any computational algorithm is classified into either one or the other. How rational is this classification?

Recall that numerical computations fundamentally require (a) definition of subsets over which mass is conserved and (b) measurement of gradients. The terms "finite difference" and "finite element" in themselves relate solely to the task of gradient evaluation. These two terms do not inherently say anything about the method of partition. It is generally assumed that the finite difference gradients are used in conjunction with explicit partitioning whereas finite element gradients are used in conjunction with weighted volume integration. However, there is nothing to preclude the use of finite-difference gradients along with weighted volume integration or the use of finite-element gradients along with explicit partitioning and surface integration.

A claim is generally made that the finite element method can handle arbitrary geometries but that the finite difference method cannot. This ambiguous claim needs to be questioned. The FEM is able to handle complex geometries not because of the finite-element gradients, but because the FEM uses an integral approach, partitioning the flow region into arbitrarily shaped subdomains. The finite-element approach to measuring gradients has nothing to do with complex geometries. In fact, it is possible to combine the integral approach with finite difference gradients and handle complex geometries with power and flexibility. This is exactly what is done in the IFDM, which is also capable of handling complex geometry.

The true power of the FEM, which is seldom properly stated, is that of measuring gradients in a very general fashion. It is precisely in those situations where generality of gradient evaluation is imperative (such as handling permeability tensor with non-zero off-diagonal terms or the general stress tensor in structural problems) that the FEM really proves superior to the IFDM and its variant, the FDM.

It appears to this writer that in identifying the FEM with weighted volume integration, a majority of the workers have not considered certain other possibilities. For one, the IFDM approach, which has positive advantages for certain problems involving isotropic materials, has not been given much attention. Also, the attractive possibility of combining explicit partition with finite element gradients has apparently not been investigated.

From the point of view of clarity therefore, it is necessary that the name of a numerical method should have two components: one to state whether the method uses an integral formulation, and the other to explain the technique of gradient evaluation. We note here that the conventional FDM is restricted to regular volume elements with sides normal to coordinate axes. Although we have treated this as a special case of the IFDM, we recognize that the restriction on regularly shaped volume elements is equivalent to directly approximating the differential equation (Zienkiewicz, 1971) since the derivation of the differential equation is tied up with the choice of regularly shaped volume elements. Hence it may be more expressive to rename the familiar numerical techniques as follows:

Current Usage

FDM: Finite difference method
FEM: Finite element method
IFDM: Integrated finite difference method

Suggested Usage

DFDM: Differential finite difference method
VIFEM: Volume integral finite element method
SIFDM: Surface integral finite difference method

Integration in Time

Turning our attention back to the discretized equations 28 to 39, we can rewrite the equations 28 to 30 into the following compact forms:

IFDM (equation 28), Lumped Mass Matrix FEM (equation 29)

$$\sum_m A_{\ell m}(\phi_m - \phi_\ell) = M_{c,\ell}\frac{\Delta\phi_\ell}{\Delta t} + Q_\ell, \qquad (32)$$

where $A_{\ell m}$ is the conductance of the ℓm-interface, given by

$$A_{\ell m} = \frac{\rho_{\ell m} K_{\ell m}\Gamma_{\ell m}}{D_{\ell m}} \qquad (33)$$

for the IFDM and

$$A_{\ell m} = \sum_e \frac{\rho K}{4\Delta_e}(b_\ell b_m + c_\ell c_m) \qquad (34)$$

for FEM with isotropic materials, and b_ℓ, b_m, c_ℓ, c_m and Δ_e are as defined in equation 24. In equation 32, Q_ℓ includes the known sources and boundary conditions.

FEM, distributed mass matrix (equation 29)

$$\sum_m A_{\ell m}(\phi_m - \phi_\ell) = \left(\sum_e \int_{V_e} N_\ell N_\ell m_{c,e}\, dV\right)\frac{\Delta\phi_\ell}{\Delta t}$$

$$+ \sum_e \int_{V_e} N_\ell N_m m_{c,e}\frac{\Delta\phi_m}{\Delta t}\, dV + Q_\ell. \qquad (35)$$

In equation 35 the second integral on the right-hand side includes summation over all m, indicating that the right-hand side includes the time derivatives of all the neighbors of ℓ with whom ℓ communicates.

In equations 32 and 35, ϕ_m and ϕ_ℓ are partly known from initial conditions, while $\Delta\phi_\ell$ and the $\Delta\phi_m$ in equation 35 are the unknowns to be evaluated. But ϕ_m and ϕ_ℓ are both functions of t and hence we have to use appropriate mean values $\bar{\phi}_m$ and $\bar{\phi}_\ell$ on the left hand side of equations 32 and 35. The criterion for choosing these mean values is as follows. Suppose we denote by $\phi_{\ell m}$ the difference in potential between subdomains ℓ and m. Then, we shall choose $\bar{\phi}_m$ and $\bar{\phi}_\ell$ in such a manner that

$$A_{\ell m}(\bar{\phi}_m - \bar{\phi}_\ell) = A_{\ell m}\int_{t_o}^{t_o+\Delta t}\frac{\partial\phi_{\ell m}}{\partial t}\, dt. \qquad (36)$$

In general let $\bar{\phi}_m = \overset{o}{\phi}_m + \lambda\Delta\phi_m$ and $\bar{\phi}_\ell = \overset{o}{\phi}_\ell + \lambda\Delta\phi_\ell$, where the superscript o denotes the known initial value at t_o and $0 \le \lambda \le 1$. If $\lambda = 0$, then all the quantities in equation 31 are known except $\Delta\phi_\ell$ and hence it is an "explicit" equation. In this case the change in potential for each subdomain is calculated independently of the others. In the case of equation 35, even if $\lambda = 0$, the right-hand side still has several unknowns, including $\Delta\phi_\ell$ and the $\Delta\phi_m$. Hence the distributed mass matrix equation is nonexplicit. We shall categorize the IFDM and the lumped mass matrix FEM as "explicit formulations" while the distributed mass matrix FEM is a "nonexplicit formulation." For want of a better expression we have used the adjective "explicit" in two different contexts: first to qualify partitioning and second, to qualify formulation. The meaning, however, should be clear from the context.

If λ is set to 0.5, the result is the well-known Crank-Nicolson or central differencing scheme while $\lambda = 1$ gives rise to the fully implicit, backward differencing scheme. From the nature of equation 35, it is obvious that λ should vary with time and should be problem dependent if one wishes to compute it exactly.

Stability of Explicit Equation

As has been shown by many (for example O'Brien and others, 1951; Dusinberre, 1961) the explicit equation will become "unstable" and yield unreasonable results if the chosen Δt exceeds some critical value. Physically the instability simply indicates that because of wrong choice of $\bar{\phi}_m$ and $\bar{\phi}_\ell$, the fluxes into a volume element are grossly over- or underestimated to such an extent that the maximum principle is violated. For a transient problem the Maximum Principle states that (for example, Neuman and Narasimhan, 1977) in the absence of sources or sinks, the maximum and the minimum potentials can occur only on the boundaries or at the initial time. As discussed by Narasimhan, stability, quantified by the so called stable time step (or time constant), is a local phenomenon (Narasimhan and others, 1978b). That is, each volume element has its own time constant expressed as the ratio of its fluid mass capacity to its surface conductance. Thus,

$$\Delta t_{stab,\ell} = \frac{M_{c,\ell}}{\Sigma_m A_{\ell m}} . \qquad (37)$$

If one wishes to use $\Delta t > \Delta t_{stab,\ell}$ for computational purposes, and get stable results, then it can be shown (Evans et al. 1954) that $0.5 < \lambda \leqslant 1$.

When $\lambda > 0$, then both equations 32 and 35 will have the unknown variable(s) occurring on both sides of the expressions, leading to a set of implicit equations. In this case, collecting all known quantities to one side (sources, boundary conditions, initial conditions) and noting that

$$\Delta\phi_\ell = \phi_\ell^{t_o+\Delta t} - \phi_\ell^{t_o} \text{ and } \Delta\phi_m = \phi_m^{t_o+\Delta t} - \phi_m^{t_o},$$

we could write these equations in a general matrix form:

$$A_{\ell m}\phi_m^{t_o+\Delta t} = Q_\ell , \quad \ell, m, = 1,2,3, \ldots, L, \qquad (38)$$

where Q_ℓ is a known vector, $A_{\ell m}$ is a matrix, and where L denotes the total number of subdomains in the system. Equations in equation 38 may be solved by direct-solution techniques or iterative techniques. These will be discussed in a subsequent section.

Estimation of Spatial Averages

In using the discretized equations over isotropic but heterogenous systems, we are constrained to use spatially averaged parameters. These are of two kinds. First, in those cases where two subdomains with different material properties have a common interface (for example in the classical finite difference method), one has to evaluate the mean permeabilities, $K_{\ell m}$, and the mean fluid densities, $\rho_{\ell m}$ (not very important) at the interface. Second, when a single subdomain of mass conservation (for example the conventional FEM) is made up of more than one material, one has to define an appropriate fluid mass capacity, $M_{c,\ell}$, for the subdomain.

First, consider mean permeability at the interface. In the simplest case, suppose each of the subdomain has a constant K and that it changes in a step-wise fashion at the interface. Then, in order to preserve continuity of flux (Edwards, 1972; Narasimhan, 1975) the correct mean value is the harmonic mean,

$$K_{\ell m} = \frac{K_\ell K_m (d_{\ell,m} + d_{m,\ell})}{K_\ell d_{m,\ell} + K_m d_{\ell,m}} . \qquad (39)$$

When extending equation 39 to those problems in which permeability is a function of potential (for example unsaturated flow), the harmonic mean may lead to severe inaccuracies (Haverkamp and Vauclin 1979). This is to be expected, since the derivation of equation 39 is based on a step-wise change of K across the interface. Obviously, when K varies over each subdomain in relation to potential, the interface $K_{\ell m}$ should be strictly evaluated based on the interface potential. The accurate estimation of interface potential, however, may be very difficult in certain problems involving sharp fronts (for example problems of infiltration in a dry soil). In this case, evaluating $K_{\ell m}$ is especially difficult. Recent empirical work by Haverkamp and Vauclin (1979) suggests that the geometric mean, $K_{\ell m} = \sqrt{K_\ell K_m}$ is a better approximation than either the harmonic mean or arithmetic mean for a one-dimensional infiltration problem studied. Another empirical study by Narasimhan and Witherspoon (1978) involving a one-dimensional infiltration problem suggested that at early times a strong upstream weighting was necessary. However, after some time had elapsed, the solution was not sensitive to changes in upstream weighting.

According to Frind (oral commun.), this problem does not arise when the exact function $K(\psi)$ is used in conjunction with volume integration as done by Frind and Verge (1978) in employing the FEM.

The spatial averaging of the capacity term arises in the conventional FEM, in which the finite elements having nodal point ℓ as a corner may be composed of different materials. In this case, the fluid mass capacity of ℓ is effectively defined as

$$M_{c,\ell} = \begin{cases} \sum_e \int_{V_e} N_\ell m_{c,e} dV, & \text{Lumped Mass Matrix} \\ \sum_e \int_{V_e} N_\ell N_{e,} m_{c,e} dV & \text{Distributed Mass Matrix} \end{cases} \qquad (40)$$

The latent assumption here is that the capacity of the subdomain is equal to the weighted sum of the capacities of the different materials contained in the subdomain. If one were to define a volume averaged fluid mass capacity for the subdomain ℓ denoted by $\langle m_c \rangle_\ell$, then for the lumped mass matrix case we will have

$$\langle m_c \rangle_\ell = \frac{\sum_e \int_{V_e} N_{\ell,} m_{c,e} dV}{\sum_e \int_{V_e} N_\ell \, dV} . \qquad (41)$$

A similar expression could be derived for the distributed mass matrix case. The physical validity of this assumption does not appear to have been investigated in the literature.

Handling Time-Dependent Parameters

In equations 32 and 35 the conductances $A_{\ell m}$ and the capacities $M_{c,\ell}$ are assumed to be constant in time. That is, the diffusion problem is governed by a linear partial differential equation.

However, in the case of nonlinear problems (for example, strongly deformable systems, saturated-unsaturated flow), both $A_{\ell m}$ and $M_{c,\ell}$ are functions of the fluid potential, or, more precisely, fluid pressure. In these problems $A_{\ell m}$ and $M_{c,\ell}$ have to be evaluated during each time step before the time derivatives of potentials can be computed.

Perhaps the simplest method of handling nonlinear coefficients is to evaluate them at some carefully estimated mean value of ϕ over the time step and to use the evaluated values as constants for the particular time step. This is the quasi-linearization approach. However, the quasilinearization scheme has to be used in conjunction with relatively small time steps, failing which integration errors will accumulate with time.

An alternate approach is the iterative approach based on the Newton-Raphson scheme (Brutsaert, 1971; Faust and Mercer, 1979). This method apparently helps take not only larger time steps than the quasi-linear scheme but is also more accurate than the latter when sharp contrasts in material properties exist over the system. A brief description of the Newton-Raphson scheme is given below.

Consider the equation

$$A_{\ell m}\phi_m^{t_o+\Delta t} = Q_\ell, \tag{42}$$

in which $A_{\ell m}$ is a function of ϕ_ℓ and ϕ_m. This means that $A_{\ell m}$ has to be evaluated at the appropriate mean values $\bar{\phi}_m$ and $\bar{\phi}_\ell$. Without elaboration, let us assume that equation 42 is transformed to

$$B_{\ell m}^{t_o+\Delta t}\phi_m^{t_o+\Delta t} - R_\ell = 0, \tag{43}$$

where $B_{\ell m}^{t_o+\Delta t}$ is the transformed conductance matrix and R_ℓ is the transformed vector of known terms. We now wish to solve for $\phi_m^{t_o+\Delta t}$ by a process of iteration. Let p denote the iteration number. Define

$$C_\ell^p = B_{\ell m}^{t_o+\Delta t,p}\phi_m^{t_o+\Delta t,p+1} \tag{44}$$

and

$$R_\ell^p = B_{\ell m}^{t_o+\Delta t,p}\phi_m^{t_o+\Delta t,p}\ . \tag{45}$$

Typically, as $p \rightarrow \infty$, $C_\ell^p \rightarrow R_\ell^p$ and hence equation 43 should be exactly satisfied. We now treat C_ℓ^p as a function of the several variables $\phi_m^{t_o+\Delta t,p+1}$ subject to the unique solution that as each estimated value of $\phi_m^{t_o+\Delta t,p}$ approaches the exact solution $\phi_m^{t+\Delta t}$, the difference $C_\ell^p - R_\ell^p \equiv 0$. In other words we are finding the zeroes of a several variable function. Now since $C^p = C^p\ (\phi_m^{t_o+\Delta t,p+1})$, we can expand C_ℓ^p as a Taylor series about the value of $C_\ell^{(p-1)}$ and neglect all derivatives higher than the first. Then,

$$C_\ell^p = \sum_m \frac{\partial C_\ell^{p-1}}{\partial\phi_m}\left[\phi_m^{t_o+\Delta t,p+1} - \phi_m^{t_o+\ t,p}\right]. \tag{46}$$

We have a new set of simultaneous equations

$$\sum_m \frac{\partial C_\ell^{p-1}}{\partial\phi_m}\phi_m^{t_o+\Delta t,p+1} = R_\ell^* \tag{47}$$

to be solved for $\phi_m^{t_o+\Delta t,p+1}$. The coefficient matrix C_p^{p-1}/m is numerically evaluated by computing C_ℓ using $\phi_m^{t_o+\Delta t,p}$ and $\phi_m^{t_o+\Delta t,(p-1)}$ and taking the difference. The iteration process repeated until $R_\ell^*\ max < \varepsilon$, where ε is the error tolerance.

Solution Strategies

Once the discretized equations are set up, the nonlinear coefficients evaluated, and the final set of simultaneous equations arrived at, the final task of numerical modeling is to solve the set of equations. In general, there are two general classes of methods available for solving a set of simultaneous equations: (1) direct solvers and (2) iterative solvers. The latter includes relaxation techniques. Choosing between the two approaches becomes especially significant since the conductance matrix is an extremely sparse one. The literature is replete with references on direct and iterative solvers and we will only attempt here to briefly discuss some of the important aspects.

There are three principal direct solution techniques: (1) solution by determinants (for example Kramer's rule), (2) solution by successive elimination of unknowns, and (3) solution by matrix inversion. Except in very special cases, it is generally understood that successive elimination of unknowns is by far the best suited direct solution technique for subsurface fluid flow problems.

In a set of equations, the unknowns can be successively eliminated either through division by leading coefficients or by Gaussian elimination. In the widely used Gaussian elimination method, the unknowns are successively eliminated by solving one equation of some unknown in terms of all the others and then substituting this value for the same unknown in all the other euqations, thereby eliminating the equation from the set. The process is reduced to a single equation in one unknown. The equation which expresses one unknown in terms of all others is the "pivotal" equation. After one unknown has been found, the remaining are found systematically by back substitution in the pivotal equations. In the Gaussian elimination method, one should always select as the pivotal equation the one that has the largest coefficient for the unknown to be eliminated.

A more recent variant for achieving Gaussian elimination is the method of matrix factorization (Remson and others, 1971; Johnson and Riess, 1977)

The factorization scheme is inspired in part by the finding that triangular systems of equations are easier to solve than square systems. Suppose the conductance matrix (A) can be factorized expressed as a product of a lower triangular matrix L and an upper triangular matrix U. Thus

$$[A] = [L][U] \tag{48}$$

and our system of equations becomes

$$[L]\{U\}\ \{d\} = \{Q\}. \tag{49}$$

We proceed to solve equation 48 in two stages. First, let [U] {ϕ} = {y}, where {y} is a dummy vector. Then the system [L] {y} = {Q} can be solved by forward substitution. Using [L] {Y} = {Q} the system [U] {ϕ} = {y} can be solved by backward substitution to get the required solution {ϕ}. The main task is therefore to decompose (A), which can be economically carried out. The LU-decomposition scheme requires (5n-4) operations in all while conventional Gaussian elimination involves $n^3/3$ operations (Remson and others, 1971).

There are three principal advantages of the direct solution techniques: (1) there is no need to make an initial guess for the solution; (2) there is no need to provide an error tolerance, and (3) convergence problems associated with iterative solvers are avoided especially when time constants vary widely among the various subdomains. The major disadvantage of the direct solver vis-a-vis sparse systems is that an excessively large number of zeroes of the sparse matrix need to be stored and operated upon.

A great deal of research effort is being expended in devising methods by which to structure matrix so as to collect all unwanted zeroes in some known region. The locations of these regions are identified through the use of suitable row and column indices. One of the simplest ways to structure the matrix is to carefully number the subdomains so that the separation between the extreme non-zero elements in a given row or column is minimal.

A very good possibility of minimizing the storage problem associated with the direct solver is the following. Recall that the stability of the explicit equation is a local phenomenon and that each subdomain has its own time constant as given by equation 37. In most problems involving irregular meshes with heterogeneous material distribution, for a given Δt some volume elements may be stable ($\Delta t < \Delta t_{stab}$) while others may not. In this case, one needs to solve a set of simultaneous equations only for those volume elements for which $\Delta t_{stab} < \Delta t$ and their immediate neighbors. If these volume elements are grouped in certain isolated regions of the flow region, then one simply has to apply the direct solver independently to one or more small portions of the large matrix with much reduced storage and computational effort. For the intervening volume elements one simply computes the potential changes in an explicit manner. Although the mixed explicit-implicit scheme has been used iteratively (Edwards, 1972; Narasimhan and others, 1978) it remains to be extended to direct solvers.

The indirect method of solving simultaneous equations includes two general types with but minor differences between them. In both these techniques the solution at a point (that is for a given volume element) is expressed in terms of its neighbors with which it communicates. Thus, for example (Remson and others, 1971),

$$\phi_\ell^{p+1} = F\left(\phi_\ell^p , \phi_m^p \right). \tag{50}$$

Now define a residual θ_ℓ by

$$r_\ell = \phi_\ell^{p+1} - \phi_\ell^p. \tag{51}$$

The aim is to minimize r_ℓ at each point by iteration until r_ℓ is acceptably small. In the relaxation method, one applies a correction to the computed values of ϕ_ℓ^{p+1} based on the magnitude of r_ℓ at each point. The iteration is then repeated. The correction applied is the "relaxation factor" or the "acceleration factor." In the iterative method, on the other hand, the latest values of ϕ_ℓ^{p+1} or ϕ_m^{p+1} are continuously substituted in equation 51 as and when they may be updated.

A wide variety of relaxation and iteration methods have been developed in the literature and one should consult recognized texts (for example Varga, 1962; Dahlquist and Bjork, 1974; Remson and others, 1971) on the subject. Some iterative schemes owe their development to petroleum reservoir engineers who are interested in a variety of single-phase as well as multi-phase systems. Among the better-known iterative/relaxation schemes one could cite: the Point-Jacobi scheme; the Gauss-Seidel scheme; the line-successive over relaxation scheme (LSOR); the alternate direct implicit scheme (ADI; Peaceman and Rachford, 1955); the strongly implicit procedure (SIP; Stone, 1968); the mixed explicit-implicit iteration procedure (Edwards, 1972; Neuman and Narasimhan, 1977; Narasimhan and others, 1978), and so on.

Some of the techniques (for example ADI) are indeed combinations of direct and iterative techniques. There is reason to believe that the efficiency of iterative schemes increases as the component of direct solver in these schemes increases. Originally developed for the classical FDM, most of these schemes scan the arrays systematically (left to right, bottom to top) during each iteration. For irregular meshes with poor structuring of nodal arrangements, these techniques may or may not be most efficient or accurate. For such problems point successive iteration methods may be more advantageous.

Moreover, most of the iteration methods operate on the entire matrix, although the stability criterion of equation 38 may only be locally violated. To avoid the extra effort involved, the mixed explicit-implicit scheme of Edwards (1972) consists in iterating only on those regions of the flow region where a group of volume elements are unstable for the time step chosen.

The chief advantage of the iteration approach is that it inherently avoids the need for storing and operating on the zeroes of a sparse matrix. Thus, iterative methods are especially economical to use for large three-dimensional problems. However, there are several disadvantages to be considered:

(1) The need for making a good initial guess. The farther the initial guess from the correct solution, the greater the solution effort.

(2) The problem of convergence. Convergence is related to the acceleration parameter, which is extremely problem dependent, and incorrect parameters can greatly increase computational effort.

(3) The need to provide error tolerance.

(4) The restrictions on time step size. In purely iterative methods convergence may pose problems with increasing time-step size.

At the present level of computational ability, it appears that for all one-dimensional problems direct solvers are preferable, while for most three-dimensional problems iterative techniques are more competitive. Depending on the nature of the particular two-dimensional problem, direct solvers may be competitive in some cases but not in others. Computationally, it is preferable to use direct solvers whenever possible, since one avoids the problem of choosing an initial guess for the solution, the error tolerance and the acceleration/relaxation parameter.

VALIDATION OF NUMERICAL METHODS

The practical outcome of using a numerical method is often a computer program or code. Once the code is written, it has to be validated to assure that it is indeed clear of programming errors and that the algorithm chosen will solve new, complex problems that fall in its fold. These programs may be either problem specific (one-dimensional consolidation problems) or general purpose.

In the logical, integral development of numerical methods presented earlier in this paper, the partial differential equation was not an essential component. Yet one of the most extensively used norms for validating numerical models is the analytic solution to the partial differential equation. In addition, models are often validated against experimental results.

In ground-water hydrology, the most widely used axisymmetric, analytic solutions are related to well hydraulics. Models may also be validated against one- or two-dimensional heat conduction problems, for which a host of interesting results are presented in Carslaw and Jaeger (1959). Validation of models for three-dimensional problems are relatively rare due to the paucity of typical three-dimensional analytic solutions.

When the problem becomes non-linear, however, the availability of analytical solutions is greatly diminished. For such problems, model validation has to depend on experimental data. As examples one could cite: (1) simulation of saturated-unsaturated flow experiments in a sand box (Vauclin and others 1975; Narasimhan and others, 1978; Narasimhan and Witherspoon, 1978); (2) simulation of free surface flow in a Hele Shaw model experiment (Marino, 1967; Narasimhan and others, 1978); and (3) simulation of the slurry consolidation experiment of Smiles and Rosenthal (1968) by Narasimhan and Witherspoon (1968).

The mathematical process of model validation includes consideration of the existence of a solution uniqueness, stability, convergence, error bounds and so on. This is a highly evolved and sophisticated field of enquiry into which we will not venture. However, in light of the logical framework presented at the beginning, we will look at what appear to be some interesting fundamental questions.

The mathematical analysis of numerical methods is heavily based on solutions to partial differential equations. In fact, the numerical method is actually treated as a solver of partial differential equations. Now, since the classical FDM seeks to approximate directly the differential equation of interest, most of the mathematical analyses have been especially devloped for the FDM. This is indeed appropriate since the explicit spatial partition of the FDM corresponds exactly with the partitioning based on which the differential equation is derived.

However, the integral approach to numerical methods shows that these methods are more versatile than being merely solvers of the differential equation. Recent work based on set theoretic concepts (Narasimhan, 1978) and volume averaging (Narasimhan, 1979) seem to suggest that the partitioning of the flow region is physically governed by certain criteria of macroscopic averaging. Subject to these criteria, one should be able to choose, for the elementary subdomain, any shape, not necessarily the regular six-sided (in three dimensions) shape chosen to derive the differential equation. Since the diffusion process is governed by specific surface (diffusion will be greater when specific surface is large and since specific surface depends on the shape of the volume element) of the volume element, it is reasonable to infer that elements with identical volume but different shapes will have different diffusion rates. Now, since the regular six-sided volume element is only one of the possible partitions that are physically justifiable based on averaging criteria, it appears that the solution of the partial differential equation is only one of the many possible solutions of the physical problem. Indeed it seems that for arbitrarily shaped volume elements for which a simple relationship does not exist between the bulk volume and the bounding surface, one may not be able to express divergence in terms of velocity gradients. In other words, a simple differential equation may not exist.

In the opinion of this writer, this reasoning raises doubts about the wisdom in our adherence to the idea that the numerical method is merely a solver of differential equations. While this is not incorrect, such a use of the numerical method becomes extremely restricted in scope. Under the circumstances, there is a definite need to look for techniques which will mathematically validate the numerical model without resorting to the analytical solution. Fundamentally, there is no reason that this should not be possible. Such validation techniques should stem from the axiomatic foundations of the theories of integration and physically based volume averaging. The time has apparently come when we have to break away from old traditions of continuum mechanics of differential equations and toward new areas of discrete systems that can now be handled with the fast computing devices which have currently become a way of life and were not available to previous generations.

FIELD APPLICATIONS

Since the early 1960s, ground-water models have been successfully applied to field problems in many parts of the world. An exhaustive summary of the field examples is clearly an impossible task. Nevertheless, to give an idea of the activity, a few references are cited below from the North American continent. A more detailed account of these examples may be found in Prickett (1975) and Narasimhan and Witherspoon (1977).

Among the earliest studies of steady-state ground-water flow is the simulation of the Gravelbourg aquifer, Saskatchewan, Canada, by Freeze and Witherspoon (1968) using the method of finite differences. Following the suggestion of Toth (1962), they used this study to identify recharge and discharge areas and evaluate regional recharge-discharge relationships.

Pinder and Bredehoeft (1968) applied a transient flow model to study the adequacy of an aquifer at the Musquodoboit village in Nova Scotia to ground-water pumpage for water supply. They treated the aquifer as a heterogeneous system with three materials and inferred that the aquifer could meet the water supply demands of the town. The numerical results compared favorably with electrical analog results.

The Prickett and Lonnquist (1971) transient model was applied to a region on a much grander scale, both in space and time. They applied their model to the Cambrian-Ordovician aquifer of Illinois, covering an area of 24,000 sq. mi. and are investigating the system response for a period of 130 years (1864 to 1995) under realistic exploitation conditions. The heavily over-pumped aquifer is partly under water table condition, and hence transmissivities were treated as a function of hydraulic head. After calibrating the model with observed data between 1864 and 1958, they went on to compute the piezometric and water table contours under expected exploitation conditions up to the year 1995.

A field problem of great practical importance in hydrology is the interaction between rainfall and the ground-water reservoir and the evaluation of the influence of the ground-water reservoir to base flow in streams. Stephenson and Freeze (1974) used a saturated-unsaturated flow finite-difference model of a two-dimensional vertical profile of a hillslope in a small watershed in Idaho. The watershed was well monitored with piezometers as well as stream gauges. After calibrating the pre-rainfall, steady-state system, they went on to simulate the system behavior over a period of 100 days during the rainy period and obtained reasonable correlations with observed piezometric levels and baseflows.

Application of numerical models to the study of soil-plant root interactions have been addressed by some workers. Feddes and others (1975) applied a two-dimensional, saturated-unsaturated flow finite-element model to study evapotranspiration in an agricultural field in De Groeve, Netherlands. The field, covered by a peaty, unsaturated soil, was underlain by a confined aquifer subject to intermittent pumping. Using the model, they computed the evolution of this system over a two week period and estimated the potential changes as well as fluid fluxes across different boundaries.

So far we have considered the classical problems of hydrogeology; those related to exploitation of the freshwater resource. Let us now pass on to consider a few applications involving the environmental consequences of groundwater exploitation.

Helm (1975) applied a one-dimensional consolidation model based on the classical Terzaghi theory of soil mechanics and successfully simulated the subsidence due to fluid production that had occurred near Pixley, California, between 1958 and 1970. The forcing function in his model was the observed hydrographs and the model helped confirm the hypothesis that the clays, subjected to frequent normal consolidation and rebound reversals, constituted the principal controller of subsidence. The same problem was also successfully simulated by Narasimhan and Witherspoon (1978) with a general purpose computer code. A similar problem, that of the spectacular subsidence around Venice, Italy, was successfully simulated by Gambolati and others (1974) using the Terzaghi theory.

More recently, the integral finite difference model was applied to simulate the land subsidence at Wairakei, New Zealand, caused by geothermal fluid extraction since the late 1950s (Narasimhan and Goyal, 1979). The three-dimensional simulation used material heterogeneity as well as preconsolidation to account for the observed off-set of the subsidence bowl and the observed plastic behavior of the subsidence-fluid pressure relation.

One of the important problems challenging hydrogeologists at the present time is that of predicting earthquakes due to induced seismicity. Hsieh and Bredehoeft (1979) have recently applied a finite-element model to study the fluid-injection versus seismicity interactions that have been so dramatically observed at the site of the Rocky Mountain arsenal in Colorado. Using an idealized model of a narrow, semi-infinite aquifer trending northwest, Hsieh and Bredehoeft were able to confirm the theory that earthquakes are triggered as the injection pressure locally exceeds a critical value.

Another problem of interest, closely related to seismicity, is hydraulic fracturing. One of the important applications of the technique of hydraulic fracturing is that of place stress estimation. Walter Palen (oral commun.) has successfully applied an integral finite difference model to simulate the hydraulic fracturing data collected in a well in South Carolina by Mark Zoback of the U.S. Geological Survey. By varying several parameters such as fracture stiffness, geometry, initial aperture, fracture toughness, and so on, and matching observed flow rates with observed injection pressures, Palen was able to estimate the least principal stress, closely matching that estimated by an independent technique.

FUTURE WORK

Over the past two decades numerical methods have witnessed a tremendous growth. We are now at a point when our ability to compute far exceeds our ability to collect field data which constitute the input for the computer models. Does this imply that numerical methods by themselves need no further improvement? The answer is no. Considerable research work remains to be done on two broad fronts. The first of these is to render the existing models more usable and efficient. Improved efficiency is to be achieved through the use of faster matrix solvers and through general computer software improvement. To make computer models more usable, input procedures, options, and terminology must be standardized.

The second and more challenging front is that of improving the conceptual base of the numerical models. There are several important problems in this regard.

Consider, for example, parameter uncertainty. All our discussions in the preceding pages centered on deterministic values of the coefficients as well as the dependent variable. However, to any one in the field of earth sciences, it is obvious that geologic phenomena in general and hydrogeologic phenomena in particular are observed on a macroscopic scale. Moreover, the scale of observation itself may very from one observer to another. If we add to this the natural spatial variability of materials in geologic systems, we can readily see that all observed parameters (which form the basis for cause-effect relationships) are subject to uncertainty due to spatial variability and variability of observations.

Since 1970 many hydrogeologists have turned their attention toward quantifying these uncertainties as either purely random processes or conditional (auto-correlated) random processes. The ideal end product of these efforts is to generate for each discrete surface segment d of a subdomain and for each subdomain itself, the conductances and fluid mass capacities as probabilistic quantities, described by some probability distribution functions. Additonally, the sources, boundary conditions, and initial conditions will also be described as probability distribution functions. The challenge now is to introduce these probability distributions in the discretized numerical equations and compute the outcomes (time-derivatives of potential) as probability distributions. Introducing these distributions greatly increases the computing effort. Probabilistic solution of realistic problems is mostly uneconomical at present, even with the availability of big computers. The increased effort stems from two considerations. The first is of course the fact that every mathematical operation has to be performed on two strings of numbers rather than just two numbers. The other more interesting one is that the algebra of distributions may not be governed by exactly the same axioms governing the algebra of real numbers. Further research in this regard appears to be necessary before generalized, probabilistic numerical models become viable computational tools.

A second major area of research is in validating numerical models. At present, analytical solutions constitute the major standard against which numerical models are validated. When such solutions are not available, as in the case of non-linear problems, many consider analytic solutions to be the ultimate standard of validation. This situation has to change if we are to realize the full power and generality of numerical methods in the context of the modern computing revolution. The integral approach followed in this paper suggests that numerical methods can be developed and validated directly from appropriate axiomatic foundations stemming from the physics of discrete phenomena as well as the theory of integration. This line of reasoning deserves the attention of active researchers. There exists a need to direct part of our attention from the popular tools of applied mathematics, such as Fourier analysis and functional analysis, and direct it to other areas of mathematics such as set theory, measure theory and topology.

A third area of research concerns the problem providing the appropriate input parameters for the numerical model. Since both laboratory generated parameters and field parameters lack physical adequacy, the overall inversion of field data to provide spatial distribution of the hydraulic parameters is being studied by many workers. This is the so called "inverse" problem. Through the use of linear and nonlinear programming, numerical models help in estimating field parameters. A basic question that arises in the case of these models is whether the parameters so generated are merely model parameters or physical parameters. Much remains to be learned in the proper use of the results generated by these sophisticated methods.

ACKNOWLEDGMENT

This work was supported by the Assistant Secretary for Conservation and Renewable Energy, Office of Renewable Technology, Division of Geothermal and Hydropower Technologies of the U.S. Department of Energy under Contract No. W-7405-ENG-48.

NOTATION

$A, A_{\ell m}$	Conductance matrix in general or a member of that matrix
$B_{\ell m}$	Modified conductance matrix or a member of that matrix
C_{ℓ}	A scalar function of rate mass accumulation in element ℓ
$d_{\ell,m}, d_{m,\ell}$	Distance from nodal points ℓ and m respectively, to the interface ℓm

Symbol	Definition	Units
$D_{\ell m}$	Distance between nodal points ℓ and m ($=d_{Pm} + d_{mP}$)	(L)
e, e_m	A finite element or the mth finite element	
g	Acceleration due to gravity	(L/T^2)
g_ℓ	Volumetric generation rate per unit volume of element ℓ	(L^3/L^3T)
G_ℓ	Volumetric generation rate from ℓ	(L^3/T)
k	Absolute permeability	(L^2)
K	Hydraulic conductivity	(L/T)
$K_\Gamma, K_{\ell m}$	Hydraulic conductivity at the surface Γ or at the interface Pm	(L/T)
L	Total number of subdomains. Also, a lower triangular matrix	
$m_{c,e}, m_{c,e_m}$	Specific fluid mass capacity of material on the finite element e or e_m	(M/L^3L)
$m_{c,\ell}$	Specific fluid mass capacity of volume element ℓ	(M/L^3L)
$M_{c,\ell}$	Fluid mass capacity of volume element ℓ	(M/L)
$n, n_\Gamma, n_{\ell m}$	Unit outer normal to surface segment Γ or interface ℓm.	
N, N_i	Weighting function	
p	Iteration number	
$\vec{q}$	Darcy velocity vector	(L/T)
Q_ℓ	General designation for a known vector or a member of a known vector	
r_ℓ, R_ℓ, R^*_ℓ	Residuals	
$\Delta t_{stab,\ell}$	Stable time step for volume element ℓ	
t	Time	(T)
U	Upper triangular matrix	
$U_{\ell m}$	Conductance between volume elements ℓ and m	(M/LT)
$U_{\ell b}$	Conductance between volume element ℓ and a boundary element	(M/LT)
V_e, V_{e_m}	Volume of finite element e, e_m	(L^3)
V_ℓ	Volume of element ℓ	(L^3)
W_ℓ	A weighting function associated with volume element ℓ	
x_1, x_2, x_3	Coordinate axes. Also used to denote x coordinates of points 1,2,3	(L)
y_1, y_2, y_3	y coordinates of points 1,2,3	(L)
Γ	Surface bounding an element	(L^2)
$\Delta\Gamma, \Delta\gamma_{e_m}, \Delta\Gamma_{\ell m}$	Discrete surface segment in general, of finite element e_m or interface between subdomains ℓ and m	(L^2)
W_e, W_{e_m}	Area of general triangular finite element or m^{th} triangular finite element	(L^2)
μ	Dynamic coefficient of viscosity	(M/LT)
ρ_ℓ	Density of water in volume element ℓ	(M/L^3)
$\rho_\Gamma, \rho_{\ell m}, \rho_{\ell b}$	Average density of water at Γ, or at the interface ℓm or ℓb	(M/L^3)
ϕ_ℓ	Average fluid potential over volume element ℓ	(L)
$\vec{\nabla}\phi_{\ell m}$	Gradient in fluid potential at interface ℓm	

REFERENCES CITED

Brustaert, Willem, 1971, A functional iterative technique for solving the Richards Equation applied to two dimensional infiltration problems: Water Resources Research, v. 7, no. 6, pp. 1583-1596.

Carslaw, H. S., and Jaeger, J. C., 1959, Conduction of heat in solids: London, Oxford University Press.

Dahlquist, G., and Bjorck, A., 1974, Numerical methods: Englewood Cliffs, Prentice Hall.

Dusinberre, G. M., 1961, Heat transfer calculations by finite differences: Scranton, Pennsylvania, International Text Book Co.

Edwards, A. L., 1972, TRUMP: A computer program for transient and steady state temperature distributions in multi-dimensional systems, Rev. II: Springfield Virginia, National Technical Information Service, National Bureau of Standards.

Evans, G. W., Brousseau, R. J., and Kierstead, R., 1954, Instability considerations for various difference equations derived from the diffusion equation: Livermore, Lawrence Livermore Laboratory, UCRL-4476.

Faust, C. R., and Mercer, J. W., 1979, Geothermal reservoir simulation: 2, Numerical solution techniques for liquid- and vapor-dominated hydrothermal systems: Water Resources Research, v. 15, no. 1, p 31-46.

Feddes, R. A., Neuman, S. P., and Bresler, E., 1975, Finite element analysis of two-dimensional flow in soils considering water uptake by rafts: II, Field applications: Proceedings, Soil Science Society of America, v.39, no. 2, p 231-237.

Felippa, C. A., 1966, Refined finite element analysis for linear and non-linear two-dimensional structures: Berkeley, University of California, Structural Engineering Laboratory.

Finlayson, B. A., and Scriven, L. E., 1967, On the search for variational principles: Journal of Heat and Mass Transfer, v. 10, p. 799.

Freeze, R. A., and Witherspoon, P. A., 1968, Theoretical analysis of regional groundwater flow: 3. Quantitative interpretations: Water Resources Research, v. 4, p. 581-590.

Gambolati, G., Gatto, P., and Freeze, F. A., 1974, Mathematical simulation of the subsidence of Venice: 2, Results: Water Resources Research, 1979, v.1, p. 563-577. 1979.

Gurtin, M. E., 1964, Variational principles for linear initial value problems: Quarterly of Applied Mathematics, v. 22, no. 3, p. 252-256.

Haverkamp, R., and Vauclin, M., 1979, A note on estimating finite difference interblock hydraulic conductivity values for transient unsaturated flow problems: Water Resources Research, v. 15, no. 1, p. 181-187.

Helm, D. C., 1975, One-dimensional simulation of aquifer system compaction near Pixley, California: 1, Constant parameters: Water Resources Research, v. 11, no. 3, p. 465-478.

Hsieh, P. A., and Bredehoeft, J. D., 1979, Reservoir analysis of the Denver earthquakes: A case of induced seismicity, submitted to Journal of Geophysical Research.

James, G., and James, R. C., 1976, Mathematics dictionary, 4th ed.: New York, Van Nostrand Reinhold Company.

Javandel, I., and Witherspoon, P. A., 1968, Application of the finite element method to transient flow in porous media: Society Petroleum Engineering Journal, v. 8, no. 3, p. 241-252.

Johnson, L. W., and Riess, R. Dean, 1977, Numerical analysis: Menlo Park, California, Addison-Wesley.

MacNeal, R. H., 1953, An asymmetric finite element network: Quarterly of Applied Mathmatics, v. 2, p. 295-310.

Marino, M., 1967, Hele-Shaw model study of growth and decay of groundwater ridges: Journal Geophysical Research, v. 72 , no. 4, p. 1195-1205.

Narasimhan, T. N., 1975, A unified numerical method for saturated-unsaturated flow [Ph.D. dissertation]: Berkeley, University of California.

___1978, A perspective on numerical analysis of the diffusion equation: Advances in Water Resources, v. 1, no. 3, p. 147-155.

___1979, A note on volume-averaging: Berkeley, Lawrence Berkeley Laboratory, LBL-8792.

Narasimhan, T. N., and Goyal, K. P., 1979, A preliminary simulation of land subsidence at the Wairakei Geothermal Field, New Zealand, Presented at the Fifth Annual Geothermal Workshop, Stanford University, Stanford, California: Berkeley, Lawrence Berkeley Laboratory, LBL-10299.

Narasimhan, T. N., and Witherspoon, P. A., 1976, An integrated finite difference method for fluid flow in porous media: Water Resources Research, v. 12, no. 1, p. 57-64.

___1977a, Numerical method for saturated - unsaturated flow in deformable porous media, 1. Theory: Water Resources Research, v. 13, no. 3, p. 657-664.

___1977b, Recent developments in modeling groundwater systems, in Proceedings, IBM Seminar on Regional Groundwater Hydrology and Modeling, Venice, Italy, May 25-26, 1976: Venice, Italy, IBM Corporation.

Narasimhan, T. N., Neuman, S. P., and Witherspoon, P. A., 1978a, Finite element method for subsurface hydrology using a mixed explicit-implicit scheme: Water Resources Research, v. 14, no. 5, p. 863-877.

Narasimhan, T. N., Witherspoon, P. A., and Edwards, A. L., 1978b, Numerical method for saturated-unsaturated flow in deformable porous media: 2, The algorithm: Water Resources Research, v. 14, 2, p. 255-261.

Neuman, S. P., 1975, Saturated-unsaturated seepage by finite elements: Journal of Hydraulic Division, American Society of Civil Engineers, v. 99, p. 2233-2250.

Neuman, S. P., and Narasimhan, T. N., 1977, Mixed explicit-iterative finite element scheme for diffusion-type problems: Part I Theory: International Journal Numerical Meth. Eng., v. 11, p. 309-323.

O'Brien, G. G., Hyman, M. A., and Caplan, S., 1951, A study of the numerical solution of partial differential equations: Journal of Mathmatical Physics, v. 29, p. 223-251.

Peaceman, D. W., and Rachford, Jr., H. H., 1955, The numerical solution of parabolic and elliptic differential equations: Journal of the Society of Industrial and Applied Mathmatics, v. 3, p. 28-41.

Pinder, G. F., and Bredehoeft, J. D., 1968, Application of the digital computer to aquifer evaluation: Water Resources Research, v. 4, p. 1069-1093.

Pinder, G. F., and Frind, E. O., 1972, Application of Galerkins Procedure to aquifer analysis: Water Resources Research, v. 8, p. 108-120.

Prickett, T. A., 1975, Modeling techniques for groundwater evaluation, in Chow, V. T., ed., Advances in hydroscience: New York, Academic Press.

Prickett, T. A., and Lonnquist, C. G., 1971, Selected digital computer techniques for groundwater resource evaluation: Bulletin 55, Illinois State Water Survey.

Remson, I., Hornberger, G. M., and Molz, F. J., 1971, Numerical methods in subsurface hydrology: New York, Wiley Interscience.

Smiles, D. E., and Rosenthal, M. J., 1968, The

movement of water in swelling materials: Australian Journal of Soil Research, v. 6, p. 237-248.

Sokolnikoff, I. S., and Redheffer, R. M., 1966, Mathematics of physics and modern engineering: New York, McGraw-Hill, p. 752.

Southwell, R. V., 1940, Relaxation methods in engineering science: London, Oxford University Press, p. 252.

Stephenson, G. R., and Freeze, R. A., 1974, Mathematical simulation of subsurface flow contributions to snow-melt runoff, Reynolds Creek Watershed, Idaho: Water Resources Research, v. 10, p. 284-294.

Stone, H. L., 1968, Iterative solution of implicit approximations of multi-dimensional partial differential equations: Journal of the Society of Industrial Applied Mathmatics, v. 5, p. 530-558.

Toth, J., 1962, A theory of groundwater motion in small drainage basins: Journal of Geophysical Research, v. 67, p. 4375-4387.

Tasny-Tschiassny, L., 1949, The triangulation of a two-dimensional continuum for the purpose of the approximate solution of second-order partial differential equations, Journal of Applied Physics, v. 20, p. 419-424.

Tyson, H. N., and Weber, E. M., 1964, Groundwater management for the nation's future--computer simulation of groundwater basins: American Society Civil Engineering Journal, Hydrology Division, 90(HY4), p. 59-77.

Varga, R. S., 1962, Matrix iterative Analysis: Englewood Cliffs, New Jersey, Prentice Hall.

Vauclin, M., Vachaud G., and Khanji, D., 1975, Two-dimensional numerical analysis of transient water transfer in saturated-unsaturated soils, in Vansteenkiste, C. C., ed., Computer simulations in water resources systems: Amsterdam, North Holland.

Winslow, A. M., 1966, Numerical solution of the quasilinear Poisson Equation in a non-uniform triangle mesh: Journal of Computer Physics, v., 1, p. 149-160.

Zienkiewicz, O. C., 1971, The finite element method in engineering science: New York, McGraw Hill.

Zienkiewicz, O. C., Mayer, P., and Cheung, T. K., 1966. Solution g anisotropic seepage by finite elements. Journal, American Society of Civil Engineers, Engineering Mechanics Division.

Manuscript Accepted by the Society October 14, 1980

Printed in U.S.A.

IV. Regional Aspects

Geological Society of America
Special Paper 189
1982

Regional Ground-Water Flow Concepts in the United States: Historical Perspective

J. D. BREDEHOEFT
W. BACK
B. B. HANSHAW
*U.S. Geological Survey, 431 National Center
Reston, Virginia 22092.*

ABSTRACT

A number of important ideas, developed during the past 100 years, form the framework of the present understanding of regional ground-water flow. The most important of these ideas are:

1. Differences in topographic elevation provide the principal driving force for regional flow.
2. Flow through confining layers forms an essential element of regional flow systems.
3. Chemical evolution within the flow systems can be used to understand the flow.
4. Moving ground water is an efficient transport mechanism for heat within the Earth.

We trace the evolution of these ideas in the United States and demonstrate their influence on the present-day understanding of flow systems with examples taken primarily from the American literature.

INTRODUCTION

A number of regional flow systems have been investigated since Darton first studied the Dakota sandstone. In this paper, we review significant ideas developed in North America that have led to the present understanding of regional ground-water flow and geochemistry, and suggest some ideas for present and future research. We do not claim that the review is complete; we accept that we may have omitted investigations or ideas that others feel are as important as those we included.

Several important ideas now provide a fundamental understanding of the phenomenon of regional flow. The most important of these ideas, which we might refer to as the principles that govern regional flow, are:

1. Differences in hydraulic head produced by topographic relief on the boundaries of the flow system are, in most instances, the driving force for the flow.
2. Most naturally occurring earth materials have finite permeability; there are no totally impermeable materials. In sedimentary deposits significant quantities of flow commonly occur through shaley confining layers.
3. Chemical reactions within the flow system occur in the moving fluid--ground water. The chemical evolution within the system can be utilized to understand the flow better.
4. Moving ground water is an efficient mechanism for the transport of heat within the Earth's crust. In areas of significant flow the natural conductive heat flux in the Earth is disturbed.
5. Regional ground-water flow plays an important role in the movement and entrapment of hydrocarbons.
6. Shales within the Earth may act as semipermeable membranes that retard the passage of charged species (ions) while allowing relatively unrestricted passage of neutral molecules (water). The pheonomenon can cause anomalously high or low osmotic pressures, electrical potential differences, and salt sieving or ultrafiltration.

MATHEMATICAL FRAMEWORK

The total concept of regional flow systems implies an understanding of the interrelations of hydrodynamics and geochemistry. In ground water we are concerned with the fluids contained in the aquifer system. That fluid is generally described by stating (1) pressure, (2) composition, and (3) energy contained in the fluid. In the general case, three coupled partial differential equations are used to describe the system:

1. A partial differential equation for pressure, which in certain simplifying instances can be reduced to an equation for hydraulic head; this is commonly referred to as the flow equation

2. A partial differential equation for chemical composition; in the case where more than one chemical constituent of the fluid is of interest, this becomes a set of partial differential equations

3. A partial differential equation for the internal energy of the fluid, generally either the temperature or the enthalpy of the fluid.

In the most general case fluid pressure, composition, and internal energy are coupled as are the equations (or sets of equations) that describe the state of the fluid. However, in many instances the coupling may be negligible and the state variables can be treated independently. For example, in many problems the change in temperature is unimportant and we consider the system isothermal, eliminating internal energy considerations.

Ground-water flow, in the most general case, is three dimensional and varies with time; to describe the system the following must be specified completely:

1. Spatial distribution of the parameters contained in the three (or more) partial differential equations mentioned above
2. Initial conditions for pressure, composition, and temperature, or enthalpy
3. Source and sinks of interest and their variation with time
4. Boundary conditions at the margins of the space of interest

Parameters of Governing Equations

Each of the governing equations contains parameters that describe certain properties of aquifers and confining layers that affect ground-water flow, chemical composition, and internal energy.

Parameters affecting flow. For flow problems in which we neglect effects of composition and temperature (internal energy) or the density, the parameters of interest are porosity; permeability; and specific storage, which reflects the combined compressibility of the fluid and the medium.

Parameters affecting chemical composition. These are related to problems of fluid composition and include dispersion coefficient, which most investigators formulate as a function of the fluid velocity and a property of the medium dispersivity; and the coefficient of molecular diffusion for the constituent in a porous medium.

Parameters affecting internal energy. Energy transport includes an additional parameter: thermal dispersivity for energy transport, which is generally thought to have a form similar to that of dispersivity for composition. The dispersivity for energy may have a different magnitude than for composition; there are, however, few empirical data. The heat capacity of the aquifer material and water is another parameter that affects internal energy, for it governs the amount of energy that can be stored within the system.

Initial Conditions

To compute transient system response, the distribution of pressure, composition, and internal energy (temperature or enthalpy) must be specified at some initial time.

Sources and Sinks

In the usual case the most important sources and sinks are:

1. Recharge or discharge, commonly through wells. Other natural recharge or discharge may also be handled as sources and sinks depending upon how the problem is formulated mathematically.
2. Chemical reactions, which either remove or add components to the fluid. This is one of the more troublesome problems since one must know what reactions are occurring as well as the rate of each reaction.

Boundary Conditions

In ground-water problems, the boundary conditions of most interest are:

1. No flux, that is, an impermeable boundary.
2. Prescribed flux, where the flow at the boundary is given some value; commonly for problems involving composition, a given composition of a particular constituent; and for energy problems, a given temperature or enthalpy.
3. Prescribed value of the state variable, where for flow the head or pressure is defined; for composition a constituent concentration is defined; and for energy either the temperature or enthalpy is defined.

In summary, each of the parameters described above, which characterize the media, can vary through space. Permeability and dispersivity for both composition and energy can have directional properties at any point in space. The value of each of the parameters must be specified throughout the region of interest if the system is to be understood quantitatively. It is from an understanding of all aspects of the geology of aquifer systems that we define the geometry and the distribution of the parameters in space, as well as define the boundary conditions.

Each of the state variables: (1) pressure or head, (2) chemical composition, and (3) temperature complement each other and can be used to investigate regional flow. As the following discusson will demonstrate, the interactions of the state variables have been used by various investigators to study regional flow.

REVIEW OF SELECTED STUDIES ON REGIONAL FLOW CONDITIONS

In 1885, T. C. Chamberlain published *The Requisite and Qualifying Conditions of Artesian Wells*, which is perhaps the first classic paper on regional flow. His discussion of "confining beds" (p. 137-138) is particularly appropriate:

> No stratum is entirely impervious. It is scarcely too strong to assert that no rock is absolutely impenetrable to water. Minute pores are well nigh all-pervading. To these are added microscopic seams, and to these

> again larger cracks and crevices. Consolidated strata are almost universally fissured. Even clay beds are not entirely free from partings. But in the study of artesian wells we are not dealing with absolutes but with availables. A stratum that successfully restrains the most of the water, and thus aids in yielding a flow, is serviceably impervious. It may be penetrated by considerable quantities of water, so that the leakage is quite appreciable and yet be an available confining stratum.

This is an important idea in our current thinking but, unfortunately, each generation of ground-water hydrologists seems to have to rediscover this thought.

One of the early studies of regional flow was that of Mendenhall (1905) in the Los Angeles area of California; Mendenhall described the regional flow over much of the San Bernardino Valley and related the flow to the topography and the alluvial geology of the valley.

The Dakota aquifer in South Dakota is the classic artesian aquifer. Many modern ideas concerning artesian aquifers stem from N. H. Darton's investigation of the Dakota aquifer during the 1890s and early 1900s. Darton recognized that the recharge to the system occurred in the Black Hills in western South Dakota while the major discharge was in eastern South Dakota, 300 to 500 km to the east. Darton (1896) wrote a preliminary report which described the system. He, and colleagues under his direction, proceeded to map systematically the area of principal development along the James River in eastern South Dakota; much of the actual mapping was done by J. E. Todd under Darton's direction. This resulted in a series of Geologic Folios, which covered the ground-water geology of the James River lowland from Nebraska to North Dakota; these folios were obviously designed to display the ground-water geology of the area under development. These are still classic examples of hydrogeological mapping. While Todd mapped the area of major development, Darton mapped the four quandrangles that encompassed the entire Black Hills, and included the areas of Dakota outcrop that he believed to be the recharge area for the Dakota aquifer in South Dakota. Darton (1909) summarized his Dakota investigations in USGS Water Supply Paper 227 in which he pointed out that potentials within the Dakota system are controlled by the elevation of the sandstone outcrops in the Black Hills. Darton (1909, p. 60) set forth his ideas concerning the mechanics of the system in the following remarks:

> The evidence of this pressure, as found in many wells in eastern South Dakota, is conclusive that the water flows underground for many hundreds of miles. Such pressures can be explained only by the hydrostatic influence of a column of water extending to a high altitude on the west. If it were not for the outflow of the water to the east and south the initial head which the waters derive from the high lands of the intake zone would continue under the entire region, but owing to this leakage the head is not maintained, and there is a gradual diminution toward the east known as "hydraulic grade," a slope sustained by the friction of the water in its passage through the strata.

Obviously, Darton fully understood that some water must flow through the overlying confining layers; he goes on to state (Darton, 1909, p. 60):

> Another factor which undoubtedly somewhat influences the hydraulic grade in the Great Plains region is a certain but unknown amount of general leakage through the so-called impermeable strata, especially when under great pressure.

Although Darton's ideas of the mechanics of the Dakota system, especially of the flow through the confining layer, were explicit, these ideas seem to have become lost in the intervening years.

Many of our modern ideas of the hydrology of artesian aquifers originate in these investigations of Darton and his colleagues. Most elementary geology texts use Darton's illustrations of the Dakota sandstone to introduce the idea of artesian flow.

The mechanics of flow in the Dakota aquifer in South Dakota became the subject of major debate in the late 1920s. A number of prominent geologists, including Meinzer, participated in the discussion. This debate, while focusing on the Dakota sandstone, became a forum for discussing the thinking of the 1920s on flow through major aquifer systems.

Russell (1928), following extensive stratigraphic study of the Dakota in the outcrop area in eastern South Dakota, questioned Darton's explanation of the artesian pressures. Russell observed that the sandstones which crop out in the Black Hills differ in age from those that comprise the aquifer in eastern South Dakota. He also pointed out the distinct variations in the chemistry of the Dakota water. Based upon this evidence, Russell suggested that the sandstones were lenticular and that the artesian pressure was the result of a reduction in pore volume within the sandstone due to sedimentary loading. Russell's paper excited considerable discussion. Terzaghi (1929) examined Russell's mechanism in light of the theory of consolidation and concluded that Russell's suggested mechanism was most improbable.Piper (1928) questioned Russell's stratigraphy and reiterated that the aquifer sandstones were probably continuous.

At the time of Russell's paper, Meinzer (1928) had a paper in press which suggested that the artesian head supports a portion of the load of the overlying rocks and that aquifers undergo small changes in pore volume as the result of changes in head. He further stated that some part of the water discharged from wells drilled in the 1890s and early 1900s to the Dakota was derived from this elastic change in pore volume. This was the first clear statement of storage in an artesian system. It is evident from pre-1928 correspondence between Meinzer and Dave Thompson, who was working in New Jersey, that Meinzer's ideas were based to a large extent upon discussions with Thompson. Thompson (1929) discussed both Meinzer's (1928) and

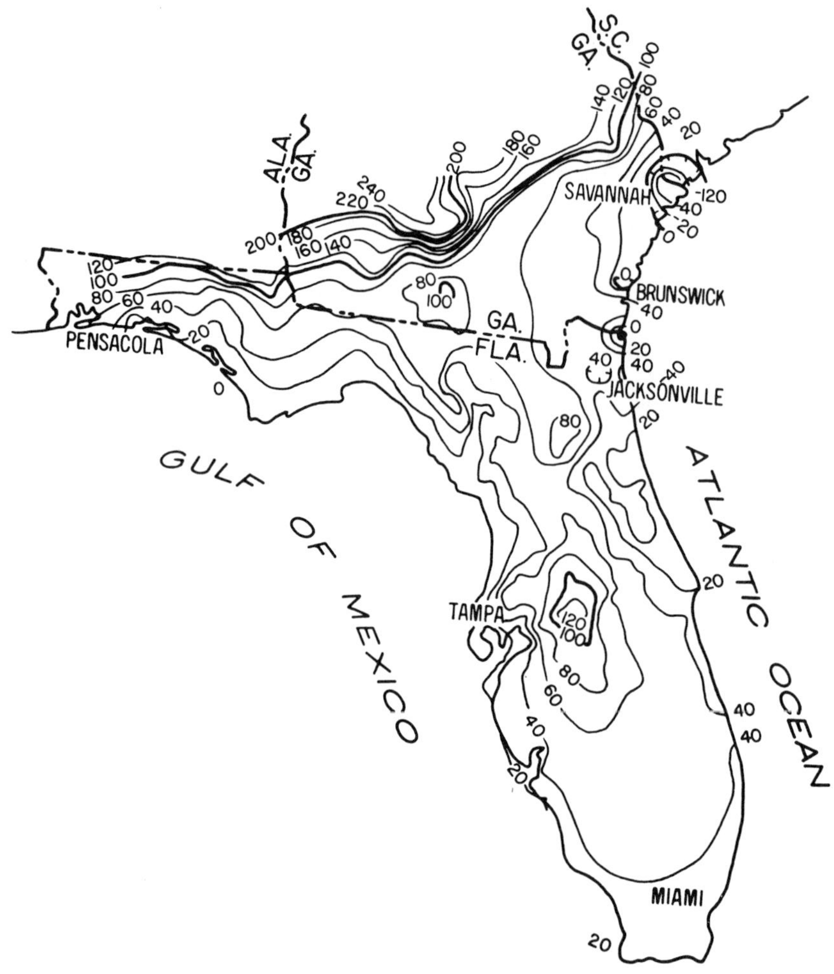

Figure 1. Potentiometric map for the principal limestone aquifer of Florida and Georgia (after Stringfield, 1966).

Russell's (1928) papers, and emphasized the importance of storage within aquifer systems.

Stringfield (1935, 1936), working in the 1930s and 1940s, made a major contribution to our understanding of regional flow by identifying the limestone aquifer of the Florida peninsula as a single system and putting together a potentiometric map for the entire Floridan aquifer (Fig. 1). Stringfield's work identified the major recharge areas as being in Florida and not Georgia as commonly believed at that time, and identified the general directions of flow. Stringfield's regional approach to hydrology of limestone aquifers, along with the work of others, has led to appreciation of regional processes of karst formation (Stringfield and others, 1977).

Lonsdale (1935) investigated the Carrizo sandstone and identified it as a major aquifer that extends across the entire coastal plain of Texas. In the area of development in southwest Texas, fresh water extends to depths of 1,600 m (5,000 ft) and is overlain by aquifers that contain salt water. The Carrizo--because of the development of ground water in the Winter Garden area as well as in other parts of Texas--has been the subject of continued investigations (Guyton, 1942; Moulder, 1957). Fresh water to a depth of 1,600 m is unusual in coastal plain areas with low topographic relief and still is not adequately explained.

Theis (1937) investigated the High Plains area of Texas and New Mexico and published an analysis of the regional flow. It was during these studies that Theis (1935) did his classic work on ground-water flow to a pumping well. Theis demonstrated that the development was so far distant from the areas of natural discharge that this discharge could not be captured. On the basis of this regional analysis, Theis showed that the High Plains development was a purely mining development.

The water supply for the large population of Long Island is almost entirely from ground water. Because of the importance of ground water, Long Island was a field laboratory for many investigations beginning in the 1930s and continuing today. The geology is typical of the coastal plain where a layered sequence of aquifers separated by confining layers make up the system. The aquifers dip gently offshore. Sea water occurs in the shallow aquifers near the coast. One of the early descriptions of the flow system was provided by

Jacob (1945). Rainfall provides the recharge; flow is away from the high area in the center of the island toward the coasts.

Sayre and Bennett (1942), working in Texas in the 1940s, identified the Edwards Limestone regional system and showed that the Edwards was a major hydrologic system which extends across much of central Texas. The system can be traced through a number of large springs; the regional flow system is primarily from west to east. Surprisingly, the anisotropy of the permeability is such that the flow is approximately parallel to the equipotential contours. This is the first place where flow of this nature was observed. Abbott (1975) pointed out that this apparent anomaly is actually a problem of scale; if wells were sufficiently close together, the detailed potentiometric gradient could be shown to be parallel to the fault trends. The structural control of cavern development permits the main transmission of ground water to occur in conduits enlarged by solution along the major faults. In considering the structural and geomorphic trends on the development of the aquifer, Woodruff and Abbott (1979) emphasize that the Edwards Limestone in the fault zone behaves like a master conduit that is a nearly homogeneous, highly porous, and permeable medium for transmission of water.

A large effort has gone into understanding the regional flow within the carbonate aquifers that underlie the Nevada Test Site and much of southeastern Nevada. Gradually in the 1950s and 1960s a major, deep Paleozoic carbonate aquifer began to be identified. Recognition of a major aquifer in such a complexly faulted and geologically complex terrain was not easy. A number of hydrologists, including Eakin (1966), Mifflin (1968), Winograd and Thordarson (1975), and Mifflin and Hess (1979) made major contributions to our understanding. Identifying and comprehending this system was one of the major hydrological breakthroughs of the past 20 years. Paleozoic carbonate rocks deposited in the miogeosynclinal portion of the Cordilleran geosyncline form an extensive regional carbonate aquifer that controls ground-water flow over much of southern Nevada (Fig. 2). The system transcends many of the mountain ranges of the area and integrates the drainage from a number of intermountain valleys. Work at the Nevada Test Site confirmed the continuity of the system. Transport of radioactive contaminants from the Nevada Test Site will ultimately be by ground water moving through the deep carbonate system. Major discharge for much of the area is at springs in Ash Meadows located in a hydrologically closed basin. Ultimately some contaminants will reach this point. Work by Winograd and Friedman (1972) suggests that the travel time from Yucca Flat--an area of much

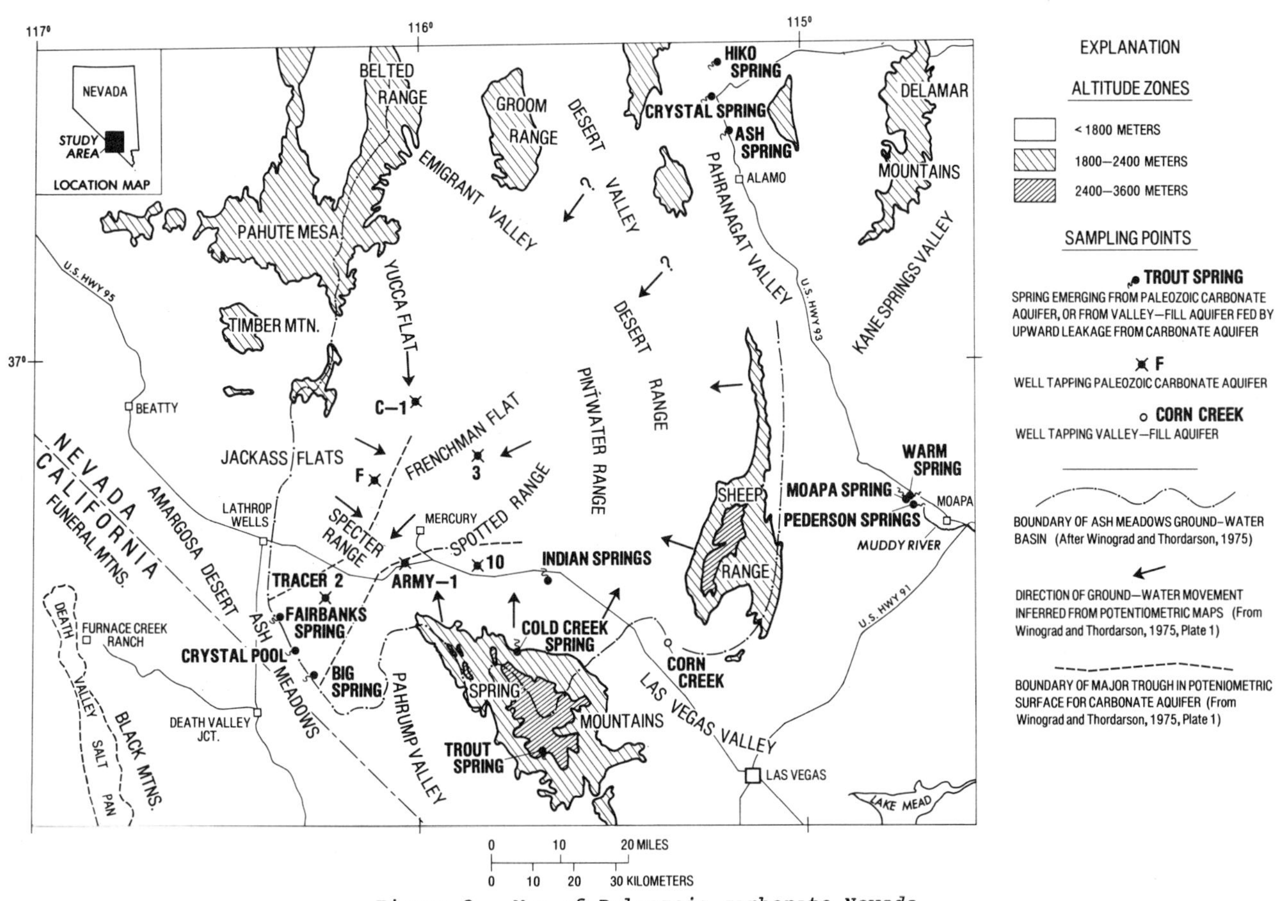

Figure 2. Map of Paleozoic carbonate Nevada system (from Winograd and Pearson, 1976).

current and past underground nuclear testing--will be of the order of 40,000 years.

Since 1975, the U.S. Geological Survey has undertaken a major investigation of the Madison Limestone regional aquifer system in northeast Wyoming, southeast Montana, and adjacent areas of South and North Dakota. Swenson (1968) promoted the idea that the Madison of this area was a major aquifer that could provide large quantities of ground water. The current project is one of the few cases in the history of ground-water studies where large amounts of money, a total of $9 million, will have been spent on deep drilling and testing for purely scientific purposes. Due to have been completed in 1980, this will be one of the most exhaustive ground-water studies ever undertaken. The following wide variety of techniques have been used in this investigation:

1. Test drilling, with appropriate coring, geophysical logging, and testing.
2. Intensive study of the subsurface geology.
3. Surface geophysics.
4. Geochemical studies of the fluids--including extensive use of stable isotopes, ^{14}C, and tritium.
5. Numerical model studies of the system.

The results to date suggest that the zones of higher permeability are controlled by regional tectonics, distribution of original lithofacies, dolomitization, dedolomitization, and by intra- and post-Mississippian karstification. The magnitude of prolonged ground-water development will depend upon induced leakage.

THE ROLE OF TOPOGRAPHY AND CONFINING LAYERS

Land elevation and relief are clearly the driving forces that determine most regional ground-water flow, as Darton elucidated. Toth (1962, 1963) examined the influence of topography upon flow. He assumed the water table to be a replica of the topography and used this as the near-surface boundary condition. His results showed the influence of this boundary condition on driving the regional flow system. The subtleties of the influence of topographic relief have become better understood in the last 15 years. Building upon the work of Toth, Freeze and Witherspoon (1966, 1967) examined the influence of the scale of the topography upon regional flow (Fig. 3). Their results demonstrated that the smaller-scale effects are damped with depth; the deeper flow reflects only the major topographic features.

For most regional flow systems, the low-permeability confining units are as important as the

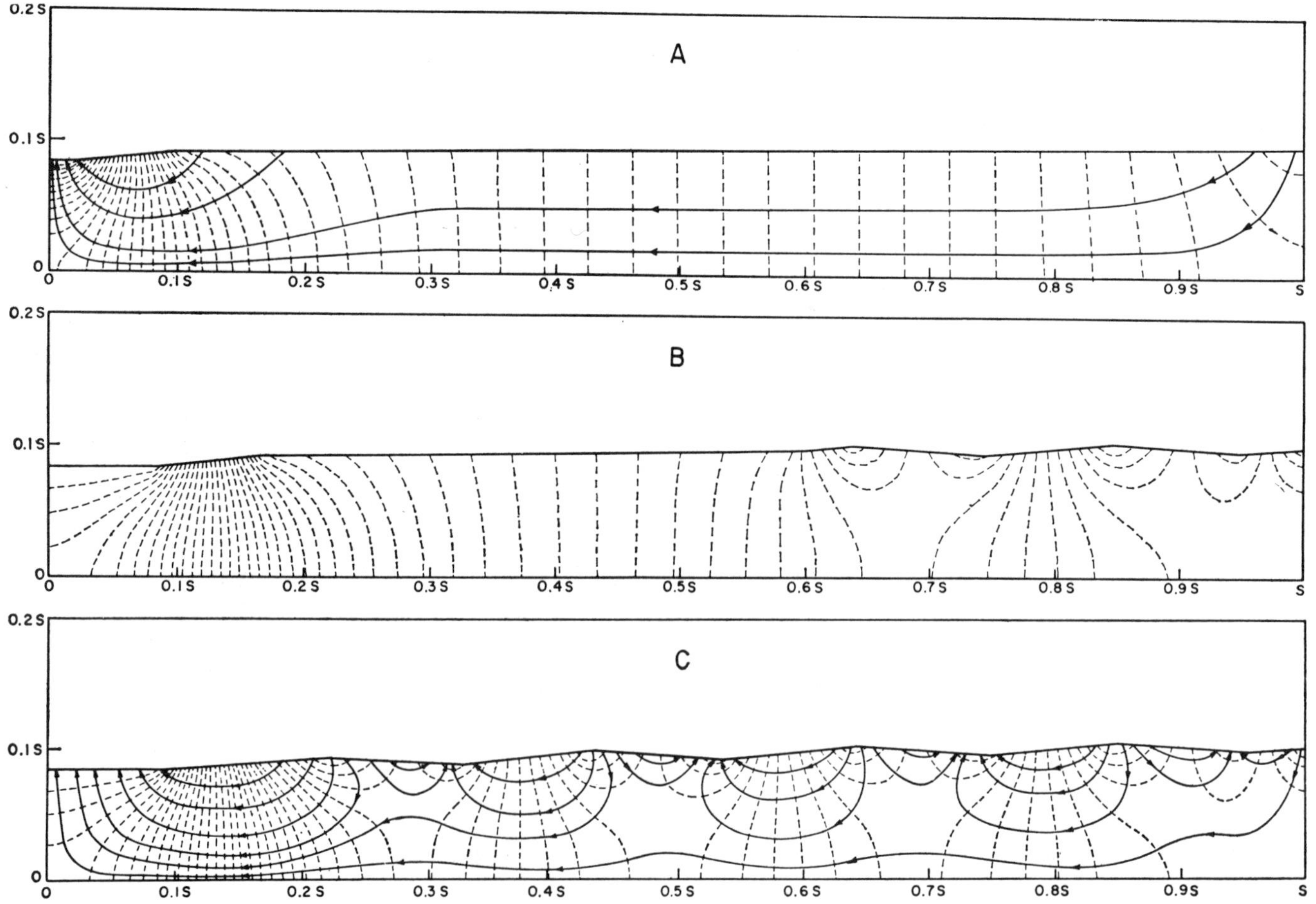

Figure 3. Effect of water-table configuration in regional ground-water flow through homogeneous isotopic media (Freeze and Witherspoon, 1967).

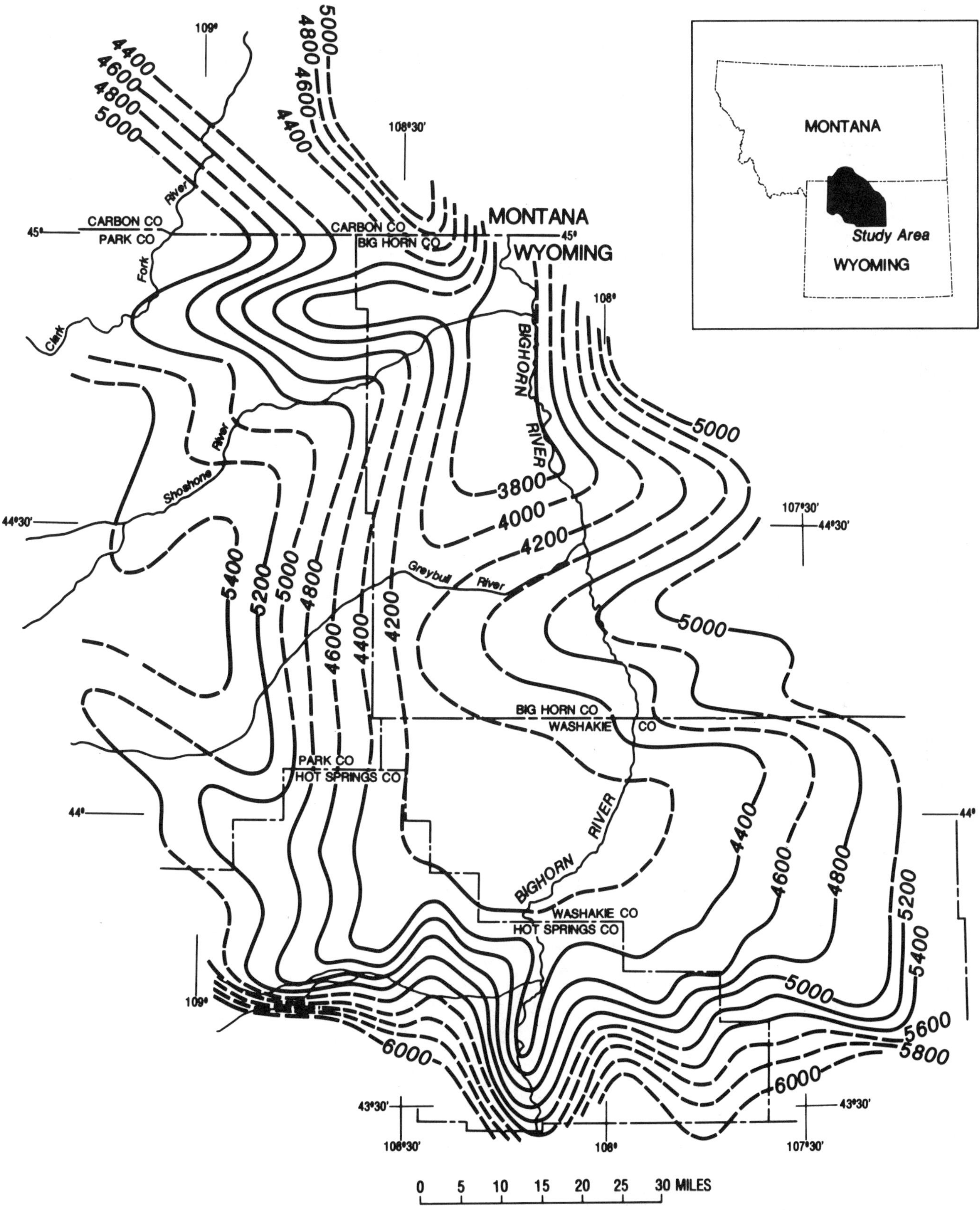

Figure 4. Potentiometric map of the Tensleep Sandstone in the Big Horn basin, Wyoming (Bredehoeft and Bennett, 1971).

more-permeable units in controlling the hydrodynamics of the system. As first stated by Chamberlin (1885), confining layers are not impermeable. The areal extent of most major systems is such that even though the permeability of the confining layers is low, significant quantities of water move through these layers. Nevertheless, not until the 1940s did ground-water hydrologists fully recognize the concept of leaky confining layers. Jacob (1946) and Hantush (1960) considered the influence of leakage on pumping and showed that in many areas of ground-water development, the magnitude of the leakage determines the ultimate size of the development. However, the permeability of the confining layers is one of the more difficult parameters to determine.

Today, our concept of regional flow allows for large quantities of both discharge and recharge to occur by movement through the confining layer. Current thinking suggests that the major discharge from the deeper permeable formations of the Illinois Basin, for instance, is by upward vertical flow through the confining layers (Bredehoeft and others, 1963; Graf, 1960). The same is true for discharge from the Tensleep Sandstone in the Big Horn Basin of Wyoming where a nearly closed depression in the potentiometric surface for the Tensleep is best explained by upward vertical movement of fluid (Fig. 4). The Carrizo sandstone also apparently discharges by means of vertical leakage (Pearson and White, 1967). Based on sulfur isotope data, Hanshaw and others (1978) have suggested that most of the water currently being produced from wells open to the Madison limestone in the Midwest area of Wyoming is leaking downward from Pennsylvanian and Permian beds.

Along the Atlantic coast between Savannah and Jacksonville the Floridan aquifer and its equivalent in Georgia contain heads near the coast that range from 10 to 20 m above sea level (see Fig. 1). This can only be explained as the consequence of a low permeability confining layer overlying a highly permeable aquifer. In studies of the "Floridan" aquifer at Brunswick, Georgia, Bredehoeft and Pinder (1973) have shown that by simulating the virgin steady-state flow system the permeability of the confining layers can be computed with reasonable precision.

Swenson (1968) reexamined the ability of the Dakota system, under existing hydraulic gradients, to transmit the quantity of water known to be produced in eastern South Dakota. He emphasized the differences in the chemistry of the Dakota water from the recharge area to the major area of production in the Missouri River valley. He suggested that cavernous limestones of the Madison group may be truncated by the lowermost sandstones of the Dakota aquifer in central South Dakota and that the bulk of Dakota water in eastern South Dakota is recharged from these underlying Madison group limestones. The suggestion that the Dakota is recharged by water from the Madison group limestones had been stated previously by Dyer and Geohring (1965) in an open-file study of the Dakota aquifer in southeastern South Dakota. Can the Dakota aquifer alone provide the quantities of water known to have been produced from the system or must one look for an additional source? This question is implied in each of the papers that followed Darton's classic of 1909. Implicit in Swenson's (1968) statements is the thought that

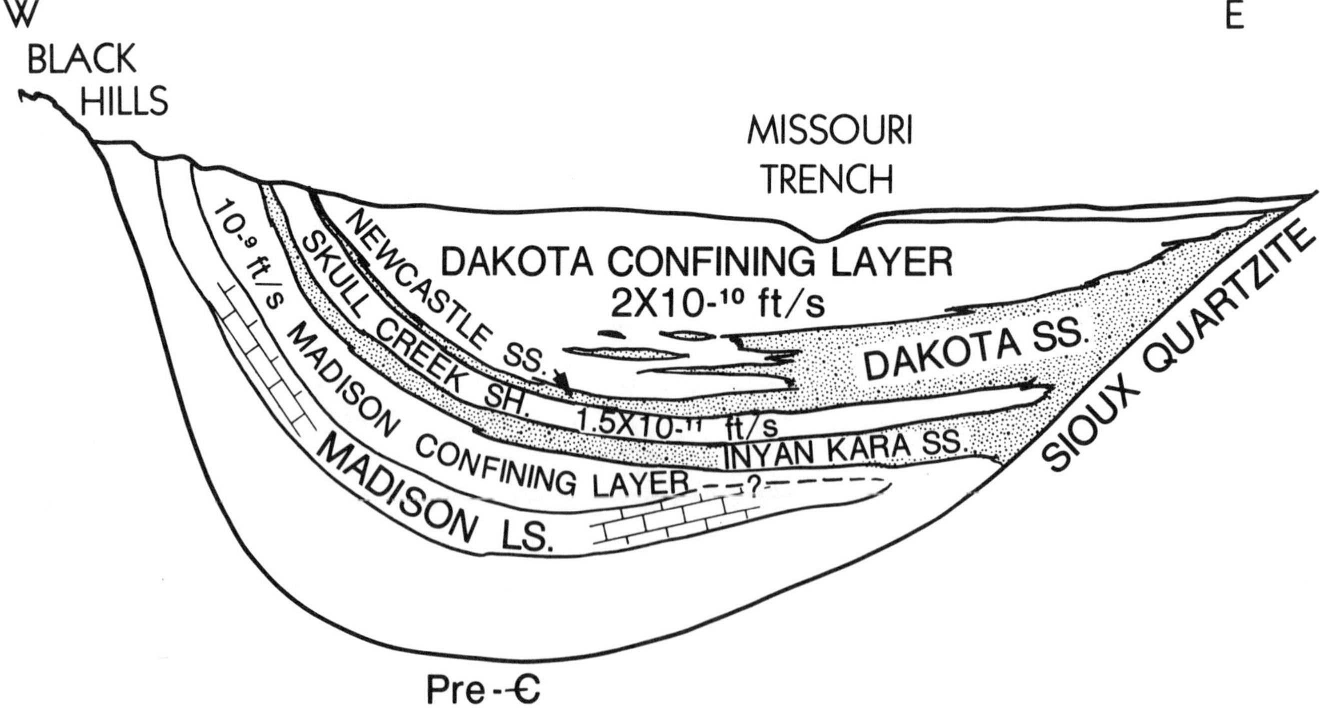

Figure 5. Schematic diagram of the Dakota aquifer system in South Dakota (after Neuzil, 1980).

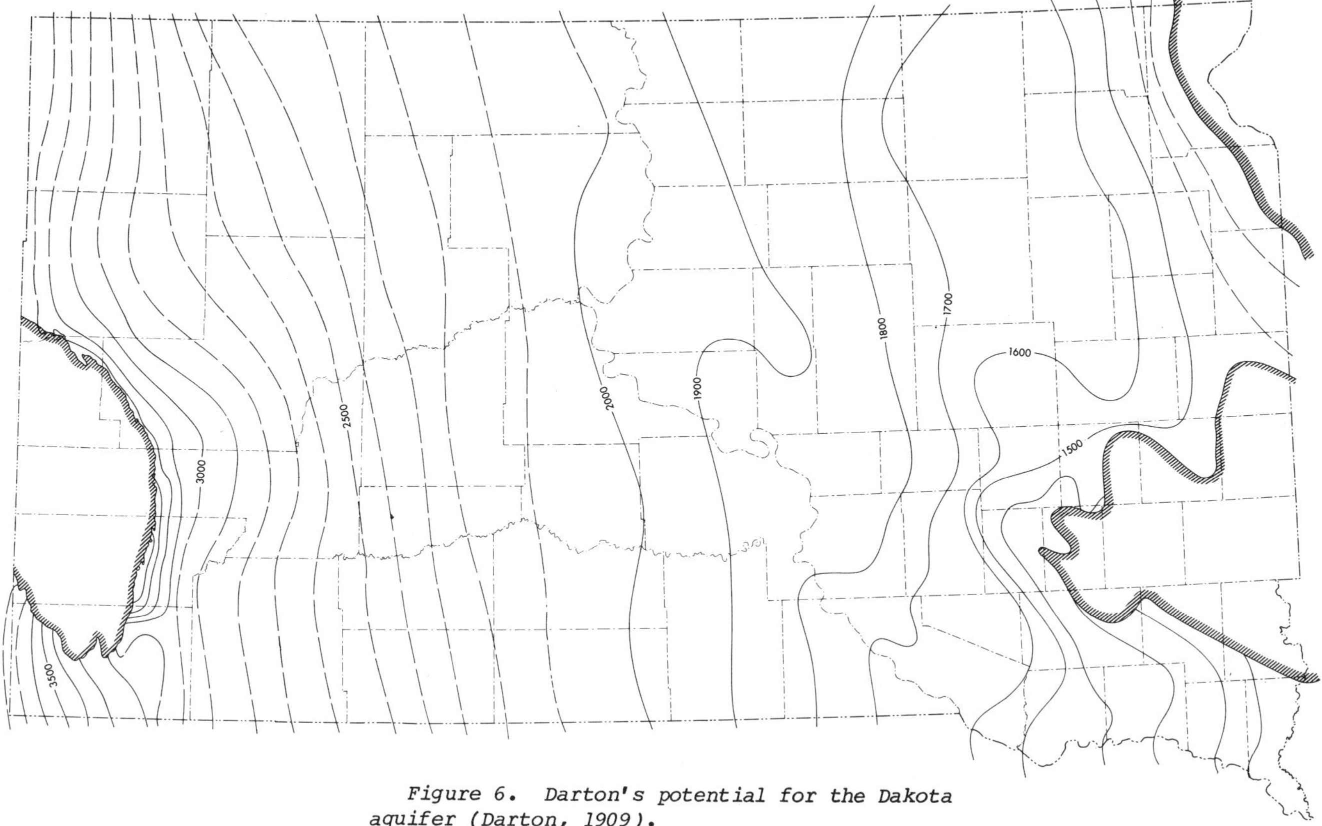

Figure 6. Darton's potential for the Dakota aquifer (Darton, 1909).

the system is nearly in steady state, that is, little additional water is available from storage. Figure 5 is a simplified schematic diagram of the Dakota aquifer system in South Dakota.

Assuming that we know: (1) the physical geometry of the system (the geology); (2) the boundary conditions for the flow system; (3) the permeability of the aquifers; and (4) the virgin head distribution in the major aquifers of the system, we can compute the permeability of the confining layers.

These computations have been made numerically for the Dakota aquifer system in South Dakota and show that the computed head in the major aquifers is sensitive to the permeability of the confining layers. Darton's potential surface (1909), reflects the influence of major development along the Missouri River and the area east of the river (Fig. 6).

Figure 7 is the computed potential surface for the Dakota aquifer and, considering the influence of pumping, the comparison is good. This procedure yields a permeability for the confining layer which is as accurate as our knowledge of the aquifer permeability.

HYDRODYNAMICS OF OIL AND GAS ACCUMULATION

M. K. Hubbert, one of the pioneers of ground-water theory (1940) became interested in regional ground-water flow and in particular the influence of hydrodynamics on the accumulation of hydrocarbons. Hubbert (1953) demonstrated that dynamic ground water flow could trap hydrocarbons in geologic structures that would not be traps in a static ground-water regime; he named the phenomenon "hydrodynamic entrapment." A number of oil companies applied Hubbert's ideas; but unfortunately, little, if any, of the resulting information was published because of its proprietary nature.

Perhaps the most active group to study regional hydrodynamics was the group organized by Hill in the late 1950s and early 1960s at Petroleum Research Corporation. With an energetic and able staff, this group compiled potentiometric maps for many of the permeable lithologic units of the intermountain basins of the Rocky Mountain region and the Colorado Plateau. The work of Petroleum Research Corporation was supported by subscription from the major oil companies. These investigations demonstrated the occurrence of regional flow systems throughout the Rocky Mountain area. Although much of the work is proprietary and remains unpublished, Hanshaw and Hill (1969) published a regional interpretation of potentiometric surfaces in the Paradox Basin.

At the same time, urged by Hubbert, the U. S. Geological Survey began an investigation of the Tensleep Sandstone in the Big Horn Basin in Wyoming. This system is particularly interesting; it contains a number of oil traps with hydrodynamic tilts of the oil-water interface. Zapp (1956) examined the discovery and early production well data and published a map of the basin indicating the tilts. Bennett, working independently from Zapp, investigated the potentiometric surface for the Tensleep

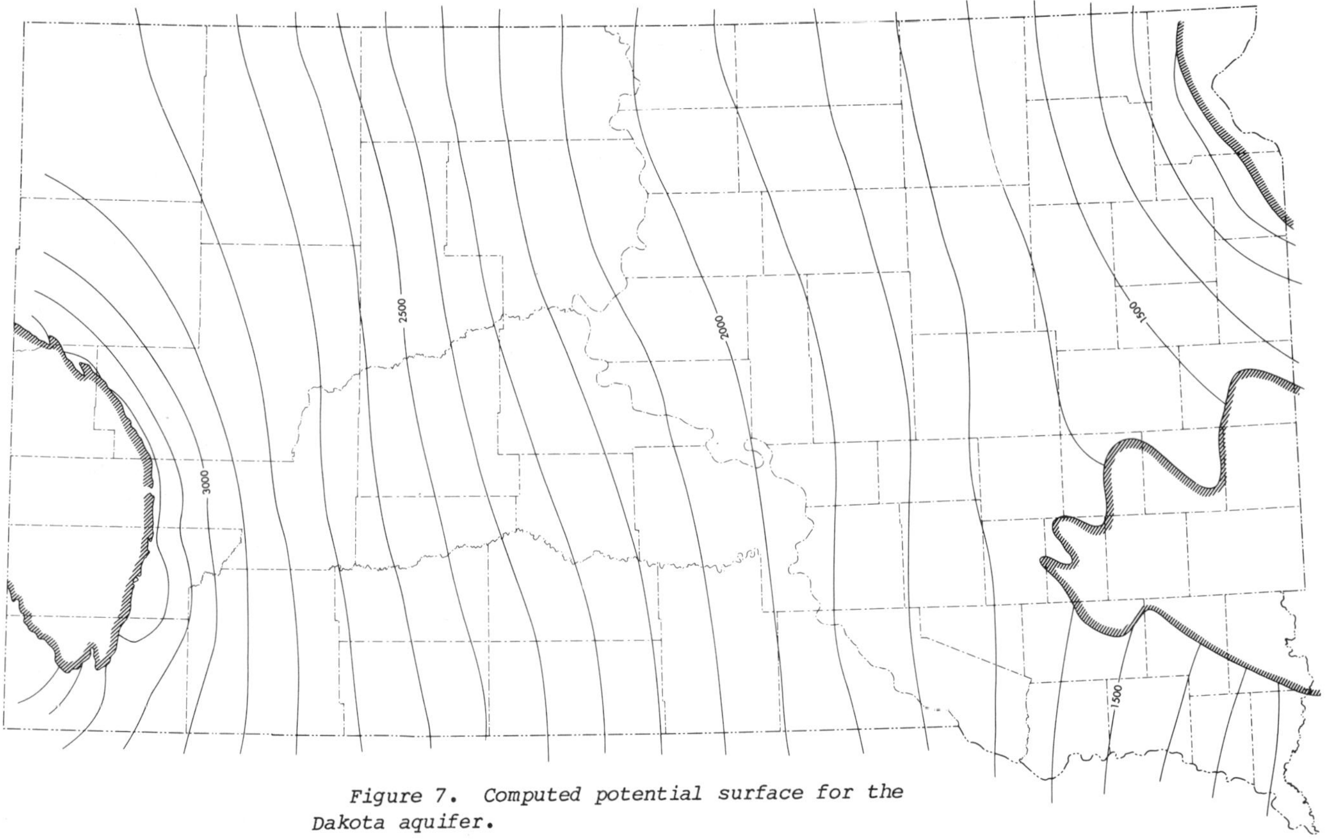

Figure 7. Computed potential surface for the Dakota aquifer.

Sandstone in the Big Horn Basin (Bredehoeft and Bennett, 1971). This map (Fig. 4) indicates that the tilts of the oil-water interface mapped by Zapp were hydrodynamically controlled. This may be the most convincing published field evidence for Hubbert's hydrodynamic entrapment theory.

GEOCHEMISTRY OF REGIONAL FLOW SYSTEMS

The geochemistry of the water in aquifers reflects both the mineralogy and the reactions occurring along the water flow path; by understanding geochemical processes controlling the water chemistry, we obtain independent information on the nature of flow systems. This understanding requires (1) data on the spatial and stratigraphic distribution of the dissolved constituents and the controlling mineralogy; (2) recognition of hydrogeologic controls on this observed distribution; (3) identification of the chemical reactions that influence this distribution; and (4) data on rates of chemical reactions and transport of dissolved constituents. The philosophical and historical development of chemical geohydrology have been discussed by Hem (1959 and 1970), Back and Hanshaw (1965), Bredehoeft and Pinder (1973), and Back and Cherry (1976).

Many of the early chemical studies of ground water were primarily descriptive and were undertaken to determine the quality of water for various uses; the earliest concern was the potential for the water to form scale in steam boilers, both stationary engines and locomotives. Later the interest turned to quality requirements for irrigation, and then for use as municipal and industrial water supplies.

One of the first studies to demonstrate the hydrologic significance of chemical reactions is the classic paper by Foster (1950) in which she explained the origin of bicarbonate water in regional aquifers of the Atlantic Coastal Plain through the use of laboratory experiments. She demonstrated that carbonaceous material in the aquifer decomposed and generated carbon dioxide which facilitated the solution of calcium carbonate; the dissolved calcium was exchanged with sodium by "base-exchange" minerals in the aquifers to produce abnormally high concentrations of bicarbonate. Two papers by White (1957a,b) stimulated a great deal of interest and additional studies; he proposed a classification of "deep" waters based on their chemical and isotopic compositions. Although he did not use the term "regional flow system," this concept was implied in his duscussion of the origin of the various types of water identified--volcanic, magmatic, connate, and metamorphic.

Geochemical studies of water took an entirely new approach in the early 1960s with the application of the principles of chemical thermodynamics. Henri Schoeller (1956) in France had previously initiated similar work but his important contributions were generally unknown to American hydrologists. The potential value of chemical thermodynamics as applied to the study of ground water was not generally recognized by hydrologists in this country until the publications of papers by Hem and Cropper (1959), which showed the application of the Nernst equation to potable water, and by Back (1961), which demonstrated the application of the law of mass action to equilibrium principles and ground-water chemistry.

Back (1966) employed the concept of hydrochemical facies he had developed earlier; this was one of the first regional studies to couple an interpretation of chemical data with the ground-water flow regime. Hydrochemical facies depict the diagnostic chemical character of water in various parts of the system. The facies reflect the effects of chemical reactions occurring between minerals within the lithologic framework and ground water. The mineralogy of an aquifer system largely determines the type of hydrochemical facies that will develop; flow patterns modify the hydrochemical facies and control their distribution. By mapping the chemical character of water in the Atlantic Coastal Plain aquifers, Back was able: (1) to demonstrate use of chemistry to identify and understand the regional flow patterns, (2) to determine the controlling chemical reactions; and (3) to delineate areas where these reactions could be studied most effectively. The primary controls on the composition of the ground water were identified as being (1) chemical character of the water as it enters the zone of saturation; (2) distribution, solubility, and sorption capacity (ion-exchange) of the minerals in the deposits; and (3) the flow rate and path of the water. Seaber (1962) made a detailed regional geochemical study of the Coastal Plain aquifers in New Jersey in which he observed geochemical patterns that correlated with the general understanding the flow system.

Interpretation of the origin of the hydrochemical facies in the Coastal Plain aquifers demonstrated that certain principles of low-temperature geochemistry that were being applied to laboratory studies at that time, primarily by R. M. Garrels, had direct application to the ground-water regime. Specifically, those ideas are: (1) that mineral equilibria studies provide a geochemically meaningful description of ground water and (2) that oxidation-reduction (redox) potential is a major control on concentration and behavior of many constituents in ground water.

The concept of regional redox zonation (Back and Barnes, 1965) was demonstrated by integrating measurements of oxidation potentials and pH with the ground-water flow pattern to explain the occurrence and concentration of iron in various parts of an aquifer. This concept was further developed in the more complex Lincolnshire limestone in England by Edmunds (1973) in which he explained not only the concentration and distribution of metals but also of sulfate, sulfide, and nitrate. Champs and others (1979) compared such redox zones in four separate aquifer systems and named three biochemical zones on the basis of redox processes: oxygen-nitrate, iron-manganese, and sulfide.

The effect of local and regional flow on the chemistry of ground water was emphasized by Thrailkill (1968) in his study of formation of caves. Water in calcareous vadose zones is often saturated or supersaturated with respect to calcite; the water may become undersaturated and thereby capable of dissolving additional calcite by being cooled or by mixing with water that has a lower partial pressure of carbon dioxide. If the vadose zone above a carbonate aquifer lacks calcareous material, the recharge water will be undersaturated with respect to calcite and the water in the regional flow system will become saturated by dissolving calcite in the aquifer. White (1977) has formalized these basic ideas to develop conceptual models for carbonate aquifers in order to explain solution and permeability distribution in various flow systems.

Recent advances by carbonate petrologists and geochemists have provided additional insights into the distribution of porosity and permeability. An understanding of (1) the origin, chemistry, mineralogy, and environments of deposition and accumulation of carbonate sediments together with (2) a comprehension of diagenetic processes that change sediments to rocks, and (3) the geochemical, tectonic, and hydrologic processes that create voids make possible the prediction of regional porosity and permeability (Hanshaw and Back, 1979). Langmuir (1979) presented an example of reactions that control the chemical character of ground water in various parts of the flow system in carbonate aquifers; he showed that the water from springs was generally more undersaturated with respect to both calcite and dolomite than was the water from wells. The springs reflect the local, shallow flow system whereas the wells tap a deeper regional flow system.

Back and Hanshaw, working with others, have used the Floridan aquifer system as their principal field laboratory to investigate the goechemistry of a carbonate aquifer system. They have extended their ideas of relating carbonate equilibrium to regional ground-water flow to other systems--the Madison Limestone in the Northern Great Plains (Hanshaw and others, 1978) and the Yucatan Peninsula (Back and Hanshaw, 1970). Based on an interpretation of chemical data, they concluded that the characteristics of the regional flow in the artesian aquifer of Florida (Fig. 8) are controlled to a large extent by the permeability of the confining layer. The lack of regional flow in the Yucatan results primarily from absence of confining beds. In Florida the chemical character of water systematically changes downgradient as minerals dissolve until the water attains equilibrium, such as is shown with respect to calcite (Fig. 9). The saturation indices (log ion activity product/equilibrium constant) indicate the water is less saturated in the recharge areas and progressively becomes more saturated toward discharge points.

As a result of their regional aquifer studies in Florida, Hanshaw and others (1971) hypothesized that dolomitization could occur within carbonate aquifers in the zone where the discharging ground water mixes with salt water. The mixing provides a continuous supply of magnesium and produces undersaturation with respect to calcite and supersaturation of dolomite. Badiozamani (1973) named this ground-water mixing model "the dorag dolomitization model" and applied it in an excellent study to explain the origin of middle Ordovician dolomites of southwestern Wisconsin.

A series of papers on geochemistry of formation

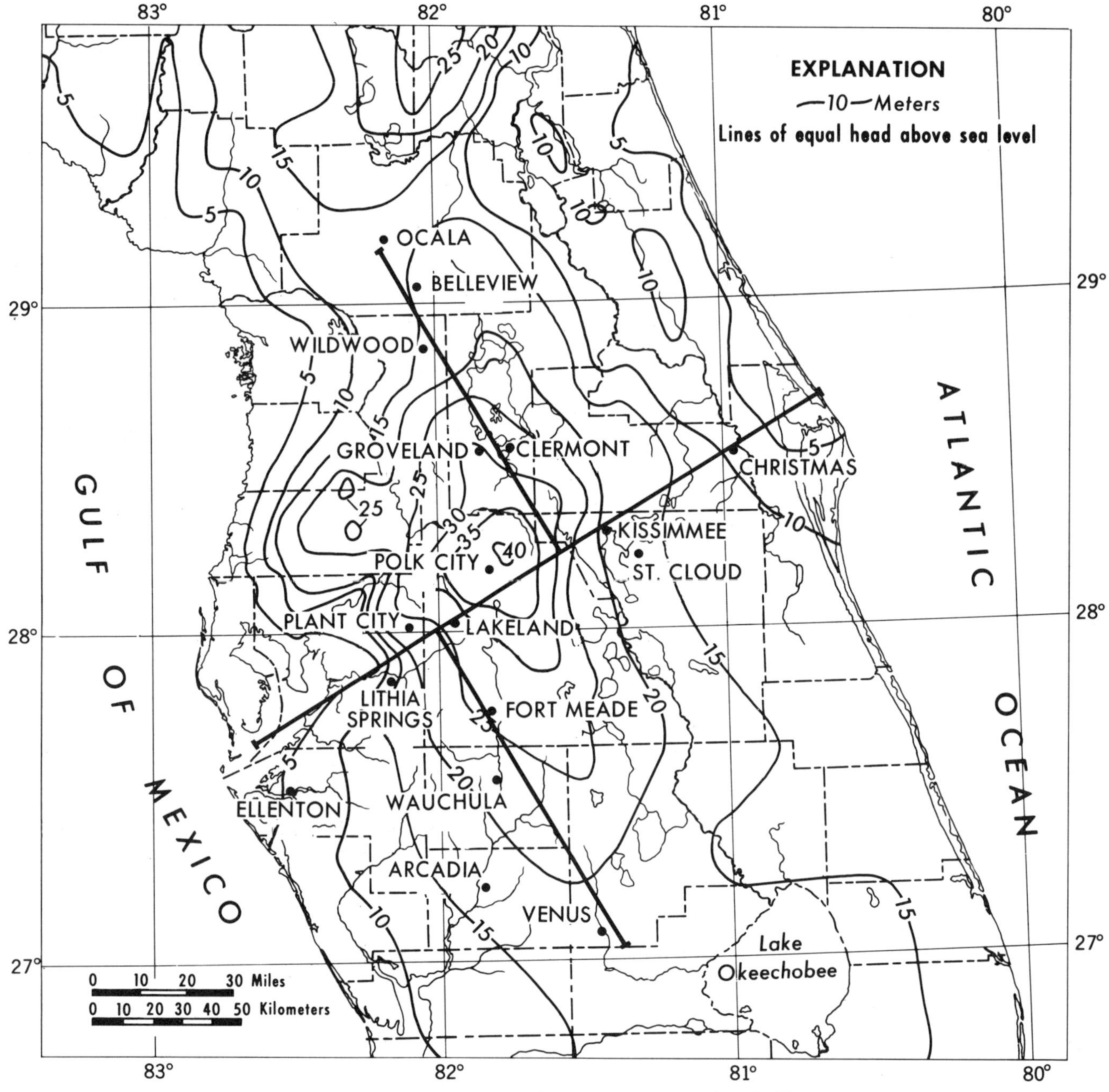

Figure 8. Map showing local and regional flow patterns in central Florida and locations of fence diagram shown in Figure 9.

waters in the western Canada sedimentary basin (Hitchon and Friedman, 1969; Billings and others, 1969; Hitchon and others 1971) conclude that these regional waters represent ancient seawater which has undergone membrane filtration, dilution of fresh-water recharge, solution of evaporites, formation of new minerals, cation exchange on clays, and desorption of ions from clays and organic matter.

In addition to studies of mineral equilibria, other chemical parameters have been used to investigate ground-water flow. Mifflin (1968) used temperature and chemical data to determine the depth of circulation and to recognize interbasin flow systems of Nevada. Three general types of flow systems--regional, local, and small-local--were delineated in the carbonate terrane using (1) the water chemistry of springs, (2) temperature, (3) variation in discharge, and (4) tritium concentrations. The relationship of sodium plus potassium to chloride plus sulfate was particularly diagnostic in that the springs issuing from regional flow systems had the highest concentrations, while springs from local systems had the lowest concentra-

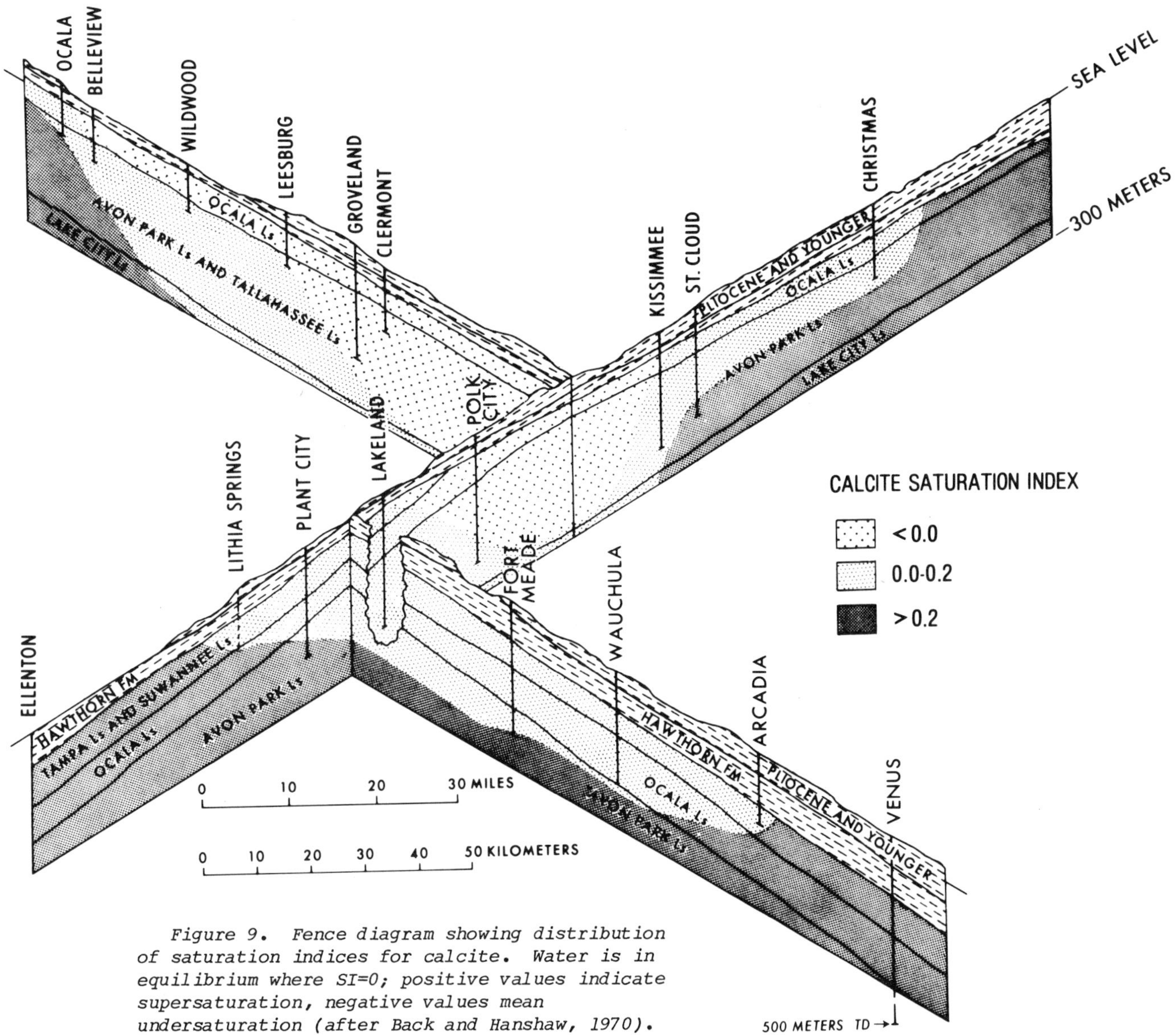

Figure 9. Fence diagram showing distribution of saturation indices for calcite. Water is in equilibrium where SI=0; positive values indicate supersaturation, negative values mean undersaturation (after Back and Hanshaw, 1970).

tions. Only the local and small-local systems contained tritium. Maxey (1968) provided a regional synthesis of flow systems in arid basins and reemphasized the chemical characteristics of springs as discussed by Mifflin.

Application of isotopic analysis of ^{14}D, ^{13}C, ^{34}S, ^{18}O, deuterium, and tritium to regional flow systems have been discussed in review articles by Back and Hanshaw (1965) and Back and Cherry (1976) and, therefore, only a few examples are mentioned here. One of the first uses of ^{14}C in regional studies was to determine ground-water velocity and residence time within the aquifer (Hanshaw and others, 1965); these data were later used to determine rates of chemical reaction and entropy production (Back and Hanshaw, 1971) and distribution of hydraulic conductivity (Hanshaw and Back, 1974).

Winograd and Pearson (1976) used chemical and isotope composition, primarily ^{14}C, of flow systems in Nevada to demonstrate regional scale flow channeling that indicate longitudinal dispersion of the order of several kilometers. Environmental isotopes have been used to analyze regional flow systems in Canada and other areas around the world (Wallick and Toth, 1976) and to deduce the flow mechanism and intergranular source of water in the chalk aquifer of the London Basin (Downing and others, 1979), and to identify the effect of sulfate-sulfide reactions on regional geochemistry (Rightmire and others, 1974).

An excellent beginning has been made in simulating geochemical patterns in regional ground-water flow (Schwartz and Domenico, 1973) in which they demonstrate that mineral dissolution, saturation constraints on dissolution, degree of saturation, partial pressure of carbon dioxide, and rates of reactions in relation to residence time of the ground water play different roles in determining the spatial distribution of dissolved constituents. Plummer (1977) has emphasized the use of mass balance relationships and mass transfer calculations to derive reaction paths that simulate observed water chemistry in regional carbonate aquifers (Plummer and Back, 1980). Evolution of ^{13}C and ^{14}C isotopes in natural waters (Wigley and others, 1978) is expressed as equations relating isotopic changes to changes in carbonate chemistry. This analysis provides an independent determination of the validity of derived reactions and thereby provides additional confidence in interpreting regional flow systems. Application of such models to other systems will be an important research activity in the 1980s.

SHALE MEMBRANE PHENOMENA

Some pressure and salinity data observed in sedimentary basins are not readily explained by purely gravitational flow or by the solution of minerals. They may, however, be explained by osmotically induced movement of water through clay and shale which serve as semipermeable membranes. Osmosis across a clay or shale layer tends to move water from an aquifer containing water with low concentration of dissolved solids to one containing water with higher concentrations. Likewise an externally applied head differential on the fluid phase may cause flow through a semipermeable membrane with resultant ion-exclusion effects. This latter phenomenon is commonly referred to as salt filtering, ultrafiltration, or hyperfiltration; it will occur only when the head difference exceeds the osmotic pressure produced by the fluids on opposite sides of the membrane.

Several authors have postulated that shales may operate as membranes. Mackay (1946) suggested that mantle-type mineral deposits may be caused by ion filtration effects of shales that act as molecular sieves and prevent hydrated ions from passing. Russell (1933) and DeSitter (1947) postulated a filtration mechanism to account for the subsurface concentration of brines. Wyllie (1955) speculated that ultrafiltration through compacting shales may explain the high salt concentration of formation water in sedimentary rocks. Bredehoeft and others (1963) and von Engelhardt and Gaida (1963) suggested a similar mechanism to account for the occurrence of brines in subsurface formations.

McKelvey and others (1957), in a discussion of hyperfiltration through synthetic membranes, suggested that compacted clays might also serve as salt-filtering membranes. McKelvey and Milne (1962) and Milne and others (1964) published the results of experiments in which salt solutions were forced at high pressure through compacted clays. They showed that compacted clays are capable of acting as membranes and suggested that the phenomenon is related to the negatively charged framework of the clay material (that is, to the ion exchange capacity).

Hanshaw (1962) also investigated the membrane properties of clays and provided the theory necessary to explain quantitatively the results of his experiments. Coplen and Hanshaw (1973) and Hanshaw and Coplen (1973) discussed the results of experiments regarding ultrafiltration by compacted clay membranes and showed that not only were dissolved solids excluded from passage through membranes but that the passage of H_2O also affected the isotopic composition of the water molecule itself.

Large differences of concentrations across compacted shales can produce enormous pressure, and it is possible to have the pressure on the liquid phase actually approach that of the lithostatic pressure (Hanshaw and Zen, 1965; Kharaka, 1971) (Fig. 10). In his thesis at the University of California, Berkeley, Hanshaw (1962) further investigated the membrane properties and ion exclusion characteristics of compacted shale systems.

In the late 1950s and early 1960s a group of scientists concerned with problems of regional flow of fluids was brought together at the Petroleum Research Corporation under the leadership of Gilman Hill. Among this group was Harold F. Walton, University of Colorado, whose perspective and theoretical approach to membrane phenomena encouraged Hill's group, which included Fred Berry and Bruce Hanshaw, to investigate effects of shale membrane phenomena in natural settings. The approach of the group was to establish geologic evidence for membrane properties of shales by examining the hydrology and geochemistry of aquifer systems throughout the world, mainly under contract to major oil companies.

Ground-water potentials that are either higher than recharge areas, or lower than discharge areas

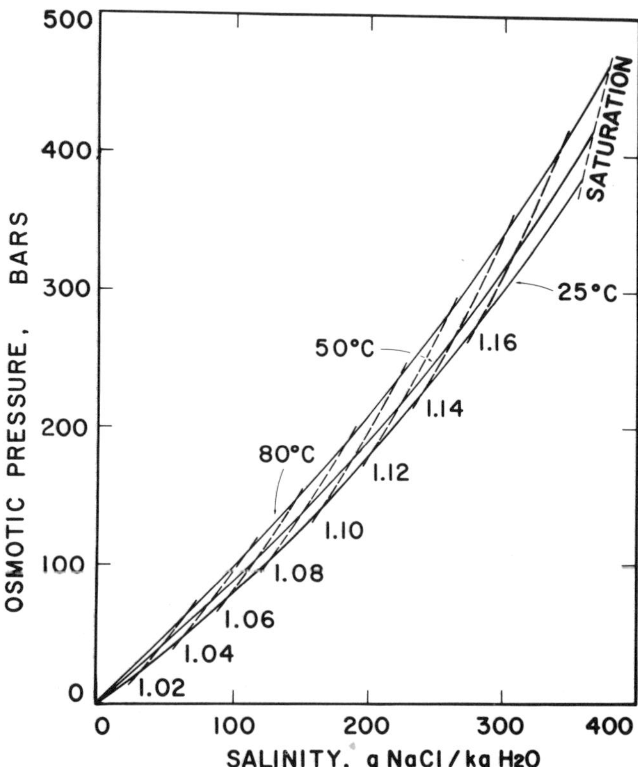

Figure 10. Calculated osmotic pressure difference across an ideal membrane (Hanshaw and Zen, 1964).

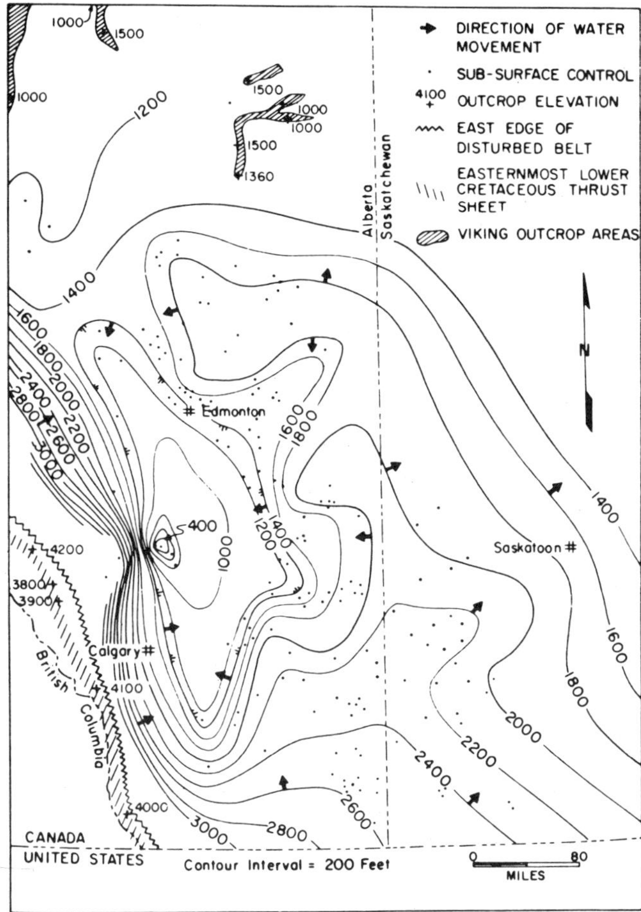

Figure 11. Potentiometric surface for the Viking Sandstone in central Alberta (Barry and Hanshaw, 1960).

or depressions in the potential surface away from high or low areas of obvious recharge or discharge are difficult to explain by normal gravitational flow. Such potentials represent anomalous hydrologic conditions that require special explanation. An example was given by Berry and Hanshaw (1960) who described a depression in the potentiometric surface contours as low as 200 m below the discharge point in the lower Cretaceous Viking Sandstone in the central Alberta Basin (Fig. 11).

Berry (1959) in a thesis at Stanford University, discussed the occurrence of depressions in potential surfaces of various formations in the Cretaceous and Jurassic systems of the San Juan Basin with potentials less than the elevation of the discharge area. Hanshaw and Hill (1969) described potentials in the Pennsylvanian and Mississippian formations in the San Juan Basin that are much higher than any of the outcrops of these units. These high and low potentials were explained by osmotic membrane processes believed to be operating between aquifers in which intervening shales act as membranes. Hitchon and Friedman (1969) discussed isotopic and chemical formation water variations in the Alberta Basin and suggested that ultrafiltration was responsible for some of the observed variations. Graf and others (1966), suggested that the highly concentrated calcium chloride brines of the Michigan Basin are the result of ultrafiltration by shale membranes. Billings and others (1969) also postulated that the geochemistry and origin of formation waters in western Canada sedimentary basins are partially the result of membrane concentration of more dilute formation water.

EXCESS PRESSURE--GULF COAST

High fluid pressures, approaching the weight of the overburden, are found in the deeper Tertiary sediments of the Gulf Coast. The phenomenon is widespread and occurs onshore and offshore across Texas and Louisiana coastal areas. The high-pressure system contains fluids: (1) at high pressures, (2) with high temperature, and (3) saturated with methane. Jones (1969) has advocated the development of this combined resource for energy. Whether or not the resource can be developed hinges upon whether enough fluids can be produced to allow high enough rates of energy extraction to make development economically feasible. The rate of fluid production, in turn, depends upon the extent, thickness, and permeability of sandstone reservoirs in the geopressured sediments (Papadopulos and others, 1975). The methane dissolved in the water may prove to be the most attractive and easily recovered portion of the resource.

A number of investigators have discussed the mechanism that produces the high fluid pressure; among the mechanisms are:

1. Continued accumulation of low permeability sediments in a subsiding basin (Bredehoeft and Hanshaw, 1968)
2. A phase change of montmorillonitic clay to ilitic clay with a consequent release of water as sediment accumulates and is buried and heated at depth (Powers, 1967)
3. The effect of thermal expansion of the fluids contained in the sediments as burial and accompanied heating proceeds (Sharp, 1978).

The debate continues over which mechanism dominates in producing the high pressures in the Gulf Coast area.

Bredehoeft and Hanshaw (1968) and Hanshaw and Bredehoeft (1968) investigated the question of how long anomalous pressures could persist in a geologic environment. The results indicate that anomalous pressures could be maintained for long periods of time, even in the geological sense; how long depends largely upon the permeability of the confining sediments. The permeability required to maintain anomalous pressures under various conditions for 100,000 years, even 1,000,000 years, is well within values that have been measured. Earth scientists are beginning to recognize systems that are not in equilibrium with present day boundary conditions. Drilling offshore near Baltimore Canyon and the North Atlantic Coast of the United States has yielded fresh water that suggests a fossil flow regime, perhaps from a period of major ice advance and lower sea level. Toth (1978) has reexamined the hydrodynamics of the Alberta Sedimentary Basin. His work suggests also that the observed potential distribution does not reflect present-day boundary conditions. These results have strong implications for understanding the accumulations of oil and gas, and perhaps other ore minerals, particularly lead and zinc.

HEAT FLOW

Stallman (1963) pointed out that moving ground water is an efficient mechanism for transporting heat. He further suggested that the earth's thermal profile would be disturbed by a small flux of ground water, and that we could analyze the earth's temperature field and obtain information on the direction and flow of ground water. This is an intriguing idea especially because temperature is one of the easier properties to measure with high accuracy. Bredehoeft and Papadopulos (1965) showed that if the heat flux and ground-water flux were both in a vertical direction, an analytical solution for the shape of the thermal profile could be obtained. From an analysis of the curvature in the thermal profile the rate of ground-water flow can be computed. Low flow rates can be detected and analyzed.

The idea of using the temperature field to determine direction and velocity of ground-water flow has probably been applied most extensively by Cartwright (1968, 1970, 1979) and by Schneider (1964a,b, 1968, 1972). Schneider (1968) mapped thermal gradients in central and northern Florida in wells tapping the principal artesian aquifer (carbonate rocks) and overlying clastic sediments. In areas where significant recharge is known to occur, gradients generally are near zero or are negative (temperature decreases with depth). On maps showing the areal temperature distribution at depths of 100 and 150 ft, "hot" regions delineate discharge areas where there is a significant upward flow component, or regions of relatively low transmissivity. "Cold" regions coincide with recharge areas or where there is a relatively high transmissivity.

Lachenbruch and Sass (1977) have examined heat flow data for much of the western United States. They obtained an analytical solution for disturbance in the vertical flux of heat caused by vertical movement of ground water using boundary conditions more appropriate for the regional heat flow problem than those used earlier by Bredehoeft and Papadopulos. Using their analytical results, they suggested that heat flow over a large area in the Basin and Range of Nevada has been significantly lowered by ground-water transport of heat. This generally fits with our understanding of a major carbonate aquifer underlying much of the region as discussed above.

FUTURE RESEARCH

As pointed out earlier (Bredehoeft, 1976) we must return to examining hydrologic systems and attempt to describe the systems in a more realistic quantitative manner. Our present knowledge of regional aquifers indicates that an increased knowledge of confining layers is critical to further understanding of the hydrodynamics of regional flow. We are becoming aware that the permeability of many confining layers depends upon fractures. Recent investigation of the Pierre Shale overlying the Dakota aquifer illustrates this point. Model analysis of the regional system indicates with a high degree of confidence that the hydraulic conductivity of the confining layer is 6 $\times 10^{-9}$ cm/s. In-place measurements in boreholes made during the summer of 1978 indicate a permeability of 5 x 10^{-11} cm/s or less. The most obvious explanation is that the confining layer is fractured. How to investigate a fractured confining layer with such low permeability challenges our imagination as researchers. Yet we must understand this problem if we are to deal intelligently with such problems as contaminant movement in these systems.

As demonstrated earlier in this paper, several studies show that chemical equilibrium does not exist in all parts of the aquifer at all times. If we are to describe quantitatively the chemical phenomena taking place within the regional flow system we must know which reactions are occurring, their progress or extent, and their rates. In reviewing hydrogeologic problems, Back and Cherry (1976) concluded in part that major scientific obstacles that currently persist in development of geochemical models for regional flow systems include (1) lack of adequate and valid geologic data, (2) deficiency of laboratory data in chemical mechanisms and kinetics of solution-precipitation reactions, and (3) difficulty of transferring chemical and geologic laboratory data to regional systems.

Assuming we understand the hydrodynamics of a system and the geochemistry we, as hydrologists, should be able to reconstruct quantitatively the flow system and the coupled chemical transport system. Although coupled flow and transport models are available, few such complete flow and geochemical reconstructions have been attempted quantitatively; yet this should be done. When this is accomplished it will lend a great deal more confidence to our understanding of the system. Transport models have reached the stage where quantitative analysis is no longer a problem, and the transport of heat by moving ground water is a promising approach by which to investigate regional ground-water flow; natural heat in the earth is a good tracer. Although the approach and ideas have been used, they probably deserve more attention.

ACKNOWLEDGMENTS

Although Robert R. Bennett has not seen this manuscript, we would like to acknowledge the great encouragement, insight, and guidance he gave us in our formative years; he taught us to appreciate and understand the importance of considering ground water from the viewpoint of regional flow systems. We have also benefited immeasurably from the stimulating discussions, continuing support, and cooperation given us by our colleagues within the U.S. Geological Survey and elsewhere. Warren Wood's thoughtful review greatly improved the manuscript.

REFERENCES CITED

Abbott, Patrick A., 1975, On the hydrology of the Edwards Limestone, south-central Texas: Journal of Hydrology, v. 24, p. 251-269.

Back, W., 1961, Calcium carbonate saturation in ground water, from routine analyses: U.S. Geological Survey Water-Supply Paper 1535-D.

___1966, Hydrochemical facies and ground-water flow patterns in northern part of Atlantic Coastal Plain: U.S Geological Survey Professional Paper 498-A.

Back, W., and Barnes, I., 1965, Electrochemical potential and iron-bearing waters related to ground-water flow patterns: U.S. Geological Survey Professional Paper 498-C.

Back, W., and Cherry, J. A., 1976, Chemical aspects of present and future hydrogeologic problems, in Advances in groundwater hydrology: American Water Research Association, p. 153-172.

Back, W., and Hanshaw, B. B., 1965, Chemical geohydrology, in Advances in hydroscience, 2: New York, Academic Press, p. 49-109.

___1970, Comparison of chemical hydrogeology of the carbonate peninsula of Florida and Yucatan: Journal of Hydrology, v. 10, p. 330-368.

___1971, Rates of physical and chemical processes in a carbonate aquifer, in Hem, J.D., ed., Nonequilibrium systems in natural chemistry: American Chemical Society, Advances in Chemistry Series 106, p. 77-93.

Badiozamani, K., 1973, The Dorag dolomitization model--Application to the Middle Ordovician of Wisconsin: Journal of Sedimentary Petrology, v. 43, p. 965-984.

Berry, F.A.F., 1959, Hydrodynamics and geochemistry of the Jurassic and Cretaceous systems in the San Juan basin, northwestern New Mexico and southwestern Colorado [Ph.D. dissert.]: Stanford University School of Mineral Sciences.

Berry, F.A.F., and Hanshaw, B. B., 1960, Geologic field evidence suggesting membrane properties of shales [abs.]: XXI International Geological Congress, 21st, Copenhagen, Rep. Session: Norden, p. 209.

Billings, G. K., Hitchon, B., and Shaw, D. R., 1969, Geochemistry and origin of formation waters in the western Canada sedimentary basin, 2. Alkali metals: Chemical Geology, v. 4, p. 211-223.

Bredehoeft, J. D., 1976, Status of quantitative groundwater hydrology, in Saleem, Z. A., ed., Advances in groundwater hydrology: American Water Resources Association, p. 8-14.

Bredehoeft, J. D., and Bennett, R. R., 1971, Potentiometric surface of the Tensleep Sandstone in the Big Horn basin, west-central Wyoming: U.S. Geological Survey Open-File Map.

Bredehoeft, J. D., and Hanshaw, B. B., 1968, On the maintenance of anomalous fluid pressures, I, Thick sedimentary sequences: Geological Society of America Bulletin, v. 79, p. 1097-1105.

Bredehoeft, J. D., and Papadopulos, I. S., 1965, Rates of vertical groundwater movement estimated from the Earth's thermal profile: Water Resources Research, v. 1, p. 325-328.

Bredehoeft, J. D., and Pinder, R. A., 1973, Mass transport in flowing groundwater: Water Resources Research, v. 9, no. 1, p. 194-210.

Bredehoeft, J. D., and others, 1963, A possible mechanism for the concentration of brines in subsurface formations: American Association of Petroleum Geologists Bulletin, v. 47, no. 2, p. 257-269.

Cartwright, K., 1968, Temperature prospecting for shallow glacial and alluvial aquifers in Illinois: Illinois State Geological Survey Circular 433, 41 p.

___1970, Ground-water discharge in the Illinois Basin as suggested by temperature anomalies: Water Resources Research, v. 6, no. 3, p. 912-918.

___1979, Measurement of fluid velocity using temperature profiles: Experimental verification: Journal of Hydrology, v. 43, p. 185-194.

Champs, D. R., Gulens, J., and Jackson, R. E., 1979, Oxidation-reduction sequences in ground-water flow systems: Canadian Journal of Earth Science, v. 16, p. 12-23.

Chamberlin, T. C., 1885, The requisite and qualifying conditions of artesian wells: U.S. Geological Survey 5th Annual Report, p. 125-173.

Coplen, T. B., and Hanshaw, B. B., 1973, Ultrafiltration by a compacted clay--1, Oxygen and hydrogen isotopic fractionation: Geochemica et Cosmochimica Acta, v. 37, p. 2295-2310.

Darton, H. H., 1896, Preliminary report on artesian waters of a portion of the Dakotas: U.S. Geological Survey, 17th Annual Report.

___1905, Preliminary report on the geology and underground water resources of the central Great Plains: U.S. Geological Survey Professional Paper 32, 432 p.

___1909, Geology and underground waters of South Dakota: U.S. Geological Survey Water-Supply Paper 227, 156 p.

DeSitter, L. U., 1947, Diagenesis in oil-field brines: American Association of Petroleum Geologists Bulletin, v. 31, p. 2030-2040.

Downing, R. A., Pearson, F. J., and Smith, D. B., 1979, The flow mechanism in the chalk based on radio-isotope analyses of groundwater in the London basin: Journal of Hydrology, v. 40, p. 67-83.

Dyer, C. F., and Geohring, A. J., 1965, Artesian-water supply of the Dakota formation, southeastern South Dakota: U.S. Geological Survey Open-File Report, 49 p.

Eakin, T. E., 1966, A regional interbasin ground-water system in the White River area, southeastern Nevada: Water Resources Research, v. 2, p. 251-271.

Edmunds, W. M., 1973, Trace element variations across an oxidation-reduction barrier in a limestone aquifer, in Proceedings, Symposium on Hydrogeochemistry and Biogeochemistry: Tokyo, Japan, p. 500-526.

Foster, M. D., 1950, The origin of high-sodium-bicarbonate waters in the Atlantic and Gulf Coastal Plains: Geochimica et Cosmochimica Acta, v. 1, p. 33-48.

Freeze, R. A., and Witherspoon, P. A., 1966, Theoretical analysis of regional ground water flow, 1, Analytical and numerical solutions to

the mathematical model: Water Resources Research, v. 2, p. 641-656.
___1967, Theoretical analysis of regional ground water flow, 2, Effect of water table configuration and subsurface permeability variation: Water Resources Research, v. 3, p. 628-634.
Graf, D. L., 1960, Geochemistry of carbonate sediments and sedimentary carbonate rocks: Part 1, Carbonate mineralogy--Carbonate sediments: Illinois State Geological Survey Circular 297, 39 p.
Graf, D. L., Meents, W. F., Friedman, I., and Shimp, N. F., 1966, The origin of saline formation waters, III, Calcium chloride waters: Illinois Geological Survey Circular 397, 60 p.
Guyton, W. F., 1942, Results of pumping tests of the Carrizo Sand in the Lufkin area, Texas: American Geophysical Union Transactions, pt. 1, p. 40-48.
Hanshaw, B. B., 1962, Membrane properties of compacted clays [Ph.D. dissert.]: Cambridge, Mass., Harvard University, 113 p.
Hanshaw, B. B., and Back, W., 1974, Determination of regional hydraulic conductivity through use of ^{14}C dating of ground water: International Association of Hydrogeology, Congress of Montpellier, v. X-1, p. 195-196.
___1979, Major geochemical processes in the evolution of carbonate-aquifer systems: Journal of Hydrology, v. 43, p. 287-312.
Hanshaw, B. B., and Bredehoeft, J. D., 1968, On the maintenance of anomalous fluid pressures, II, Source layer at depth: Geological Society of America Bulletin, v. 79, p. 1107-1122.
Hanshaw, B. B., and Coplen, T. B., 1973, Ultrafiltration by a compacted clay membrane, II, Sodium ion exclusion at various ionic strengths: Geochimica et Cosmochimica Acta, v. 37, p. 2311-2327.
Hanshaw, B. B., and Hill, G. A., 1969, Geochemistry and hydrodynamics of the Paradox basin region, Utah, Colorado, and New Mexico: Chemical Geology, v. 4, p. 263-294.
Hanshaw, B. B., and Zen, E-an, 1965, Osmotic equilibrium and overthrust faulting: Geological Society of America Bulletin, v. 76, p. 1379-1386.
Hanshaw, B. B., Back, W., and Rubin, M., 1965, Radiocarbon determinations for estimating groundwater flow velocities in central Florida: Science, v. 148, p. 494-495.
Hanshaw, B. B., Back, W., and Deike, R. G., 1971, A geochemical hypothesis for dolomitization by ground water: Economic Geology, v. 66, p. 710-724.
Hanshaw, B. B. Busby, J., and Lee, R., 1978, Geochemical aspects of the Madison aquifer system, in Williston Basin Symposium: Montana Geological Society, 24th Annual Conference, p. 385-389.
Hantush, M. S., 1960, Modification of the theory of leaky aquifers: Journal of Geophysical Research, v. 64, p. 1043-1052.
Hem, J. D., 1959, Study and interpretation of the chemical characteristics of natural water: U.S. Geological Survey Water-Supply Paper 1473.
___1970, Study and interpretation of the chemical characteristics of natural water: U.S. Geological Survey Water-Supply Paper 1473, second ed., 363 p.
Hem, J. D., and Cropper, W. H., 1959, Survey of ferrous-ferric chemical equilibria and redox potentials: U.S. Geological Survey Water-Supply Paper 1459-A.
Hitchon, B., Billings, G. K., and Klovan, J. E., 1971, Geochemistry and origin of formation waters in the western Canada sedimentary basin, III, Factors controlling chemical composition: Geochemica et Cosmochemica Acta, v. 35, p. 567-598.
Hitchon, B., and Friedman, I., 1969, Geochemistry and origin of formation waters in the western Canada sedimentary basin, 1, Stable isotopes of hydrogren and oxygen: Geochimica et Cosmochimica Acta, v. 33, p. 1321-1349.
Hubbert, M. K., 1940, The theory of groundwater motion: Journal of Geology, v. 48, p. 785-944.
___1953, Entrapment of petroleum under hydrodynamic conditions: American Association of Petroleum Geologists Bulletin, v. 37, p. 1954-2026.
Jacob, C. E., 1945, Correlation of ground-water levels and precipitation on Long Island, N.Y.: New York Department of Conservation, Water Power and Control Commission Bulletin, GW 14.
Jacob, C. E., 1946, Radial flow in a leaky artesian aquifer: American Geophysical Union Transactions, v. 27, pt. 2, p. 574-586.
Jones, P. H., 1969, Hydrodynamics of geopressure in the northern Gulf of Mexico basin: Journal of Petroleum Technology, v. 21, p. 803-810.
Kharaka, Y. K., 1971, Simultaneous flow of water and solutes through geological membranes [Ph.D. dissert.]: Berkeley, University of California, 263 p.
Lachenbruch, A. H., and Sass, J. H., 1977, Heat flow in the United States and the thermal regime of the crust: American Geophysical Union Geophysical Monograph 20, p. 626-675.
Langmuir, Donald, 1971, The geochemistry of some carbonate ground waters in central Pennsylvania: Geochimica et Cosmochimica Acta, v. 35, p. 1023-1045.
Lonsdale, J. T., 1935, Geology and ground-water resources of Atascosa and Frio Counties, Texas: U.S. Geological Survey Water-Supply Paper 676, 90 p.
Mackay, R. A., 1946, The control of impounding structures on ore deposition: Economic Geology, v. 41, p. 13-46.
Maxey, George B., 1968, Hydrogeology of desert basins: Ground Water, v. 6, no. 5, p. 10-22.
Meinzer, O. E., 1928, Compressibility and elasticity of artesian aquifers: Economic Geology, v. 23, p. 263-291.
Mendenhall, W. C., 1905, The hydrology of San Bernardino Valley, California: U.S. Geological Survey Water-Supply Paper 142.
Mifflin, M. D., 1968, Delineation of ground-water flow systems in Nevada: Reno, Desert Research Institute, Hydrology and Water Resources Publication 4, 111 p.
Mifflin, M. D., and Hess, J. W., 1979, Regional carbonate flow systems in Nevada: Journal of Hydrology, v. 43, p. 217-238.

Milne, I. H., McKelvey, J. G., and Trump, R. P., 1964, Semi-permeability of bentonite membranes to brines: American Association of Petroleum Geologists Bulletin, v. 48, p. 103-105.

Moulder, E. A., 1957, Development of ground water from the Carrizo Sand and Wilcox Group in Dimmit, Zavala, Maverick, Frio, Atascosa, Medina, Bexar, Live Oak, McMullea, La Salle, and Webb Counties, Texas: U.S. Geological Survey Open-File Report.

McKelvey, J. E., and Milne, I. H., 1962, The flow of salt solutions through compacted clays: Clays and Clay Minerals Proceedings, p. 248-259.

McKelvey, J. G., Spiegler, K. S., and Wylie, M.R.J., 1957, Salt filtering by ion-exchange grains and membranes: Journal of Physical Chemistry, v. 61, p. 174-178.

Neuzil, C. E., 1980, Fracture leakage in the Cretaceous Pierre Shale and its significance for underground waste disposal [Ph.D. dissert.]: Baltimore, Johns Hopkins University.

Papadopulos, S. S., and others, 1975, Assessment of onshore geopressured-geothermal resources in the northern Gulf of Mexico basin: U.S. Geological Survey Circular 726, p. 125-146.

Pearson, F. J., Jr., and White, D. E., 1967, Carbon 14 ages and flow rates of water in Carrizo Sand, Atascosa County, Texas: Water Resources Research, v. 3, no. 1, p. 251-261.

Piper, A. M., 1928, Discussion of The origin of artesian pressure by W. L. Russell: Economic Geology, v. 23, p. 686-696.

Plummer, L. N., 1977, Defining reactions and mass transfer in part of the Floridan aquifer: Water Resources Research, v. 13, p. 801-812.

Plummer, L. N., and Back, William, 1980, The mass balance approach: Application to interpreting the chemical evolution of hydrologic systems: American Journal of Science, v. 280, p. 130-142.

Powers, M. C., 1967, Fluid-release mechanisms in compacting marine mudrocks and their importance in oil exploration: American Association of Petroleum Geologists Bulletin, v. 51, p. 1240-1254.

Rightmire, C. T., and others, 1974, Distribution of sulphur isotopes of sulphates in groundwaters from the principal artesian aquifer of Florida and the Edwards aquifer of Texas, United States of America, in Isotope techniques in groundwater hydrology 1974: Vienna, International Atomic Energy Agency, v. 2, p. 191-207.

Russell, W. L., 1928, The origin of artesian pressure: Economic Geology, v. 23, p. 132-157.

___1933, Subsurface concentration of chloride brines: American Association of Petroleum Geologists Bulletin, v. 17, p. 1213-1228.

Sayre, A. N., and Bennett, R. R., 1942, Recharge, movement, and discharge in the Edwards Limestone reservoir, Texas: American Geophysical Union Transactions, pt. 1, p. 19-27.

Schneider, R., 1964a, Geologic and hydrologic factors related to artificial recharge of the carbonate rock aquifer system of central Israel: Symposium of Haifa Publication No. 72, p. 37-45.

___1964b, Relation of temperature distribution to ground-water movement in carbonate rocks of central Israel: Geological Society of America Bulletin, v. 75, p. 209-216.

___1968, An interpretation of the geothermal field associated with the carbonate-rock aquifer system of Florida abs.: Geological Society of America Special Paper 101, Abstracts for 1966, p. 191-192.

___1972, Distortion of the geothermal field in aquifers by pumping: U.S. Geological Survey Professional Paper 800-C, p. 267-270.

Schoeller, H., 1956, Geochemie des eaux souterraines: Paris, Editions Techniques.

Schwartz, F. W., and Domenico, P. A., 1973, Simulation of hydrochemical patterns in regional ground-water flow: Water Resources Research, v. 9, no. 3, p. 707-720.

Seaber, P. R., 1962, Cation hydrochemical facies of ground water in the Englishtown Formation, New Jersey: U.S. Geological Survey Professional Paper 450, article 51.

Sharp, J.A., 1978, Energy and momentum transport model of the Ouachita basin and its possible impact on formation of economic mineral deposits: Economic Geology, v. 73, p. 1057-1066.

Stallman, R., 1963, Computation of groundwater velocity from temperature data, in Bentall, R., ed., Methods of collecting and interpreting groundwater data: U. S. Geological Survey Water-Supply Paper 1544-H, p. 36-46.

Stringfield, V. T., 1935, The piezometric surface of the artesian water in the Florida peninsula: American Geophysical Union Transactions, p. 524-529.

___1936, Artesian water in the Florida peninsula: U. S. Geological Survey Water-Supply Paper 773-6, p. 115-195.

___1966, Artesian water in Tertiary limestone in the southeastern states: U.S. Geological Survey Professional Paper 517.

Stringfield, V. T., LeGrand, H. E., and LaMoreaux, P. E., 1977, Development of karst and its effects on the permeability and circulation of water in carbonate rocks with special reference to the southeastern states: Geological Survey of Alaska Bulletin, Part G, 94, 68 p.

Swenson, F. A., 1968, New theory of recharge to the artesian basin of the Dakotas: Geological Society of America Bulletin, v. 79, p. 163-182.

Terzaghi, C., 1929, Discussion of The origin of artesian pressure by W. L. Russell: Economic Geology, v. 24, p. 91-100.

Theis, C. V., 1935, The relation between the lowering of the piezometric surface and the rate and duration of discharge of a well using groundwater storage: American Geophysical Union Transactions, v. 18, pt. 2, p. 522.

___1937, Amount of ground-water recharge in the southern High Plains: American Geophysical Union Transactions, v. 18, pt. 2, p. 564-568.

Thompson, D. G., 1929, Discussion of The origin of artesian pressure by W. L. Russell: Economic Geology, v. 24, p. 758-771.

Thrailkill, John, 1968, Chemical and hydrologic factors in the excavation of limestone caves: Geological Society of America Bulletin, v. 79, p. 19-46.

Toth, J., 1962, A theory of groundwater motion in stream drainage basins in central Alberta, Canada: Journal of Geophysical Research, v. 67, p. 4375-4387.

___1963, A theoretical analysis of groundwater flow in small drainage basins: Journal of Geophysical Research, v. 68, p. 4795-4812.

___1978, Gravity-induced cross-formational flow of formation fluids, Red Earth region, Alberta, Canada: Analysis, patterns, evolution: Water Resources Research, v. 14, p. 805-843.

von Engelhard, W., and Gaida, K. H., 1963, Concentration changes of pore solutions during the compaction of clay sediments: Journal of Sedimentary Petrology, v. 33, p. 919-930.

Wallick, E. I., and Toth, J., 1976, Methods of regional groundwater flow analysis with suggestions for the use of environmental isotopes, in Proceedings of an advisory group meeting, Vienna, 27-31 Jan. 1975: Interpretation of environmental isotope and hydrochemical data in groundwater hydrology: Vienna, International Atomic Energy Agency, p. 37-64.

White, D. E., 1957a, Thermal waters of volcanic origin: Geological Society of America Bulletin, v. 68, p. 1637-1658.

___1957b, Magmatic, connate, and metamorphic waters: Geological Society of America Bulletin, v. 68, p. 1659-1682.

White, W. B., 1977, Conceptual models for karst aquifers: Revisited, in Dilamarter, R. R., and Csallany, S. C., eds., Hydrologic problems in karst regions: Western Kentucky University, p. 176-187.

Wigley, T.M.L., Plummer, L. N., and Pearson, F. J., Jr., 1978, Mass transfer and carbon isotope evolution in natural water systems: Geochimica et Cosmochimica Acta, v. 42, p. 1117-1139.

Wilson, W. E., and Gerhardt, J. M., 1979, Simulated changes in potentiometric levels resulting from groundwater development for phosphate mines, west-central Florida: Journal of Hydrology, v. 43, p. 491-515.

Winograd, I., and Friedman, I., 1972, Deuterium as a tracer of regional groundwater flow, southern Great Basin, Nevada and California: Geological Society of America Bulletin, v. 83, p. 3691.

Winograd, I. J., and Pearson, F. J., 1976, Major carbon-14 anomaly in a regional carbonate aquifer: Possible evidence for megascale channeling, south-central Great Basin: Water Resources Research, v. 12, p. 1125-1143.

Winograd, I. J., and Thordarson, W., 1975, Hydrogeologic and hydrochemical framework, south-central Great Basin, Nevada-California, with special reference to the Nevada Test Site: U.S. Geological Survey Professional Paper 712-C, 109 p.

Woodruff, C. M., Jr., and Abbott, P. L., 1979, Drainage-basin evolution and aquifer development in a karstic limestone terrain in south-central Texas, U.S.A.: Earth Surface Processes, v. 4, p. 319-334.

Wyllie, M.R.J., 1951, An investigation of the electrokinetic components of the self-potential curve: American Institute of Mining Engineers Transactions, v. 192, p. 1-18.

___1955, Role of clay in well-log interpretation, in Proceedings, First National Conference on Clays and Clay Technology: California Division of Mines Bulletin, v. 169, p. 282-305.

Zapp, D. D., 1956, Structure contour map of the Tensleep Sandstone in the Big Horn basin, Wyoming and Montana: U.S. Geological Survey, Oil and Gas Investigations, Map 182.

Manuscript Accepted by the Society October 14, 1980

Printed in U.S.A.

Geological Society of America
Special Paper 189
1982

Hydrogeology of Induced Seismicity and Tectonism: Case Histories of Kariba and Koyna

DAVID T. SNOW
Groundwater and Geological Engineering
Colorado Springs, Colorado 80907

ABSTRACT

This paper represents a search for a set of conditions necessary and sufficient to enhance seismicity upon reservoir filling. In the small number of reservoirs that have clearly been seismogenic, criticality may have been attained beforehand by recent stress changes. These can be deduced from the erosion of landforms, the deposition of sediments, crustal uplift and geostatic cooling or subsidence and heating, as well as from pore-pressure changes resulting from river incision and alluviation. Geomorphic evolution has produced effective-stress changes greater in magnitude than the superposed stresses from reservoir filling. Increased total stresses due to reservoir weight extend to lesser distances than the increased fluid pressures imposed by changes of the regional ground-water flow system. Thus the field of reduced effective stress, wherein failure by fault slippage can occur, depends upon the hydrogeologic setting, which is systematically examined in this report. Widely spaced stiff fractures in unweathered basement rock are essential to reduction of effective stresses. Consolidation shortens the fluid-transient period. Field evidence suggests that Poisson's ratio for extensive rock masses exceeds 0.33, a parameter needed for estimating horizontal stress changes from vertical changes.

Application of theoretical stress changes in different tectonic environments suggests that natural and man-made seismicity may be influenced by surficial loading. Because the cumulative changes of principal stresses are the result of compensative stabilizing and destabilizing effects of total load and fluid pressure, each case needs to be analyzed individually. Even in thrust-fault regions, reservoir filling might stimulate seismicity.

Kariba Reservoir in Africa exemplifies induced seismicity in a normal-fault environment. Faults in the Karoo-filled, ancient mid-Zambezi rift have been inactive since Cretaceous time, but minor pre-reservoir seismicity attests to criticality, which was maintained by erosion. The river lowered the valley floor without faulting, whereas in other rifts, topographic similarity and basement stresses were maintained by fault displacements. The existence of faults, artesian aquifers, and thermal springs submerged by Kariba Reservoir imply that upon filling, fluid pressures were increased over distances as much as several reservoir widths away from the site and to depths of several kilometres. Whereas post-reservoir seismicity was produced by decrease of effective stress, subsidence was responsive to increase of total stress.

Koyna Reservoir in India exemplifies induced seismicity in a wrench-fault environment. The site is topographically uplifted and close to the western fault-boundary of the Indian peninsula. The flat-lying Deccan lava flow contacts, together with vertical north-south wrench faults, promote extensive ground-water circulation from the reservoir to the coastal lowland. Uplift and erosion maintained criticality and minor prior seismicity. Reservoir filling renewed faulting by increasing the pore pressure in the system. A major earthquake swarm, culminating in a disastrous M = 6.7 event, propagated down the hydraulic gradient.

Progressive decrease of effective stress is inadequate as a criterion for seismogenicity of reservoirs because it is also satisfied by many nonseismic reservoirs. Strength and pore-pressure measurements deep beneath reservoir shores are not recommended; the former are too imprecise, the latter too small for utility. Seismic prediction may depend in the future on techniques of structural

Dedicated to my professor and friend, Paul A. Witherspoon, on the happy occasion of his sixtieth birthday. For maintaining at Berkeley a place for his graduates to return for renewal of inspiration in hydrology, I am grateful.

discrimination. Heterogeneity of the crust is the main problem: large areas are stressed subcritically, probably because the weakest faults limit the stresses transmitted. Among the factors needing study are differences in fault materials, contact geometry and continuity, and fault orientation with respect to stresses. If analysis of historic and predicted stress changes satisfy necessary but not sufficient criteria for failure, then until means are available to identify criticality, the information can find practical application in adjustment of probabilistic measures of historic seismicity.

INTRODUCTION

In a few dozen cases of reservoir loading (National Academy of Sciences, 1972), the coincidence in place and time between reservoir filling and enhanced seismicity has left little doubt that a causal relationship must exist. But for the majority, consisting of thousands of reservoirs, no relationship is apparent (Simpson, 1976). Microseismicity is probably more common than macroseismicity around reservoirs, and in all likelihood numerous cases of induced microseismicity have gone unnoticed. A few reservoirs with high-gain seismic networks have proven quiet (Johnson and others, 1977; Daley and others, 1977). The search for an understanding of the underlying processes continues, with the objective of eventually attaining predictive capabilities.

The seismicity observed after filling Lake Mead (Carder, 1945) was the first suggestion of a relationship. In 1965-1966, earthquake swarm activity coincident with injection of fluid into fractured basement rocks 3650 to 3662 m deep in the Denver Basin (Evans, 1966) convinced many that increased fluid pressures could trigger releases of stored elastic energy by reducing resistance to fault slippage. It was startling to realize that the Laramide revolution was not over. M. Goguel noted (oral presentation, Colloquium on Seismic Effects of Reservoir Impoundment, London, 1973) that the few incontrovertible cases of induced seismicity show that mountain-building stresses are not relieved by creep as previously believed, but rather persist for extremely long periods below failure levels.

The most likely condition in a seismic or recently tectonic region is that the state of stress is close to critical for slip-failure along some fault of optimum orientation and/or of weakest materials (Snow, 1972). A threshold of strength may exist if the history of load changes has produced stresses in the past more favorable for failure than now exist. Presumably, aseismic blocks situated between critical weak surfaces may be subjected to adverse load changes, such as increased pore pressure, with an indefinite degree of safety.

Post-Pliocene stripping of about 600 m of overburden from the Denver Basin progressively reduced confinement and therefore fault strength (Snow, 1968a). Earthquakes may have manifested stress decreases throughout the unloading period. The intensity VII quake of November 7, 1882, near Denver (Hadsell, 1968) was illustrative. The man-made swarm of 1965-1968 suggested (Snow, 1972) that circumstances might be found that would allow discrimination between seismic and aseismic reservoirs by showing that in seismic cases a history of changes has kept effective stresses critical, thus prone to trigger failure by small increases of pore pressure. My studies of Lakes Kremasta, Kariba, Koyna, and others have shown that in each case the filling of seismogenic reservoirs has been preceded by a period of diminished effective stresses. Thus a necessary condition for reservoir induced seismicity may be a history of decreasing effective stresses. The converse--that nonseismic reservoirs have a built-in safety threshold by reason of prior increases of effective stress--has not been substantiated. There has been a lack of stimulus to devote equal time to the geomorphic and hydrogeologic reconstruction of effective stress histories for quiet reservoirs. It is evident that in provinces of uniform geology and stress history, such as the Appalachian Piedmont, there are few seismogenic reservoirs; in fact, the majority are aseismic (Severy and others, 1975). For discrimination between potentially active and inactive reservoir sites, the inconsistencies demand admission of factors other than a progressive decrease of effective stress.

No new factors being apparent, efforts have been directed toward refined computations of stress changes. Gough and Gough (1970a) emphasized the changes of the total stress field and the subsidence resulting from filling Lake Kariba in eastern Africa. Withers (1977) and Bell and Nur (1978) applied reservoir loading as a change of boundary conditions to the Biot consolidation equations, computing the changes of pore pressure and effective stresses in a semi-infinite, homogeneous, compressible porous medium. Withers assumed soil-like behavior of rock under compression; that is, instantaneous compression was assumed to be small compared to the eventual compression of rock drained of excess pressure, implying that compression of voids by fluid expulsion is large compared to mineral compression. That is presumably correct for core samples of intact crystalline rock in a laboratory. Deformability of a fractured rock mass has two parts: that due to the rock matrix is $1/E$, where E = Young's modulus; normal deformability of a set of parallel fractures is $1/K_n d$, where K_n is normal stiffness and d is the average spacing between fractures. Snow (1968a) found that in one near-surface, fractured, weathered rock mass of 100 m dimensions, deformation resulted largely from the closure of fractures upon increase of effective stresses (drawdown by a well). The ratio of rock to fracture moduli, $E/K_n d$, was about 300. However, laboratory-derived measures of normal stiffness for unweathered fractures or saw cuts, together with field measures of spacing, suggested that for deep, hard, unweathered rock masses, fractures at high stress levels contribute a minute portion of the compression ($E/K_n d = 2 \times 10^{-3}$ to 2×10^{-2}) (Snow, 1972). Field deformation data are needed for larger and deeper unweathered masses that include faults and shears. In any event, the consolidation effects on pore pressures are confined to the immediate vicinity of a reservoir, whereas perfect fluid coupling, high hydraulic diffusivity, and extensive boundary-potential changes (water table) are responsible for more widespread and therefore more significant regional pore-pressure increases when a reservoir fills.

An unrealistic assumption is that the reservoir effect is applied solely at the reservoir bottom, so that the water table is unaffected elsewhere. Precipitation, infiltration, and continuity of discharge toward a filling reservoir require that the water table rise in the banks and piezometric levels increase below, extending the reservoir's effect by diminishing effective stresses over a broad region compared to the "bulb of pressure" under a (solid) reservoir.

An extreme case is that of a reservoir in karstic terrain: the water table is nearly planar, coincident with streams that appear and disappear along their courses. If a reservoir is successfully filled in karstic terrain, the regional water table will rise nearly as much as the reservoir. Kremasta Reservoir, Greece (Snow, 1972), produced a widespread rise recorded in observation wells as far away as 7 km; spring flows increased as far away as 13 km. There are also permeable volcanic terrains whose flat water tables do not reflect topography. The filling of Teton Reservoir, Idaho, produced a widespread rise of the water table in fractured tuffs (Independent Panel, 1976).

To evaluate the effects of reservoirs impounded on sedimentary terrains or certain volcanic terrains, basin hydraulics and circulations need to be analyzed. These are governed largely by bedding, faults, and topography. Fault zones in underlying metamorphic and granitic basement rocks respond to the widespread potential changes at the lower limits of the basin. Basin circulations are governed largely by the distribution of hydraulic potential, $\phi = p/\gamma + z$, where p is pressure, $\gamma = \rho g$ is the unit weight of water, and z is elevation.

Hydrogeologic thinking has been dominated by the erroneous notion that the basement is impermeable. In Denver's Rocky Mountain Arsenal Well, the first 60 m of fractured and weathered metamorphics proved a better reservoir than the overlying 3,602 m of Paleozoic to Cenozoic sediments (Scopel, 1964).

The basement underlying continental sediments has a preserved regolith of subhorizontal attitude which may be a significant aquifer compared to either the underlying unweathered rock or the overlying sediments. As will be shown, the movement and depth of hypocentral activity west of Koyna Reservoir, India, suggests propagation of a pressure front along the top of granite underlying the Deccan volcanics. In many cases, it is important to evaluate combinations of bedded aquifers and intersecting fault zones, as may be seen in the case of Kariba Reservoir, Zambia-Zimbabwe.

Hydraulic potential in the crystalline rocks under the Denver Basin is about 300 m subhydrostatic, possibly reflecting outlet potentials at the nearest Precambrian outcrops in Minnesota and Texas, hundreds of kilometers away. At least 1200 m of potential difference exists between fractured rock beneath the Denver Basin and the corresponding rocks that crop out immediately west of the thrust zone bounding the Rocky Mountain Front Range. The geometry of a fault system is clearly significant in determining the effective-stress changes at hypocentral depths in the basement, particularly when the contrast between faults and their surroundings render the faults prominent aquifers in some places and aquicludes in others.

An increase of pore pressure resulting from surface loading of a reservoir floor that propagates no fluid pressures represents an extreme (no coupling). The elasticity-consolidation-diffusion theory would then suggest (Withers, 1977) increased stability everywhere except directly beneath a filling reservoir. Greater coupling actually results in transient and greater increases of pore pressure farther from the reservoir. These realistic circumstances suggest that increases in instability should be expected to migrate away from the reservoir, where decreases of effective stress are possible. Any immediate pore-pressure increase due to surface loading speeds the attainment of a steady-state potential distribution. Full coupling can usually be expected, since ground water discharges at rivers and must continue to discharge unless the regional water table is lower than the reservoir. At hypocentral depths in fault zones communicating with reservoirs, the effects of the boundary changes are superimposed on the original potential distribution. If the saturated portion of a fault zone is bounded by a water table undergoing a regional change, effective stresses at locations distant from the elastically compressed reservoir region may decrease by a significant proportion of the reservoir pressure. In pervious karstic or volcanic terrains with low water tables, full coupling is predictable, and large effective-stress changes may occur (Bell and Nur, 1978). However, where clay aquicludes overlie alluvial or fractured aquifers--a common situation in desert basins of southwestern United States--coupling is not to be expected (Snow, 1975).

The position of a steep conductive fault relative to the reservoir can significantly influence the distribution of changes in hydraulic potential. Figure 1 represents a homogeneous isotropic terrain, or the plane of a near-vertical fault zone striking across-valley. Upon reservoir filling, the water table rises in the banks to remain tangent to the shore, and water continues to discharge. If the fault zone is homogeneous or depth-dependent with respect to hydraulic diffusivity (K/S_s, the ratio of hydraulic conductivity to specific storage), then throughout most of the zone of earthquakes, the region of deepest circulation must experience, on the average, about half the change of exit potential, that is, pressures corresponding to half the depth of the reservoir where the fault crosses.

If a fault lies along an inundated valley, coupling is complete and penetration is enhanced. River courses are commonly at least partly controlled by erodible faults. With a valley fault originally saturated to river level, potential change upon submergence is equal to the depth. As the submerged fault length is often great compared to the hypocentral depth, the change of potential there may approach the depth of the overlying reservoir. At Kariba, most epicenters were in the deepest basin, where not only the greatest surface load was applied, but, perhaps more significantly, where

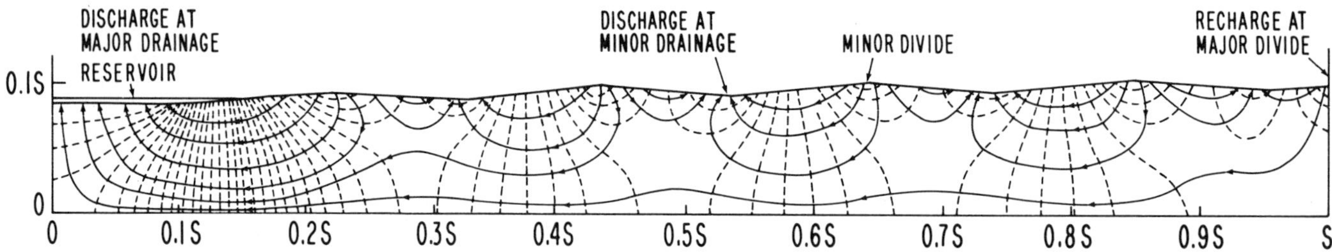

Figure 1. Regional flow net; vertical section in homogeneous material from major divide to major drainage. Dashed lines equipotentials, solid lines flow lines, S scale of drainage pattern of rivers. Depth of flow is assumed to be 0.1S (Freeze and Witherspoon, 1967).

major longitudinal faults (Deka fault and others parallel to it) were subject to the maximum potential change. Where a dam occupies a faulted valley, a sharp discontinuity of boundary potential change results in steep gradients and underseepage. Symmetry requires that the potential change beneath the dam be about half the maximum reservoir depth. A narrow reservoir is associated with total stress changes that diminish rapidly with depth, especially near load boundaries, such as the dam. Because pore pressures are more widely distributed, the region near the lower end of a reservoir will undergo diminishing effective stress upon filling, as exemplified by Koyna.

Faults parallel to a reservoir but not intersecting it are subjected to a range of effective-stress changes, depending upon position within the compressive domain under the surface load and upon pore-pressure conducting features, such as cross faults continuous with the reservoir. A system of intersecting longitudinal, transverse, or diagonal faults may produce conductivity approaching isotropy in the transverse plane (Snow, 1969). Change of potential distribution then resembles the isotropic case of Figure 1.

The prediction of the distribution of effective-stress changes clearly requires that hydrogeologic knowledge be obtained for each reservoir. It requires more than casual study of the region, with attention to hydrologic contrasts between lithologic units, to the relation between fault patterns and to the reservoir, and to whether different fracture sets act as aquifers or aquicludes. Boundary conditions and the changes they have undergone in the past guide judgment of future effects of reservoir filling.

Perspective on the hydrogeology of an area may be derived from its history. It becomes clear that hydraulic load changes are not the only ones to have produced an effective stress history, but also that erosion of terrain, alluviation of valleys, fault displacements, and epeirogeny may have in the past produced effective stress changes of magnitude greater than reservoir changes. Some guides to numerical translation of surface conditions to subsurface changes of effective stress were developed in a prior publication (Snow, 1972). Subsequent application to six reservoirs has shown the need to expand the scope of generalizations, introducing idealized effects of isostatic crustal uplift and subsidence resulting from erosion and deposition with concurrent cooling and heating, much as suggested by Haxby and Turcotte (1976).

STRESS CHANGES DUE TO RESERVOIRS AND OTHER LOADS

Because reservoirs produce maximum changes of only a few bars, they must be imposed on areas already close to failure in order to be seismogenic. The combinations of principal stress changes leading toward failure in different tectonic environments are best envisioned with the help of Mohr's representation (Figs. 2, 3, and 8), where by failure is indicated by any stress circle tangent to the failure envelope of slope ϕ. The effective-stress friction angle for faults may vary from 25°

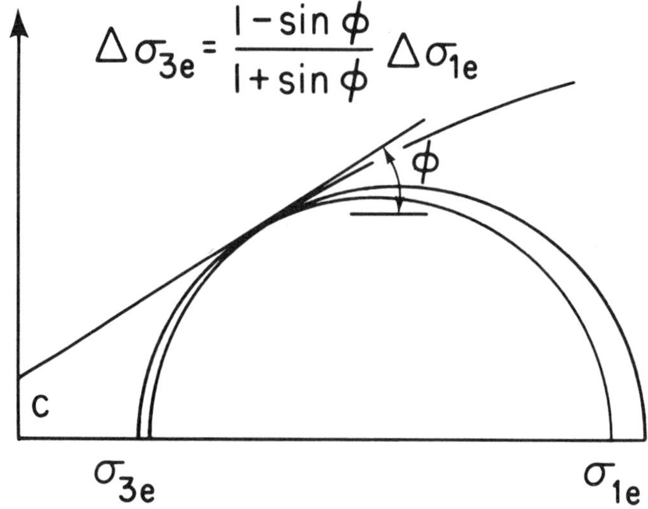

Figure 2. Mohr's circles of effective stresses are tangent to failure envelope of slope ϕ and lie below envelope for stable stress conditions. For stability to be maintained when stresses are critically close to failure, principal stress changes of the same sign must be

$$\frac{\Delta\sigma 3}{\Delta\sigma 1} > \frac{1 - \sin\phi}{1 + \sin\phi}, \text{ if positive,}$$

or

$$\frac{\Delta\sigma_3}{\Delta\sigma_3} < \frac{1 - \sin\phi}{1 + \sin\phi}, \text{ if negative,}$$

such that the circle moves away from the envelope.

to 45°, with lower values for gouge and higher values for more granular or brecciated materials or with more angular asperities. Established faults are assumed to have effective-stress friction angles close to 30°, such that the criterion of induced failure is about one-third. Absolute stresses are unknown but increase with depth. Therefore, only cumulative changes of stress are plotted, relative to a failure envelope of unknown position. The origin of shear and compressive stress coordinates may be assumed equal to some original (tectonic) state of stress. Because stresses may be noncritical initially, Withers (1977, p. 212) measured relative stability for different stresses by the distance between the change-of-effective-stress circle and an assumed failure envelope (Fig. 3). For any loading history before or after reservoir filling, a sequence of circles describes a locus of closest points of approach, indicating by proximity to the envelope the states of relative instability.

If a reservoir were filled instantaneously, surface loadings would be imposed normal to the reservoir bottom equal to the depth times the unit fluid weight. Compressive total stresses develop instantaneously in the underlying and surrounding rock. Gough and Gough (1970a) and Withers and Nyland (1976) have computed stresses and elastic deflections for such impoundments as Lake Kariba. Withers (1977) adopted material properties for a porous, permeable, homogeneous-isotropic half-space and computed the distribution of pore pressures induced by the surface loads and consolidation of saturated rock. He first assumed noncoupling by the reservoir (that is, a solid lake, like ice, but initially with atmospheric pore pressure at the contact). Pore pressures produced by the compression diffuse away, so that effective stress changes approach total stress changes. He also computed the fully coupled effect (with no change of adjacent water table), but preferred to compute reservoir effects on the assumption that most of the changes are dissipated close to the reservoir (0.25 coupling). In reality, coupling is complete because

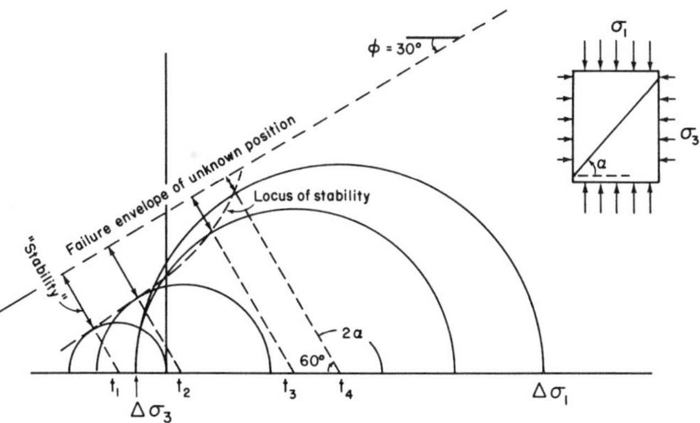

Figure 3. Mohr's circles of cumulative effective-stress changes $\Delta\sigma_1$ and $\Delta\sigma_2$, centered at $(\Delta\sigma_1 + \Delta\sigma_3)/2$. Relative distance from a hypothetical failure envelope defines stability. The locus of closest point of approach of these circles depicts the stability history (Withers, 1977).

reservoirs occupy valleys, which are areas of ground-water discharge; many, such as Kariba and Koyna, have hot springs indicating deep meteoric circulation. Furthermore, the water table is changed in the reservoir banks, in some cases to remote distances from them. Any initial pore pressure produced by consolidation under the reservoir only speeds adjustment of pressures for the entire flow region. Whereas it would be of interest to compute the transient hydraulics for changing reservoir and water table boundaries incorporating excess pressures due to the surface loads, the hydrogeologic implications of aquifers, faults, barriers, and boundaries will have more far-reaching impacts on the steady-state distribution of effective-stress changes than does the consolidation, which has a temporary effect, mainly near the reservoir.

Steady-state conditions for various loadings are emphasized in this paper, and only qualitative effects of transient conditions are mentioned. Reservoir geometries lie between the extremes of an infinite reservoir of constant depth and a point reservoir of negligible volume, like a well. The latter may raise the water table locally, or in the limit, everywhere by the same amount. Simpson (1976, p. 144-146) discusses effects of a finite reservoir, which may be to propagate failure upon filling.

Material properties of fractured porous media are embodied in the modulus of the solids, E; the normal stiffness, K_n; spacing, d; and aperture, 2b, of fractures. In the defining sketch (Fig. 4),

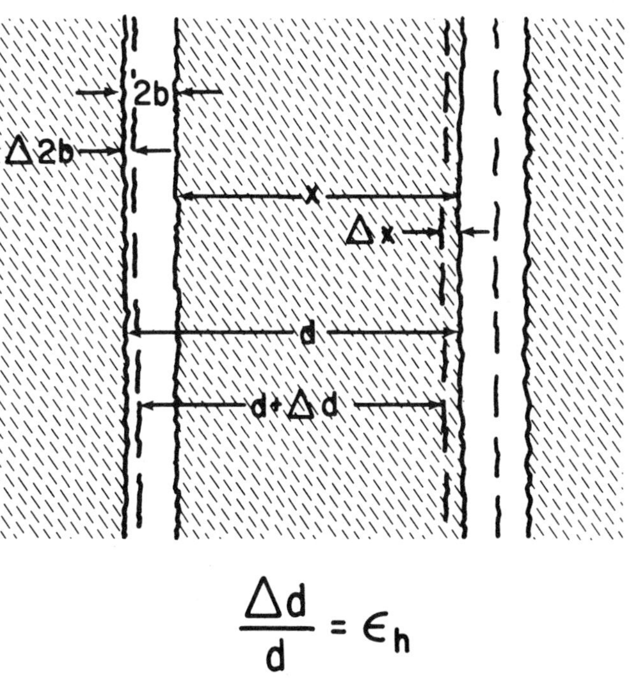

$$\frac{\Delta d}{d} = \epsilon_h$$

$$\frac{\Delta x}{x} = \frac{\Delta x}{d} = \epsilon_x$$

Figure 4. Deformation of a rock mass with deformable vertical fractures.

strain of the mass normal to a set of parallel fractures is:

$$\varepsilon_h = \frac{\Delta d}{d},$$

strain of the solids is

$$\varepsilon_x = \frac{\Delta x}{x} \cong \frac{\Delta x}{d}$$

and

$$\Delta d = \Delta x + \Delta 2b.$$

For plane-strain conditions, horizontal strain vanishes,

$$\varepsilon_h = \frac{\Delta x + \Delta 2b}{d} = 0,$$

so that upon increase of effective stress, as by a drop of pore pressure, the solid blocks expand as the deformable fractures close:

$$\Delta x = -\Delta 2b = \frac{1}{K_n} \Delta\sigma_{ex} \quad \text{and} \quad \varepsilon_x = \frac{1}{K_n d} \Delta\sigma_{ex}.$$

Deformability is linear in a limited range of effective stresses, but greater stiffness values apply at increasing loads and depths. With compression and extension positive, Hooke's Law for total stresses is:

$$\varepsilon_x = -\frac{1}{E}\Delta\sigma_{tx} + \frac{\nu}{E}\Delta\sigma_{ty} + \frac{\nu}{E}\Delta\sigma_{tz}, \qquad (1)$$

where σ_{tx}, σ_{ty}, σ_{tz} are total stresses in the x, y, z directions and ν is Poisson's ratio. For the solid blocks, if symmetric in x and z,

$$\varepsilon_x = -\frac{1}{E}[(1 - \nu)\Delta\sigma_{tx} - \nu\Delta\sigma_{ty}].$$

The steady-state condition after filling an infinite reservoir (Fig. 5) is such that the total vertical load change and pore pressure change due to the unit weight and height of water is $\Delta\sigma_{ty} = \gamma h$ and the effective vertical change $\Delta\sigma_{ey} = 0$. The total horizontal stress may be partitioned into effective stress and fluid pressure:

$$\Delta\sigma_{tx} = \Delta\sigma_{ex} + \gamma h.$$

Thus

$$\varepsilon_x = -\frac{1}{E}[(1 - \nu)(\Delta\sigma_{ex} + \gamma h) - \nu\gamma h] = \frac{1}{K_n d}\Delta\sigma_{ex},$$

$$\Delta\sigma_{ex} = \frac{2\nu - 1}{\frac{E}{K_n d} + 1 - \nu}\gamma h.$$

The same result is obtained in two steps: the elastic response of an infinite uncoupled reservoir, $[\nu/(1 - \nu)]\gamma h$, minus the steady-state pore-pressure change, γh. Field evidence showed (Snow, 1968a) that weathered, near-surface, fractured metamorphic rock has $K_n d \cong 10^3$ bars and rock has $E \cong 10^5$ bars. Therefore, an infinite reservoir produces no steady-state effective-stress changes in the regolith, because the fractures are deformable compared to the rock. A rock cut by numerous gouge-filled faults might behave similarly. Field evidence is offered later to suggest that a high value of Poisson's ratio, close to $\nu = 0.35$, is appropriate.

At depth, where faults may be rare, fractures fresh, and loads high, fault deformability may be akin to laboratory breaks and saw cuts of stiffness $K_n \cong 10^5$ bars/cm, and spacing is of the order of 10^2 cm. Therefore, $K_n d \cong 10^7$ and $E/K_n d$ is small.

If the water table is raised uniformly in fractured rock, as in Figure 6, then $\Delta\sigma_{ty} = 0$, since porosity is nil (Snow, 1968b). From equation 1,

$$\varepsilon_x = -\frac{1}{E}(1 - \nu)(\Delta\sigma_{ex} + \gamma h) = \frac{1}{K_n d}\Delta\sigma_{ex},$$

$$\Delta\sigma_{ex} = \frac{\nu - 1}{\frac{E}{K_n d} + 1 - \nu}\gamma h.$$

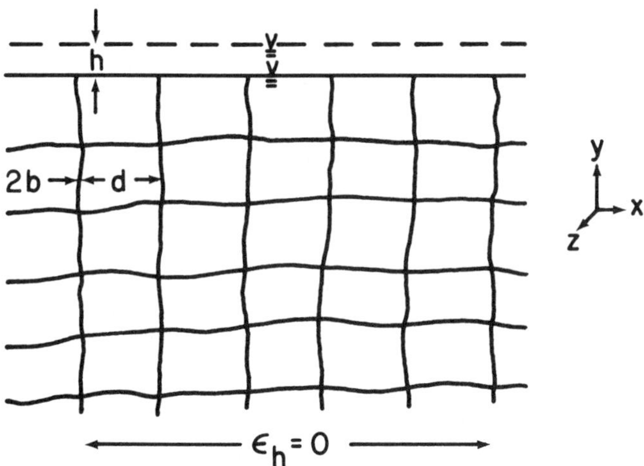

Figure 5. Vertical section of a fractured rock mass beneath an infinite reservoir.

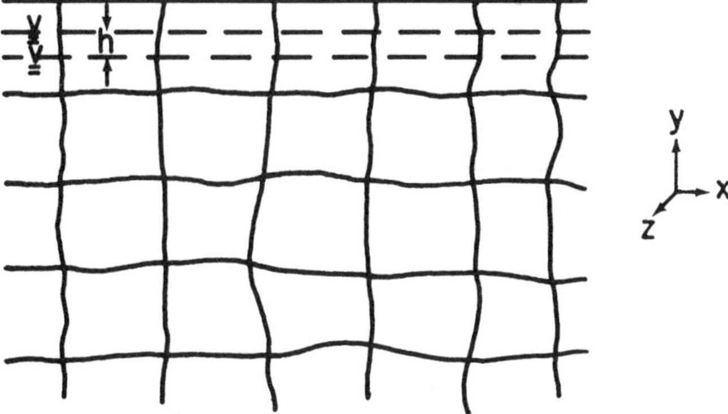

Figure 6. Fractured rock mass subject to change of water table.

Accordingly, when the water table is raised, then near the surface or anywhere that $E/K_n d$ is large, there is no change of effective horizontal stress, only a vertical one: $\Delta\sigma_{ex} = 0$ and $\Delta\sigma_{tx} = \gamma h$; $\Delta\sigma_{ey} = -\gamma h$. At depth or where $E/K_n d$ is small, $\Delta\sigma_{ex} = -\gamma h$.

The transient loading, before a pore-pressure front arrives at hypocentral depths, is the dead weight of the infinite reservoir,

$$\Delta\sigma_{ex} = \frac{\nu}{1-\nu}\gamma h,$$

whereas no effect is felt before the transient arrives following a rise of water table, because the total stress change is nil.

Stress changes produced by alluviation or erosion can be deduced similarly. Deposition of alluvium by running water is accompanied by a rising water table,

$$\Delta\sigma_{ty} = \gamma_{sa} h.$$

The unit weight of saturated alluvium γ_{sa} with void ratio 0.5 is about 2.1γ, or for saturated rock, such as lava, $\gamma_{sa} \cong 2.6\gamma$. Changes of vertical effective stress are $\Delta\sigma_{ey} = 1.1\gamma h$ for alluvium, or $1.6\gamma h$ for rock. From equation 1,

$$\Delta\sigma_{ex} = \frac{3.1\nu - 1}{\frac{E}{K_n d} + 1 - \nu}\gamma h \quad \text{or} \quad \frac{3.6\nu - 1}{\frac{E}{K_n d} + 1 - \nu}\gamma h,$$

respectively. If the basement rocks are weathered, $\Delta\sigma_{ex} = 0$; if fresh, $\Delta\sigma_{ex} = 0.13\gamma h$ for alluvium or $\Delta\sigma_{ex} = 0.40\gamma h$ for rock, small changes that depend on $K_n d$ and ν.

Erosion, accompanied by a fall of the water table, will produce the same magnitudes of change as alluviation. Usually, planation processes remove lithified, dense rock rather than alluvium. As Haxby and Turcotte (1976) have suggested, erosion is attended by isostatic uplift, and if it is radial to the earth, circumferential extension decreases horizontal stresses. Figure 7 shows a crustal block subjected to erosion e, measured relative to the strata. Unloading of the Denver Basin, for example, can be computed by projection of the missing Ogallala Formation. Assuming a reasonable density contrast, isostatic uplift is about 82% of the erosion. In other cases, uplift is directly measurable from elevations of successive peneplains, pediments, stream terraces or beaches with respect to sea level or graded streams. Beneath the Atlantic coastal plain are formation contacts that were once subhorizontal but are now tilted seaward at slopes that are steeper for older surfaces than younger. Projection landward measures successive Piedmont uplifts (Snow, 1977). If changes of topographic elevation can be assessed, rather than the positions of stratigraphic markers, the uplift was about 5.6 times the elevation change. Similarly, Fig. 7 shows that isostatic subsidence is about 62% of the depth of alluviation. The changes in total horizontal stress due to the horizontal extension or contraction caused by

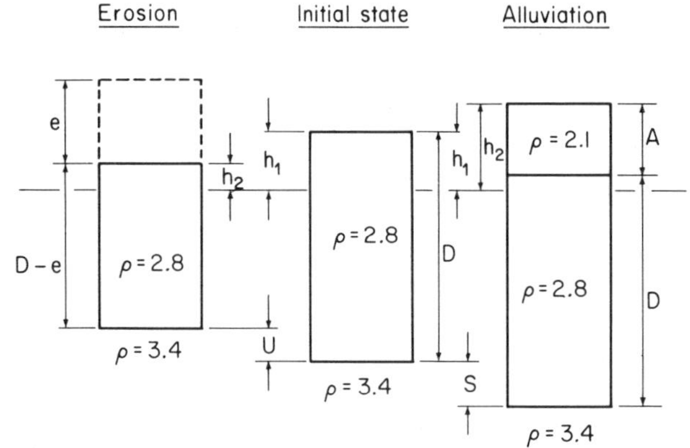

Let D = thickness of earth's crust, of density = 2.8, buoyed by a mantle of density = 3.4. Let erosion = e, alluviation = A, producing uplift = U or subsidence = S. Equating the mass of the columns, before and after geomorphic change, we have for erosion

$$2.8D = 2.8(D - e) + 3.4U,$$
$$U = 0.82e,$$

and for alluviation

$$2.8D + 3.4S = 2.8D + 2.1A,$$
$$S = 0.62A.$$

Changes of elevation produced are

$$h_2 - h_1 = U - e = -0.22U$$

and

$$h_2 - h_1 = A - S = 0.61S.$$

Both the vertical and circumferential strains produced by vertical (radial) uplift are U/a, where a is the radius of the earth. Applying Hooke's Law for uplift alone, $\Delta\sigma_y = 0$ and $\Delta\sigma_x = \Delta\sigma_z$, gives

$$\epsilon_x = \frac{\nu - 1}{E}\Delta\sigma_x,$$

$$\Delta\sigma_x = -\frac{E}{1-\nu}\frac{U}{a}$$
$$= -\frac{10^5}{0.65}\frac{U}{6.38 \times 10^6}$$
$$= -2.4 \times 10^{-2} U$$
$$= 2.4 \times 10^{-2} S.$$

U e a

Figure 7. Alluviation and erosion; isostatic subsidence and uplift and accompanying stress changes.

uplift or subsidence, respectively, are shown to be about 2.4×10^{-2} bars per metre.

As erosion lowers the ground surface, warm rock is uplifted, exposed and ultimately cooled; and because the process is slow enough to be essentially steady-state, the thermal profile is retained. Cooling of a half-space by a uniform temperature change ΔT produces a stress change computed in two steps: unrestrained strains are $\epsilon_x = -\epsilon_y = \alpha\Delta T$, according to the expansion coefficient α; applying

Hooke's Law to restore the strains gives

$$\varepsilon_x = -\frac{1}{E}\Delta\sigma_x + \frac{\nu}{E}\Delta\sigma_y + \frac{\nu}{E}\Delta\sigma_z,$$

but $\Delta\sigma_x = \Delta\sigma_z$ and $\Delta\sigma_y = 0$. The sign of restoring strain is opposite to the thermal strain, so that

$$-\alpha\Delta T = \frac{\nu - 1}{E}\Delta\sigma_x.$$

Taking $\alpha = 8 \times 10^{-6}/°C$, $E = 10^5$ bars, and $\nu = 0.35$ (for reasons explained later), then $\Delta\sigma_x =$ 1.23 bars/°C. If the normal geothermal gradient is -1°C/40 m, then $\Delta\sigma_x = -3.08 \times 10^{-2}$ b/m of uplift, and vice versa. As an inspection of Mohr diagrams will show, erosion of mountain masses perpetuates faulting there, in part because uplift and cooling decrease effective stresses; conversely, basins and fault troughs receiving alluvium from nearby mountains are stabilized, partly because subsidence and heating increase effective stresses.

Table 1 summarizes effective stress changes due to a variety of causes. Clearly, other causes also act, such as relief by faulting, formation of crustal structures or changes in mantle convection.

TABLE 1. COEFFICIENTS OF EFFECTIVE STRESS CHANGE, $\Delta\sigma_e = Ch$

Causes	Coefficient C (bars/m): Vertical change ($\Delta\sigma_{ey}$)	Horizontal change ($\Delta\sigma_{ex}$)
Fill an Infinite Reservoir		
transient (no hydraulic coupling)	0.10	0.054
steady (complete coupling)	0	-0.046
subsidence accompanying filling	0	0.007
Raising Water Table		
through porous materials (for example, tuff, alluvium of 20% porosity)		
transient	0.020	0.011
steady	-0.080	-0.089
through fractured rocks		
transient	0	0
steady	-0.10	-0.10
Alluviation (with equal water table rise) by porous materials (for example, gravel, silt, ash)	0.11	0.013
attendant subsidence	0	0.016
attendant heating	0	0.019
by rock (for example, lava)	0.16	0.040
attendant subsidence	0	0.021
attendant heating	0	0.023

Notes: (1) h>0 for filling a reservoir, raising the water table, alluviation, subsidence, heating, and vice versa.
(2) $\nu = 0.35$; for C in psi/ft, multiply by 4.30.
(3) stress changes of seismic interest are in fractured rocks: tabulated coefficients are for fresh rock at depth and high stress ($E \ll K_n d$). For weathered or highly faulted rock ($E \gg K_n d$), $\Delta\sigma_{ex} = 0$ for all cases.
(4) subsidence or uplift, measured independently, or changes of elevation, can be translated into depths of alluviation or erosion, for application of above coefficients.

TECTONIC ENVIRONMENTS

Figure 8 illustrates the stress changes leading to failure in normal, wrench, and thrust-fault environments. In normal-fault regions (σ_1 vertical, $\Delta\sigma_2 = \Delta\sigma_3$), filling of an infinite reservoir that is h metres deep produces transient compressive loads

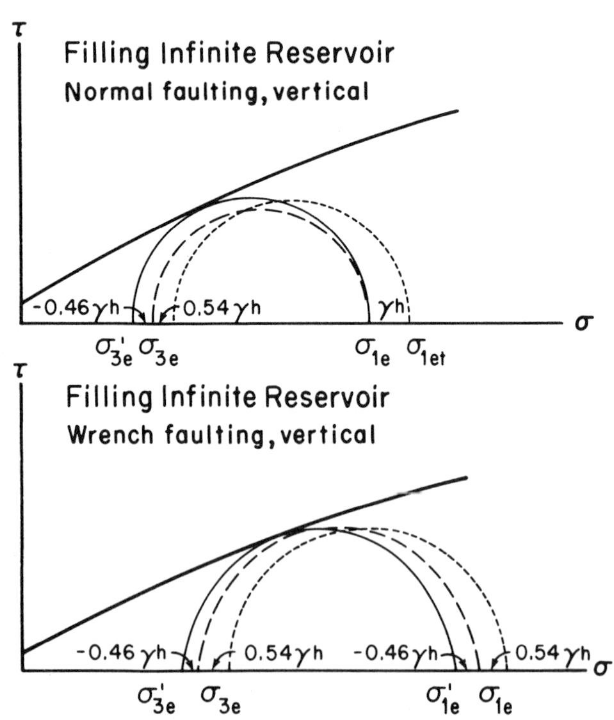

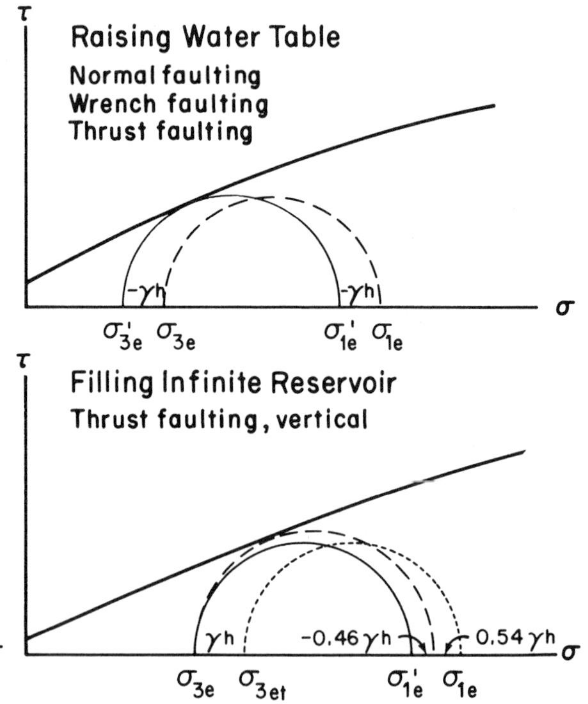

Figure 8. Effective stress changes in unweathered, hard rocks with wide-spaced fractures. Initial circle shown dashed, transient circle dotted, final circle solid.

$\Delta\sigma_1 = 0.10h$ and $\Delta\sigma_3 = 0.054h$ bars, but failure is not likely to be induced, because the stress circle moves away from the 30° envelope: $\Delta\sigma_3/\Delta\sigma_1 = 0.54 > 1/3$. Conversely, emptying a wide reservoir could lead to failure. Low tide conditions have been correlated with times of high coastal seismicity (Berg, 1967). Klein (1976) has shown good correlation of induced quakes with tidal loadings.

When pore pressures from an infinite reservoir penetrate to hypocentral depths, the steady-state effective stresses both drop by 0.10h, shifting the Mohr envelope left to possible failure. Conversely, the long-term effect of lowering a wide reservoir is to stabilize a normal-fault region.

Subsidence due to a reservoir's weight may increase σ_3 by 0.007h bars, tending to stabilize. Reservoirs are probably too small in dimensions to benefit from isostatic subsidence, since the observed subsidences at Lake Mead (18 cm, Carder, 1945) and at Lake Kariba (22 cm, Gough and Gough, 1970a) are consistent with elastic compressions under total-stress changes. Conceivably, induced seismicity may ultimately die out by subsidence, stress flow, or failure.

Raising the water table, as may occur upon filling a narrow canyon reservoir, by injection at one or more wells, by irrigation projects, or by climatic change will have minor transient effects at depth, but the steady-state effect is a reduction of all effective stresses. Movement of stress circles of essentially unchanged diameter toward the failure envelope is unfavorable in all tectonic environments, normal, wrench, or thrust-fault. Correlation of seismicity with river stage (McGinnis, 1963) in the Mississippi Valley is consistent with this finding, because water tables rise and fall with the rivers and with infiltration. Conversely, a lowered water table stabilizes. Huntley (1976) envisioned the current seismic quiescence of the north end of San Luis Valley, Colorado (the extremity of the Rio Grande graben), where young scarps exist, to be due to piezometric decline caused by irrigation pumpage.

For realistic material constants, it is shown in Figure 8 that failure of critically stressed normal-fault regions is fostered by decreasing effective stresses, not by increasing them. For example, if an area were buried in lava but the water table remained unchanged, $\Delta\sigma_1$ and $\Delta\sigma_3 > 0$ and elastically, $\Delta\sigma_3/\Delta\sigma_1 = \nu/(1 - \nu)$. In Figure 2 it was shown that if

$$\frac{\Delta\sigma_3}{\Delta\sigma_1} = \frac{1 - \sin\phi}{1 + \sin\phi},$$

the stress circle slides tangent to the envelope. In Figure 9, the line

$$\frac{\nu}{1 - \nu} = \frac{1 - \sin\phi}{1 + \sin\phi}$$

separates the combinations of ν and ϕ, satisfying the criteria of failure expressed in Figure 2. Appropriate properties for hard, fractured rock masses lie in the upper field, predicting normal-fault failure only for decreasing effective stresses.

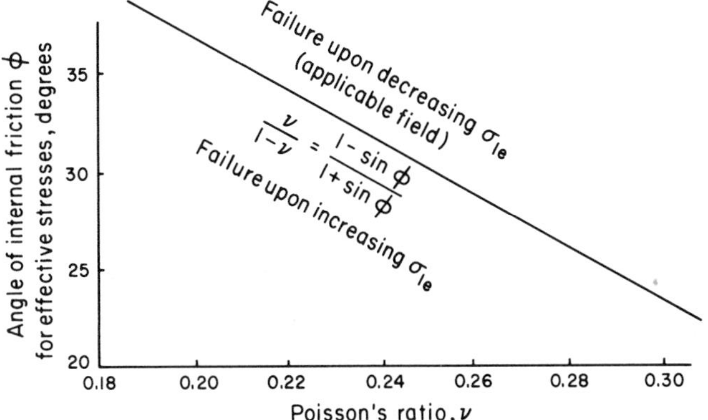

Figure 9. Combinations of friction angle and Poisson's ratio producing failure upon increase or decrease of effective stress.

If Gough and Gough (1976, p. 213) are correct in interpreting the initial seismicity at Kariba to have been triggered by changes in total (uncoupled) stresses, the fact that the failure took place upon increasing stresses would require an unusually low Poisson's ratio, $\nu < 0.25$, or an unexpectedly low friction angle, $\phi < 30°$.

The seismically active portion of the Basin and Range province is its western extremity, where high, geomorphically youthful, fault-bounded mountain masses are separated by deep, narrow, alluviated valleys. In contrast, the quiescent central and eastern portions feature old, narrow, corroded mountains separated by extensive alluvium-filled basins burying the inactive border faults. For alluviation to arrest normal faulting, Poisson's ratio must be relatively high. For each metre of saturated alluvium, $\Delta\sigma_1 = 0.11$ bars and $\Delta\sigma_3 = [(3.1\nu - 1)/(1 - \nu)](0.1) + 0.015 + 0.018$, the latter terms being subsidence and heating effects. For stability in the presence of a variety of faults of friction angle $\phi = 30°$, $\Delta\sigma_3/\Delta\sigma_1 \geq 0.33$ (see Fig. 2), which requires $\nu \geq -0.33$, according to the above relation. But if subsidence is only partially effective on the scale of intermountain basins ν must be slightly greater, close to 0.35.

It is appropriate that a young, narrow graben, like Death Valley, initially shows active faulting, because isostatic subsidence is slow until the width of the alluviating valley approaches the crustal thickness. Thereafter, continued alluviation and subsidence may shift epicenters to faults in the adjacent mountain block.

Conversely, where h metres of rock erosion in a normal-fault region takes place, stability develops initially as the stress circle shrinks, $\Delta\sigma_1 = -0.16h$ and $\Delta\sigma_3 = -0.04h$. When uplift occurs to attain isostatic balance, the horizontal stress falls, $\Delta\sigma_3 = -0.021h$, and by cooling, $\Delta\sigma_3 = -0.023h$, for a cumulative $\Delta\sigma_3 = -0.084h$. Thus $\Delta\sigma_3/\Delta\sigma_1 = -0.084/-0.16 = 0.53 > 1/3$, which is unstable for decreasing stresses.

A normal-fault region that can be demonstrated to be subsiding in absence of alluviation should be stabilizing by increases of σ_3; the converse would hold for uplift. An increase of elevation during alluviation connotes subsidence and stabilization, whereas a decrease of elevation during erosion is consistent with uplift and instability in a normal-fault environment. Thus a storage reservoir on a young mountain block in the Basin and Range province could be seismogenic, whereas one in a nearby intermontane valley is less likely to be.

In *wrench-fault regions*, σ_2 is vertical and $\Delta\sigma_1 = \Delta\sigma_3$ in response to vertical load changes. Filling an infinite reservoir has a temporary stabilizing effect, but upon arrival of the pore-pressure change, the stress circle (Fig. 8) shifts toward the failure envelope, fostering failure. Conversely, a sudden lowering of a stabilized lake may trigger failure, but eventually increases the stability. Subsidence beneath a reservoir h meters deep may produce $\Delta\sigma_1 = \Delta\sigma_3 = 0.007h$, helping to stabilize the site.

The steady-state effect of a general change in the water table is to produce instability while rising and stability while falling.

Alluviation in wrench-fault terrain has a stabilizing effect, since both principal stresses σ_1 and σ_3 increase equally. In cases of widespread filling--for instance, by basalt flows of the Deccan or the Columbia Plateau--subsidence and heating both augment the stresses by $\Delta\sigma_1 = \Delta\sigma_3 = 0.021$ and 0.023, respectively, so that filling tends to stabilize these regions. Erosion produces destabilizing effects.

Regions of coincident wrench and normal faulting are known, such as in Dixie Valley, Nevada (Shawe, 1965), where σ_3 is west-northwest but $\sigma_1 = \sigma_2$. Alluviation without subsidence first produces $\Delta\sigma_v = 0.11h$ and $\Delta\sigma_h = 0.013h$, stimulating normal faulting, since $\Delta\sigma_3/\Delta\sigma_1 = .12 < 1/3$, but not wrench faulting, whose stability is enhanced by equal increases of σ_1 and σ_3, and also by the effect of increased confinement, $\Delta\sigma_2 = 0.11h$ on a three-dimensional Mohr envelope. As a young graben matures, normal faulting may replace wrench faulting, then as subsidence and heating occur, both are stablized ($\Delta\sigma_3/\Delta\sigma_1 = 0.048/0.11 = 0.44 > 1/3$).

The Denver Basin, a wrench-fault province (Healy and others, 1968) was subjected to about 600 m of post-Pliocene erosion without a fall of basement piezometric level (being subhydrostatic, controlled by exit levels elsewhere). Uplift and cooling further diminished stresses:

$$\Delta\sigma_1 = \Delta\sigma_3 = -0.125h - 0.018h - 0.020h$$

$$= -0.163h \cong -98 \text{ bars.}$$

The Laramide revolution has been kept active by progressive movement of the stress circles toward instability. Natural quakes occurred, such as the November 7, 1882, event, and the region was ripe for induced seismicity when basement pore pressures were augmented by injection at the Rocky Mountain Arsenal Well (up to 90 bars at the well; Healy and others, 1968).

In *thrust-fault regions*, σ_3 is vertical. The filling of an infinite reservoir produces a transient, $\Delta\sigma_1 = 0.054h$ and $\Delta\sigma_3 = 0.10h$, stably shrinking and shifting the circle. In Pakistan, Tarbella Reservoir's filling was marked by decreased seismicity in that part of the Baluchistan arc (Jabob and Ambruster, 1976). The steady state, whose cumulative effect is a change of $\Delta\sigma_1 = -0.046h$, is likewise stable, and any lake-produced subsidence, $\Delta\sigma_1 = 0.007h$, does not nullify it. Conversely, lowering a reservoir on thrust terrain could stimulate seismicity. The Prince William Sound (Good Friday Quake) thrust event of 1964 occurred at low tide (Berg, 1966).

Raising the water table may lead to instability in thrust regions, as elsewhere. A swarm of quakes around Delhi, India, was possibly related to a general water table rise in alluvium on a quartzite basement following establishment of irrigation practice (P.C. Chatterji, oral commun., 1968). Conversely, drainage may stabilize a thrust region, or any other.

When a thrust-fault area is alluviated by a thickness of h metres with a concurrent water-table rise, then $\Delta\sigma_1 = 0.013h$ and $\Delta\sigma_3 = 0.11h$ bars, so that a marked tendency to stabilize develops initially. As subsidence and heating occur, the horizontal stresses increase to a cumulative $\Delta\sigma_1 = 0.048h$, but always in a stable manner, because the vertical loading change $\Delta\sigma_3$ is greater. The same is true for more dense fills, which might approach $\Delta\sigma_1 = 0.084h$ and $\Delta\sigma_3 = 0.16$.

Conversely, erosion of rock decreases confinement, $\Delta\sigma_3 = -0.16h$, and $\Delta\sigma_1 = -0.04h$, driving thrusts toward failure. When uplift and cooling effects are added, the cumulative change is $\Delta\sigma_1 = -0.084h$, which also promotes instability.

If S metres of subsidence occurs without deposition, the induced stresses $\Delta\sigma_1 = 0.024S$ bars and $\Delta\sigma_3 = 0$ could lead to thrust-fault instability, and uplift to stability.

Table 2 summarizes the sense of changes predicted for each of the various events discussed, including those of geologic history recorded in landforms or deposits, as well as events imposed by man's works.

NORMAL FAULTING AT LAKE KARIBA, EASTERN AFRICA

Kariba Reservoir occupies 250 km of the Zambezi valley in eastern Africa. Pronounced seismicity followed filling in 1958. The relationships between faults and geomorphology are essential evidence for construction of a chronology of stresses in the region.

Geologic Setting

The geology of the Kariba Lake area has been described by earlier workers (Bond, 1953; Hitchon, 1958; Knill and Jones, 1965; Drysdall and Weller, 1966; Kirkpatrick and Robertson, 1968; Gough and Gough, 1970b). As seen in Figure 10, the lake occupies a northeast-trending structural and topographic trough, wherein bedrock is mostly the Permo-Carboniferous Karoo Series of sediments. The Zambezi River has partly exhumed the trough, locally exposing Precambrian metamorphics where the river has been superposed across horst blocks, as at Kariba damsite. In Zambia and Zimbabwe, northwest and southeast, respectively, from the northeast-flowing Zambezi River, coarse-grained

TABLE 2. STABLE (S) AND UNSTABLE (U) EVENTS FOR CRITICALLY STRESSED TECTONIC ENVIRONMENTS

Event	Region		
	Normal fault	Wrench fault	Thrust fault
Raise water table: canyon lake, injection wells, irrigation, increase precipitation	U	U	U
Lower water table	S	S	S
Fill infinite reservoir -			
Transient, high tide	S	S	S
Steady	U	U	S
Empty Infinite Reservoir -			
Transient, low tide	U	U	U
Steady	S	S	U
Subsidence due to filling	S	S	U
Rebound due to emptying	U	U	S
Alluviation with water table rise,	U	S	S
plus subsidence and heating	S	S	S
Erosion with water table fall,	S	U	U
plus uplift and cooling	U	U	U
Subsidence alone, without deposition	S	S	U
Uplift alone, without erosion	U	U	S

Note: Since some of the above isolated effects never act alone, a numerical summation is necessary to predict stability.

gneissic rocks with steep foliation strike northeast. At Kariba, the gorge exposes muscovite gneiss and quartz gneiss cut by Precambrian pegmatites and Jurassic dolerites.

A nonmetamorphic Precambrian series, considered part of the basement cover by most geologists (Pallister, 1971), was faulted during the pre-Karoo rifting, and may be buried beneath parts of the Karoo trough (Bond, 1953). Bond (written commun., 1972) calls it the Sijarira Series for exposures in the Sijarira horst block southwest of Lake Kariba. About 600 m thick, the series includes conglomerates at the base, overlain by quartzites with pale mauve, greenish, or chocolate shales, clay-pellet conglomerates, and current-bedded sandstones.

Faults and fractures of the pre-Karoo rift strike parallel to the Precambrian foliation. Numerous pre-Karoo graben formed up and down east Africa by processes apparently similar to the Tertiary rifting that formed the modern valleys. Dixey (1956), McKinlay (1965), and Bishop and Trendall (1967) considered the Karoo to have been deposited in schist-belt valleys, partly erosional, partly faulted. Younger Karoo beds overlap onto the basement, indicating considerable relief before deposition. Jurassic and Cretaceous faulting subsequently enhanced the basement irregularities, likewise following the persistent northeast grain.

Drysdall and Weller (1966) say that the Karoo beds were deposited in downwarped troughs that were precursors of the graben that coincided with the Tertiary rifts. Definite faulting did not begin until late in Karoo times. The sequence, up to 2,800 m thick, is entirely continental. The Dwyka Tillite and outwash gravels are found in some basal exposures; coarse colluvial conglomerates and sandstones are basal elsewhere. Above these are the Ecca coal measures and siltstones. These two units total at most 100 m, whereas the overlying Madumabisa Shale, containing lacustrine sandstone and siltstone, is 450 to 640 m thick. It contains some marly limestones and has calcareous cement throughout. Overlying the Madumabisa is the Escarpment Grit, 365 m of coarse sandstone, a prominent cliff former. It overlaps the trough-bounding faults, locally lying directly upon Precambrian rocks. It is unconformable on older, inward-tilted Karoo beds, suggesting contemporaneous subsidence. Next above is the Stormberg Series, 120 m of pebbly arkoses, mudstones, and basaltic lavas. This sequence of Karoo lithologies is distributed throughout several isolated basins in east Africa, in extensive basins of south Africa, and in southern Brazil, for the series was formed before the breakup of Gondwanaland.

The stratigraphy dates the breakup of Gondwanaland as mid-Jurassic to early Cretaceous. Jurassic sediments deposited near the east African coast were tilted and warped, since Cretaceous beds deposited upon them are discordant. The Gondwana erosion surface, cut during late Jurassic time across rocks of varied lithology, was buried by marine Cretaceous sediments.

Numerous northeast-trending normal faults of Jurassic and Cretaceous age dip 50° to 65° both southeast and northwest. These are well exposed in the benches of the Momba Colliery in Zambia and in the Wankie coal mines of southern Zimbabwe. In some other rifts, these faults have been reactivated during Miocene, Pliocene, and Pleistocene times (Pallister, 1971). No Tertiary faulting is evident in the mid-Zambezi trough, though the modern seismicity at Kariba is apparently related to this same fault system, as judged by (first-arrival) fault-plane solutions. The northwest flank of the asymmetric mid-Zambezi trough is an exhumed fault contact. The southeast flank, originally a monoclinal flexure, was broken during Cretaceous time to form graben and horst structures. Sijarira is a prominent horst east of Binga (Figs. 10 and 11). There is no evidence of post-Cretaceous faulting, though prominent faultline scarps have formed by exhumation.

Streams originating on the flat, widespread post-African surface fall 500 m to the Zambezi valley, a region of low Karoo hills where drainage is locally subsequent to the layered, tilted (up to 30°) red Karoo sandstones and shales, or to the strike faults. The escarpment zone on the Zambian side is deeply dissected, exposing Precambrian rocks. Its breadth (10 to 20 km) indicates long quiescence since latest fault movement, whereas on the northwest side of the Luangwa trough, northeast of

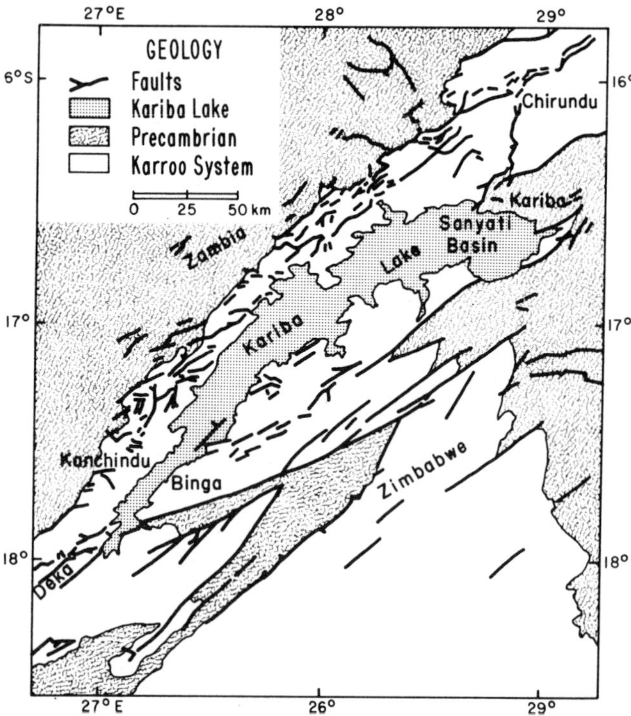

Figure 10. Geology and fault pattern in the environs of Lake Kariba (Gough and Gough, 1970b).

Kariba, the scarps show little retreat. Movement there may have only shortly preceded recent rifting along Lakes Nyasa, Tanganyika, and Mweru (Drysdall and Weller, 1966, p. 66).

Cretaceous continental sedimentation was also rift-controlled. Most of the Gondwana erosion surface was destroyed. Remnants of post-Gondwana valleys are now found only on the eastern highlands and on the Great Dyke. Uplift and a new post-Gondwana base level resulted in deep troughs carved and faulted along the earlier lines. The Zambezi River course was established at this time, consequent to rifts preceding the Cretaceous valley fills. Beds bearing dinosaur bones are unconformable on Karoo rocks in the mid-Zambezi rift, the Luangwa rift, and in the Livingstone area. Cretaceous beds are found at Kanchindu on the Zambian shore of Lake Kariba and on the south side of the Kariba gorge (Hitchon, 1958).

There was continued faulting during the Cretaceous in Malawi; wrench faults displace Karoo rift faults and dinosaur beds. The normal fault on the northwest flank of Luangwa Valley also displaces dinosaur beds. King (1962) noted that rifting lowered the Karoo beds along the mid-Zambezi trough, because Cretaceous beds filled it practically to the brim. No known Cretaceous continental deposits lie outside the rift valleys (Pallister, 1971).

From Cretaceous to Miocene time, erosion bevelled the vast, smooth African pediment surface across Precambrian to Cretaceous rocks. A deep laterite developed, and Zambezi tributaries flowed consequent to the surface.

In the Miocene, rifting resumed elsewhere in eastern Africa, but the mid-Zambezi faults were not reactivated. Vulcanism began in Kenya. Epeirogeny developed during Miocene time, uplifting the African surface and initiating new erosion cycles. The maximum uplift, about 600 m, occurred along a north-south axis through eastern Zimbabwe. The surface tilted down to the coast in Mozambique and down to the interior across Zimbabwe and Zambia into the Kalahari Basin.

Two post-Miocene pedimentation cycles passed through the Kariba area and upstream of Victoria Falls, forming post-African surfaces (1) and (2). Much of Zimbabwe displays surface (1), which also lies behind the escarpment north of the Zambezi, becoming extensive in southwest Zambia, where it is covered by Kalahari sands at about 1,000 m elevation. Southeast of Kariba Lake, surface (1) is found at 1,070 m on the Chizarira Range. Nearby, surface (2) is at 840 m. The latter tilts downstream to the northeast, falling about 60 m in 250 km (Bond, written commun., 1972).

Pliocene faulting occurred at the south end of the Tanganyika trough (Baker, 1971). The northeast boundary fault of Luangwa Valley may have been reactivated in the Pliocene. Mio-Pliocene normal faults in Kenya were both en echelon and stepped.

Post-African surface (3) was eroded in late Pliocene, possibly in response to tilting of the older surfaces, rejuvenating the Zambezi River system. The surface is at the lip of Victoria Falls, and forms an extensive plain believed equivalent to the 610-m surface of Bond (written commun., 1972) and the 550-m surface of Lester (written commun., 1972).

After surface (3) formed, a major stream rejuvenation caused 150 m or more of late Pliocene Zambezi River incision, producing prominent gorges where the river was superimposed on crystalline basement rocks. The gorges were filled with gravel and red Kalahari-type (dune) sands, then locally cemented by chalcedony. The Zambezi River was raised by alluviation above the 610 m level, as indicated by red sand deposits upon surface (3). The chalcedony is banked against the Binga fault and against the Sanyati fault near Chipeso Hot Spring (Fig. 11). Unfortunately, the deposit is useless for age-dating, since the mobile dune sands composing it have been oft-remobilized during the Tertiary into younger and younger surface covers and channel-fillings.

Another rejuvenation, possibly by tilting, lowered the rivers and developed, upon stillstand, the End-Tertiary surface of Bond (written commun., 1972), at elevation 480 m. Batoka Gorge (head of Lake Kariba) was eroded through gneisses, and the wide Sanyati Basin was opened up on Karoo beds.

Pleistocene rifting was very active: the east and west rift zones of Kenya, Tanganyika, Uganda, and Ethiopia developed most of their morphology then, but the mid-Zambezi trough remained dormant as far as geomorphology indicates. Bond (oral commun., 1972) noted faults cutting cemented Kalahari sands in the Wankie Game Preserve, south of Kariba Lake, but since the sands are of indefinite age, the faults cannot be reliably dated. Lester (oral commun., 1972) believes they are Quaternary faults. Along Lake Tanganyika, there is active faulting, there being several

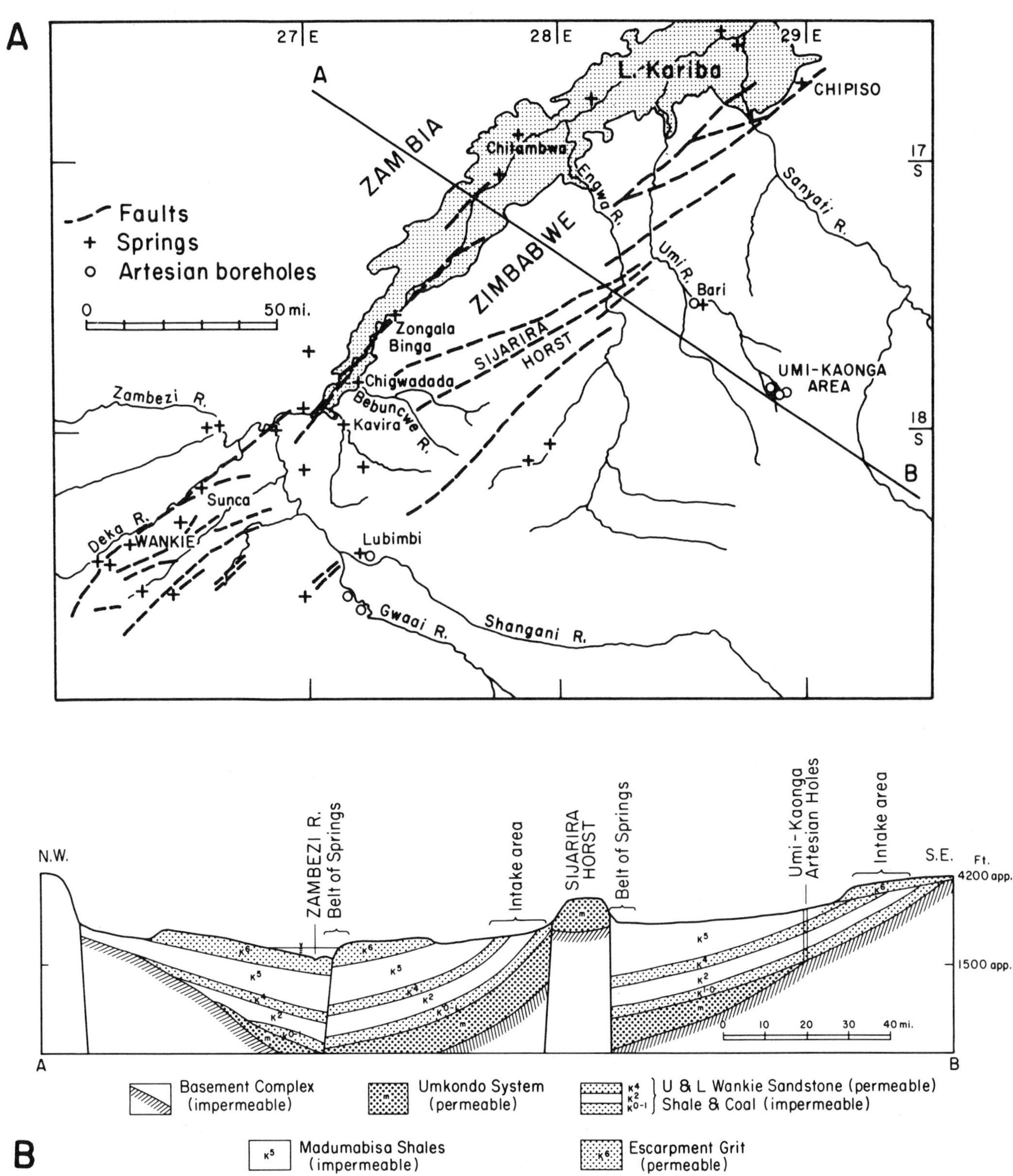

Figure 11. Map (A) and diagrammatic section (B) of the Zambezi Basin, showing artesian structures controlling hot springs (Bond, 1953).

metres of vertical and horizontal displacement of roadways (R. Mäckel, oral commun., 1972).

Broad continental warping affected much of eastern Africa during the Pleistocene. Parts of the Zambezi River were incised, other parts were abandoned, and huge swamps (such as Okavanga) formed. In the mid-Zambezi reach, a series of topographic steps formed in response to the epeirogeny, continuing to a historic (pre-Kariba reservoir) level of about 435 m. Climatic changes attending different glacial stages might also have influenced stream regimen. Data on sedimentation and water level from the great rift lakes would bear on this topic.

The principal evidence for modern, pre-impoundment tectonism is Bond's testimony (oral commun., 1972) that while sitting on alluvium about 25 km northeast of Binga, close to the Binga fault-line scarp, he and others in his company felt a quake. No stations recorded it, though Bond timed its occurrence. Archer and Allen (1969, p. 132), and Archer (oral commun., 1972) indicate that some prior events were recorded at Binga. None originated at the northeast end of the lake, where most of the induced seismicity developed. Mann (1967) says that the Deka fault, which extends from Binga southwesterly 330 km into Botswana, is active. Quakes also occur in the Okavanga Swamp (I. Gough, oral commun., 1973).

Archer (oral commun., 1972) mentioned minor seismic activities, remote from Lake Kariba, that suggest modern tectonism: occasionally, there are quakes along an east-west trend through Namwalla, Kafue, Mozobuku and Churundu, passing 55 km north of Lake Kariba; at the opposite extreme, a quake stripped the rip-rap off Antelope Dam, south of Bulawayo, 385 km away from Kariba Lake.

The Deka fault has no mylonite, little brecciation, and no cataclasis of quartz grains (Mann, 1967). Apparently, no great stresses acted normal to the fault. Mild thermal metamorphism is suggested by chloritization of minerals and distillation of oil from coal (Mann, 1967, p. 27). At least 305 m of vertical displacement is deduced from stratigraphic throw, down on the northwest. These attributes are consistent with normal faulting. Joints in the country rock are steep, predominantly in two sets striking N40°E and N80°W. The former is parallel to the faults, the latter at an acute angle, suggesting to Mann a dextral wrench system. The possibility of wrench-fault stresses (σ_2 vertical) in the region is supported by evidence of such faults in southern Malawi. Nevertheless, the recurrence of normal faulting during the past 300 million years and the consistent deduction of normal displacements from first-motion studies (Sykes, 1967; Gupta and others, 1972) suggest that other alternatives are unlikely.

Hydrology

Ground water in eastern Africa is not plentiful, since most rock formations there are crystalline. But at the caves at Sinoia, Zimbabwe, it can be seen that beneath the flat, dry veld there is a water table that persists throughout a variety of metamorphic Precambrian rocks. Perennial streams rise in gorges tributary to the Zambezi River. Whereas the laterized veld exposures tell little of the bedrock pore structure, outcrops in gorges reveal an abundance of joints and faults, at least some of which probably have hydraulic significance at depths of thousands of metres (Secor, 1965). Though fractures become less frequent with depth and finer in aperture (Snow, 1968c), pore-pressure transients are propagated, and their influence on strength is still great.

Oldest of the sediments are the poorly cemented, coarse, pervious Sijarira Series penetrated by some artesian wells in Zimbabwe. Bond (1953) believes they underly the Karoo in the mid-Zambezi trough. Karoo rocks are important to the reservoir loading behavior. The basal Dwyka beds are likely aquifers, though there has been some cementation. The Ecca coal measures are also permeable, as I have observed at the Momba Colliery in Zambia. Money (oral commun., 1972) described an inrush of water to an exploration shaft driven close to the coal. Artesian conditions probably existed there in the Gwembe coal seam. Whereas the Madumabisa Shale is probably impervious, its carbonate cement suggests that faults cutting it may be open, brecciated, or dissolved. The Escarpment Grit, locally cemented, is evidently pervious as a whole. Money (oral commun., 1972) encountered artesian flows in drill holes penetrating Escarpment Grit in the Barotseland Basin of southwest Zambia.

Faults in the mid-Zambezi trough are aquifers, since numerous hot springs rise along them (Maufe, 1933) southeast of Lake Kariba, especially near Wankie. Using borehole thermistry to obtain geothermal gradients, Bond (1953) estimated from spring temperatures the depth of meteoric water circulation in various parts of the mid-Zambezi Basin. Figure 11a, modified from Bond (1953), locates the springs, the faults, and the reservoir. Figure 11b shows likely paths of circulation, 1 to 3 km deep.

Zongala Hot Spring emerges at the Binga fault. Bond (1953) reports a geyser 60 to 90 cm high at 96.7°C. In August 1972, the jet was 40 cm high at 97°C, discharging about 1 ℓ/s (liter/second). Maufe (1933) describes a history of gradually declining activity. Other, more copious hot springs discharge along the same fault. Near Binga, there are about 610 m of Karoo shale, 350 m of sandstone, and about 880 m of Sijarira quartzite. Using 17.5 m/°C as the measured gradient in shale and 33.0 m/°C as that in sandstone, Bond calculated a possible rise of 72°C above the mean annual surface temperature of 23°C, which approaches the observed spring temperature. Thus the spring waters are probably of meteoric origin.

Several copious springs emerge from the Deka fault. Sunga spring discharges 26 ℓ/s at a "warm" temperature, which Bond attributes to the thin stratigraphic section there, the Sijarira being absent. Near the Gwaai River is Lubimbi spring, a lush area (about 1 km^2) of evapotranspiring grasses. In 1972, the hottest eye was at 80°C, a temperature in accord with Bond's report that 600 to 900 m of Sijarira rocks underly the Karoo in that region. Chipiso Spring, in the Sanyati Basin near Kariba, lies on a fault line (Maufe, 1933) and discharges about 0.5 ℓ/s at 82°C, as measured in 1972.

Hot springs of meteoric origin attest to hydraulic continuity to considerable depths. Since the Kariba-area springs occur at faults, they show these faults to be capable of increasing pore pressure in the basin and probably deeper when their exits are submerged. Since the temperatures are consistent with the depth of sediments, it indicates that little heat is lost, either by dilution or conduction, as the spring waters rise along the steep faults (Fig. 11b), which must have high fluid conductivity compared to the rest of the circulation path (the confined basal aquifers). When the exit potential ($p/\gamma + z$) is increased by submergence, potential must change by nearly the same amount where the faults intersect the deep-basin aquifers, such as the Dwyka or Sijarira formations. Furthermore, the changes of potential must be distributed regularly, perhaps linearly within uniform aquifers, with nearly full reservoir change at the faulted deep end and no change at the recharge end. Fractured basement metamorphic rocks in contact with the aquifers, though not very permeable, may be presumed to transmit potential changes deeply through faults and fractures, quickly in some faults, slowly in the fine, tighter, or gouge-filled conduits. Potential changes may be propagated to depths of several kilometres, especially along dead-ended, steeply inclined faults. Thus the effect of surface potential changes must reach to hypocentral depths, even remote from a reservoir, if faults are continuous conduits to those depths. Induced seismicity proves such continuity exists, for in absence of pore-pressure conduction, effective stresses must rise beneath a reservoir, and hence stabilize faults. The hydrology of some deep mines, production from fractured-basement oil fields, and the result of injecting fluids into basement rocks at the Rocky Mountain Arsenal Well (Healy and others, 1968) all support the supposition that faults, and perhaps other fractures, are hydraulically significant conductors to depths of several kilometres. Especially along recently active faults that have not been resealed by mineralization, it is likely that pore-pressure changes are propagated to great depths beneath a filling reservoir. Observed time lags between filling and seismicity are evidence of hydraulic transient phenomena not explicable by elasticity.

Increases of hydraulic potential resulting from reservoir filling decrease the vertical and horizontal effective stresses both near the surface reservoir and far from it. Rising potentials drive stresses toward failure. Such regional effects were deduced from geology, piezometry, and springflow observations near Kremasta, Greece (Snow, 1972). Similarly, effective stresses were diminished in the mid-Zambezi Basin by impoundment of Lake Kariba, probably quite quickly at basal aquifer depth, but more slowly deep in the basement rocks, whose hydraulic diffusivity controls the delays.

Subsidence

Gough and Gough (1970a) computed the distribution of total stress changes produced by Lake Kariba; computed elastic deflections were consistent with subsidence observed near the lake (Sleigh, 1969), especially along a profile east of the Sanyati Basin (Fig. 12). The agreement shows that elastic deflections are computable from lake bathymetry, assuming an isotropic, continuous half-space having Young's modulus of 0.85×10^6 bars.
The deflections are the same for a porous basement as for a nonporous basement. A saturated mass having fractures that are stiff and widely spaced ($K_n d >> E$) has bulk deformability K nearly equal to the rock-grain deformability K_s, thus it strains vertically according to

$$\varepsilon_y = \frac{1}{E}(\Delta\sigma_{ty} - \alpha\gamma h) \simeq \frac{\Delta\sigma_{ty}}{E},$$

where $\alpha = 1 - K/K_s$ (Nur and Byerlee, 1971). Deformation of the joint blocks aggregates more than the joints. For an infinite reservoir, there is negligible change of vertical contact (effective) stress between blocks, and with stiff, widely spaced fractures, $E/K_n d$ is small, so that $K \simeq K_s$. If the reservoir is narrow, or if only the water table is raised in fractured rock, there is negligible change of total stress (porosity being nil), so there can be little subsidence. With stiff fractures, decreased effective stress produces negligible heave by fracture opening. The argument of Nur and Byerlee is sustained by the Kariba experience, for in the case of an infinite reservoir, there is a vertical strain even though there is no change of effective stress; conversely, in the case of a rising water table, no strain would result even upon change of effective stress.

For strength, effective stress is paramount in importance; application of Nur and Byerlee's equation for effective pressure, $\langle P \rangle = P_c - \alpha P_p$, in terms of confining and pore pressures would lead to error. In fractured rocks, deformability of the voids is unimportant; shear strength depends upon normal effective stress according to $\sigma_e = \sigma_t - \gamma h$, in terms of total stress and pore pressure.

The continuum stress distribution computed by Gough and Gough (1970a) is qualitatively correct for a saturated, fractured medium at depths where the fractures are so stiff (K_n) and widely-spaced (d) that the product greatly exceeds the modulus of the solid grains, $K_n d >> E$, properties appropriate for deeply-buried, fresh, crystalline basement rocks, but neither for sedimentary rocks like those of the Karoo nor for weathered crystalline rocks. If the voids are relatively deformable, the increase in total horizontal stress must essentially equal the increase in pore pressure because effective stress can change but little ($K_n d << E$). Crude approximations of the total and effective stress changes may be all that are warranted in circumstances of heterogeneity and uncertain hydraulic and elastic properties, such as exist in the Karoo sequence.

Seismicity

Kariba Reservoir extends 250 km along the Zambezi River, which borders Zambia and Zimbabwe.

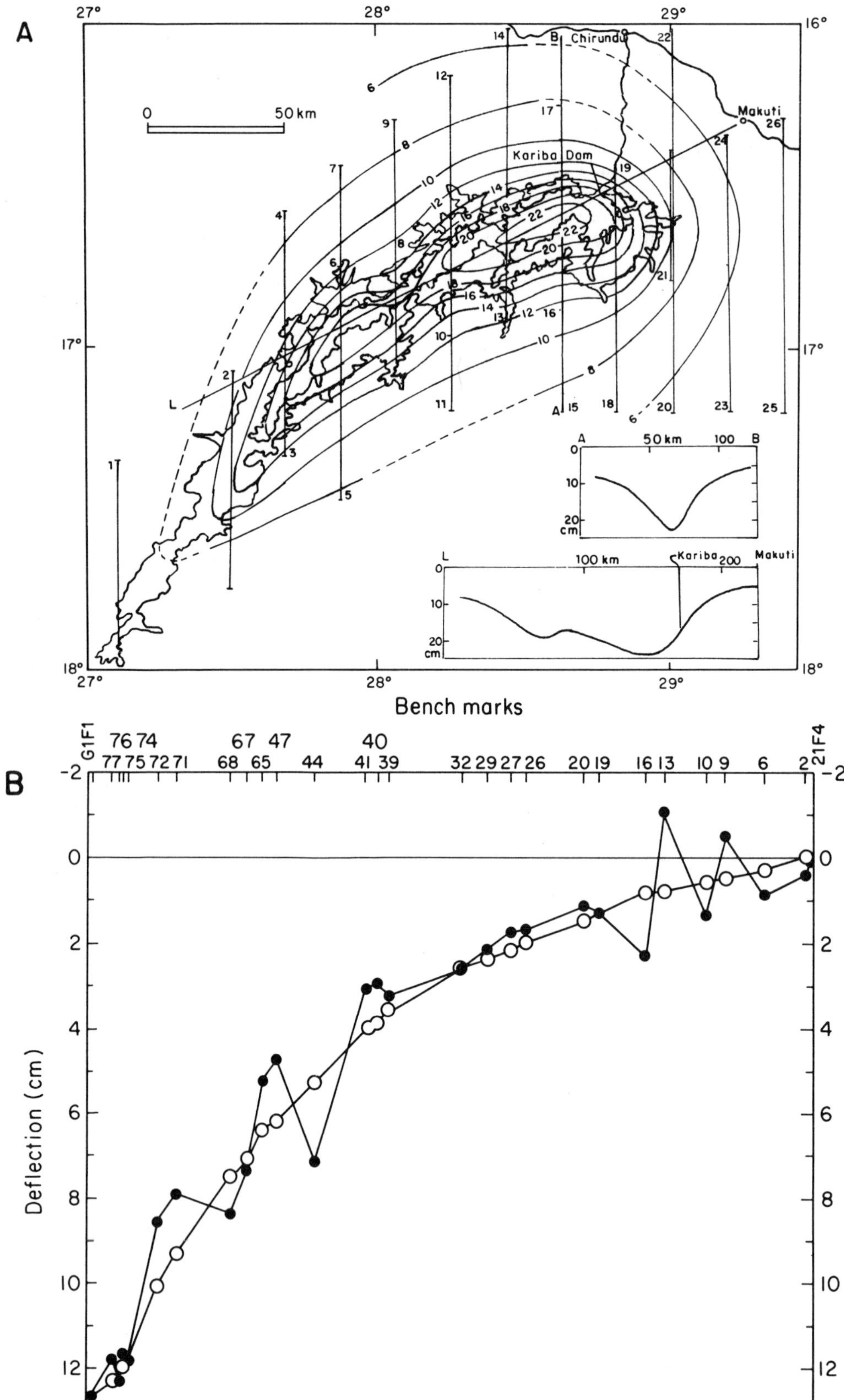

Figure 12. (A) Computed contours of subsidence around Lake Kariba (Gough and Gough, 1970a). (B) Measured (dots) and calculated (circles) deflections along the crooked road from Kariba (left) to Makuti (right) relative to Makuti (Gough and Gough, 1970a).

An arch dam 168 m high in Kariba gorge was closed in December 1958. The seismic activity that began about June 1959 and grew to a climax of two M = 5.5 quakes and one M = 5.8 quake in September 1963 has been described by Gough and Gough (1970b). The following comments reflect many of the observations and deductions of those pioneering authors.

In a normal-fault environment, the minimum tectonic stress is horizontal and the maximum is vertical (the overburden stress). For Kariba, Sykes' (1967) fault-plane solutions for two of the major shocks of 1963 corroborate normal displacements. Gupta and others (1972) agreed. Archer (1972, oral commun.) notes that epicenters determined from different remote stations match poorly, azimuthal differences of velocity being consistent with shallow hypocenters. No detailed studies of Kariba quakes have been completed for lack of close-in stations.

Figure 13 is a plot of lake levels compared with the frequency of earthquakes. There is evidence of a time-lag on the order of one year, for seismicity continued long after the fall of levels that occurred from March 1963 through February 1964. The increase in seismic activity starting in June 1963 might correlate with prior lake levels, which may have increased hydraulic pressure in the hypocentral region, several kilometres below the lake. There was a delay of a month and a half between peak lake level and peak activity.

Since 1968, there have been continued but low general levels of activity until September-November 1971, when a burst occurred with earthquake frequency higher than it had been since September 1964 (Archer, 1972). One quake was of magnitude 3.8. The record of Lake Kariba levels shows that annual fluctuations have been about 2.5 m, and that the 1971 burst followed three months after the June to July peak of 484.5 m, which was no different from peaks of other years when seismicity was low.

The Kariba tremors may be subdivided (Gough and Gough, 1970b) on the basis of time and location. One group occurred near Binga, perhaps along the Binga fault, during early reservoir filling in 1961-1962. A seismographic station was temporarily located at Binga. Local seismicity observed by Bond (oral commun., 1972) preceded lake filling, so only small stress changes (0.4 bars maximum) were needed to trigger a swarm of at least 192 small quakes ($M_L \leq 3.2$), 180 of which were in a four-day burst. The reservoir had risen very little above the river level at Binga. It is apparent from this swarm that an active fault can be sensitive to stress changes of only a few tenths of a bar. There seems to be no critical reservoir depth required, such as the 100-m criterion of Rothé (1969).

A group of epicenters was recorded downstream as far as 65 km from the lake, between Kariba and Chirundu. These had magnitudes of $M_L \leq 3.6$, occurring in a burst during 1961. The local total stress changes due to the lake were 0.06 bar for the farthest and 0.3 for the closest of the Chirundu group. To explain these remote epicenters, Gough and Gough invoked triggering by incremental total stress. The preexistence of critical stresses is suggested by continued seismicity (up to 1972) along the trend from Namwalla to Chirundu, as noted by Archer (oral commun.). Thus seismicity in the Chirundu region in 1961 could have been unrelated to Lake Kariba, or triggered by small stress changes. If the focal-plane mechanism of quakes on the Namwalla-Chirundu trend is normal, then increases of total stress are not expected to trigger failure. Rather, remote pore pressure increases cannot be wholly discounted because they decrease the total stresses. From Sanyati Basin to Chirundu there are at least five normal faults that bound the Precambrian horsts. When one end of a conduit is subject to an increase in hydraulic potential of about 160 m, the change in potential will extend over as large an area of the plane as is affected by flow and interconnection, even to many kilometres away from the reservoir. Deep ground-water circulation paths as much as 200 km in length have been described in the Basin and Range province of Nevada (Mifflin, 1968), with transit times up to 20,000

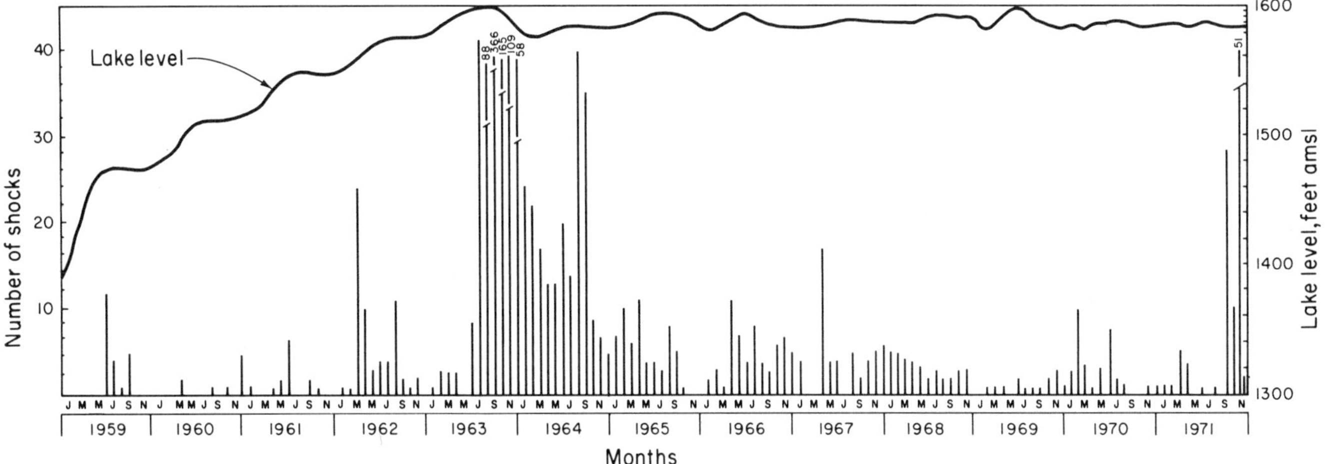

Figure 13. Monthly earthquake frequency and water levels of Lake Kariba during 1969-1971. Water levels courtesy of African Power Corporation. Seismicity from the data of Archer (1968, 1972) (after Gupta and others, 1972).

years. In karstic regions, such as near Kremasta, Greece, spring-flow fluctuations have indicated potential changes at least 13 km from a reservoir (Snow, 1972). Changes of potential at the distal end of a conduit, as at Chirundu, must be of vanishing magnitude.

The greatest grouping of earthquakes and the largest shocks originated beneath and near the Sanyati Basin, the widest and deepest part of the reservoir (Fig. 10). The reservoir trends east-west at that end, and is crossed by numerous northeast-trending normal faults, including those that bound several Precambrian horst blocks. Hot springs, such as Chipiso (Fig. 11), just east of the Sanyati Basin, indicate hydrologic continuity along at least some of these normal faults to depths of 2 km or more. Since these faults are recharged tens of kilometres away at higher elevations, the discharge ends of extensive circulation cells are subject to the full increase of reservoir potential when submerged. Depending upon the heterogeneity of fluid conductivity in such cells, the increase in hydraulic potential at the base of the sedimentary sequence near the reservoir could be 0.7 to 0.9 times the depth of submergence. Faults must penetrate the basement, since Precambrian horsts are exposed. Pore-pressure conduction below the sediments is therefore likely. Open faults at depths of several kilometres are mechanically possible (Secor, 1965) and are certainly within the experience of the mining profession (see the paper by Cook in this volume). Where there is circulation, there is dissipation of potential along flow lines in fault zones penetrating the basement. But in open, dead-ended, noncirculating fractures that penetrate to great depths, the steady potential change throughout that pore space depends only upon the potential change at the upper end. Faults terminating at the base of the Karoo or Sijarira sequence could conduct a major portion of the reservoir head to great depths as well as do faults cutting the sediments.

Gough and Gough (1970b) observed no earthquakes at the eastern extremity of the lake, even though there are known faults there. Total stress and pore-pressure changes were presumably small where the reservoir is shallow. If parallel faults are loaded unequally, those that undergo the greatest pore-pressure changes are driven more toward failure than are faults receiving small changes.

Gough and Gough (1970b) speculated on possible mechanisms of rock failure as causes of the Kariba earthquakes:

(i) By the direct effect of shear stress caused by load across fault planes,

(ii) By an indirect effect of the added stress in triggering larger initial stress, and

(iii) By the effects of increased fluid pressure in the groundwater.

Gough and Gough (1970b, p. 98) rejected (i), since a cohesionless fault has more shear strength than the 2 bars imposed at Kariba. Indeed, unstressed ground at hypocentral depth is nonexistent, nor can cohesionless, faulted ground support effective tensions. If stresses are near-critical for normal faulting (Fig. 8a), failure requires changes of the effective stresses such that Mohr's circle approaches the envelope. As has been illustrated (Fig. 9), in order to cause failure by increasing the vertical stress, Poisson's ratio and the friction angle on faults would have to have unrealistic values. It is this fact more than any other that supports hypothesis (iii) versus (ii).

Gough and Gough (1970b) computed the progressive growth of total shear stress as the reservoir filled, seeking correlation with the seismicity. Shear stress is an incomplete criterion of failure for materials having internal friction. The difference between principle stresses is particularly nondiagnostic for a normal-fault environment, whereas the ratio of principle effective-stress changes is more useful. If elastic-induced total stresses were primarily responsible, there should have been no periods of activity during low levels of shear stress, such as during the period October 1966 to January 1967. Delayed response is more likely to attend an effective-stress history, since permeability to and storativity of ground water produce transients.

Gough and Gough (1970b) noted that a high value of b' in the magnitude-frequency relationship for Kariba earthquakes is consistent with the range of b' for aftershocks. The implication is that the quakes occur in a region of low stress. Comparison of the radiated seismic energy with the computed decrease of potential energy stored elastically during subsidence of the loaded region indicated that the seismicity could be produced at low efficiencies. That finding is consistent for a normal-fault environment: vertical stress drives the motion. It does not, however, support the notion that small differences in total stress produced by the weight of water will reactivate old faults if stress levels are low. High b' values also attend reservoir-induced seismicity in wrench-fault terrain, where stresses are high.

The Kariba evidence favors the interpretation that pore-pressure penetration and reduced effective stresses following rises in lake levels triggered the earthquakes. It will be shown that the loading history fostered a critically stressed environment before the reservoir filling began.

Loading History

The geomorphic history of the Kariba area has been translated into a history of changes of effective stresses. The changes refer to some arbitrary depth in the crust subject to changes of pore pressure as well as overburden load. The deducible stress-producing events are enumerated in Table 3, and the cumulative stress changes from Jurassic to present are graphed in Figure 14. Two major river incisions have occurred, one of which took place in the Cretaceous, before dinosaur beds were deposited not far above today's level of the Zambezi River. After subsequent filling of the Cretaceous valleys and continental-scale planation, a series of erosional steps were developed, so that the Zambezi has only recently cut deeper than the Cretaceous thalweg. Before the reservoir was filled, the

TABLE 3. EFFECTIVE-STRESS CHANGES AT KARIBA

Event	Bars			
	Vertical		Horizontal	
	$\Delta\sigma_1$	$\sum\Delta\sigma_1$	$\Delta\sigma_3$	$\sum\Delta\sigma_3$
0 Datum, Jurassic Gondwanaland rifting, stresses critical				
1. Post-Gondwana surface cut in late Jurassic, possibly 270 to 300 m below datum, possibly at 1,560 m elevation. Zambezi drainage established (?)	- 45.4	- 45.4	- 28.5	- 28.5
2. Degradation of mid-Zambezi V. to elevation 533 m at Kanchindu and in other rift valleys; Cretaceous beds. Eroded 1,027 m in trough	-164.3	-209.7	-102.7	-131.2
3. Cretaceous valley fill in all rifts, possibly to 1,433 m at Tundazi Peak., Sijarira Range. Faulting. Alluviate 900 m in trough	+ 97.6	-112.1	+ 54.0	- 77.2
4. Cutting of African Surface by Miocene time. Alluviate 76 m to elevation 1,509 m.	+ 8.4	-103.7	+ 4.6	- 72.6
5. Erosion to post-African surface (1), to elevation 1,070 m at Chicarira Range. Eroded 439 m of alluvium	- 48.3	-152.0	- 26.3	- 98.9
6. Erosion to post-African surface (2), to elevation 838 m. Eroded 232 m of alluvium.	- 25.5	-177.5	- 13.9	-112.8
7. Erosion to 610 m surface, 228 m of alluvium.	- 25.1	-202.6	- 13.7	-126.5
8. Zambezi Canyon incision, 150 m into surface, lowered water table, but negligible aereal denudation.	+ 15.0	-187.6	+ 15.0	-111.5
9. Filled Zambezi (and tributary) canyons 150 m, raised water table only.	- 15.0	-202.6	- 15.0	-126.5
10. Porous Kalahari sand covered 610 m surface, raised topography to 622 m without change of water table.	+ 2.5	-200.1	+ 1.1	-125.4
11. Erosion to 480 m, end-Tertiary surface of Bond (1960), narrow in upper reservoir area, wider downstream. Erosion 12 m, dry; 77 m alluvium, wet; 53 m rock, wet.	- 2.5 - 8.5 - 8.5	-219.6	- 1.1 - 4.6 - 5.3	-136.4
12. Incision (by steps) to 434 m pre-Kariba Lake level. Local earthquake. Erode 46 m rock.	- 7.4	-227.0	- 4.6	-141.0
13. Filling Lake Kariba to 487 m maximum. Infinite equivalent.	0	-227.0	- 3.1	-144.1

stresses were sufficiently critical to manifest themselves seismically, so that the environment was ripe for reservoir-induced activity. Progressive historic decreases of "stability" are consistent with pre-reservoir criticality.

The loading history was appropriate for earthquake initiation in a critically-stressed normal-fault environment. The Kariba Basin topography below lake level featured subdued hills, thus the water table was close to the surface. When an extensive reservoir inundates such an area, pore-pressure changes are nearly equal to the vertical total-stress changes. But since filling decreases the horizontal effective stress, σ_{3e}, by nearly half the water load, γh, the circle of stresses shifts left toward the failure envelope (Fig. 8a). Under such seismic conditions as existed (at Binga) before filling, the stress circle was already tangent (at least for some faults). Binga lies on a normal fault believed reactivated by the lake.

Other rift faults were activated in the Sanyati Basin, the region of deepest, widest submergence.

At least 1,000 m of erosional unloading took place from the level of the Miocene African surface down to that of the modern river. Figure 14 illustrates the changes in σ_{1e} and σ_{3e} and the progressive shift of the stress circles toward a 30° envelope. There is no evidence for post-Cretaceous faulting in the Kariba area, though some parts of the east African rift system have had active faulting during Cretaceous, Miocene, Pliocene, Pleistocene, and Holocene times. In the mid-Zambezi trough, neither surfaces nor post-Cretaceous deposits have been displaced. Yet modern prelake seismicity at the end of the denudation suggests that criticality must have been attained.

One circumstance that accords with the maintenance of nearly critical stresses is that "the rift valleys traverse regions which have been uplifted epeirogenically to form part of the 'basin and swell' structure of Africa" (Baker, 1971, p. 544). In the well-studied rifts of Kenya, fault displacement and uplift have been of similar magnitude. We may speculate that there may have been but slight changes in stress under the rift-valley floors, which maintained nearly constant elevation as the surrounding country was uplifted. It may imply that crustal stresses are limited by the strength of critical features such as rifts.

Northeast of and in line with the mid-Zambezi rift is the fault-bounded Luangwa Valley, showing a more youthful scarp than borders the mid-Zambezi trough. Northeast of Lake Nyasa, along the same trend, is the Ruhuhu Basin. It shows late-Tertiary rejuvenation of its boundary faults. All three Karoo valleys are topographically similar; their floors are at comparable elevations. The southernmost, the mid-Zambezi, has maintained its level by erosion, with the Zambezi River the agent. Its tributary, the Luangwa River, was unable to keep up with epeirogeny; part of the Luangwa valley-lowering was produced by downfaulting. In the northernmost, the Ruhuhu, faulting was the dominant mechanism, erosion playing only a minor part. Stresses were maintained alike in all three valleys. It may be supposed that if the Zambezi had coursed elsewhere, the mid-Zambezi trough might have had fault activity like the more active rifts. It is a corollary of this reasoning that differences in land-forming processes can dictate the tectonism of a locale, since stresses may differ only by small amounts from the state critical for failure. The sudden filling of a lake can therefore easily trigger seismicity.

It appears that a rift valley is a self-perpetuating zone of structural failure. Topographically lower than the surrounding high pediments, it is characterized by low stresses because of low strength. Erosion and weakness of normal faults have perpetuated the depression since Precambrian time.

In the absence of established rift valleys, epeirogeny leads to stress changes that promote normal faulting. Uplift decreases σ_3, fostering instability. The post-Miocene African surface, remarkable for its perfect planation across diverse rock types, tilts due west from an altitude of

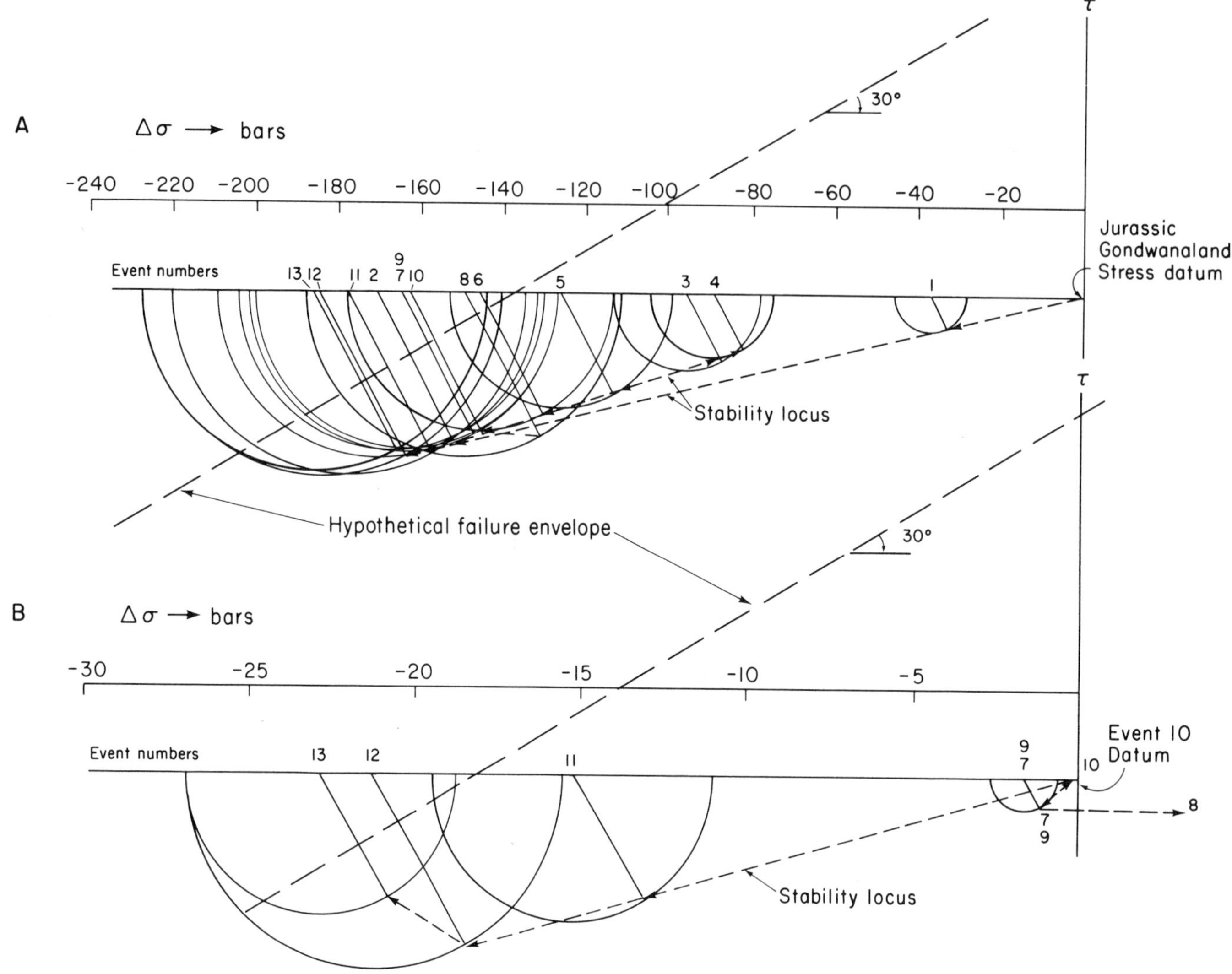

Figure 14. Stability locus, Kariba loading history. (A) Jurassic to present, showing diminishing stability. (B) End of Tertiary (610-m surface) to present, at different scale, showing progressive weakening since Kalahari sands deposited (10); lake filling aggravates instability (13). Note that if the actual stress datum were known and depicted as a large circle in the positive stress quadrant, subsequent stress circles would describe the same locus as the above circles of change relative to the datum.

2,000 m in the Eastern Highlands of Zimbabwe to 1,675 m at Marandellas, 1,520 m at Salisbury, 1,465 m at Gwelo, 1,310 m at Lusaka, Zambia and 1,250 m at Bulawayo. Farther west in Zambia it is covered by sands where it dips into the Kalahari Basin. The most active seismicity has been along the axis of maximum uplift in the Eastern Highlands and in neighboring Mozambique, northward into the Shire Valley of Malawi and the troughs of Lakes Nyasa and Tanganyika. The mid-Zambezi valley, with its flanking position, has been largely aseismic.

WRENCH FAULTING AT KOYNA, INDIA

Koyna Dam in Maharashtra impounds Shivaji Sagar Lake, an elongate reservoir with north-south trend, parallel to and only 5 to 10 km east of the Indian continental divide, which is only 50 km from the west coast. A tunnel and penstock through basalt deliver the water to a power plant on the west side, taking advantage of a prominent west-facing escarpment.

Geologic Setting

The Koyna region (Fig. 15) seemingly has such simple geology that conventional mapping was done only after a 1967 earthquake nearly demolished Koyna Dam. For many kilometres around, the rocks are Deccan basalt flows (traps), deeply covering Paleozoic metamorphic and granitic basement rocks. Nevertheless, important aspects of the geology, hydrology, and tectonics remain uncertain.

The nearest exposed basement rocks are 150 km

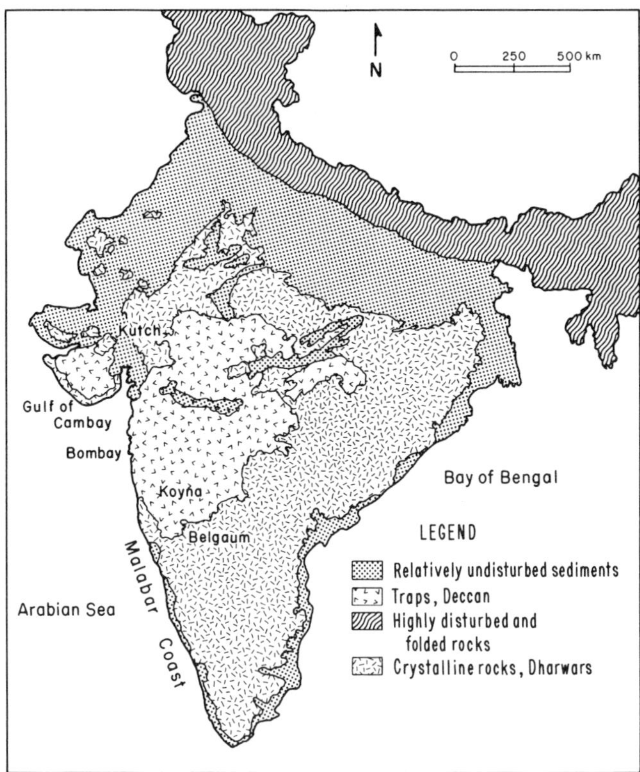

Figure 15. General geology of India, showing the location of Koyna (after Housner, 1969).

south of Koyna. Near Dudhganga Damsite (Fig. 18), the basalt is unconformable on Paleozoic shale and quartzite, oxidized red but lacking an appreciable regolith. Elsewhere, the basement under the traps consists of Archean metasedimentary gneisses, schists, and granites (Krishnan, 1960). Most of the periphery of the Deccan exposes granite, so geophysical interpretations assume a granite basement. According to Gupte (oral commun., 1973), a regolith is present at Belgaum, where the contact has at least 300 m of relief (Committee of Experts, 1968). Interpretation of derivatives of the gravity field indicate (Guha and others, 1974a, Fig. 22) a basement high just south of Koyna, with about 1 km of Deccan cover.

A regression of the sea took place in Kutch in mid-Cretaceous time, with Deccan eruptions following. The entire Deccan region may have been eroded since late Cretaceous time. The volcanics are older than overlying Paleocene sediments north of Bombay.

Near the coast, the basalt attains maximum thickness of a few kilometres. The Western Ghats (west-facing escarpments) expose over 1,000 m of Deccan flows, though faulting may have repeated the strata.

The presence of widespread flows and dikes suggests that the traps are vent eruptives like the Columbia basalts, but Gupte (1971) finds their structure indicative of central crater and vent eruptions. Numerous subhorizontal pipe-like vents exist, filled with breccia and surrounded by radial columnar joints. At least six are exposed in the Koyna quarry. The flow units are discontinuous, forming chaotic piles of overlapping individuals with slightly different textures, separated by sharp contacts and dipping in different directions. Rocks on one hill therefore seldom correlate with rocks on adjacent hills at the same elevation. In the northern Deccan are basalt dikes in northeast-trending swarms and radial groups, absent near Koyna. Composition differences suggest that the dikes postdate the flows and do not represent feeder vents (Agashe and Gupte, 1971).

The relation of Deccan eruptions to seafloor spreading is unknown. Krishnan (1968, p. 9) believes that the linear southwest coast of India (Malabar) separated from Madagascar during Late Cretaceous time, after the Deccan volcanism. There is no evidence of submarine emplacement: pillow structures and marine interbeds are absent. Rather, the base and many interflow contacts were apparently subaerial. Steamtubes at the base of flows are common, indicating wet ground.

Soft, easily weathered horizons called red boles are commonly observed. Once thought to be soil horizons, they are actually basalts that have each been hydrothermally altered and reddened by the succeeding flow. The so-called intertrappean sediments do not exist, at least in areas studied by Gupte (oral communucation, 1973) or by me. The red boles are hard rocks in the subsurface that rapidly disintegrate to red clay upon exposure. The succession of lavas allowed little if any time for weathering; contacts are tight in most cases.

Compact, fine-grained, occasionally porphyritic basalts usually overlie the red boles. Crude columnar joints and extensive curved fractures are common to these homogeneous, hard, brittle rocks. The most voluminous flows are amygdaloidal, grading upward to vesicular or scoreaceous textures. They commonly overlie the compact flows. The amygdules are zeolites, often vuggy. Since cavities are not interconected and joints are usually absent in vesicular flows, they are watertight compared to the compact fractured flows. Most of the rare fractures are cemented by calcite. Seepages along horizontal contacts have been observed occasionally in penstock excavations, and grass stays green where some horizons crop out on hillslopes; thus some contacts are aquifers. Though water supplies are meager in trap rocks, occasional springs on steep mountain slopes indicate that saturated, continuous conduits do exist.

The attitude of the flows is essentially horizontal. Some authors have suggested dips of 1° to 2° toward the west, but in view of the irregular contacts observed by Gupte, it is uncertain whether any systematic inclination exists, either original or due to tilting or folding.

The nature of faulting in the Deccan is controversial. Many geologists in India believe that normal faulting predominates, a style suggested by the step-like escarpments of the Western Ghats, but inconsistent with continental drift. The faults dip 80 to 90°, unlike most normal faults. There is no evidence for rifting along the western coast, which is more likely to be a wrench fault.

India was joined to Africa before the Late Jurassic breakup of Gondwanaland. Krishnan (1968) notes Precambrian units common to India and Africa and to India and Australia. Jurassic sedimentary rocks of the Kutch region north of Bombay may have been contemporaneous with similar rocks found along the African coast.

The post-Cretaceous history of the Indian Ocean has been deduced from the paleomagnetic stratigraphy (McKenzie and Sclater, 1973). In 75 million years, India drifted northward about 2,000 km on a crustal plate bounded by two transform faults whose sea-bottom vestiges are the Chagos-Laccadive Plateau and the Ninety East Ridge (Fig. 16). About 45 million years ago, the plate margins separated in a northeast-southwest direction along a myriad of northeast transform faults recognizable in the many northeast-trending ridges between the Mascarine Plateau and the Chagos-Laccadive Plateau. India rotated about 30° counterclockwise in the process and collided with Asia about 20 million years ago. Guha (1970, p. 14) noted the correspondence between seismic and geologic evidence for northerly continental drift. Horizontal displacements require a compressive mechanism. Wrench-fault tectonics have probably prevailed from Cenozoic to Holocene time, with marginal Himalayan thrusting after the Miocene collision. Topographic uplift along steep north-south faults can be ascribed to minor dip-slip components of motion along wrench faults of great displacement. The horizontal strata give no clues to the strike-slip components. As described below, wrench faulting is compatible also with the recent ground breakage at Koyna and with seismic evidence.

Koyna lies just east of the continental divide, the crest of the Western Ghats, a steep, west-facing basalt escarpment only 50 to 70 km from the Arabian Sea. In the first few kilometres, west-flowing streams fall about 1,000 m nearly to sea level, then flow to the coast in valleys cut across a low-lying hilly region called the Konkan. Remnants of flat erosional surfaces reflect a history of uplift of the Konkan, as does the elevated interior plateau east of the Ghats.

The escarpment parallels the linear coast for about 1,400 km, from the Gulf of Cambay to the tip of the peninsula. The corroded crest winds sinuously (Fig. 19) over a zone that is only about 2 km wide in the Koyna region, but as much as 10 km wide near Bombay. The Committee of Experts (1968) found little evidence of fault control of the Ghats, except for unconfirmed fault contacts of the basalt against the Precambrian at the southerly edge of the Deccan near Belgaum (Figure 15). No single master fault is known; instead, there may be a

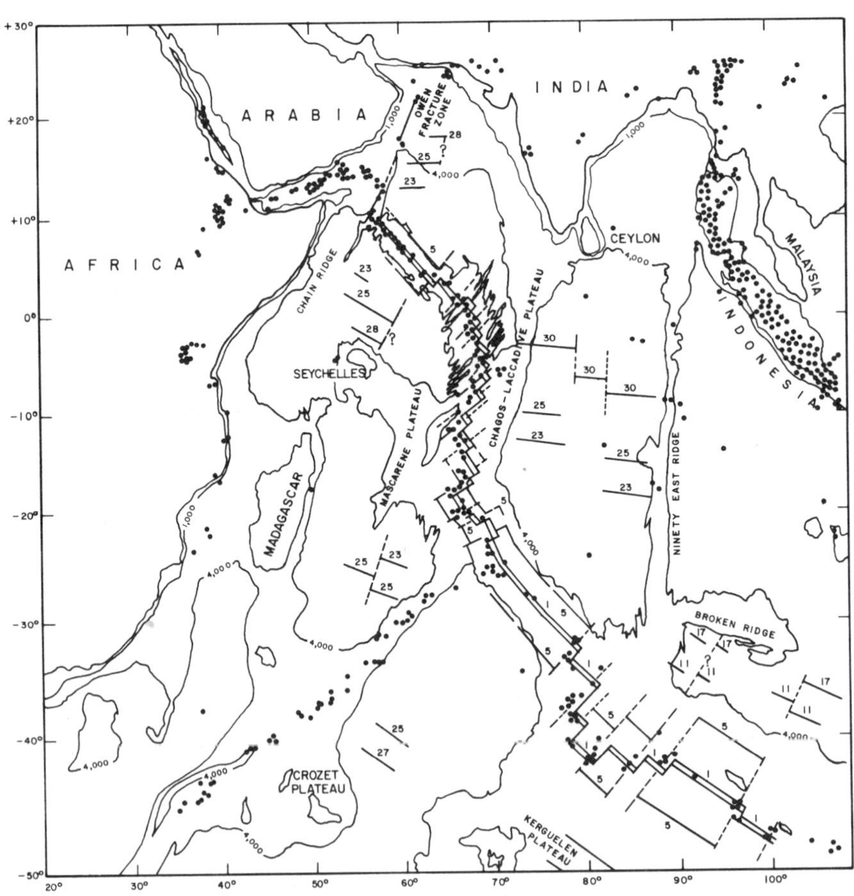

Figure 16. Magnetic anomalies in Indian Ocean are represented by straight lines. The numbering of the lines starts with 1 on the ridge axis (two closely spaced lines) and extends to 33, which identifies molten rock that solidified about 80 million years ago. Earthquakes recorded over a recent 7-year period are indicated by black dots. The major features on the floor of the Indian Ocean are the three large ridges that run north and south. India, once a part of Gondwanaland, was attached to Madagascar and Antarctica (McKenzie and Sclater, 1971).

multitude of faults responsible for the offset of erosion surfaces.

The linear coastline is probably fault-controlled, but geophysical evidence is inconclusive (Committee of Experts, 1968). The crest of the Ghats has probably not retreated 50 to 70 km from a coastal fault: near the Vaitarna River, 2 km west of and below the escarpment, there are flat erosional remnants at elevations of 320 m and 150 m on valley walls whose floor is at 65 m. These are remnants of flat-bottomed valleys; their formation implies that watershed areas once lay west of the escarpment. Therefore, in spite of the steepness of the 1,035-m scarp, the crest cannot have retreated eastward more than 1 km since the 320-m surface formed. If the crest retreated only 1 km while the river incised 255 m, it is unlikely to have retreated 50 km from the coast while the river incised the first 715 m. The scarp must have its origin at one or more north-south faults fairly close by.

Thirty-three thermal springs (Gupta and others, 1967) along a 360-km north-south line 20 to 25 km west of the Ghats suggest a fault zone (Fig. 17). All the springs issue from Deccan rocks or surficial deposits, none from Precambrian rocks. In both of these areas, the geology is monotonous, relatively unmapped, and concealed by vegetation and laterization. However, at least one of the hot springs (Aravali) lies on the extension of a prominent north-south valley, as do some cool springs. Many north-south streams in the Konkan may be structurally controlled. By intersecting the west-flowing consequent streams, they form an imperfect trellised drainage pattern.

A history of the uplift is consistent with gravity data for the western Indian peninsula. Glennie (1932) and Guha and others, (1974a, appendix) depict a strong gravity gradient beneath the Konkan (Fig. 18). Offshore, and locally in the Konkan north of Bombay, there is a strong positive Bouguer anomaly (up to 80 mgal). From about the Western Ghats eastward across the peninsula, there is a strong negative Bouguer anomaly (to -100 mgal). The steep gradient probably corresponds to the contact of a low-density granitic continental plate against a high-density basaltic oceanic plate in the Arabian Sea, juxtaposed by transform faulting. It implies neither that the Deccan sequence is thicker offshore nor that a source of eruptions once lay there. Gupte's (1971) evidence for numerous vents and

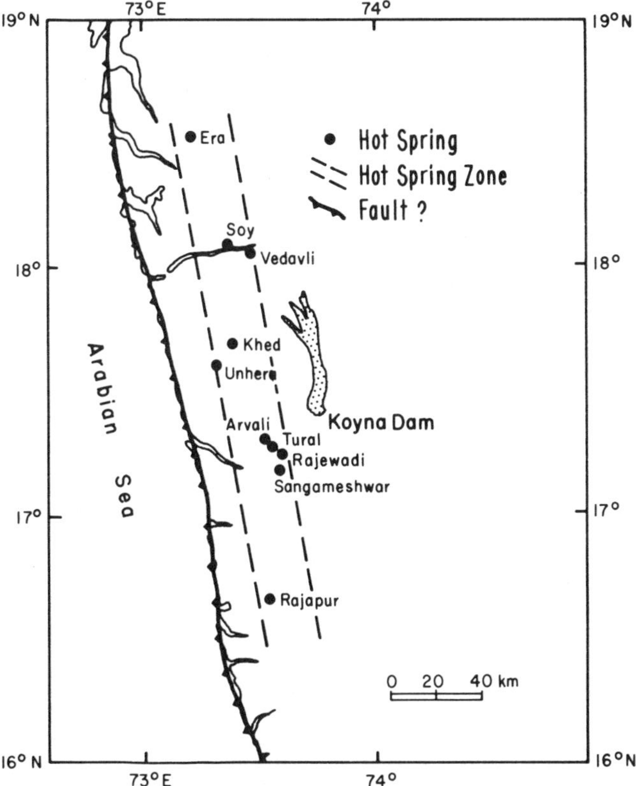

Figure 17. Location of thermal springs in a zone parallel to the Western Ghats (Guha and others, 1974a).

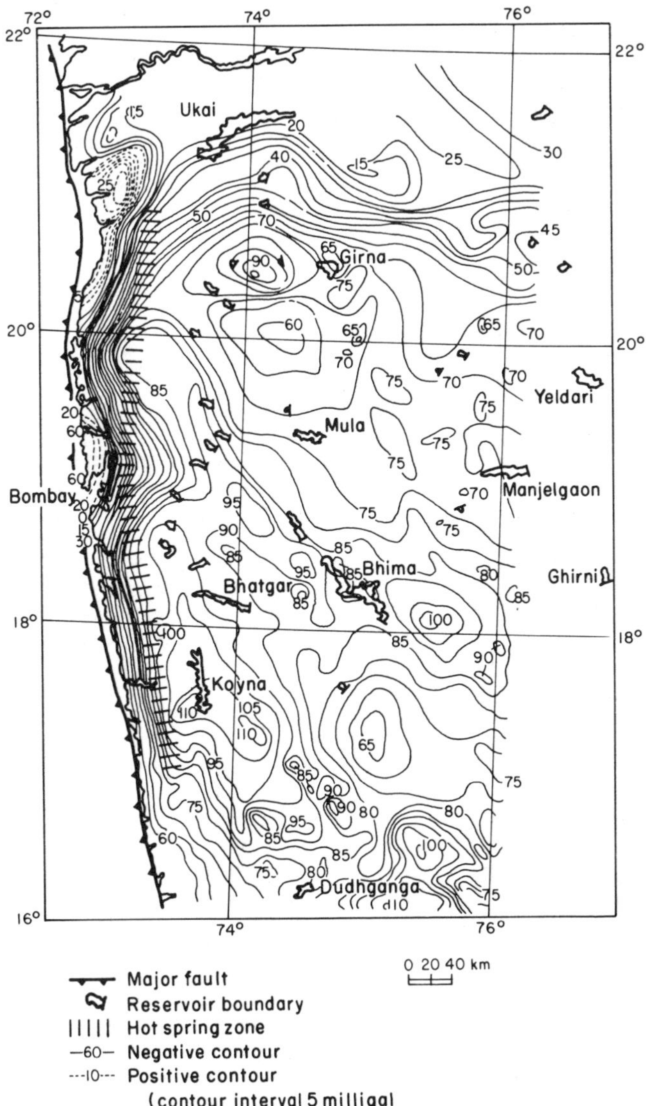

Figure 18. Bouguer gravity anomaly and reservoirs in West Maharashtra, India (Guha and others, 1974a).

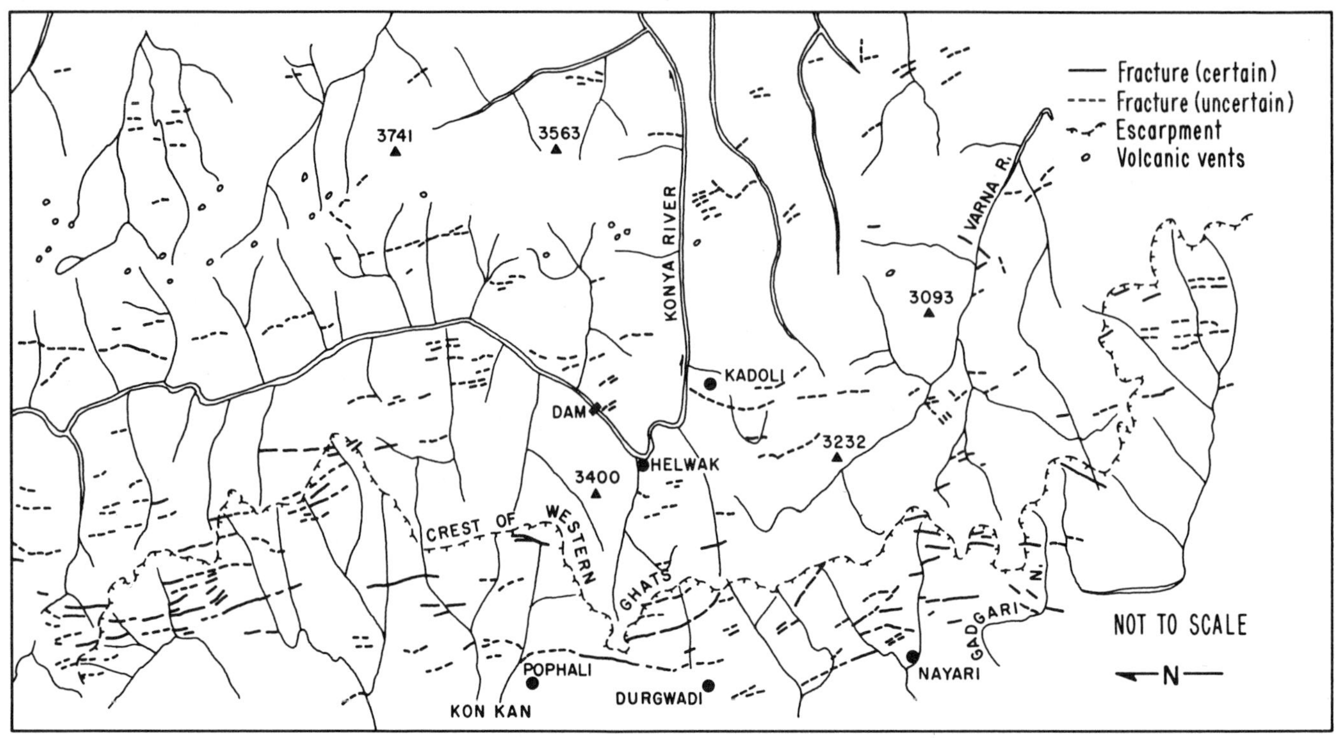

Figure 19. Linear terrain features indicating fractures in the Koyna area (Deshpande and Jagtap, 1971).

central eruptions rather than fissure sources implies that numerous fault conduits existed in the late Cretaceous rather than a single rupture. Accordingly, the continental margin may also be diffuse--a zone of parallel faults.

Using new gravity data and derivatives of the field, Guha and others (1974a) reinterpreted the basement configuration. It plunges to depths of 6 or 7 km near the coast, rises to a ridge 1 km deep near the Koyna River, then falls eastward to 3 km. The Western Ghats coincide with the inferred granitic ridge, and eastward-tilted erosion surfaces mirror the basement dip. Guha speculates that the composite bedrock geometry produces a stress concentration at the ridge sufficient to make the seismic response to reservoir loading more effective at Koyna than at other reservoirs. However, Figure 8 shows that failure cannot be produced by an increase of effective stress, even in the presence of concentrations. The pattern of topographic uplift has provided some gravitational compensation. When friction is dissipated by lateral slip during wrench faulting, vertical displacements are facilitated, with negative gravity areas rising relative to positive areas.

The frequency of steep, nearly north-south fractures cutting the Deccan increases westward (Fig. 19). Distinct from the curving, diversely oriented cooling cracks in compact lava flows, the north-south fractures are planar, systematically oriented, and cross-cutting. Continuous individuals or zones cut the cliffs, laterite caps, and, particularly, excavation faces. Excavations show many minor and some major north-south water-bearing zones. Crossing Koyna Dam is a vertical zone 1 to 3 m wide that had to be excavated and back filled with concrete. In the tailrace tunnel, Phase III Koyna Project, three near-vertical, north-northeast to north-northwest zones pinch and swell, with maximum widths as great as 10 m. They contain weakly permeable, sheared rock and clay gouge. One cathedraled to the river bed, which was fortunately blanketed with concrete. A sink hole developed where a weak zone crossed the tunnel alignment. Project geologist Sharma (oral commun., 1973) noted that lavas on opposite sides of that zone are offset 3 m. At Station 1080 in the tailrace canal, a horizontal red bole is offset 1.2 m (east side down). Thus it is consistent to regard all steep, continuous fractures and zones that strike N15°W to N15°E strike as faults. Few show vertical offsets, but horizontal motion of any magnitude may have occurred. The sheared materials in these zones suggest that at least some are important wrench faults. A continental plate that has drifted thousands of kilometres relative to other plates is bounded by transform faults parallel to the continental motion. As in California's San Andreas fault zone, a multitude of India's Malabar wrench faults occupy a zone many kilometres wide and undergo the same motion as the master transform.

East of the Western Ghats are several planar erosional surfaces, sloping gently southeast (Fig. 21a). Streams flowing southeast to the Bay of Bengal (700 km away) have flatter gradients than do streams flowing west to the Arabian Sea (60 km away). The highest may be called the Mahabaleshwar surface after a town near the headwaters of the Koyna, Krishna, and Gavitri rivers. It is Eocene to Pliocene in age. The surface is an extensive terrain with 100 m of relief, proof that it is not structurally controlled by a lava flow. Laterite, locally over 25 m thick, mantles bedrock. Unweathered trap rocks are exposed at the edge of the surface. At Arthur's Seat, a 1200-m precipice faces the west-flowing Gavitri River. The Krishna River meanders east for several kilometres on the Mahabaleshwar surface, then plummets 80 m into a box gorge. The Koyna River falls steeply from the Mahabaleshwar surface to a one-sided valley truncated by a west-facing 700-m precipice, at the bottom of which is the west-flowing Sivitri River. The divide may be a fault-line scarp or the lower Sivitri may be capturing the Koyna headwaters by scarp retreat. The Koyna River is unique in that it flows south for 70 km parallel to and only 5 to 10 km east of the continental divide, before turning abruptly east just downstream of Koynanagar (Fig. 19). All other streams flow east or southeast, consequent to the oldest erosion surface. Many of their headwaters have been captured by the Koyna River.

Suspected structural control of the north-south reach of the Koyna River has been reexamined since the 1967 earthquake (Das and Roy, 1969). Near the reservoir are triangular ridge facets and north-south lineaments, though no obvious fault zone controls the river. The lake is not in a graben, nor is there normal faulting. The shear zone cutting Koyna damsite extends about 1 km downstream.

Younger erosion surfaces lie about 215 and 365 m below the Mahabaleshwar surface. Both form shoulders on the valley walls in the headwaters and hilltops farther downstream. In the 100 m between that third surface and the river, benches record successive stillstands.

The surfaces have been profiled using Survey of India topography. Figure 20 shows the locations of the profiles that appear in Figure 21. In profile 1-1', ridge-top remnants of the Mahabaleshwar surface indicate that it tilts 1:200 to the south. The second surface has similar tilt, but the third is flatter, like the Koyna River, sloping 1:1,100. In profile 2-2', flat hilltops on the continental divide define the Mahabaleshwar surface, but lower surfaces are absent. The slope of the summit surface from Mahabaleshwar at 1,340 m to Dharmsala

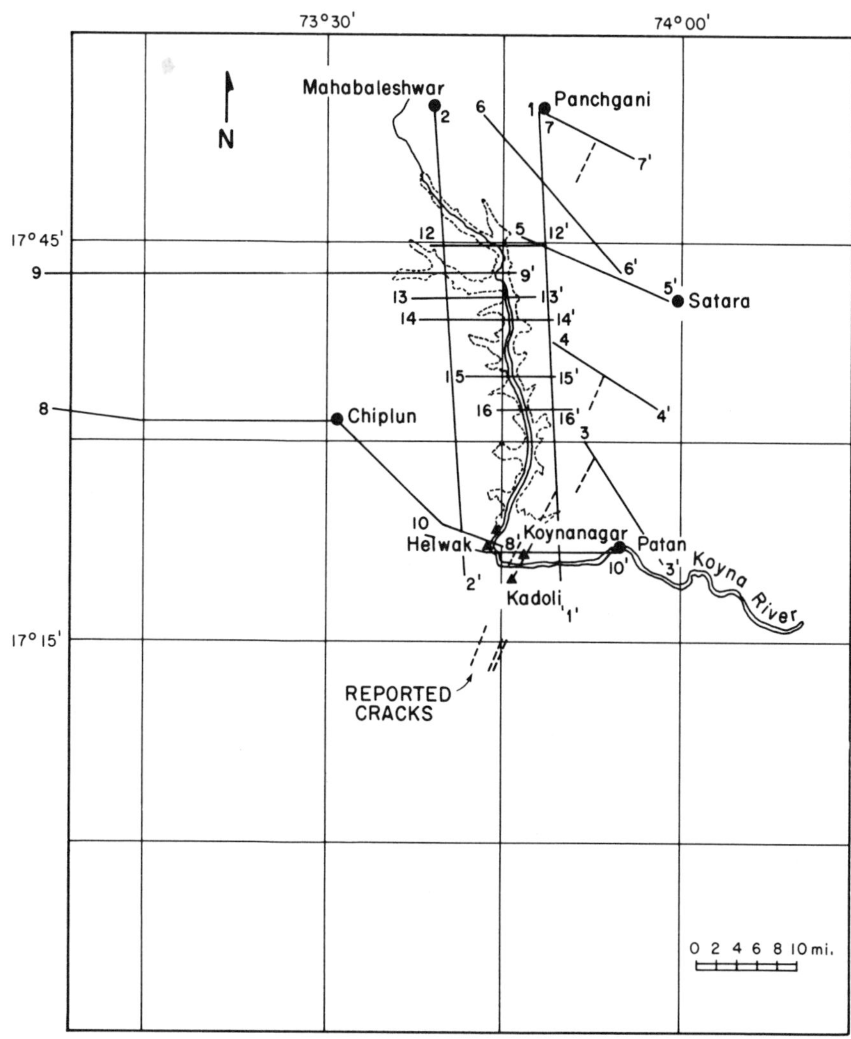

Figure 20. Location of topographic profiles given in Figure 21.

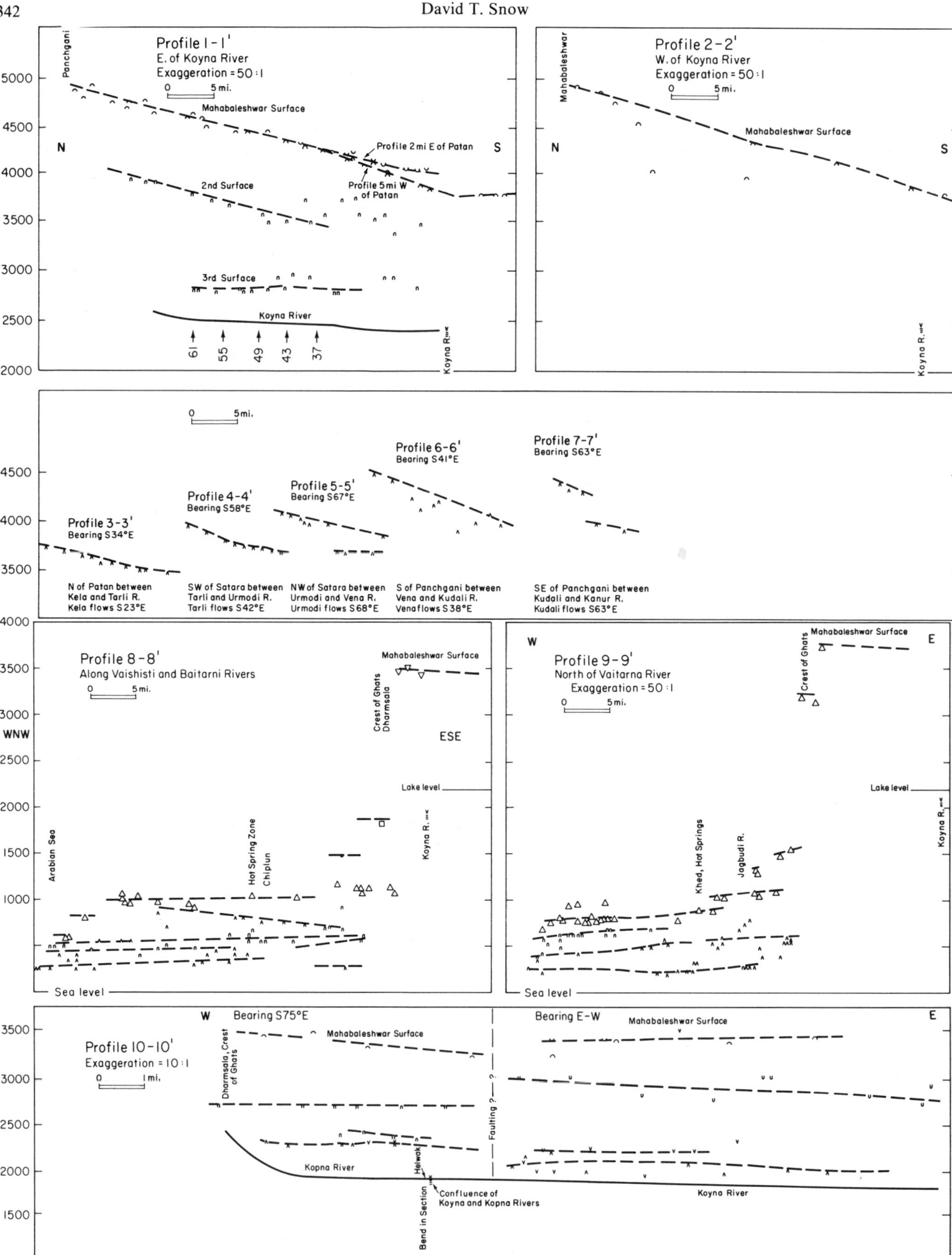

Figure 21. Profiles of geomorphic surfaces, Koyna Reservoir area.

near Koynanagar at 1,035 m probably represents tectonic tilt of a peneplane developed near sea level. Sections 3-3' through 7-7' show the Mahabaleshwar surface on each of five ridges trending from S34°E to S67°E and profile 10-10' trends east-west. By stereographic projection, a great circle describes a tilt of about 1:200 to the south-southeast.

Profile 10-10' shows displacement of the two oldest surfaces near Koyna Reservoir, indicating downfaulting to the west. Profile 1-1' indicates that the Koyna River was controlled by faulting before tilting began, because the second surface parallel to the Mahabaleshwar surface was cut on the flanks of the Koyna valley. Therefore, rivers such as the Kela, Tarli, Urmodi, Venna, Kudali, and Kanur, which flow between S23°E and S28°E, consequent to the earliest tilt, must also have predated most of the tilting.

The Koyna River has gravel terraces downstream from Helwak, where its direction of flow changes from south to east, but gravel fill and terraces are absent upstream from the bend. Gupte and others (1972) say that about 25 m of alluvium is found at most Deccan damsites situated on east-flowing rivers. Recurrent southerly tilting, such as occurred after the second surfaces formed, is a more likely explanation than faulting for nondeposition in the rejuvenated north-south reach.

Profiles 8-8' and 9-9' show the Western Ghats. The Mahabaleshwar surface is above 1,000 m. On the Konkan are remnants of erosion surfaces below 300 m. Intermediate vestiges are near the Ghats. The highest Konkan surface may be the Mahabaleshwar surface, for both are deeply laterized. Displacements on faults probably explains formation of the Ghats, since scarp retreat cannot.

Faults cannot be located by profiling the thalweg of a stream because a knick-point formed by offset retreats from the trace. The Jagbudi River (profile 9-9') coincides with a striking lineament that trends N15°E for about 27 km and even crosses the continental divide. In homogeneous rock, such a trace must be a fault. Yet erosion has so thoroughly leveled the surface across the fault that its existence is concealed. Many faults trending N15°W to N15°E lie in a zone 10 to 20 km wide, as demonstrated by exposures, by topography, and by the zone of hot springs.

Ground breakage occurred during the December 10, 1967, earthquake. The breaks were reported in a zone 45 km long (Fig. 20) and trending N23°E, parallel to computed focal-plane mechanisms (Guha and others, 1970). The minimum observed extent of ground breakage is 7 km. The active fault is about 6 km downstream from Koyna Dam, traversing the villages of Donachiwadi and Kadoli, north and south of the Koyna River, respectively. Most of the trace cuts laterite, but on the rim south of Kadoli, bedrock is visible where landslides have peeled away from the scarp. The ground breakage forms a sinistral en echelon pattern (Fig. 22). North of Danachiwadi, an earthen retaining wall is offset 30 cm left-laterally. Erosion between 1967 and my inspection in May 1973 obliterated many details. North-northeast of Donachiwadi, the trace is oblique to the sloping ground, giving an unnatural aspect to the erosional gully formed along the fault. An older gully, 2.5 m deep, is parallel to the 1967 trace (N20°E) and about 10 m east of it. Serving no irrigation purpose, it is probably a fault trace predating Koyna Reservoir. Koyna and Kremasta, Greece, may be the only seismogenic reservoirs to have initiated ground breakage, hence it is important to recognize that breakage also preceded reservoir filling.

Fracture orientations in the Koyna region vary from N25°W to N25°E, constituting one set or two. The westerly trends may have formed while India drifted along between transform faults parallel to the Chagos-Laccadive zone. The more easterly fractures may be more recent, formed during northeast drifting along Indian Ocean wrench faults. The modern north-northeast faulting (the Koyna earthquake and attendant ground breakage) is

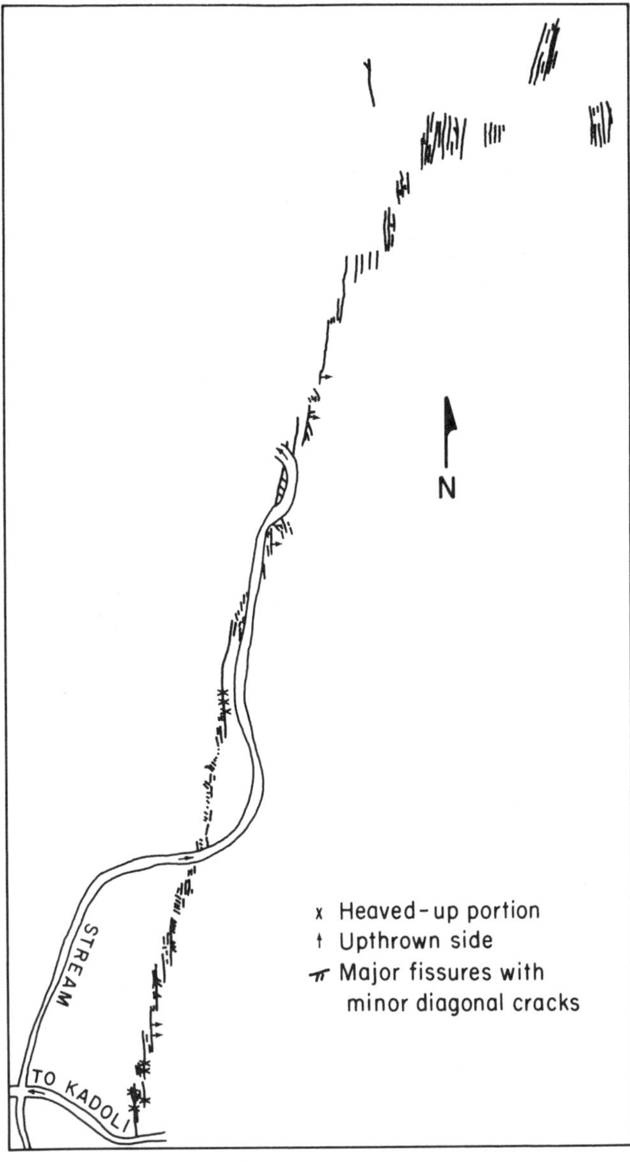

Figure 22. Fissures and cracks south of Koyna River (Das and Roy, 1969).

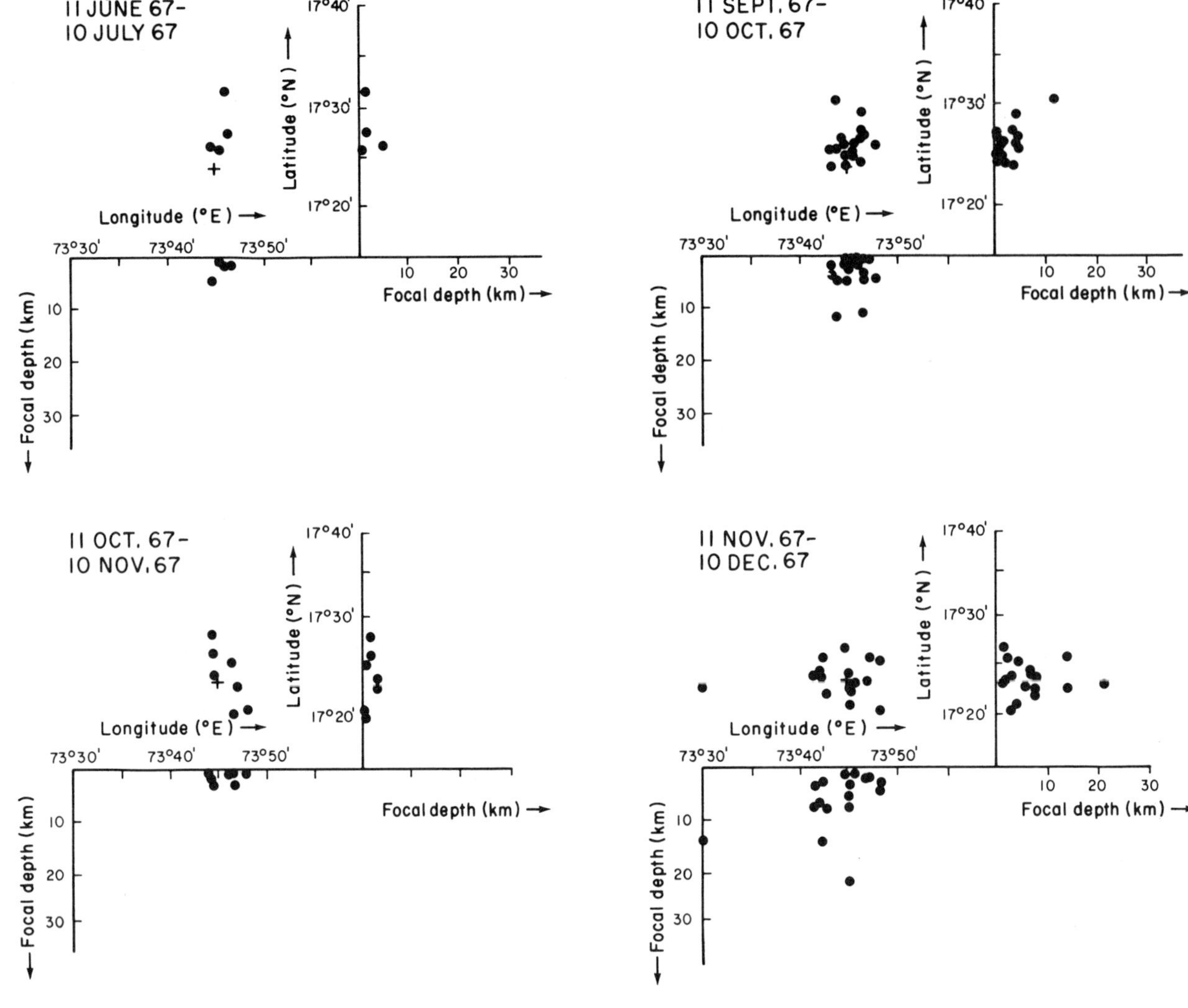

Figure 23. (Above) Projections on different planes of foci of foreshocks in the Koyna Project area. (Facing page) Projections on different planes of foci of aftershocks in the Koyna Project area. Note the migration of hypocenters southwesterly and to greater depths (Guha and others, 1970).

consistent with the tectonic setting. The recent seismicity is merely another event whose time of occurrence, but not cause, was determined by the hydrologic change brought about by reservoir filling.

Hydrology

In Maharashtra, nearly all the rain falls during the June to September monsoons, and drought prevails the rest of the year. In the Koyna region, near the continental divide, normal annual rainfall is about 7.5 m, abnormal as little as 4.0 m. Following deficient monsoons, ground-water storage is depleted and base flow disappears. Since trap rocks and laterites are poor reservoirs, storage lakes are vital for agriculture and power--the main purpose of the Koyna Project. From Koyna Reservoir (elevation 660 m), a 3.7 km tunnel through the divide delivers water to penstocks and a power cavern at Pophali on the Vaitarna River (elevation 170 m). The declivity of the Western Ghats makes the project a unique hydroelectric scheme. Likewise, there is a steep hydraulic gradient for ground waters in the underlying trap rocks, a fact bearing on the reservoir-induced seismicity.

During monsoons, Koyna Reservoir rises quickly, about 55 m, producing a greater annual load fluctuation than any other seismic reservoir; the effects should be noticeably periodic. Guha and others (1974a) found that the seismicity had an annual return period with a 6- to 8- week lag behind peak reservoir levels. The correlation was best from the time of the major quake of 1967 until 1969. There are suspected hydrologic reasons for such behavior.

When the rains come, the entire terrain is recharged. At Mahabaleshwar, dug wells 5 m in diam-

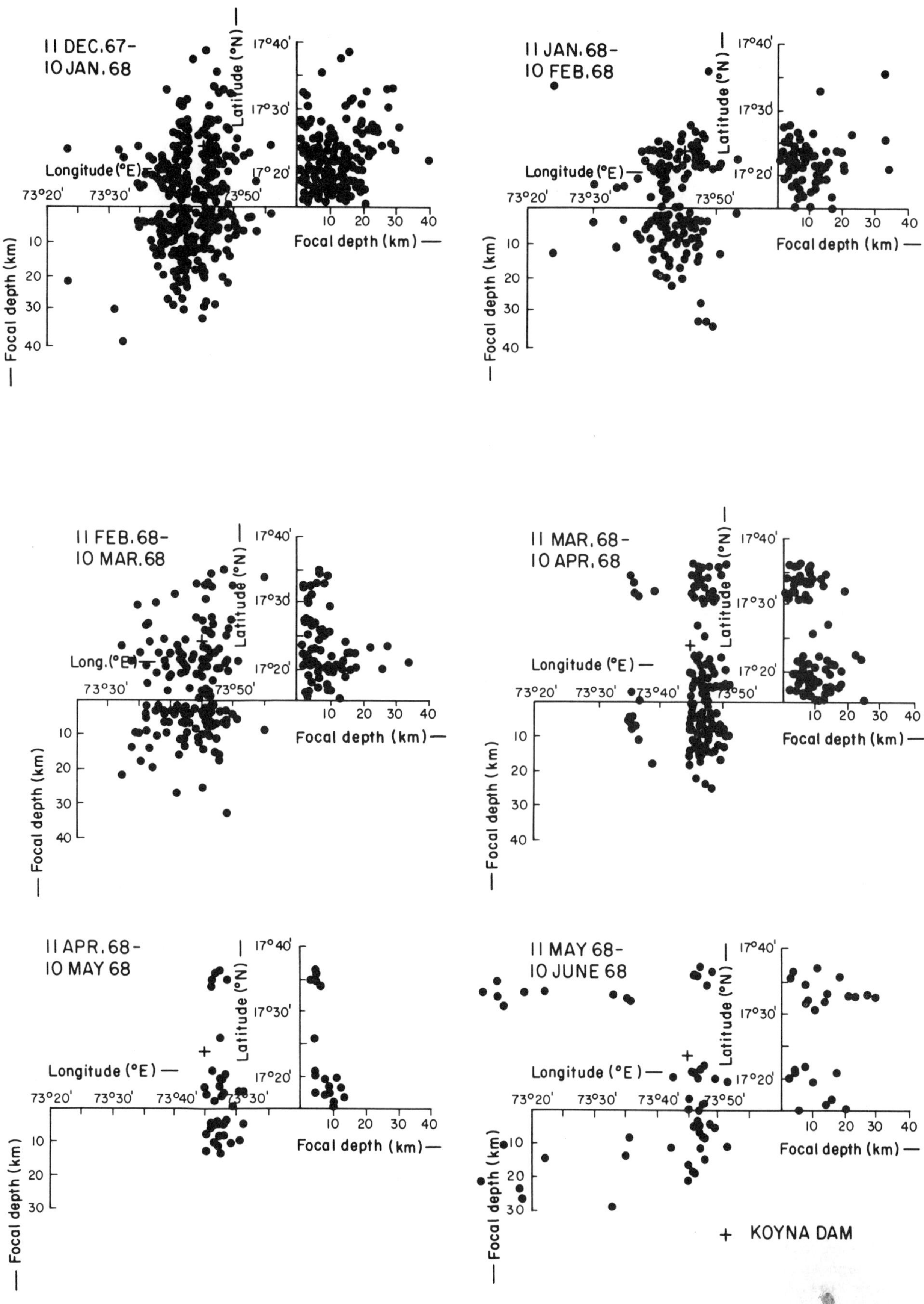
11 DEC.67-
10 JAN.68
Latitude (°N)
Longitude (°E)
Focal depth (km)
11 JAN.68-
10 FEB.68
11 FEB.68-
10 MAR.68
Long.(°E)
11 MAR.68-
10 APR.68
11 APR.68-
10 MAY 68
11 MAY 68-
10 JUNE 68
+ KOYNA DAM

eter penetrate 25 m of laterite to unweathered, less permeable basalt. In droughts, water is drawn from the laterite until it dries up. During the rains, the wells fill to the surface in response to ground-water recharge. At the edge of the flat topographic surfaces, a few springs emerge from the base of the laterite. On the rolling Mahabaleshwar surface, some springs emerge above the base, the ground water being perched on caliche layers. Because laterite mantles young erosion surfaces near the rivers, as well as old, high surfaces, seasonal fluctuations of pore pressure (1 to 3 bars) must invade the basalt from all recharge boundaries. Rivers like the Koyna and the Vaitarna, which are drainage boundaries for the entire ground-water province, fluctuate only a few metres, whereas reservoirs fluctuate many metres. If permeability distribution were known, changes in hydraulic boundary conditions could be translated into changes of effective stress throughout the region.

The Deccan basalt is nearly impermeable, as indicated by meager well yields and nearly dry excavations. Ground water constitutes a marginal resource. Whereas compact basalt has continuous, curving fractures, they appear watertight in underground excavations. Amygdaloidal basalt is essentially unfractured, and cavities are seldom interconected, thus its permeability is even lower. Yet there are defects that contain and convey small amounts of water. The red boles are probably impermeable in situ, but some flow contacts are permeable, as indicated by rare seeps in excavations and horizons of grass on cliff faces. Rare springs occur in gullies crossing permeable contacts. Meager discharges (up to a few litres per minute) support a dry-farming economy. Some springs continue to flow long after the rivers disappear.

The permeability of fractured rock generally diminishes with depth, partly because fractures close with increasing effective stress (Snow, 1968b). In lavas, fractures fill with minerals, thus close with depth. Nevertheless, miners know that faults can conduct copious amounts of water to depths of over 1 km. Pore pressure penetrates far deeper (> 3.5 km in the Rand mines). Nearly vertical north-south faults near Koyna are hydraulically significant. Typical of these are the zones cut by the Stage III tailrace tunnel. Leakage along these faults was about 0.1 l/s where the tunnel crosses them 6 m under the Vaitarna River. Gouge makes them very poor water conductors but capable of transmitting pore pressures, presumably to great depths. Lesser faults and fractures may be better conductors if clean than are the wide, gouge-filled ones. Vertical fractures outnumber the permeable horizons, but the vertical permeability may not exceed the horizontal when measured in east-west planes. However, the geometry of the two-set system assures anisotropy: maximum permeability is horizontal and north-south, possibly twice that on east-west planes (Snow, 1969). Therefore, pore-pressure changes must propagate more rapidly southward of Koyna Reservoir than downward or to either east or west. The distributions of hypocenters after the December 10, 1967, quake (see Figs. 23 and 26) are consistent with that inference.

Hot springs of the Konkan (Fig. 17), Kariba, and Kremasta indicate deep circulation. Aligned in a north-south band near the foot of the Western Ghats, they are probably fault-controlled.

There are reports (Gupta and others, 1967; Agashe and Gupte, 1971) that the temperature and discharge of springs changed at the time of the major quake of December 10, 1967. Guha and others (1974a) have noted that reliable data are absent for the pre-quake period. Increases of temperature and discharge (Table 4) are consistent with the effects of major earthquakes noted elsewhere (Sherard, 1966, p. 34). Presumably, stress drops open fractures, augmenting flow at fixed gradients. More rapid rise of water results in less cooling.

Hot water discharge proves that steep conduits exist, whether the heat is due to deep meteoric circulation or to volcanism (Agashe and Gupte, 1971; Gupta et al., 1972). Mild chemical content and modest temperatures (< 60°C) suggest a meteoric origin. Ground-water circulation in and below the traps is at least 1 km deep (depending on dilution).

Like the springs, Koyna Reservoir overlies one or more faults, including some that may have controlled the erosion of the north-south reach of the river. If the hot springs are at the discharge end of a complex U-tube conduit system (Fig. 24) and the Koyna River is at the recharge end, then the raising of Koyna Reservoir should eventually increase spring flows and pore pressures. Absence of seismicity at other reservoirs of the Western Ghats could relate to absence of such conduits, limiting the regional hydraulic effect. The north-south orientation of Koyna Reservoir parallel to the fault system probably produced a greater, more pervasive pore-pressure change than have reservoirs trending east-west across the faults. The extent is governed largely by the scale of the ground-water flow system affected by the reservoir. Since most reservoirs are elongate, their effect on potential distribution may best be envisioned in transverse sections. But a longitudinal section may also be necessary to reveal certain important

Table 4. THE TEMPERATURES OF KONKAN HOT SPRINGS BEFORE AND AFTER THE DECEMBER 10, 1967 EARTHQUAKE.

	Temperature (°C)		
Hot Spring	1940	1956	Post quake
Unhera, Pali	41.5	43.3	46
Sow, Mahad	41.5	40	43
Aravali	40	42.2	43
Rajwadi	54	47.8	54
Tural	61	60	65
Rajapur	42	41.1	44
Agni Kund *	58	60	58
Anasuya Kunda *	56	50	54
Bhimeshar Kund *	51	46.1	52
Kothavalas' Spring *	52	46.7	50
Surya Kund *	50	47.8	50
Chandra Kund *	50	44.4	49

Source: Agashe and Gupta, 1971.
*Vajreshwari.

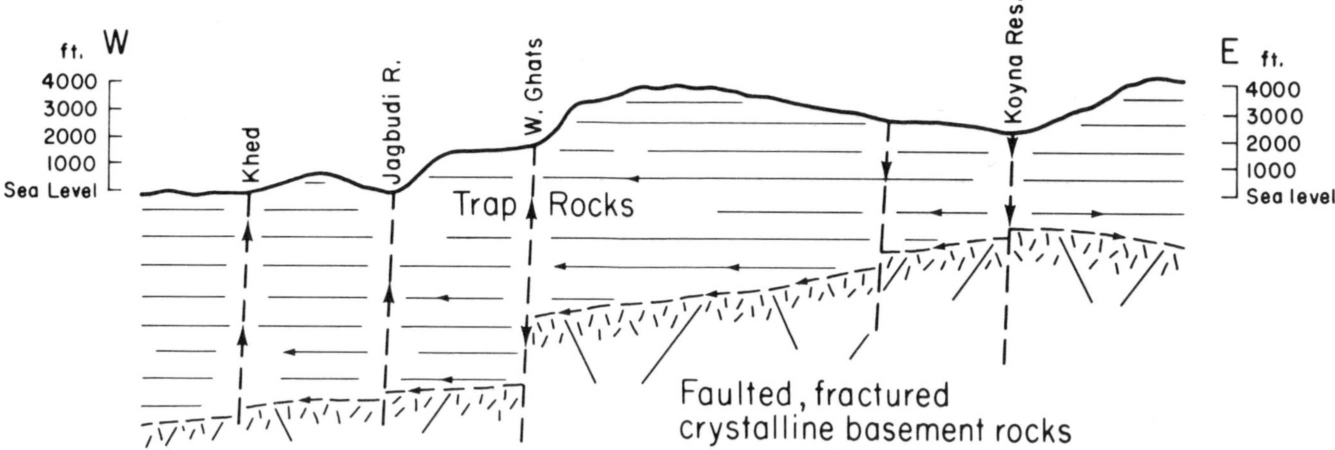

Figure 24. Schematic geologic section (exaggeration 5X) illustrating possible U-tube ground-water circulation in faults and contact surfaces of Deccan trap rocks and in faults and the fractured regolith on the buried basement granitic (?) rocks. Pore pressures in faults older than Cretaceous are governed by the potential distribution in the ground-water flow system.

effects. If a river is one of many flowing off a highland, the regional flow pattern at considerable depths is parallel to the rivers. Because Koyna is parallel to the continental divide, it probably raised the potentials on a regional scale, as well as on the local scale of its own valley. This may be a cause of the pronounced induced activity at Koyna.

Koyna promotes recharge high on the eastern slope. As depicted in Figure 24, flow is downward along vertical north-south faults, horizontal westward 8 to 10 km along contacts, then upward along faults to discharge at low positions in the Konkan. The distribution of boundary potential change is controlled by contrasts of conductivity along the seepage paths. If the distal end (near the springs, in this case) is resistive compared to the proximal, then most of the reservoir potential change applies along the horizontal portions. The deep-seated flow segments are most important, for basement faults may not extend into the cover (trap rocks), but probably communicate with the unconformity. The hydraulics of a surficial layer can influence pore pressures in much deeper zones, especially via essentially dead-ended faults extending to hypocentral depths. Because total-stress changes from the surficial load of a reservoir are trivial (and of wrong sign) at depths of 5 to 10 km, pore-pressure transmission to such depths is a plausible cause of induced seismicity.

A reservoir produces a three-dimensional hydraulic influence. The sharp discontinuity in charge of boundary potential at the dam results in steep gradients and flow downstream. When potential rises downvalley, remote from the surficial deadweight, it develops a region of maximum decrease of effective stress and may stimulate seismicity. Such was the case at Koyna (Figs. 23a and b) after December 10, 1967. At Hsinfengkiang Reservoir in China (Wang and others, 1976), most hypocenters were beneath the canyon reaches, whereas few were under the wide body of the reservoir. In wrench-fault environments (India and China), decreasing effective stress can develop induced seismicity, whereas increasing effective stress stablizes.

Geophysical Observations

Guha and others (1970) catalogued earthquakes for the period A.D. 1341 to the present (Fig. 25). Before the M = 5.6 event of September 1967 at Koynanagar, there were 163 recorded in peninsular India, none exceeding an estimated Richter magnitude of 6.5. Along the western coast and offshore, events up to M = 4.0 have occurred. Minor quakes have been slightly more frequent on the east coast than the west, but the interior has been so quiescent that engineers have generally considered the peninsula to be a stable aseismic block. The distribution of coastal quakes may represent breakage along plate margins, whereas the infrequency of interior quakes suggests continuity of stresses there, like those of the margins. The linear west coast and the strong associated gravitational gradient (Fig. 18) signify an important tectonic discontinuity. Guha (1970, p. 14) attributes the seismicity of the peninsula to plate tectonics--specifically, to a coastal transform fault. Focal-plane solutions and ground breakage are consistent with left-lateral wrench faulting.

Attributes of normal faulting are doubtful. Uplift of Deccan erosion surfaces may have accompanied greater wrench-fault components. The Committee of Experts (1968) and Guha and others (1970) speculated on the relationship of the earthquakes to a coastal area fold between Bombay and Kutch, called the Panvel Flexture. Gupte (oral commun., 1973) questions its existence: in the welter of different initial dips of basalt flows, there is no discernible systematic dip.

After the first filling of Koyna Reservoir during the period July to September 1962, small (felt) quakes at Koyna became common, increasing after the 1963 monsoon. The rattle of windows was accompanied by rumbling sounds, such as have also been reported from Kremasta, Monteynard, Lake Mead,

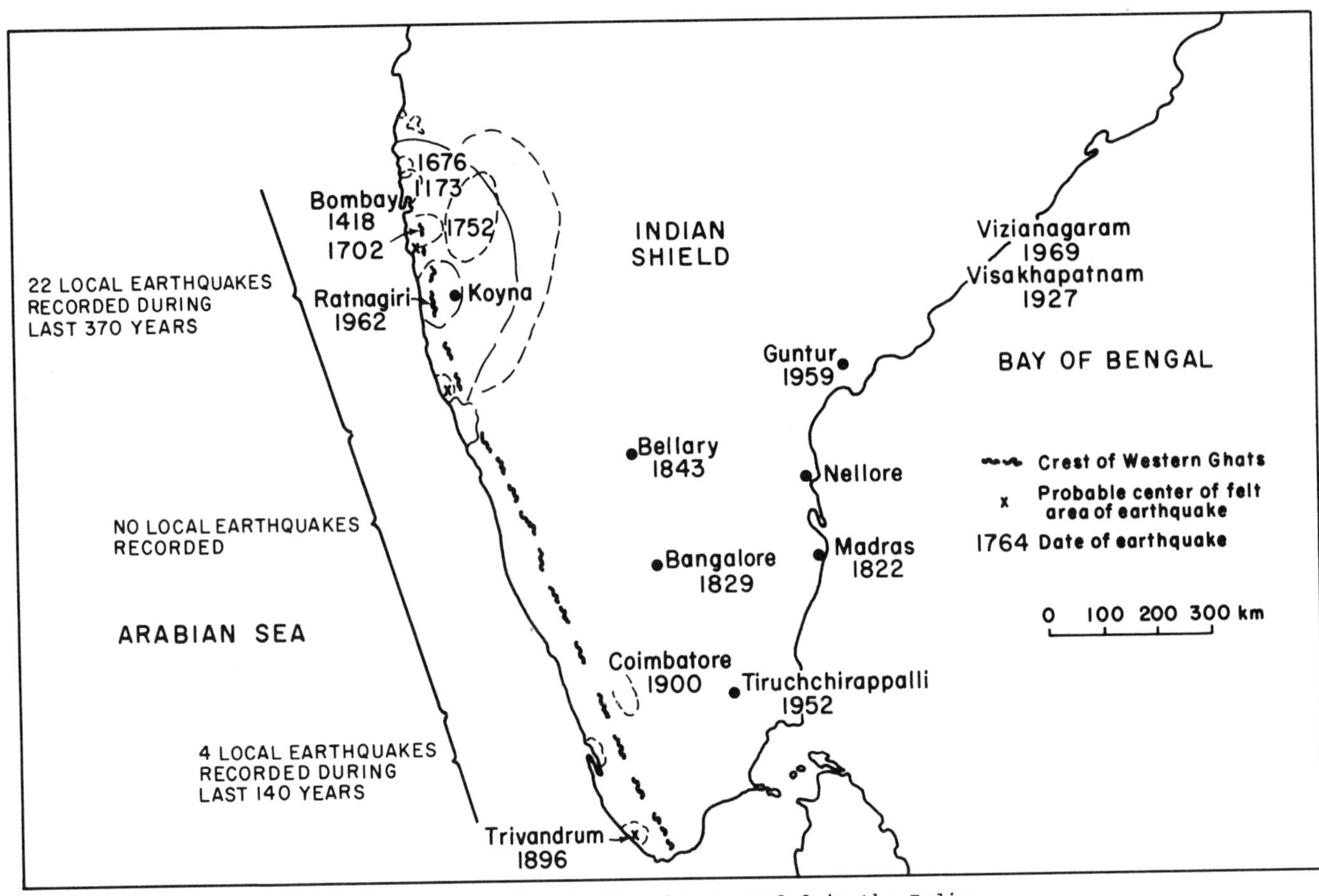

Figure 25. Earthquakes recorded in the Indian Peninsula.

and other seismic reservoirs in the world. The quakes have been attributed to very shallow foci.

The first quake causing concern for safety occurred at Koynanagar on September 13, 1967, with magnitude 5.8, felt to a radius of 120 km. It had intensity VII, falling off sharply nearby. Its focus was at 3 to 7 km, probably below the basalt. Figure 26 shows isoseismals of the September 13 and December 10, 1967, quakes, both centered close to the dam. The September 13 event stimulated instrumentation efforts by the Koyna Project management and the Central Water and Power Research Station. The December 10 quake made the work assume grave importance. No other structure has been so nearly destroyed by man-made earthquakes as was Koyna Dam. With many power projects in planning and construction stages, Indian geophysicists and engineers are "leaving no stone unturned" in efforts to understand the phenomenon. Koyna is currently the best-documented seismic reservoir in the world.

The magnitude 7.0 earthquake of December 10, 1967, had its apparent epicenter about 10 km north of the dam, as computed from remote stations, but the isoseismals centered about 6 km southeast of the dam, coincident with the ground breakage. It was the greatest quake in peninsular India's history. The damage exceeded that of any other man-made quake, shutting down the hydroelectric plant, cracking the concrete gravity dam from face to face below waterline, totally destroying two villages, razing many of the construction facilities at Koynanagar, killing about 200 people and injuring over 2,000. Masonry home construction was a prevalent cause of loss. Many rockfalls and a few landslides were initiated. Ground breakage occurred intermittently along the fault zone, trending N20°E through the center of greatest intensity and parallel to the long axis of the elliptical isoseismals. Fault-plane solutions for both the September and December quakes (Lee and Raleigh, 1969; Sykes, 1970; Tandon and Chaudbury, 1971) were consistent with the pattern of en echelon cracking (Fig. 22), indicating left-lateral displacement in offsets of about 30 cm. Fractures were reported elsewhere (see Fig. 26b) also oriented north-northeast. In common with other large, shallow, reservoir-associated quakes, the region of high intensity was local, whereas the felt area (radius 700 km) was large. Guha and others (1970) theorized that the energy was channeled in low-velocity rock underlying the basalt. For aftershocks, both Pg and Sg velocities decrease with depth of focus, consistent with the underlying granite, whose existence was anticipated by Glennie (1932). Velocity of Pg was an appropriate 5.6 km/s. The December 10 focus was undoubtedly shallow, judged by the intensity pattern. Foreshocks were all less than 4 km deep, whereas aftershocks were generally deeper.

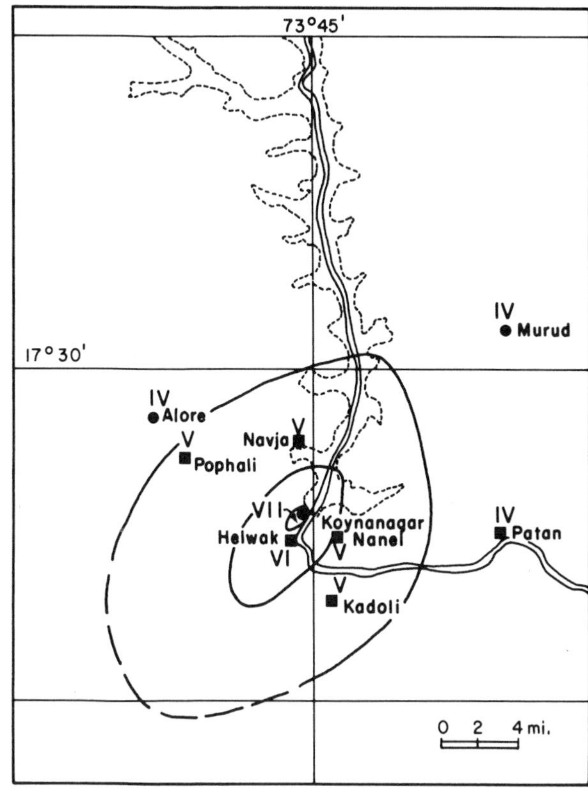

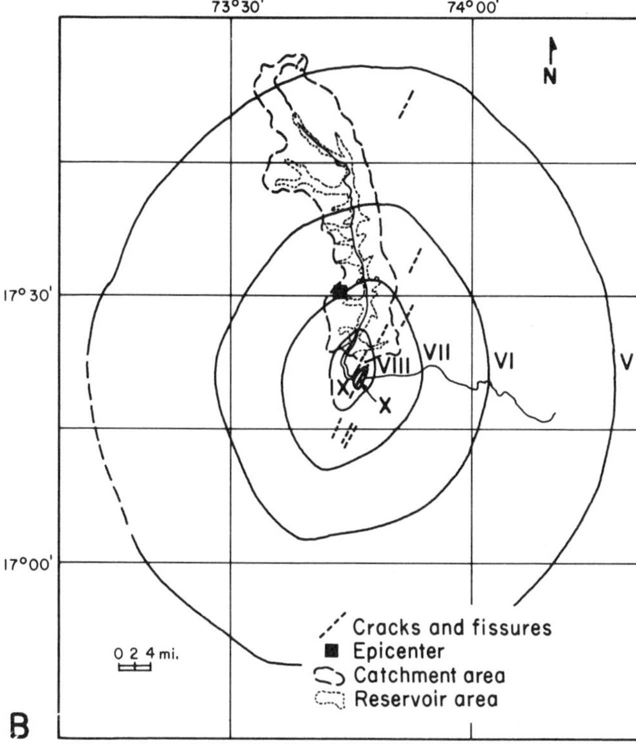

Figure 26. (A) Isoseismals of the earthquake of September 13, 1967, Koynanagar. (B) Isoseismals of the main earthquake of December 10, 1967 (Guha and others, 1970).

Most foreshocks had epicenters upstream from the dam, but most aftershocks were downstream, to the south and west (compare Figs. 23a and 23b). Quakes of larger magnitude had deeper foci. Seismicity that began in the basalts may have propagated into the granite basement, consistent with a hydraulic transient inducing failure upon attainment of sufficient pore pressure.

Because of power failure, no record was made by the seismograph at Koyna Dam, but an accelerograph and pendulum tiltmeters produced valuable records close to the epicenter of a major quake. A 16-s tilt immediately preceded the earthquake and recorded about 1 m of subsidence near the dam. Some ground disturbance seems to occur whenever a quake of $M \geq 4.0$ is recorded. The proportion of the deflections due to ground movement or to the dam deflecting its foundation is unknown. Carlson strainmeters at the base of the dam recorded sharp changes that do not generally coincide with the major shocks (Fig. 27), though their sense is believed consistent with the stress drop of the

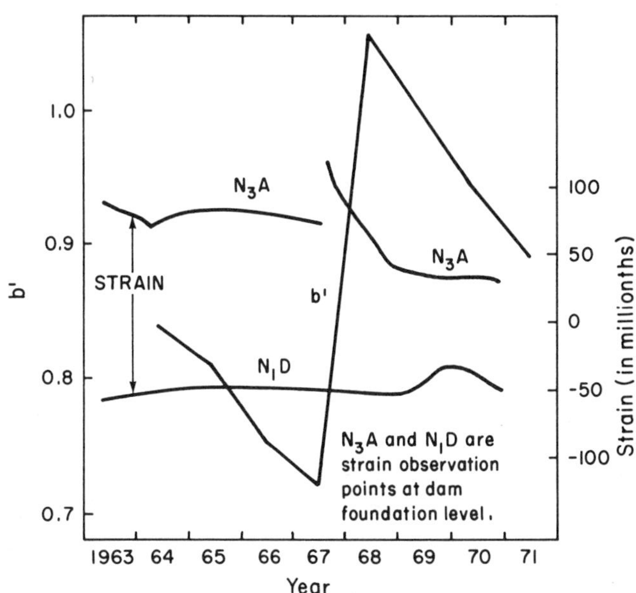

Figure 27. Changes of the magnitude-frequency slope parameter b', related to the time of the December 10, 1967, quake (Guha and others, 1974a).

main 1967 event. Stress changes of about 12 bars (40 microstrains) compare well with the stress drop computed from other field data and the change in slope b' in the frequency-magnitude relationship (Fig. 27). Hourly magnetic observations employed two proton-precession magnetometers, one near Koyna and another 120 km away at Poona. Total field changes in the reservoir area could be attributed to aftershocks. Changes were large compared to earthquake responses elsewhere, perhaps due to the high piezomagnetic susceptibility of basalt. Telluric observations showed no significant changes attending any of the $M > 4$ aftershocks.

A system of permanent seismic observatories (Fig. 28) was established in 1968. By 1973, over 25,000 $M > 1.0$ quakes had been registered, a unique record of a man-made swarm (Guha and others 1974b).

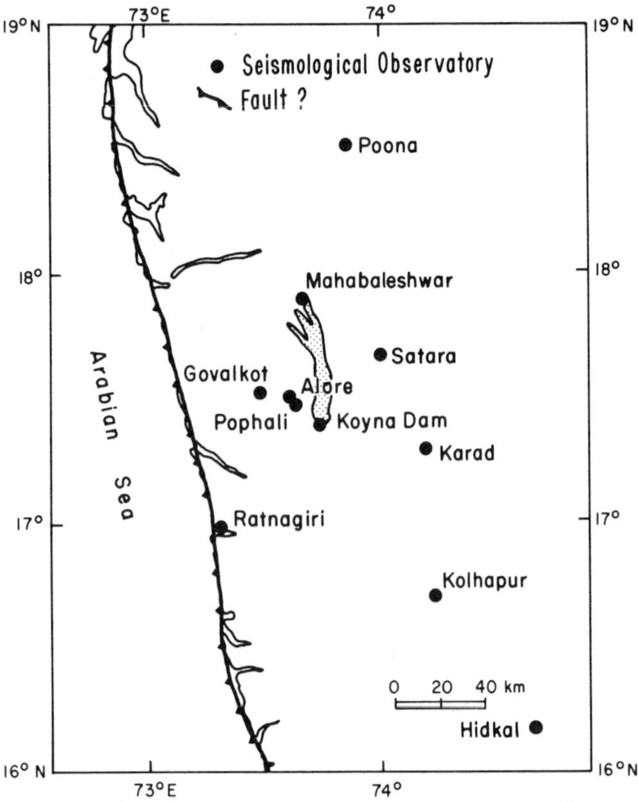

Figure 28. Seismological observations of West Maharashtra (Guha and others, 1974a).

From the aftershocks, Guha and others (1974a) have determined exponential decay laws for aftershock frequency and energy. Deviations from the law, expressed by the ratio of observations to expectations, reveal seasonal variations. Good correlation between frequency and lake level can be shown (Fig. 29) for 1968 to 1971, with a time lag of 3 to 4 months. As at Lake Mead (Carder, 1968), Kremasta (Comninakis, and others 1968), Kurobe (Hagiwara and Ohtake, 1972), and Denver (Simon, 1969), the correlation deteriorated during later cycles.

Epicenters are spatially related to the reservoir, clearly clustered near the deep end (Fig. 30). They form half an ellipsoid, with major axes plunging about 20° to the N20°E, parallel to the pattern of isoseisms and the December 10, 1967 fault. The center lies about 4 km downstream from the dam, so that most of the hypocenters are well out of the reservoir's total-stress field. The plunge reflects diminishing total-stress changes at increasing depth. Eccentricity of the hypocentral pattern and scatter toward the west of the reservoir (Fig. 23b) are believed to be explained by the hydraulic anisotropy.

There were changes in the frequency-magnitude relationship: as seen in Figure 27, the slope b' decreased regularly during 1964-1967, increased abruptly after the main 1967 shock, and has decreased regularly since. By installing microseismic networks at new reservoirs, Guha and others (1974a) hope to predict soon after filling whether danger exists according to changes in b'. Damage might be minimized by this method, though indications would be available too late for design modifications. Nine sites in the peninsula interior and in the seismically active Himalayan region have shown no reservoir-induced enhancement. One in the interior and two on the western margin have had only minor isolated shocks, consistent with background for the area. One reservoir on the eastern margin and three (besides Koyna) on the western margin may have produced seismic disturbances. The

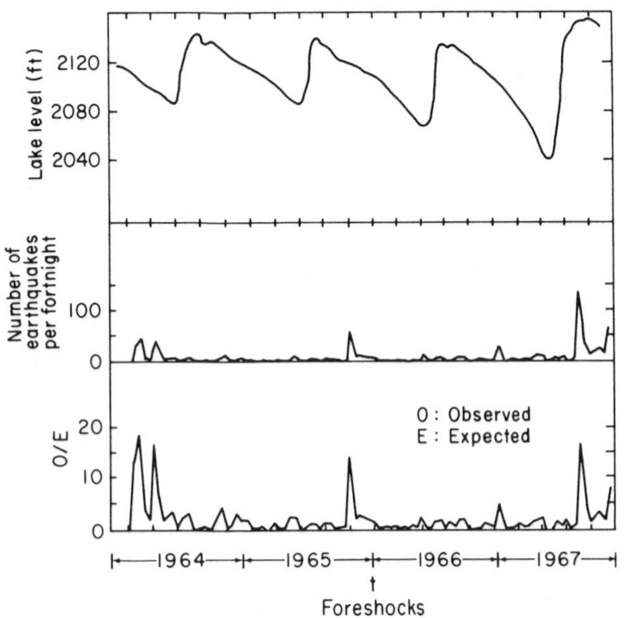

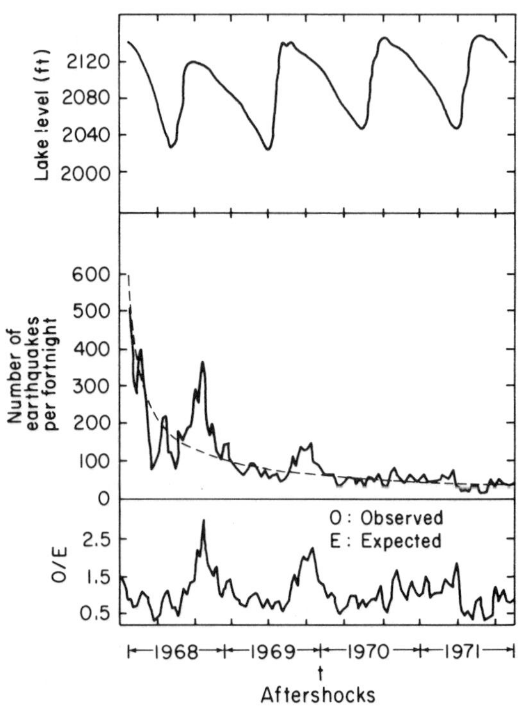

Figure 29. Correlation of earthquakes with lake levels (Guha and others, 1974a).

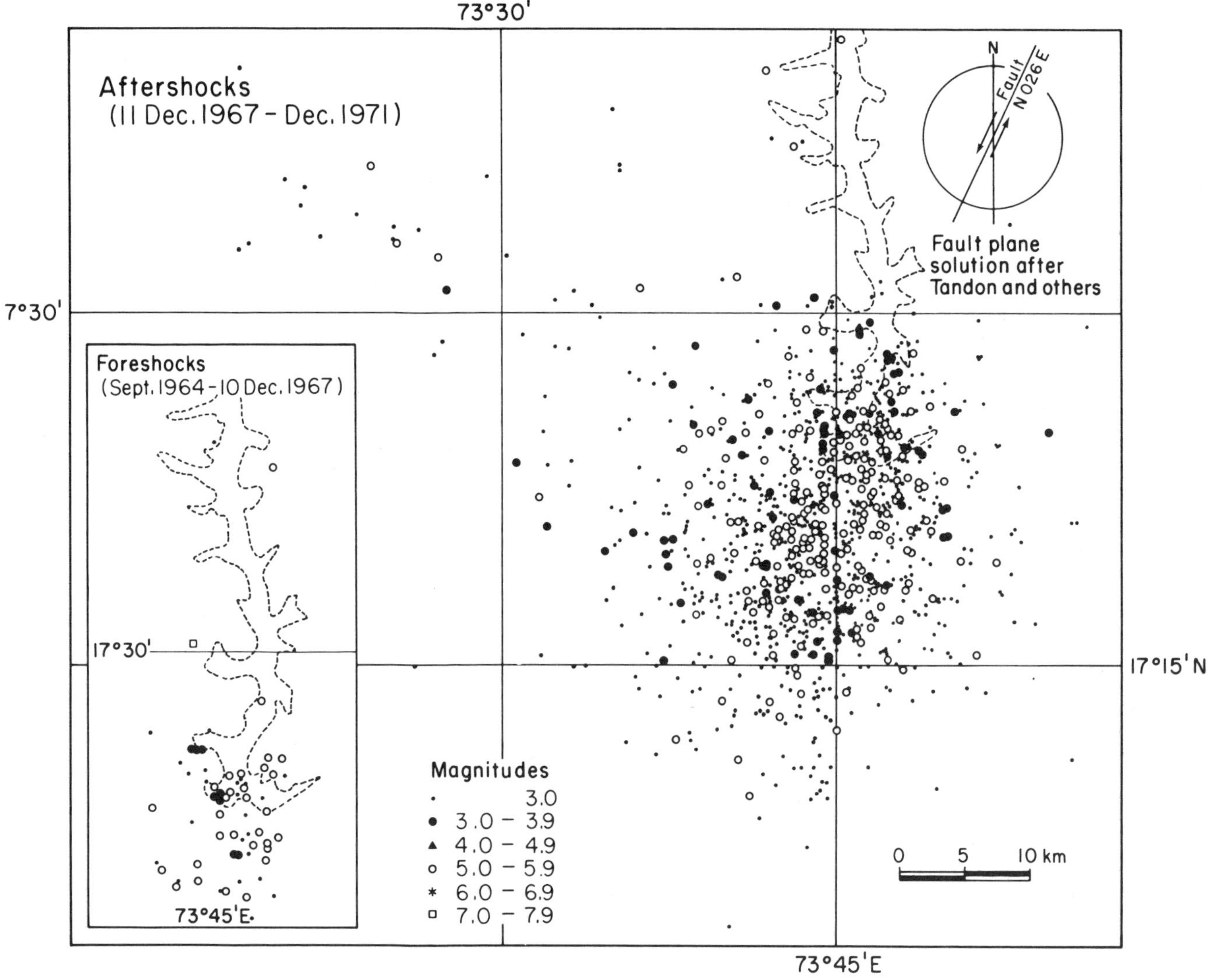

Figure 30. Distribution of epicenters for foreshocks and aftershocks (Guha and others, 1974a).

seismic history of these localities seems too short and uneventful to make a strong case for reservoir triggering. However, a brief microearthquake swarm (M < 0) occurring during first filling of Mula Reservoir shows a convincing relationship to reservoir level. These data are similar to observations of microseismicity at Hendrick Verwoerd Reservoir, South Africa (Green, 1973). Microseisms may attend the filling of many reservoirs, and doubtless others will be found throughout the world as sensitive instruments are deployed. High values of b', as found at Mula Reservoir, may suggest safety in some cases.

Loading History

The objective of these case histories is to find special circumstances that discriminate between seismic reservoirs and aseismic reservoirs. A new tool is the relation of the reservoir's induced stress changes to the history of pre-reservoir stress changes (Snow, 1972). The history from the time of last faulting is most significant, but an even longer record deduced from geomorphology also proves instructive.

Though pre-reservoir seismicity in the Indian peninsula was subdued, the long history of rare events suggests that rock stresses have been critical somewhere, sometimes: continuity suggests that stresses have been high everywhere--that is, close to failure. At Koyna, stresses were critical before reservoir filling, as proven by the gullied fault trace parallel to the 1967 fracture northeast of Donachiwadi.

The filling of a reservoir in hydraulic continuity with faults will always diminish effective stresses nearby, especially in the horizontal plane. When critical conditions for wrench-fault slippage preexist, a shift of Mohr's circle toward the origin should stimulate motion. Thus, for Koyna, reservoir-induced seismicity is consistent with that hypothesis; the problem is that at other Deccan reservoirs, most of which are definitely aseismic, similar circumstances may also have prevailed, though perhaps to lesser degree.

The following analysis of stress history from the time of development of the Mahabaleshwar surface (Miocene?) to the present suggests that conditions for faulting must have recurred from time to time. Stress changes (Table 5) are attributable to geomorphic development:

1. Uplift of 810 m recorded in the elevation and tilt of the Mahabaleshwar surface decreased total horizontal stresses elastically (-0.024 b/m; Fig. 7) in the hypocentral region 1 to 3 km beneath Koyna, and cooling had an additional effect (-0.031 b/m). Simultaneously, the discharge end of the west-flowing ground-water circulation cell (Fig. 24) was lowered relative to the intake. In the cell, it is assumed that 1/3 the hydraulic potential is dissipated in tight vertical faults below the Koyna intake area, 1/3 is dissipated in vertical faults below the Konkan discharge area, and 1/3 is dissipated along the more pervious, but 40-km-long horizontal paths. At the bottom of the cell under Koyna, or nearby in any deeper faults connected to the cell at the regolith, hydraulic potential decreased about a third of the uplift, namely 270 m, increasing vertical and horizontal effective stresses.

2. The valleys of the Koyna and other east-flowing rivers were eroded. Several stillstands are recorded in terrace surfaces (Fig. 21). Stress changes were computed in two artificial steps whose sum is equivalent to the actual history: first, the Koyna River cut a narrow chasm, lowering the water table to its pre-reservoir configuration; then the solids formerly filling the valleys were eroded away. Lowering the river to elevation about 599 m, 564 m below the Mahabaleshwar surface, decreased hydraulic potential in the hypocentral region by 2/3, that is, 375 m, reducing pore pressure and increasing all effective stresses. Erosion has been evaluated by planimetering five east-west topographic sections across Koyna Valley (Fig. 31). Removal of 63% of the slab between horizontal planes at the river and the Mahabaleshwar surface resulted in an average unloading of 356 m of rock with density 2.85. Assuming Poisson's ratio to be 0.35, the decrease of horizontal effective stresses was computed. The net effect of valley formation was a decrease of stresses, 63.9 bars vertically and 17.1 bars horizontally.

3. Events equivalent to a uniform uplift resulted in stream incision in the Konkan. The uplift (285 m) and attendent cooling reduced horizontal effective stresses. In actuality, uplift was doubtless concurrent with the incision of Konkan rivers as well as with that of the Koyna and other east-flowing rivers. The west-flowing rivers cut into the peneplaned Konkan to depths below present sea level, since estuaries now extend 45 km from the coast. But these rivers now run on bedrock about where they cross the hot spring line (that is, near Chiplun and Khed), so that any coastal emergences and submergences had no effect on relative hydraulic potentials within the ground-water cell. Likewise, canyon-cutting and subsequent alluvial filling that occurred in lower-altitude reaches of the east-flowing Deccan rivers (Gupte, 1972) have had no influence on potentials in the cell since the Koyna River flowed on bedrock in the reservoir reach. Measured by upper surface elevations of 292 m and 274 m at Chiplun and Khed, respectively, uplift, Konkan incision, and average lowering of the discharge potential was about 285 m. Assuming 1/3 that decrease in the hypocentral region (95 m), all effective stresses increased 9.5 bars.

4. Filling Koyna Reservoir to 77.6 m imposed a deadweight and raised hydraulic potentials. Assuming that 2/3 of the hydraulic potential increase at the cell inlet is propagated to hypocentral depth, the reservoir decreased effective stresses about 5.2 bars. The deadweight, computed from cross sections, was equivalent to 12.5 m of water over the entire area (concentrated loads are spread out at depth). The corresponding increase of vertical effective stress is computed elastically. The net changes produced by filling the reservoir were

Table 5. EFFECTIVE STRESS CHANGES AT KOYNA

Event	Stresses (bars)			
	Vertical		Horizontal	
	Change $\Delta\sigma_v$	Cumulative $\Sigma\Delta\sigma_v$	Change $\Delta\sigma_h$	Cumulative $\Sigma\Delta\sigma_h$
1 Uplift of Mahabaleshwar surface, 810 m relative to Konkan	0	0	-44.5	
Lower potential 270 m due to circulation to the Konkan, 1/3 of uplift	+ 27.0	+27.0	+27.0	-17.5
2 Lower potential at Koyna River 564 m by river incision, 2/3 effective at hypocenter	+37.5		+37.5	
Erode Koyna Valley, equivalent to 356 m dry rock, ρ = 2.85	-101.4	-36.9	-54.6	-34.6
3 Lower potential 285 m at Konkan, 1/3 effective at hypocenter	+9.5		+9.5	
Uplift of peninsula 285 m, cooling	0	-27.4	-15.6	-40.7
4 Raise potential by filling Koyna Reservoir 77.6 m	-5.2		-5.2	
Add equivalent 12.5 m infinite reservoir deadweight	+1.3	-31.3	+.7	-45.2

Note: For wrench faulting, cumulative changes were $\Sigma\Delta\sigma_1 = \Sigma\Delta\sigma_3 = \Sigma\Delta\sigma_h$.

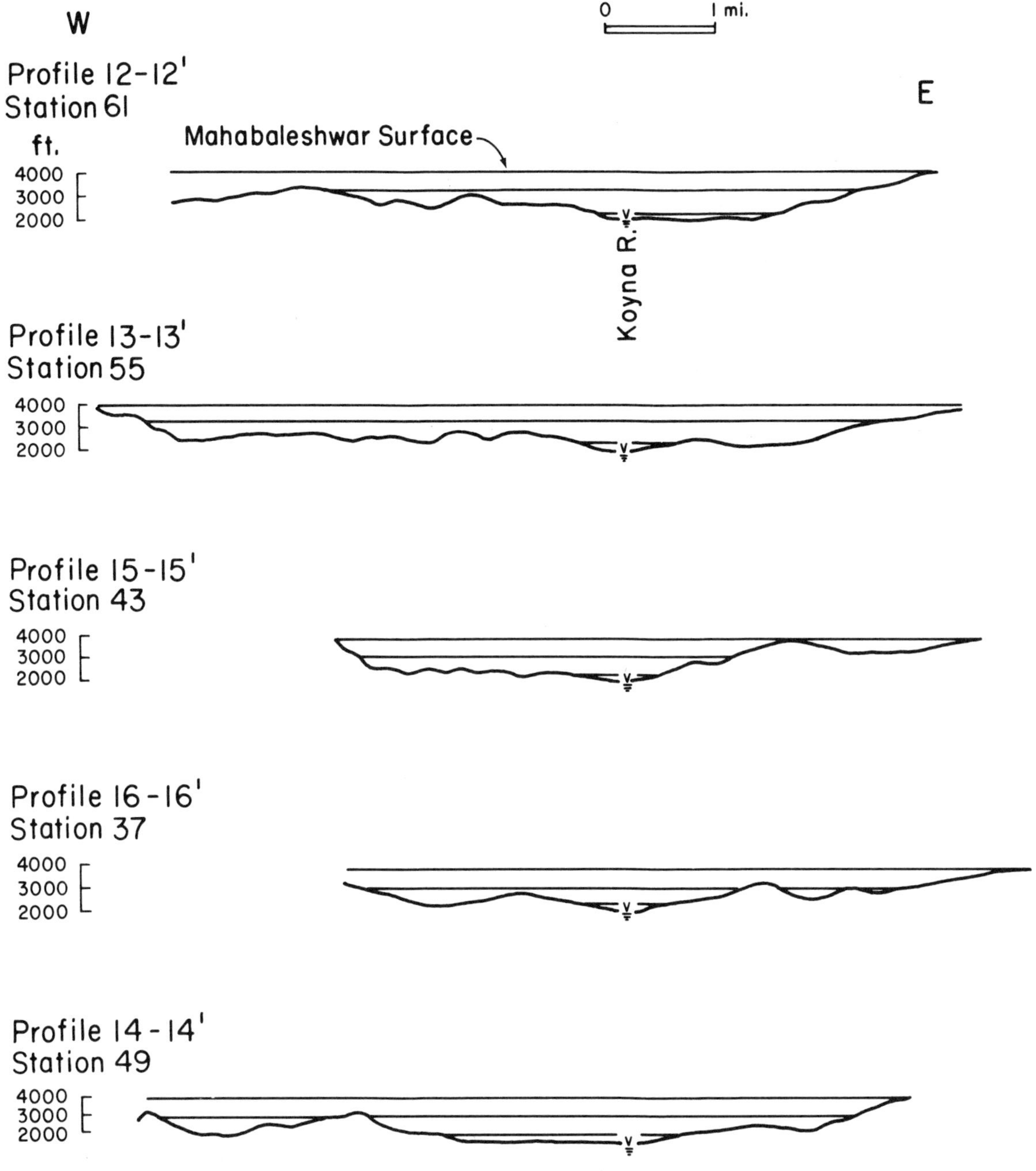

Figure 31. Topographic profiles across Koyna River valley.

decreases of 4.0 bars vertical effective stress and 4.5 bars horizontal effective stress.

Mohr's circles in Figure 32 summarize the stress changes. The inital circle (centered at 0) is some prehistoric state of tectonic stresses at an indefinite depth. Because $\Delta\sigma_1 = \Delta\sigma_3$, the circles maintain constant diameter, and with progressively decreasing effective stresses, 1 to 3, the stability locus moves horizontally toward the failure envelope (dashed to dotted circles). Because faulting occurred at Donachiwadi before the reservoir project was conceived, it is presumed that stresses attained criticality, as shown by tangency of circle 3. Further adverse changes were produced by reservoir filling; faulting took place to diminish σ_1 or increase σ_3 or both, maintaining tangency.

DISCUSSION

It is encouraging to find that seismogenic reservoirs can be explained in terms of effective-stress changes that prepared the site and then triggered failure, but stress history may never be routinely diagnostic. Many aseismic reservoirs would also prove to have a history of decreasing "stability."

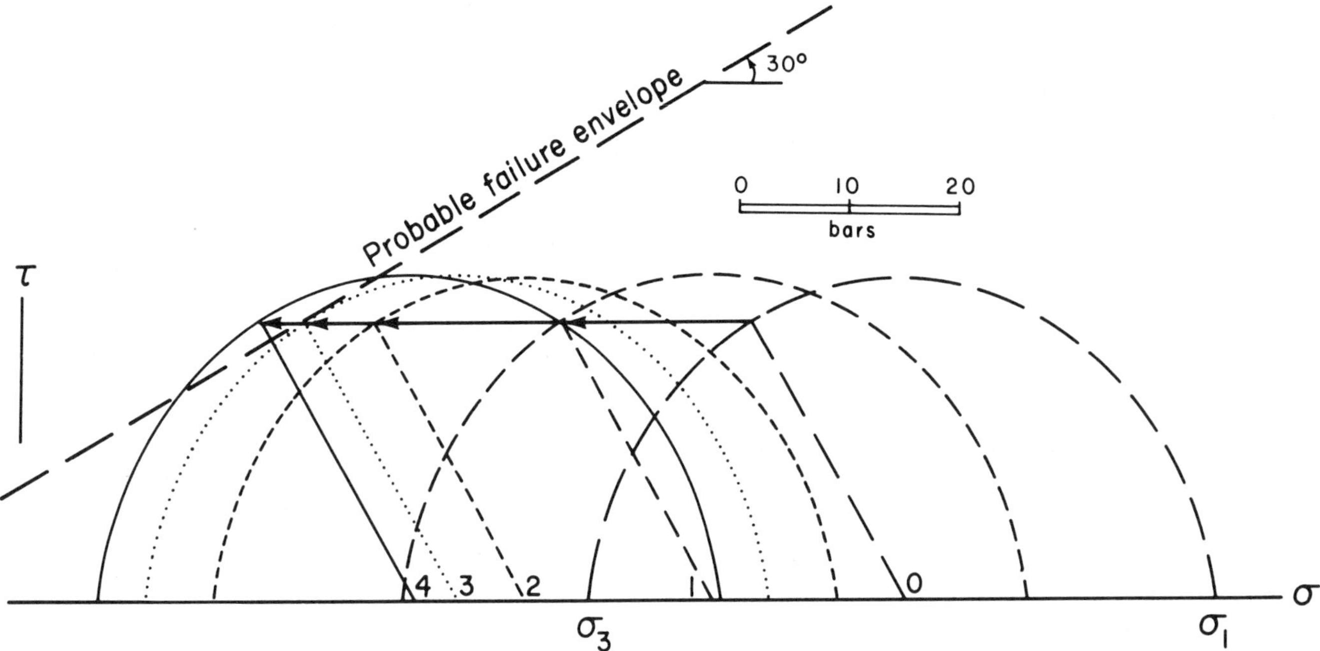

Figure 32. Mohr's representation of horizontal ($\Delta\sigma_1 = \Delta\sigma_3$) compressive effective stress changes at Koyna Reservoir since earliest Tertiary time. The dashed circle (center 0) schematically represents the original (Miocene?) stresses before uplift began. The size of the circles is arbitrary, since magnitudes and deviator stresses vary with depth. Circles 1 to 3 show historic changes, the effects of uplift, erosion, and changes in piezometric level up to the prereservoir condition of criticality (active faulting). Circle 4 indicates reservoir-induced movement toward further instability relieved by faulting.

Uplift, erosion, and rising piezometric head are the main causes of decreased effective stress, processes common to reservoirs in most mountainous areas of the world. Typical aseismic reservoirs should be studied in as much detail as the obviously seismic reservoirs. Handin and Nelson's (1973) study of Lake Powell is a unique investigation of the lack of seismicity at a major impoundment. At the First International Symposium on Induced Seismicity, Banff, Alberta, 1975, I predicted a reservoir to be aseismic because it was to be impounded (Palo Verde Reactor cooling ponds) above an aquiclude, therefore raising effective stresses in the bedrock.

To elucidate causes of reservoir-induced seismicity, it may seem necessary to demonstrate pore-pressure propagation to hypocentral depths. Costly peizometers, installed at depths of 2 to 4 km to record the effects of filling, would enhance the credibility of effective-stress-failure theory. Diffusivity determinations would allow propagation rates to be estimated elsewhere. Without piezometry we cannot know if such tardy events as the August 1, 1975 quake at Oroville, eight years after filling (Bell and Nur, 1978) or the swarm at Clark Hill Reservoir, South Carolina, 21 years after filling (Talwani, 1976), may be related to reservoir filling (Snow, 1977).

Expenditures on deep holes have not been routinely justifiable. The U. S. Geological Survey holes at Monticello Reservoir, South Carolina (M. Zoback, written commun., 1979) are the only known holes designed to test reservoir-induced seismicity at hypocentral depth (2 km). Apparently, conductive faults or fractures can be found there that register reservoir pore pressures. Hydrofracturing can determine approximate stresses, directions, and threshold values, but criteria for reservoir-induced seismicity probably cannot be attained, because strength properties can be no more precisely defined than in-situ stress magnitudes. Reservoir-caused changes are small compared to the uncertainties. Hydrofracture stress measurements have successfully reconciled the onset of seismicity at the Rangeley injection experiment (Raleigh and others, 1976), where 71-106 bars fluid pressure was imposed. Resolution is inadequate if pressure changes are of only a few bars, as in reservoir loadings. There is no inherent deficiency in fluid-pressure measurement, but there is a scatter of strengths from randomly placed tests.

Heterogeneity is the underlying impediment to successful reservoir-seismicity prediction. Noncritical stresses obviously exist at most reservoir sites; strength differs significantly from place to place, even in a nearly uniform stress field at a given depth. Stress in rock masses (up to a scale of hundreds of kilometres) may not exceed the strength of the weakest throughgoing fault. In a system of parallel faults, each may have slightly different frictional properties. Earthquakes would tend to be localized by slip on the weakest, all others being subcritical. Because real faults occur in more complex intersecting arrangements, stresses are redistributed upon failure and finite displacement, so that numerous faults of more than one orientation participate in producing seismicity.

Seismic gaps and aseismic regions indicate that large bodies of rock have subcritical stresses for unstable (stick-slip) sliding or that they undergo stable (aseismic) sliding. Unstressed (gravity-only) regions like the Black Hills (E. Hoskins, oral commun., 1970) are rare. With effective stresses at shallow hypocentral depths being on the order of 200 to 2000 bars, even modest strength differences, say 1 to 2° ϕ-angle, give higher-strength faults some 10 to 100 bars threshold to be overcome before abrupt slip can occur. Reservoirs develop effective-stress changes generally less than 10 bars, thus their placement is fortuitous. The proportion of reservoirs that are seismogenic is an indication of the scale of structural domains bounded by weak stress-governing faults. Domains with dimensions of hundreds of kilometres may be subcritical, since reservoirs rarely pressurize the bounding faults that have critical stresses. Predictive capabilities depend, therefore, not so much on the precision of estimating stresses or stress changes, but rather on structural discrimination beyond our current capabilities. We need to identify systematic differences between faults that may govern strength. Among them are fault-zone materials, contact geometry, fault orientations, and continuity or extent.

Long (1974) attributes seismic differences to stress concentrations in geologic units of different elastic modulus. It is unresolved what the relative importance of material elasticity versus fault strength may be in localizing earthquakes. Few rock contacts other than those of young plutons are welded; rather, with age they tend to become pseudoplanar faults. If fault weakness determines regional stress, classification of fault strength is applicable to prediction of induced seismicity.

Microseismic swarm activity, as developed beneath Jocassee Reservoir, South Carolina (Talwani, (1979), would identify an attractive place for deep hydrofrac research boreholes. At a prospective reservoir site, several holes could be drilled and used to test several faults for criticality. Pressures could be applied far in excess of those that the reservoir would impose to define some measure of the variable-strength threshold possessed by faults in that region.

In some provinces, such as the California Coast Ranges, geomorphic and stratigraphic criteria can discriminate between active and inactive faults if economic needs justify the expense, as is often the case for nuclear reactors. But in many regions, all faults seem geologically inactive, though earthquakes suggest modern tectonism. For example, the most recently active fault known among the many identified in the Atlantic Piedmont is the Belaire fault (Prowell and others, 1975), though its latest movement may have been in the Eocene. Seismicity suggests that others lacking surface evidence of recency, such as the Modoc fault, may be more active than the Belaire. For reactor and reservoir studies, injection testing with microseismic detection could be more diagnostic of criticality than are fault-dating studies.

Some crustal regions have pore pressures far in excess of hydrostatic, whether of tectonic, osmotic, dehydration, or magmatic origin. Where pore pressures approach lithostatic (total overburden), a surficial increase of piezometric elevation will not alter the existing state of (vanishing) effective stress at depth. California Coast Range reservoirs may therefore be safe even though the region is already seismic. In contrast, regions of modest seismicity and hydrostatic pore pressure (for example, the Piedmont) can be sensitive to small reservoir-induced changes.

A mapping of principal stress directions and magnitudes might disclose systematic differences between structural domains, the loading experience showing which are critical and which are safe for loading. Ege (1977) has been the first to measure stresses at enough points in a region (seven stations in Rainier Mesa, Nevada Test Site) to be able to contour stress. Deep-well stress data from oil-field hydrofracture has been studied (Kehle, 1964). If hydrofracture tests were done in mineral exploration holes, it would yield basement-rock data for mountainous terrain more applicable to reservoir engineering. A mining industry motivated to conduct stress measurements may be that involved in in-situ leaching in hard, fractured rocks.

The magnitudes of induced earthquakes may depend on hydrogeology, since earthquake energy is related to the dimensions of faulted surfaces. If only local pore-pressure changes occur, only small fault areas may be mobilized. Thus a reservoir in a graywacke province whose streams have developed a finely textured drainage pattern may induce only minor quakes that get lost in the background. In karst regions with regional flow systems, such as at Kremasta, Greece, big events may occur. Lake Bileča in the Dalmatian karst region (Grančarevo Dam, Yugoslavia, Roksandić, 1970) is microseismic, perhaps because it does not influence the regional flow system. The lake lies on an anticline of relatively nonkarstic Triassic dolomite, whereas the Cretaceous limestones are cavernous downdip. A main-stem reservoir will influence the major cells of a fault-zone circulation system, whereas an off-channel storage reservoir, or one high on the watershed, produces a more local potential perturbation. Kariba exemplifies a reservoir in a major regional discharging reach in a deep topographic basin. Cherry Creek Reservoir (Hetch Hetchy Project) in California and Rampart Reservoir in Colorado are examples of aseismic, high-level, off-channel storage reservoirs. Deep reservoirs are usually at the discharging end of major ground-water circulation cells, and the transient pulse in hydraulic head propagates upstream. Koyna is an exception, being situated at the upstream, recharging end of the circulation cells, both east and west.

Though prediction of induced seismicity is beyond our current means, engineering use can still be made of the technology described herein. We can establish whether necessary conditions for failure exist even if sufficient conditions for failure remain unknown. Seismic prediction is already probabilistic, so that the findings of a study of the historical and projected stress changes, favorable or unfavorable, can be applied by shifting any magnitude-frequency curve toward lower or higher frequencies, respectively. Furthermore, its slope might be changed by using higher b' values when induced seismicity is likely. This work offers no guidance either to the amount of shift justified or to the change of slope. Larger changes in quake statistics should be applied to reservoirs of large dimensions and depths, critical position in ground-water circulation cells, active faults, instrumental evidence of local seismicity, as well as a stress history suggestive of maintained criticality.

Other guidelines may be developed by determining such characteristics of faults as fault-zone materials, fault-surface morphology, fault length, interconnectedness, rectilinearity, and orientations. Hydrofracture tests might be conducted at several depths in several holes to estimate the existing threshold strength.

ACKNOWLEDGMENTS

Field work and much of the conceptual effort was done under contract 14-08-0001-13079 with the U. S. Geological Survey, Earthquake Generation Program, funded by the Advanced Research Projects Agency.

Sincere thanks are given to the many hospitable people in eastern Africa who gave advice and assistance, maps and publications during 1972. An incomplete list of benefactors includes A. R. Drysdall of the Zambian Geological Survey and S. Ward and R. Mäckel at the University of Zambia; Leomba Leomba, my field assistant; J. W. Wiles, J. G. Stagman, and I. M. Kirkpatrick of the Rhodesian Geological Survey; G. Bond and L. Lester of the University of Rhodesia; R. W. Sleigh, Dept. of the Surveyor General; C. F. Allison, Central African Power Board; and C. B. Archer, Department of Meteorological Services.

I gratefully acknowledge the help of the many people who assisted me in India during 1973. Among them are C. V. Gole, S. K. Guha, A. Ansari, and S. P. Agarwal of the Central Water and Power Research Station; V. R. Deuskar and A. Y. Daffle of the Koyna Project, Maharashtra State; and P. B. Gupte of the College of Engineering, Poona, who introduced me to important aspects of Deccan geology. The use of maps, air photos, literature, housing, and transportation, as well as the hospitality so generously provided, greatly facilitated my work.

Typing, drafting, and editorial assistance were provided by Lawrence Berkeley Laboratory, University of California.

NOTATION USED

Symbol	Definition	Units
A	= depth of alluvium	L
a	= radius of earth	L
α	= coefficient of thermal expansion	$L/°C$
	= inclination of failure plane to axis of minimum compression	
	= ratio of bulk to grain compressibility, K/K_s	
b	= bars prssure	M/LT^2
$2b$	= fracture width or aperture	L
b'	= slope of earthquake magnitude - log frequency relationship	
D	= thickness of earth's crust	L
d	= average separation between parallel fractures	L
E	= Young's modulus	M/LT^2
e	= erosion depth	L
ε	= strain	
g	= acceleration of gravity	L/T^2
γ	= unit weight of water, ρg	M/L^2T^2
γ_{sa}	= unit weight of saturated alluvium	M/L^2T^2
γ_r	= unit weight of saturated rock	M/L^2T^2
h	= head of water	L
	= height above datum	L
	= depth of fill, scour	L
	= subscript for "horizontal"	
K	= hydraulic conductivity	L/T
	= bulk deformability	L^2/MT^2
K_n	= normal stiffness of fracture	M/L^2T^2
K_s	= bulk deformability of grains (rock blocks)	L^2/MT^2
ν	= Poisson's ratio	
ϕ	= hydraulic potential, unit weight basis	L

ϕ	= angle of internal friction	
P	= pressure intensity	M/LT^2
ρ	= density	M/L^3
S_s	= specific storage	1/L
S	= subsidence of earth's crust	L
σ	= stress (positive in compression)	M/LT^2
σ_1	= maximum principle stress	M/LT^2
σ_3	= minimum principle stress	M/LT^2
σ_e	= effective stress	M/LT^2
σ_t	= total stress	M/LT^2
t	= time	
u	= uplift of earth's crust	L
v	= subscript for "vertical"	
z	= elevation	L

REFERENCES

Agashe, L. V., 1971, Mode of eruption of the Deccan trap basalts: Bulletin Volcanologique, v. 35, no. 3, p. 591-601.

Agashe, L. V., and Gupte, R. B., 1971, The Koyna earthquake and igneous activity, in Symposium on Koyna Earthquake: Indian Journal of Power and River Valley Development, p. 61-62.

Archer, C. B., 1972, A supplementary catalogue of earthquakes in the Lake Kariba area, 1969-1971: Salisbury, Goetz Observatory Report, Meteorological Services.

Archer, C. B., and Allen, N. J., 1969, A catalogue of earthquakes in the Lake Kariba area, 1959-1968: Salisbury, Goetz Observatory Report, Meteorological Services, Salisbury.

Baker, B. H., 1971, Explanatory note on the structure of the southern part of the African rift system: in Tectonics of Africa, Paris, UNESCO, p. 543-548.

Bell, M. L., and Nur, A., 1978, Strength changes due to reservoir-induced pore pressure and stresses and their application to Lake Oroville: Journal of Geophysical Research, v. 83, no. 39, p. 4469-4483.

Berg, E. 1966, Triggering of the Alaskan earthquake of March 28, 1964, and major aftershocks by low ocean tide loads: Nature, v. 210, p. 893-896.

Bishop, W. W., and Trendall, A.F., 1967, Erosion surfaces, tectonics, and volcanic activity in Uganda: Geological Society of London Quarterly Journal, v. 122, no. 488, p. 385-420.

Bond, G., 1953, The origin of thermal and mineral waters in the middle Zambezi Valley and adjoining territory: Rhodesian Geological Society Bulletin 23, p. 1-18.

Brace, W. F., 1972, Pore pressure in geophysics: American Geophysical Union Monograph 16, p. 265-273.

Carder, D. S., 1945, Seismic investigation in the Boulder Dam area 1940-1941 and the influence of reservoir loading on local earthquake activity: Seismological Society of America Bulletin, v. 35, no. 4, p. 175-192.

___1968, Reservoir loading and local earthquakes, in Proceedings Sixth Annual Engineering Geology and Soils Symposium, p. 225-241.

Chandler, R., and Msaki, P., 1972, Annual Bulletin Number 1: Seismograph Station, School of Natural Science, Lusaka (Zambia), 12 p.

Committee of Experts, 1968, Koyna earthquake, December 11, 1967: New Delhi, Resolution No. E1. 1.21 (16)/67, v. 1, 75 p.

Comninakis, P., and others, 1968, Foreshock and aftershock sequences of the Cremasta earthquake, and their relation to the water loading of the Cremasta artificial lake: Annali di Geofisica, v. 21, p. 39-71.

Das, S. R., and Roy, A. K., 1969, Report on the photo-geological study (with limited field check) of the Koyna region and the coastal tract in Satara, Ratnagiri and Coloba district, Maharashtra: Geological Survey of India, 20 p.

Daly, W., Judd, W., and Meade, R., 1977, Evaluation of seismicity at U.S. reservoirs: U.S. Commission on Large Dams, 27 p.

Deshpande, B. G., and Jagtap, P. N. 1971, Interpretation of aerial photographs of Koyna region, in Symposium on Koyna Earthquake: Indian Journal of Power and River Valley Development, p. 25-26.

deSwardt, A. M., 1965, Rift faulting in Zambia, in UNESCO Seminar on E. African Rift System: Nairobi, University College, p. 105-114.

Dixey, F., 1956, The east African rift system: Oxford, Colonial Geological Mineral Resources Bulletin Supplement no. 1.

Drysdall, A. R., and Weller, R. K., 1966, Karoo sedimentation in northern Rhodesia: Geological Society of South Africa Transactions, v. 69, p. 39-69.

Ege, J. R., 1977, In situ stress measured at Rainier Mesa, Nevada, and a few geologic implications [Ph.D. Thesis]: Golden, Colorado School of Mines, 172 p.

Evans, D. M., 1966, The Denver area earthquakes and the Rocky Mountain Arsenal disposal well: Mountain Geologist, v. 3, no. 1, p. 23-36.

Freeze, R. A., and Witherspoon, P. A., 1967, Theoretical analysis of regional groundwater flow. 2. Effect of water-table configuration and subsurface permeability variations: Water Resources Research, v. 3, no. 2, p. 623-634.

Glennie, E. A., 1932, Gravity anomalies and the structure of the earth's crust: Survey of India Professional Paper 27.

Gough, D. I., 1969, Incremental stress under a two-dimensional artificial lake: Canadian Journal of Earth Science, v. 6, p. 1067-1075.

Gough, D. I., and Gough, W.I., 1970a, Stress and deflection in the lithosphere near Lake Kariba: Royal Astronomical Society Geophysical Journal, v. 21, pp. 65-78.

___1970b, Load-induced earthquakes at Lake Kariba:

Royal Astronomical Society Geophysical Journal, v. 21, p. 79-103.
___1976, Time dependence and trigger mechanisms for the Kariba (Rhodesian) earthquakes, in Milne, W. G. ed., First International Symposium on Induced Seismicity: Engineering Geology, v. 10, p. 211-217.
Green, W. R. E., 1973, Seismic activity observed at the Hendrick Verwoerd Dam, in International Colloquium on Seismic Effects of Reservoir Impounding: Royal Society of London Summaries, 2 p.
Guha, S. K., and others, 1970, Recent seismic disturbances in the Shivajisagar Lake area of the Koyna Hydroelectric Project, Maharashtra, India: Khadakwasla, Poona, India, Central Water and Power Resources Station, 129 p.
Guha, S. K., and others, 1974a, Case histories of some artificial crustal disturbances: Engineering Geology, v. 8, p. 59-77.
___1974b, Koyna earthquakes (Oct. 63- Dec. 73): Poona, Central Water and Power Research Station, 340 p.
Gupta, H. K., 1972, Some discriminatory characteristics of earthquakes near the Kariba, Kremasta, and Koyna artificial lakes: Seismological Society of America Bulletin, v. 62, no. 2, p. 493-507.
Gupta, H. K., and Combs, J., 1976, Continued seismic activity at the Koyna reservoir site, India, in Milne, W. G., ed., First International Symposium on Induced Seismicity: Engineering Geology, v. 10, p. 307-313.
Gupta, H. K., and Rastogi, B. K., 1976, Dams and earthquakes: Amsterdam, Elsevier, 229 p.
Gupta, H. K., Ramamurthy, D., and Purandare, S.C., 1967, Report on the geophysical investigations of thermal springs of Ratnagiri district, Maharashtra State: Geological Survey of India, 9 p.
Gupta, H. K., Rastogi, B. K., and Narain, H., 1972, Common features of the reservoir-associated seismic activities: Seismological Society of America Bulletin, v. 62, no. 2, p. 481-492.
Gupte, R. B., 1971, Engineering aspects of the Deccan trap basalts: Maharashtra Public Works Journal, v. 7, nos. 3-4, p. 7-20.
Gupte, R. B., and Rajguru, S. N., 1972, Late Pleistocene geomorphological history of rivers of western Maharashtra: Bulletin Volcanologique, v. 35-3, p. 686-695.
Hadsell, F. A., 1968, History of earthquakes in Colorado: Colorado School of Mines Quarterly, v. 68, p. 57-72.
Hagiwara, T., and Ohtake, M., 1972, Seismic Activity associated with the filling of the reservoir behind the Kurobi Dam, Japan, 1963-1970: Tectonophysics, v. 15, p. 241-254.
Handin, J., 1969, On the Mohr-Coulomb failure criterion: Journal of Geophysical Research, v. 74, no. 22, p. 5343-5348.
Handin, J., and Nelson, R. A., 1973, Why is Lake Powell aseismic?, in International Colloquium on Seismic Effects of Reservoir Impounding (COSERI): Royal Society of London Summaries, 3 p.
Haxby, W. F., and Turcotte, D. L., 1976, Stresses induced by the addition or removal of overburden and associated thermal effects: Geology, v. 4, p. 181-184.
Healy, J. N., and others, 1968, The Denver earthquakes: Science, v. 161, p. 1301-1310.
Hitchon, B., 1958, The geology of the Kariba area: Northern Rhodesia Geological Survey Report 3, 41 p.
Housner, G. W., 1969, The seismic events at Koyna Dam: Proceedings, Eleventh Symposium on Rock Mechanics, Berkeley: American Institute of Mining, Metallurgical, and Petroleum Engineers, p. 541-558.
Huntley, D., 1976, Groundwater recharge to the aquifers of northern San Luis Valley, Colorado [Ph.D. thesis]: Golden, Colorado School of Mines, 247 p.
Independent Panel, 1976, Failure of Teton Dam, in Report to Department of Interior and State of Idaho: U.S. Government Printing Office.
Jacob, K. G., and Ambruster, J., 1975, Tarbella Reservoir, Pakistan: A region of compressional tectonics with reduced seismicity upon initial filling: Paper presented at first International Symposium on Induced Seismicity, Banff, Canada, Sept. 15-19, 1975.
Johnson, S. J., and others, 1977, Reservoirs and induced seismicity at Corps of Engineers projects: Vicksburg, Waterways Experiment Station Miscellaneous Paper S-77-3, 14 p.
Kehle, O. K., 1964, The determination of fracture stresses through analysis of hydraulic well fracturing: Journal of Geophysical Research, v. 69, p. 259-273.
King, L. C., 1951, South African scenery (Second edition): Edinburgh, Oliver and Boyd, 379 p.
___1962, Morphology of the earth: Edinburgh, Oliver and Boyd, 699 p.
Kirkpatrick, I. M., and Robertson, I.D.M., 1968, A geologic reconnaissance of the Makuti-Kariba road: Rhodesian Geological Survey, 16 p.
Klein, F. W., 1976, Tidal triggering of reservoir-associated earthquakes, in Milne, W. G., ed., First International Symposium on induced Seismicity: Engineering Geology, v. 10, p. 197-210.
Knill, J. L., and Jones, K. S., 1965, The recording and interpretation of geological conditions in the foundations of the Rosiers, Kariba, and Latiyan Dams: Geotechnique, v. 158, no. 1, p. 94-124.
Krishnan, M. S., 1960, Geology of India and Burma (Fourth edition): Madras, Higginbothams, 604 p.
___1968, The evolution of the Indian Peninsula: Indian Geoscience Association Journal, v. 8 p 1-11.
Lee, W.H.K., and Raleigh, C. B., 1969, Fault-plane solution of the Koyna (India) earthquake: Nature, v. 223, p. 172-173.
Long, L. T., 1974, Speculations on gravity anomalies, stress amplifications and earthquakes: Seismological Society of America Earthquake Notes.
Mäckel, R., 1971, The physiographic regions of Zambia--A proposal for discussion: Zambia Geographic Association Magazine, v. 19, 11 p.
Maini, Y.N.T., 1971, In situ hydraulic parameters in jointed rocks--Their measurement and interpretation [Ph.D. thesis]: London, Imperial College, 312 p.

Major, M. W., and Simon, R. B., 1968, A seismic study of the Denver (Derby) earthquakes: Colorado School of Mines Quarterly, v. 63, no. 1, p. 9-56.

Mann, A. G., 1967, Investigations of jointing associated with the Deka fault, Rhodesia: Publisher unknown, p. 24-28.

Maufe, H. B., 1933, A preliminary report on the mineral springs of Southern Rhodesia: Salisbury Geological Survey of Southern Rhodesia Bulletin No. 23.

McGinnis, L. D., 1963, Earthquakes and crustal movement as related to water load in the Mississippi Valley region: Illinois Geological Survey Circular 344, 20 p.

McKenzie, D. P., and Sclater, J. G., 1971, The evolution of the Indian Ocean: Scientific American, v. 224, no. 5, p. 62-72.

McKinlay, A. C. M., 1965, The coal fields and coal resources of Tanzania: Geological Survey of Tanzania Bulletin 38.

Mifflin, M. D., 1968, Delineating groundwater flow systems in Nevada: Reno, Center for Water Resources Research, Technical Report Series H-W, Publication no. 4, 54 p.

National Academy of Sciences, 1972, Earthquakes related to reservoir filling: Washington, D.C., 24 p.

Nur, A., and Byerlee, J. D., 1971, An exact effective stress law for elastic deformation of rock with fluids: Journal of Geophysical Research, v. 76, no. 26, p. 6414-6419.

Pallister, J. W., 1971, The tectonics of East Africa in Tectonics of Africa: Paris, UNESCO, p. 511-542.

Prowell, D. C., O'Connor, J. B., and Rubin, M., 1975, Preliminary evidence for Holocene movement along the Belaire fault zone near Augusta, Georgia: U.S. Geological Survey Open-File Report 75-680, 8 p.

Raleigh, C. B., and others, 1976, An experiment in earthquake control at Rangeley, Colorado: Science, v. 191, p. 1230-1237.

Roksandic, M. M. 1970, Influence de la charge d'un reservoir sur l'activité seismique: Proceedings, Second Congress of the International Society for Rock Mechanics, v. 3, p. 483-488.

Rothé, J. P., 1969, Earthquake and reservoir loadings: Santiago: Proceedings, Fourth World Conference on Earthquake Engineering, v. 1, p. 28-38.

Scopel, L. J., 1964, Pressure injection disposal well, Rocky Mountain Arsenal, Denver, Colorado: Mountain Geologist, v. 1, p. 35-42.

Secor, D. T., 1965, The role of fluid pressure in jointing: American Journal of Science, v. 203, p. 633-646.

Seraphim, J. L., and del Campo, A., 1965, Interstitial pressures on rock foundations of dams: Journal of the Soil Mechanics and Foundations Division, American Society of Civil Engineers, v. 91, no. 5, p. 65-85.

Severy, N. I., Bollinger, G. A., and Bohannon, H. W., Jr., 1975, A seismic comparison of Lake Anna and other Piedmont Province reservoirs in the eastern U.S.A.: Presented at First International Symposium on Induced Seismicity, Banff, Alberta, 24 p.

Shawe, D. R., 1965, Strike-slip control of Basin-Range structure indicated by historical faults in western Nevada: Geological Society of America Bulletin, v. 76, p. 1361-1378.

Shehata, W. M., 1971, Geohydrology of Mt. Vernon Canyon area, Jefferson County, Colorado [Ph.D. thesis]: Golden, Colorado School of Mines, 164 p.

Sherard, J. L., 1966, A study of the influence of the earthquake hazard on the design of embankment dams: California Department of Water Resources, 339 p.

Simon, R. B., 1969, Seismicity of Colorado: Consistency of recent earthquakes with those of historical record: Science, v. 165, p. 897-899.

Simpson, D. W., 1976, Seismicity changes associated with reservoir filling, in Milne, W. G., ed., First International Symposium on Induced Seismicity: Engineering Geology, v. 10, p. 123-150.

Sleigh, R. W., 1969, How Kariba dented the earth: Rhodesian Science News, v. 3, no. 5, p. 129-131.

Snow, D. T., 1968a, Fracture deformation and changes of permeability and storage upon changes of fluid pressure: Colorado School of Mines Quarterly, v. 63, no. 1, p. 201-244.

___1968b, Rock fracture spacings, openings, and porosities: American Society of Civil Engineers, Soil Mechanics and Foundations Division Journal, v. 94, no. 1, p. 73-91.

___1968c, The frequency and apertures of fractures in rock: International Journal of Rock Mechanics and Mineral Science, v. 7, p. 23-40.

___1969, Anisotropic permeability of fractured media: Water Resources Research, v. 5, no. 6, p. 1273-1289.

___1972, Geodynamics of seismic reservoirs: Proceedings, Symposium on Percolation through Fissured Rocks: Stuttgart, German Society of Soil and Rock Mechanics, p. T2J 1-19.

___1975, Predicted aseismicity of Palo Verde Reactor Ponds: Presented at First International Symposium on Induced Seismicity, Banff, Alberta.

___1977, Induced seismicity at Richard B. Russell Reservoir: Geological and seismological evaluation of earthquake hazards at Richard B. Russell Project, Corps of Engineers, Savannah District, 50 p., reviews and reply.

Sykes, L. P., 1967, Mechanism of earthquakes and nature of faulting on the mid-oceanic ridges: Journal of Geophysical Research, v. 72, p. 2131-2153.

___1970, Seismicity of the Indian Ocean and a possible nascent island arc between Ceylon and Australia: Journal of Geophysical Research, v. 75, no. 26, p. 5041-5055.

Talwani, P., 1976, Earthquakes associated with the Clark Hill Reservoir, South Carolina--A case of induced seismicity: Engineering Geology, v. 10, p. 239-253.

___1979, Induced Seismicity and Earthquake prediction studies in South Carolina: Final Report, U.S. Geological Survey Contract 14-08-0001-14553, 36 p.

Tandon, A. N., and Chaudbury, H. M., 1971, Seismometric study of the Koyna earthquake of December

11, 1967, in Symposium on Koyna Earthquake: Indian Journal Power and River Valley Development, p. 103-118.

Wang, M. Y., and others, 1976, Mechanism of the reservoir impounding earthquakes at Hsinfengkiang and a preliminary endeavor to discuss their cause: Engineering Geology, v. 10, p. 331-351.

Withers, R. J., 1977, Seismicity and stress determination at man-made lakes [Ph.D. thesis]: Edmonton, University of Alberta, 241 p.

Withers, R. J., and Nyland, E., 1976, Theory for the rapid solution of ground subsidence near reservoirs on layered and porous media, in Milne, W. G., ed, First International Symposium on Induced Seismicity: Engineering Geology, v. 10, p. 169-185.

Wolters, R., and others, 1972, Observations on the openings of fissures, their arrangement and extension, in Proceedings, Symposium on Percolation through Fissured Rock: Stuttgart, German Society of Soil and Rock Mechanics, p. T1J 1-13.

Manuscript Accepted by the Society October 14, 1980

Printed in U.S.A.

Geological Society of America
Special Paper 189
1982

Comments on Isotope Dating of Ground Waters in Crystalline Rocks

P. FRITZ
Department of Earth Science, University of Waterloo
Waterloo, Ontario, Canada N2L 3G1

ABSTRACT

Age dating of ground water from crystalline rocks, shales, and clayey sediments is an essential component in assessing hydrogeologic suitability of such rocks for long-term containment of toxic materials. Isotope techniques available for this task are discussed and it is noted that, at present, only three tools can be used for these investigations. They are based on tritium, ^{14}C, and elements of the uranium decay series. There is little hope that others will be developed in the near future except for applications where it is possible to collect extremely large volumes of water within a reasonable time. The limitations of the available tools are discussed and directions in which major research efforts should proceed in future are indicated. It is also emphasized that no single tool, including isotope analyses, can provide absolute water ages. Therefore, these techniques must be combined with geochemical, hydrological, and mathematical studies before it will be possible to arrive at reasonable age estimates of ground waters in low-permeability rocks.

INTRODUCTION

The determination of the age of a ground-water body is among the most difficult problems encountered in hydrologic investigations because estimates based on hydraulic data are often just that: estimates. This is especially true if very large or deep systems are considered in which ground-water residence times have to be measured in tens of thousands or even hundreds of thousands of years. Physically based age calculations must include hydraulic gradients gathered from modern well installations. It is often quite impossible to estimate these for prehistoric times, especially in areas that have been glaciated during the most recent geologic history or where major climatic shifts have occurred. Direct dating techniques based on environmental isotopes and decay products of radioactive decay series are, theoretically, not limited by such events and could be used for ground-water age determinations.

This was first suggested by Brinkman and others (1959), who proposed that ground water could be dated through ^{14}C analyses on the dissolved inorganic carbon. They tested the method on ground waters in Germany and then proceeded to "date" ground-water samples from oases in the Libyan deserts. Knetsch and others (1962) reported ^{14}C ages (uncorrected) between 20,000 and 30,000 years and also noted that the ^{18}O and ^{2}H contents of these waters were respectively 3 °/oo and 23 °/oo lower than those known for central Europe. This suggested that the desert waters formed in Europe, in a climate even cooler than today's, and that fossil water from a previous pluvial period was preserved as ground water and was now actively discharging in artesian wells.

From the beginning it was noted, however, that many of the calculated ^{14}C ages were much too high because no adequate corrections for a secondary dilution or loss of ^{14}C in the ground water were available. As a result, a great number of studies were published proposing methods to transform measured ^{14}C activities into absolute water ages. Most ^{14}C corrections require a knowledge of processes which determine the geochemical characteristics of the ground water and which must be obtained from detailed hydrogeologic investigations. Unfortunately these investigations were often bypassed for lack of cooperation with hydrogeologists and/or lack of suitable field installations. However, results were still published and there is now a wealth of data in the literature which have only historic value.

This was recognized by several groups, notably within the U.S. Geological Survey, the International Atomic Energy Agency (IAEA) in Vienna and some researchers at North American and European universities. Efforts were made to promote a "combined approach," which relies on the parallel use of environmental isotope techniques (with stable and radioactive isotopes), detailed geochemical analyses, and extensive hydrogeologic field investigations. This effort was successful, and environmental isotope techniques are now widely accepted by hydrogeologists. This approach also demonstrated that it is possible to make very concrete statements about origin, age, and geochemical history of a water mass. Much of this research was done on ground waters from aquifers in permeable, sedimentary rocks and was summarized in the 1970, 1974 and 1978 proceedings of symposia organized by the IAEA.

This presentation focuses on possibilities and limitations of determining the age of ground water in low-permeability rocks, an area which has received little attention because little interest existed. Yet low-permeability rocks are now considered for the disposal or storage of toxic materials for almost geologic time spans. Under these conditions even a "dry" cavern will eventually be filled with water, and it is important to predict the flow path the waters will follow once they begin to flow from a sealed repository. Age determinations done in conjunction with studies in fracture hydrology would provide a time frame that could permit an extrapolation from the past to the future.

A number of isotopic tools suitable for research in low-permeability environments are available or are under development. The most important of these are based on ^{3}H, ^{14}C, and the uranium decay series. Others might be developed in the future, although there are some limitations, especially for those tools which at present either require extremely long preparation procedures and/or very large samples. Such is the case of ^{32}Si ($t_{1/2} \sim 500$ years; Lal and others, 1970), ^{39}Ar ($t_{1/2} = 269$ years; Oeschger and others, 1974) or ^{81}Kr ($t_{1/2} = 2.6 \times 10^{5}$ years; Oeschger and Gugelmann, 1974). These tools have or will have their applications in other hydrogeologic investigations and will be essential for the eventual intercalibration of different isotope-dating tools.

New analytical possibilities, such as the development of accelerator techniques, will reduce sample size, which is especially important for ^{14}C ($t_{1/2} = 5730$ years) and ^{36}Cl ($t_{1/2} = 3.08 \times 10^{5}$ years) (Elmore and others, 1979). The latter isotope is under intensive study by researchers at the University of Arizona and the University of Rochester and results show that it should soon be possible to proceed to field testing of the ^{36}Cl-tool (S. Davis, 1979, oral commun.). The interpretation of the result will require, however, a close look at the geochemistry of chlorine, which may be more complex than usually assumed. Even in crystalline terrains chlorine might be added in significant amounts to the water, thus diluting the original ^{36}Cl concentrations. Ion exchange may also have some effect.

^{85}Kr is also potentially useful. Its abundance in the atmosphere is essentially controlled by man's nuclear activities and it can be extracted and analyzed from only a few hundred liters of water (Rozanski and Florkowski, 1979). However, its halflife is only 10.76 years and, therefore, it competes directly with tritium, which can be analyzed on much smaller samples. Furthermore, the analytical techniques required for ^{85}Kr are much more complicated than the commercially available liquid scintillation counters which suffice for most tritium determinations. However, ^{85}Kr plays a unique role in studies based on rare gas analyses because it indicates whether air contamination has occurred.

A summary of dating tools under development and the research required before they can be applied for absolute or relative ground-water dating has been presented by Davis (1978). His presentation does not build on these discussions but deals with tools already employed for dating ground waters in low-permeability environments: tritium, ^{14}C, and the uranium decay series. Each will be treated separately with special emphasis on the experiences gained at the Stripa, Sweden, test site (Fritz and others, 1979).

TRITIUM DATING

Tritium is a radioactive hydrogen isotope with a half-life of 12.3 years. It is part of the water molecule and, therefore, the only direct dating tool available. Analyses are done on 1-l samples if tritium enrichment precedes the counting in gas or liquid scintillation counters [detection limit <1 T.U. where 1 tritium unit (T.U.) represents one tritium atom for 10^{18} hydrogen atoms]. Less than 15 cm^{3} of water are needed if direct counting is done with detection limits of about $\pm$ 10 T.U.

Young vs. Old Ground Water

Physical testing of aquifer parameters in wells installed in low-permeability rocks is very difficult and a tritium analysis on as little as 10 cm^{3} water can be very important. Such analysis may permit the recognition of residual drill water or give a first indication of the degree of rock impermeability. Fracture permeability may be much greater than anticipated.

On a large scale, flow through fractured crystalline rocks was studied by Rauert and Stichler (1974) and Fontes and others (1979). The first study dealt with samples collected from drips and "springs" with a discharge of up to 10 l/s in a tunnel through the Tuxer Hauptkamm in the Austrian Alps. In the second study, similar samples were collected in the Mont Blanc tunnel in the French-Italian Alps. The observations made were similar in both studies: stable isotope data (^{2}H and ^{18}O) show a dependence on the depth below ground surface, reflecting decreasing heavy isotope contents with increasing altitude of recharge. This is shown in Figure 1, where, at the entrance to the Mont Blanc Tunnel, the $\delta^{2}H$ values are 20 to 30°/₀₀ higher than in the central part. This preservation of altitude effects in the tunnel samples suggests that flow in the fracture systems is essentially vertical, probably because of the large sink in the form of the tunnel.

However, it is important that a difference in

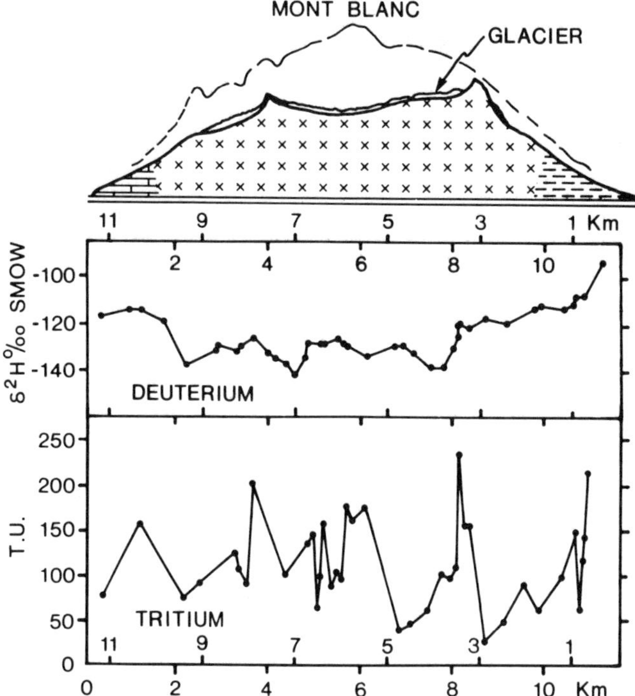

Figure 1. ^2H and ^3H concentration in drip waters collected in the Mont Blanc road tunnel. Note that δ^2H decreases in the center of the tunnel but no systematic pattern is observed in the tritium contents (after Fontes and others, 1979).

tritium contents was observed at the two localities. Whereas Rauert and Stichler (1974) found low tritium contents in the center of the tunnel and high values in seeps closer to the entrance (reflecting longer residence times for the water without tritium), there is no logical pattern visible for the samples from the Mont Blanc Tunnel (Fig. 1). All samples contain high tritium and are, therefore, very young. However, to actually date these waters, samples would have to be collected for several years at regular intervals. The results would then be compared with tritium fallout curves, which have been highly variable for the past 20 years. Bomb testing and, locally, nuclear power plants, have augmented background tritium levels of 10 to 20 T.U. by several orders of magnitude. The results show, however, that all Mont Blanc samples have bomb tritium and, in part or in toto, are younger than 25 years (prebomb water would show less than 5 T.U.).

There is a similar situation in the Sudbury mining district of Canada, where actively discharging ground waters were sampled from a number of mines (Fritz and Reardon, 1978). Even those samples collected approximately 1,200 m below ground contained significant amounts of tritium, demonstrating that in an environment disturbed by mining activities, young ground water can be channelled to great depth in a relatively short time. Because the premining hydraulic gradients are not known, no comments can be made on the premining situation.

The presence of young water at those depths is disturbing if one considers deep crystalline rocks for waste repositories. Thus, results from the Stripa granite have special significance. This granite is adjacent to a 400-year-old iron ore mine, and presumably all fracture systems in the granite have been dewatering into this mine. However, a number of boreholes drilled from mine levels and terminating 300 to 900 m below ground surface were flowing wells whose water did not contain measurable amounts of tritium. Tritium was only found in old mine workings and in the drill fluids used.

A monitoring program at sites such as Stripa is therefore very important, because it remains to be seen whether the mining activity associated with the construction of the test facilities did not activate "fossil" systems. If this were the case, and if these systems have a limited storage capacity--as might be expected for fracture systems in a granitic rock mass--then we might see increasing tritium contents in the ground waters as soon as the old water has "run out." If we do not observe this, the existence of regional flow systems in these shield environments would have to be investigated.

Tritium Loss by Diffusion

The intergranular porosity of a crystalline rock is usually so small that diffusive losses of tritium into this matrix can be neglected. This is not so if water flow through clayey material is investigated. Fractured tills and clays can transmit significant amounts of water (for example, Grisak and Cherry, 1975; Grisak and others, 1976), but unfortunately only a few isotope data are available from such systems. Mathematical models designed to describe the transport of rapidly decaying radionuclides through fractured clays are also not yet fully operational.

Only some preliminary results can be presented. Day (1977) investigated a number of piezometer sites installed in the Lake Agassiz clay deposits in south-central Manitoba. Visual evidence had previously indicated that these clays were fractured. Because they overlie an important carbonate aquifer, some concern arose over their ability to protect this aquifer from surface pollution.

Laboratory hydraulic conductivities for this glaciolacustrine clay were determined to be between 10^{-8} and 10^{-10} cm/s. Field values determined through single-well response tests were at least one and frequently two orders of magnitude greater. Values as high as 3×10^{-7} cm/s were obtained in the field. This difference is indicative of a medium in which water flow is largely controlled by fractures.

Depth profiles for tritium and ^{18}O contents were obtained from multilevel installations. As an example, Figure 2 shows data from one piezometer test in an area with significant vertical downward hydraulic gradient. Day (1977, p. 84, 85) and Cherry and others (1979) point out that the tritium depth profiles appear to indicate much slower vertical movements through the aquitard than would be expected if continuous, well-connected fractures controlled the bulk hydraulic conductivity of the deposits. Figure 2 shows the tritium front (indicating bomb tritium from 1953 onward) to be at a depth of about 9 m, suggesting a maximum rate of movement on the order of 0.5 m per year. A rate between 20 and 150 cm per day has been predicted from hydraulic considerations at this site.

At this site ^{18}O data provide another indication for the age of a water sample because the carbonate aquifer contains water that is at least 10,000 years old and has up to 10 ‰ less ^{18}O than modern water (Fritz and others, 1974). At present, this aquifer yields up to 10^6l of water per day for municipal and industrial purposes in the Winnipeg area. This pumping resulted in the creation of a sizeable drawdown cone with pronounced downward hydraulic gradients (Render, 1970). Before pumping started, however, hydraulic gradients were upwards, permitting the penetration of these old waters into the clay deposits. As the hydraulic gradient became reversed, modern water began to flush these old waters from the aquitard. This explains the ^{18}O gradients shown in Figure 2 and provides an additional tracer, which indicates where the modern recharge front is located. Tritium and ^{18}O show a remarkable agreement, with the ^{18}O front being slightly lower. The tritium inputs measurable today commenced only about 25 years ago whereas the ^{18}O-rich waters began to enter the clays as soon as the hydraulic gradients permitted--roughly 100 years ago.

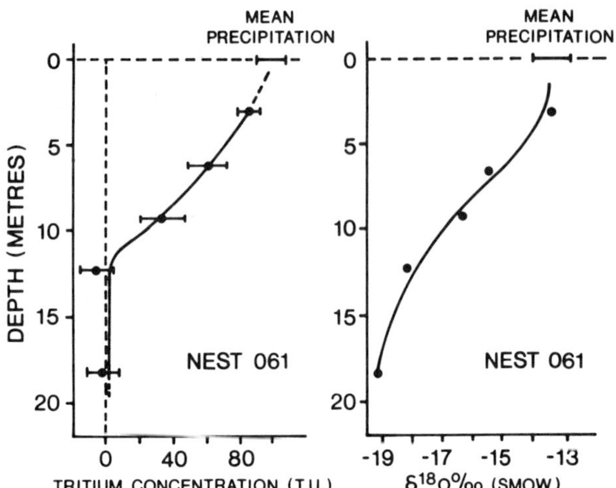

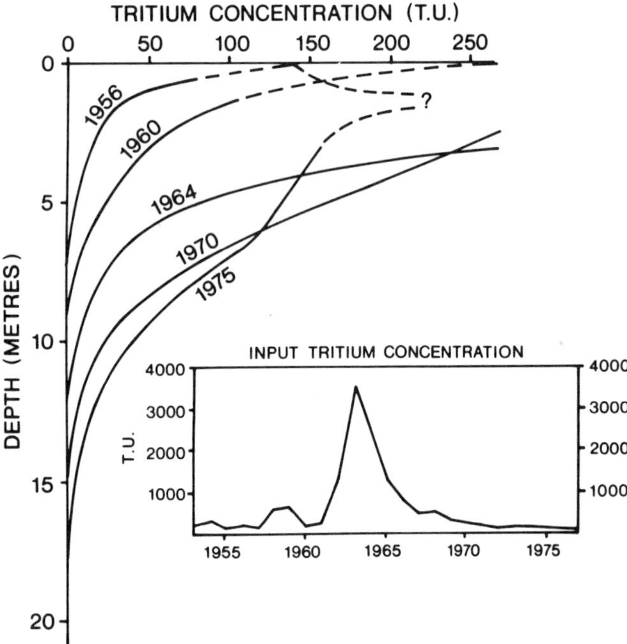

Figure 2. TOP: Tritium and ^{18}O contents in waters sampled at the specified depth in Lake Agassiz clay. The abundance of both isotopes shows a marked decrease with depth as a result of diffusive loss from fractures into the clay matrix. BOTTOM: simulated tritium concentrations using the given tritium input function for model calculations. Tritium loss by decay and diffusion into the matrix (that is, addition of tritium-free water from the matrix) were assumed (after Day, 1977).

The discrepancy between the water movements indicated by isotopic data and hydraulic data had to be solved. G. Allison (oral commun.) and Day (1977) proposed that although fracture flow dominates water movements in the aquitard, the depletion of isotopic contents of ground water in the fractures is attributed to interaction with water in the clay matrix. Molecular diffusion is thought to be controlling and would occur in response to concentration gradients. In time such gradients would become different for the two isotopes: the matrix will be filled up with the stable ^{18}O but a tritium gradient would be maintained because of its decay. Thus tritium would never be found below a certain depth.

The input functions for the two isotopes are also quite different. The ^{18}O contents would be essentially constant from one year to the next and reflect the annual average isotopic composition of precipitations. Tritium input was constant before bomb testing started but has varied considerably since the early 1950s. Figure 2 also shows the local tritium input function and a number of calculated profiles based on diffusion from fractures into the matrix. Diffusion within the fractures was neglected because of the relatively high flow velocities. Modeling for ^{18}O with a simple isotopic difference between input and water in the clay matrix (constant input) yielded similarly satisfying results. In both cases the observed profiles could be reproduced with fracture spacings as low as 5 cm.

Similar observations have since been made at other localities (Cherry and others, 1979) and all demonstrate that reasonable descriptions of flow paths and determination of water ages are only possible through a combined consideration of isotopic and hydrogeologic data. Velocities determined from hydrogeologic testing alone are misleading because they do not consider actual water movements. If chemical analyses are added, such studies allow detailed discussions on potential contaminant migration and the suitability of clay deposits to retain waste.

^{14}C DATING

Since the introduction of ^{14}C as a ground-water dating tool (Brinkman and others, 1959) it has received more attention than any other dating tool with the exception of tritium. Analyses are done on the carbon of the dissolved inorganic carbonate in a water. Thus only indirect water ages are obtained and one must assume that the carbonate travels with the speed of the water. Counting of ^{14}C activities can be done by gas or liquid scintillation counters, for which 1 to 5 g of carbon are required. This signifies that 50 to several hundred litres of water must be processed for a conventional analysis.

This limitation should be overcome once accelerator-based ^{14}C determinations are possible with the required accuracy. Sample sizes would be reduced to one litre or less.

Comparison With Other Dating Tools

Great efforts have been undertaken to develop ^{14}C dating into a practical and universally applicable tool in hydrogeology. Unfortunately too many publications leave the impression that the basic problems have been solved and that little is needed--usually only some aquifer specific information--before reliable age determinations are possible. Not so. This is not to say that reasonable age estimates have not been possible, but the hydrology and geochemistry of most aquifers with long residence times are so complex that simple model considerations can only describe a few aspects of the system. Only where all aspects of hydrogeologic investigations are combined is it probable that the final interpretation is close to reality. One such case is the Floridan aquifer system, which, during the past decade, has been investigated with all possible tools--despite the fact that it is not the best-instrumented system and that most wells are open holes with ill-defined sampling points (Wigley and others, 1978).

What is ultimately necessary, however, is a comparison of the ^{14}C tool with other dating techniques. These are available but most have very serious restrictions because of sample size or complex preparation procedures. Of special significance, therefore, are the studies done by Loosli and Oeschger (1979) who compared ^{14}C measurements with ^{39}Ar determinations. ^{39}Ar is produced by cosmic rays, and its natural atmospheric concentrations are 0.112 ± 0.012 dpm (decays per minute) ^{39}Ar per litre of Argon. The half-life of ^{39}Ar is only 269 years and it is thus not useful for dating very old waters. Sample size required is close to 15 m^3--an amount not usually available from low permeability rocks. The ^{14}C-^{39}Ar comparisons were therefore possible only in major aquifer systems and results from 11 wells are shown graphically in Figure 3. Agreement does not exist. All argon ages indicate that the waters should be younger than 750 years, even those with a ^{14}C age of greater than 25,000 years. Figure 3 shows a simple mixing line between fossil and modern waters. Mixed samples should either fall on this line or to the right (that is, below it)--depending on the nature of the mixing process. The graph shows that some samples might be explained by such mixing but not the ^{14}C old samples.

It is difficult to find an explanation for the discrepancy. Loosli and Oeschger (1979) show that ^{39}Ar can be produced from ^{39}K through a $^{39}K(n,p)^{39}Ar$ reaction. In one of their samples (not shown in Fig. 3) they measure an enrichment by a factor of 3.8 over atmospheric levels. However, this occurred in a thermal spring and is probably an exception rather than a rule, although no data are available to support the assumption.

Loosli and Oeschger then questioned the ^{14}C ages, but they felt that the ages could not be corrected to the young age dates indicated by the argon data. It was also argued that ^{18}O data support the high ^{14}C ages because the older waters had slightly lower ^{18}O contents. This, however, is not a valid argument because these lower ^{18}O contents could simply reflect a slightly cooler climate that existed during the "little ice age" of the Middle Ages--an interpretation that would support the argon data.

Loosli and Oeschger (1979) also compared ^{14}C with ^{32}Si ages and gave results for two samples: compared with ^{14}C ages (based on A_0 = 85 pmC) close to 5,000 and 6,000 years, the ^{32}Si ages range from 300 to 1,200 years, depending on the choice of the initial conditions. Again the ^{14}C age appears to be much too old. Although no geochemical corrections were applied, such corrections would probably reduce the age determination by another few thousand years.

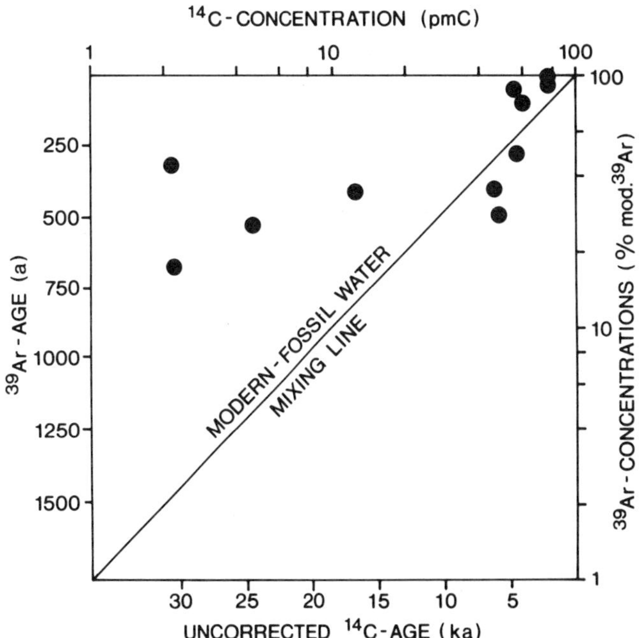

Figure 3. Correlation of ^{39}Ar and ^{14}C water ages determined on ground waters from aquifers in central Europe. If simple mixing of fossil and modern water had occurred, the data points should fall on the mixing line shown. Other mixing models would move the points to the right (after Loosli and Oeschger, 1979).

Thus, the first serious effort to compare water ages calculated by different age-dating tools did not yield the agreement one would have hoped for, although the comparison was done on aquifers that have been investigated extensively. No explanation for the differences were given but the need for an interdisciplinary approach in these studies was emphasized.

^{14}C as a Dating Tool in Low-Permeability Rocks

Diffusion and Isotope Exchange. A calibration of the ^{14}C age-dating tool through other isotope techniques also requires a closer look at how physico-chemical processes might affect the ^{14}C activity of dissolved inorganic carbon. As far as physical processes are concerned, both diffusive and dispersive loss must be considered as well as subsurface production of ^{14}C. In a porous medium the dispersion coefficients usually reflect both molecular diffusion and mechanical dispersion. They may also be used to describe how the distribution of ^{14}C-bearing carbonate species is modified in the aqueous phase. The importance of this has not been tested but will have to be assessed in shales or clayey sediments.

In a fractured medium, the connectivity of the fracture systems determines how much mechanical "dispersion" or "spreading" is possible. Molecular diffusion in the direction of transport will only be important if water moves slowly in fractures. However, diffusion into the matrix may well occur in sedimentary rocks. The process would be the equivalent of the one described above for tritium, although in a crystalline rock matrix intergranular diffusion is less important than in sedimentary rocks. However, total porosities between 1 and 2% do exist (see Gale, this volume) and some ^{14}C loss may occur (Davison and others, 1979).

In environments where there is a small ratio of carbon atoms in the aqueous carbonate to the carbon in the minerals in contact with the water (that is, thin calcite-coated fractures and marls), solid-state diffusion may be important. Thilo and Münnich (1970) suggest that such "exchange" takes place. Figure 4 shows the magnitude of retardation and "exchange" measured by Thilo and Münnich (1970) in column experiments using ^{14}C-enriched Na_2CO_3 solutions. Irreversible exchange was assessed by determining the recovery of ^{14}C after flushing the column with ^{14}C-free water following the uptake experiment. Figure 4b shows that the "permanent" loss of ^{14}C becomes increasingly significant with higher temperatures. Thilo and Münnich attribute this to "some kind of diffusion" into the solid phase. However, if it were a simple diffusion process, then this uptake should be largely reversible. This is not the case and, instead, one could suggest that chemical interaction between aqueous and solid phase (recrystallization) permanently modified the concentration relationships by approaching thermodynamic equilibrium between the two phases (see also Anderson, 1968). The distinction is important.

Both diffusion and exchange will not only affect the distribution of ^{14}C but also of ^{13}C between the two phases. However, because of the small

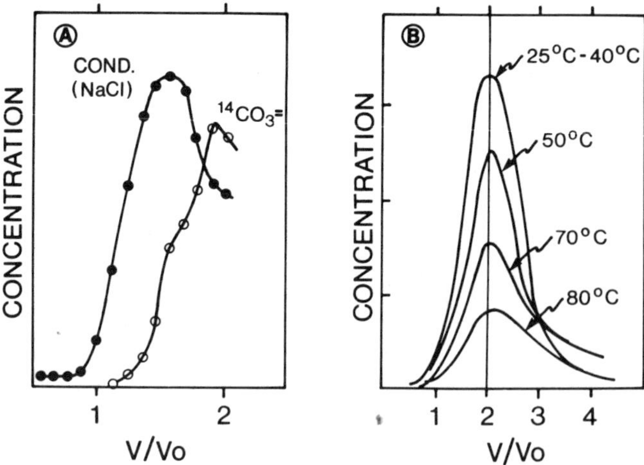

Figure 4. A: Retardation of a ^{14}C pulse in loss soil. B. Increasing loss of ^{14}C from ^{14}C-spiked water moving through a column packed with calcite particles. V describes the volumes of water used to elute the injected ^{14}C (volume V_o). In both cases a ^{14}C loss shows that exchange with the solid carbonate in the columns has taken place (after Thilo and Münnich, 1970).

concentration gradients between solid and aqueous phase for ^{13}C as compared with the steep ^{14}C gradients, the effects will be different for the two isotopes. (There is only a few parts per thousand difference between solid and aqueous ^{13}C concentrations; crystals could be devoid of ^{14}C and decay would maintain the gradients). In general, one would expect that simple diffusive processes could be important for ^{14}C but would have a small or even negligible effect on ^{13}C. If, however, chemical exchange takes place, then both isotopes will be affected and isotopic equilibrium between the aqueous and solid phases would be established.

According to Thilo and Münnich's (1970) experiments, exchange processes can quickly lead to an almost total loss of ^{14}C from the aqueous phase, and if the above-mentioned experiments also reflect natural situations then "the reliability of high ^{14}C ages from calcareous aquifers would have to be questioned." There is little direct evidence to show that this is the case. A study by Anderson (1968) indicates that only surface layers are affected. Rye and O'Neil (1968) reported that the carbon in the aqueous phase of a fluid inclusion in a calcite was out of equilibrium with calcite carbon by several parts per mil, although the assigned age of the inclusion was greater than 10^7 years.

A similar lack of isotopic equilibrium is also noted for the fracture calcites in the Stripa granite. Their $\delta^{13}C$ values vary from -15 to +15 ‰ whereas the aqueous carbonate shows a much smaller range, between about -14 and -18 ‰ (Fritz and others, 1979). However, the data available to date are limited, and further verification of the importance of these processes is needed before they can be considered insignificant. Therefore, no quantification in the form of a ^{14}C correction is possible at present .

Ground Water Chemistry and ^{14}C Contents. Purely chemical effects causing dilution or loss of ^{14}C from ground-water systems have been studied extensively. Although no concensus exists on how to correct for these interferences and arrive at actual water ages, the basic processes are understood and can be described in detail if sampling throughout the flow system is possible (for example, Wigley and others, 1978).

Chemical changes are often paralleled by modifications of the ^{13}C content of the dissolved inorganic carbon because reactants and products may have variable isotopic compositions. Thus mixing of different reservoirs and/or exchange between them can cause $\delta^{13}C$ shifts. These shifts, however, have to be known if ^{13}C is to be used to assess modification of ^{14}C content. Figure 5 summarizes in simplified form the most important processes, which will be discussed in the following section in more detail.

Calcium carbonate is present in all rock types and, therefore, after its dissolution ^{14}C dilutions have to be expected. The distinction between open and closed system dissolution, as discussed by Deines and others (1974), and Wigley and others (1978) is important. As the ground-water geochemistry evolves, calcite saturation is reached and the precipitation of calcite may become significant.

If dissolution of ^{14}C-free calcite took place under closed system conditions--recognizable in the geochemical evolution of a ground water--then it would be possible to quantify the ^{14}C dilution. These calculations are based on chemical and isotopic mass balances and require as input a knowledge of ground-water geochemistry, field pH

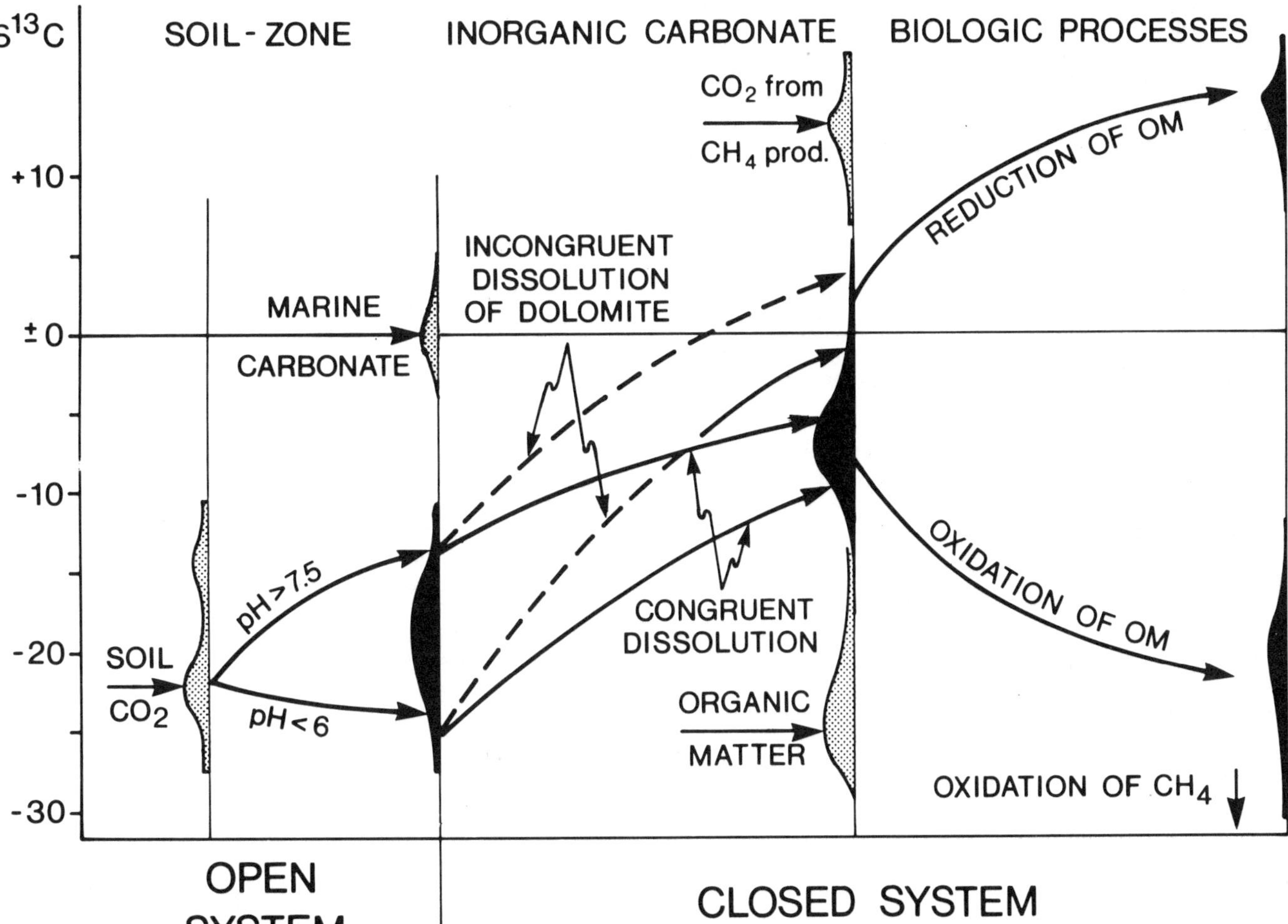

Figure 5. Modifications of the $\delta^{13}C$ of dissolved inorganic carbon by different processes. $\delta^{13}C$ values of sources are shown as dark areas. In the recharge areas of a ground-water system, the dissolution of calcite and isotope exchange with soil carbon dioxide controls much of the chemistry and ^{13}C contents of the newly forming ground water. Depending on pH and the $\delta^{13}C$ of the soil CO_2 different "initial" $\delta^{13}C$ values can be attained. If this is followed by a closed system dissolution of calcite or dolomite then usually these $\delta^{13}C$ values increase considerably. Following the simple dissolution of calcite, typical $\delta^{13}C$ values fall between -12 and -8. If incongruent dissolution of dolomite takes place, marine carbonate values can be reached. Biologic processes can further modify the $\delta^{13}C$ of the dissolved inorganic carbon and the direction of change will depend on whether oxidation or reduction of organic matter takes place.

of the water, and soil-CO_2 and carbonate mineral carbon isotopic compositions (Reardon and Fritz, 1978; Fontes and Garnier, 1979).

Precipitation of calcite is usually discussed only in the context of the incongruent dissolution of dolomites, but it also occurs without it. For example, fracture coatings in crystalline rocks typically contain calcite. Furthermore, geochemical data gathered at Stripa (Fritz and others, 1979) document that calcite must precipitate in the fractures. Figure 6 is a plot of total inorganic carbon (TIC) contents versus the saturation index of calcite. It shows that once calcite saturation has been reached, carbonate is lost, evidently through precipitation of calcite. The total dissolved inorganic carbon content decreases from about 10^{-3} moles/l in the shallow ground waters to less than 10^{-4} moles/l in the deepest waters. Calcite saturation is maintained because the pH rises continuously and at depth reaches values above 9.5.

This simple removal of carbonate from solution causes some isotope fractionation whereby the residual aqueous carbon will be depleted in heavy isotopes. The change in isotopic composition will follow a "Rayleigh distillation" with isotope effects that will depend on the pH of the water at which this precipitation occurs. These fractionation effects are shown in Figure 7 and are close to 2.5 ‰ for waters whose pH is > 7.5, but may exceed 10 ‰ at lower pH values. The effects of this removal on the residual carbonate are small if less than 50% of the original carbon is removed, but may become significant thereafter. The $^{13}C/^{12}C$ ratios in the aqueous carbonate of the Stripa ground waters does decrease with depth and 2 ‰ to 3 ‰ difference is observed. Assuming that the isotope effect is close to 2.5 ‰ ($\alpha_{calc.-H_2O} = 1.0025$), then this decrease represents a removal of close to 80% of the aqueous carbon reservoir. This agrees with the chemical data and supports the model proposed.

A knowledge of these effects is necessary if ^{13}C is used to correct for dead carbon contributions. ^{14}C contents, however, are

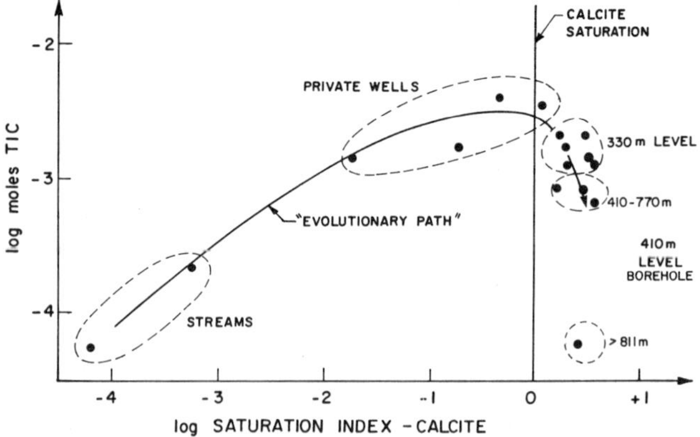

Figure 6. Saturation index (SI) for calcite versus total inorganic carbon contents from ground waters at Stripa, Sweden. Note the loss of aqueous carbonate with depth and the preservation of calcite saturation close to SI = 1.

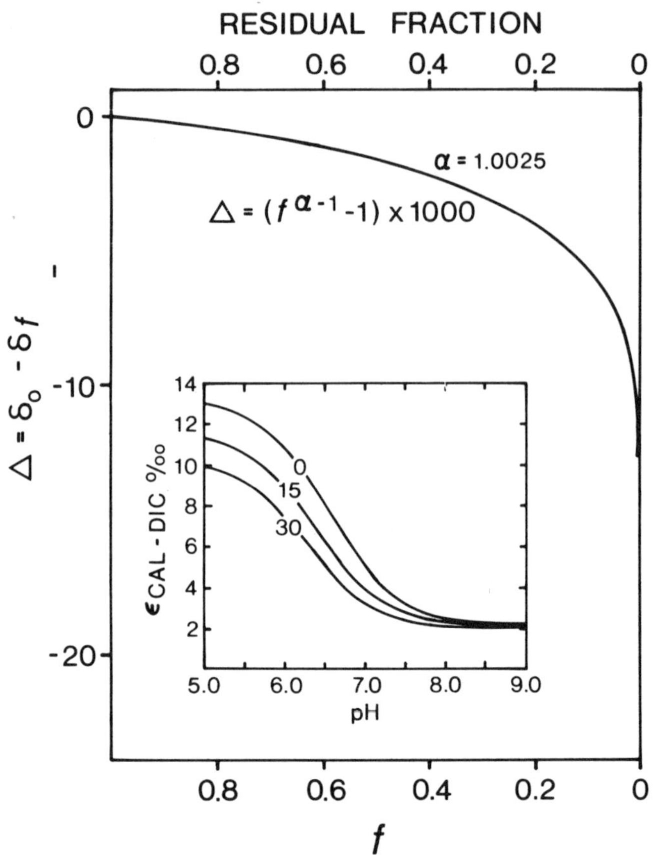

Figure 7. INSET: Carbon isotope fractionation effects between solid calcite precipitated from a solution with given $\delta^{13}C$. The effects increase with decreasing pH because of the greater importance of H_2CO_3, which has less ^{13}C than HCO_3^- or $CO_3^{=}$ (after Wigley and others, 1978). Outer diagram shows carbon isotope effects as they can occur during precipitation of calcite from a finite carbon reservoir (dissolved inorganic compound). Plot δ_o-δ_f vs f, the residual fraction of carbonate in solution.

affected only a little because a 3 ‰ shift in ^{13}C contents corresponds to only 0.6% ^{14}C fractionation and is thus within the error usually made on ^{14}C determinations.

In carbonate aquifers the incongruent dissolution of dolomite is one of the most important processes that can affect the geochemistry of a ground water. This occurs because it is accompanied by calcite precipitation. Not only can ^{14}C be lost in this calcite but rock carbonate can also be added in much greater amounts than is possible under congruent dissolution conditions. The ^{14}C activity of a ground water may become arbitrarily small even in the absence of radioactive decay (Wigley and others, 1978). It is very difficult to model this evolution because one has to rely on increasing Mg^{2+} contents in the water as an indicator for the amount of dolomite-carbon added. Although this may be an acceptable approach in sedimentary rocks, in crystalline environments, where primary silicates transform into clays and some ionic compounds,

Mg^{2+} is lost in the clays. In the ground waters in the Stripa granite, Mg^{2+} concentrations are always below 10^{-4} moles/l despite increasing salinities with depth. Similar observations are made on deep ground waters on the Canadian Shield.

Fortunately, dolomite is not a dominant fracture mineral or primary constituent of crystalline rocks. Nor is it abundant in shales and clay deposits, unless it has been deposited as a detrital component.

Studies on ground waters of the Canadian Shield have shown that biogenic carbon dioxide can have a profound effect on the carbon isotopic composition of the aqueous carbonate. This is not so much due to the oxidation of organic matter as to methane-producing bacteria, which are common. "Contamination" of ground waters with biogenic CO_2 can be expected where the waters are active (Barker and others, 1978; Barker, 1979). Such CO_2 contributions are immediately reflected in the ^{13}C contents, and $\delta^{13}C$ values as high as +16 °/°° have been measured at pH > 8. If such water precipitated calcite, similarly high $\delta^{13}C$ values would be noticeable in the precipitates. Such is the case for some fracture calcites in the Stripa granite where $\delta^{13}C$ values close to +15 °/°° have been measured. The effects of biogenic carbonate on the ^{14}C content will depend on the food source. In crystalline environments this could be either organic or inorganic carbon in the ground water because the aquifer rocks do not contain organic matter. If organic carbon were decomposed, it would have approximately the age of the ground water, whereas the inorganic carbon would have a mixed origin. Therefore, in the latter case no additional ^{14}C contamination takes place. The biogenic CO_2 would partially reset the ^{14}C clock, that is, increase the ^{14}C of the aqueous carbonate in the water. No ^{14}C data are yet available on methane-contaminated systems, but work is in progress and corrections can be proposed for systems in which a quantification of these processes is possible. This has to be done by studying the geochemical evolution of a ground water body. The correction procedures are essentially mass balance considerations (Barker and others, 1979).

^{14}C adjustment procedures proposed in the literature rely either on simple carbonate dissolution models where

$$H_2CO_2 + CaCO_3 \rightleftarrows 2H_2CO_3^- + Ca^{2+}$$

or complex consideration based on the total chemistry of ground waters from various sampling points in a flow system. Fontes and Garnier (1979) have summarized these procedures and propose "correction" procedures that involve a combination of ^{13}C and chemical mass balances with an approximation for isotope exchange. The ^{14}C "corrections" based on this approach can be very significant and can exceed by a considerable margin one ^{14}C-halflife, which is the maximum correction derived from the simple carbonate dissolution model. Unfortunately these considerations do not include silicate dissolution-precipitation. However, they dominate the geochemical evolution of most ground waters in crystalline rocks and, therefore, these model calculations are normally not applicable. Instead it is necessary to assess each geochemical process separately. For example, in the case of the Stripa ground waters, the following conclusions were drawn from ^{14}C and chemical analysis of ground water in the granite.

The geochemistry of the shallow ground waters is determined by carbonate dissolution and ion exchange. Silicate dissolution takes place but does not dominate. Therefore, geochemical correction procedures outlined above could be applied. With these it is estimated that about 50% of the initial ^{14}C activity of these ground waters is "lost" because of dilution with ^{14}C-free carbonate carbon.

In the deeper ground waters, after calcite saturation has been reached, silicate geochemistry dominates the water, and calcite precipitation does occur. Dolomite has not been recognized and ^{13}C contents of the aqueous carbon show that these ground waters were not affected by organic carbon produced by bacterial activities. Thus no further ^{14}C dilution took place, and decreasing ^{14}C contents with depth almost certainly reflect loss of ^{14}C by decay and thus increasing age.

Exchange with carbonate minerals and diffusive losses into the matrix of the granite (dead-end pores, etc) could further decrease the ^{14}C contents of the aqueous carbon. However, ^{13}C analyses on fracture calcite show that with few exceptions, no isotopic equilibrium between aqueous and solid carbonate exists and, therefore, it was tentatively assumed that exchange was negligible. Similarly, fracture-hydrology data and preliminary calculations show that diffusive loss probably had little influence on the ^{14}C activities of the samples.

GROUND WATER DATING WITH THE URANIUM DECAY SERIES

Uranium Activity Ratios

Uranium dating based on $^{234}U/^{238}U$ ratios has a potential advantage over the ^{14}C dating method because existing technology requires much smaller samples for uranium dating. Furthermore, the dating range extends well beyond that of ^{14}C and theoretically, ages between 50,000 and 1.5 x 10^6 years can be determined. This method is based on the fact that the activity ratio $^{234}U/^{238}U$ varies in a given ground water system as a function of age and geochemistry. Normally both have to be considered simultaneously.

If a geochemical system is in secular equilibrium then the activity ratios of parent and daughter isotopes are equal to one. Such is also the case for the uranium decay series and the $^{238}U/^{234}U$ activity ratio:

$$^{238}_{92}U \xrightarrow[4.5 \times 10^9\ a]{\alpha} {}^{234}_{90}Th \xrightarrow[24d]{\beta} {}^{234}_{91}Pa \xrightarrow[1\ min]{\beta}$$

$$^{234}_{92}U \xrightarrow[2.5 \times 10^5 a]{\alpha} {}^{230}_{90}Th.$$

However, the decay of ^{238}U in minerals has a recoil effect. If ground water leaching occurs, ^{234}U is preferentially dissolved. This is probably the principal reason that the activity ratios of most ground waters are 1 or greater than 1 and only rarely smaller (Osmond, 1979).

The activity ratio of a ground water can be modified through the preferential decay of ^{234}U. With time the ratio will again approach 1. If no uranium is added and no other quantifiable geochemical processes modify the ratio, then it is possible to utilize the changing activity ratio of a flow system for age determinations.

Brines from a salt dome in New Mexico (drill hole ERDA, No. 6) have been dated with this method by Barr and others (1979), using the following relationships:

$$t = [(A_m - 1)/(A_o - 1)]/-\lambda_{234}.$$

A_o and A_m are the initial and measured activity ratio respectively and λ_{234} is the decay constant of ^{234}U (= ln 2/248,000). To use this formula, initial activity ratios had to be deducted from comparisons with data collected elsewhere. With these data, one finds that the age of this brine, (its inclusion into salt) must have occurred between 500,000 and 1,300,000 years ago. This assumes that no interaction between brine and host rock took place and there is evidence that indeed chemical equilibria between rock and fluid have not been reached. The resulting simple relationship between activity ratios and age is graphically shown in Figure 8. The intercept at time = 0 is the measured activity, A_m, and the most probable initial ratio for the ERDA-6 sample is also shown. Note that if leaching occurs, then the ^{234}U contents in the aqueous phase will increase. The ages calculated using this equation are then too young.

In Stripa, uranium isotope data were also used as age indicators. Nothing is known about rock-water interaction because no fracture mineral data are yet available. Therefore, it was assumed that rock-water interactions were not important. Age differences between sampling points can be calculated, and data are plotted in Figure 8. They indicate about 4 x 10^5 years age difference between Stripa 16 and Stripa 29. Stripa 17 and 29 are respectively the top and bottom of the same drillhole and an age difference of about 1.5 x 10^5 years is suggested.

It is difficult to say how realistic these calculated ages are. In the Stripa ground waters the uranium contents decrease with increasing depth probably in response to loss of carbonate through calcite formation (Fig. 8). We have no information about what the isotope effects may be in this case--be it during precipitation or subsequent preferential leaching of ^{234}U.

Barr and others (1979) show that for a closed system, such as the brine included in a salt mass, it is possible to assess the intensity of rock-water interactions because the release of ^{234}Th and ^{234}U from the matrix into the fluid can be estimated. However, in ground-water systems leaching is not the only important process and

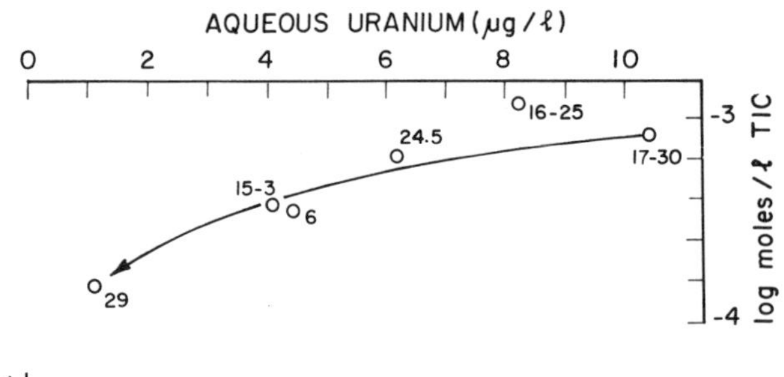

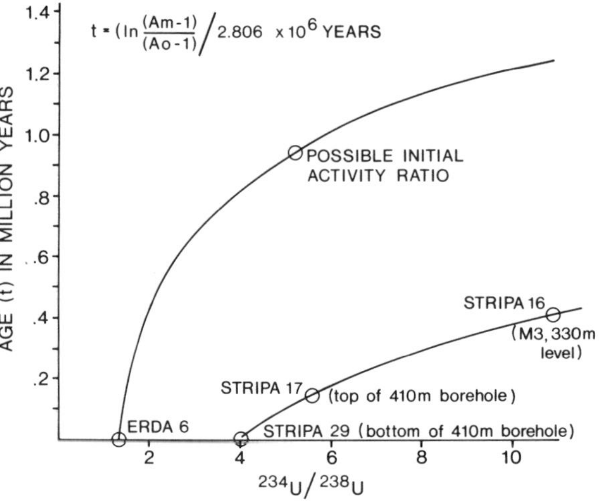

Figure 8. Uranium disequilibrium ages as a function of changing $^{234}U/^{238}U$ activity ratios in ground waters. Result of ERDA 6 (Barr and others, 1979) and its estimated age assuming initial conditions. Three Stripa samples are also shown indicating that the age difference between them could be as high as 4 x 10^5 years provided no interaction with rock uranium has occurred. The upper part of this figure shows that at Stripa, loss of uranium from the aqueous phase parallels decreasing aqueous carbonate concentrations. Because of this, no rock-water interaction was assumed, although this is probably not entirely correct and results shown are more likely to be maximum ages.

considerably more complex model considerations will be necessary. This is shown by Osmond and Cowart (1976) and Osmond (1979) who summarized the various fractionating processes affecting the uranium activity ratios and document applications in the tracing of water masses, mixing of water, and aquifer-water interactions. This type of information will provide the basis for detailed field investigations, which must emphasize the recognition of the evolutionary paths a ground water has undergone. As for ^{14}C, water age dating based on a single number or with a single ratio is simply not possible. Furthermore, again as for ^{14}C, no unique correction procedure can be proposed; each ground-water system has to be assessed on its own. Mineral analyses are as essential as information about regional flows in fractured rock, and neither can be obtained cheaply.

Rare Gas Analyses: Helium

Where detailed field investigations are planned,

the analytical work should also include rare gas analyses with special emphasis on helium. In a closed system, the amounts of helium dissolved in the water depend on the quantity of uranium in the aquifer minerals and the rate at which uranium is released from them. In any given system this release will be constant in time. The amounts accumulated will thus be proportional to the time the system was under closed system conditions.

The rate at which the ^4He increases is given by the equation:

$$^{4}He_{\text{rate of solution}} = \rho . \phi^{-1} [1.19 \times 10^{-13} (U)$$

$$+ 0.288 \times 10^{-13} (Th)] \ cm^3 STPy^{-1} cm^{-3} H_2O,$$

where ρ = bulk density of the rock and ϕ = fractional porosity. U and Th are uranium and thorium contents in ppm. It is possible to utilize this relationship to determine ground-water ages from helium concentration, provided uranium contents of the rocks (thorium can be assumed to be about four times as abundant as uranium) and porosity are known.

Marine (1976) used helium concentrations to date the ground water encountered in wells drilled into crystalline metamorphic rocks underlying the Savannah River Plant. Making assumptions on uranium and thorium contents in the rocks, fracture spacing, porosity, and helium generation rate, Marine calculated water ages close to 840,000 years.

A similar approach was taken in Stripa where helium concentrations in the ground water are about 5 orders of magnitude above atmospheric saturation levels. The total porosity of the granitic rock was measured to be close to 10^{-2} and uranium contents in the mine area were determined from borehole samples to be between 30 and 160 ppm with an average value close to 45 ppm. With this average value and the measured ^4He concentrations, the calculated water ages would be close to 1.4×10^5 years at the 330 m hole (~ 900m below gound surface). These are considerably higher age estimates than those derived from the ^{14}C data, even without assuming that some helium might have been lost from the waters during their long subsurface history.

As crude as all these estimates may be, they show the potentials of this method. In the context of nuclear waste disposal, however, they also show that apparently ground water systems in undisturbed crystalline rocks exist under closed-system conditions and that even the very mobile helium is retained.

SUMMARY

The age dating of ground waters in low-permeability rocks poses a special challenge because sample sizes will be small. The most widely applied tools are based on tritium, ^{14}C, and the uranium decay series. However, presentations made at a 1978 workshop on the age dating of old ground waters (University of Arizona) and the 1974 and 1978 symposia volumes of meetings on isotope hydrology (I.A.E.A., Vienna, Austria) show that a number of other isotope tools have great potential for such studies in normal aquifers. Because of sampling or analytical problems it is very unlikely that they can be used in the forseeable future for age determinations on much smaller samples. The only exceptions may be ^{36}Cl, for which little is known about isotope geochemistry, and ^{85}Kr, which competes directly with tritium.

The limitations encountered in applying the more conventional techniques in low-permeability rocks have also not been fully explored because virtually no interest in flow systems in such rocks existed. This suddenly changed when the potential for the storage of toxic waste in crystalline rocks, clayey sediments, or shales required urgent assessment. The disposal concepts for nuclear waste storage in a number of countries were conceived assuming suitability of crystalline rocks for the long-term containment of byproducts from the nuclear fuel cycle. These rock types have been selected because it was assumed that ground-water movements in them would be limited. To verify this, a combination of physically based tools, isotope-geochemical techniques, and mathematical models will have to be employed. This "combined approach" is essential because none of these techniques alone is able to provide definite answers in terms of water ages.

This applies to the most simple isotope tool, tritium, although the presence of tritium in any water is a clear indicator that young water is present. In crystalline rocks it should also be possible to relate tritium to ground-water velocities, provided uncontaminated samples can be obtained. In clayey sediments, however, this is not necessarily the case if isotopic differences exist between the water in the fractures and in the matrix. This will be the case whenever water movements and modern recharge or infiltration takes place. Diffusive loss of tritium can then occur and water movements (plus contaminants?) can no longer be directly related to tritium concentrations in samples collected from installations in such sediments.

Carbon-14 and "uranium" dating of ground waters are tools that are difficult to use unless one can demonstrate that the geochemical evolution of a ground water follows simple processes whose influence on the carbon and uranium budget can be assessed. An integration of these isotope investigations with geochemical and hydrogeological studies is not only recommended but necessary.

In the foregoing, an attempt was made to summarize the geochemical interferences that may influence the carbon geochemistry and thus the ^{14}C budget of ground waters in crystalline environments. It is evident that no simple correction formula can be proposed and for a given ^{14}C concentration it is impossible to give a unique solution in the form of a water age.

To carry any discussion beyond a general level will require very detailed geochemical data. A knowledge of the nature of possible geochemical reactions and associated mass-transfers is a prerequisite for a comprehensive evaluation of the possible magnitude of ^{14}C dilutions. Yet, to apply the results of such consideration, it is necessary to know a priori which processes are dominant in the system under consideration. This is usually not the case, and simplified model considerations will have to be used to eliminate those reactions that are impossible or unlikely. This task is facilitated in crystalline rocks because--on the basis of the data available to us--one can show that the most complicating process, namely the incongruent dissolution of dolomite, is of little importance.

This does not eliminate the need for detailed investigations of the geochemical evolution of a ground water before acceptable model consideration will provide us with appropriate ^{14}C corrections. The necessity for such investigations was dramatically emphasized by a comparison of ^{39}Ar and ^{14}C ages done on major aquifers in different sedimentary rocks. No agreement between the two existed and no reasonable explanations for the discrepancies have yet been found, although detailed geochemical studies on these waters had not yet been done.

Despite apparent problems, case histories seem to indicate that it is possible to obtain reasonable age estimates from ^{14}C analyses. This has not yet been proven for dating based on elements of the uranium decay series. Age determinations have been attempted using the activity ratio $^{234}U/^{238}U$, as well as helium and radon accumulations resulting from uranium decay. The results seem to prove the usefulness of such analyses for the dating of very old waters. However one has to have a very detailed knowledge of the geochemical history of a ground water and its physical setting before actual ages can be calculated from model considerations. The geochemical history is needed to identify those processes which induce isotope fractionation and thus modify activity ratios. A knowledge of the physical setting is required to assess the possibility of helium and radon migration from or to sampling points.

In conclusion, no isotope tool exists (nor any other tool) that can be applied to ground water age dating on its own. Absolute age dating, as is done for rocks and other solid matter, is not possible. However, detailed site investigations such as those carried out in Sweden and, hopefully, at several localities in North America should provide the background data necessary to assess the limits of these methods and the reasonableness of a water age calculated from a measured activity of a given isotope or decay product.

REFERENCES CITED

Anderson, T. F., 1968, Surface measurements in calcite grains by isotopic exchange with ^{14}C labelled carbon dioxide: Geochimica et Cosmochimica Acta, v., 32, p. 1177-1186.

Barker, J., 1979, Methaner in groundwater [Ph.D. dissertation]: Waterloo, University of Waterloo, 309 p.

Barker, J., Fritz, P., and Brown, R. M., 1979, ^{14}C measurements in aquifers with methane, in proceedings, Symposium Isotope Hydrology, Neuherberg, June 1978: International Atomic Energy Agency, Symposium 228/33, v. II, p. 661-678.

Barr, G. E., Lambert, S. J., and Carter, J. A., 1979, Uranium isotope disequilibrium in ground waters of southeastern New Mexico and implications regarding age dating of waters, in Proceedings, Symposium Isotope Hydrology, Neuherberg, June 1978: International Atomic Energy Agency, Symposium 228-28, v. 2, p. 645.

Brinkman, R., Münnich, K. O., and Bogel, J. C., 1959, ^{14}C Altersbestimmungen von Grundwasser: Naturwissenschaften, v. 46, p. 10-12.

Cherry, J. A., and others, 1979, Hydrogeologic properties and pore water origin and age: Clayey till and clay in South Central Canada, in Proceedings, Workshop on Low Flow, Low Permeability Measurements in Largely Impermeable Rocks. Paris, Organization for Economic and Commercial Development and International Atomic Energy Agency, p. 31-48.

Davis, S. N., 1978, Workshop on dating of old ground water: Office of Waste Isolation, Union Carbide, Rept. Y/OWI/SVB-78/55412, 138 p.

Davison, C., C., Grisak, G. E., and Williams, D. W., 1979, Field permeability and hydraulic potential measurements in crystalline rock and solute transport through finely fractured media, in Proceedings, Workshop on Low Flow and Low Permeability Measurements in Largely Impermeable Rocks: Paris, OECD-IAEA, p. 139-156.

Day, M. J., 1977, Analysis of movement and hydrochemistry of groundwater in the fractured clay and till deposits of the Winnipeg area, Manitoba (M.S. thesis): Waterloo, University of Waterloo.

Deines, P., Langmuir, D., and Harmon, R. S., 1974, Stable carbon isotope ratios and the existence of a gas phase in the evolution of carbonate groundwater: Geochimica et Cosmochimica Acta, v. 38, p. 1147-1164.

Elmore, D, and others, 1979, Analysis of ^{36}Cl in environmental water samples using an electrostatic accelerator: Nature, v. 277, p. 22-25.

Fontes, J. C., Bortolami, G. C., and Zuppi, G. M., 1979. Hydrologic isotopique du Massif du Mont-Blanc, in Proceedings, Symposium Isotopic Hydrology, Neuherberg, June 1978: International Atomic Energy Agency Symposium 228-22.

Fontes, J. C., and Garnier, J. C., 1979, Determination of initial ^{14}C activity of the total dissolved carbon: Water Resource Research, v. 15, p. 399-413.

Fritz, P., Barker, J., and Gale, J. E., 1979, Isotopic and geochemical characteristics of groundwaters in the Stripa granitic pluton: Berkeley, Lawrence Berkeley Laboratory, LBL-8285, 120 p.

Fritz, P., Drimmie, R. J., and Render, F. W., 1974, Stable isotope contents of a major prairie aquifer in central Manitoba, Canada, in Proceedings, Symposium Isotopic Hydrology: International Atomic Energy Agency, Symposium 182-22, I, p. 379-398.

Fritz, P., and Reardon, E. J., 1978, Isotopic and chemical characteristics of mine water in the

Sudbury area: Waterloo Research Institute, Report 609-06.

Fritz, P., and others, 1978, The carbon isotope geochemistry of a small watershed in northeastern Ontario: Water Resources Research, v. 14, p. 1059-1067.

Grisak, G. E., and Cherry, J. A., 1975, Hydrologic characteristics and response of fractured till and clay confining a shallow aquifer: Canadian Geotechnology Journal, v. 12, p. 23-43.

Grisak, G. E., and others, 1976, Hydrogeologic and hydrochemical properties of fractured till in the Interior Plains region: Royal Society of Canada Special Publication no. 12, p. 303-335.

International Atomic Energy Agency, 1970, Proceedings, Symposium on Isotope Techniques in Ground Water Hydrology: Vienna, International Atomic Energy Agency.

___1974, Proceedings, Symposium on Isotope Techniques in Ground Water Hydrology: Vienna, International Atomic Energy Agency.

___1978, Proceedings, Symposium on Isotope Techniques in Ground Water Hydrology: Vienna, International Atomic Energy Agency.

Knetsch, G., and others, 1962, Untersuchungen an Grundwassern der Ost-Sahara: Geologische Rundschau, v. 52, p. 587-610.

Lal, D., Nijampurkar, V. N., and Rama, S., 1970, Silicon-32 hydrology, in Proceedings, Conference on Isotope Hydrology: International Atomic Energy Agency, Symposium 129-54, p. 847-868.

Loosli, H. H., and Oeschger, H., 1979, ^{39}Ar, ^{14}C and ^{85}Kr measurements in groundwater samples, in Proceedings, Symposium on Isotopic Hydrology, Neuherberg, June 1978: International Atomic Energy Agency, Symposium 228.

Marine, W. I., 1976, Geochemistry of groundwater at the Savannah River Plant: Aiken, South Carolina, DuPont De Nemours and Co., No. DP 1356 (report to ERDA), p. 102.

Oeschger, H., and Gugelmann, A., 1974, Ausblick auf die Wasserdatierung mit Hilfe von Edelgasisotopen: Oester. Wasserwirtsch., v. 26, p. 51-56.

Oeschger, H., and others, 1974, ^{39}Ar-dating of groundwater, in Proceedings, Conference on Isotopic Hydrology: International Atomic Energy Agency, Symposium 182-37, v. I, p. 179-190.

Osmond, K., and Cowart, J.B., 1976, The theory and uses of natural uranium isotopic variations in hydrology: Atomic Energy Review, v. 14, p. 589-752.

Osmond, K., 1979, Uranium disequilibrium in hydrologic studies, in Fritz, P., and Fontes, J. C., eds., Handbook on environmental isotope geochemistry: Amsterdam, Elsevier, v. 1.

Rauert, W., and Stichler, W., 1974, Groundwater investigations with environmental isotopes, in Proceedings, Symposium on Isotopic Hydrology: International Atomic Energy Agency, Symposium 182-24, p. 431-443.

Reardon, E. J., and Fritz, P., 1978, Computer modelling of groundwater ^{13}C and ^{14}C isotope compositions: Journal of Hydrology, v. 36, p. 201-224.

Render, F. W., 1970, Geohydrology of the metropolitan Winnipeg area as related to groundwater supply and construction: Canadian Geotechnical Journal, v. 7, p. 243-274.

Rozanski, K., and Florkowski, T., 1979, ^{85}Kr dating of groundwater, in Proceedings, Symposium on Isotopic Hydrology, Neuherberg, June 1978: International Atomic Energy Agency Symposium, 228/51.

Rye, R. O., and O'Neil, J. R., 1968, The O^{18} content of water in primary fluid inclusions from Providencia, North-Central Mexico: Economic Geology, v. 63, p. 232-238.

Thilo, L., and Münnich, K. O., 1970, Reliability of carbon-14 dating of groundwater effect of carbonate exchange, in Proceedings, Symposium Isotopic Hydrology: International Atomic Energy Agency, Symposium 129/17, p. 259-270.

Wigley, T. M. L., Plummer, L. N., and Pearson, F. J., 1978, Mass transfer and carbon isotope evolution in natural water systems: Geochimica et Cosmochimica Acta, v. 42, p. 1117-1139.

Manuscript Accepted by the Society October 14, 1980

Printed in U.S.A.

Geological Society of America
Special Paper 189
1982

Geothermal Resource Exploration

HAROLD A. WOLLENBERG
Earth Sciences Division
Lawrence Berkeley Laboratory
University of California
Berkeley, California 94720

ABSTRACT

Exploration for geothermal resources in the Basin and Range geomorphic province of the western United States is much akin to mineral resource exploration. In this region, potential geothermal resources are located at or near zones of intersection of normal faults, where fracture permeability permits deep circulation of meteoric water into high-temperature areas afforded by high regional heat flow.

Exploration encompasses regional surveys, covering several thousand square kilometres, followed by more detailed studies of several hundred square kilometres to identify sites for deep drillholes. The most effective methods for regional exploration are geologic studies based on areal multispectral surveys, heat-flow drilling, chemical geothermometry of warm- and cold-water sources, and age-dating of rock. Detailed heat-flow-hydrologic surveys, gravity and active-seismic surveys, and geologic mapping are most effective in locating targets for deep confirmatory drilling. Geochemical-halo surveys may also be very effective in both regional and areal exploration.

Results of geochemical, radiochemical, and geophysical surveys combine to furnish a comparison of long-term flow rates of geothermal systems, thereby permitting preliminary evaluation of the relative viability of the systems.

INTRODUCTION

The occurrences of geothermal resources have much in common with those of hydrothermal mineral deposits. Present-day geothermal systems are depositing minerals that will form the hydrothermal orebodies of the future. In this respect, then, exploration for geothermal resources follows much of the rationale used for mineral exploration.

Most geothermal systems explored to date are found in consolidated rocks. Therefore, the phenomenon that most often governs the occurrence of a geothermal resource is fracture permeability. Fracture systems permit deep circulation of meteoric water and the development of hydrothermal convection systems. As with mineral prospects, geothermal exploration targets occur in areas with deep, pervasive fracturing, often at the intersections of prominent fault zones or lineaments. An example is the geothermal site near Beowawe, Nevada, illustrated in the high-altitude photograph in Figure 1. Here the east-northeast-trending Malpais escarpment is cut by a north-northwest-trending fault a few hundred metres east of a strongly altered zone encompassing the hot springs area.

The phenomenon that creates most geothermal resource areas in the Basin and Range geomorphic province of the western United States is described by Hose and Taylor (1974). Meteoric water penetrates deeply (3 to 5 km) along permeable zones in regions of relatively high crustal heat flow (Fig. 2). In such regions the geothermal gradient may be 40 °C/km to > 60 °C/km, with temperatures at depth 4 to 5 km in excess of 200 °C. Water percolating downward is heated and rises through the fractured, permeable rock on the upward limb of a convection cell. Where a small portion of this system "daylights," hot springs occur, often at the fault-zone margin of a Basin and Range valley. Field surveys in this region indicate that there are hidden hydrothermal systems, circulating beneath a cover of Tertiary or Quaternary sediments, that do not have hot springs. These hidden systems may be manifested on the surface by zones of alteration, geochemical halos, and electrical or thermal anomalies.

Although heat sources may be different in other terranes, circulation of meteoric water at depth also occurs in geothermal areas associated with the deep sedimentary basin of the tectonically active Salton Trough of California and Mexico; in active volcanoes and caldera occurrences; and in the strongly faulted, recently volcanically active Geysers/Clear Lake region of northern California.

The goals of geothermal exploration are (1) to discern the anomalies associated with a circulating

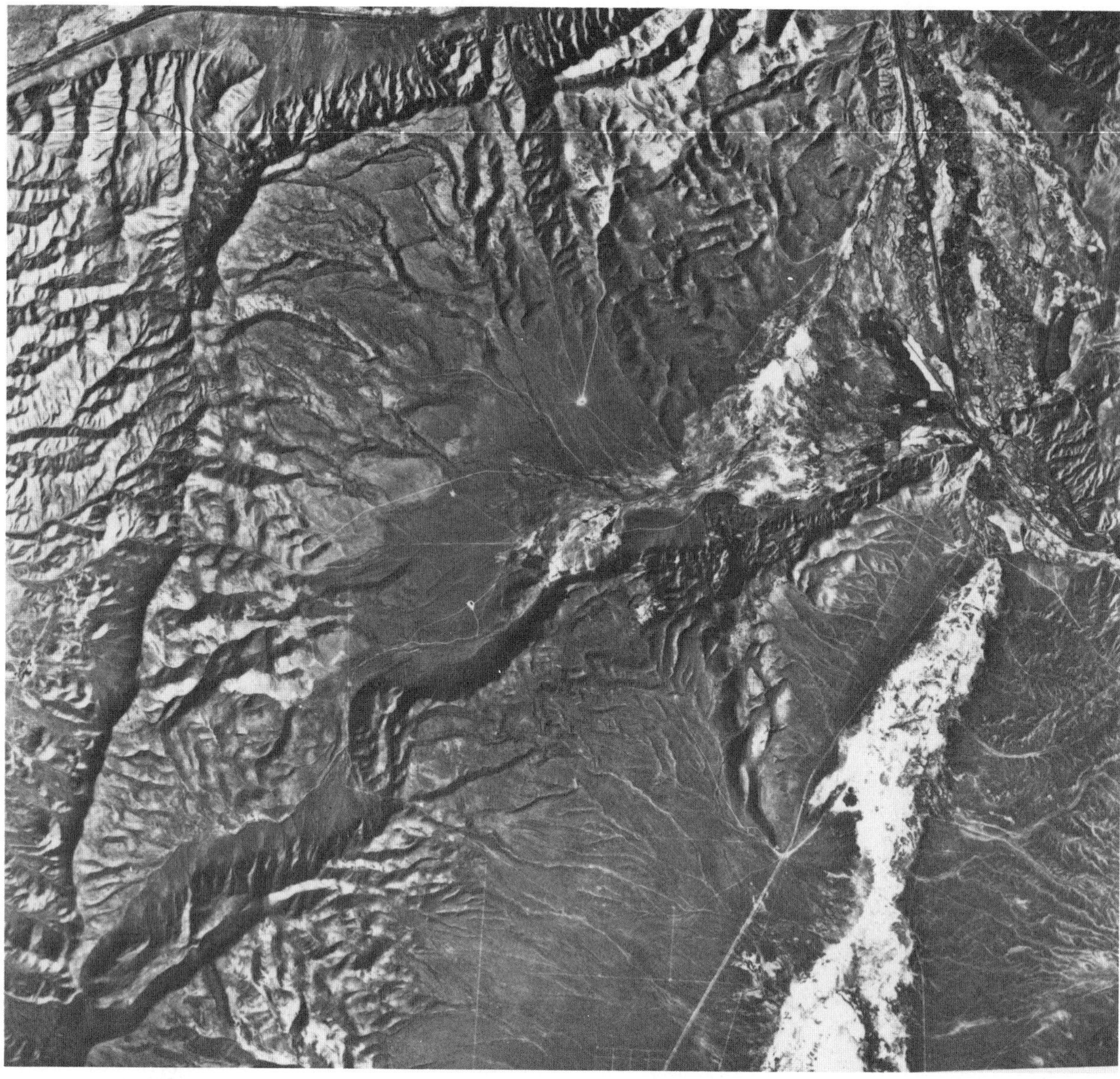

Figure 1. Vertical aerial photograph of Whirlwind Valley region, Nevada, from 18 km above surface, showing association of geothermal area (center) near intersection of east-northeast-trending Malpais escarpment and north-northwest-trending zone of en echelon faults. Width of field is approximately 27 km.

hydrothermal system by means of geological, geochemical, and geophysical surveys and (2) to locate the sites for holes that, when drilled and tested, will confirm the presence of the system and permit evaluation of the system's resource potential.

Under the leadership of Paul A. Witherspoon, Lawrence Berkeley Laboratory's Earth Sciences Division has conducted such surveys in varied terranes. It is this experience, primarily in the Basin and Range province, that has formed the basis for the following description and evaluation of geothermal exploration techniques.

REGIONAL STUDIES

Exploration programs for Basin and Range geothermal resources have been described by Keller and others (1979), Ward and others (1978), Mabey and others (1978), and Wollenberg and Goldstein (1977). Goldstein (1977) evaluated the various surface geological, geochemical, and geophysical techniques and ranked them as to their relative effectiveness in reconnoitering large regions of several thousand square kilometres, and in more detailed exploration, of areas of several hundred square kilometres. The regional techniques evaluated are listed (roughly in decreasing order of importance) in Table 1.

Along with an extensive survey of the existing geologic and hydrologic data and literature on the region, geological studies are considered most effective in the reconnaissance phase. These studies incorporate mapping, which relies on areal multispectral surveys. High- and medium-altitude low-sun-angle photography is needed to discern

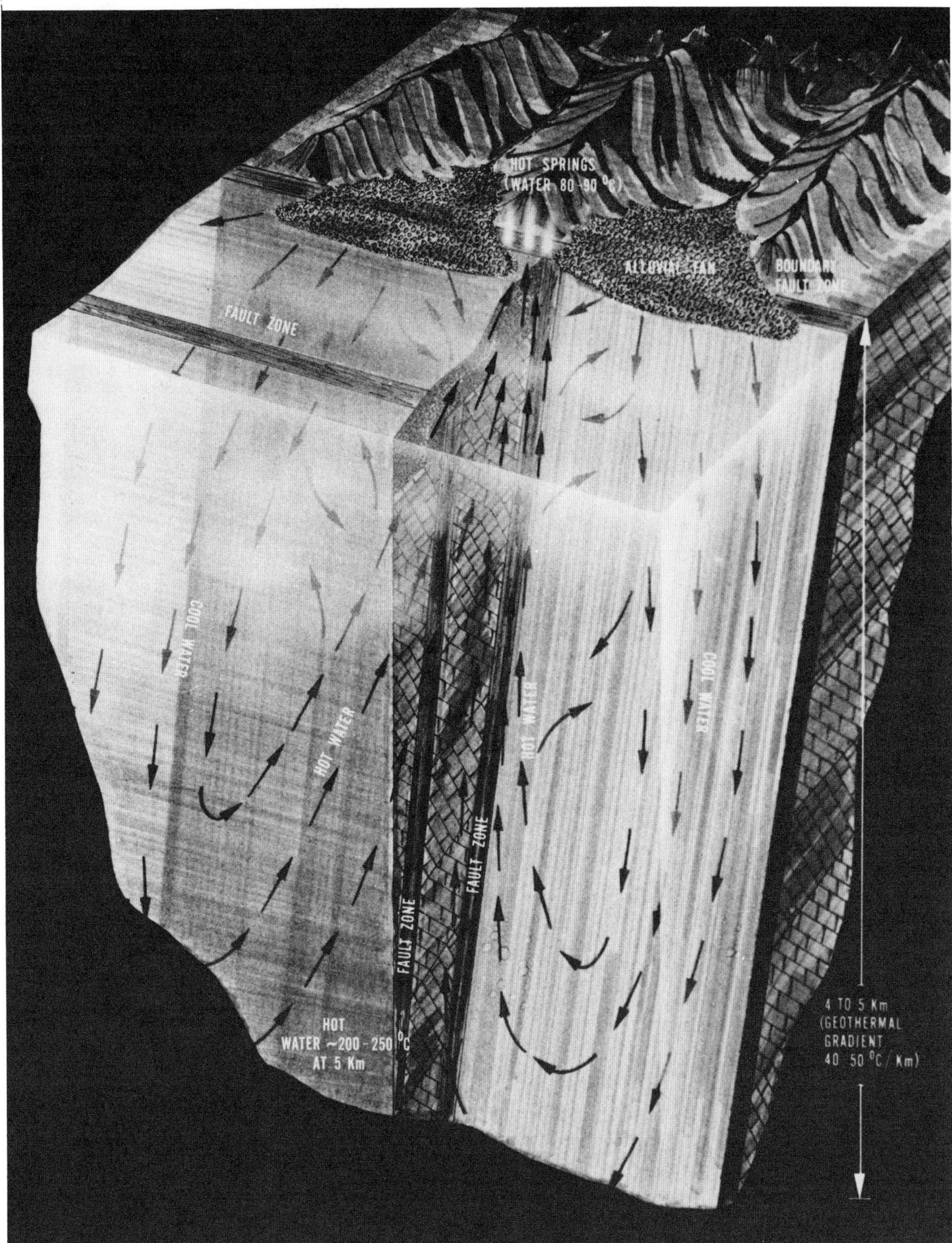

Figure 2. Cutaway schematic diagram of a geothermal system in Basin and Range geomorphic province.

structural patterns, and thermal-infrared imagery is used to disclose unknown thermal springs, zones of anomalously high heat flow, and, in some cases, evidence of fault zones concealed beneath alluvium.

Age dating of extrusive rocks is another important aspect of geological studies. Geochemical methods, especially chemical geothermometry based on silica (Fournier and Rowe, 1966) and alkali-element ratios (Fournier and Truesdell, 1973) of thermal waters, are also considered very effective in reconnaissance. Of nearly equal

TABLE 1. REGIONAL TECHNIQUES USED IN GEOTHERMAL RESOURCE EXPLORATION

1. Geology
   a. Aerial multispectral surveys
      i. High-altitude photography at low sun angles
      ii. Medium-altitude color and color-infrared photography
      iii. Thermal infrared imagery
   b. Geologic mapping to supplement existing data
2. Heat-flow surveys
3. Age dating of rocks
4. Geochemical surveys
   a. Chemical geothermometry
   b. Regional geochemical halos
5. Seismotectonic studies
6. Microearthquakes
7. Aeromagnetics
8. Gravity
9. Hydrology

importance are regional heat-flow surveys based on temperature profiles, and measurements of the thermal conductivity of cores and cuttings from new or existing drillholes. Broad-scale regional structural patterns may be disclosed by airborne magnetometric and regional gravity surveys.

On the basis of their studies in the northwestern Basin and Range province, Keller and others (1979) considered the most useful geophysical methods in geothermal exploration to be those that define regional heat flow and disclose the subsurface configuration on the intermountain basins. The techniques most likely to be useful include heat-flow, gravity, seismic-reflection and electrical-resistivity surveys.

AREAL STUDIES

When promising areas have been delineated from regional surveys, detailed studies incorporating several disciplines are employed; these are listed in Table 2. The more effective studies are heat-flow surveys, active and passive seismic surveys, detailed geologic mapping, and gravity, electrical-resistivity, geochemical-halo, and trace-element surveys. This paper briefly describes some of these techniques and gives an example that combines geophysical, hydrologic, and geochemical data to estimate the long-term viability of a Basin and Range geothermal system.

HEAT FLOW AND HYDROGEOLOGIC STUDIES

Heat flow and hydrogeology are closely related. This is exemplified by experiences in northern Nevada (Sass and others, 1977; Olmstead and others, 1975) and in Long Valley, California (Sorey and Lewis, 1976). In their Long Valley studies, Sorey and Lewis combined spring-discharge and chemical-mixing models to estimate the total convective heat flow from hot springs. The estimated heat flow, 4.3×10^7 cal/s, could result from a conductive heat flow of 10 $\mu cal/cm^2 \cdot s$ over the caldera area of 450 km^2, a value compatible with measurements by Lachenbruch and others (1976).

Surveys of conductive heat flow revealed and delineated convecting geothermal systems in Grass Valley, Nevada (Sass and others, 1977). Eighty-two holes, whose depths ranged from 18 to 400 m, were drilled over an area of 200 km^2. Initially, six holes, spaced a few kilometres apart, were drilled to depths of 150 to 200 m in alluvium and Tertiary sedimentary rocks to obtain background conductive heat flow. Holes were logged by radiation and electrical methods. Cores and cuttings were collected to determine thermal conductivity. Nearby, 15 holes were drilled to depths below the water table in the discharge area of Leach Hot Springs to detail the local heat-flow pattern and to obtain hydrologic data (Olmstead and others, 1975). Thirty intermediate-depth heat-flow holes (50 to 100 m) were drilled to furnish additional detail in areas between the initial deep holes. From these intermediate holes, Sass and others (1977) determined that, at several locations, heat flow is strongly related to shallow-depth (15 to 20 m) temperatures. As a result, 31 shallow holes were drilled to detail the heat flow in zones of specific interest. The conductive heat-flow pattern is shown in Figure 3. Besides the obvious thermal anomaly associated with Leach Hot Springs, a conductive thermal anomaly is evident in the vicinity of Panther Canyon and another in the valley, cen-

TABLE 2. AREAL TECHNIQUES USED IN GEOTHERMAL RESOURCE EXPLORATION

1. Heat-flow surveys
2. Active seismic surveys
3. Detailed geologic mapping
4. Hydrogeologic studies
   a. Mass heat flow
   b. Mixing models
   c. Flow rates
5. Geochemical halos
6. Gravity surveys
7. Electrical resistivity surveys
8. Passive seismic surveys
   a. Teleseismic P-wave
   b. Microearthquake and ground noise
9. Self potential
10. Ground magnetics

tered approximately 5 km southwest of the hot springs. These anomalies have no obvious surface manifestations and may be conductive indicators of convecting geothermal systems. Such anomalies are likely targets for deep, confirmatory drilling.

Olmstead and others (1975) investigated the hydrogeologic settings of several thermal spring areas in Nevada. In their study of subsurface temperature distributions, they evaluated total heat flux and fluid discharges of individual spring systems. To accomplish this they (1) compiled geologic data on the areas, supplemented in several cases by more detailed geologic mapping; (2) repeatedly measured spring flows and temperatures; and (3) drilled grids of test wells in the vicinity and away from thermal springs, measured temperature gradients in the wells, and monitored water-surface elevations. From these surveys they determined the nature of fluid discharge of the thermal systems and estimated the heat- and water-discharge fluxes. Total heat discharge from the convecting thermal systems included that from spring flow and from associated conductive heat flow. Total water discharge included flow from spring orifices (including steam), evapotranspiration, and subsurface outflow. Water discharge was also calculated by the heat-budget method, which incorporates the estimates of reservoir temperature and heat discharge.

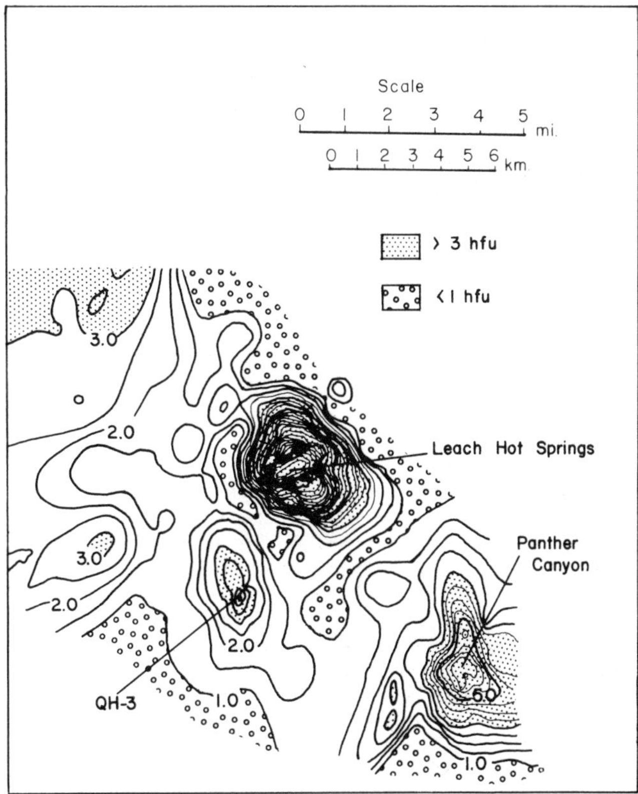

Figure 3. Contours of conductive heat flow ($cal/cm^2/s$) in the vicinity of Leach Hot Springs, Grass Valley, Nevada. (After Sass and others, 1977.)

GEOCHEMICAL SURVEYS

Trace-element contents of stream water, stream sediments, and soils have been used extensively in mineral exploration to discern chemical halos, which may surround concealed orebodies. From analyses of drill cuttings, Bamford (1978) demonstrated that the areal zonation of some trace elements centers on known geothermal sites. At both the vapor-dominated systems of The Geysers, California, and the liquid-dominated system at Roosevelt Hot Springs, Utah, areal zoning was evident in the distributions of arsenic, antimony, lead, zinc, manganese, boron, and tungsten (Fig. 4).

Silica contents of warm-and-cold spring waters within a hydrologic region have been used in mixing models (Fournier and Truesdell, 1974) to estimate the temperature of unmixed hot water at depth within a geothermal system and the proportion of nearer-surface cold water mixing with the hot water.

Radioelement and trace-element contents of warm-and cold-spring waters furnish estimates of the age of a geothermal system, and, when combined with geophysical data (to be discussed below), provide estimates of the long-term flow rate of a system. Oxygen and hydrogen isotope ratios of warm and cold waters, combined on a diagram (Fig. 5) developed by Craig (1961), indicate the amount of contact a water has had with rocks at depth in passage from its meteoric source to the sampling site. The amount of "oxygen shift" (the displacement from the meteoric water line for a given hydrogen/deuterium ratio) in general is greater for waters of high temperature or waters that have undergone little mixing with near-surface cold water.

Soil-gas emanations may delineate fault zones along which hot waters are circulating. A survey in Buffalo Valley, Nevada, using alpha-track detectors, disclosed sharp anomalies in ^{222}Rn associated with the hot springs (Wollenberg, 1976). A radon survey by Nielson (1978) in the vicinity of Roosevelt Hot Springs, Utah, indicated anomalies in alluvium overlying fault zones, as did a mercury survey in the same area by Capuano and Bamford (1978).

SEISMIC SURVEYS

Active seismic methods incorporating reflection and refraction techniques and passive methods including teleseismic, microearthquake, and ground-noise monitoring have been applied to geothermal exploration.

Reflection and refraction techniques have proved valuable in delineating suballuvial structures. In Grass Valley, Nevada, reflection surveys using a "Vibroseis" source (Majer, 1978) disclosed the faulting associated with Leach Hot Springs, as illustrated in Figure 6. Because the fault pattern determines the plumbing system controlling the circulation of water at depth, and because such faults are often concealed by an alluvial veneer, seismic-reflection surveys are considered very

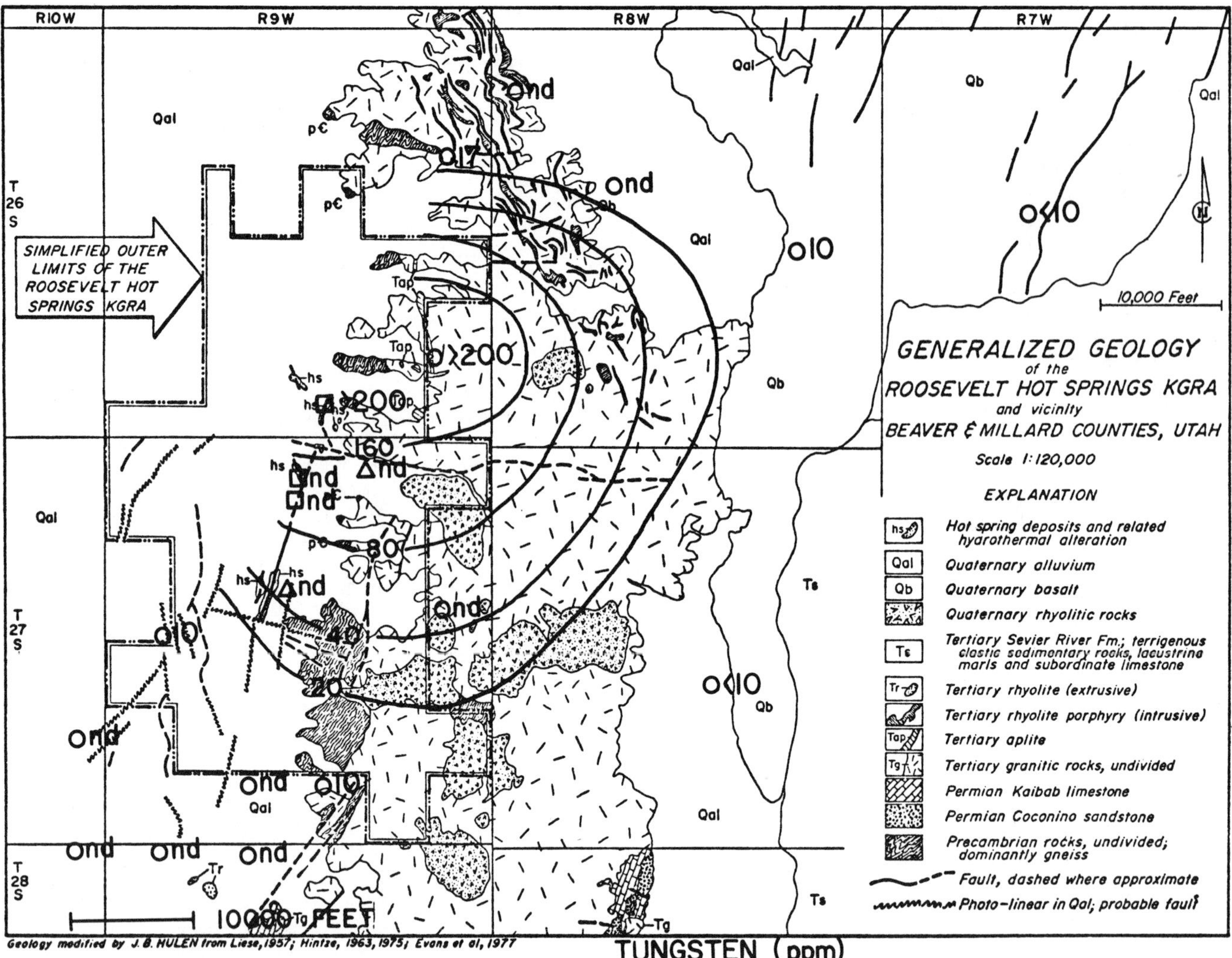

Figure 4. Contours of tungsten contents of samples of upper 100 ft of rock or soil, superimposed on a generalized geologic map of Roosevelt Hot Springs region, Utah. (After Bamford, 1978).

effective tools in detailed exploration of suspected geothermal resource areas.

Monitoring the arrival of compressional (P) waves from distant seismic events may indicate the presence of zones of densification associated with circulating hydrothermal systems. In Grass Valley, Majer (1978) discerned such a zone by slightly earlier arrivals of teleseismic P waves in the vicinity of Leach Hot Springs (in comparison with arrivals at stations away from the springs), indicating densification of sediments by deposition of material from the hydrothermal system (Fig. 7). The densification was corroborated by a detailed gravity survey (Goldstein and Paulsson, 1977).

In assessing surface geophysical techniques in geothermal exploration, Goldstein and others (1978) pointed out that most microearthquake and ground-noise surveys to date have not been definitive because of the limited capabilities of seismic instrumentation and data processing. More refined field systems, incorporating large numbers of seismometers and on-line data processing, should permit these surveys to become more useful in geothermal exploration.

ELECTRICAL METHODS

Electrical methods in geothermal exploration have emphasized measurements of electrical resistivity. Circulating thermal water, or zones of alteration resulting from such circulation, provide a more electrically conductive environment than warm-dry rock or cold-wet rock. On the other hand, vapor-dominated zones and areas of subsurface hydrothermal deposition may be more resistive. Fault zones with circulating thermal water and areas of elevated heat flow may also be discerned by measurements of electrical self-potential (Corwin and Hoover, 1978). Goldstein and others (1978) described and compared electrical geophysical methods that are used in geothermal exploration.

Both natural-field and controlled-source methods are used to measure electrical resistivity. Controlled-source methods rely on either injection

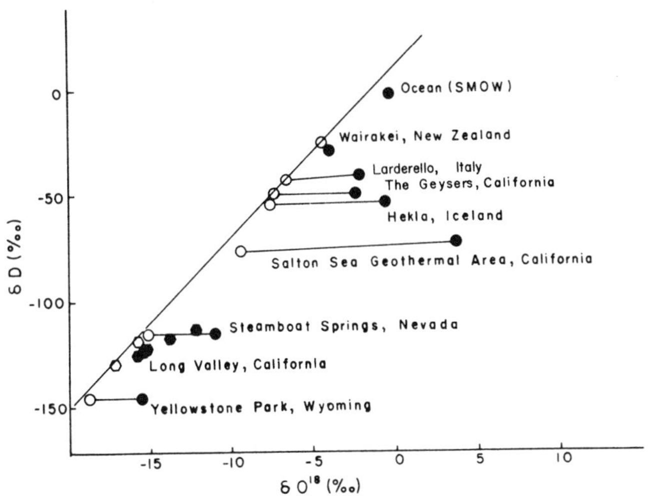

Figure 5. Oxygen- and hydrogen-isotopic compositions of some major geothermal areas. Solid symbols represent thermal springs; open symbols represent fresh waters. (Source: Mariner and Willey, 1976; reprinted by courtesy of American Geophysical Union.)

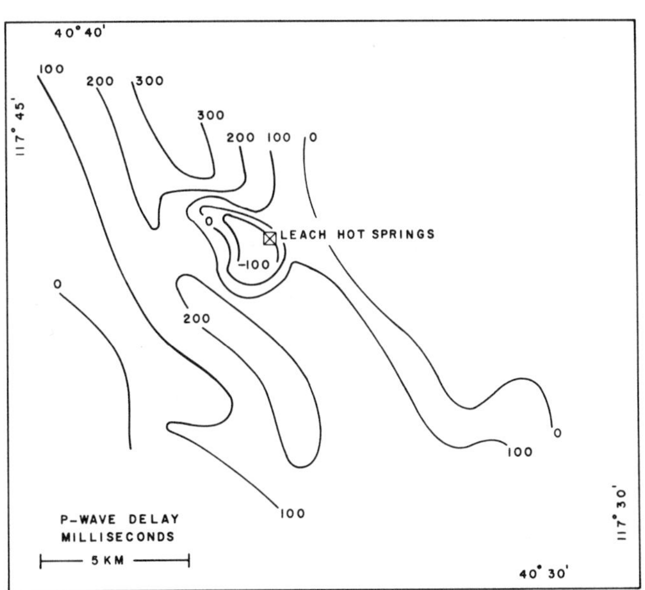

Figure 7. Contours of P-wave delay, in milliseconds, in Grass Valley, Nevada. Note negative delay (or advance) near Leach Hot Springs. (After Majer, 1978.)

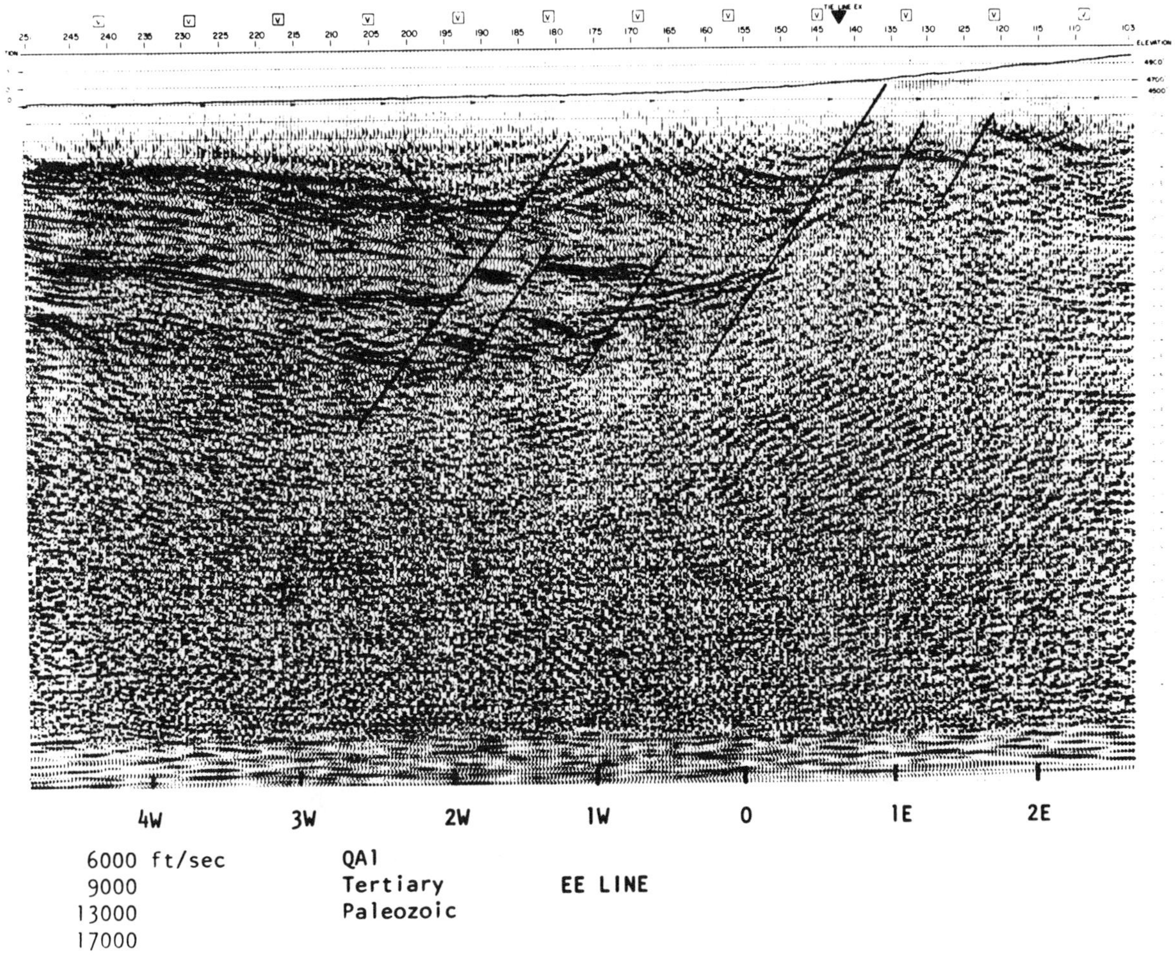

4W 3W 2W 1W 0 1E 2E

6000 ft/sec QA1
9000 Tertiary EE LINE
13000 Paleozoic
17000

Figure 6. Seismic-reflection profile across eastern portion of Grass Valley, Nevada. Solid triangle at top of diagram indicates location of Leach Hot Springs. Vertical and horizontal scales are in kilometres, as shown along bottom of diagram. (After Majer, 1978.)

of current into the earth through a pair of electrodes or induction of current within the ground by means of an oscillatory current in a loop of wire set on the surface. In the former method, resistivities are calculated from voltages measured between a pair of electrodes at varying distances from the current dipole. In the case of the oscillatory current source (the electromagnetic or EM method) perturbations in the Earth's magnetic and electrical fields are measured at varying distances from the current loop and with varying current frequencies in the loop. Natural-field methods rely on (1) that measurements of differences in small voltages result from differences in the Earth's electrical field caused by the presence of a conductor at depth (telluric method) or (2) a combination of magnetic and electrical-field measurements (magnetotelluric methods).

A detailed evaluation of electrical resistivity and telluric methods was made by Beyer (1977), on the basis of surveys in potential geothermal resource areas in northern Nevada. (A composite diagram of the results of several geophysical surveys is shown in Fig. 8.) Beyer concluded that the roving-dipole (or bipole-dipole) resistivity method was least useful in exploring these geologic settings, whereas the dipole-dipole method provided a more definitive assessment of the resistivity structure but required extensive interpretation. In a modeling study comparing resistivity methods, Beyer concluded that the dipole-dipole method was most effective in discerning the resistivity structure of an area containing sharp, lateral discontinuities, as in a faulted Basin and Range setting, and that the Schlumberger method was applied best to delineation of flat-lying geothermal targets in a regularly layered setting. Beyer also concluded that the electrical-field-ratio telluric method furnished a relatively rapid electrical reconnaissance of a large region, indicating areas that could be covered in greater detail by the more expensive dipole-dipole technique.

A large, controlled-source (60 W) electromagnetic system is presently being developed and tested (Morrison and others, 1978, and may offer advantages over conventional galvanic (conduction) types of surveys in areas where it is difficult to lay out long wires or to inject current into the ground. The EM system has greater flexibility in rugged terrain but at present the data are no easier to interpret than dipole-dipole resistivity data.

In a recent article summarizing their earth-science activities at Roosevelt Hot Springs, Ward and others (1978) described the application of several electrical-geophysical techniques to discern the resistivity structure of this promising geothermal area. They employed active electrical methods, including dipole-dipole, Schlumberger, and electromagnetic soundings, and determined that the geo-electric model resulting from these surveys was valid for depths to 500 m; below that depth, the resolution of active electrical methods was not adequate for the particular geologic setting. Ward and others (1978) indicated that one-dimensional models based on data from passive magnetotelluric and audio-frequency magnetotelluric (AMT) surveys did not yield realistic distributions of resis-

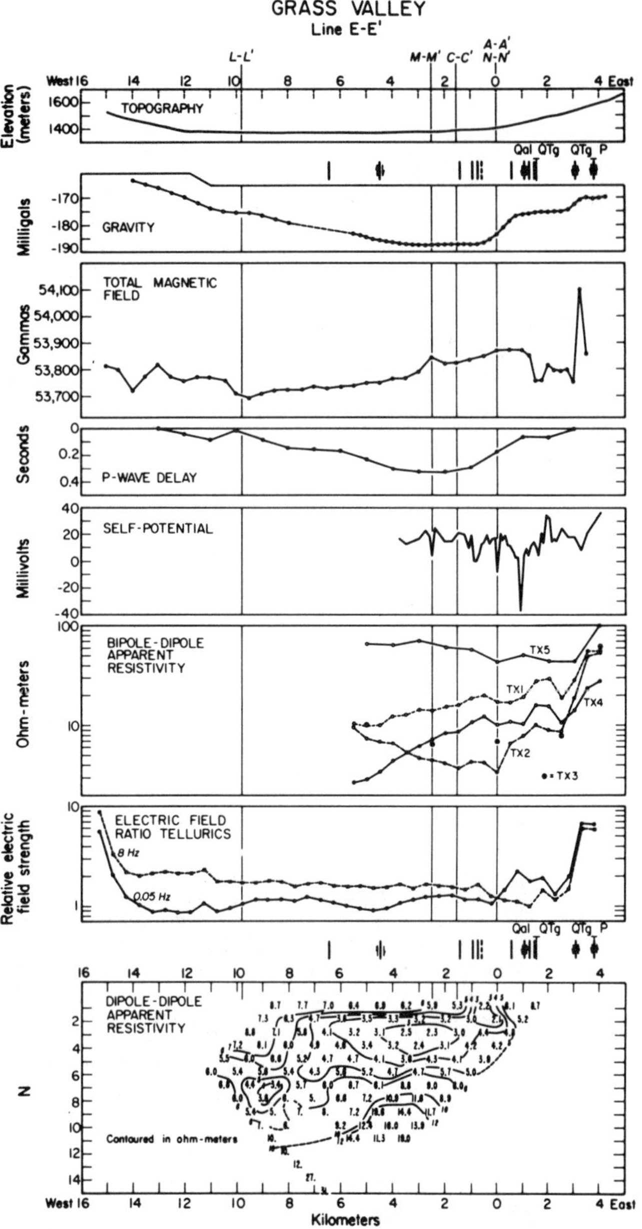

Figure 8. Composite of geophysical data on a vertical section transecting Grass Valley, Nevada. Leach Hot Springs roughly coincides with position "0." (After Beyer and others, 1976, p. 132.)

tivities at depth. Three-dimensional modeling in progress might furnish a more believable pattern. However, Mabey and others (1978) found that at the Raft River geothermal areas in Idaho, a resistivity anomaly discerned by telluric current and AMT surveys coincided with the location of the geothermal reservoir.

In summary, electrical techniques that have proved most useful in disclosing and delineating Basin and Range geothermal systems are the self-potential method, which detects circulating ground water, and the dipole-dipole and Schlumberger resistivity methods, from which models of the configuration of zones of electrical resistivity at depth can be constructed. Electrical-field-ratio telluric surveys are useful for reconnaissance of large areas and for disclosure of areas that may merit more detailed surveys by dipole methods.

AGES AND FLOW RATES OF HYDROTHERMAL SYSTEMS

Geochemical and geophysical data can be combined to estimate the ages and flow rates of geothermal systems. The subsurface flow rates, which may be substantially greater than surface-spring discharge rates, are indicators of the system's viability.

The age of a geothermal system is estimated from the concentrations of uranium and its daughter products in hot-spring waters and in cold ground waters of the hydrologic regions surrounding the hot-spring systems. In the northern Nevada studies conducted by Lawrence Berkeley Laboratory, trace-element geochemistry (Wollenberg and others, 1977) indicated that hot waters analyzed were very low in uranium--less than a few tenths of a part per billion (ppb). Cold-spring waters contained from 1 to more than 3 ppb (Fig. 9). Though they contain little uranium, the hot-spring waters have abundant daughter products of the ^{238}U decay series. This abundance was also noted by O'Connell and Kaufmann (1976) and Cadigan and Felmlee (1978) in their studies of radioelement contents of hot-spring systems. The marked difference in uranium contents of hot and cold springs and the presence of uranium daughters in the cold springs suggest that if the cold-spring water is water recharging the hydrothermal system, then uranium is concentrating at depth within the system. This concentration may be due to reduction of the uranyl (U^{6+}) ion in uranium carbonate to the less mobile U^{4+} ion. This reduction would be accomplished in the reducing environment of a geothermal system, which is afforded by the presence of hydrogen sulfide. Of equal importance in the deposition of uranium is the retrograde solubility of uranyl carbonate, which precipitates out at high temperatures. (The equilibria and solubility of actinide compounds in aqueous environments are presently under study at Lawrence Berkeley Laboratory, with impetus from uranium-resource assessment and nuclear-waste isolation programs.)

On the basis of evidence that uranium is concentrating in the hot-spring systems, a knowledge of the uranium content of cold- and hot-spring waters permits one to estimate the age of a hot-spring system. The methods of calculation and limiting assumptions have been described in detail by Wollenberg and others (1977). In its simplest form, the equation to calculate the age of a system is

$$T = Ra/U\lambda_{Ra},$$

where T is the apparent age of the system, Ra is the ^{226}Ra content observed in the hot-spring water, U is the difference between uranium measured in the cold and hot waters, and λ_{Ra} is the radioactive decay constant of radium-226. If secular equilibrium does not exist in the uranium decay series in the cold or hot waters, a more complex equation must be applied to correct the apparent age determined from the above equation.

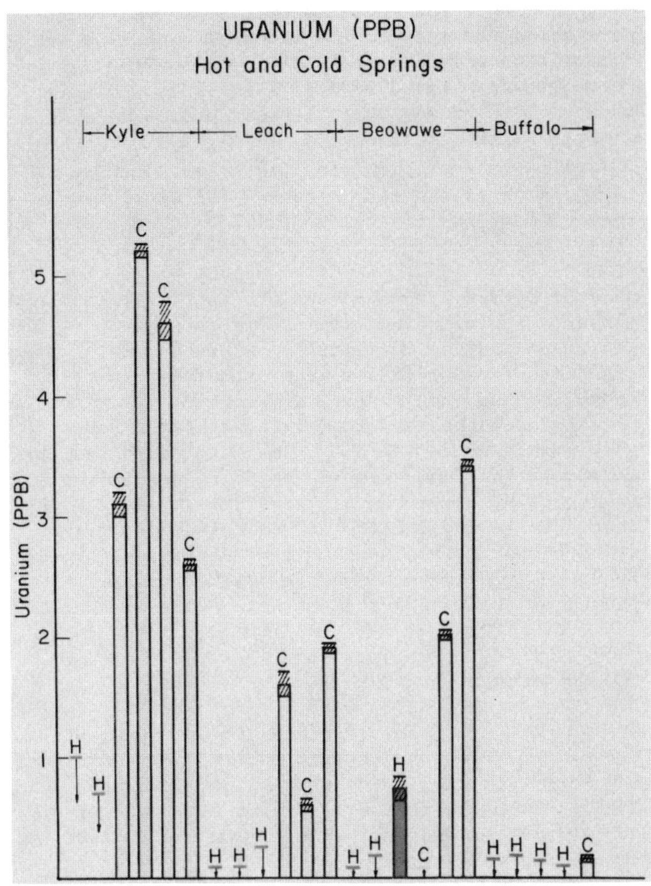

Figure 9. Contents of uranium in cold- and hot-spring waters in four geothermal areas in northern Nevada. C = cold springs; H = hot springs; tails of vertical arrows indicate detection limits. (After Wollenberg and others, 1977.)

Using this concept, the age of the Leach Hot Springs system was estimated to be 310,000 yr, the system at Kyle Hot Springs 78,000 yr.

The calculated ages can be used with spring-water chemistry and geophysical data to estimate the long-term flow rates of the spring systems. This method is based on the concept that the mass of material deposited by a convecting system is the product of the system's age, flow rate, and the concentration of chemical constituents of the hot water (total dissolved solids). Kyle Hot Springs is an example. Gravity surveys in Buena Vista Valley, in which the springs are located, permitted Goldstein and Paulsson (1977) to estimate the excess mass of material deposited by the convecting system at Kyle as 7×10^8 metric tons. If this material were deposited uniformly over the life of the system (apparent age 7.8×10^4 yr), the deposition rate would be about 8.97×10^3 tons/yr, or 2.83 g/s. The sum of major chemical constituents of Kyle Hot Springs water totals 1.6 g/l. Assuming that this is the chemical content of the upward-convecting hot water and that the concentration has remained essentially unchanged over the lifetime of the system, thus furnishing the mass of the material deposited, then the long-term steady-state flow rate of the Kyle Hot Springs system would be 1.77×10^2 l/s. This is more than 500 times the observed surface discharge at Kyle (Mariner and others, 1974).

Similar calculations for Leach Hot Springs, based on an excess mass of 2.5×10^8 metric tons (Goldstein and Paulsson, 1977), an apparent age of 3.1×10^5 yr, and total chemical constituents of 3.23×10^{-1} g/l, yielded a steady-state flow rate of 80 l/s. Olmsted and others (1975) combined measured surface flow with estimates of evaporation and transpiration to calculate a present-day surface discharge rate of 15 l/s, approximately one-fifth the estimated steady-state flow rate.

Estimated ages and flow rates at Leach and Kyle are summarized in Table 3. Chemical geothermometry indicates that temperatures at depth within the Leach and Kyle systems are similar. However, even though surface discharge at Leach is much greater than that at Kyle, the flow-rate estimates suggest that the Kyle system may have greater resource potential than the system at Leach Hot Springs.

ACCUMULATION OF URANIUM

Assuming that the difference in uranium content between cold- and hot-spring waters (ΔU) represents the amount of uranium depositing at depth, and combining this with the long-term flow rate of a geothermal system, one can estimate the rate of uranium accumulation:

Uranium accumulation rate (g/s)

$$= \text{flow rate (l/s)} \times \Delta U \text{(g/l)}.$$

This estimate requires that the flow rate be based on an age of the geothermal system by a method that does *not* incorporate the system's uranium content, such as isotope dating of spring deposit material.

If uranium accumulated steadily at the rate of several kilograms per year over the age of a geothermal system, several hundred to more than a thousand metric tons might be deposited in systems of the age of those at Leach and Kyle Hot Springs.

SUMMARY

Geothermal exploration incorporates a number of techniques, ranging from high-altitude (~18 km) aerial surveys, to heat-flow and hydrogeologic drilling to depths of 300 m, to electrical soundings penetrating several kilometres into the earth. Regional geological and geophysical surveys cover broad regions of several thousand square kilometres with the goal of finding target areas of a few hundred square kilometres. These are investigated in detail by geological, geophysical, and geochemical techniques to locate sites of a few tens of hectares for drilling deep test wells to examine a potential geothermal reservoir.

In the Basin and Range geomorphic province of the western United States, geothermal resource areas are localized by zones of fracture permeability, afforded by intersections of faults, which permit percolation of meteoric water to depths where temperatures reach more than 200 °C. In these respects, geothermal exploration targets are similar to those of hydrothermal mineral deposits. Exemplified by the apparent concentration of uranium in geothermal systems, today's geothermal resources may be tomorrow's ore deposits.

The U.S. Department of Energy, in cooperation with industry, is doing deep drilling to examine targets identified by surface surveys and shallow drilling. The resulting case studies will furnish definitive

TABLE 3. ESTIMATES OF AGE AND FLOW RATE

	Kyle Hot Springs system	Leach Hot Springs system
Apparent age*	7.8×10^4 yr	3.1×10^5 yr
Excess mass†	7.8×10^8 metric tons	2.5×10^8 metric tons
Total dissolved solids	1.6 g/l	0.3 g/l
Long-term flow rate	177g/s	80g/s

* From radio-geochemical data and calculations (Wollenberg and others, 1977).
† From gravity studies (Goldstein and Paulsson, 1977).

evaluations of the usefulness of the exploration techniques in locating geothermal resources.

ACKNOWLEDGMENTS

I wish to thank Beverly Strisower for her valuable assistance in gathering information and Ellen Diamond for improving the text. I appreciate discussions with Norman E. Goldstein on the applicability of geophysical techniques to geothermal exploration. This work was supported by the Assistant Secretary for Conservation and Renewable Energy, Office of Renewable Technology, Division of Geothermal and Hydropower Technologies of the U.S. Department of Energy under Contract No. W-7405-ENG-48.

REFERENCES CITED

Bamford, R. W., 1978, Geochemistry of solid materials from two U.S. geothermal systems and its application to exploration: University of Utah Research Institute, IDO/77.3.2, 196 p.

Beyer, H., 1977, Telluric and dc resistivity techniques applied to the geophysical investigation of Basin and Range geothermal systems [Ph.D. thesis]: Berkeley, University of California and Lawrence Berkeley Laboratory, LBL-6325 (3 parts), 461 p.

Beyer, H., and others, 1976, Preliminary open-file report on geological and geophysical studies in Grass Valley, Nevada: Berkeley, Lawrence Berkeley Laboratory, LBL-5262, 144 p.

Cadigan, R., and Felmlee, K., 1978, Radioactive springs geochemical data related to uranium exploration: Journal of Geochemical Exploration, v. 8, p. 381-395.

Capuano, R. M., and Bamford, R. W., 1978, Initial investigation of soil mercury geochemistry as an aid to drill-site selection in geothermal systems: University of Utah Research Institute, Earth Sciences Laboratory, IDO/78-1701.b.3.3, ESL-13, 32 p.

Corwin, R. F., and Hoover, D. B., 1978, The self-potential method in geothermal exploration: Berkeley, Lawrence Berkeley Laboratory, LBL-7075, 56 p.

Craig, H., 1961, Isotopic variations in meteoric waters: Science, v. 133, p. 1702.

Fournier, R. O., and Rowe, J. J., 1966, Estimation of underground temperatures from the silica content of water from hot springs and wet-steam wells: American Journal of Science, v. 264, p. 685-697.

Fournier, R. O., and Truesdell, A. H., 1973, An empirical Na-K-Ca geothermometer for natural waters: Geochimica et Cosmochimica Acta, v. 37, p. 1255-1275.

___1974, Geochemical indicators of subsurface temperature, Part II, Estimation of temperature and fraction of hot water mixed with cold water: U.S. Geological Survey Journal of Research, v. 2, no. 3, p. 263-270.

Goldstein, N. E., 1977, Northern Nevada geothermal exploration strategy analysis: Berkeley, Lawrence Berkeley Laboratory, LBL-7012, 52 p.

Goldstein, N. E., and Paulsson, B., 1977, Interpretation of gravity surveys in Grass and Buena Vista Valleys, Nevada: Berkeley, Lawrence Berkeley Laboratory, LBL-7013, 43 p.; and Geothermics (in press).

Goldstein, N. E., Norris, R. A., and Wilt, M. J., 1978, Assessment of surface geophysical methods in geothermal exploration and recommendations for future research: Berkeley, Lawrence Berkeley Laboratory, LBL-6815, 166 p.

Hose, R. K., and Taylor, B. E., 1974, Geothermal systems of northern Nevada: U.S. Geological Survey Open-File Report 74-271, 26 p.

Keller, G. V., Grose, L. T., and Crewdson, R. A., 1979, Geophysical exploration for geothermal prospects north of Gerlach, Nevada. Submitted to Geophysics.

Lachenbruch, A. H., and others, 1976, Geothermal setting and simple heat conduction models for the Long Valley Caldera: Journal of Geophysical Rsearch, v. 81, no. 5, p. 769-784.

Mabey, D. R., and others, 1978, Reconnaissance geophysical studies of the geothermal system in southern Raft River Valley, Idaho: Geophysics, v. 43, no. 7, p. 1470-1484.

Majer, E. L., 1978, Seismological investigations in geothermal regions [Ph.D. thesis]: Berkeley, University of California and Lawrence Berkeley Laboratory, LBL-7054, 223 p.

Mariner, R. H., and Willey, L. M., 1976, Geochemistry of thermal waters in Long Valley, Mono County, California: Journal of Geophysical Research, v. 81, no. 5, p. 792-800.

Mariner, R. H., and others, 1974, The chemical composition and estimated thermal reservoir temperatures of the principal hot springs of northern and central Nevada: U.S. Geological Survey Open-File Report 74, 32 p.

Morrison, H. F., and others, 1978, Description, field test, and data analysis of a controlled-source EM system (EM-60): Berkeley, Lawrence Berkeley Laboratory, LBL-7088, 142 p.

Nielson, D. L., 1978, Radon emanometry as a geothermal exploration technique: Theory and an example from Roosevelt Hot Springs KGRA, Utah: University of Utah Research Institute, Earth Science Laboratory, IDO/78-1701.b.1.1.2, ESL 14, 31 p.

O'Connell, M., and Kaufmann, R., 1976, Radioactivity associated with geothermal waters in the western United States: Basic Data, U.S. Environmental Protection Agency Technical Note ORP.LV-75-8A.

Olmsted, F. H., and others, 1975, Preliminary hydrogeologic appraisal of selected hydrothermal systems in northern and central Nevada: U.S. Geological Survey Open-File Report 75-56, 267 p.

Sass, J. H., and others, 1977, Application of heat-flow techniques to geothermal energy exploration, Leach Hot Springs area, Grass Valley, Nevada: U.S. Geological Survey Open-File Report 77-762 and Lawrence Berkeley Laboratory, LBL-6809, 125 p.

Sorey, M. L., and Lewis, R. E., 1976, Conductive heat flow from hot springs in the Long Valley caldera, Mono County, California: Journal of Geophysical Research, v. 81, no. 5, p. 785-791.

Ward, S. H., and others, 1978, A summary of the geology, geochemistry, and geophysics of the Roosevelt Hot Springs thermal area, Utah: Geophysics, v. 43, no. 7, p. 1515-1542.

Wollenberg, H. A., 1976, Radioactivity of geothermal systems, in Proceedings, Second United Nations Symposium on the Development and Use of Geothermal Resources, San Francisco, May 1975, Volume 2: Washington, D.C., U.S. Government Printing Office (Lawrence Berkeley Laboratory; University of California), p. 1283-1292.

Wollenberg, H. A. and Goldstein, N. E., 1977, Evaluation of geothermal exploration techniques in northern Nevada, in Proceedings, American Nuclear Society, Topical Meeting on Energy and Mineral Resource Recovery: U.S. Department of Energy conference 770440, p. 551-561.

Wollenberg, H. A., and others, 1977, Geochemical studies at four northern Nevada hot spring areas: Berkeley, Lawrence Berkeley Laboratory, LBL-6808, 69 p.

Manuscript Accepted by the Society October 14, 1980

V. Engineering Aspects

Geological Society of America
Special Paper 189
1982

Hydrogeology of Radioactive Waste Isolation: The Challenge of a Rational Assessment

STANLEY N. DAVIS
Hydrology and Water Resources Department
University of Arizona, Tucson, Arizona 85721

ABSTRACT

Plans to isolate radioactive waste raise more questions than can be answered within the lifetime of today's researchers. The existence of a multitude of unanswered questions, however, is not a unique phenomenon related to waste isolation, but it is a general characteristic of all science. Although most questions related to the isolation of radioactive waste are interesting scientifically, answers to only a small but critical fraction of the questions are vital to the practical aspects of the problem.

Almost all hydrogeologic evaluations of repository sites convey an excessively negative view with respect to geologic processes. Repositories do not necessarily become more hazardous with time. Many slow, normal geologic processes will seal and cover repository sites, making radionuclide migration less likely in the future than at present. Furthermore, several more rapid geologic processes, which are viewed with great apprehension, will not always have negative effects. Faulting could block ground-water circulation near the repository and consequently slow the outward migration of radionuclides; volcanism could cover repository sites, with only a slight chance of a volcanic vent contacting the radioactive waste; and in most of the northern part of the Midwest, future glacial activity would probably plaster the ground surface with layers of till, thus helping to isolate and protect any potential deep repository.

Safety of waste repositories will depend on multiple barriers to migration of radionuclides. Further research focused on each barrier, however, is required before repository safety is assured.

INTRODUCTION

In this paper I discuss some of the more general hydrogeologic aspects of the isolation of radioactive waste. Although the escape of gas, erosion of the land surface by water and wind, uptake by plant roots, excavation by explosions, and penetration by animals and man can all cause the undesired spread of radioactive waste, the most likely agent for this spread is commonly assumed to be ground water that has passed through the waste.

The required time of isolation for hazardous wastes may vary from a few days, for certain pathogenic organisms and many radionuclides with short half-lives, to an infinite length of time for many inorganic poisons and a few radionuclides. For all waste, and particularly for radioactive waste, the time required for isolation is also dependent on the physical surroundings of the waste disposal site, which in turn will control the rate of partial release and resulting concentrations of noxious materials in the biosphere. Despite the fact that required time of isolation is highly site specific, it is customary to speak of a set time frame for a particular category of radioactive waste. Stated values generally vary from 10^2 years for wastes containing primarily activation products to 10^7 years or more for wastes containing significant amounts of transuranics. The process of choosing an isolation time ranges from one bordering on the occult to an almost infinitely complex process of scenario construction based on field data of highly variable quality. Unfortunately, some of the resulting numbers have been proposed as official guidelines without taking into account any of the site characteristics that must be understood in order to estimate potential radiation doses to man. Nevertheless, I have generally assumed isolation times of a few hundred years for low-level wastes not containing significant amounts of transuranic elements. For high-level wastes and transuranic wastes, I have assumed isolation times of 10^4 to 10^5 years, even though many of the radioactive constituents have much longer half-lives.

Unless otherwise noted, the following

discussion relates to deeply buried high-level waste. Many of the ideas presented, however, apply equally well to transuranic and low-level waste buried at shallow depths. Also, I have made no attempt to be comprehensive, which would duplicate summary publications such as Bredehoeft and others, 1978; Bradshaw and McClain, 1971; Cohen, 1977; KBS, 1977a; Martinez, 1977; NAS Panel on Bedrock Disposal, 1972; NAS Panel on Geologic Site Criteria, 1978; Papadopulos and Winograd, 1974; U.S. Atomic Energy Commission, 1974; U.S. House of Representatives, 1976; Wingorad, 1974; and Witherspoon and Degerman, 1978.

BACKGROUND

The management of radioactive waste has been a growing problem for more than 35 years. Significant amounts of radioactive military wastes began accumulating in the mid-1940s. At this writing, the largest volume of waste in a noncommunist country at any single location is presently at the Hanford Reservation in the state of Washington, where waste began to be generated in 1944. Disposal of military waste in the United States has been largely in surface and near-surface tanks, cribs, pits, ponds, trenches, and landfills. High-level military waste is now being stored in large steel tanks buried at a shallow depth, primarily at the Hanford and Savannah River (South Carolina) reservations. At Hanford, ^{90}Sr and ^{137}Cs have been separated from much of the high-level waste. These radionuclides are stored temporarily within steel cylinders that have been placed in a large cooling bath. At Oak Ridge, Tennessee, waste material rich in ^{137}Cs was mixed with cement and injected into shale under a pressure high enough to split the shale horizontally (Weeren and others, 1978). The final result is a series of horizontal sheets of solid radioactive cement enclosed in a nonpermeable shale several hundred feet below the surface. Early in the history of waste disposal in the United States, large amounts of low-level waste were placed in casks and barrels and dumped at sea. Although sea disposal is still practiced by some foreign countries, it was phased out in the United States during the decade from 1960 to 1970. Shallow burial in special landfills has generally replaced sea disposal in the United States. The first of these landfills for commercially generated waste was opened in 1962 near Beatty, Nevada.

In the mid-1950s, concomitant with the development of commercial nuclear reactors, the magnitude of the future waste-disposal problem became apparent to most informed scientists. The U.S. Atomic Energy Commission (AEC) responded to this problem by supporting a Committee on Geologic Aspects of Radioactive Waste Disposal within the National Academy of Sciences (P. Durbin, 1976, unpub. report). The initial agreement between the AEC and the Academy was signed on February 28, 1955. In September 1955, a preliminary conference was held at Princeton University. The official committee that subsequently was formed eventually issued a report in September 1957. Among other things, the first committee report concluded that permanent storage of high-level waste might best be accomplished by placing liquid wastes in cavities in salt. The next most promising method was considered to have been stabilization of the waste in slag or ceramic material and in dry mines, surface sheds, or large cavities.

The Academy Committee on nuclear waste has continued to function but is at present called the Committee on Radioactive Waste Management. This parent committee supervises numerous panels, each of which is dedicated to a separate topic. Committee members are not paid and come from various professional and academic organizations. The Academy employs a small full-time staff to assist the committee and various panels. This staff also includes a small number of scientists.

Some of the earliest hydrogeologic work through government agencies was done by Arthur M. Piper and C. V. Theis, both of the U.S. Geological Survey. Most of their early work, which began about 1950, is described in short unpublished reports. Starting in the mid-1960s and increasing in frequency up to the present, most of the national laboratories dealing with nuclear energy, as well as numerous governmental agencies, have assembled advisory committees concerned with the isolation of nuclear waste. These committees, together with full-time professionals in contracting and subcontracting organizations, account for several hundred individuals at present concerned with the hydrogeological aspects of the problem alone.

Much of the federally funded work in the United States on the management of commercially generated radioactive waste has been summarized in three publications by the U.S. Department of Energy (1979a). These publications are in ten ponderous volumes that defy systematic study by their overwhelming bulk. They do, nevertheless, provide an up-to-date picture of the entire commercial-waste problem. Problems on the management of defense waste are covered in numerous publications for each of the major locations where this waste is stored.

The most concise and well-written proposals thus far for the possible disposal of commercially generated waste have come from Sweden. These reports (KBS, 1977a; 1978) represent summaries of more than 100 separate studies on almost all phases of the problem. The studies were required as a result of the "stipulation law" passed by the Swedish parliament in April 1977. This law stipulated that new nuclear power units could not be put into operation unless the owner were able to show that the radioactive waste problem had been solved in a completely safe way. The matter was still politically unresolved as of December 1979, despite the fact that the KBS reports proposed extremely conservative disposal plans. (A national plebiscite on the matter of nuclear energy was held in Sweden during the spring of 1980.)

One of the recent complications to develop in the waste-management field has been the problem of reprocessing commercial fuel rods. On April 7, 1977, President Carter announced a policy of nonproliferation of nuclear weapons, which included a decision that deferred indefinitely the reprocessing of spent fuel from commercial reactors. This decision has complicated the disposal of commercial waste by introducing an awkward geometry to waste containers if the fuel rods are not cut into pieces or otherwise reconstituted, by introducing large quantities of radionuclides with very long half-lives, by increasing the chances of future human intrusion in search of uranium and plutonium and by increasing the possibility of criticality if uranium and plutonium are not removed by reprocessing (Allen, 1978). Before the President's decision, isolation of high-level waste for 1,000 yr was generally considered adequate. At present, suggestions have been made for time periods exceeding 1,000,000 yr.

RESEARCH IN AN INFINITE SEA OF QUESTIONS

The safe isolation of nuclear waste has been the topic of a burgeoning number of technical and nontechnical publications. This rapid growth in literature has largely been due to the many questions that continue to arise concerning the safety of nuclear power generation. Power engineers, who are faced with the task of providing the public with an ever-increasing amount of electricity, view waste isolation as an expensive and difficult problem, but one with solutions well within the reach of present engineering technology. In contrast, the theoretical scientist feels especially concerned about the need to make reliable, very long-term predictions on a system whose characteristics are extremely ill-defined. As a result, he feels it imperative to find answers to many basic questions which may overshadow engineering or economic considerations. Some might feel that a single well-trained researcher will always generate many more questions than can be answered even within several lifetimes. Out of the many questions that are so generated, it is essential that research be directed only toward those that are critical to the solution of the practical problem. Clearly, the ability to obtain significant answers to all questions will require investments which far exceed available time and resources.

The answer to the problem of an almost infinite number of questions is an effective administration of research money designated for problems of radioactive waste. First, questions should be judged as to whether they are based on sound scientific principles. Second, questions should be viewed in the context of the entire waste-disposal effort. This requires not only a careful consideration of the phenomenon postulated, but a judgment concerning the relevance of the phenomenon in the final assessment of the long-term safety of a waste repository. Considerable progress in the required systems approach has already been made, but as yet few proposed research projects have been screened for relevancy except by subjective judgment. Finally, administrators must be forceful and well-enough informed to say no and be able to convince politicians, potential researchers, and the public in general that research has been rejected for sound reasons.

In conclusion, theoretical science will always generate more questions than it can possibly answer. Therefore, a strong and enlightened administrative control is imperative to channel research funds toward answering questions of significance. This is all the more important since the nuclear-waste disposal problem has very strong sociopolitical implications apart from its extreme technical complexity. Undue insistence on a full investigation of irrelevant or remote questions tends to confuse or excite the public, rendering a difficult problem to be a well-nigh impossible one. Here we find ourselves at the sensitive and ill-defined interface between the fields of science and the humanities. The hydrogeologist is thus challenged to reach beyond mere technical tools to solve the problems of nuclear-waste isolation.

Most studies on nuclear-waste containment make several "conservative" assumptions which may possibly be necessary for safety analyses but which present a negative image for the general public. The difficulty is that the nontechnical person may be told that an event may happen, but the fact that the chance of that event taking place may be only one in a billion is not particularly meaningful to him. Two fundamental assumptions, rarely expressed but always implied, are first that future civilizations will be ignorant of locations and dangers of nuclear wastes and second that they will still retain the skills to locate and exhume wastes from nuclear repositories. The combination of these assumptions appears to be quite extreme, if not illogical.

Within hydrogeology, a negative image is always conveyed when future geologic events are discussed. The passage of time with accompanying geologic activity, however, need not always degrade the integrity of nuclear-waste repositories. If for nothing else than public reassurance, the positive aspects of geologic processes should always also be presented. Furthermore, in many discussions of geologic criteria, specialists incorrectly assume that certain events or processes are automatically to be avoided (NAS Panel on Geologic Site Criteria, 1978). Three common examples follow.

Glaciation

If an area is potentially subject to glaciation within the next few thousand years, it is usually considered unsuitable for the location of a repository. Nevertheless, in areas of continental glaciation in the north-central part of the United States, glacial scour is usually limited and deposition of till having very low permeability is widespread within a few hundred kilometres of the glacial margin. Thus, in locations such as Iowa, north-central Missouri, central Illinois, and central Ohio, a rather firm prediction could be made that future glaciation would deposit yet another layer of low-permeability material to further isolate hydrogeologically the underlying Paleozoic rocks. Even in the unlikely possibility that a meltwater channel would intersect an area of interest, past events would suggest that future vertical scour from such a channel would be considerably less than 100 m, compared with the total depth of more than 300 m for repositories for high-level radioactive wastes.

Faulting

In dense, brittle rocks that have high compressive strengths, faulting can produce brecciated zones and open fractures having high permeabilities. However, in soft rocks and nonindurated sediments, faulting can greatly reduce the local permeability, as in many of the alluvial valleys of California, which are widely known to have fault barriers to ground-water movement. Furthermore, if faulting produces earthquakes, severe damage to properly engineered bedrock repositories is not likely (Pratt and others, 1978). The hazards of faulting have been generally overemphasized except where the repository could be actually intersected by a newly formed fault. Under special circumstances, therefore, faults could even add to the safety of a repository. This would be particularly true in Gulf Coast sediments and in alluvial basins of the southwestern United States, provided the fault is hydrogeologically upgradient from the repository.

Volcanism

Although volcanism is generally considered to have a highly negative influence on a repository site, it actually could most commonly add to the safety of the site. Ash, tuff, and mudflows from volcanic activity would tend to blanket the land surface with material having a low permeability and a high capacity to adsorb radionuclides. Dikes in the subsurface would form ground-water barriers. Lava flows would protect the land surface from effects of rapid land erosion. Inasmuch as an area of active volcanism would be evacuated during eruptions, the direct intersection of a volcanic vent or dike with a repository, as unlikely as this would be, would probably not create a major catastrophe. Most of the solid waste material could be assimilated in the lava or encased in volcanic rock and could become immobilized after the event.

Negatively oriented presentations not only may convince nonscientists that repositories for nuclear waste will automatically become less safe with time, but these presentations have also added great confusion to the choice of the best medium for a geologic repository. In order to establish a viewpoint and even perhaps to obtain research funding, researchers have sometimes enlarged on the dangers of particular geologic materials. Because salt has generally been the favored medium, it has also been the popular target of criticism. Unfortunately, in order for salt to have become a primary candidate in the past, other geologic materials were either ignored or in some instances even downgraded. The obvious result is that today, as scientific opinion shifts away from salt, the public is left with the incorrect impression that salt is unsafe and that repositories must now be located in less-than-desirable rock types. Had the positive aspects of various media been emphasized at first, the ebb and flow of hydrogeological opinions would not have left the problem of choosing the best repository medium in its present state of confusion.

MULTIPLE BARRIERS

The approach to containment of nuclear waste that has been developed by the Department of Energy (DOE) is commonly referred to as the multiple-barrier approach. This is a sound approach as long as practical objectives are kept in mind. The central idea is to provide as many barriers to waste migration as possible. Theoretically, each barrier will give significant protection to the biosphere, and two or three barriers taken together should provide an acceptable degree of protection.

A chemically stable waste form can be the first barrier to radionuclide migration. Spent fuel or waste from fuel reprocessing may be mixed with cement or incorporated in ceramics, glass, or other material of low solubility. Recently, Ringwood (1978) advocated an artificial rock made of minerals known to have low solubilities under temperature and salinity conditions expected in high-level repositories. The artificial rock would contain between 2% and 10% by weight of nuclear waste. Whether or not this will prove feasible remains to be determined. If the minerals prove to be as stable as predicted, then all other barriers may be redundant.

Waste containers and packing materials can also be significant barriers, although the tentative plans for waste handling in the United States do not at present include extra-resistant containers such as those proposed by Swedish planners (KBS, 1977b). Thick copper, nickel-steel, or lead containers may be packed in a mixture of quartz sand and clay. The metal is relatively insoluble, the quartz would add mechanical stability and increased thermal conductivity, and the clay would prevent radionuclide migration because of its low permeability and high sorptive capacity for ions in the liquid phase. The Swedish "gold-plated" solution to questions of waste-container stability, unfortunately, will require leaving several hundred million dollars worth of lead and copper in the subsurface. This deposit will, therefore, invite future human intrusion if the dangers of radioactivity are ignored.

In most plans for geologic isolation of waste, the host rock for the repository is considered one of the most important barriers. This is particularly true if the dimensions of the host rock are relatively large and the equivalent bulk permeability is very low. If thick salt is present and if the salt retains its integrity, then migration of water through the host rock will not be significant over periods of millions of years. Unfortunately, the integrity of the salt is difficult to predict, particularly where it has been disturbed by construction of a repository and heat accumulation from the waste.

The last barrier to the outward migration of radionuclides from a repository is the migration path of the radionuclides before it reaches artificial or natural exits to the biosphere. In some hydrogeological settings, such as the Nevada Test Site, the potential migration path is measured from many tens to more than a hundred kilometres. In other localities, such as near the burial site for low-level radioactive waste at Maxey Flats, Kentucky, the potential migration path is measured in a few hundred metres or less. If available aquifers have low permeabilities and hydraulic gradients are small and nearly horizontal, travel times for water may be measured in hundreds of years per kilometre. Radionuclides with strong sorptive tendencies will have travel times measured in tens of thousands of years per kilometre or more. Under such circumstances, the travel path itself would provide enough transit time so that the biosphere would be effectively protected from these radionuclides.

RESEARCH NEEDS

Some of the present needs for hydrogeological research related to high-level radioactive-waste isolation are given below. I emphasize applied research needs because the urgency of most of the problems does not allow us to wait the several decades that would be necessary for a thorough investigation of all fundamentals.

The Waste Form and Related Barriers

Research is needed to define the manner in which radioactive waste could enter into the ground water. At present, the rates at which radionuclides would most likely enter groundwater in the event of containment failure are largely unknown; in published literature, these rates appear to be largely speculative. Despite the large number of variables such as waste form, container construction, water temperatures, groundwater flux, and mineralogy of surrounding rocks, some reasonable bounding values of radionuclide concentrations in ground water resulting from containment failure should be obtained. In the absence of such information, hydrogeologists have arbitrarily assumed values that could be too high by one or more orders of magnitude. For example, in one case, de Marsily and others (1977) assumed a removal of radionuclides by dissolution of a glass waste form during a period of 5,000 yr, with a ground-water flux of only 6.34×10^{-2} cm/yr, in a repository having radioactive glass with a mass concentration of 3×10^{3} mg/cm^{2} of repository (taken normal to the assumed direction of flow). Unless the aging and devitrification of the glass causes all the radioactive material to be selectively ejected from the glass matrix into the water, the above rate of dissolution would require continuous SiO_2 concentrations in the ground water of more than 9,400 mg/l. This is about two orders of magnitude higher than would be expected even if all the silica in the ground water were derived from the glass and not from the surrounding rock, a situation which is quite unlikely.

Host-Rock Barrier

Most of the waste isolation research to date has been related to the host rock. Containment failure, if it occurs in the future, will be along tubular features (drill holes and solution cavities) or planar features (faults, joints, and modern fractures). Moreover, these features will be widely and irregularly spaced. Studies of rock-matrix permeabilities, distribution coefficients of granular materials, and fluid-flow models that assume spatially averaged hydraulic conductivities, for example, have little value in hydrogeologic studies of most host rocks. Any rock having a hydraulic conductivity high enough and evenly enough distributed to be measured reliably in the field before repository excavation will probably not be a satisfactory host rock for that repository.

As unsettling as it may seem, a prospective deep geologic repository should probably not be investigated thoroughly by drilling from the surface prior to large-scale subsurface excavations. The drillholes themselves will generally create a greater future permeability in the host rock than exists in nature, so that a highly detailed investigation of any potential repository by drilling will automatically destroy the practical usefulness of the host rock. Reliance must be placed instead on standard geoscience techniques such as (1) comparisons between incomplete information of the host rock at the repository site and the more-detailed information of the same rock unit at distances of several kilometres from the site; (2) surface geophysics; (3) detailed mapping of surface geology; (4) detailed studies of rock characteristics at the repository level by exploratory drifts and horizontal drill holes; and, most important, (5) detailed and continuing hydrogeologic mapping and investigations during excavation and operation of the repository. Most of these techniques are site specific, and large research projects on these techniques without geographic focus would have minimal benefits.

As implied above, the matter of borehole plugging is a critical factor in evaluating the long-term hydrogeologic conditions in the host rock. Techniques developed for production and storage of petroleum products (Herndon and Smith, 1976) are probably satisfactory, but field demonstrations of borehole plugging in the host rock are needed. Robert F. Walters (1977, written commun.) has suggested that initial exploratory boreholes should be positioned where the access shafts will be located. These preliminary holes would then be plugged, after which the shaft would be drilled, and large-diameter cores from the shaft that contain the small-diameter plugged holes would be taken to the laboratory for inspection and testing. Although the final closure of the shafts may not take place for several decades after the repository opens, careful planning is needed so that initial shaft construction and location will be compatible with eventual plugging after the repository is no longer in use.

Most plans for deep repositories include the construction of a test shaft and drifts before final construction of major roads or buildings. It is essential that this be done as early as possible to allow time for various in situ tests similar to those conducted by the Swedish and American hydrogeologists in granitic rock (Witherspoon and Degerman, 1978). Although much of the existing information will be useful for general planning of repositories in granitic rocks regardless of their geographic locations, site-specific tests will always be needed.

Two types of hydrogeological research are needed most for massive rocks such as granite, limestone, thick basalt flows, and nonwelded tuff.

The first is to develop reasonable radionuclide transport models for individual fracture zones. Observations of relatively sound massive rock suggest that water-bearing fractures at depths of about 100 m tend to be vertical and are spaced 20 to 200 m apart or more. Moreover, only a few of these fractures would probably have significant hydraulic conductivities. It is likely that only hydraulic modeling, taking into account the individual fractures, would have meaning in the host rock within a few hundred metres of the repository. The probability of a given fracture being present in the rock mass as a whole could, of course, be expressed statistically if information concerning spatial distributions of various types of fractures can be obtained. If the other natural variables appear to be too difficult to define in general terms, however, extensive field measurements and experiments may be required at each repository site. This may be particularly true with regard to the transport delay factors or so-called distribution coefficients for various dissolved constituents moving within the rock fractures.

The second general category of hydrogeologic research would be a very general study of the statistical distributions of water-bearing fractures in various rock units in order to develop methods of repository evaluation. Preliminary work (Davis and Turk, 1964) suggests that strong hydrogeologic similarities exist among specific rock types in various regions at moderate depths. Whether or not meaningful correlations can be established among hydrogeologic characteristics at depths in excess of 100 m remains to be demonstrated. A careful study of records from existing deep tunnels, mines, and drillholes might help answer this important question. As indicated above, the medium itself should not be extensively perforated by test holes in order to obtain the desired information before excavating a repository. Nevertheless, the probability of encountering significant water-bearing fractures per given volume of repository is a critical value needed for preliminary planning purposes.

Most studies of postulated containment failures of repository host rocks assume simplistic flow systems with normal hydraulic gradients, which are maintained for hundreds of thousands of years. Indeed, many analysts bypass the flow problem and find it more convenient to assume zero transit time for all radionuclides to flow through the host rock from a leaky repository into some overlying aquifer. Such assumptions are hardly justified. Commonly, the worst imagined case is considered to be a repository in salt destroyed by subsidence collapse. Even in this case, the water entering the salt will become saturated with salt, which in turn will increase the water density well above the overlying fresh water. The dense brine will remain puddled in the collapsed repository and will move out very slowly into adjacent rocks. Because of the density of the brine, it will spill out into the deepest parts of adjacent permeable beds, and radionuclides in the brine will eventually reach the surface only through very slow dispersion. Research to formulate realistic models of this scenario is needed.

An additional argument for using a soluble form of silica for backfilling in a repository would be to slow the dissolution of glass if it is used as a medium to stabilize the waste. Water that is nearly saturated with silica before reaching the radioactive glass would have little corrosive effect on the glass, thus greatly adding to the safety of the disposal method. As the silica-rich backfill around the waste canisters is altered by circulating ground water, a clay-like residue may be formed. This residue should have a low permeability and should also have a significant exchange capacity, which would in turn help to immobilize a large number of radionuclides.

If temperatures rise sufficiently high next to waste canisters in granitic, basaltic, or tuffaceous materials, and if gas pressures in a repository are low to moderate, the small amount of water that seeps into the repository will vaporize and not reach the canisters. This thermal barrier can provide an important safety factor during the early history of waste isolation (Cohen and Steinborn, 1978). Hydrogeologic research on this thermal barrier effect is needed.

Migration Path Barrier

Once the host rock is breached, migrating water will eventually carry some radionuclides to the land surface. Fortunately, distances traversed by the contaminated water will generally be so large and migration rates will be so slow that the pathway itself becomes a protective barrier between the repository waste and the biosphere. Conventional ground-water modeling becomes very useful at this stage for evaluating numerous possible repository sites. Although additional research is needed to adapt fluid-flow and mass transport models to various natural settings, and although a number of phenomena related to fluid transport are poorly understood, most uncertainties at present in practical applications are related to testing and calibrating the models. This is particularly true for models of flow through fractured materials.

Extensive field testing related to fracture flow is needed in particular. Relatively small-scale field tests in fractured rock are required first. These tests would involve drilling wells and observation holes, measuring hydrogeologic parameters, injecting tracers and observing their migration, and finally selecting the model that best describes the field data. Testing of ground-water models by data collected over periods of time similar to migration times of radionuclides--that is, hundreds to thousands of years--is of course impossible, so scaling up from the proposed field tests to an actual repository remains a problem.

A partial solution to this scaling problem is to try to date the existing ground water in the vicinity of a proposed repository. Various methods of dating exist (Davis, 1978, unpub. report), but more field-oriented work is needed, particularly for very old ground water. Indeed, if ground water can be demonstrated by dating to be virtually static in and around a proposed repository and if the regional flow is nearly horizontal and the hydrogeologic regimen will not change materially in the future, then the problem of ground-water migration during the next few thousand years can be ignored. Ground-water dating could, therefore, make most of the calculations of regional migration unnecessary in some repository settings. If dating of water at various points shows that water is moving actively through aquifers, then past water velocities might be estimated. Furthermore, if a single sample of ground water from a single point in an aquifer indicates various dates by different methods, the degree of large-scale hydrodynamic dispersion in an aquifer might possibly be estimated. Field data on both ground-water dispersion and velocities are essential for proper calibration of models of natural systems.

The Final Barrier--Water Potability

In many of the analyses of containment failure of proposed repositories in salt, the contaminated water is assumed to arrive directly at the surface to be consumed forthwith by humans. This assumption, however, ignores the fact that water leaving the salt repository would actually be a brine and would need to be diluted more than two orders of magnitude before it is even remotely fit for human consumption. Indeed, this fact suggests an interesting possibility for applied research. Might there be some denaturing agent which is not a poison but which could be added to the backfill or to the waste itself in a repository so that any ground water that moved through the repository would have a disagreeable taste even if it is diluted by several orders of magnitude? This would add a final barrier to the potential movement of radionuclides from a repository to humans.

CONCLUDING STATEMENT

I have asked often in various committee meetings why particular esoteric research has been undertaken or is proposed in connection with repositories for radioactive waste. A very common answer is that public safety is not really at stake but that an apprehensive public must be reassured by the research. For this reason, it seems questionable whether the difficult program of waste isolation should be founded on such a basis.

Radioactive waste is hazardous, the problem is very serious, and numerous studies are needed; nevertheless, an unending demand for more time and more research money will eventually backfire. Some proposals for the solution of the waste problem involve spending several million dollars to prevent the loss of a single life at some indefinite date in the far future. As noble as this may seem, the practical question may be asked "How many *hundreds* of lives might be saved in the *near* future if this amount of money were spent for other safety-related activities?" Unless such practical questions are addressed, it would take only a few strategically placed articles in popular national magazines or a few television exposés, together with the emergence of other national issues competing for funds, to divert the object of public concern from radiation dangers to the ever-present problems of inflation and taxes. Once political support is lost, a responsible solution to the waste problem will be difficult, and the nuclear-waste program may return to the situation of the 1950s, when meager amounts of money were available only for interim storage of high-level waste. Hydrogeologists share an important responsibility in seeing that research moves ahead rapidly and that the badly needed practical effort is not diverted toward noncritical issues.

ACKNOWLEDGMENTS

Although I owe a debt to many individuals with whom I have worked in problems of nuclear waste during the past 20 years, they are in no way associated with the rather freewheeling discussion given here. I would like to add, however, that if parts of my presentation have scientific merit, there is a good chance that inspiration for those parts came from Arthur Piper and, in the last few years, from Paul Witherspoon and his research team. Lewis Cohen and T. N. Narasimhan were kind enough to read the present manuscript, which has been improved greatly by their critical comments.

REFERENCES CITED

Allen, E. J., 1978, Criticality analysis of aggregations of actinides from commercial nuclear waste in geological storage: Oak Ridge National Laboratory, ORNL/TM-6458, 153 p.

APS Study Group, 1977, Report to the APS by the study group on nuclear fuel cycles and waste management: Reviews of Modern Physics, v. 50, no. 1, part II, 185 p.

Bradshaw, R. L., and McClain, W. C., 1971, Project salt vault: A demonstration of the disposal of high-activity solidifed wastes in underground salt mines: Oak Ridge National Laboratory, ORNL-4555, 356 p.

Bredehoeft, J. D., and others, 1978, Geologic disposal of high-level radioactive wastes--Earth-science perspectives: U.S. Geological Survey Circular 779, 15 p.

Cohen, B. L., 1977, The disposal of radioactive wastes from fission reactors: Scientific American, v. 236, no. 6, p. 21-31.

Cohen, J. J., and Steinborn, T., 1978, High temperature emplacement of nuclear waste in geologic media, in Science underlying radioactive waste management, Symposium A, Abstracts: Boston Material Research Society Annual Meeting, p. 44.

Davis, S. N., and Turk, L. J., 1964, Optimum depth of wells in crystalline rocks: Groundwater, v. 2, no. 2, p. 6-11.

de Marsily, G., and others, 1977, Nuclear waste disposal: Can the geologist guarantee isolation?: Science, v. 197, p. 519-527.

Herndon, J., and Smith, D. K., 1976, Plugging wells for abandonment, a state-of-the-art study and recommended procedures: Duncan, Oklahoma; Halliburton Services (prepared for Union Carbide Corp., Office of Waste Isolation), 57 p.

KBS, 1977a, Handling of spent nuclear fuel and

final storage of vitrified high-level reprocessing waste, Part II, Geology: Stockholm, Karn-Bransle-Sakerhet (Nuclear Safety Project), 81 p.

____1977b, Handling of spent nuclear fuel and final storage of vitrified high-level reprocessing waste, Part IV, Safety analysis: Stockholm, Karn-Bransle-Sakerhet (Nuclear Safety Project), 125 p.

____1978, Handling and final storage of unre processed spent nuclear fuel: Stockholm, Karn-Bransle-Sakerhet, 2 vols.

Martinez, J. D., ed., 1977, An investigation of the utility of Gulf Coast domes for the storage or disposal of radioactive wastes: Institute for Environmental Studies, Baton Rouge, Louisiana State University, Y/OWI/SUB-4112/37, 475 p.

NAS Panel on Bedrock Disposal, 1972, An evaluation of the concept of storing radioactive wastes in bedrock below the Savannah River Plant site: Washington, D.C., National Academy of Sciences, 86 p.

NAS Panel on Land Burial of Low-Level Waste, 1976, The shallow land burial of low-level radioactively contaminated solid waste: Washington, D.C., National Academy of Sciences, 150 p.

NAS Panel on Geologic Site Criteria, 1978, Geological criteria for repositories for high-level radioactive wastes: Washington, D.C., National Academy of Sciences, 19 p.

Papadopulos, S. S., and Winograd, I. J., 1974, Storage of low-level radioactive wastes in the ground: Hydrogeologic and hydrochemical factors: U.S. Geological Survey Open-File Report 74-344, 49 p.

Pratt, H. R., Hustruld, W. A., and Stephenson, D. E., 1978, Earthquake damage to underground facilities: Savannah River Laboratory Publications DP-1513, 84 p.

Ringwood, A. E., 1978. Safe disposal of high-level nuclear reactor wastes: A new strategy: Canberra, Australian National University Press, 64 p.

U.S. Atomic Energy Commission, 1974, High-level radioactive waste management alternatives: U.S. Atomic Energy Commission, WASH-1297, 94 p.

U.S. Department of Energy, 1979a, Management of commercially generated radioactive waste; draft environmental impact statement: U.S. Department of Energy Publication DOE/EIS-0046-D, 2 vols.

U.S. House of Representatives, 1976, Low-level radioactive waste disposal: Hearings before a subcommittee on government operations: U.S. House of Representatives, Ninety-Fourth Congress, Second Session, 434 p.

Weeren, H. O., Moore, J. G., and McDaniel, E. W., 1978, Waste disposal by shale fracturing at Oak Ridge National Laboratory, in Science underlying radioactive waste management, Symposium A, Abstracts: Material Research Society Annual Meeting, Boston, p. 22.

Winograd, I. J., 1974, Radioactive waste storage in the arid zone: EOS (American Geophysical Union Transactions), v. 55, no. 10, p. 884-894.

Witherspoon, P. A., and Degerman, O., 1978, Swedish-American cooperative program on radioactive waste storage in mined caverns: Lawrence Berkeley Laboratory Report LBL-7049, SAC-01, 25 p.

Manuscript Accepted by the Society October 14, 1980

Geological Society of America
Special Paper 189
1982

Ground-Water Problems in Open-Pit and Underground Mines

NEVILLE G. W. COOK
Professor of Mining Engineering
Department of Materials Science and Mineral Engineering
University of California
Berkeley, California 94720

ABSTRACT

Ground water constitutes a significant operational problem and potential hazard in open-pit and underground mining. The *pressure* of ground water in geologic discontinuities adversely affects the safety of height and slope angles of open-pit mines, justifying measures to reduce ground-water pressures in their vicinity. The principal hazard in underground mining and tunneling arises from *massive* inflows of ground water when the mine unexpectedly intersects large, water-bearing geologic discontinuities; *normal* inflows are on the order of 2.5 tons of water per ton of rock mined. In tunneling, pilot drilling ahead of the advancing face is used to detect such discontinuities, which can then be sealed by ring drilling and cover grouting. Extensive deformations resulting from large areas of unsupported roof make such grouting ineffective in many mine excavations.

Surprisingly few data are available with which to calculate the permeability of rock masses at depth. Two sets of data quoted yielded permeabilities of 0.8×10^{-9} m/s for tunnels, and 1.6×10^{-6} m/s for mined excavations, at a depth of about 1.5 km below surface and before the permeability of the rock mass is enhanced by deformations induced by mining. The difference between these values may be attributed to local reductions in the permeability of the rock mass by the grouting of major discontinuities in tunneling, which is not done ordinarily in mining.

Where precautions such as regional dewatering, installation of watertight bulkheads, and adequate spare pumping capacity are not sufficient to cope with potential inflows in mining, the extent of the unsupported roof in mines must be broken by regularly spaced pillars, so that pilot drilling and cover grouting can be done.

INTRODUCTION

Ground water may constitute a significant operational difficulty and potential hazard in open-pit and underground mining and in tunneling. Problems related to ground water range from preventing flooding in shafts sunk through water-bearing rocks, to environmental pollution after mining. Shaft sinking, on the one hand, is a highly specialized activity in mining and tunneling, and methods for coping with ground water by cement grouting, freezing, and other techniques have been documented well (Atherton and Garrett, 1959; Marsh, 1959; Weehuisen, 1959). On the other hand, environmental pollution of ground water by mining activities has been the subject of recent debate and legislation (Federal Water Pollution Control Act, 1972). Accordingly, neither of these two topics is discussed in this paper. Rather, the purpose of this paper is to examine the nature and magnitude of operational and hazardous aspects of ground water in open-pit and underground mining and tunneling. Of the few published sources of data sufficient to enable calculation of the permeability of a rock mass at depths of more than 1 km, some are presented. These data are used to estimate values of permeability when the major water-bearing discontinuities are sealed by grouting and when they are not. The effects of deformation, caused by mining, on the permeability of the rock mass are shown by reference to the rate of water inflow.

OPEN-PIT MINES

Although the presence of ground water does cause significant operational problems in open-pit

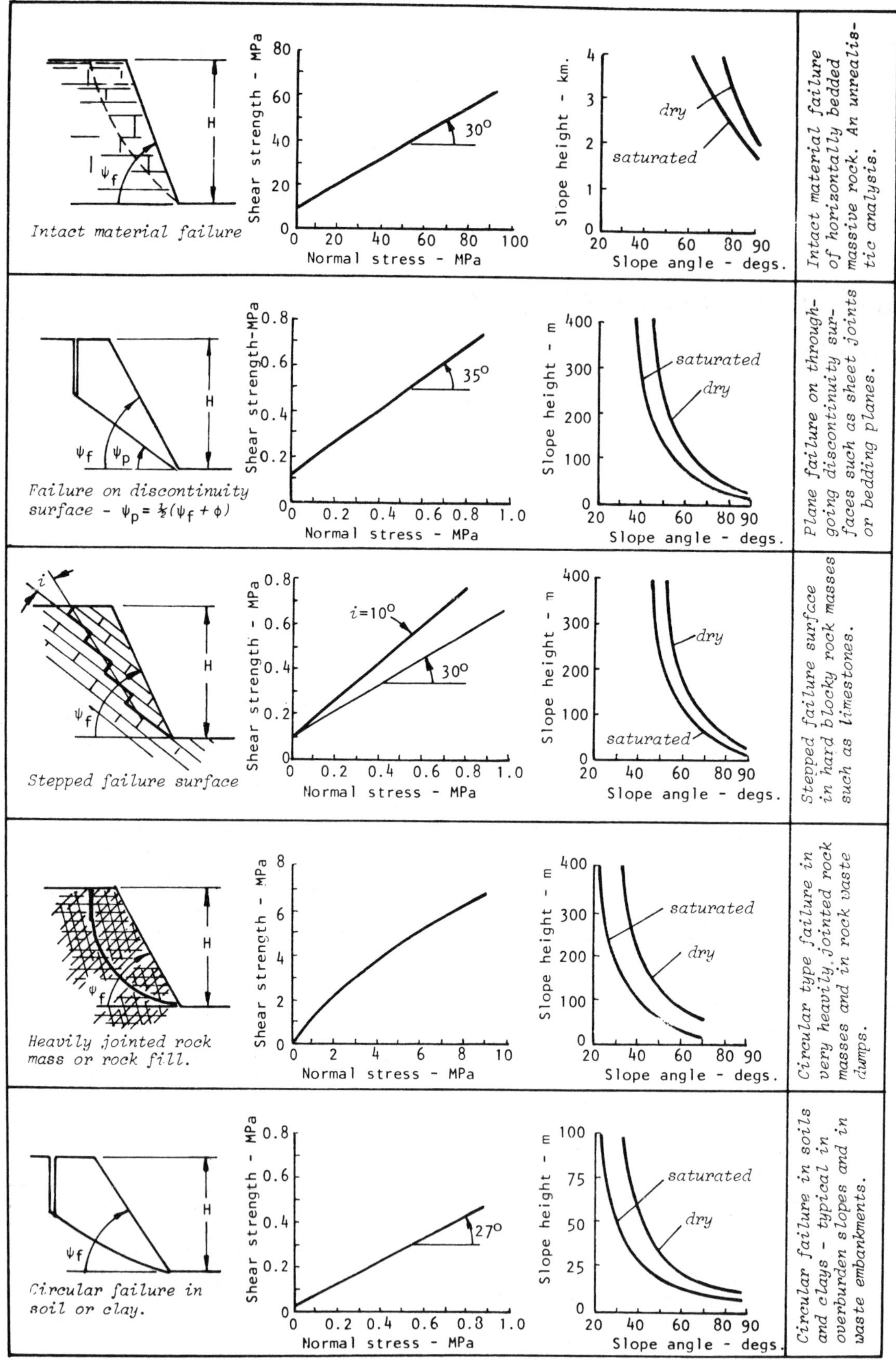

Figure 1. Diagrams illustrating five modes of potential slope failure for materials with different properties showing the adverse effects of ground water (saturation) on slope angle and slope height. (After Hoek and Bray, 1977.)

mines (such as the need to use waterproof blasting agents and accessories in wet blast holes), the major concern is with the effects of ground-water pressures rather than quantities on the stability of pit slopes. These effects have been studied extensively, and a comprehensive appraisal of them has been made by Hoek and Bray (1977).

The stability of open-pit slopes is maintained by the cohesion of the slope material and its frictional resistance to sliding. Several forms of slope failure can occur, depending upon the nature of the slope material and its geologic structure. Hydraulic pressure from ground water tends to promote all the different forms of slope failure. This happens by two separate mechanisms. First, the pressure of ground water reduces the value of the effective stress in the slope material, thereby reducing the frictional resistance to sliding. Second, ground-water pressure in back of a slope may generate a significant force toward the open pit, tending to promote sliding. It is important to understand that both mechanisms are results of ground-water *pressure* rather than *quantity*. Ground-water pressure can have a devastating effect on the stability of any open-pit slope.

One of the most meaningful ways to demonstrate stability of open-pit slopes is by showing the graphic relationship between slope height and slope angle. Such relationships are sketched in Figure 1 for five principal forms of slope failure, for both dry and saturated conditions. These forms of failure fall into two categories: plane failure and circular failure. Both have been analyzed in detail by Hoek and Bray (1977). Circular failure is the classic form of failure in soil mechanics (Terzaghi, 1943; Taylor, 1948; Bishop, 1955). Wedge failures can be regarded as a more complicated, three-dimensional plane failure (Londe, 1965; Goodman and Taylor, 1967). A different mode of slope failure, *toppling*, has been recognized recently (Muller, 1968; Goodman and Bray, 1976, Hoek and Bray, 1977). However, toppling is not yet understood as well as other forms of slope failure, and the effect of ground water on toppling has not been taken into account.

The effects of ground-water pressure on the stability of open-pit slopes have been excellently illustrated by Hoek and Bray (1917) and are reproduced here in Figure 1. For each of the five principal forms of slope failure, the maximum height of the slope for any given slope angle decreases drastically from the dry to the saturated condition. Alternatively, for any slope height a significant reduction in slope angle is necessary to maintain stability under saturated conditions compared with dry conditions. There are great economic advantages to maintaining the steepest slope angle possible within the limits of safety, as this reduces the amount of waste rock that must be mined to extract the ore. These advantages justify substantial efforts to achieve and maintain dry rather than saturated conditions in the mine (Moffit and others, 1971; Stewart and Kennedy, 1970). This involves careful hydrogeological studies and engineered remedies. These remedies range from simple procedures, such as installing surface drainage to divert precipitation from the proximity of the open pit, to drilling wells and developing drifts adjacent to the slopes in order to reduce ground-water pressures in their vicinity (Hoek and Bray, 1977; Durston, 1979; Lopaschuk, 1979; Pentz, 1979). The measurement and analysis of ground-water flow in the vicinity of open-pit slopes, the evaluation of the effects of ground water on slope stability, and the drainage of slopes are described fully in the Pit Slope Manual (Canada Center for Mineral and Energy Technology, 1977).

TUNNELS AND UNDERGROUND MINES

Although some underground mines and tunnels are excavated in rock strata with relatively homogeneous porosity and permeability, many are not. The continuity of most rock is interrupted in some degree by bedding planes, joints, fractures, and faults. In massive and stratiform rocks with low hydraulic permeabilities, these discontinuities provide the principal conduits for the movement of ground water.

In rock with homogeneous but not necessarily isotropic hydraulic properties, the effects of ground water on tunneling and mining can be predicted with comparative ease and certainty. Unfortunately, this situation is the exception rather than the rule; most mines and tunnels are excavated in rocks where the dominant permeability arises from discontinuities. For example, Trexler (1979) reported the primary source of ground water in an old, deep mine as being natural fractures intersected by drillholes and drifts; of about 2,340 holes and drifts, only 115 holes and 2 drifts had significant inflows (26 to 33 l/s). Significant inflows of ground water into the workings of mines and tunnels are common. In general, it is necessary to install pumps with capacities large enough to remove this water. For example, in 1958, Charbonnage de France pumped 2.4 tons of water for each ton of coal mined; in England and Wales in 1962, 2.44 tons of water were pumped for each ton of coal mined (Orchard, 1969). Gold mines of the Witwatersrand System pump about 1 ton of water per ton of rock mined at a weighted average depth of 1.5 km (Whillier, 1978). Unfortunately, other measurements, which would be needed to infer information such as the permeability of the rock mass, are not complete, so that the above data are not sufficient for such calculations.

There is even less information about catastrophic inflows than there is about normal inflow, so that few, if any, meaningful calculations concerning this hazard can be made. Major disasters or potential disasters from sudden, unexpected inflows or inrushes of water are fortunately rare, but have occurred throughout the history of mining. In July 1847, the sea broke into Workington Colliery in Great Britain, causing the loss of 27 lives. This coal mine was being worked with headings from 3.7 m (12 ft) wide to 4.6 m (15 ft) wide and pillars 13.7 m (45 ft) by 9.1 m (30 ft) wide at depths from 27.4 m (90 ft) to 45.7 m (150 ft) beneath the sea floor (Orchard, 1969). In October 1968, the world's richest gold mine, West Driefortein, on the Witwatersrand System, South Africa, experienced an inrush of 3,700 l/s that took 26 days to bring under control, and which put the mine out of full production until August 1969 (Tress, 1974). The information pertaining to the source of this inflow is a hurriedly written note which reads as follows: "Stope 46B on 6 level west of 4 shaft has broken open and water is running 6 inches deep into the shaft" (Cartwright, 1970). Cummins and Given (1973) reference eight cases of disastrous mine flooding from sudden, unexpected inflows of water.

New gold mines in the Orange Free State portion of the Witwatersrand System were opened during the 1950s. Because of the significant costs of pumping large quantities of water from underground mines, the new mines were used to make measurements of the rates of ground-water inflow as a function of primary mine development by tunneling and subsequent mining of the stratiform gold reef (Venter, 1969). These data are virtually unique in that sufficient information concerning the rate of inflow, the length and size of tunnels developed, the area mined, and the height of the water table was obtained to enable simple calculations to be done on the permeability of the stratiform rock mass within which these excavations were being made.

The flow of ground water through geologic discontinuities has been compared with the laminar flow of fluid between closely spaced parallel plates (Huitt, 1956; Snow, 1968). The volume rate of flow through such a discontinuity can readily be shown as

$$q = \frac{g(2b)^3 i}{12\eta}, \qquad (1)$$

where q = the volume rate of flow per unit length of the discontinuity in a direction normal to that of the flow ($m^3/m/s$);
$2b$ = the aperture of the discontinuity (m);
g = the gravitational acceleration (9.81 m/s^2);
i = the hydraulic gradient (m/m); and
η = the kinematic viscosity of water (m^2/s).

A similar expression for the volume rate of flow through such a discontinuity into a circular hole from the outside of a concentric cylinder (Narasimhan and Witherspoon, 1976) is:

$$Q = \frac{2\Pi(2b)^3 g}{12\eta(\ln r/r_h)}(H-H_h), \qquad (2)$$

where Q = the volume rate of flow through the discontinuity into the hole (m^3/s);
r = the radius to the cylinder (m);
r_h = the radius of the hole (m);
H = the hydraulic head outside the cylinder (m);
H_h = the hydraulic head in the hole at the discontinuity (m); and the other symbols are as defined above.

The volume rate of flow into a horizontal, subsurface tunnel can be found using the theory of images (Ferris and others, 1962) from an equation similar to equation 2, namely:

$$Q = \frac{2\Pi(2b)^3 g}{12\eta(\ln 2r/r_t)} H, \qquad (3)$$

where 2θ = twice the distance to the water table, that is, the separation between the tunnel and its image (n);
r_t = the radius of the tunnel (m);
H = the hydraulic head of the water table (m); and the other symbols are as defined above.

These equations demonstrate the overriding effect of the aperture of the discontinuity on the rate of ground-water flow. The volume rates of flow through discontinuities with various spacings are shown as a function of aperture in Figure 2.

Figure 3 illustrates the cumulative distribution of exposed fractures in granite as a function of aperture at Pikes Peak, Colorado (Snow, 1970). This shows that the hydraulically dominant fractures with large apertures are much less frequent than those with small apertures. Accordingly, the flow of large quantities of ground water into underground excavations is likely to be a capricious phenomenon, resulting from the infrequent intersection of discontinuities with large apertures. Ground-water inflows during tunneling have been studied with the use of physical and numerical models by Goodman and others (1965), but these studies assume a knowledge of hydraulic permeability or transmissivity along the tunnel axis. Even the most careful site investigations before excavation commonly fail to detect the presence of some of the most hazardous hydrogeologic features and certainly do not quantify their properties (Olivier, 1970). Snow (1968) has shown

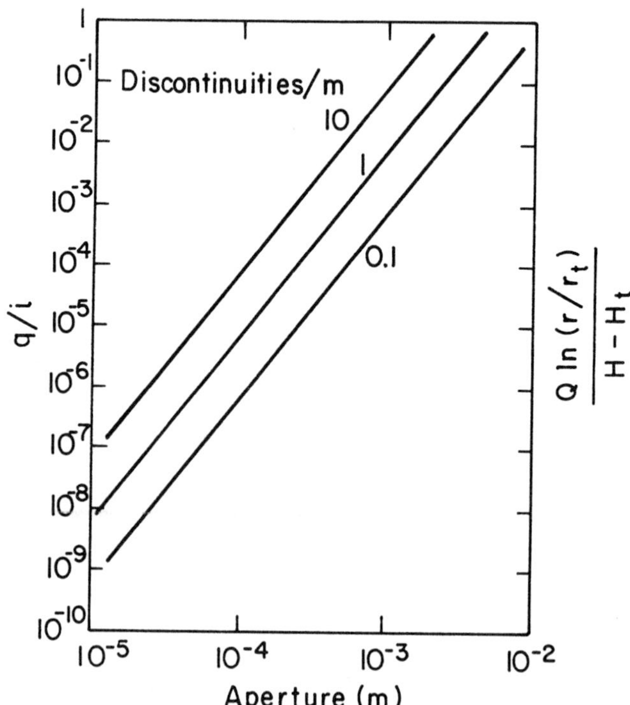

Figure 2. Volume rates of linear and radial flow through discontinuities as a function of their aperture for three different discontinuity spacings.

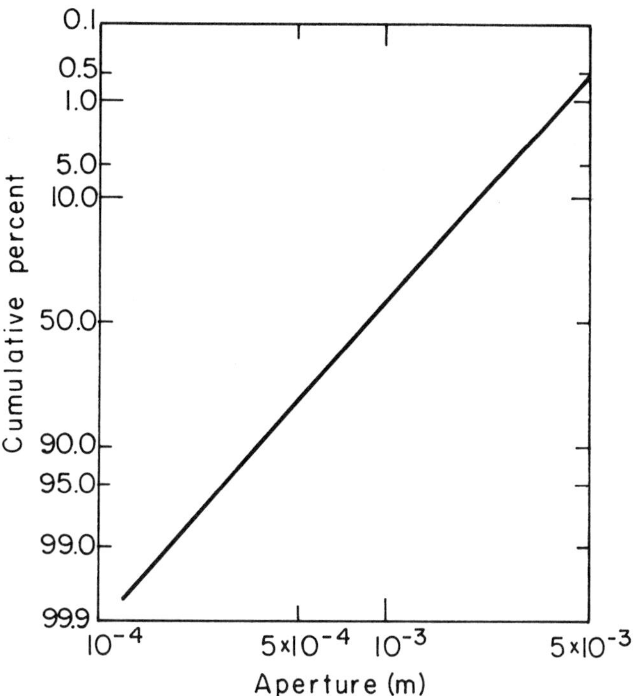

Figure 3. Cumulative distribution of exposed fractures with various apertures in granite from Pikes Peak, Colorado. (After Snow, 1970.)

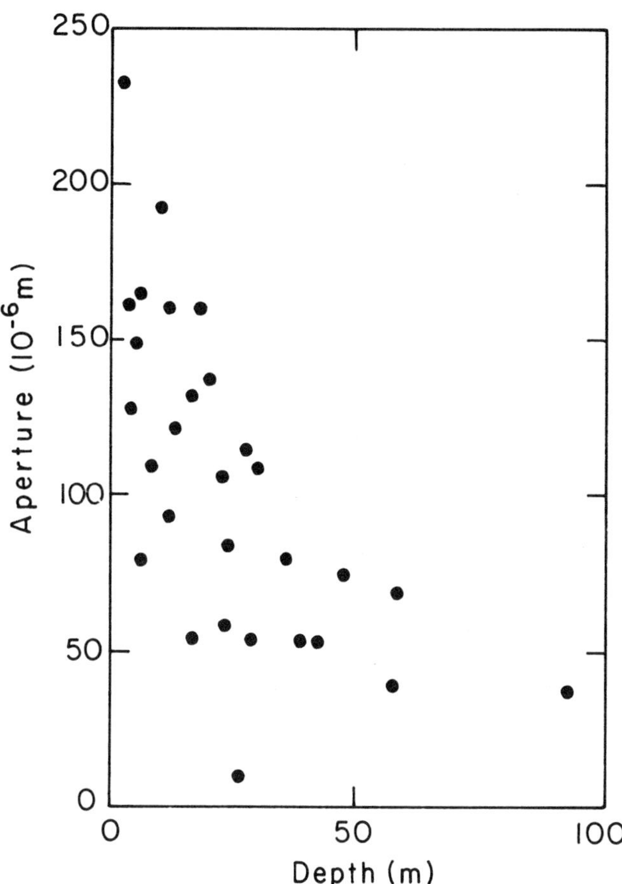

Figure 4. Changes in the apertures of fractures as a function of depth. (After Snow, 1968.)

that the aperture of fractures decreases with increasing depth (compressive stress), as shown in Figure 4. The extent to which the excavation of underground openings reduces the compressive stress across discontinuities is therefore also likely to affect their aperture and hydraulic transmissivity.

Horizontal and vertical tensile stresses are induced in the roofs and floors of underground openings as a result of changes in the distribution of stresses in the rock adjacent to them. These changes can be calculated by analytical and numerical methods (Obert and Duvall, 1967; Jaeger and Cook, 1976). Figure 5 shows the height above a tubular underground opening at which the total vertical stress reverts from tension to compression, as a function of the ratio between the depth and minimum horizontal dimensions of the unsupported roof. The extent of total horizontal tension is generally less than that of total vertical tension.

Figure 6 shows the rate of pumping from a new gold mine in the Witwatersrand System as a function of the cumulative length of the tunnels and cumulative area mined (Venter, 1969). Note that the rate of pumping, or water inflow, increased initially as a linear function of the cumulative length of the tunnels, as would be expected from equation 3. Soon after mining began on the stratiform gold reef, the rate of inflow accelerated rapidly. If the rate of inflow is examined as a function of the cumulative area mined, we see that initially the rate of inflow increased linearly with the area mined and then at a much greater rate. A similar pattern for an adjacent mine is shown in Figure 7. The ratio of the depth to the minimum horizontal dimension of the area mined at which the

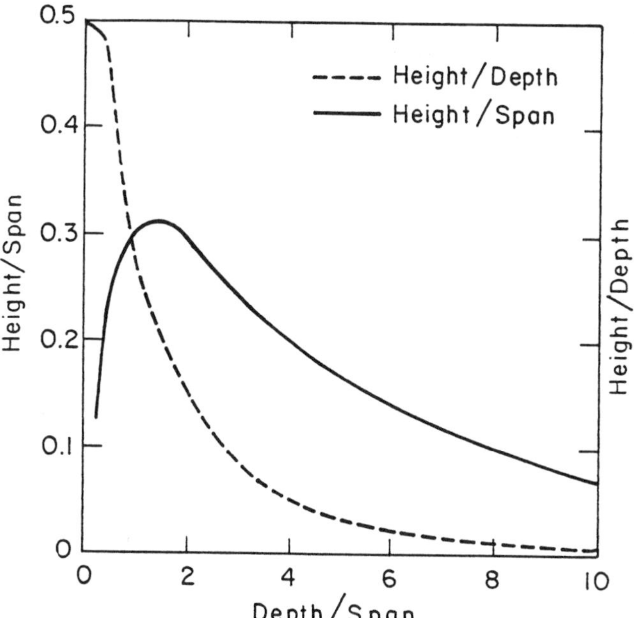

Figure 5. Height above a mine roof to which the zone of vertical tension extends as a function of the ratio between the depth of the roof below surface and its unsupported span or minimum horizontal dimension.

inflow accelerated, D/S, is about 8 in the first example and about 5 in the second example.

Figure 5 shows that at values similar to these, the height at which vertical tension is induced begins to increase rapidly. Presumably, the sudden increase in inflow occurs when the induced vertical tension extends to the height of water-bearing discontinuities in rocks with little permeability other than that caused by discontinuities. Both mines are located in the Orange Free State gold field of the Witwatersrand System. This system comprises alternating groups of argillaceous and arenaceous sediments with an age of about 2.5 b.y. In the Orange Free State these sediments are overlain by lava of the Ventesdorp System and by impermeable sandstones, mudstones, and shales of the Karoo System (Haughton, 1969).

Using equations 1 and 3 and the linear portions of the data plotted in Figure 6, the permeability can be estimated. This gives a value of 0.8 x 10^{-9} m/s from the inflow to the tunnels and 1.6 x 10^{-6} m/s for the initial inflow into the mined area. This disparity may be accounted for by the practice of pilot drilling and cover grouting during excavation of shafts and tunnels. For safety, pilot boreholes are drilled ahead of the advancing end of the tunnel to detect discontinuities that carry large volumes of water. If necessary, such discontinuities are sealed by the injection of cement grout through rings of cover holes before tunneling is continued. Injection grouting has been found to have little effect in the case of mining, because the extensive changes in stress and deformation associated with large mine excavations destroy the effect of grouting. Thus, the largest discontinuities that account for the greatest inflow of water are sealed as tunneling proceeds but not as mining proceeds. Where it is necessary to prevent large inflows through the disturbance of discontinuities as a result of mining, the extent of the exposed mine roof must be

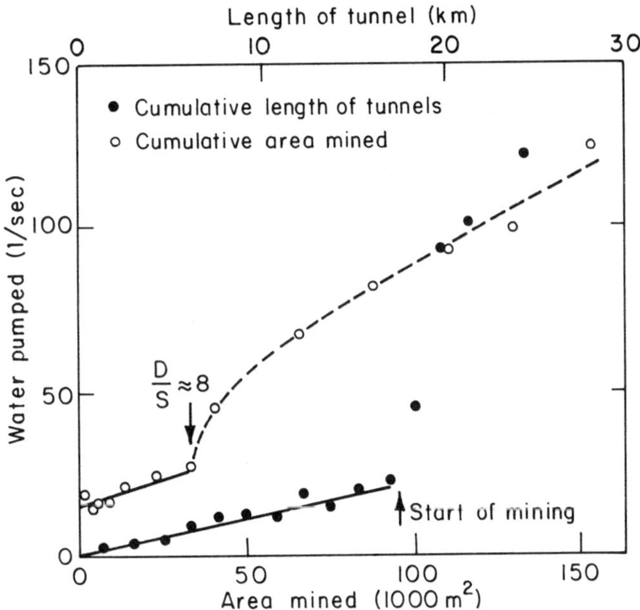

Figure 6. The rate of pumping from a new mine as a function of the cumulative length of tunnels developed and the cumulative area mined. (After Venter, 1969)

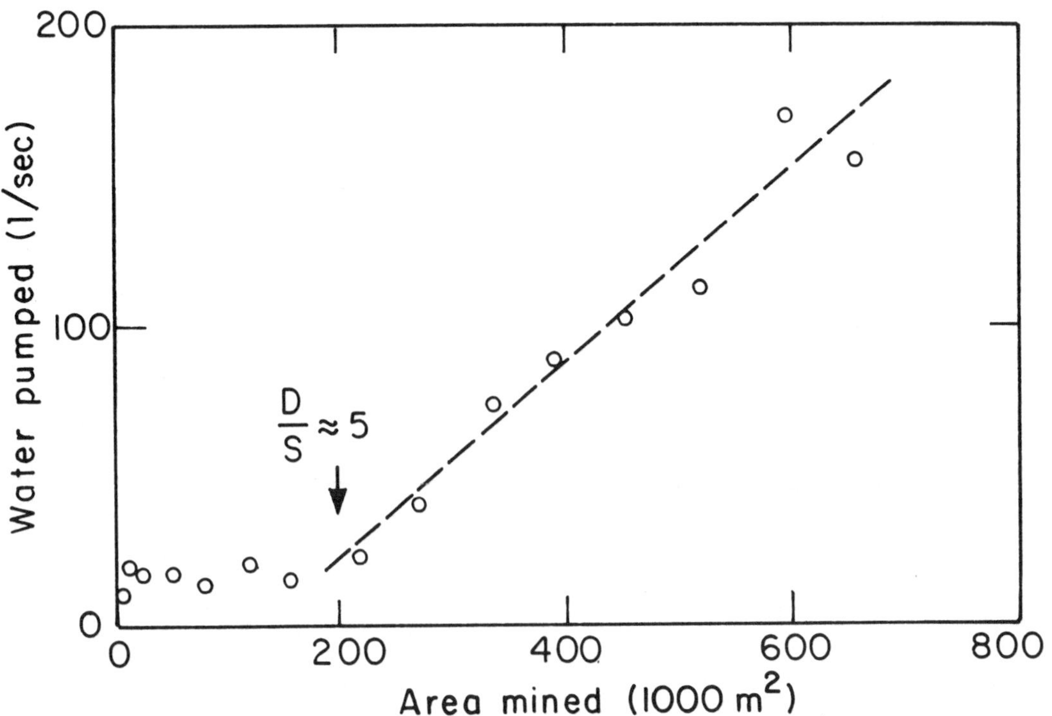

Figure 7. Rate of pumping from a new mine as a function of the cumulative area mined. (After Venter, 1969.)

limited by leaving regular and adequate pillars, as is done in room and pillar coal mining. Such procedures, though they sterilize the ore in the pillars, may be necessary when mining beneath large volumes of ground water or beneath oceans and lakes, where dewatering is impractical. Also, pilot drilling and cover grouting can then be used to reveal and seal major discontinuities.

MODELING

The effects of ground water in open-pit and underground mines and tunnels are twofold. First, there are those resulting from ground-water pressure on slope stability, particularly open-pit slopes. Second, there are those effects related to the quantity of ground-water inflow, which may be manageable or hazardous.

A number of numerical models exist with which to simulate the flow of ground water into mines, although they have not been used widely for this purpose. In general, there are too few field data available to make meaningful use of these models. For example, Brown (1979) stated that simple hydrological analyses are usually most appropriate for mining problems, because the information available is limited and the details of complex analyses are difficult to evaluate in terms of mining practice. Ignoring the unexpected consequences of an undetected, large, water-bearing discontinuity and the effects of major deformations of unsupported mine roofs, these models can be used to evaluate ground-water inflows to underground workings and ground-water pressure distributions in the vicinity of open-pit slopes. For example, in deep mines under saturated conditions, the finite-difference models of Pinder and Bredehoeft (1968) and Prickett and Lonnquist (1971) or the finite-element models of Gupta and Tanji (1966) and Javandel and Witherspoon (1968) can be used. For shallow or open-pit mines, free-surface models such as those of Jeppson (1968) and Neuman and Witherspoon (1970) can be used. If information is available concerning variable saturation, models such as those developed by Freeze (1971), Neuman (1973), and Narasimhan and Witherspoon (1977) may be more applicable.

CONCLUSION

Ground water constitutes one of the principal operational difficulties and potential hazards in mining and tunneling. Several numerical models have been developed to simulate ground-water flow and pressures. These models can be applied to the study of ground-water problems in open-pit and underground mines to a limited extent only. In general, meaningful simulation of practical problems with the use of these models requires more field data than usually exist. In particular, the hazard of major water-bearing discontinuities arises because they are difficult to detect, and the effects on the permeability of major deformations of unsupported mine roofs are not understood.

In open-pit mines, the pressure of ground water in discontinuities has a major deleterious effect on the stability of slopes. Measures to depressurize this ground water to maintain slopes at greater angles than would otherwise be safe are often economically very rewarding.

In tunnels and underground mines unexpected inflows constitute the major hazard. Few mines and tunnels are completely dry. The ventilation air removes about 10 g of water per cubic metre of air; rates of airflow in tunnels are typically 10 m^3/s and in mines, 500 m^3/s. Most underground mines have pumps to handle the ordinary inflow of ground water, which is normally in the range of 100 to 2500 l/ton of ore mined. Flooding commonly occurs when an unanticipated major water-bearing discontinuity is exposed. In tunneling and in mine-tunnel development, pilot drilling ahead of the advancing tunnel face, with appropriate precautions, is used to guard against flooding. If a major discontinuity is located it can be sealed by cover drilling and cement grouting. Such discontinuities have proved very difficult to locate by means other than pilot drilling. A single discontinuity can result in an inflow of about 1,000 l/s. Grouting has not proved to be effective in sealing such discontinuities in the vicinity of excavations with extensive unsupported roofs made in mining the ore; thus the hazard of flooding is present wherever mining is done below the ground-water table. Precautions against such flooding include pilot drilling, regional dewatering, the provision of water-tight bulkheads that can be used to seal portions of a mine, and adequate spare pumping capacity. Where such precautions are inadequate, the extent of the unsupported mine roof must be limited by regularly spaced pillars, so that pilot drilling and cover grouting can be used to detect and seal major discontinuities, as in tunneling.

REFERENCES CITED

Atherton, F. G., and Garrett, W. S., 1959, The history of cementation in shaft sinking, *in* Symposium on Shaft Sinking and Tunnelling: London, The Institution of Mining Engineers.

Bishop, A. W., 1955, The use of the slip circle in the stability analysis of earth slopes: Geotechnique, v. 5, p. 7-17.

Brown, A., 1979, Drainage excavation and control for underground oil shale projects, *in* Proceedings, First International Mine Drainage Symposium: San Francisco, Miller-Freeman Publishing Co.

Canada Centre for Mineral and Energy Technology, 1977, Pit slope manual: Ottawa, Printing and Publishing, Supply and Services.

Cartwright, A. P., 1970, West Driefontein--Ordeal by water, *in* Goldfields of South Africa Limited: Johannesburg, J. G. Ince & Son (Pty.).

Cummins, A. B., and Given, I. A., 1973, SME mining engineering handbook: New York, American Institute of Mining, Metallurgical, and Petroleum Engineers.

Durston, K. J., 1979, Open pit dewatering at the Pine Point operations of Cominco Limited, *in* Proceedings, First International Mine Drainage Symposium: San Francisco, Miller-Freeman Publishing Co.

Ferris, J. G., and others, 1962, Theory of aquifer tests, ground water hydraulics: U.S. Geological Survey Water-Supply Paper 1536-E.

Freeze, R. A., 1971, Three dimensional transient saturated-unsaturated flow in a groundwater basin: Water Resources Research, v. 7, p. 347.

Goodman, R. E., and Bray, J. W., 1976, Toppling of rock slopes, in Proceedings, Specialty Conference on Rock Engineering for Foundations and Slopes: New York, American Society of Civil Engineers, p. 201-234.

Goodman, R. E., and Taylor, R. L., 1967, Methods of analysis of rock slopes and abutments: A review of recent developments, in Fairhurst, C., ed., Failure of breakage of rocks: American Institute of Mining Engineers Transactions, p. 303-320.

Goodman, R. E., and others, 1965, Ground water inflows during tunnel driving, Engineering Geology, v. 2, p. 39-56.

Gupta, S. K., and Tanji, K. K., 1976, A three dimensional Galerkins finite element solution flow through multi aquifers in Sutter Basin, California: Water Resources Research, v. 7, p. 347.

Haughton, S. H., 1969, Geological history of southern Africa: Johannesburg, Geological Society of South Africa.

Hoek, E., and Bray, J., 1977, Rock slope engineering: London, Institution of Mining and Metallurgy.

Huitt, J. L., 1956, Fluid flow in simulated fractures: American Institution of Chemical Engineering Journal, v. 2, p. 259-264.

Jaeger, J. C., and Cook, N.G.W., 1976, Fundamentals of rock mechanics: New York, John Wiley and Sons.

Javandel, I., and Witherspoon, P. A., 1968, Application of the finite-element method for transient flow in porous media: Society of Petroleum Engineers Journal, v. 8, p. 241.

Jeppson, R. W., 1968, Seepage from channels through layered porous media: Water Resources Research, v. 4, p. 435.

Londe, P., 1965, Une méthode d'analyse à trois dimensions de la stabilité d'une rive rocheuse: Paris, Annales des Ponts et Chaussées, p. 37-60.

Lopaschuk, R., 1979, Experience with drainage control at Cyprus Anvil Mining Company, Yukin, in Proceedings, First International Mine Drainage Symposium: San Francisco, Miller-Freeman Publishing Co..

Marsh, F., 1959, Shaft sinking in Great Britain since 1947, in Symposium on Shaft Sinking and Tunnelling: London Institution of Mining Engineers.

Moffit, R. B., Friese-Greene, T. W., and Lillico, T. M., 1971, Pit slopes--Their influence on the design and economics of open pit mines, in Proceedings, Second Symposium on Stability in Open Pit Mining: New York, American Institute of Mining Engineers.

Muller, L., 1968, New considerations of the Vajont Slide, Felsmechanik und Engenieurgeologie, v. 6, no. 1, p. 1-91.

Narasimhan T. N., and Witherspoon, P. A., 1976, An integrated finite-difference method for analyzing fluid flow in porous media: Water Resources Research, v. 12, p. 57-63.

___1977, Numerical model for saturated-unsaturated flow in deformable porous media, part I, theory: Water Resources Research, v. 13, p. 657-664.

Neuman, S. P., 1973, Saturated-unsaturated seepage by finite elements: American Society of Civil Engineers Proceedings, Hydraulic Division Journal (HY12), p. 99.

Neuman, S. P., and Witherspoon, P. A., 1970, Finite-element method of analyzing steady flow into a free surface: Water Resources Research, v. 6, p. 889.

Obert, L., and Duvall, W. I., 1967, Rock mechanics and the design of structures in rock: New York, John Wiley and Sons.

Olivier, H. J., 1970, Geological-engineering exploration for the Orange-Fish Tunnel, in The Technology and Potential of Tunnelling (Proceedings, The South African Tunnelling Conference): Johannesburg, Associated Scientific and Technical Societies of South Africa.

Orchard, R. J., 1969, The control of ground movements in undersea working: The Mining Engineer, v. 128, p. 259-268.

Pentz, D. L., 1979, Case examples of open pit mine drainage, in Proceedings, First International Mine Drainage Symposium: San Francisco, Miller-Freeman Publishing Co.

Pinder, G. F., and Bredehoeft, J. D., 1968, Application of the digital computer for aquifer evaluation: Water Resources Research, v. 4, p. 1069.

Prickett, T. A., and Lonnquist, C. G., 1971, Selected digital computer techniques for ground water resource evaluation: Illinois State Water Survey Bulletin 55.

Snow, D. T., 1968, Rock fracture spacings, openings and porosities: American Society of Civil Engineers Proceedings, Soil Mechanics and Foundations Division Journal, SM 1, p. 73-91.

___1970, The frequency and apertures of fractures in rock: Journal of Rock Mechanics and Mining Science, v. 7, p. 23-40.

Stewart, R. M., and Kennedy, B. A., 1970, The role of slope stability in the economics, design and operation of open pit mining, in First Symposium on Stability in Open Pit Mining: New York, American Institute of Mining Engineers.

Taylor, D. W.,1948, Fundamentals of soil mechanics: New York, John Wiley.

Terzaghi, K., 1943, Theoretical soil mechanics: New York, John Wiley.

Tress, P. W., 1974, West Driefontein Gold Mining Company Limited, The flooding and the recovery; October, 1968 to January, 1972, in Association of Mine Managers of South Africa, Papers and Discussions 1972-73: Johannesburg, Chamber of Mines of South Africa, p. 617-686.

Trexler, B. D., 1979, Hydrogeology of a lead zinc mine, in Proceedings, First International Mine Drainage Symposium: San Francisco, Miller-Freeman Publishing Co.

Venter, P. P., 1969, The problem of underground water in mines, in Proceedings, Symposium on Groundwater in Southern Africa: Johannesburg, Chamber of Mines of South Africa.

Weehuisen, J. M., 1959, New shafts of the Dutch State Mines, in Proceedings, Symposium on Shaft Sinking and Tunnelling: London, Institution of Mining Engineers.

Whillier, A., 1979, Recent advances in the cooling of deep mines in South Africa, in Proceedings, Eleventh Commonwealth Mining and Metallurgical Congress, Hong Kong, 1978: Institution of Mining and Metallurgy.

Manuscript Accepted by the Society October 14, 1980

Printed in U.S.A.

Geological Society of America
Special Paper 189
1982

Prediction of Chemical Problems in the Reinjection of Geothermal Brines

O. WERES and J. A. APPS
Earth Sciences Division
Lawrence Berkeley Laboratory
University of California
Berkeley, California, 94720

ABSTRACT

Amorphous silica and silicates, and the carbonates and sulfates of calcium, strontium, and barium are the solids most commonly deposited by geothermal brines. The phenomenology and kinetics of their precipitation are reviewed in this paper. Practical methods of avoiding or reducing reinjection well damage are discussed, with emphasis on the important and well-researched problem of removing colloidal amorphous silica from spent geothermal brines before reinjection. Need for further research is also discussed. It is concluded that the interaction of the reinjected brine with the reservoir rock and the brine's effect on rock properties are the areas that most need further research.

INTRODUCTION

Geothermal resources have received worldwide attention as alternative or supplementary sources of energy. The overwhelming majority of these resources are reservoirs of hot water saturating the pores and fractures of rock formations. The exploitation of such resources involves extracting the hot water from the reservoir, recovering the heat, and disposing of the residual water by returning it via an injection well to the rock formation adjacent to the reservoir.

A major problem with this process is that sometimes the injected fluids precipitate dissolved solids that could either seal off the injected formation (and necessitate expensive treatment to restore well injectivity), or, in extreme cases, lead to the abandonment of the well.

The large investment associated with any injection well makes it particularly important to anticipate problems due to declining formation permeability, and to take appropriate corrective measures to maintain adequate injectivity over the life of the plant. To do this requires a semi-quantitative, if not quantitative, predictive capability with regard to when, where, how much, and what kind of precipitates are likely to occur. This presupposes an adequate understanding of the chemical properties of geothermal fluids. It is the purpose of this paper to suggest how this might be achieved and indicate where additional research is required to achieve this objective.

The nature and extent of problems likely to arise during injection of geothermal fluids depends on the initial chemical compositions and temperatures of the fluids recovered, fluid mixing, the method by which heat is recovered from the mixed fluid, the addition of makeup water or other discharge streams, treatment processes prior to injection, and finally the physical and chemical characteristics of the rock into which the fluid is injected. It is almost impossible to generalize the problems arising from so many variables. Furthermore, each geothermal resource will be exploited in a manner which best suits the nature of the resource, the economic needs of the community, and the environmental constraints of the locality. Experience from operating sites throughout the world does give some insight into the problems that may be encountered, and a careful evaluation of the impact of each variable on the injection problem can help us anticipate serious problems and avoid costly design mistakes.

Geothermal fluids occur with a wide range of temperatures and salinities. The distribution of temperatures and salinities at some major geothermal resource areas in the United States and Mexico is displayed in Table 1. In some cases the temperature approaches that of the critical point of pure water, as at Cerro Prieto, Mexico; and the salinity may approach the saturation solubility of sodium chloride, as at Niland, California.

Although not completely understood, it appears that the chemical composition of geothermal fluids is determined by reaction of modified meteoric waters with the host rocks. The high salt content of some geothermal fluids may be due to the dissolution of evaporite beds in sedimentary formations, or to processes involving hyperfiltration, boiling, or diffusion under established thermal gradients. It is probable that partial to complete thermodynamic equilibrium is established between the geothermal fluid and the exposed surfaces of minerals composing the host rock. This is particularly true for secondary salts such as calcite, barite, anhydrite, and possibly fluorite. At temperatures greater than 200 °C, equilibrium with respect to quartz, feldspars, micas, and other secondary silicates such as epidote and chlorite is commonly assumed to have been achieved, and this assumption is used to estimate geothermal reservoir temperatures (Fournier and Truesdell, 1970, 1973). However, equilibrium with respect to all silicates over the temperature range of interest has not been proven (Ellis and Mahon, 1977). Reservoir rocks show partial alteration, the formation of metastable phases, compositional inhomogeneities in single phases, and isotope disequilibrium, all signs of incomplete reaction and the nonattainment of thermodynamic equilibrium.

It is difficult to confirm that limited thermodynamic equilibrium has been achieved. To do so, we must know the chemical composition of the exposed mineral surfaces and the aqueous phase, and have precise information on the thermodynamic properties of solid phases and the geothermal fluid. Neither compositional nor thermodynamic information is adequate at this stage to confirm equilibrium in more than a few limited cases. Fortunately, the concentrations of major scaling components of a geothermal fluid can usually be reconciled with the assumption of thermodynamic equilibrium with respect to the more soluble minerals in the host rock. Minor components in the geothermal fluid, such as aluminum, magnesium, and ferrous ion which are essential to the computation of solubility products of many silicates are rarely significant in considering injection problems, and do not concern us here.

An accurate and reliable chemical analysis of the geothermal fluid is most important. Such an analysis is a prerequisite for any predictive estimates of scaling rates. Unfortunately, there are few truly reliable analyses of representative fluids from geothermal reservoirs.

Geothermal plants are constructed either for

TABLE 1

CLASSIFICATION OF IMPORTANT NORTH AMERICAN GEOTHERMAL RESOURCES BY TEMPERATURE AND SALINITY†

Down Hole Fluid Temperature *	Very Low Salinity 0-2 x 10^3 ppm	Low Salinity 2 x 10^3-2 x 10^4 ppm	Intermediate Salinity 2 x 10^4- 2 x 10^5 ppm	High Salinity > 10^5 ppm
Very High Temperature > 240°C	Yellowstone, Wyoming	Puma, Hawaii Valles Caldera, New Mexico Roosevelt Hot Springs, Utah	Cerro Prieto, Mexico	Salton Sea (Niland) California
High Temperature 150°C-240°C	Mono-Long Valley, California Beowawe, Nevada	East Mesa California Heber, California Steamboat Springs, Nevada Roosevelt Hot Springs, Utah	Brawley, California	
Intermediate Temperature	Raft River, Idaho Fly Ranch Hot Springs, Nevada	Brady Hot Springs, Nevada		
Low Temperature 40°C-90°C	Klamath Falls, Oregon			

* Temperature is based on actual well measurements.
† From Apps, 1977.

electric power generation or for heating purposes. Higher temperature resources are more desirable for electric power generation, but tend to have more problems involving fluid disposal because of the normally higher concentration of dissolved salts in the geothermal fluids. Of the great variety of systems proposed for power generation, most fall into two broad categories: those that employ steam flashed from the geothermal fluid, (flash steam cycles), and those that employ a secondary working fluid heated by the geothermal fluid (binary cycles). Flash steam cycles cause partition of volatiles into the vapor (steam) phase and concentration of the residual brine by removal of up to 30% of the water. Binary cycles usually leave the residual brine with approximately the same bulk composition. Each system affects the scaling behavior of a given geothermal fluid in a different manner and can cause precipitation of solids that differ in composition, quantity, physical character, and location within the plant and subsequent disposal stream. Thus, problems likely to be encountered during injection of the residual brines depend not only on the composition and temperature of the reservoir fluid, but also on the type of plant being operated.

Operating experience at geothermal power plants around the world suggests that the scales most likely to cause injection well damage are those of amorphous silica, calcium carbonate (calcite and aragonite), and sulfates, such as gypsum or hemihydrate and anhydrite, and barite or baryto-celestite. Other scales, such as amorphous silicates, galena, iron sulfides (pyrrhotite and pyrite), and chalcopyrite, sometimes occur, but usually in insignificant amounts due to the small concentration of heavy metals in most geothermal fluids. The exceptionally hot and saline brines in the Salton Sea area, however, do contain substantial amounts of iron, lead, and other heavy metals, and there such "exotic" phases precipitate in significant amounts and may cause practical problems.

Injection of the residual brines may be accompanied by addition of makeup water from other sources. This makeup water may be needed to prevent or minimize subsidence, to recharge the geothermal reservior, or to dilute a potentially supersaturated brine. The chemical composition of such water could be quite different from that of the residual geothermal fluid. Depending on the location and availability, the makeup water could be fresh, irrigation runoff, saline lake, or even seawater. Care should be taken in the selection and use of such waters, because they may contain dissolved constituents that would cause supersaturation when mixed with geothermal fluids or when reheated in the subsurface geothermal reservoir. Soluble constituents such as sodium bicarbonate and sodium or magnesium sulfate, for example, could readily produce carbonate and sulfate precipitates.

Finally, the physical characteristics and temperature of the formation receiving the fluids will have some bearing on the extent and character of precipitation. In most cases, the extent of precipitate formation expected as a result of reheating, mixing with formation waters, and reaction with host rock minerals can be estimated theoretically. However, the impact over time of such precipitation on permeability, and the methods of restoring permeability after well damage due to precipitation will require laboratory experiments and field evaluation. Several investigators have addressed these problems (Grubbs and others, 1972; Lund and others, 1976; Lund and Fogler, 1976; Warner, 1966) and the subject has been surveyed by Warner and Taft (1965).

In attempting to determine whether precipitation of a given solid may result in well damage, the investigator has at his command a number of valuable tools that will enable him to form an intelligent estimate of the potential for serious problems arising under any given set of circumstances. These are:

1. The solubility of the material in question as a function of temperature, pressure and composition of the aqueous phase.

2. Its rate of precipitation onto seed crystals as a function of temperature.

3. The rates of homogeneous and heterogeneous nucleation of particles of the material in the aqueous phase as a function of temperature.

The kinetic factors given in items 2 and 3 can be markedly affected by the presence of inhibiting or catalyzing ions that may lead to widely varying rates of precipitation. Nevertheless, such factors can be studied in the laboratory under controlled conditions and the predictive capability extended to include the presence of such ions.

Of greater hindrance to developing a predictive capability is the effect that foreign ions have on crystal habit, and the significance of precipitation or removal of suspended solids on the permeability of the formation aquifers penetrated by the injection well. Much will depend upon current experience and confirmatory laboratory and field tests. Once the investigator has evaluated the potential problems arising from a given field situation, he will be in a position to recommend strategies to avoid or minimize them. These strategies may include treatment with inhibitors to retard or minimize precipitation, or treatment to remove scale-forming components from the aqueous phase prior to injection.

In the following sections, we shall discuss the current state of knowledge with respect to the most important scale formers, recommend where additional work is required, and suggest techniques for overcoming serious scaling problems where they arise. Greater emphasis is placed on problems relating to silica scaling and precipitation because this subject has been extensively studied at the Lawrence Berkeley Laboratory (Weres, Yee and Tsao, 1980).

CHEMISTRY OF SILICA IN GEOTHERMAL BRINES

Quartz is the most stable and most common form of silica in nature. The rock matrices of most geothermal reservoirs contain quartz, and the solubility of quartz under the given conditions usually determines the concentration of dissolved silica in the brine. The predominant form of this dissolved silica is monosilicic acid, $Si(OH)_4$.

When the brine is cooled in the course of

production and energy extraction, it becomes supersaturated with quartz. The deposition of quartz and the other crystalline silica minerals from aqueous solution is very slow. The rate is negligible under most conditions of interest to geothermal practice.

The precipitation of silica from geothermal brines usually becomes noticeable only after supersaturation is attained relative to amorphous silica, and the usual product is amorphous silica.

There are two basic pathways for the precipitation of amorphous silica:

1. Homogeneous nucleation and growth of colloidal silica particles in the solution phase, which then coagulate and precipitate or adhere to solid surfaces.

2. Molecular deposition upon solid surfaces to give a dense, compact silica glass.

The rate of homogeneous nucleation depends very strongly upon the saturation ratio; that is, the ratio of silica concentration to the equilibrium solubility under the given conditions. Roughly speaking, it is very slow when the saturation ratio is less than two, and rapid when the saturation ratio is greater than three. At saturation ratios below about three, there is a certain "lag time" during which the dissolved silica concentration does not change noticeably. This "lag time" increases very rapidly with decreasing silica concentration, and decreases with increasing pH and salinity.

When conditions are such that homogeneous nucleation is unimportant (for example, when the saturation ratio is less than 2, or the flow velocity is high enough for nucleation to be limited to points further downstream), the molecular deposition mechanism is dominant, and silica scale may be deposited by molecular deposition alone. The rate of molecular deposition on solid surfaces is usually very slow, but may become significant under conditions of high silica concentration and temperature. What appears to be scale deposition by this mechanism has been reported by Harrar and others (1979d).

Amorphous silica precipitation is favored by high initial brine temperature, because cooler brines do not contain enough dissolved silica for homogeneous nucleation to occur even after they have been flashed down to atmospheric pressure. In practice, this means that silica precipitation is likely to be a problem only if the reservoir temperature is above about 240 °C. This is illustrated in Table 2.

The kinetics of homogeneous nucleation and molecular deposition have been studied in detail, both experimentally and theoretically (Weres, Yee, and Tsao, 1980). This report also contains a review and discussion of earlier work in this area.

TYPES OF SILICA DEPOSITS

The deposition of vitreous silica on solid surfaces by the molecular deposition process is slow, but the material is hard to remove once formed. Physical methods like cavitation and hydroblasting may be effective in surface equipment, but slow chemical dissolution with hot hydrofluoric acid or caustic may to be the only way to remove it downhole.

The first step in the conversion of colloidal silica to solid or semisolid deposits is coagulation and settling, or chemically equivalent adhesion to solid surfaces.

The rate and extent of coagulation of colloidal silica is determined mostly by pH and salinity. Above about pH 6, the colloid becomes less stable with increasing pH. The rate and extent of flocculation also increase with salinity. Di- and polyvalent cations are particularly effective in this regard, especially at pH values above about 7.5. In practice, calcium is usually by far the most important of these. Larger particles are more strongly flocculated by divalent cations (Iler, 1975) even though this is probably not true of flocculation by univalent cations.

Unfortunately, available data are inadequate to allow silica coagulation to be predicted quantitatively. However, good semiquantitative data regarding "critical coagulating concentrations" at room temperature are available in the literature. (See Allen and Matijević, 1969, 1970, 1971; Matijević, 1973; Iler, 1975).

The conversion of flocculated colloidal silica to an adherent gel or solid scale involves cementation of the particles by molecular deposition of dissolved silica between them. The rate of this chemical cementation process depends upon the concentration of dissolved silica (among other factors). When the colloidal particles are approximately in equilibrium with dissolved silica, cementation is slow. Under these conditions, cementation involves dissolution of silica from the particles and its redeposition at the interparticle contact points. The only driving force for it is the difference of curvature of the silica surface; the curvature near the interparticle contact point has a large negative element, and this reduces the solubility of the solid silica there. Massive adherent deposits will form only at points where hydrodynamic conditions allow the floc to settle out and remain undisturbed.

Occasionally, colloidal silica deposits may be cemented by the deposition of calcium carbonate or iron sulfide among the silica particles.

Fairly hard solid deposits may form at a moderate rate (centimeters per year) where both colloidal silica and substantially supersaturated dissolved silica are present. Cementation occurs relatively fast under these conditions. Sometimes, when the dissolved silica concentration is high enough and the deposit remains in good contact with the brine, cementation can proceed to the point of producing a glasslike material. (This is known to occur in some of the equipment at Cerro Prieto).

POSSIBLE POST-REINJECTION SILICA PLUGGING MECHANISM

Brine injection can damage the receiving formation if the brine is supersaturated with silica or contains colloidal silica. Any brine that is supersaturated relative to amorphous silica will deposit vitreous silica on solid surfaces at a rate determined by the dissolved silica concentration and temperature. The rate may be neglible (the usual case at moderate concentrations and temperatures), or it may not be. What constitutes a negligible rate will depend on the physical properties of the reservoir. For example, 20 $\mu M \cdot y^{-1}$ would be negligible when the injected fluid is going into a 4-mm fracture. It would be catastrophic if the rock has pore permeability only and an average pore size of 10 μm.

Note that the rate of molecular deposition is determined *only* by the temperature, concentration, and other chemical variables, and is not affected by the hydrodynamic state of the fluid. This means that it will continue until the supersaturation is reduced by some combination of concentration decline and reheating.

If brine that contains a floc is reinjected, the floc will accumulate in any place where the brine enters pores or cracks fine enough to filter out the floc. These places will eventually be plugged. If pore permeability is dominant, the plugging will take place right at the wellbore where it will do the most harm most rapidly. (However, damage at the wellbore is also the easiest to undo, by treating the well with sodium hydroxide or hydrofluoric acid.) In a fractured reservoir, the floc may finally come to rest so far from the injection well that no significant injectivity decline will ever be observed. Also, if the floc is not deposited by the time the brine is reheated, it will partially or completely redissolve.

If, in addition to floc, the brine also contains enough dissolved silica to cause rapid cementation, the spatial distribution of the damage will be similar but the damage will be worse and harder to undo. Fortunately, significant supersaturation cannot persist for long in the presence of large colloid loadings, and this fact will tend to reduce the extent of cementation in the formation.

It must be considered that reaction of the brine with reservoir materials could destabilize the colloid and accelerate deposition. For example, injecting brine that is undersaturated with calcite into a calcite-bearing formation would cause the pH and calcium content of the brine to increase, thereby destabilizing the colloid.

Finally, a colloid-free brine with high enough supersaturation could undergo nucleation after reinjection. Whether nucleation actually occurs would be determined by whether or not competing processes such as molecular deposition or reheating can reduce the silica supersaturation rapidly enough to prevent it. Nucleation in the formation could involve rapid cementation of the colloid produced because of the inevitability of passing through a stage of simultaneous substantial colloid loading and substantial silica supersaturation. The extent of damage would again depend on rate and the physical properties of the reservoir. Greater supersaturation would cause nucleation nearer to the wellbore, and therefore, be more likely to cause serious damage. It remains to be considered whether or not flow through a porous medium can inhibit nucleation by removing newly nucleated particles from the brine which adhere to the rock surfaces before they grow to significant size.

BRINE TREATMENT FOR SILICA REMOVAL AND SCALE CONTROL

A salient characteristic of geothermal energy conversion processes is that they involve large mass flows. Typically, about 10 kg of steam is required per kWh generated, and 20 to 40 kg of spent brine must be disposed of. This means that a brine treatment cost of more than 10^{-5} $\$ \cdot kg^{-1}$ of brine is significant, and a cost of more than about 10^{-4} $\$ \cdot kg^{-1}$ is unlikely to be acceptable. This means that brine treatments must be simple, and that chemicals added in concentrations greater than a few parts per million must be very cheap. Also, the sheer volume of brine to be treated can render a process impractical because of unacceptably large brine storage, chemical delivery or handling requirements. Finally, to be commercially viable, the overall energy extraction and conversion system must be as cheap, reliable, and as easy to operate as other types of electric power plants.

With the above considerations in mind, we proceed to review possible approaches to avoiding silica-related problems in reinjection.

Five distinct approaches have been developed for the control of silica precipitation from geothermal brines, all of which are discussed below.

1. Avoidance or Minimization of Silica Supersaturation

Supersaturation can be avoided by preventing brine cooling to below the amorphous silica saturation temperature for the given dissolved silica concentration. This is the only completely safe, simple approach available and is a matter of clever energy conversion process design rather than actual brine treatment. The substantial difference between the solubility of quartz, which determines the initial silica concentration in the brine, and that of amorphous silica makes the goal attainable in many cases (see Table 2). A successful practical demonstration of this approach will be discussed in the following section.

Under some circumstances a saturation ratio somewhat in excess of unity may give nucleation and molecular deposition rates low enough to be practically insignificant. This is particularly likely at temperatures below 100 °C. This possibility should be carefully considered using available kinetic data (Weres, Yee, and Tsao, 1980) when designing for a saturation ratio of unity or lower appears impractical for some reason.

2. Kinetic Inhibition of Molecular Deposition and Nucleation

The rates of all silica polymerization and precipitation processes decrease with decreasing pH. Therefore, one way to slow down silica precipitation, although not to stop it, is to acidify the brine (Owen, 1975). Alternatively, the pH may be kept from rising by avoiding flashing

TABLE 2. Silica Concentrations, Saturation Ratios, and Deposition Rates Resulting from Flashing Water Initially Saturated with Quartz

	Init. t(°C)	200	220	240	270	300
	Solub.Quartz(ppm)	297	375	440	521	598
Final t(°C)	Solub. Am.Sil.(ppm)					
70	245	1.315*	1.387	1.469	1.619	2.004
		390†	520	646	844	1,198
		1.594‡	2.123	2.638	3.443	4.891
		0.010§	0.039	0.11	0.36	1.65
100	364	1.238	1.303	1.377	1.514	1.859
		368	489	606	789	1.112
		1.010	1.342	1.664	2.167	3.054
		0.0003	0.027	0.10	0.39	1.94
130	509	1.164	1.224	1.292	1.416	1.728
		346	459	568	738	1.033
		0.679	0.902	1.117	1.449	2.030
		-	-	0.036	0.30	1.88
160	680	1.093	1.148	1.210	1.324	1.607
		325	430	532	690	961
		0.477	0.633	0.783	1.014	1.413
		-	-	-	0.016	1.29

* Ratio of concentration by flashing
† Final concentration of SiO_2(ppm)
‡ Final saturation ratio over solid AS
§ Molecular deposition rate in micrometers/day) at pH=7.0 and $[Na^+]$=0.069.

Note: One-stage steam separation has been assumed for simplicity.

This Table quoted from Weres, Yee and Tsao (1980), where it is designated "Table 4.1."

the brine, as in a binary cycle with downhole pumped brine production. Another effect of lowering pH is to inhibit the coagulation of colloidal silica.

Lowering pH or keeping it from rising may also decrease or eliminate the precipitation of carbonates and sulfides.

Most discussions of treatment by acidification assume the use of hydrochloric acid. Sulfuric acid is cheaper, but may cause or aggravate the precipitation of calcium, strontium or barium sulfate. However, this risk should be carefully evaluated on a case-by-case basis, and the use of sulfuric acid should not be rejected out of hand because of it. The cheapest acid of all is carbon-dioxide-rich geothermal condenser off-gas. This additive is available at no cost, but its use would require large and potentially complicated and expensive gas-liquid contact equipment. The use of condenser off-gas would probably introduce some hydrogen sulfide and atmospheric oxygen into the brine. (Condensers usually have minor leaks which cause air to be entrained. In the case of contact condensers, air dissolved in the cooling water also ends up in the off-gas.) Oxygen could aggravate corrosion problems or cause the precipitation of ferric hydroxide or similar substances.

Workers at Lawrence Livermore Laboratory have extensively evaluated the use of hydrochloric acid addition for scale control at Niland, California (Owen, 1976; Grens and Owen, 1976; Austin and others, 1977). It was concluded that adding about 200 ppm of HCl to the brine could completely eliminate scaling throughout the system at a chemical cost of 1 or 2 mils/kWh. Special alloys would have to be used to keep corrosion down to acceptable levels. Also, it was found that pumping the acidified brine through sandstone cores dissolved the calcite in the matrix, which led to serious plugging with loosened matrix material (Piwinskii and Netherton, 1977). This means that the brine pH would have to be raised again before reinjection by adding base. All in all, brine acidification to the point of complete elimination of scale proved to be a complicated and costly proposition. The addition of a smaller amount of acid, which would reduce but not completely stop scale deposition, would probably be more practical in that case.

Harrar and others (1979e) discussed the possibility of adding hydrogen peroxide to Niland brine to decrease the pH. The hydrogen peroxide oxidizes the ferrous iron in the brine to ferric iron. The ferric iron hydrolyzes, and this reaction releases protons. Overall, four protons are released per molecule of hydrogen peroxide consumed and this actually makes the addition of hydrogen peroxide cheaper than the addition of an equivalent amount of hydrochloric acid. The use of hydrogen peroxide or other oxidizing agents to lower brine pH in this way is limited to the few iron-rich brines like those at Niland. Also, the ferric hydroxide precipitate that is formed may cause undesirable side effects.

Working at Wairakei and Broadlands, New Zealand, Rothbaum and others (1979) found that lowering the pH of the brine to 4 or 5 reduced the rate of scale deposition by up to a hundredfold. Working with simulated Cerro Prieto brines in their laboratory, Weres, Tsao and Iglesias (1980) found that even small pH changes (less than one unit) could significantly reduce the rate of scale deposition.

3. Aging the Brine

"Aging" the brine to convert dissolved silica to colloidal silica has been used in practice at the Otake plant in Japan (Yanagase and others, 1970). It has been experimented with on a smaller scale at Ahuachapán in El Salvador (Cuéllar, 1975), and at Wairakei and Broadlands in New Zealand (Rothbaum and Anderton, 1975; Rothbaum and others, 1979).

At Otake and Ahuachapán the desired goal was to reduce the rate of scaling in surface waste brine disposal pipes and canals. In both places, untreated brine delivered to these conduits apparently contained both colloidal silica and a substantial excess of dissolved silica. Rapid accumulation of hard scale of the cemented colloidal aggregate type was observed. "Aging" the brine by retaining it in a suitable tank for a period of time was tried in both cases in a conscious attempt to reduce the dissolved silica concentration by allowing time for its conversion to colloidal silica.

The experiments succeeded in both cases, and a commercial-scale "brine aging pond" was actually constructed and put into routine operation at Otake. In both places, the rate of scale accumulation in the brine disposal system was greatly diminished. Scanning electron micrographs of scale specimens from Otake clearly show that the scale at the outlet of the "aging pond" is a fluffy, weakly cemented floc-like material. Cuéllar comments on the accumulation of about 14.5 metric tons of scale in the aging pond, but notes that: "Almost all deposits formed in the tank are very light, porous, and easily removed" (Cuéllar, 1975, p. 1346).

In both cases, only a fraction of the colloidal silica in the brine was deposited in the "aging pond" or disposal system. Most of it was apparently carried away by the brine in metastable or weakly coagulated, nonadhesive form.

In New Zealand, brine aging was initially experimented with as the first step in a more complicated course of brine treatment. The primary purpose was to convert dissolved silica, which cannot be removed by coagulation, to colloidal silica, which can. This goal was achieved (Rothbaum and Anderton, 1975). Brine aging did not significantly decrease the rate of scale deposition but it did cause the scale that was deposited to be softer and easier to remove (Rothbaum and others, 1979).

The time needed for the dissolved silica concentration to drop into the range of slowly-decreasing "near steady state" values will vary widely with the chemical parameters of the brine. In the cited instances, adequate aging times were found to be as follows: Broadlands, 30 min; Ahuachapan, 45 min; Otake, 1 h, and Wairakei, 2 1/2 h.

In general, the time required will decrease with increasing initial supersaturation, temperature, pH and salinity. The reader is referred to the data presented by Weres, Yee and Tsao (1980), and Weres, Tsao, and Iglesias (1980) in this regard.

A variety of different physical configurations for the brine aging vessel seem to work well. The various configurations all provide approximately plug flow reactor conditions. We refer the reader to the original papers and to the review by Phillips and others (1977) for further details. Our only firm recommendation is that the aging vessel have a cover of some sort in order to prevent contact with air and to minimize heat loss.

4. Chemical Inhibition of Adhesion

Inhibiting the adhesion of colloidal silica to solid surfaces causes the colloidal silica to remain suspended in the brine rather than being deposited to form solid scale.

Harrar and his co-workers have systematically searched for chemical additives that have this effect when added to Niland brine in small concentration (Harrar and others, 1979a, b, c, d, e, and f). They have succeeded in identifying several commercial products that are effective. They have not yet reported whether or not adding these substances to the brine hinders subsequent colloidal silica removal by flocculation. (Such an effect is to be expected, because the principle involved here is colloid stabilization, which is the opposite of induced flocculation.)

5. Removal of Colloidal Silica by Coagulation and Settling

In most cases, it will probably be necessary to remove any colloidal silica that may be present in the brine before reinjection. Similar needs are common in other circumstances, and a sophisticated body of technology exists and is available for this purpose. Technical evolution in this area has largely consisted of careful application of existing knowledge to the particular circumstances of geothermal development in specific areas.

In New Zealand, the goal was to develop a practical treatment which would yield brine innocuous enough to deliver to nonelectrical users or to dispose of in fresh surface waters without environmental risk. Primarily, this required the removal of colloidal silica and arsenic, and secondly, some reduction of dissolved silica to below the levels attainable by aging alone. The addition of 400 to 700 ppm of calcium oxide (unslaked lime; CaO) and about 10 ppm of sodium hypochlorite (NaClO) to the aged brine followed by settling was found to remove all of the colloidal silica and most of the arsenic and dissolved silica (Rothbaum and Anderton, 1975). The product is a high surface area mixture of colloidal amorphous silica and amorphous calcium silicate which might have some practical use.

Removal of colloidal silica alone is not explicitly discussed by Rothbaum and Anderton, but the data they present suggest that about 100 ppm CaO may be adequate for this purpose. This relatively large amount appears to be needed because the New Zealand brines are of rather low salinity and, in particular, contain little calcium. Therefore, a large increase in pH and/or calcium ion concentration is needed to flocculate the silica in them.

We note that increasing the pH or calcium ion concentration may cause or aggravate calcium carbonate or sulfate precipitation either at the surface or after reinjection. This is particularly to be expected in the case of a very low-calcium and high-bicarbonate brine like that at Broadlands. This possibility must be carefully considered on a case-by-case basis if lime addition is being considered.

Colloidal silica removal from the brine at Niland (California) has been demonstrated by Quong and others (1978) on a 38 $l \cdot min^{-1}$ pilot plant scale. The suspended solids in the flashed brine at Niland consist mostly of amorphous silica, but contain smaller amounts of other substances as well. However, the silica dominates their overall behavior. This brine is very saline and has a high calcium content, but the pH is only 5.7 to 6.0. This low pH value causes the decline in dissolved silica to be slow, and the rate and extent of coagulation of colloidal silica to be small. Raising the pH helps matters considerably, but causes the supernatant brine to cloud with ferric hydroxide subsequent to precipitation. This behavior is due to the very high iron concentration of up to 2 g $\cdot l^{-1}$ in Niland brines. (This effect might be avoidable by isolation from the air, but this is not discussed in the cited paper.) All in all, Niland is a rather difficult case.

Quong and others (1978) have demonstrated that reaction clarification followed by filtration is an effective brine treatment method in the case of Niland.

First, a few ppm of the coagulant is added to the brine and it is rapidly stirred in a reaction vessel for 5 min. Then it enters the clarifier. This device has a large amount of slowly stirred coagulated silica sludge at the bottom. Sludge is continuously recycled from the bottom of the clarifier into the reaction vessel. The brine slowly flows through the sludge, and as it does, the very high density of flocculated colloidal silica in the sludge rapidly lowers the dissolved silica concentration in the brine, as well as coagulating and retaining most of colloidal silica. The brine then flows up and out of the sludge blanket at about 0.25 $mm \cdot s^{-1}$. This is the maximum velocity consistent with good separation of the clarified brine from the sludge through gravitational settling of the latter. The clarifier overflow still contains an unacceptably high concentration of suspended solids (44 ppm in the best performance reported). However, a final sand bed filtration reduces this to below 2 ppm, which is considered good enough for reinjection.

A pilot-sized commercial reactor-clarifier (114 $l \cdot min^{-1}$), which incorporates the reaction vessel described above within the clarifier, was successfully tested at Niland by Featherstone and others (1979). Subsequently, a larger-scale pilot test utilizing a 6080 $l \cdot min^{-1}$ throughput rate was carried out by the San Diego Gas & Electric/Magma Power Company at Niland. The larger system used an Eimco reactor-clarifier and a granular medium gravity filtration system. System upsets due to inadvertent contamination of the brine with air reduced the quality of the processed brine with respect to injectability. However, elimination of air entrainment would probably result in production of a highly-injectable effluent.

On the basis of results obtained working with simulated Cerro Prieto brines, Weres, Tsao and Iglesias (1980) have proposed a much simpler treatment for Cerro Prieto: 5 to 20 min. aging, addition of 20 to 30 ppm CaO with vigorous stirring, and separation of floc from brine in a clarifier. The resulting floc settles at about 1 $mm \cdot s^{-1}$ and the "clarified brine" contains less than 4 ppm suspended solids.

It is possible that adding lime to the brine will cause or enhance precipitation of calcium or barium carbonate from the brine. It was not possible to address this question in the laboratory but we did examine it theoretically. We concluded that, if precipitation of carbonates is indeed encountered in practice and causes practical problems, adding a modest amount of acid before the final stage of steam separation will eliminate this problem at reasonable cost, as well as reduce silica scaling in the low-pressure steam separator (Weres, Tsao and Iglesias, 1980).

That such a simple process appears adequate for Cerro Prieto is a fortuitous happenstance. When flashed down to 100 °C, the brine is in a chemical state conducive to very fast colloidal silica nucleation and dissolved silica decline. The resulting colloid is only moderately coagulated in the unmodified brine, and does not begin to settle until the aging process has run its course. However, it is near to the pH range of strong coagulation. Increasing the pH by about 1/2 unit by adding a small amount of base (CaO is the cheapest) causes very rapid and essentially complete coagulation.

The silica removal processes reviewed here are all variations on the same basic concept, but differ greatly in its concrete realization. These variations are largely due to the different brine chemistries in the different areas. This demonstrates that an "off-the-shelf" approach is not appropriate in the area of geothermal chemistry. Rather, the process must be tailored to the particular characteristics of the given brine. This is particularly true if commercial flocculants or similar additives are to be used, because the effectiveness of such products can vary greatly with what seem to be minor variations in brine composition (Weres, Tsao and Iglesias, 1980).

Field Experience with Geothermal Reinjection

Large-scale geothermal brine reinjection experiments at Ahuachapán, Otake, and Niland have been reported on in the literature.

At Ahuachapán, the brine was initially saturated with quartz at about 240 °C, and was flashed down to about 155 °C, and reinjected at 152 °C. Therefore, it was slightly undersaturated with amorphous silica at injection (see Table 2). As predicted, there was no permeability reduction or scaling of any sort observed over the duration of the injection experiments, which totaled almost 2 x 10^6 m^3 of brine. This is the classic demonstration of the supersaturation-avoidance approach to silica precipitation control (Einarsson and others, 1975; Cuéllar, 1975.)

Routine reinjection of spent brine from the geothermal development at Otake began in 1972. As of 1975, 8 x 10^6 tonnes had been injected into three wells. The receiving interval was a highly-permeable stratum of tuff breccias 300 m thick. Over the three years, a slow decrease in injectivity of the three wells was observed. For example, with the water level at the wellhead, reinjection well No. 1 took 310 $tonnes \cdot hr^{-1}$ at the beginning of injection, but only 120 $tonnes \cdot hr^{-1}$ in 1975. Scale was deposited throughout the length of this well. At the wellhead, it was about 25 mm thick after three years and consisted mostly of SiO_2 (Kubota and Aosaki, 1975).

Unfortunately, Kubota and Aosaki say nothing about the preinjection treatment that the brine received (if any), or even about the brine temperature. However, Yanagase and others (1970) reported that a full field capacity brine aging pond was in routine operation as of 1970. Therefore, it seems likely that the injected brine was being treated in this way. This means that the brine probably contained about 200 ppm of metastable colloidal silica and was supersaturated with amorphous silica by a few percent. If so, the reported injectivity decrease was probably due to slow deposition and cementation of colloidal amorphous silica both near to and at some distance from the wellbore.

The injectability of the spent brine at Niland has been extensively studied at that site by laboratory methods as well as some actual injection tests (Morse 1977, 1978; Netherton and Owen, 1978; Tewhey and others, 1978; Hasbrouck and others, 1979; Owen and others, 1979; Quong and others, 1978). Injection tests showed that untreated brine quickly and completely plugged up the pores of the receiving formation right at the wellbore. However, injection was not stopped by this plugging because a large fracture zone continued to take the brine. The membrane filter test method commonly used in petroleum practice to evaluate the injectability of waters (Barkman and Davidson, 1972) was adapted to geothermal use and successfully applied. Its results agreed well with those of the injection experiments. Also, membrane filtration experiments clearly demonstrated the mechanism of deposition of cemented colloidal aggregates from brine that contained both colloidal and supersaturated dissolved silica. Core-flushing test methods and equipment for evaluating injectability were developed, applied and described as well.

Outstanding Research Needs

A basic need in all areas of geothermal chemistry is for carefully collected and exhaustively analyzed field data of various sorts. This does not necessarily mean that grandiose new experimental programs need to be started. Rather, ongoing test and operational activities should be planned with the participation of chemists, and provisions should be made for adequate and well-defined sampling, analysis, and description. The ultimate product of this work should be a series of well-documented case studies that describe and explain what actually happens. Thorough characterization of all precipitated solids must be included in such studies, and every effort should be made to compare and reconcile the predictions of theoretical and laboratory studies with field experience.

The major outstanding need in the area of silica chemistry is a capability to quantitatively explain and predict the colloidal stability and coagulation behavior of colloidal silica, and its cementation following coagulation. Combined with existing knowledge of the chemical kinetics of amorphous silica precipitation, this information would go a long way toward systematizing the design of preinjection brine treatment processes.

A related need is for a better understanding of the deposition of amorphous silicate materials. Work in this area should start with careful characterization of observed precipitates, and move toward the sort of predictive capability that exists for pure amorphous silica.

Chemistry of Calcium Carbonate Precipitation

Calcium carbonate, $CaCO_3$, exists in three polymorphic forms, calcite, aragonite and vaterite. Three hydrates, mono-, tri- and hexa-hydrocalcites, have been synthesized but are of no importance to problems involving the injection of geothermal fluids. Calcite is thermodynamically most stable under the pressures and temperatures encountered in a geothermal system. However, aragonite can be kinetically favored in the presence of many other ions, particularly magnesium and strontium.

Calcium carbonate possesses retrograde solubility (Berendsen, 1971; Holland, 1967). Therefore, the operation of a binary geothermal system, in which the geothermal fluid is not flashed, is unlikely to cause precipitation unless imcompatible fluids from different sources are mixed or carbon dioxide is permitted to escape from the fluid.

Flashing of the geothermal fluid, even if minimal, will lead to separation of carbon dioxide from the liquid phase, and can easily induce supersaturation with respect to calcium carbonate. This is due to a shift in the speciation in the aqueous phase and is conventionally described by the following reactions:

$$2HCO_3^- \rightarrow CO_3^{2-} + H_2O + CO_2\uparrow,$$

$$CO_3^{2-} + Ca^{2+} \rightarrow CaCO_3(s).$$

Thus a reduction in the concentration of dissolved carbon dioxide results in the precipitation of calcium carbonate. Further flashing of the geothermal fluid will increase the residual concentration of calcium in solution and increase ionic strength, both of which can lead to enhanced precipitation (see especially Holland, 1967, p. 408, Fig. 9.22) but this is more than offset by the solubility increasing effect of brine cooling.

Precipitation of calcium carbonate is therefore thermodynamically favored if the geothermal fluid is initially saturated with respect to calcite and undergoes flashing in the geothermal well or in the plant. A perturbation in the flow of a producing geothermal well due to changes in casing diameter can trigger flashing, and induce immediate precipitation of calcium carbonate adjacent to the point of disturbance. Much of the precipitated calcium carbonate may remain suspended in solution as particulates. The distribution between suspended particulate and adhering scale should be studied under field conditions and the size distribution and physical properties of the particulates evaluated. Such information could be used to determine separation procedures and evaluate the impact suspended solids might have on permeability degradation in the formation adjacent to the injection well.

Supersaturation in the residual brine may be enhanced by the addition of makeup water containing dissolved calcium or carbonate ions. Even if undersaturated at the point of injection, reheating the brine or mixing with formation waters can easily lead to supersaturation in the formation adjacent to the injection well. Thus it is extremely important that semiquantitative if not quantitative estimates be made of the rate and nature of calcium carbonate precipitation, and of how it may ultimately affect the operation of an injection well. In the following discussion, an appraisal is made of what is currently known about the kinetics of calcium carbonate precipitation, and what more needs to be done to predict brine injection problems due to carbonate precipitation.

The kinetics of precipitation and dissolution of calcium carbonate polymorphs has been investigated on many occasions. Unfortunately, no satisfactory model yet exists to explain all the phenomena observed, and no quantitative predictions are possible regarding the rate of dissolution or precipitation, or the polymorph(s) likely to form under a given set of conditions. Nevertheless, much valuable research has been done that provides a basis for interpreting field observations, and which may, with some effort, provide guidelines for resolving the current uncertainties.

In common with silica, and other salts, calcium carbonate polymorphs can in principle form from solution either by homogenous nucleation, by heterogenous nucleation on particles of a substrate, or by crystal growth. Wollast (1971), using classical nucleation theory, attempted to calculate the degree of supersaturation of an aqueous solution with respect to calcite at 30 °C before homogenous nucleation would occur spontaneously. He found that a product of the activities of the ions Ca^{++} and CO_3^{--} equal to 2000 times the product of the concentration of Ca^{++} and CO_3^{--} at the equilibrium value (K_{dp}) is insufficient to cause the nucleation of calcite. In contrast, a similar calculation with respect to heterogeneous nucleation on a substrate showed that supersaturations of only four times K_{sp} could be tolerated before precipitation would occur. In natural environments, microcracks and grain boundaries of rock-forming minerals could act as preferred nucleation sites where crystal growth could start when the geothermal fluid is even slightly below true saturation.

Although the classical theory used by Wollast leaves much to be desired, it does demonstrate quite clearly that homogeneous nucleation due to flashing of geothermal fluids requires very high levels of supersaturation. Under unusual circumstances, it is possible that such levels might be achieved. However, it is more likely that "pseudo-homogeneous" nucleation will occur due to the presence of dissolved impurities, or that simple heterogeneous nucleation will occur before such levels of supersaturation can be attained.

Under flashing conditions, in either producing geothermal wells or steam separators, calcium carbonate precipitation can be exceedingly rapid due to the elevated temperature and the high level of supersaturation frequently achieved. Under injection conditions, supersaturation may still prevail in the injected fluid, due to the residual effects of flashing or to mixing with makeup water. Supersaturation may also be reestablished as a result of reheating or mixing of the fluid in the subsurface environment. However, the levels of supersaturation will not normally be high, and precipitation will occur either through pseudohomogeneous or heterogeneous nucleation or orderly crystal growth.

In practice, it has been found that only modest levels of supersaturation (3 to 30 times) will lead to what appears to be homogeneous nucleation (Packter, 1968; Packter and Uppaladini, 1974; Packter, 1975a) as distinct from heterogeneous nucleation. According to studies by Packter (1968), this form of nucleation will vary in rate of nuclei formation over approximately five orders of magnitude for sparingly soluble alkaline-earth metal, and lead salts with induction times varying from 1 to 10^5 seconds at 22 °C. Furthermore, he found that where the saturation ratio was greater than about 10, the rate of nucleation varied linearly as some power of the metal salt concentration, thus

$$R_n = K_n C^p,$$

where R_n is the rate of nucleation, in nuclei$\cdot l^{-1}\cdot s^{-1}$ k_n is the nucleation rate number, and p is the reaction rate order. For calcium carbonate, p = 4.2.

Experimental observations of "homogeneous" nucleation are therefore quite inconsistent with the theoretical treatment of Wollast, and the ranges of observed "homogeneous" nucleation fall within supersaturation levels that might easily be encountered during the mixing of residual brines with makeup water during injection. Because induction times at relatively low levels of supersaturation may be quite long, it is feasible for the observed form of homogeneous nucleation to cause precipitation after injection and in the rock formation adjacent to the wellbore. Thus, this form of nucleation cannot be ignored in any treatment of calcium carbonate precipitation kinetics. A possible explanation for the inconsistency between theory and observation regarding homogeneous nucleation of calcium carbonate is that the nucleation is in fact heterogeneous, but is induced by the presence of foreign ions or complexes too small to be considered nuclei in the accepted sense. Because such nuclei may be present only in minute concentrations, it will probably be exceptionally difficult either to prove the existence of this mechanism, or to identify the controlling species.

During most laboratory experiments to study nucleation phenomena in the aqueous phase, (Klein and Fontal, 1964; Packter and Sahay, 1974; Packter 1975a) heterogeneous nucleation is observed prior to homogeneous nucleation due to the presence of many suspended minute particles. Geothermal fluids, after passing through a geothermal power plant will inevitably contain many particulates including suspended clay particles, products of abrasion, and small crystallites of sparingly soluble salts. Their importance must necessarily be assessed in terms of their ability to act as nuclei that produce suspended crystallites of such size as to plug the pores of the rock formation accepting the injected fluid.

Calcium carbonate crystal growth or dissolution in the aqueous phase at near ambient temperatures has been studied many times, particularly as it pertains to natural systems where calcium carbonate is a major component of the affected rock or sediment. In particular, great interest has been shown in the effect of foreign ions on the formation of calcite or aragonite in seawater.

Calcium carbonate crystal growth or dissolution depends upon both a forward reaction in which ions containing the Ca^{2+} and CO_3^{2-} components are transferred to the surface of the crystal, and a back reaction in which the same components are detached from the crystal surface and dispersed. At equilibrium, of course, the two processes proceed at the same rate, and there is no net loss or gain of mass by the participating phases. Experimental studies are undertaken with the hope that the mechanism(s) involved in this mass transfer process will be identified. Until recently, such attempts have yielded mixed results, but two important papers by Plummer and others (1978, 1979) provide a very clear insight into the complexity of the mechanism of precipitation and dissolution of calcite.

These two papers and an earlier paper by Plummer and Wigley (1976) report the results of a series of experiments in which semioptical-grade Iceland spar crystals of predetermined surface area were dissolved in dilute solutions as a function of pH, P_{CO_2}, and temperature. Studies were made both with a pH-stat, and by free drift (of pH) when the reaction was near equilibrium. The pH range studied was from 2 to 7, P_{CO_2} from 0.0 to 0.97 atm, and temperatures ranging from 5 °C to 60 °C. According to Plummer and others (1979), the net rate of reaction of calcite between 5 °C and 60 °C can be described by the equation

$$R = k_1 a_{H^+} + k_2 a_{H_2CO_3^*} + k_3 a_{H_2O} - k_4 a_{Ca^{2+}} a_{HCO_3^-},$$

where k_1, k_2, k_3, and k_4 are temperature-dependent rate constants and,

$$H_2CO_3^* = H_2CO_3^\circ + CO_2^\circ.$$

The degree to which either $a_{H_2CO_3^*}$ or a_{H_2O} affects the dissolution rate is a function both of pH, the partial pressure of CO_2 and temperature. Plummer and others (1978) show that at 25 °C, a_{H^+} contributes primarily to the dissolution process below pH 4.5 when P_{CO_2} is less than 1 atm. Above P_{CO_2} = 0.1 atm and at 5.5 < pH < 6.5, the activity of $H_2CO_3^*$ controls the dissolution rate, whereas below P_{CO_2} = 0.1 atm and at pH > 6.5, it is the activity of water, aH_2O, that controls the dissolution rate. Separating these regions are regions where the dissolution rate is controlled by two or more mechanisms, as shown in Figure 1. The back (precipitation) rate constant k_4 is less precisely determined than the forward rate constants. Plummer and others (1979) describe a relationship for k_4, thus:

$$k_e = \frac{K_2}{K_c}\left\{k_1' + \frac{1}{a_{H^+(s)}}[k_2 a_{H_2CO_3^*(s)} + k_3 a_{H_2O(s)}]\right\},$$

where K_2 and K_c are equilibrium constants for the reactions:

$$HCO_3^- = H^+ + CO_3^{2-}$$

and

$$K_c = CaCO_3(calcite) = Ca^{2+} + CO_3^{2-}.$$

The subscript, (s), refers to ions bound in the adsorption or surface layer of the calcite.

By replacing the theoretical k_1' with the experimentaly determined k_1' the investigators were able to show that the above expression showed the correct temperature dependence when plotted as function of P_{CO_2} and compared with experimental values, implying that $k_1' \approx (10\text{-}20)k_1$. However, the temperature range compared was limited to 5 to 35 °C.

Reddy and Nancollas (1971) and Nancollas and Reddy (1971) have investigated the crystal growth of calcite from agitated supersaturated solutions of sodium bicarbonate in a supporting calcium chloride electrolyte between 10 °C and 40 °C. The pH of the solution used was approximately 8.5 to 8.8 and specially synthesized seed crystals were used as the growth substrate. The investigators found that their data agreed with an empirical kinetic equation in which the rate of crystallization is

$$\frac{dT_{Ca^{2+}}}{dt} = - k_G\, s(m_{Ca^{2+}}\, m_{CO_3^{2-}} - K_{SP}/f_2^2),$$

where $dT_{Ca^{+2}}$ = the total change of Ca^{+2} in solution,

k_G = rate constant for crystal growth,

s = the surface area of the crystals,

$m_{Ca^{2+}}$ and $m_{CO_3^{2-}}$ = the molarities of Ca^{2+} and CO_3^{2-} ions in solution,

K_{SP} = the solubility product for calcium carbonate, and

f_2 = divalent activity coefficient.

The activation energy for crystal growth was found to be 11.0 ± 1.0 kcal·mole^{-1} over the range of temperatures studied. Because of the apparent lack of dependence of the growth rate on stirring rate, it was assumed that calcite growth is a surface-controlled process.

Plummer and others (1979) reviewed this work and found that, for the conditions of the experiment, their equation expressing the net rate of reaction reduced to the same form as given above but that close experimental agreement was not obtained, either because their model was wrong, or because surface P_{CO_2} values differed from those in the bulk fluid. The latter explanation appears reasonable, but emphasizes the difficulties involved in reconciling experiments by different investigators.

The work so far accomplished is still insufficient to predict the impact of calcium carbonate precipitation when supersaturated fluids are injected into permeable rock formations. Dissolution or precipitation studies have been made over too restricted a temperature range, and the precipitation and dissolution mechanisms have not yet been unambiguously resolved. For example, Lund and others (1973) found that the acid dissolution of dolomite appeared to be surface reaction controlled between 25 °C and 50 °C, whereas at 100 °C, dissolution exhibited diffusion control. Similar observations were made by Lund and others (1975) where, at 800 psig and -15.6 °C, calcite dissolution in acid was limited by both mass transfer and surface reaction rates, whereas at 25 °C, dissolution was entirely diffusion-controlled.

The effect of foreign ions has yet to be more clearly quantified for the range of compositions, salinities, and temperatures normally encountered during the brine disposal process. Calcite precipitation or dissolution rates near equilibrium appear to be strongly affected by the presence of some kinds of foreign ions. The prevailing interpretation for this behavior is that the ions are adsorbed on the surface of the growing crystal and inhibit further growth or dissolution by blocking the active sites. As an example, ground waters may be found to be undersaturated with respect to coexisting calcite, when equilibration should have been attained within a few years at most (Plummer and Wigley, 1976). Similar observations have been made by Kitano and Hood (1965) with respect to the presence of organic ions in solution. On other occasions, when waters are supersaturated with respect to calcite, the metastable polymorphs aragonite or vaterite are formed, presumably because the kinetics of their growth are faster than those of calcite, and less sensitive to interference by foreign ions. Typical examples of such aragonite precipitation from seawater are given by Cloud (1962) and Bathurst (1968). Aragonite also precipitates from hot spring waters (Kitano, 1962b, Berendsen, 1971) so it is important to consider whether aragonite rather than calcite may precipitate during injection of residual geothermal fluids, and under what conditions its formation would be favored. Furthermore, the whole subject of the value of inhibitors hinges on their effectiveness in controlling not only calcite precipitation, but aragonite precipitation as well.

Experimental studies relating to this subject can be broken down into two classes: (1) Studies that involve the pseudohomogeneous or heterogeneous nucleation of calcite or aragonite from supersaturated solutions. In these studies the chemical and phase composition of the resulting precipitate is examined as a function of the initial solution composition. (2) Studies in which the rate of growth of calcite is observed in the presence of a solution of predetermined composition saturated with respect to and in the presence of aragonite. Because of the relevance of the first-class studies to reinjection conditions, this subject will be considered first.

Kitano (1962a, b) made an extensive study of the precipitation of calcium carbonate from supersaturated solutions of calcium bicarbonate as a function of the concentration of various inorganic ions in solution, and as a function of temperature. Of the many observations he made, those of particular significance to the injection of residual brines are:

1. Alkali chlorides tend to inhibit aragonite formation.
2. Increasing sodium carbonate concentration favors the precipitation of aragonite up to approximately 0.01 $eq \cdot l^{-1}$. At greater concentrations, calcite is favored.
3. Increasing strontium bicarbonate favors the precipitation of aragonite up to 2 $mg \cdot l^{-1}$. At greater concentrations, calcite is favored.
4. With the addition of sodium hydroxide, only calcite is formed.
5. Magnesium ions very strongly favor aragonite formation. In particular, magnesium chloride is proven more effective than either magnesium bicarbonate or magnesium sulfate. Magnesium was not observed to coprecipitate with calcium carbonate under the conditions of the study.
6. Small amounts of barium favor calcite formation.
7. Sulfate ions favor aragonite formation.
8. The presence of $NaHCO_3$, Na_2CO_3, or $Mg(HCO_3)_2$ drastically increases the rate of calcium carbonate precipitation.
9. Alkali chlorides decrease slightly the rate of calcium carbonate precipitation.
10. At temperatures greater than 40 °C, aragonite tends to be favored. However, at higher CO_2 pressures (about 1 atm), and where the pH is somewhat lower, calcite is favored. More basic waters, where the partial pressure of CO_2 is lower, tend to favor aragonite.
11. The polymorph vaterite was observed to form at temperatures generally greater than 80 °C. However, this polymorph is so unstable that it is not expected to exist without inversion to either aragonite or calcite.

The two last observations by Kitano should be compared with the observations of Kinsman and Holland (1969) who studied the precipitation of calcium carbonate from seawater at temperatures between 3 °C and 80 °C by the slow addition of dilute Na_2CO_3 solutions. At 30 °C, aragonite was the only phase precipitated, but at 16 °C calcite containing as much as 8 to 10 mol % $MgCO_3$ also precipitated from solution. At 3 °C vaterite or calcium carbonate monohydrate were the dominant phases.

Kitano reports that several of his observations are in conflict with earlier studies that showed that the presence of alkali chlorides favored aragonite formation, and that aragonite is favored in solutions with a pH greater than 8. Indeed, some of the conclusions reached from his own studies appear, at first glance, to be in conflict. In view of the complexity of solution chemical composition and temperature effects on the competing kinetics of reaction, and the potential of foreign ions to modify the relative rates of precipitation, it is clear that opportunities still remain to develop an internally consistent theory that will reconcile the wide range of disparate observations.

A more recent study by Moller and Rajagopalan (1975) does shed some light on the problem of precipitation in the presence of magnesium, an ion that appears to inhibit calcite formation, except at elevated temperatures in the presence of chloride ion (Kitano, 1962b). By varying the Mg/Ca ratio in solution, Moller and Rajagopalan showed that when the Mg/Ca ratio reached 4, a sudden drop in both the nucleation rate and rate of precipitation of calcium carbonate was observed, and the phase precipitating changed from calcite to aragonite.

Moller and Rajagopalan attribute the nucleation rate differences between calcite and aragonite to different orientations of the triangular carbonate ion in the two lattices as suggested by Lippmann (1973). However, changes in nucleation rate within each polymorph as a result of increasing magnesium ion concentration may be attributed to direct interaction of this ion on the surface, such as adsorption at growth centers or Ca^{2+} - Mg^{2+} ion exchange. The preferred explanation of Lippmann (1973) and others that the observed decrease in nucleation rate is due to the large dehydration energy of the magnesium ion is discounted in favor of the concept that the small dehydrated magnesium ions in subcritical nuclei distort the lattice to such an extent that formation of the calcite nucleus becomes impossible when Mg/Ca>4. The study was conducted at ambient temperature and in a "homogeneous solution" consisting only of calcite powder dissolved in carbonic acid and magnesium chloride to make up the selected Mg/Ca ratios.

Observations relating to the transformation of aragonite to calcite in the presence of the aqueous phase are generally consistent with nucleation and growth studies of these phases from solution. Bischoff and Fyfe (1968) made quantitative studies of the effect of alkali and alkaline earth halides, CO_2, KOH, and $KAl(SO_4)_2$, on this transformation at 108 °C. They also found that calcite nucleation occurred on the aragonite surface, and that the rate-determining step involved calcium and bicarbonate ions. It was catalyzed by the increasing ionic strength of the aqueous phase. Solutions high in calcium and bicarbonate ion were considered to be most conducive to calcite formation. Magnesium and sulfate ion inhibited calcite formation. These conclusions were subsequently evaluated further by Bischoff (1968) who showed that the ionic strength of the solution as determined by varying concentrations of NaCl between 0.02 N and 0.3 N had a very marked effect in decreasing the nucleation time of calcite in the presence of differing concentrations of magnesium chloride, whereas increasing concentration of magnesium chloride had the opposite effect. Observations similar to those of Bischoff and Fyfe (1968) and Bischoff (1968) were made by Yoshinaga and Tarutani (1975) who studied the conversion of aragonite and strontianite aragonite to calcite in various salt solutions at 20 °C.

McLester and others (1970) studied the effect of differing concentrations of alkaline earth metal ions on the aragonite/calcite transformations at 50 °C. They found that inhibition followed the sequence Be>Mg>Sr>Ba and was consistent with either a surface or solution effect.

What are the implications of these studies in terms of the problems of reinjection? Most spent geothermal fluids will have moderate to high ionic strength, will be quite alkaline, and will contain relatively low concentrations of such ions as Mg, Sr, and Ba. Sulfate would also be low unless makeup water from agricultural runoff, saline lake water, or seawater were to be incorporated with the spent fluid. Injection temperatures could be anywhere from ambient to about 70 °C or above. Under these conditions, calcite would be the most likely phase to precipitate from the injected fluid. The induction time for nucleation and growth could be heavily dependent on the ionic strength and small amounts of magnesium, strontium, or barium. The fate of these alkali earth metals during passage through the geothermal plant could have a marked effect on subsequent calcium carbonate precipitation. Increasing magnesium ion content causes a progressive decrease in both the calcite and aragonite nucleation rates.

Chemistry of Sulfate Precipitation

Sulfates are unlikely to represent as serious a potential problem to the injection of spent geochemical fluids as either silica or calcium carbonate. However, it is entirely possible that such phases as gypsum ($CaSO_4 \cdot 2H_2O$), hemihydrate ($CaSO_4 \cdot 1/2H_2O$), anhydrite or barite ($BaSO_4$) and barytocelestite ($(Ba,Sr,CA)SO_4$) will form in an injection well and the surrounding formation, particularly if makeup water containing sulfate is mixed with residual brines prior to injection. Information on the solubility of calcium sulfate phases is well known (Holland, 1967; Collins 1975). Gypsum shows a solubility maximum at about 35 °C, whereas both hemihydrate and anhydrite show strong retrograde solubility.

In the presence of water, at one atm pressure, gypsum becomes thermodynamically less stable than anhydrite above 40 °C. Hemihydrate is metastable with respect to anhydrite at all temperatures in the presence of water at 1 atm. It does, however, become more stable than gypsum at temperatures above 105 °C. The relative stability of these phases is affected by the ionic strength of the solution.

Data on the solubility of barite and celestite are scarce. Available information has been summarized by Holland (1967), who shows that barite solubility in water as a function of temperature is prograde up to at least 150 °C, while for celestite, it is strongly retrograde. The presence of some electrolytes, such as sodium chloride, causes the solubility of both barite and celestite to increase (Holland, 1967; Collins, 1975). At 25 °C, maximum solubility is reached for the celestite at about 3 N NaCl. The effect of increased solubility due to NaCl appears to be more pronounced at elevated temperatures. A 5.0-M NaCl solution at 150 °C dissolves approximately 40 times as much $BaSO_4$ as water alone. Furthermore, the effect of variable solubility as a function of ionic strength at different temperatures must be modeled precisely, using an equation that accurately reflects the thermodynamic properties of the electrolyte. Until recently, little progress had been made in that area, although it now appears electrolyte models developed by Pitzer and his co-workers (Pitzer, 1979) and employed with success by Harvie and Weare (1979) may be effective in resolving this problem.

It is possible that most barium sulfate precipitates will contain significant quantities of strontium and minor amounts of calcium. Baryto-celestite containing up to 40% strontium has been observed to precipitate from the geothermal fluids at East Mesa (Taylor and others, 1979). Barite oil well scales also contain significant amounts of strontium (Gates and Caraway, 1965; Weintritt and Cowan, 1967).

At all temperatures up to 100 °C, gypsum precipitation from supersaturated solutions is favored kinetically. Above 105 °C, the hemihydrate is kinetically favored over gypsum (Nancollas and others, 1973). At even higher temperatures, (>120 °C), anhydrite appears to be kinetically favored and is observed as a scalant in heat exchangers (Comeaux, 1977). However, Austin and others (1975) report both anhydrite and hemihydrate formation at temperatures between 125 °C and 150 °C. The formation of any one or more of these phases can also be affected by such factors as ionic strength, the presence of other ions in solution, and degree of supersaturation.

Gypsum nucleation and precipitation has been studied many times. Nancollas and others (1973) cite studies prior to 1977. More recent studies include those by Konak (1974) and Packter (1975b). Only Schierholtz (1958) and Konak (1974) indicated that a diffusion mechanism controls the nucleation and growth of gypsum. The other studies concluded that the rate-determining step involved a surface reaction.

Liu and Nancollas (1970) measured the crystal growth rate of gypsum on seed crystals between 15 °C and 45 °C. They found that the precipitation reaction obeyed an equation of the form:

$$-\ dm/dt = kn(m-m_o)^2,$$

where m = $[Ca^{2+}] = [SO_4^{2-}]$, the concentration of free ions,

m_o = solubility value at a given ionic strength, and

n = the number of growing sites.

This reaction was found to have an activation energy of 15 ± 0.5 kcal·mole^{-1} within the range of temperatures measured. Subsequently, Nancollas and others (1973) extended crystal growth studies on gypsum to between 60 °C and 100 °C. They found that the reaction kinetics were the same as at lower temperatures, with a similar activation energy of 14 ± 2 kcal/mole^{-1}. In addition, Lui and Nancollas (1973) studied the linear growth of gypsum crystallites between 25 °C and 55 °C using a microscopic method, and found that the fastest growing faces had an activation energy 16± 1.5 kcal·mole^{-1}, in substantial agreement with earlier studies.

According to Liu and Nancollas (1971), and Nancollas and Gill (1979) gypsum dissolves according to a first-order mechanism, thus

$$\frac{dC_{Ca}}{dt} = k''s\ (C^o_{Ca} - C_{Ca}),$$

where C_{Ca} = total calcium concentration in solution,

C^o_{Ca} = solubility value at a given temperature,

k'' = first order rate constant, and

s = constant proportional to the number of growth sites on the added seed crystals.

The dissolution mechanism is markedly affected by stirring rate, and therefore is probably diffusion-controlled. It is about 50 times faster than the growth rate of gypsum (Nancollas and others, 1978). Furthermore, it is considerably faster than the dissolution rate of barium sulfate, and has, according to Liu and Nancollas (1971), an activation energy of 10 ± 1.5 kcal·mole^{-1} over the temperature range between 10 °C and 30 °C. This value is rather high for a diffusion-controlled process.

The hemihydrate, $CaSO_4 \cdot 1/2H_2O$, appears in two crystalline modifications known as the α and β forms respectively. Nancollas and others (1978) studied the growth of the α form between 195 °C and 142 °C and dissolution between 90 °C and 140 °C using solutions containing equivalent concentrations of calcium and sulfate ions. This work was subsequently extended by Nancollas and Gill (1979) with solutions containing nonequivalent concentrations of calcium and sulfate ions at 120 °C. Both the crystal growth and dissolution of the hemihydrate followed similar kinetic equations to those of gypsum. However, the activation energies for these two processes have not been evaluated. It should be noted that the formation of hemihydrate in geothermal systems is likely to be a transient phenomenon. According to Nancollas and Gill (1979) as precipitation temperatures approach 120 °C, the transformation of hemihydrate to anhydrite becomes so rapid that only a pesudo-equilibrium would be approached with respect to hemihydrate.

Nancollas and Gill (1979) studied anhydrite growth on seed crystals between 120 °C and 150 °C. They found that crystal growth was linearly dependent on the square of the supersaturation corresponding to the equation

$$-\frac{dC_{Ca}}{dt} = ks(C_{Ca} - C^o_{Ca})^2,$$

where the parameters used are as as given above. The activation energy for this reaction was computed to be 13.9 ± 2.2 kcal·mole^{-1}.

Further studies on the dissolution of anhydrite showed it was 40 times faster than the corresponding crystallization reaction and that a reaction second order in undersaturation was followed, thus

$$\frac{dC_{Ca}}{dt} = k^{\ddagger}s(C^o_{Ca} - C_{Ca})^2,$$

in common with other sparingly soluble sulfates such as barite (Nancollas and Liu, 1975) and celestite (Campbell and Nancollas, 1969). This is in contrast to the first-order mechanism observed for the hydrated sulfates. The rate constant for this reaction was found to be 14.6 ± 2.4 kcal·mole^{-1}.

The nucleation of barite and barytocelestite in geothermal fluids will be due to the presence of suspended nuclei. Mealor and Townshend (1966) show that the homogeneous nucleation of barium sulfate requires a supersaturation factor close to 1000. This is greatly in excess of the supersaturations attained under experimental conditions, and implies that all nucleation phenomena involving barite precipitation observed in the laboratory must be either pseudohomogeneous or heterogeneous. Under geothermal conditions, it is doubtful whether supersaturations greater than 10 will be reached, so it is to be expected that all such nucleation reactions will be heterogeneous.

Evidence is presented by Turnbull (1953) that at 25 °C, and at supersaturations up to 16, the mechanism of precipitation is probably due to heterogeneous nucleation because the rate depends upon the choice of reagent solution. Subsequent crystal growth appears initially to be limited by the transfer of ions across the crystal solution interface. As precipitation progresses, the rate of crystal growth becomes slower and is assumed to be diffusion-limited.

Collins and Leineweber (1956) reached similar conclusions. They believed that the mechanism of nucleation was heterogeneous, even after cleaning reagents, and supersaturations of up to 32 times were attained for initial conditions. Collins and Leineweber also developed a model to describe the growth of crystallites in terms of a diffusion-controlling mechanism in which the least concentrated ion controlled the rate of growth. This model is consistent with their experimental results and again suggests that during the early stages of crystallite growth, a diffusion-controlling mechanism predominates.

Nancollas and Liu (1975) have investigated both the crystal growth and dissolution kinetics of barite seed crystals. At approximately seven times saturation, the precipitation kinetics appeared to obey a surface-rate-controlled reaction described by the equation:

$$(\frac{-dm}{dt} = k_c s(m - m_o)^2,$$

where m and m_o represent the barium sulfate solutions at time, t, and at equilibrium; k_c is the rate constant for crystallization; and s is a linear function of the number of active growth sites in the seed crystals per unit volume of solution. The negligible effect of stirring on the crystal growth rate, and the activation of energy of the reaction, 10 kcal·mole^{-1}, both support the interpretation of surface rate control for crystal growth.

Studies on dissolution of barite crystal were found to obey a similar rate equation, thus:

$$(dm/dt) = k_d s(m_o - m)^2,$$

where k_d is the appropriate dissolution rate constant. The dissolution reaction was also insensitive to stirring rate. In another study by Campbell and Nancollas (1969) on the crystal growth and dissolution of strontium sulfate, similar rate equations were found to be consistent with experimental observations.

In summary, it seems possible to generalize on what is known to date about nucleation and precipitation of sulfates in geothermal systems. It is almost certain that nucleation under all geothermal conditions will be controlled by the presence of nuclei suspended in the aqueous phase. Initial growth of the nuclei may be diffusion-controlled, but thereafter growth changes to a reaction that is surface-rate-controlled and second order with respect to supersaturation of the reacting species. Nancollas and Reddy (1974) believe the latter is due to the slow dehydration of the lattice ions prior to their incorporation in the crystal structure but it may also be due to adsorption or surface diffusion (Nancollas and Gill, 1979).

BRINE TREATMENT TO PREVENT SCALING BY CARBONATES AND SULFATES

If scaling in the well and adjacent formation during injection appears to be a problem, and if the selections of geothermal plant type and makeup water fail to alleviate the problem, then three possible courses of action remain open. If the problem is not likely to be severe, periodic scale removal by mechanical means (redrilling or reaming) or acidization of the injected fluid may be all that is required to maintain the injection well in good condition. With anticipated higher accumulation rates, inhibitors could be added to the injected fluid. Finally, if scaling rates are expected to be severe, the fluid could be treated to remove the precipitate prior to injection. The second approach has much appeal, particularly when the inhibitor requirements are small and therefore economical. Vetter (1976), in discussing the problem of scale removal from wells by mechanical means, points out that such an approach is expensive, and is not always effective because the technique does not permit removal of scale deposits from the adjacent formation. Chemical methods of scale removal are therefore preferred.

Chemical methods of scale removal, both in the well and adjacent formation, usually entail acid dissolution. If the scale is a carbonate, such as calcite, this is easily accomplished using hydrochloric acid. Under these conditions, where the pH is less than 5, the rate of dissolution is directly proportional to the hydrogen ion activity, as shown by Lund and others (1975) and by Plummer and others (1979). Other scales, such as those of barium sulfate, are much more difficult to remove. According to Vetter, various solvents, such as distilled water, sodium chloride solution, and even chelating agents such as ethylenediaminetetraacetic acid (EDTA) and nitriletriacetic acid, have been tried by the oil industry in attempting to redissolve barite. Such methods are slow and Vetter recommends they be resorted to only in specific cases.

The use of inhibitors in controlling scaling rates of calcite, gypsum, hemihydrate and barite has been studied with respect to scaling problems in heat exchangers, evaporators, boilers, cooling towers and wells, and review articles have been written surveying the problem as it relates to wells (Nancollas and Reddy, 1974; Vetter, 1976; Vetter and Kandarpa, 1979).

Calcite precipitation and growth is inhibited by organic phosphonates and phosphates. Thus Reddy and Nancollas (1973) and Nancollas and Reddy (1974) report that as little as 0.5 ppm or 1×10^{-6} mole$\cdot$l^{-1} of the phosphonate ENTMP will almost completely stop calcite precipitation in the presence of seed crystals at 25 ^{o}C. These authors found other phosponates to be even more effective. ENTMP also inhibits nucleation of calcium carbonate and affects both morphology and the proportion of the calcium carbonate polymorphs (vaterite and aragonite) that may form. Reitemeier and Buehrer (1940) and Buehrer and Reitemeier (1940) studied the inhibiting effect of sodium hexametaphosphate at concentrations of the order of 1 ppm and found that it prevented precipitation of calcium carbonate from ammoniacal bicarbonate solutions.

The inhibition of gypsum nucleation and crystal growth has been studied by McCartney and Alexander (1958), Crawford and Smith (1966), Nestler (1968), Smith and Alexander (1970), Nancollas and Reddy (1974) and by Nancollas and others (1978). Nancollas and others (1978) have also studied hemihydrate growth inhibition, and Nancollas and Gill (1979) have studied the inhibition of both hemihydrate and anhydrite nucleation and crystal growth. McCartney and Alexander concluded that the most effective materials in decreasing nucleation of gypsum were colloidal materials having regularly spaced, ionized carboxyl groups in a chain structure. Among these was polyacrylic acid that prevented crystallization of gypsum at 30 ^{o}C for up to five hours, even at concentrations as low as 1.5 ppm. These investigators also noted that inhibition was more effective with this additive at higher rather than lower pH. Crawford and Smith investigated the sorption of polyacrylic acid at a concentration equal to 2×10^{-4} mole$\cdot$l^{-1} acrylic acid on calcium sulfate at 30 ^{o}C, and found that its uptake closely paralleled the removal of calcium sulfate from solution. The presence of polyacrylic acid increased the induction period, and markedly decreased the rate of precipitation by a factor of about 500.

Nestler (1968) continued investigations of polyacrylic acid adsorption on gypsum. He found an increase in adsorption capacity with increasing molecular weight, and that adsorption equilibrium was attained within 10 min. for the highest molecular weight (22,200) studied. Smith and Alexander verified the effectiveness of polyacrylic acid, and confirmed and extended previous observations regarding its pH dependence and molecular weight in inhibiting gypsum crystallization.

Nancollas and Reddy (1974) have surveyed calcium sulfate inhibition, citing gelatin in concentrations from 1 to 100 ppm as being effective. Phosphonates will inhibit calcium sulfate growth only temporarily, and their effectiveness to do even that declines with increasing temperature. Nancollas and others (1978) found phytic acid at 7×10^{-7} mole$\cdot$l^{-1} to completely inhibit gypsum crystallization for about 24 hours at 70 ^{o}C, whereas the phosphonic acid ENTMP at 1×10^{-5} mole$\cdot$l^{-1} at 120°C and 132 times supersaturation inhibited hemihydrate growth for at least seven days. Nancollas and Gill (1979) also found that polyacrylic acids and ENTMP were effective in retarding the growth of not only hemihydrate but also anhydrite, at concentrations as low as 10^{-7}mole$\cdot$l^{-1}. They noted, however, that at higher temperatures, organic phosphonates accelerated the transformation of hemihydrate to anhydrite. Finally, Nancollar and Liu (1975) and Leung and Nancollas (1978) have shown respectively that the phosphonic acids TENTMP and NTMP will markedly retard barium sulfate crystal growth, and that growth is almost completely inhibited at an NTMP concentration of 5×10^{-7} mole$\cdot$l^{-1}.

It is generally believed that inhibitors such as polyacrylic acid, phosphonates, and other naturally-occurring inorganic and organic inhibitors act through surface adsorption and blocking of further crystal growth. Their effectiveness under industrial plant conditions has been demonstrated. However, the application of inhibitors to the problems of subsurface goethermal fluid disposal requires care. Inhibitors may be preferentially adsorbed by rock-forming minerals, may decompose or coprecipitate. Removal by any one of these processes could lead to enhanced local precipitation in the vicinity of the wellbore. The choice of inhibitor should therefore be determined by verification of its effectiveness under simulated or actual field conditions.

Should the effectiveness of inhibitors prove unsatisfactory, preinjection treatment would have to be used. The choice of technique would depend upon the availability of raw materials, space for facilities, economic costs, and a fundamental understanding of the mechanisms involved. Because the nature of the problem will be heavily dependent on the conditions pertaining to each site, only general comments on treatment procedures can be suggested. A general review of problems encountered in treating waste waters prior to injection is given by Sadow (1972).

Treatment processes involve three considerations. The first concerns the precipitation of the offending scale-former. The second requires that the injected fluid no longer continue to precipitate a solid phase, or that precipitation rates become so slow that no harmful effects over the life of the injection well would be anticipated. Finally, suspended materials must be removed to ensure that physical plugging of the well is prevented.

The precipitation of the offending scale-formers can be achieved by additives, such as slaked lime in the case of calcium carbonate or calcium sulfate. Lime causes gross supersaturation of the scale-former, and the rate of precipitation is markedly accelerated. The recycling of seed crystals or other nuclei, and/or the use of fluidized beds or other techniques that enhance the surface area of the precipitating phase will also accelerate the rate of removal and decrease the holding time, and hence, the volume of fluid that must be impounded before injection. Because most reactions involve surface rate control, excessive agitation may be unnecessary, but care should be taken to insure that the rates and mechanisms of precipitation are known and understood should diffusional transport become rate-controlling at some stage during treatment.

Because of the large volume of fluids, efforts should be made to insure that the precipitant forms crystals or crystal aggregates that can be filtered or separated by conventional means. Voluminous floc-like or fine powdery precipitates not only entrap large volumes of water, but rapidly fill holding ponds and present additional costly disposal problems. Ideally, the precipitant should be marketable and if this is so, consideration may have to be given to avoiding coprecipitation of undesirable minor constituents of the geothermal fluid. For example, if gypsum is produced and used as a soil amendment, the presence of coprecipitated lead and arsenic may be unacceptable.

After removing excess precipitant, the spent fluid may require further treatment to prevent continued precipitation in the transfer lines and injection well. This can be accomplished by the use of inhibitors or through pH adjustment with acid such as HCl or CO_2 to eliminate the supersaturated state. In all cases, consideration must be given to cost, because of the large volumes of fluid requiring treatment.

CONCLUSIONS AND RECOMMENDATIONS

The review of studies conducted so far on the precipitation kinetics of silica, carbonate, and sulfates indicates that it should be feasible to anticipate and control or overcome a wide variety of scaling problems arising from the injection of geothermal fluids in subsurface formations. Research on the precipitation kinetics of calcium carbonate requires additional study to evaluate the effects of contaminating ions. Further study is also required to confirm the mechanism of precipitation at temperatures to 150 °C. With regard to both carbonates and sulfates, further study is required to clarify the mechanism as well as rates of pseudohomogeneous nucleation. The mechanisms of precipitation of sulfates appear to be fairly well-known, but confirmatory studies over a wider temperature range (between 25 °C and 150 °C) should be undertaken.

Finally, research should be carried out to evaluate the effect of precipitation of silica, carbonates, and sulfates in a variety of rock types, in order to relate progressive precipitation and changes in porosity to permeability changes. Only if this is accomplished will the full benefits of the past and present efforts to understand and quantify the mechanisms of precipitation of potential scale-formers be obtained.

ACKNOWLEDGMENT

This work was supported by the Assistant Secretary for Conservation and Renewable Energy, Office of Renewable Technology, Division of Geothermal and Hydropower Technologies and by the Director, Office of Energy Research, Office of Basic Energy Sciences, Division of Engineering, Mathematics, and Geosciences of the U.S. Department of Energy under Contract No. W-7405-ENG-48.

REFERENCES CITED

Allen, L. H., and Matijević E., 1969, Stability of colloidal silica. I. Effect of simple electrolytes: Journal Colloid Interface Science v. 31, p. 287-296.

Apps, J. A., 1977, The definition of engineering development and research problems relating to the use of geothermal fluids for electric power generation and nonelectric heating: Berkeley, Lawrence Berkeley Laboratory, LBL-7025, 57 p.

Austin, A. E., and others, 1975, Precipitation of calcium sulfate from sea water at high temperatures: Destination, v. 16, p. 331.

Austin, A. L., and others, The LLL geothermal energy program status report January 1976-January 1977: Livermore, Lawrence Livermore Laboratory, UCRL-50046-76, p 173.

Bathurst, R. G. C., 1968, Precipitation of oids and other aragonite fabrics in warm seas, in Muller, C., and Friedman, G. M., eds., Recent developments in carbonate sedimentology in central Europe: New York, Springer-Verlag, p. 1-10.

Barkman, J. H., and Davidson, D. H., 1972, Measuring water quality and predicting well impairment: Journal Petroleum Technology, p. 865-873.

Berendsen, P., 1971, The solubility of calcite in CO_2-H_2O solutions, from 100° to 300°C, 100 to 1000 bars, and 0 to 10 weight per cent CO_2, and geologic applications [Ph.D. dissertation]: Riverside, University of California, 203 p.

Bischoff, J. L., 1968, Kinetics of calcite nucleation: Magnesium ion inhibition and ionic strength catalysis: Journal of Geophysical Research, v. 73, no. 10, p. 3315-3322.

Bischoff, J. L., and Fyfe, W. S., 1968, Catalysis, inhibition, and the calcite-aragonite problem, American: I, The aragonite-calcite transformation: Journal of Science, v. 266, p. 65-79.

Campbell, J. R., and Nancollas, G. H., 1969, The crystallization and dissolution of strontium sulfate in aqueous solution: Journal of Physical Chemistry, v. 73, no. 6., p. 1735-1740.

Cloud, P. E., Jr., 1962, Behaviour of calcium carbonate in sea water: Geochimica et Cosmochimica Acta, v. 26, p. 867-884.

Collins, A. G., 1975, Geochemistry of oilfield waters: New York, Elsevier, 496 p.

Collins, F. C., and Leineweber, J. P., 1956, The kinetics of the homogeneous precipitation of barium sulfate: Journal of Physical Chemistry, v. 60, p. 389-394.

Comeaux, R. V., 1977, Calcium sulfate anhydrite scaling of water cooled exchangers: The National Association of Corrosion Engineers (Corrosion/77), Paper no. 64, 22 p.

Crawford, J. E., and Smith, B. R., 1966, Adsorption of polyelectrolytes during crystallization of inorganic salts: Journal of Colloid and Interface Science, v. 21, p. 623-625.

Cuéllar, G., 1976. Behavior of silica in geothermal waste waters, in Proceedings, Second United Nations Symposium on the Development and Use of Geothermal Resources, San Francisco,

20-29 May 1975 Volume 2,: Washington, D. C., U. S. Government Printing Office, v. 2, p.1349-1363.

Einarsson, S. S., Vides, A. R., and G. Cuéllar, 1975, Disposal of geothermal waste water by reinjection, in Proceedings, Second United Nations Symposium on the Development and Use of Geothermal Resources, San Francisco, 20-29 May 1975: p. 1349-1363.

Ellis, A. J., and Mahon, W. A. J., 1977. Chemistry and geothermal systems: New York, New York Academic Press, 392 p.

Fournier, R. O., and Truesdell, A. H., 1970, Chemical indicators of subsurface temperature applied to hot spring waters of Yellowstone National Park, Wyoming, U.S.A., in Proceedings, Symposium, Development and Utilization, Geothermal Resources, Pisa: Geothermics, Special Issue 2, v. 2, Pt. 1, p. 529-535.

Fournier, R. O., and Truesdell, A. H., 1973, An empirical Na-K-Ca geothermometer for natural waters: Geochimica et Cosmochimica Acta, v. 37, p. 1255-1275.

Gates, G. L., and Caraway, W. H., 1965, Oil well scale formation in water-flood operations using ocean brines, Wilmington, California: U.S. Bureau Mineral Investigation, no. 6658, 28 p.

Grens, J. Z., and Owen, L. B., 1976, Inhibiting deposition of siliceous scale: Conference on scale management in geothermal energy development, Chicago Operations Office, COO-2607-4, p. 161-168.

Grubbs, D. M., Haynes, C. D., Hughes, T. H., and Stow, S. H., 1972, Compatibility of subsurface reservoirs with injected liquid wastes: National Resources Center, University of Alabama, 89 p.

Harrar, J. E., and others, 1979a, Studies of brine chemistry, precipitation of solids, and scale formation at the Salton Sea geothermal field: Livermore, Lawrence Livermore Laboratory, UCRL-52640.

Harrar, J. E., and others, 1979b, Preliminary results of tests of proprietary chemical additives, seeding, and other approaches for the reduction of scale in hypersaline geothermal systems: Livermore, Lawrence Livermore Laboratory, UCID-18051.

Harrar, J. E., and others, 1979c, On-line tests of organic additives for the inhibition of the precipitation of silica from hypersaline geothermal brine: Livermore, Lawrence Livermore Laboratory, UCID-18091.

Harrar, J. E., and others, 1979d, On-line tests of organic additives for the inhibition of the precipitation of silica from hypersaline geothermal brine: II, Tests of nitrogen-containing compounds, silanes, and additional ethoxylated compounds: Livermore, Lawrence Livermore Laboratory, UCID-18195.

Harrar, J. E., and others, 1979e, On-line tests of organic additives for the inhibition of the precipitation of silica from hypersaline geothermal brine: III, Scaling measurements and tests of other methods of brine modification: Livermore, Lawrence Livermore Laboratory, UCID-18238.

Harrar, J. E., and others, 1979f, Field tests of organic additives for the control of scale at the Salton Sea geothermal field: Geothermal Resources Council, Transactions, v. 3, p. 295-298.

Harvie, C. E., and Weare, J. H., 1979, The prediction of mineral solubilities in natural waters: The Na-K-Mg-Ca-Cl-SO_4-H_2O system from zero to high concentration at 25°C: Geochim. et Cosmochim. Acta (in press).

Hill, J. H., and others, 1979, Apparatus and techniques for the study of precipitation of solids and silica from hypersaline geothermal brine: Livermore, Lawrence Livermore Laboratory, UCRL-52799.

Holland, H. H. 1967, Gangue Minerals in ore deposits, in Barnes, H. L., ed., Geochemistry of hydrothermal ore deposit: New York, Holt Rinehart and Wilson, p. 382-432.

Iler, R. K., 1975, Coagulation of colloidal silica by calcium ions, mechanism, and effect of colloid size: Journal Colloid Interface Science, v. 53, p. 476-488.

Kinsman, D. J. J., and Holland, H. D., 1969, The co-precipitation of cations with $CaCO_3$: IV, The co-precipitation of Sr^{2+} with aragonite between 16° and 96°C: Geochimica et Cosmochimica Acta, v. 33, p. 1-17.

Kitano, Y., 1962a, The behavior of various inorganic ions in the separation of calcium carbonate from a bicarbonate solution: Bulletin Chemical Society Japan, v. 35, no. 12, p. 1973-1980.

Kitano, Y. 1962b, A study of the polymorphic formation of calcium carbonate in thermal springs with an emphasis on the effect of temperature: Bulletin Chemical Society Japan, v. 35, no. 12, p. 1980-1985.

Kitano, Y., and Hood, D. W., 1965, The influence of organic material on the polymorphic crystallization of calcium carbonate, Geochimica et Cosmochimica Acta, v. 29, p. 29-41.

Klein, D. H., and Fontal, B., 1964, Heterogeneous and homogeneous nucleation of barium sulphate: Talanta, v. 11, p. 1231-1237.

Konak, A. R., 1974, The catalysed and uncatalysed nucleation of gypsum crystals from aqueous solution: Kristallu and Technik, v. 9, no. 7, p. 779-787.

Kubota, K., and Aosaki, K. 1975, Reinjection of geothermal hot water at the Otake geothermal field, in Proceedings Second United Nations Symposium on the Development and Use of Geothermal Resources, San Francisco, 20-29 May 1975, Volume 2: Washington, D. C., U. S. Government Printing Office, p. 1379-1383.

Leung, W. H., and Nancollas, G. H., 1978, Nitrilotri (methylenephosphonic acid) adsorption on barium sulfate crystals and its influence on crystal growth: Journal of Crystal Growth, v. 44, p. 163-167.

Lippmann, F., 1973, Sedimentary carbonate minerals: New York, Springer Verlag, 228 p.

Liu, S. T., and Nancollas, G. H., 1970, The kinetics of crystal growth of calcium sulfate dihydrate: Journal of Crystal Growth, v. 6, p. 281-289.

Liu, S. T., and Nancollas, G. H., 1971, The kinetics of dissolution of calcium sulfate dihydrate: Journal Inorganic Nuclear Chemistry, v. 33, p. 2311-2316.

Liu, S. T., and Nancollas, G. H., 1973, Linear crystallization and induction period studies of the growth of calcium sulfate dihydrate crystals: Talanta, v. 20, p. 211-216.

Lund, K., and Fogler, H. S., 1976, Acidization-V: The prediction of the movement of acid and permeability fronts in sandstone: Chemical Engineering Science, v. 31, p. 381-392.

Lund, K., Folger, H. S., and McCune, C. C., 1973, Acidization-I: The dissolution of dolomite in hydrochloric acid: Chemical Engineering Science, v. 28, p. 691-700.

Lund, K., and others, 1975, Acidization-II: The dissolution of calcite in hydrochloric acid: Chemical Engineering Science, v. 30, p. 825-835.

Lund, K., Fogler, H. S., McCune, C. C., and Ault, J. W., 1976, Acidization-IV: Experimental correlations and techniques for the acidization of sandstone cores: Chemical Engineering Science, v. 31, p. 373-380.

McCartney, E. R., and Alexander, A. E., 1958, The effect of additives upon the process of crystallization: I, Crystallization of calcium sulfate: Journal of Colloid Science, v. 13, p. 383-396.

McLester, M. E., Martin, D. F., and Taft, W. H., 1970, Effects of alkaline-earth metal ions on the transformation of aragonite to calcite in aqueous solution: Journal Inorganic Nuclear Chemistry, v. 32, p. 391-399.

Mealor, D., and Townshend, A., 1966, On homogeneous nucleation in barium sulphate precipitation from homogeneous solution: Talanta, v. 13, p. 1191-1193.

Moller, P., and Rajagopalan, G., 1975, Precipitation kinetics of $CaCO_3$ in presence of Mg^{2+} ions: Zeitschrift für Physikalische Chemie Neue Folge, v. 94, p. 297-314.

Nancollas, G. H., and Gill, J. S., 1979, The formation and dissolution of high temperature forms of calcium sulfate scales: The influence of inhibitors: International Symposium on Oilfield and Geothermal Chemistry, Society of Petroleum Engineers SPE-7861, p. 7-14.

Nancollas, G. H., and Liu, S. T., 1975, Crystal growth and dissolution of barium sulfate, Society Petroleum Engineers Journal, v. 15, no. 6, p. 509-516.

Nancollas, G. H., and Reddy, M. M., 1971, The crystallization of calcium carbonate: II, Calcite growth mechanism: Journal of Colloid and Interface Science, v. 37, no. 4, p. 824-830.

Nancollas, G. H., and Reddy, M. M., 1974, The kinetics of crystallization of scale-forming minerals: Society Petroleum Engineers Journal, p. 117-126.

Nancollas, G. H., Reddy, M. M., and Tsai, F., 1973, Calcium sulfate dihydrate crystal growth in aqueous solution at elevated temperatures: Journal of Crystal Growth, v. 20, p. 125-134.

Nancollas, G. H., Eralp, A. E., and Gill, J. S., 1978, Calcium sulfate scale formation: A kinetic approach: Society of Petroleum Engineers Journal, p. 133-138.

Netherton, R., and Owen, L. B., 1978, Apparatus for the field evaluation of geothermal effluent injection, _in_ Geothermal Energy - Novelty Becomes Resource: Transactions, Geothermal Resources Council Annual Meeting, Hilo, Hawaii, 25-27 July 1978, v. II, p. 487-490.

Nestler, C. H., 1968, Adsorption and electrophoretic studies of poly(acrylic acid) on calcium sulfate: Journal of Colloid and Interface Science, v. 26, p. 10-18.

Packter, A., 1968, The precipitation of sparingly soluble alkaline-earth metal and lead salts: Nucleation and growth orders during the induction period: Journal Chemistry Society (A), p. 859-862.

Packter, A., 1975a, The precipitation of calcium carbonate powders from aqueous solution with slow development of supersaturation: Induction periods crystal numbers and final sizes: Kristall und Technik, v. 10, no. 2, p. 111-121.

Packter, A., 1975b, Die Ausfallung von Calciumsulfat-Dihydrat aus wassri-gen Losungen. Kinetik des Kristallwachstums und der endgultigen Kristallugrossen: Zement-Kalk-Gips, no. 2, p. 88-92.

Packter, A., and Sahay, S., 1974, The precipitation of sparingly-soluble alkaline-earth metal and lead salts with slow development of supersaturation (slow anion and hydroxyl ion addition): Kinetics of nucleation and induction periods: Kristall und Technik, v. 9, no. 5, p. 485-496.

Packter, A., and Uppaladini, S. C., 1974, The precipitation of alkaline-earth metal carbonate powders from aqueous solution: Crystal numbers and final sizes: Kristall und Technik, v. 9, no. 9, p. 983-994.

Phillips, S. L., Mathur, A. K., and Doebler, R. E., 1977, A study of brine treatment: Berkeley, Lawrence Berkeley Laboratory, LBL-6371.

Pitzer, K. S., 1979, Characteristics of very concentrated aqueous solutions: Berkeley, Lawrence Berkeley Laboratory, LBL 9660.

Piwinskii, A. J., and Netherton, R., 1977, An experimental investigation of the permeability of Kayenta and St. Peter sandstones to hypersaline brine in the temperature interval 70 to 90°C at 10.3-MPa confining pressure: Livermore, Lawrence Livermore Laboratory, UCRL-52382.

Plummer, L. N., and Wigley, T. M. L., 1976, The dissolution of calcite in CO_2-saturated solutions at 25°C and 1 atmosphere total pressure: Geochimica et Cosmochimica Acta, v. 40, p. 191-202.

Plummer, L. N., Wigley, T. M. L., and Parkhurst, D. L., 1978, The kinetics of calcite dissolution in CO_2-water systems at 5° to 60°C and 0.0 to 1.0 atm CO_2: American Journal Science, v. 278, p. 179-216.

Plummer, L. N., Wigley, T. M. L., and Parkhurst, D. L., 1979, Critical review of the kinetics of calcite dissolution and precipitation, _in_ Jenne, E. A., ed., Chemical modelling in aqueous systems: ACS Symposium Series, v. 93, p. 537-573.

Quong, R., Schoepflin, F., Stout, N. D., Tardiff, G. E., and McLain, F. R., 1978, Processing of geothermal brine effluents for injection, _in_ Geothermal Energy - Novelty Becomes Resource: Transactions, Geothermal Resources Council Annual Meeting, Hilo, Hawaii, 25-27 July 1978, v. II, p. 551-554.

Reddy, M. M., and Nancollas, G. H., 1971, The crystallization of calcium carbonate: I, Isotopic exchange and kinetics: Journal of Colloid and Interface Science, v. 36, no. 2, p. 166-172.

Reddy, M. M., and Nancollas, G. H., 1973, Calcite crystal growth inhibition by phosphonates: Desalination, v. 12, p. 61-73.

Reitemeier, R. F., and Buehrer, T. F., 1940, The inhibiting action of minute amounts of sodium hexametaphosphate on the precipitation of calcium carbonate from ammoniacal solutions I: Quantitative Studies of the Inhibition Process: Journal Physical Science, v. 14, p. 535-551.

Rothbaum, H. P., and Anderton, B. H., 1975, Removal of silica and arsenic from geothermal discharge waters by precipitation of useful calcium silicates, in Proceedings, Second United Nations Symposium on the Development and Use of Geothermal Resources, San Francisco, 20-29 May 1975: Washington, D. C., U. S. Government Printing Office, p. 1417-1425.

Rothbaum, H.P., and others, 1979, Effect of silica polymerisation and pH on geothermal scaling: Geothermics, v. 8, p. 1-20.

Sadow, R. D., 1972, Pretreatment of industrial waste waters for subsurface injection, in Cook, T. D., Underground waste management and environmental implications: American Association Petroleum Geologists, Memoir 18, p. 93-101.

Schierholtz, O. J., 1958, The crystallization of calcium sulphate dihydrate: Canadian Journal Chemistry, v. 36, p. 1057-1063.

Smith, B. R., and Alexander, A. E., 1970, The effect of additives on the process of crystallization: II, Further studies on calcium sulphate (1): Journal of Colloid and Interface Science, v. 34, no. 1, p. 81-90.

Taylor, R. L., and others, 1979, Section 5: Geochemistry, in Howard, J. H., and others, Geothermal resource and reservoir investigations of U. S. Bureau of Reclamation leaseholds at East Mesa, Imperial Valley, California: Berkeley, Lawrence Berkeley Laboratory, LBL-7094, p. 165-233.

Turnbull, D., 1953, The kinetics of precipitation of barium sulfate from aqueous solution: Acta Metallurgica, v. 1, p. 684-691.

Vetter, O. J., 1976, Oilfield scale - can we handle it?: Journal of Petroleum Technology, p. 1402-1408.

Vetter, O. J., and Kandarpa, V., 1979, Scale inhibitor evaluation for oilfield and geothermal operations, in International Symposium on Oil Field Geothermal Chemistry: Society of Petroleum Engineers, SPE 7864, p. 31.

Warner, D. L., 1966, Deep well waste injection-reaction with aquifer water: Journal of the Sanitary Engineering Division, Proceedings of the American Society of Civil Engineers, v. 92, p. 45-69.

Warner, D. L., and Taft, R. A., 1965, Deep well injection of liquid waste (Environmental Health Series, Water Supply and Pollution Control):, U.S. Dept of Health, Education and Welfare, Public Health Service Pub. 999-WP-21, 55p.

Weintritt, D. J., and Cowan, J. C., 1967, Unique characteristics of barium sulfate scale deposition: Journal Petroleum Technology, v. 19, p. 1381-1394.

Weres, O., Tsao, L., and Iglesias, E., 1980, The chemistry of silica in Cerro Prieto brines: Berkeley, Lawrence Berkeley Laboratory, LBL-10166.

Weres, O., Yee, A., and Tsao, L., 1980, Kinetics of silica polymerization: Berkeley, Lawrence Berkeley Laboratory, LBL-7033.

Wollast, R., 1971, Kinetic aspects of the nucleation and growth of calcite from aqueous solutions, in Bricker, O. P., ed., Carbonate cements: The John Hopkins University Studies in Geology, no. 19, p. 265-273.

Yanagase, T., Suginohara, Y., and Yanagase, K., 1970, The properties of scales and methods to prevent them, in Proceedings, United Nations Symposium on the Development and Utilization of Geothermal Resources: Geothermics, Special Issue 2, v. 2, Pt. 2, p. 1619-1623.

Yoshinaga, Y., and Tarutani, T., 1975, Transformation of strontian aragonite to calcite in aqueous solutions: Kyushu University, Memoirs of the Faculty of Science, Ser. C., v. 9, no. 2, p. 231-234.

Manuscript Accepted by the Society October 14, 1980

Printed in U.S.A.

Geological Society of America
Special Paper 189
1982

Aquifer Thermal Energy Storage: A Survey

CHIN FU TSANG AND DEBORAH L. HOPKINS
Lawrence Berkeley Laboratory, University of California, Berkeley, California 94720

ABSTRACT

The disparity between energy production and demand in many power plants has led to increased research on the long-term, large-scale storage of thermal energy in aquifers. Field experiments have been conducted in Switzerland, France, the United States, Japan, and the People's Republic of China to study various technical aspects of aquifer storage of both hot and cold water. Furthermore, feasibility studies now in progress include technical, economic, and environmental analyses, regional exploration to locate favorable storage sites, and evaluation and design of pilot plants.

Several theoretical and modeling studies are also under way. Among the topics being studied using numerical models are fluid and heat flow, dispersion, land subsidence or uplift, the efficiency of different injection/withdrawal schemes, buoyancy tilting, numerical dispersion, the use of compensation wells to counter regional flow, steam injection, and storage in narrow glacial deposits of high permeability.

Experiments to date illustrate the need for further research and development to ensure successful implementation of an aquifer storage system. Areas identified for further research include shape and location of the hydrodynamic and thermal fronts, optimal flow rate and formation permeability, thermal dispersion, natural regional flow, land subsidence or uplift, thermal pollution, water chemistry, wellbore plugging and heat exchange efficiency, and control of corrosion.

INTRODUCTION

The need for energy storage arises from the disparity between energy production and demand. The development of viable storage methods will play a significant role in our ability to implement alternative energy technologies and use what is now waste heat. The ability to provide heat at night and during inclement weather is a key factor in the development of solar energy. Conversely, winter cold, in the form of melted snow or water cooled to winter air temperatures, can be used as a coolant or for air conditioning. Practical storage systems would also allow us to capture the heat that occurs as a by-product of industrial processes and power production. Industrial plants and electric utilities generate tremendous amounts of waste heat, which is usually dissipated through an expensive network of cooling towers or ponds to avoid thermal pollution. Because periods of heat demand do not generally coincide with electricity generation or industrial production, a viable storage method is essential if this heat is to be used. Such a method would not only provide for the use of what is now waste heat, but would significantly decrease the necessary investment in cooling and backup heating systems.

In recent years, aquifers have been studied as a very promising means for the long-term, large-scale storage of thermal energy. Aquifers are porous underground formations that contain and conduct water. Confined aquifers are bounded above and below by impermeable clay layers and are saturated by water under pressure. They are physically well suited to thermal energy storage because of their low heat conductivities, large volumetric capacities (on the order of 10^9 m^3) and ability to contain water under high pressures. Aquifers are also attractive storage sites because of their widespread availability.

Aquifer storage is not a new concept. Since the 1950's aquifers have been used to store fresh water, oil products, natural gas, and liquid wastes. However, their use for thermal energy storage was not suggested until the 1970's. Initial studies were conducted by Rabbimov and others (1971), Meyer and Todd (1972), Kazmann (1971), and Hausz (1974).

At present, there are several projects throughout the world in which the technical, economic, and environmental aspects of aquifer storage are being studied. We shall survey the status of these projects, which are divided into three broad categories: field experiments, feasibility studies, and theoretical and modeling studies. The paper concludes with a discussion of key problems that warrant further research.

FIELD EXPERIMENTS

Initial field experiments have been described by Werner and Kley (1977), Kley and Nieskens (1975), and Despois and Nougarede (1977 unpub. report). Generally, field projects to date have been performed on a relatively small scale and have used water of moderate temperatures (not greater than 55 °C or less than 5 °C). Most of these experiments focused on obtaining pressure and temperature data with the objectives of understanding heat and fluid flow in the aquifer and validating numerical models. There is a need both to extend the temperature range of the investigations and to examine the concept of energy storage on a larger scale. It is also important to look more carefully at other facets of energy storage in aquifers, including the effects of regional flow, land uplift or subsidence, water treatment, and economic feasibility.

A number of experiments have been carried out in recent years. Some of the major projects that are either under way or were recently completed are summarized below. First-cycle data for most of these experiments are summarized in Table 1.

Switzerland

In Switzerland, district heating accounts for 50% of total energy consumption. The amount of yearly consumption that must be stored ranges from 30% for continuous production to as much as 50% for solar production. At present, underground heat storage is being studied as a possible solution to the problem of seasonal storage of thermal energy (Mathey and Menjoz, 1978).

In 1974, the University of Neuchâtel conducted an experiment in which 494 m^3 of hot water was injected into a shallow, phreatic aquifer (Mathey, 1975). Details of the experiment are outlined in Table 1. The high permeability of the aquifer and the small injection volume caused the lighter hot water to rise and resulted in a large heat loss through the upper unconfined layer (Fig. 1). After pumping 16,370 m^3 of water, an amount over 30 times the injection volume, only 40% of the heat was recovered.

France

In France, several theoretical calculations have indicated that thermal energy storage in aquifers is both economically and technically feasible. Field experiments were designed and carried out during the three-year period from 1976 to 1978. Participating agencies included the Bureau de Recherches Géologiques et Minières, Orléans; Centre d'Études Nucléaires de Grenoble; and École des Mines de Paris.

TABLE 1. FIRST CYCLE DATA FROM AQUIFER THERMAL ENERGY STORAGE FIELD EXPERIMENTS

Site	Ambient temp. (°C)	Injection temp. (°C)	Amount injected (m^3)	Injection period (days)	Storage period (days)	Amount withdrawn (m^3)	Energy recovery ratio (%)	Comments and references
University of Neuchatel, Switzerland (1974)	11	51	494	9	122	16,370	~ 40	Shallow phreatic aquifer large buoyance convection (Mathey, 1975)
Bonnaud, France (1976-77)	12	40	1,400	20	122	3,000	30	Evidence of strong thermal dispersion (de Marsily, 1978)
Campuget Exp. Gard, France (1977-78)	14	33.5	20,200	88	42 32	17,000	20	Shallow phreatic aquifer (Cormary and others, 1978)
Auburn Univ., Alabama, U.S.A. (1976)	20	36.4	7,688	17	44	14,260	68	Injection well clogging (Molz and others, 1978)
Auburn Univ., Alabama, U.S.A. (1978-79)	20	55	54,784	80	51	55,345	65	Evidence of ground surface movement (see Molz and others, 1979)
Yamagata Basin, Japan (1977-78)	16	23.7	8,843	64	96	9,930	~ 40	Doublet wells too close to each other (Yokoyama and others, 1978)
Texas A & M Univ., College Station, Texas U.S.A. (1978-79)	21	8.9	31,800	92	60			In progress (Reddell, personal communication)

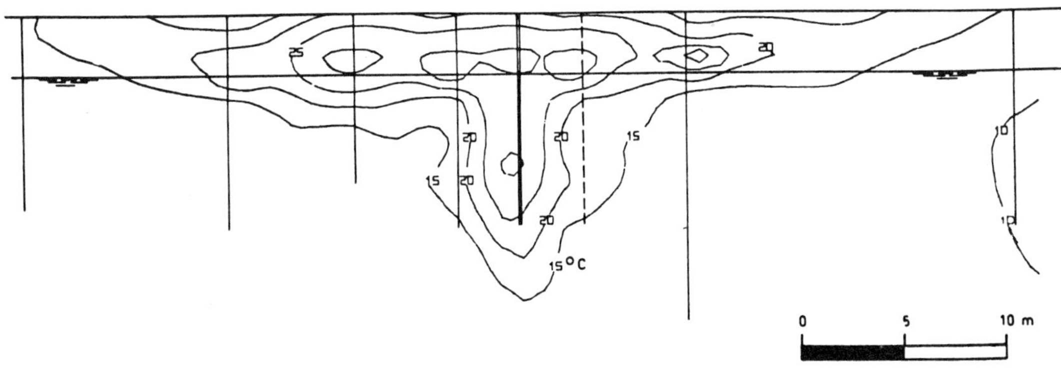

Figure 1. Experiment by the University of Neuchâtel: extension of a thermal disturbance in a gravelly aquifer 16 d after a 223-h injection of 51 °C water, at 37 l/min, into the Colombier-Robinson well (Mathey and Menjoz, 1978).

The Bonnaud experiments (1976 to 1977) involved the injection of hot water into a confined, shallow aquifer, 3 m thick (Fabris and Gringarten, 1977). Eleven observation wells were used to obtain detailed data from two series of heat storage experiments. A fence diagram of the experimental well field is shown in Figure 2. The first series of experiments consisted of three successive injection and production cycles and was followed by the second series (1977) consisting of four cycles. Results of the first cycle are shown in Table 1. More detailed results are shown in Table 2, which also lists results of experiments using normal-temperature (19 to 20 °C) water for comparison. A temperature log for one of the observation wells is shown in Figure 3. The low thermal recovery of the first cycle is attributed in part to the small scale of the experiment and the substantial thermal dispersion which was observed throughout the course of the study.

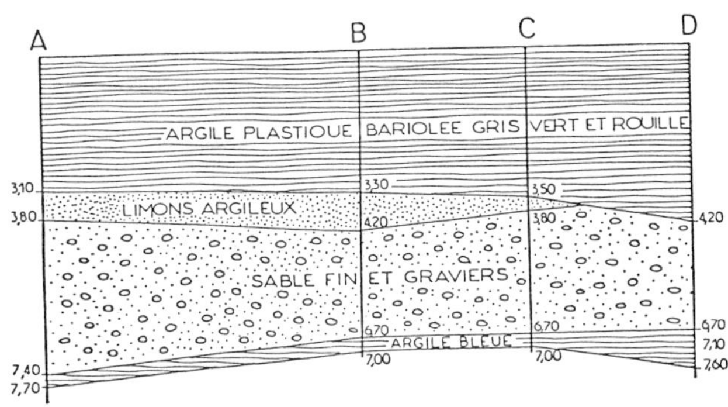

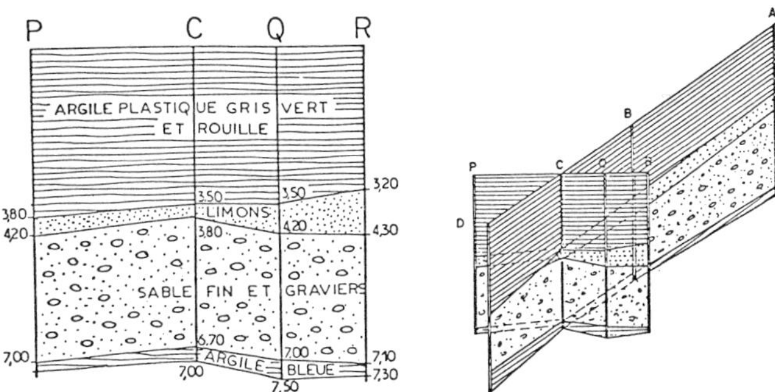

Figure 2. Fence diagram of the Bonnaud experimental well field (Fabris and Gringarten, 1977).

The Campuget experiment (1977 to 1978) was a full-scale storage project with the goal of storing enough heat to meet the needs of 100 housing units (Cormary and others, 1978). Research included: careful monitoring of a phreatic aquifer used for large-scale interseasonal storage, measuring the efficiency of heat recovery, performing a numerical simulation, and studying applications to space heating using existing greenhouses. Over a three-month injection period, 20,200 m^3 of water, heated by simple solar collectors and heat pumps, was injected into a shallow, unconfined aquifer (Table 1). The experiment consisted of two waiting and withdrawal periods. The change in temperature during each period is shown in Table 3.

During pumping, a significant decrease in efficiency was observed between the first and second withdrawal periods, from 54% to 10%. As in the case of the Swiss experiment, the water was stored in a shallow, phreatic aquifer, resulting in a large heat loss through the unsaturated zone (Fig. 4). Contributing to the loss of efficiency was a decrease in air temperature from 16 °C in October 1977 to 6 °C in January 1978 and an accumulation of 700 mm of rainfall between October and March, which caused a decrease in the thickness of the

TABLE 2. APPARENT THERMAL CONDUCTIVITY VALUES* FOR SEVERAL HOT WATER STORAGE EXPERIMENTS

Research Institute	Aquifer thickness (m)	Injection temperature (°C)	Injection Q_i (m^3/hr)	Injection t_i duration (days)	Production Q_p (m3/hr)	Production t_p duration (days)	Thermal storage radius (m)	Apparent conductivity $\tilde{\lambda}$ (cal/ms^{-1}/°C)	$\tilde{\lambda}/\lambda$
ENSMP † BURGEAP ‡ at Neuilly 1974	10	19	15	3	30	3	8	4.5	7.5
ENSMP BURGEAP at Noisy 1974	30	20	115	3	130	3.7	12	1.8	4
BRGM § at Bonnaud 1977	2.5	34	3.4	6 3 6 6	3.4	6 3 6 6	10 7 10 10	12	20

* The apparent thermal conductivity, denoted by $\tilde{\lambda}$, is derived from field temperature data. The actual thermal conductivity of the aquifer is denoted by λ.
† Ecole Nationale Supérieure des Mines de Paris, Paris, France
‡ Bureau de Géologie Appliquée, Neuilly, France
§ Bureau de Recherches Géologiques et Minières, Orléans, France

Source: Sauty and others, 1979, Table 1

unsaturated zone from 3 to 1 m from October 1977 to January 1978.

Throughout the experiments, precipitation of carbonates was neither observed nor detected through chemical analysis. Local bacterial growth did develop but was successfully treated with chlorine injections and did not result in clogging.

United States

Auburn University. A series of field experiments is being conducted by the Water Resources Research Institute of Auburn University to test the concept of storing large quantities of hot water in confined aquifers and to provide data for calibrating mathematical models. For a preliminary experiment initiated in June 1975 (Molz and others, 1978), a well field was constructed consisting of a central injection-production well surrounded by 10 observation and 3 boundary wells (Fig. 5). An additional observation well was placed just above the upper confining layer. Warm water (averaging 35 °C) was obtained from the effluent discharge canal of a power plant and injected into a confined aquifer where the ambient temperature of the formation water was about 20 °C (Table 1). Clogging, thought to be primarily due to a high level of suspended solids in the injected water, was a serious problem throughout the injection period. Following a storage period of approximately 41 days, 14,260 m^3 of water was produced from the aquifer. Figure 6 shows production rate and temperature as functions of time. In view of the relatively small volume of injected water and a partially penetrating injection well, an energy recovery ratio of 68% was encouraging.

For a second set of experiments begun in March 1978 (Molz and others, 1979), water obtained from a shallow semi-confined aquifer at the experimental site was heated and injected into the deeper confined aquifer (Table 1). After a period of storage, the water was produced until its temperature fell to 33 °C which was 13 °C above the ambient water temperature in the aquifer. During this first 6-month cycle, clogging problems were encountered but were much less severe than in the preliminary experiment. The improvement is attributed to the use of ground water rather than surface water and to some backwashing of the injection well. The loss of permeability is probably due to clay particle swelling, dispersion, and migration in the storage formation. A second problem encountered was some type of convection or mixing in the observation wells, which caused erroneous temperature readings. The problem was corrected early in the experiment by backfilling the wells with coarse sand.

A second 6-month cycle has just been completed (Molz and Parr, 1979). Injection volume and temperature were similar to that in the first cycle. The aquifer was still "warm" from the previous cycle where production had stopped 13 °C above ambient temperature. For the injection period, the average specific capacity of the injection well was found to be significantly greater than in the previous experiment. This indicates that the degree of clogging may stabilize at an acceptable level when water is low in suspended solids and suitable backwashing procedures are used during injection. Measurement of ground surface elevation changes indicated a rise of 4 mm near the injection well during injection. The ground subsided to its original elevation during production. An energy recovery ratio of 73% at 33 °C has been obtained,

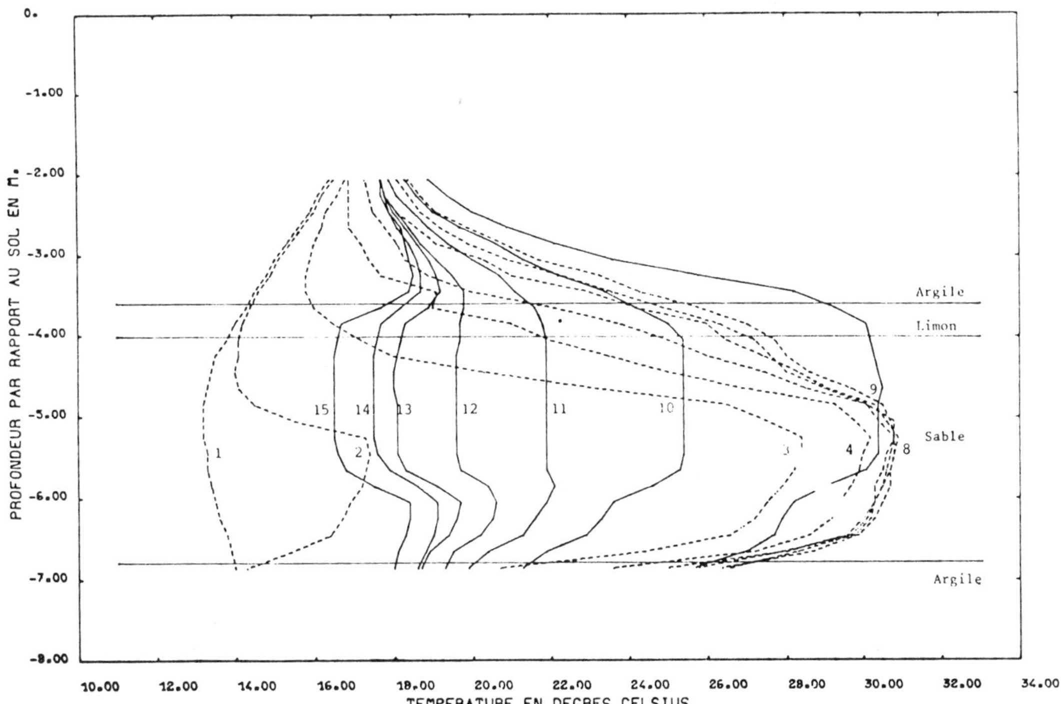

Figure 3. Bonnaud experiment (first cycle) temperature log in well P2 for the period from August 12 to 14, 1977 (Fabris and Gringarten, 1977).

which is significantly better than that obtained for the first cycle.

Texas A & M University. The objective of the experiment at Texas A & M University is to demonstrate the economic and technical feasiblity of cold water storage in aquifers. In particular, Reddell and others (1978) are developing a prototype system in which water is chilled by winter cold and stored in an aquifer for use in summer for air conditioning. Chilled-water storage is of special interest in Texas, where air conditioning accounts for one-third of the residential energy load. Aquifer storage is considered technically feasible based on several years' experience with injection, storage, and recovery of surface water, and is especially attractive in that 80% of the land in Texas is underlain by aquifers.

The storage site is located in the alluvial floodplain of the Brazos River. Water quality is poor, having high concentrations of sodium, chloride, and iron. The site consists of 2 wells that can be used for either injection or production and 12 observation wells, which are used to monitor water levels and temperature profiles. In the winter of 1978 to 1979, water was pumped from a shallow aquifer into a 5000 ft^2 (464.5 m^2) cooling pond. When wet-bulb temperatures dropped below 50 ^{o}F (10 ^{o}C), water was pumped through a spray system and cooled to the air temperature. The chilled water was reinjected into the aquifer and stored until summer. Currently, cold water is being withdrawn and used in a heat-exchange process for air conditioning while warm waste water is being reinjected into the aquifer. Initial data are shown in Table 1.

Preliminary calculations indicate that the efficiency of heat recovery should improve with each injection-pumping cycle, up to three to five cycles. This could be verified by a long-term study over a period of more than 3 years. A long-term study would also allow for statistical variation in weather and would thus permit some evaluation of long-term efficiency.

TABLE 3. TEMPERATURES MEASURED AFTER EACH PHASE OF THE CAMPUGET EXPERIMENT

Phase	Length of period (d)	Temperature at end of period* (^{o}C)
Continuous injection of 20,200 m^3 of hot water	88	33.5
First waiting period	42	30.0
Withdrawal of 5,000 m^3 of water	43	21.0
Second waiting period	32	19.0
Withdrawal of 12,000 m^3 of water	55	14.0

Source: Cormary and others, 1978

* Initial aquifer temperature was 14 ^{o}C

Japan

In the temperate zone, Japan has a climate characterized by warm summers and cold winters and is well suited to the development of a total

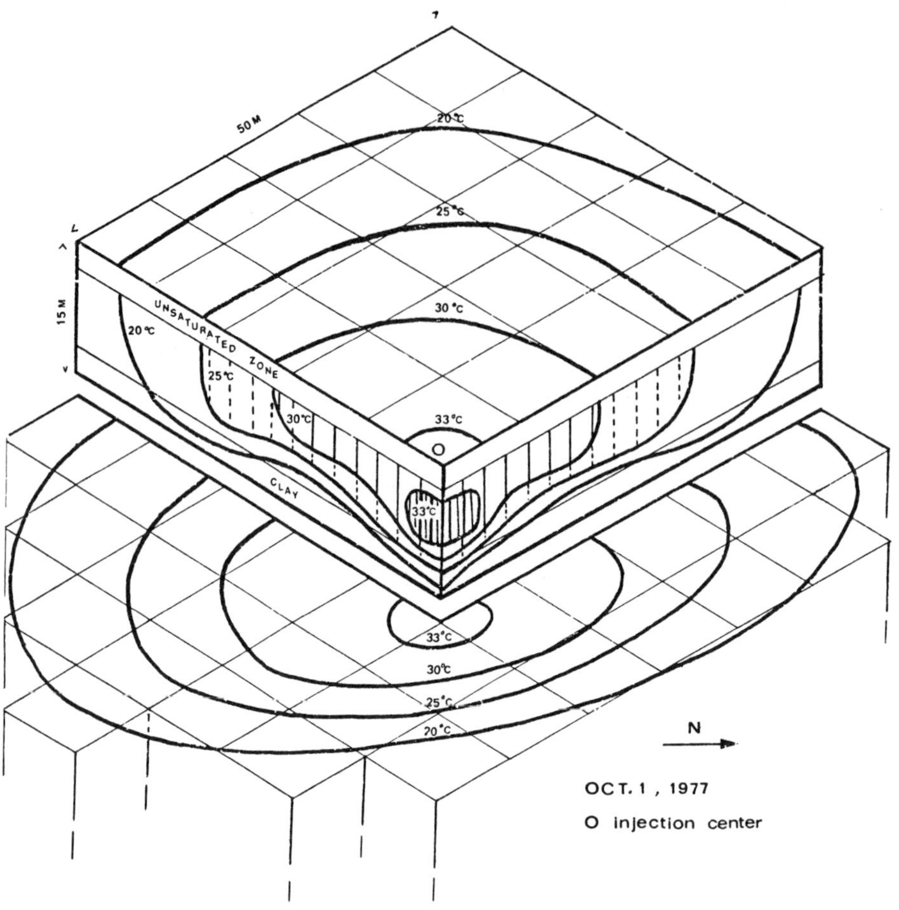

Figure 4. Campuget experiment: temperature distribution in the aquifer at the end of the storage period, October 2, 1977 (Cormary and others, 1978).

energy system. The average temperature in August is 24.5 °C, compared with an average temperature in January of -1.2 °C. Average yearly precipitation is 1,200 mm, of which 300 mm is snow. In 1977, a field project was initiated to study seasonal storage of thermal energy: winter storage of cold water for summer use and summer storage of warm water for winter use (Yokoyama and others, 1978). Applications to air conditioning, fish breeding, and agriculture are being studied.

The test site, consisting of two dual-purpose wells and one observation well, is located in Yamagata Basin, which is underlain by alluvial deposits. Experiments are being conducted using a confined, sand-and-gravel aquifer, 19 m thick. During summer, cold water is withdrawn and used to air-condition a commercial building while warm waste water is sprinkled on the 300-m^2 roof for heat collection. After passing through a filtration tower, the water is further heated by a heat exchanger and recharged through a second well. The process is reversed during winter when warm water is withdrawn and used to melt snow.

For the first experiment, warm water was injected between July 16 and September 18, 1977 (Table 1). Researchers were able to maintain a constant water level throughout the experiment, even with daily discharging. There was no evidence of clogging, and permeability appeared to remain constant.

Between January and March 1978, some 9,430 m^3 of water from melted snow with an average temperature of 5.3 °C was injected into the aquifer. The water was withdrawn 4 mo later at an average temperature of 14 °C. Failure to get cooler water may be attributed to the small volume of injected water and the short distance between injection and production wells, which resulted in the mixing of warm and cold water. At present, a second cold-water storage experiment is under way. Cold water stored during winter 1979 will be withdrawn during summer and used to cool a commercial building.

In addition to field experiments, numerical methods have been used to simulate aquifer storage. A three-dimensional model that takes thermal convection into account has been developed.

China

In China, use of aquifers for thermal energy storage largely developed out of reinjection experiments designed to reduce subsidence and raise ground-water levels (City of Shanghai Hydrogeological Team, 1977). During the late 1950's, in

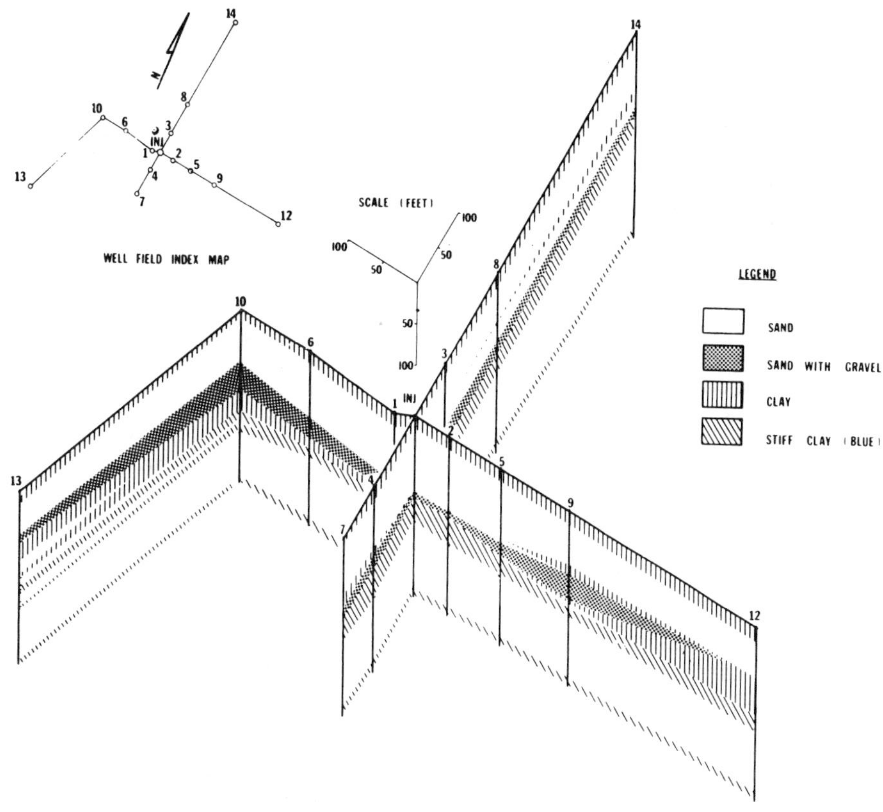

Figure 5. Experiment by Auburn University: fence diagram of the experimental well field. The injection well is labeled INJ (Molz and Warman, 1978).

the city of Shanghai, widespread use of groundwater by a number of factories led to subsidence and a significant drop in the ground-water level. In an attempt to remedy these problems, several factories began experimenting with reinjection of cold water from air-conditioning systems. Experiments over the next few years were generally successful in restoring ground-water levels and increasing output from production wells. In addition, reinjection and well construction methods were continually improved through experiments with different techniques, volumes, and injection periods.

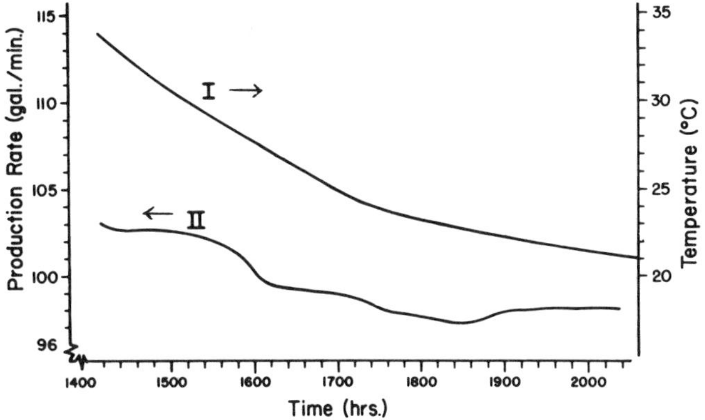

Figure 6. Experiment by Auburn University: temperature (curve I) and production (curve II) as functions of time during the recovery portion of the experiment (Warman and others, 1977).

During the spring and summer of 1965, the Shanghai Cotton Mill Factory initiated a large-scale artificial recharge experiment using four water sources: deep well water, industrial waste water, filtered industrial waste water, and tap water. Researchers also experimented with continuous versus intermittent withdrawal and different reinjection/shut-in cycles. Temperature changes and water quality were monitored both before and after injection. Results indicated that there was little water flow in the aquifer and that there were only small changes in water temperature. These experiments became the basis for later projects, which used winter injection of cold water for summer use and summer injection of hot water for winter use.

During the same period, the Shanghai Water Company conducted extensive experiments using a variety of reinjection methods and three specially designed reinjection wells, 95 m deep, to study changes in ground-water level, water quality, and temperature. Their experiments yielded relatively complete quantitative records, which confirmed the effectiveness of using gravity recharge and underground production methods to raise ground-water levels.

Based on these large-scale experiments and their own studies, the City of Shanghai Hydrogeological Team concluded that by using reinjection, they could effectively control subsidence and ground-water levels and that it was possible to store cold water in winter for summer use in air conditioning. These conclusions led to a city-wide reinjection program in which 70 factories used 134 deep wells for simultaneous recharge. As a result of the program, the ground-water level increased by more than 10 m.

Ground water produced during summer had a very low temperature and thus became a new source of chilled water for industrial use. At the conclusion of summer pumping there was a net average increase in the land level of 6 cm--the first time in several decades of continuous subsidence that any surface uplift had been observed. The program grew in subsequent years so that there are now several hundred wells in use. Production and injection methods have been greatly improved and the program has been expanded to include summer injection of hot water for use in winter. Because of the success at Shanghai, many industrial cities and large villages have adopted similar reinjection and thermal energy storage programs.

FEASIBILITY STUDIES

Current feasibility studies of energy storage in aquifers range from general economic and systems analyses to evaluation and design of pilot plants. The technical work being done entails field and laboratory investigations of the hydrogeological, chemical, and biological aspects of aquifer storage. Several programs call for regional geologic surveys to locate possible storage sites and many studies are aimed at specific applications. Feasibility studies in progress are summarized in Table 4. The following are brief descriptions of several of these projects.

United States

Desert Reclamation Industries/Port Authority of New York. The objective of a study being conducted by the New York State Energy Research and Development Administration is to assess the feasibility of converting the air-conditioning system at the John F. Kennedy Airport in New York City from a conventional refrigeration machine to a system using cold water stored in an aquifer under the airport (Hibshman, 1978). The project would be equivalent in magnitude to providing central air conditioning to every home in a city of 25,000.

Water would either be drawn directly from near-freezing Jamaica Bay, or be chilled by winter air. Three ways of capturing cold from winter air have been considered: cooling towers, dry coolers, and cooling ponds. Use of cooling ponds has been rejected because of space limitations and the dangers of attracting birds and creating fog.

General Electric Tempo. The GE Tempo Center for Advanced Studies is conducting a study to estimate the value of annual-cycle thermal energy storage for a proposed hot water district heating system in the St. Paul/Minneapolis urban area of Minnesota (Meyer, 1980). The proposed system, based on cogeneration of power and heat by the Northern States Power Company, would use coal-fired cogeneration for baseload power and oil-fired boilers for peak and standby needs. Unlike most large district-heating systems in the United States, which produce steam, this system would send out hot water, which is the common practice in Europe.

Economic and environmental benefits of a system incorporating thermal energy storage are being assessed and compared with a cogeneration system that does not use energy storage. Preliminary results indicate that the use of heat storage wells may allow cogeneration capacity to be reduced by about 20%. This reduction in capital requirements is considered to be a measure of the expenditure that could be made for thermal energy storage.

University of Texas. The University of Texas, with Subsurface Disposal Corporation and Bovay Engineers, Inc., is in the second year of a study assessing the feasibility of deep aquifer storage of high-pressure hot water [343 °C, 18.6 MPa (2,700 psi)] and deep cavern storage of hot oil (Collins and others, 1978; Collins, 1979). A study already completed has indicated that underground storage of high-temperature, high-pressure fluid is geologically feasible in approximately 80% of the continental United States.

To study aquifer storage, mathematical modeling and computer simulation are being used to evaluate thermal losses, pumping requirements, and solution transport of minerals. Additional studies of drilling procedures, well design, and cavern leaching have been performed and will be used to make cost estimates. On the basis of preliminary results, it appears that aquifer storage of hot water at temperatures above 149 °C may not be feasible because of downhole pumping requirements and problems associated with silica dissolution and reprecipitation.

A mathematical model has also been developed to study heat storage in solution caverns in massive salt deposits. Salt formations are promising storage sites because of their low porosity and permeability and their semi-plastic properties, which allow small fractures and openings to close. At present, the most serious problem facing salt-cavern storage appears to be possible deformation of the cavern due to creep or plastic flow. A possible solution is to fill the cavern with gravel or coarse sand.

Results just completed for a 10-MWhe system indicate that a hot-oil, gravel-filled, two-well thermocline system, providing 80 MWhe of storage capacity, could be built for about $3.4 million. Operating costs would amount to about 10% of energy transferred (during pumping) with thermal losses of less than 1% when the system is operated on a 24-h storage-recovery cycle at a temperature of 343 °C. These figures indicate that such storage systems would be very cost effective for power generating systems.

Bellingham, Washington. Rocket Research Company directed a study of the feasibility of using industrial waste heat in a planned district heating system in Bellingham, Washington. The source of energy is the Intalco Aluminum Company,

TABLE 4. SUMMARY OF CURRENT FEASIBILITY STUDIES IN AQUIFER THERMAL ENERGY STORAGE

Research institute	Program objectives	Scope of study
General Electric-TEMPO Center for Advanced Studies, Santa Barbara, California, United States (Meyer, 1980)	Assess the feasibility of installing a new district heating system in the Minneapolis/St. Paul metropolitan area based on cogeneration of power and heat and use of aquifers for storage	System design analysis; environmental analysis; economic analysis including estimates of a reasonable time scale, capital requirements, and fuel savings achievable
University of Texas, Houston, United States (Collins and others, 1978)	Assess the feasibility of deep aquifer storage of high-pressure hot water and deep cavern storage of hot oil	Mathematical modeling and computer simulation to study thermal losses, pumping requirements, solution and transport of minerals, and thermomechanical stresses
Tennessee Valley Authority (TVA), Jackson, Tennessee, United States (Eissenberg, 1979)	Survey the potential of thermal energy storage in aquifers in the TVA service area	Parametric modeling of aquifers and aquifer storage; development of criteria for determining the suitability of an aquifer for thermal energy storage
Oak Ridge National Laboratory, Oak Ridge, Tennessee, United States (Eissenberg, 1979)	Perform an environmental impact analysis of aquifer thermal energy storage	Survey environmental and economic effects of aquifer storage; economic analysis of chilled water storage at JFK airport
New York State Energy Research and Development Administration, Desert Reclamation Industries, Plainfield, New Jersey, and the Port Authority of New York and New Jersey, United States (Hibshman, 1978)	Assess the feasibility of converting the air conditioning system at the John F. Kennedy airport in New York City from a conventional refrigeration machine system to a system using cold water stored in an aquifer under the airport	Economic and technical analyses; study of cooling towers, dry coolers, and cooling ponds as methods of capturing winter cold.
Rocket Research Company, Redmond, Washington, United States (Eissenberg, 1979)	Assess the feasibility of using waste heat from an industrial plant in a planned district heating system in Bellingham, Washington	Market and technical analyses; review of alternative heat source options; provision of a conceptual design and cost estimate; large-scale demonstration project
Hooper and Angus Associates Ltd. and Hydrology Consultants Ltd., Toronto, Canada (1979)	Provide a preliminary assessment of using aquifers for chilled water storage or as a source of chilled water.	Technical and economic analyses; analysis of using heat pumps in conjunction with aquifers for heating and cooling applications
Weizmann Institute of Science, Rehovaot, Israel (Nir and Schwarz, 1978)	Study the concept of a total energy system that would utilize aquifer storage to provide cold water for cooling a power plant and warm water for agricultural uses	Environmental, economic, and technical analyses; review of possible pilot operations
Faculte Polytechnique de Mons, Belgium (Brych, 1979)	Evaluate the feasibility of several methods of underground thermal energy storage.	Technical and economic evaluation; testing of promising methods; site selection and implementation of a full-scale storage system
RISØ National Laboratory, Roskilde, Technical University of Denmark, Lyngby, and the Danish Geological Survey, Denmark (Qvale, 1978)	Locate favorable sites for warm water storage and implement a test facility	Nationwide geological and hydrological survey to locate favorable sites; development of mathematical models; design, construction, and operation of a demonstration plant
Kernforschungsanlage Julich, Messerschmitt-Bolkow-Blohm, Munchen, Bundesanstalt fur Geowissenschaften und Rohstoffe, Berlin, Kraftanlangen Heidelberg, and Univ. of Stuttgart (Jank, 1978)	Conduct a comprehensive review of existing information on energy storage systems and provide an advanced systems analysis of operating a storage system within a regional and national system of energy supply	Analysis of potential savings from incorporating thermal energy storage into district heating systems; analysis of using artificial lakes, aquifers, and aquifers filled with artificial bulk material for energy storage; experimental investigation of chemical transport, corrosion and biological processes in an aquifer; design and construction of a small-scale pilot system
Swedish Board for Energy Source Development, Stockholm, Sweden (Mathey, personal commun.)	Determine necessary technical, economic, environmental, and institutional conditions for storage of heat from various sources	Investigation of several types of accumulators including oil tanks and ground water basins
University of Neuchâtel and the Institute de Production d'Energie de l'Ecole Polytechnique Federale, Lausanne, Switzerland (Mathey, personal commun.)	Study various aspects of aquifer storage of hot water with temperature as near as possible to 100 °C	Research of biochemical, thermal, and hydraulic aspects of hot water storage including field and lab tests and numerical modeling; determination of optimal sites and management schemes
École des Mines de Paris, Fontainebleau, France (Iris and de Marsily, 1979)	Study the technical and economic feasibility of a space heating system for the Paris area using aquifers for heat storage, solar captors for heat production, and heat pumps for energy transformation	Determination of in situ parameters necessary to predict the efficiency of a storage system; definition of optimal conditions for a storage site; optimization of a global system; environmental analysis including a study of bacteriological pollution

which generates 93.3 °C waste heat at a rate of 4.85 x 10^8 KJ/h (4.6 x 10^8 Btu/h). An initial study has demonstrated the feasibility of using this waste heat for space heating in 12,000 homes in Bellingham. The average return of investment is estimated at 63%.

Research plans call for market and technical analysis, a review of alternative heat-source options, and provision of a conceptual design and cost estimate. A large-scale demonstration project is also planned.

Canada

At present, approximately 3% of Canada's annual energy expenditure is for mechanical refrigeration for air conditioning. It is believed that an annual storage cycle of chilled water in aquifers can reduce this expenditure. Hooper and Angus Associates Ltd., in association with Hydrology Consultants Ltd. (1979), is working under a contract with the Ministry of Energy, Mines and Resources, Canada, to provide a preliminary technical and economic assessment of using aquifers for chilled-water storage or as a source of chilled water. Specific objectives of the study are to identify the most promising applications and systems and to determine areas for further research and development. In addition to studies of chilled-water extraction and storage, the study will include an examination of using heat pumps in conjunction with aquifer storage for heating and cooling applications.

Israel

In Israel, aquifer storage is being studied as part of a total energy system that would provide water for both heating and cooling purposes. (Nir and Schwarz, 1978). In semi-arid zones, inland location of power plants is difficult because of limited water resources for direct or wet tower cooling. Finding year-round users of waste heat is also a problem in this region. The only potential use identified thus far is for winter agriculture where heat would be used in greenhouses and soil heating. A total energy system using aquifers for storage offers a possible solution to the problems associated with seasonal demand for heat and the large amounts of cooling water required by power plants.

A preliminary study of using aquifer storage in conjunction with power production has been undertaken in southern Israel. The project under study calls for the power plant to withdraw cold water from the aquifer and return warm water, in a closed cycle, to a warm region of the aquifer. During the cold season there would be an additional cycle in which warm water would be withdrawn and delivered to users.

The following topics have been identified for further study: crop selection to make maximum use of available heat; heat dissipation in excess of agricultural demand; plant response to heat input in surrounding soil, water, and air; heat dissipation from soils; and materials and configurations for efficient heat transfer from water to soil. Pilot operations under consideration include: recharge into specified geological formations; aquifer operation with controlled storage and recovery; control of greenhouse and uncovered soil temperatures using warm water and heat sources; and a power plant condenser operation with high temperatures and variable water quality.

THEORETICAL AND MODELING STUDIES

Table 5 summarizes current theoretical and modeling studies of energy storage in aquifers. As the table shows, a number of numerical models are under development although their details are yet to be reported. We shall comment on only three of these projects, conducted by Lund University, Sweden; Bureau de Recherches Géologiques et Minières (BRGM), France; and Lawrence Berkeley Laboratory, U.S.A.

At Lund, a two-dimensional finite-difference model was specifically developed to study the storage of hot water in eskers (long and narrow glacial deposits of high permeability). In addition to the esker project, a number of theoretical studies were made using semi-analytic methods to examine several related topics in thermal storage, including buoyancy tilting of a vertical thermal front, entropy analysis of numerical dispersion, and effects of temperature-dependent viscosity in a two-well extraction-injection system (Hellstrom, 1978; Claesson and others, 1978).

A number of numerical models were developed at BRGM, France, for the study of fluid and heat flow, including a two-dimensional, steady-flow semi-analytical model; a layered, two-dimensional, finite-difference model; and a program to calculate dispersion effects (Gringarten and others, 1977). Some of these techniques were used to model the Bonnaud experiment, which was described in the first section.

A number of numerical models have been developed over the last 6 yr at Lawrence Berkeley Laboratory to study single- and two-phase fluid and heat flow in porous media (Tsang and others, 1978b). Among these models is the program CCC (conduction, convection and compaction), which was chosen for the aquifer thermal energy storage (ATES) studies. CCC employs the integrated finite-difference method and is a fully three-dimensional model incorporating the effects of complex geometry, temperature-dependent fluid properties, gravity, and land subsidence or uplift. This code has been validated against a number of semi-analytic solutions and is currently used to model the Auburn field data (Molz and others, 1978). Extensive generic studies of the ATES concept have been made using CCC. Some of the results are illustrated in Figures 7, 8, and 9 which show the thermal front diffusion during hot-water storage in an inhomogeneous aquifer (Fig. 7), for a well partially penetrating the aquifer (Fig. 8) and for a two-well extraction-and-injection system (Fig. 9). For the particular case of a low-permeability aquifer with a 10^6 kg/d flow rate, calculated energy balances for successive production-storage cycles are listed in Table 6.

On the whole, a substantial amount of modeling

TABLE 5. NUMERICAL MODELING STUDIES IN AQUIFER THERMAL ENERGY STORAGE

Research institute	Project
Technical University of Denmark, Denmark (Qvale, 1978)	One- and two-dimensional finite-element models; study of using compensation wells for countering regional flow
Lund University, Sweden (Hellstrom, 1978)	Two-dimensional, doublet, semi-analytic model; two-dimensional finite-difference program developed to study storage in eskers
University of Neuchâtel, Switzerland (Mathey, 1977; Mathey and Menjor, 1978)	Two- and three-dimensional finite-element models
École des Mines de Paris France (de Marsily, 1978)	Two-dimensional, radial, finite-difference model, two- and three-dimensional finite element models
Bureau des Recherches Géologiques et Minières (BRGM), France (Gringarten and others 1977; Sauty and others, 1979)	Layered, two-dimensional, finite difference model, modeling of the Bonnaud experiment, dispersion modeling studies
University of Yamagata, Japan (Yokoyama and others, 1978)	Finite difference method using a complex potential function
United States Geological Survey, United States (Papadopulos and Larson, 1978)	Intercomp model (finite-difference scheme) used to model the Auburn (1976) experiment
Lawrence Berkeley Laboratory, United States (Tsang and others, 1978)	Three-dimensional integrated finite difference model for conduction, convection, and consolidation; extensive generic studies; modeling of the Auburn (1978) experiment
University of Texas, United States (Collins and others, 1978)	Model to study steam injection into permeable earth strata (two-phase program)

work is being done. However, there is a great need to have these models properly validated and then carefully applied to help us understand the processes underlying the ATES concept.

FUTURE RESEARCH

In this section we put forth what we consider to be the key technical problems that need to be addressed in current or future ATES projects. On the one hand, based on experiences in petroleum engineering, hydrology, and geothermal energy development, the ATES concept is expected to be technically feasible. On the other hand, significant research and development are needed to ensure successful implementation of the ATES system. These areas of needed research include the following.

Shape and Location of the Hydrodynamic and Thermal Fronts. The hydrodynamic front may be tracked by chemical tracers. It is expected to arrive at any observation point before the thermal front. Thus, proper monitoring may yield information about the stored thermal bubble before its arrival at any observation well. The thermal front may be tracked by temperature measurements made in observation wells or by surface geophysics. The latter is a new method proposed by LBL which may provide an economical way to monitor the hot-water bubble during injection, storage, and production stages. Because most of the heat loss is through the hot-cold water interface, a knowledge of the shape and location of the thermal front is quite important for implementation of the ATES concept.

Optimal Flow Rate and Formation Permeability. The Neuchâtel and Campuget experiments indicated that if the flow rate is slow and vertical formation permeability high, buoyancy convection of the lighter hot water will be significant, leading to high heat losses and low energy recovery. Studies are needed to set design guidelines concerning the

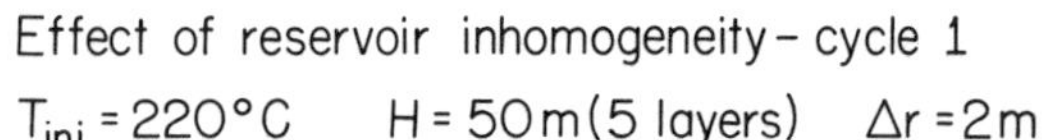

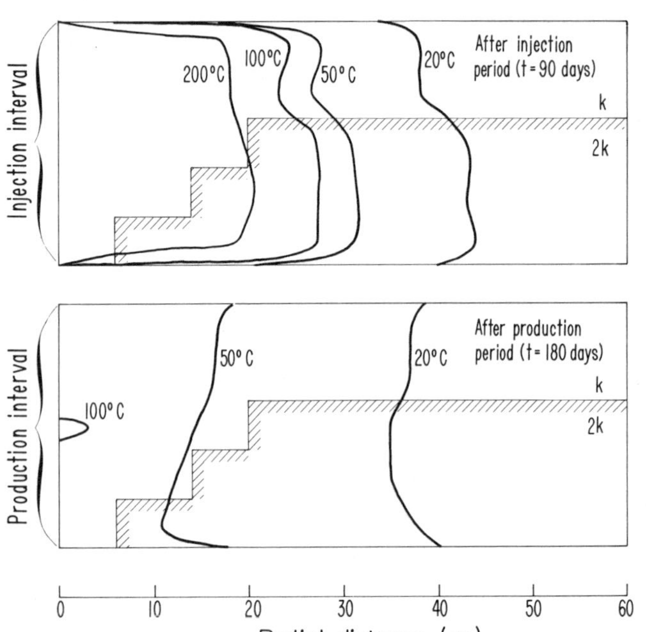

Figure 7. Effects of reservoir inhomogeneity after 90-d injection and production periods (Tsang and others, 1979).

CYCLE I: Partial Penetration

T_{inj} = 220°C ; ΔR = 2m ; H = 100m

Injection Zone

220° 150° 100° 40° 20°

Isotherms after 90 Days Injection (t=90)

Production Zone

150° 100° 40° 20°

Isotherms after 90 Days Injection (t=180)

0 10 20 30 40 50 60 70 80

Radial Distance (m)

Figure 8. Calculated isotherms for partial penetration after 90-d injection and production periods in the first cycle (Tsang and others, 1979).

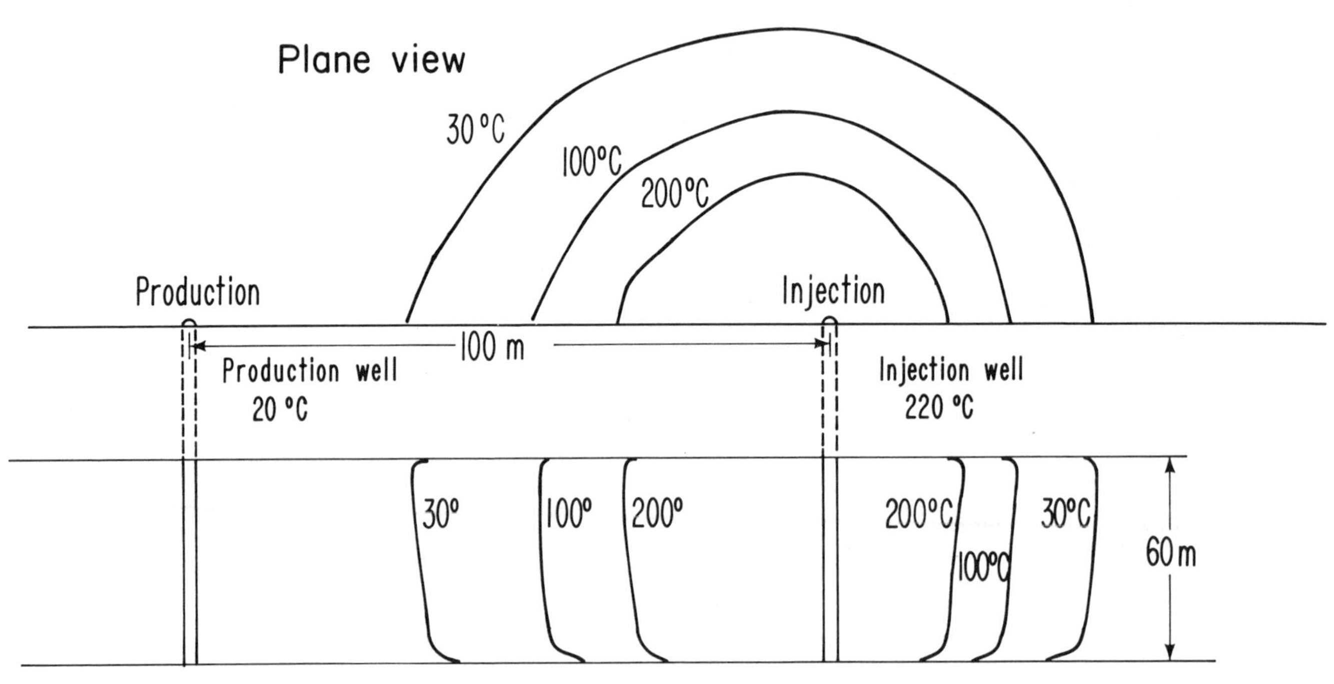

Figure 9. Calculated isotherms for a two-well system after 90 d of injection; plane and cross section views (Tsang and others, 1979).

TABLE 6. COMPUTED ENERGY BALANCE FOR A LOW-PERMEABILITY AQUIFER

Cycle	Energy injected (J x 10^{13})	Energy recovered (J x 10^{13})	Energy loss from aquifer (J x 10^{13})	Energy diffused to heat the aquifer (J x 10^{13})	Energy recovered (%)	Prod. temp. at end of cycle (°C)
1	5.71	4.96	0.053	0.71	86.8	124
2	5.71	5.09	0.068	0.55	89.2	139
3	5.71	5.14	0.077	0.49	90.0	147
4	5.71	5.18	0.084	0.45	90.7	151
5	5.71	5.20	0.091	0.42	91.1	155

Source: Tsang, and others, 1978, Table 2

*Calculations are for the first five cycles for the case of 90-day injection (flow rate of 10^6 kg/day) and 90-day production periods. Injection and ambient temperatures are 220 and 20 °C, respectively; the aquifer is 100 m thick.

choice of aquifers with optimal permeability values for a given storage rate. Such guidelines may help to increase energy recovery and make a given ATES system economically feasible.

Thermal Dispersion. Due to the nonhomogeneity of the aquifer porous media, fingering or extra dispersion will occur at the thermal front. This was observed in the Bonnaud experiment and it tends to significantly decrease the energy recovery. Hence, theoretical and experimental studies are required to estimate its effects and, if possible, to design a way to minimize them.

Natural Regional Flow. A substantial regional flow will move the hot water bubble away from the storage site. Studies on countering such flow by means of compensation wells were done in Denmark and in the United States, at Louisiana State University. Further work is needed in this area.

Land Subsidence or Uplift. Highly accurate land-level surveys should be made on the surface, and vertically within wells, in order to detect land movements during an ATES operation and to evaluate the accuracy of subsidence models. These surveys will yield information necessary for environmental impact statements and developing guidelines for injection and production operations.

Thermal Pollution. Proper accounting of the heat left in the aquifer at each storage-recovery cycle should be made. The rate of dissipation of the heat into the surroundings has to be investigated to ensure minimal effects on the environment.

Water Chemistry. Experience should be gained in the careful analysis of the compatibility of the aquifer water and the hot/cold storage water. The interaction between the injected water and the porous rock medium should also be investigated. Both laboratory experiments and in-situ studies are much needed to evaluate any adverse effects.

Wellbore Plugging and Heat Exchange Efficiency. Scaling and biological growth result in reduced efficiency in the heat exchangers above ground and in well plugging below ground. Specific studies should be made to determine general factors responsible for these adverse effects.

Corrosion. An area of general concern is corrosion. Advanced techniques should be used to measure corrosion rates through the system. Corrosion control techniques need to be developed for general application over a wide range of conditions.

In any pilot or demonstration project, it is crucial to both adequately address the key problems listed above and demonstrate the economic feasibility of the ATES concept. In this way we will obtain a proper understanding, a data base, and the necessary working experience that will allow the general commercial implementation of aquifer thermal energy storage systems.

ACKNOWLEDGMENTS

We wish to thank Cheri Plumlee, Lois Armetta, and Goran Hellstrom for their contribution to the publication of this paper.

This work was supported by the Director, Office of Energy Research, Office of Basic Energy Sciences, Division of Engineering, Mathematics, and Geosciences of the U.S. Department of Energy under Contract No. W-7405-ENG-48.

REFERENCES CITED

Brych, J., 1979, Aquifer storage efforts in Belgium, in ATES newsletter: Berkeley, Lawrence Berkeley Laboratory, PUB-294, v. 1, no. 2.

Bureau de Recherches Géologiques et Minières and others, 1976, Etude experimentale d'un doublet hydrothermique, in Rapport scientifique annual: Orleans, France, Service Geologique National, 9 p.

City of Shanghai Hydrogeological Team, 1977, Artificial recharge into groundwater systems (in Chinese, Summary translated by C. F. Tsang), in ATES newsletter; Berkeley, Lawrence Berkeley Laboratory, PUB-294, v. 1, no. 3.

Claesson, J., and others, 1978, Theoretical analysis and computer simulation of solid-fluid heat storage systems in the ground; extractions of earth heat, in Interim report, part I and II: Lund, Sweden, Lund Institute of Technology, Departments of Mathematical Physics and Building Science, 190 p.

Collins, R. E., 1979, High temperature underground thermal energy storage, in ATES newsletter: Berkeley, Lawrence Berkeley Laboratory, PUB-294, v. 1, no. 2.

Collins, R. E., and others, 1978, High temperature underground thermal energy storage, in Proceedings, Thermal Energy Storage in Aquifers Workshop: Berkeley, Lawrence Berkeley Laboratory, LBL-8431, p. 62-66.

Cormary, Y., and others, 1978, Heat storage in a phreatic aquifer, Campuget experiment (Gard, France), in Proceedings, Thermal Energy Storage in Aquifers Workshop: Berkeley, Lawrence Berkeley Laboratory, LBL-8431, p. 88-93.

de Marsily, G., 1978, Peut-on stocker de l'energie dans le sol?: Annales des Mines, v.1978, no. 4, p 11-24.

Eissenberg, D. M., 1979, A summary, in ATES newsletter: Berkeley, Lawrence Berkeley Laboratory, PUB-294, v. 1, no. 2.

Fabris, H., and Gringarten, A. C., 1977, Etude experimental de stockage et de transfert de chaleur en aquifère, paper RS 598 in Resumé des principaux resultats scientifiques et techniques due Service Géologique National pour 1977: Orleans, France, Service Géologique National, Supplement to the Bulletin of the Bureau du Recherches Géologiques et Minières.

Gringarten, A. C., and others, 1976, Etude des possibilités de stockage d'eau chaude en proven ance de centrales thermiques: Rapport 76, Societe Geologique National 389, AME-BRGM.

____1977, Etude théorique du stockage d'eau chaude par puits unique aquifère en l'absence d'écoulement regional (programme Esther), paper RS-576, in Resumé des principaux resultats scientifiques et techniques du Service geologique national pour 1977: Orleans, France, Service Geologique National, Supplement to the Bulletin of the BRGM.

Hausz, W., 1974, Heat storage wells, Presented at Symposium on efficient use of fuels in process and manufacturing industries, Chicago: Santa Barbara, Institute of Gas Technology (GE-Tempo paper P-655).

Hellstrom, G., 1978, Aquifer storage projects in Sweden, in Proceedings, Thermal Energy Storage in Aquifers Workshop: Berkeley, Lawrence Berkeley Laboratory, LBL-8431, p. 67-70.

Hibshmant, H. J., 1978, Air conditioning Kennedy Airport with winter cold, in Proceedings, Thermal Energy Storage in Aquifers Workshop: Berkeley, Lawrence Berkeley Laboratory, LBL-8431, p. 56-61.

Hooper and Angus Assoc. Ltd. and Hydrology Consultants Ltd., 1979, Study to examine the technical and economic feasibility of using aquifers as a chilled water source or storage medium for building air conditioning systems: Toronto, Energy Mines and Resources Canada, 2 vols.

Iris, P., and de Marsily, G., 1979, Campuget experiment; interpretation by models--development of the research, in ATES newsletter: Berkeley, Lawrence Berkeley, Laboratory, PUB-294, v. 1, no. 3.

Jank, R., 1978, Aquifer storage efforts in Germany, in Proceedings, Thermal Energy Storage in Aquifers Workshop: Berkeley, Lawrence Berkeley Laboratory, LBL-8431, p. 71-74.

Kazmann, R. G., 1971, Exotic uses of aquifers: Journal of Irrigation Drainage Division, American Society Civil Engineers, v. 97, IR3, p. 515-522.

Kley, W., and Nieskens, H. G., 1975, Möglichkeiten der warmespeicherung in einem Porengrundwasserleiter und technische probleme bei einer Rückgewinnung der Energie. (Practical methods of heat storage in a porous aquifer and technical problems of energy recovery): Zeitschrift der Deutschen Geologischen Gesellschaft, v. 126, no. 2, p. 397-409.

Larson, S. P., Papadopulos, S. S., and Mercer, J. W., 1976, Transport of water and heat in an aquifer used for hot water storage; digital simulation of field results, (abstract), in Program, Symposium on Uses of Aquifer Systems for Cyclic Storage of Water, Fall Annual Meeting of the American Geophysical Union, San Francisco: Tulsa, American Geophysical Union.

Mathey, B., 1975, Le stockage de chaleur dans les nappes souterraines (application a l'énergie solaire): 2me Symposium de la Societé Suisse, Lausanne EPFL.

____1977, Development and resorption of a thermal disturbance in a phreatic aquifer with natural convection: Journal of Hydrology, v. 34, p. 315-333.

Mathey, B., and Menjos, A., 1978, Underground heat storage; choice of a geometry and efficiency, in Proceedings, Thermal Energy Storage in Aquifers Workshop: Berkeley, Lawrence Berkeley Laboratory, LBL-8431, p. 80-87.

Meyer, C. F., 1978, Large-scale thermal energy storage for cogeneration and solar systems, in Proceedings, 5th Energy Technology Conference and Exposition, Washington, D.C.: Government Institute, Inc.

____1980, Potential benefits of thermal energy storage in the proposed Twin City district heating cogeneration system, in Proceedings, International Assembly on Energy Storage, Dubrovnik, Yugoslavia: National Academy of

Science (USA) and Council of Academies (Yugoslavia).

Meyer, C. F., and Hausz, W., 1978, Energy management objectives and economics of heat storage wells, in Proceedings, Thermal Energy Storage in Aquifers Workshop: Berkeley, Lawrence Berkeley Laboratory, LBL-8431, p. 17-28.

Meyer, C. F., and Todd, D. K., 1973, Heat storage wells: Water Well Journal, v. 27, no. 10, p. 35-41.

Meyer, C. F., Todd, D. K., and Hare, R. C., 1972. Thermal storage for eco-energy utilities: GE-Tempo Report GE72TMP-56.

Molz, F. J., and Parr, A. D., 1979, Confined aquifer experiment--heat storage, in ATES newsletter: Berkeley, Lawrence Berkeley Laboratory, PUB-294, v. 1, no. 3.

Molz, F. J., and Warman, J. C., 1978, Confined aquifer experiment--heat storage, in Proceedings, Thermal Energy Storage in Aquifers Workshop: Berkeley, Lawrence Berkeley Laboratory, LBL-8431, p. 46-50.

Molz, F. J., Warman, J. C., and Jones, T. E., 1978 Aquifer storage of heated water, Part I, a field experiment: Ground Water, v. 16, no. 4, p. 234-241.

Molz, F. J., and others, 1979, Thermal energy storage in confined aquifers. Auburn, Alabama, Water Resources Research Institute, Auburn University, (preprint), 29 p.

Nir, A., and Schwarz, J., 1978, Survey of thermal energy storage in aquifers coupled with aqricultural use of heat under semi-arid conditions, in Proceedings, Thermal Energy Storage in Aquifers Workshop: Berkeley, Lawrence Berkeley Laboratory, LBL-8431, p. 77-79.

Noyer, M. L., 1977, Simulation des transferts thermique dans les aqui feres: conditions de validité des solution analytiques: Orleans, France, Department Geothermie, BRGM.

Papadopulos, S. S., and Larson, S. P., 1978 Aquifer storage of heated water, Part II, Numerical simulation of field results: Ground Water, v. 16, no. 4, p. 242-248.

Qvale, E. B., 1978, The Danish seasonal aquifer warm-water storage program, in Proceedings of Thermal Energy Storage in Aquifers Workshop: Berkeley, Lawrence Berkeley Laboratory, LBL-8431, p. 75-76.

Manuscript Accepted by the Society October 14, 1980

Printed in U.S.A.

Conversion Tables

TABLE 1. VOLUME (L^3)

	m^3	litre	bbl	gallon (U.S.)	gallon (Imp.)	ft^3
m^3	1	10^3	6.289	2.642×10^2	2.20×10^2	35.315
litre	10^{-3}	1	6.289×10^{-3}	0.2642	0.220	3.5315×10^{-2}
bbl	0.1590	1.590×10^2	1	42.0	34.97	5.6146
gallon (U.S.)	3.7854×10^{-3}	3.7854	2.381×10^{-2}	1	0.8327	0.13368
gallon (Imp.)	4.546×10^{-3}	4.546	2.860×10^{-2}	1.201	1	0.16054
ft^3	2.832×10^{-2}	28.32	0.17811	7.481	6.229	1

TABLE 2. FLOW RATE: (L^3/t) OR (M/t)

	m^3/sec	litres/sec	litres/min	bbl/day	gallons/min (U.S.)	gallons/min (Imp.)	ft^3/sec	klb/hr (ρ_w=1.0)	klb/hr (ρ_w = 0.9)	acre-ft/day
m^3/sec	1	1.0×10^3	6×10^4	5.434×10^5	1.585×10^4	1.320×10^4	35.315	7.94×10^3	7.14×10^3	70.045
litres/sec	1.0×10^{-3}	1	6×10^1	5.434×10^2	1.585×10^1	1.320×10^1	3.5315×10^{-2}	7.94	7.14	7.0045×10^{-2}
litres/min	1.667×10^{-5}	1.667×10^{-2}	1	9.057	0.2642	0.220	5.885×10^{-4}	1.32×10^{-1}	1.19×10^{-1}	1.1674×10^{-3}
bbl/day	1.840×10^{-6}	1.840×10^{-3}	1.104×10^{-1}	1	2.917×10^{-2}	2.429×10^{-2}	6.498×10^{-5}	1.46×10^{-2}	1.31×10^{-2}	1.2889×10^{-4}
gallons/min (U.S.)	6.31×10^{-5}	6.31×10^{-2}	3.785	34.29	1	0.8327	2.2280×10^{-3}	0.50	0.45	4.4192×10^{-3}
gallons/min (Imp.)	7.58×10^{-5}	7.58×10^{-2}	4.546	41.18	1.201	1	2.676×10^{-3}	0.60	0.541	5.3072×10^{-3}
ft^3/sec	2.8317×10^{-2}	2.8317×10^1	1.699×10^3	1.539×10^4	4.488×10^2	3.737×10^2	1	2.24×10^2	2.02×10^2	1.9835
klb/hr (ρ_w=1.0)	1.26×10^{-4}	1.26×10^{-1}	7.56	68.5	2.00	1.67	4.46×10^{-3}	1	0.900	8.8256×10^{-3}
klb/hr (ρ_w=0.9)	1.40×10^{-4}	1.40×10^{-1}	8.42	76.2	2.22	1.85	4.95×10^{-3}	1.11	1	9.8062×10^{-3}
acre-ft/day	1.4276×10^{-2}	1.4276×10^1	8.5658×10^2	7.7584×10^3	2.2629×10^2	1.8842×10^2	0.5042	1.1331×10^2	1.0198×10^2	1

TABLE 3. PRESSURE ($ML^{-1}t^{-2}$)

	N/m^2 (Pascals)	kg_f/m^2	lb_f/in^2 (psi)	bars	atm	ft of water (at 68°F)	m of water (at 68°F)	$dynes/cm^2$	mm of Hg	mbars	in. of Hg
N/m^2 (Pascals)	1	$1.0197x10^{-1}$	$1.4504x10^{-4}$	10^{-5}	$9.8692x10^{-6}$	$3.3512x10^{-4}$	$1.0214x10^{-4}$	10	$7.50x10^{-3}$	10^{-2}	$2.953x10^{-4}$
kg_f/m^2	9.80665	1	$1.4223x10^{-3}$	$9.8067x10^{-5}$	$9.6787x10^{-5}$	$3.287x10^{-3}$	$1.002x10^{-3}$	$9.8067x10^{1}$	$7.353x10^{-2}$	$9.8067x10^{-2}$	$2.895x10^{-3}$
lb_f/in^2 (psi)	$6.8948x10^{3}$	$7.031x10^{2}$	1	$6.8948x10^{-2}$	$6.8046x10^{-2}$	2.311	0.7044	$6.8947x10^{4}$	$5.1715x10^{1}$	$6.8948x10^{1}$	2.0360
bars	10^{5}	$1.0197x10^{4}$	14.504	1	0.98692	33.512	10.214	10^{6}	750	10^{3}	$2.953x10^{1}$
atm	$1.0133x10^{5}$	$1.0332x10^{4}$	14.696	1.01325	1	33.956	10.349	$1.0133x10^{6}$	760	$1.0133x10^{3}$	29.921
ft of water (at 68°F)	$2.984x10^{3}$	$3.042x10^{2}$	0.4327	$2.984x10^{-2}$	$2.945x1^{-2}$	1	0.3048	$2.984x10^{4}$	$2.2382x10^{1}$	$2.984x10^{1}$	$8.812x10^{-1}$
m of water (at 68°F)	$9.790x10^{3}$	$9.980x10^{2}$	1.419	$9.790x10^{-2}$	$9.662x10^{-2}$	3.281	1	$9.790x10^{4}$	$7.3435x10^{1}$	$9.790x10^{1}$	2.891
$dynes/cm^2$	10^{-1}	$1.0197x10^{-2}$	$1.4504x10^{-5}$	10^{-6}	$9.8692x10^{-7}$	$3.3512x10^{-5}$	$1.0214x10^{-5}$	1	$7.50x10^{-4}$	10^{-3}	$2.953x10^{-5}$
mm of Hg	$1.3332x10^{2}$	$1.36x10^{1}$	$1.9337x10^{-2}$	$1.3332x10^{-3}$	$1.3158x10^{-3}$	$4.4679x10^{-2}$	$1.362x10^{-2}$	$1.3332x10^{3}$	1	1.3332	$3.937x10^{-2}$
mbars	10^{2}	$1.0197x10^{1}$	$1.4504x10^{-2}$	10^{-3}	$9.8692x10^{-4}$	$3.3512x10^{-2}$	$1.0214x10^{-2}$	10^{3}	$7.50x10^{-1}$	1	$2.953x10^{-2}$
in. of Hg	$3.3864x10^{3}$	$3.454x10^{2}$	$4.9116x10^{-1}$	$3.3864x10^{-2}$	$3.3421x10^{-2}$	1.135	$3.459x10^{-1}$	$3.3864x10^{4}$	25.40	$3.3864x10^{1}$	1

TABLE 4. ENERGY ($ML^{2}t^{-2}$)

	ergs	Joules (N-m)	cal	kWh	Btu	hp-h	ft-lb wt.
ergs	1	10^{-7}	$2.3885x10^{-8}$	$2.7778x10^{-14}$	$9.4782x10^{-11}$	$3.7251x10^{-14}$	$7.3756x10^{-8}$
Joules (N-m)	10^{7}	1	$2.3885x10^{-1}$	$2.7778x10^{-7}$	$9.4782x10^{-4}$	$3.7251x10^{-7}$	$7.3756x10^{-1}$
cal	$4.1868x10^{7}$	4.1868	1	$1.1630x10^{-6}$	$3.9683x10^{-3}$	$1.5596x10^{-6}$	3.0881
kWh	$3.6x10^{13}$	$3.6x10^{6}$	$8.5986x10^{5}$	1	$3.4122x10^{3}$	1.3410	$2.6552x10^{6}$
Btu	$1.055x10^{10}$	$1.055x10^{3}$	$2.5200x10^{2}$	$2.9307x10^{-4}$	1	$3.9301x10^{-4}$	778.16
hp-h	$2.6845x10^{13}$	$2.6845x10^{6}$	$6.4119x10^{5}$	$7.457x10^{-1}$	$2.5445x10^{3}$	1	$1.98x10^{6}$
ft-lb wt.	$1.3558x10^{7}$	1.3558	$3.2383x10^{-1}$	$3.7662x10^{-7}$	$1.2851x10^{-3}$	$5.0505x10^{-7}$	1

TABLE 5. COMPRESSIBILITY (Lt^2M^{-1})

	m^2/N (Pascals)$^{-1}$	m^2/kg_f	in.^{2}/lb$_f$ (psi)$^{-1}$	bars^{-1}	atm^{-1}	(ft of water)$^{-1}$ at 68°F	(m of water)$^{-1}$ at 68°F	cm^2/dyne	mm^{-1} Hg	mbar^{-1}	in.$^{-1}$ Hg
m^2/N (Pascals)$^{-1}$	1	9.80665	6.8948x10^{3}	10^{5}	1.0133x10^{5}	2.984x10^{3}	9.790x10^{3}	10^{-1}	1.3332x10^{2}	10^{2}	3.3864x10^{3}
m^2/kg_f	1.0197x10^{-1}	1	7.031x10^{2}	1.0197x10^{4}	1.0332x10^{4}	3.042x10^{2}	9.980x10^{2}	1.0197x10^{-2}	1.36x10^{1}	1.0197x10^{1}	3.454x10^{2}
in.^{2}/lb$_f$ (psi)$^{-1}$	1.4504x10^{-4}	1.4223x10^{-3}	1	14.504	14.696	0.4327	1.419	1.4504x10^{-5}	1.9337x10^{-2}	1.4504x10^{-2}	4.9116x10^{-1}
bars^{-1}	10^{-5}	9.8067x10^{-5}	6.8948x10^{-2}	1	1.01325	2.984x10^{-2}	9.790x10^{-2}	10^{-6}	1.3332x10^{-3}	10^{-3}	3.3864x10^{-2}
atm^{-1}	9.8692x10^{-6}	9.6787x10^{-5}	6.8046x10^{-2}	0.98692	1	2.945x10^{-2}	9.662x10^{-2}	9.8692x10^{-7}	1.3158x10^{-3}	9.8692x10^{-4}	3.3421x10^{-2}
(ft of water)$^{-1}$ at 68°F	3.3512x10^{-4}	3.287x10^{-3}	2.311	33.512	33.956	1	3.281	3.3512x10^{-5}	4.4679x10^{-2}	3.3512x10^{-2}	1.135
(m of water)$^{-1}$ at 68°F	1.0214x10^{-4}	1.002x10^{-3}	0.7044	10.214	10.349	0.3048	1	1.0214x10^{-5}	1.362x10^{-2}	1.0214x10^{-2}	3.459x10^{-1}
cm^2/dyne	10	9.8067x10^{1}	6.8948x10^{4}	10^{6}	1.0133x10^{6}	2.984x10^{4}	9.790x10^{4}	1	1.3332x10^{3}	10^{3}	3.3864x10^{4}
mm^{-1} Hg	7.50x10^{-3}	7.353x10^{-2}	5.1715x10^{1}	750	760	2.2382x10^{1}	7.3435x10^{1}	7.50x10^{-4}	1	7.50x10^{-1}	25.40
mbar^{-1}	10^{-2}	9.8067x10^{-2}	6.8948x10^{1}	10^{3}	1.0133x10^{3}	2.984x10^{1}	9.790x10^{1}	10^{-3}	1.3332	1	3.3864x10^{1}
in.$^{-1}$ Hg	2.953x10^{-4}	2.895x10^{-3}	2.0360	2.953x10^{1}	29.921	8.812x10^{-1}	2.891	2.953x10^{-5}	3.937x10^{-2}	2.953x10^{-2}	1

TABLE 6. DYNAMIC VISCOSITY ($ML^{-1}t^{-1}$)

	poises (g/cm-s)	Pa-s (kg/m-s)	lbm/ft-s	lbf-s/ft^2	centi-poises	lbm/ft-h	lbf-s/in.^{2}	kgf-s/m^2	dynes-s/cm^2
poises (g/cm-s)	1	10^{-1}	6.7197x10^{-2}	2.0886x10^{-3}	10^{2}	2.4191x10^{2}	1.4504x10^{-5}	1.0197x10^{-2}	1
Pa-s (kg/m-s)	10	1	6.7197x10^{-1}	2.0886x10^{-2}	10^{3}	2.4191x10^{3}	1.4504x10^{-4}	1.0197x10^{-1}	10
lbm/ft-s	1.4882x10^{1}	1.4882	1	3.1081x10^{-2}	1.4882x10^{3}	3.600x10^{3}	2.1584x10^{-4}	1.5175x10^{-1}	1.4882x10^{1}
lbf-s/ft^2	4.788x10^{2}	4.788x10^{1}	32.174	1	4.788x10^{4}	1.1583x10^{5}	6.9447x10^{-3}	4.8824	4.788x10^{2}
centipoises	10^{-2}	10^{-3}	6.7197x10^{-4}	2.0886x10^{-5}	1	2.4191	1.4504x10^{-7}	1.0197x10^{-4}	10^{-2}
lbm/ft-h	4.1338x10^{-3}	4.1338x10^{-4}	2.7778x10^{-4}	8.6336x10^{-6}	4.1338x10^{-1}	1	5.9956x10^{-8}	4.2153x10^{-5}	4.1338x10^{-3}
lbf-s/in^2	6.8948x10^{4}	6.8948x10^{3}	4.6331x10^{3}	1.44x10^{2}	6.848x10^{6}	1.6679x10^{7}	1	7.031x10^{2}	6.8948x10^{4}
kgf-s/m^2	9.8067x10^{1}	9.8067	6.59	2.0482x10^{-1}	9.8067x10^{3}	2.3723x10^{4}	1.4223x10^{-3}	1	9.8067x10^{1}
dynes-s/cm^2	1	10^{-1}	6.7197x10^{-2}	2.0886x10^{-3}	10^{2}	2.4191x10^{2}	1.4504x10^{-5}	1.0197x10^{-2}	1

TABLE 7. PERMEABILITY $(L^2)^*$

	cm^2	m^2	ft^2	Darcy	cm/s	ft/s	ft/yr	$1/s\text{-}m^2$	gpd(U.S.)/ft^2 (Meinzer)	ft/day
cm^2	1	10^{-4}	$1.076x10^{-3}$	$1.013x10^{8}$	$9.8067x10^{4}$	$3.217x10^{3}$	$1.015x10^{11}$	$9.8067x10^{5}$	$2.079x10^{9}$	$2.7798x10^{8}$
m^2	10^4	1	$1.076x10^{1}$	$1.013x10^{12}$	$9.8067x10^{8}$	$3.217x10^{7}$	$1.015x10^{15}$	$9.8067x10^{9}$	$2.079x10^{13}$	$2.7798x10^{12}$
ft^2	$9.290x10^{2}$	$9.290x10^{-2}$	1	$9.4134x10^{10}$	$9.1107x10^{7}$	$2.989x10^{6}$	$9.433x10^{13}$	$9.1107x10^{8}$	$1.933x10^{12}$	$2.5826x10^{11}$
darcy	$9.869x10^{-9}$	$9.869x10^{-13}$	$1.062x10^{-11}$	1	$9.678x10^{-4}$	$3.175x10^{-5}$	$1.002x10^{3}$	$9.678x10^{-3}$	$2.053x10^{1}$	2.7434
cm/s	$1.0197x10^{-5}$	$1.0197x10^{-9}$	$1.0976x10^{-8}$	$1.033x10^{3}$	1	$3.281x10^{-2}$	$1.035x10^{6}$	10	$2.120x10^{4}$	$2.8346x10^{3}$
ft/s	$3.108x10^{-4}$	$3.108x10^{-8}$	$3.346x10^{-7}$	$3.149x10^{4}$	$3.048x10^{1}$	1	$3.156x10^{7}$	$3.048x10^{2}$	$6.463x10^{5}$	$8.64x10^{4}$
ft/yr	$9.849x10^{-12}$	$9.849x10^{-16}$	$1.060x10^{-14}$	$9.980x10^{-4}$	$9.659x10^{-7}$	$3.169x10^{-8}$	1	$9.659x10^{-6}$	$2.048x10^{-2}$	$2.738x10^{-3}$
$1/s\text{-}m^2$	$1.0197x10^{-6}$	$1.0197x10^{-10}$	$1.0976x10^{-9}$	$1.033x10^{2}$	10^{-1}	$3.281x10^{-3}$	$1.035x10^{5}$	1	$2.120x10^{3}$	$2.8346x10^{2}$
gpd(U.S.)/ft^2 (Meinzer)	$4.810x10^{-10}$	$4.810x10^{-14}$	$5.173x10^{-13}$	$4.872x10^{-2}$	$4.716x10^{-5}$	$1.547x10^{-6}$	$4.883x10^{1}$	$4.716x10^{-4}$	1	$1.3368x10^{-1}$
ft/day	$3.5973x10^{-9}$	$3.5973x10^{-13}$	$3.8721x10^{-12}$	$3.6452x10^{-1}$	$3.5278x10^{-4}$	$1.1574x10^{-5}$	$3.6525x10^{2}$	$3.5278x10^{-3}$	7.4804	1

* $\rho_w = 1$, viscosity = 1 centipoise.

TABLE 8. THERMAL CONDUCTIVITY $(MLt^{-3}T^{-1})$

	$g\text{-}cm/s^3\text{-}K$ (ergs/s-cm-K)	$kg\text{-}m/s^3\text{-}K$ (W/m-K)	$lbm\text{-}ft/s^3\text{-}^oF$	$lbf/s\text{-}^oF$	cal/s-cm-K	$Btu/h\text{-}ft\text{-}^oF$
$g\text{-}cm/s^3\text{-}K$ (ergs/s-cm-K)	1	10^{-5}	$4.0183x10^{-5}$	$1.2489x10^{-6}$	$2.3885x10^{-8}$	$5.778x10^{-6}$
$kg\text{-}m/s^3\text{-}K$ (W/m-K)	10^5	1	4.0183	$1.2489x10^{-1}$	$2.3885x10^{-3}$	$5.778x10^{-1}$
$lbm\text{-}ft/s^3\text{-}^oF$	$2.4886x10^{4}$	$2.4886x10^{-1}$	1	$3.1081x10^{-2}$	$5.9438x10^{-4}$	$1.4379x10^{-1}$
$lbf/s\text{-}^oF$	$8.0068x10^{5}$	8.0068	$3.2174x10^{1}$	1	$1.9124x10^{-2}$	4.6263
cal/s-cm-K	$4.1868x10^{7}$	$4.1868x10^{2}$	$1.6824x10^{3}$	$5.2291x10^{1}$	1	$2.4191x10^{2}$
$Btu/h\text{-}ft\text{-}^oF$	$1.7307x10^{5}$	1.7307	6.9546	$2.1616x10^{-1}$	$4.1337x10^{-3}$	1

TABLE 9. SPECIFIC HEAT ($L^2t^{-2}T^{-1}$).

	J/kg-K	kcal/kg-K (cal/g-K)	kWh/kg-K	Btu/lbm-°F
J/kg-K	1	2.3885×10^{-4}	2.778×10^{-7}	2.3885×10^{-4}
kcal/kg-K (cal/g-K)	4.1868×10^{3}	1	1.163×10^{-3}	1
kWh/kg-K	3.6×10^{6}	8.6×10^{2}	1	8.6×10^{2}
Btu/lbm-°F	4.1868×10^{3}	1	1.163×10^{-3}	1

TABLE 10. DIFFUSIVITY (L^2t^{-1})

	cm^2/s (stokes)	m^2/s	ft^2/h	centistokes	ft^2/s
cm^2/s (stokes)	1	10^{-4}	3.8750	10^{2}	1.0764×10^{-3}
m^2/s	10^{4}	1	3.875×10^{4}	10^{6}	1.0764×10^{1}
ft^2/h	2.5807×10^{-1}	2.5807×10^{-5}	1	2.5807×10^{1}	2.7778×10^{-4}
centistokes	10^{-2}	10^{-6}	3.875×10^{-2}	1	1.0764×10^{-5}
ft^2/s	9.29×10^{2}	9.29×10^{-2}	3.6×10^{3}	9.29×10^{4}	1

Printed in the U.S.A. by Malloy Lithographing, Inc., Ann Arbor, Michigan 48106